HEAT TRANSFER IN CONVECTIVE FLOWS

presented at
THE 1989 NATIONAL HEAT TRANSFER CONFERENCE
PHILADELPHIA, PENNSYLVANIA
AUGUST 6–9, 1989

sponsored by
THE HEAT TRANSFER DIVISION

edited by
R. K. SHAH
GENERAL MOTORS CORPORATION

THE AMERICAN SOCIETY OF MECHANICAL ENGINEERS
United Engineering Center   345 East 47th Street   New York, N.Y. 10017

ISBN No. 0-7918-0349-X

PREFACE

HEAT TRANSFER IN TURBULENT FLOWS

Richard H. Pletcher, Iowa State University

D. W. Pepper, Marquardt Company

Turbulent flow heat transfer occurs in a wide range of important applications. Despite the importance of turbulent convection and the relatively intense level of research activity in this field, much remains unknown. It is still not possible to accurately predict turbulent heat transfer in many configurations of practical interest. The objective of this session was to provide a forum for investigators to present, discuss, and compare results of research in this important field. The twelve papers on this topic are representative of the broad range of issues and problems under investigation in turbulent convection. The papers deal with both forced and mixed convection, and address issues of turbulence modeling, heat transfer augmentation, and laminar to turbulent transition. The efforts of the authors, reviewers, and the session co-chairman, Darrell W. Pepper, are gratefully acknowledged.

HEAT TRANSFER IN RECIRCULATING FLOW

M. E. Crawford, The University of Texas at Austin

N. K. Anand, Texas A&M University

R. S. Amano, The University of Wisconsin-Milwaukee

The broad topic of recirculating flow occurs in internal and external geometries when the streamlines within some part of the flowfield deviate from a nearly parallel direction and turn back on themselves. This typically occurs in internal flows in the vicinity of expansions and contractions, flow through corrugated channels, and flow within enclosures. External recirculating flow occurs when the flow encounters surface discontinuities such as backward facing steps, flow in regions of strong surface curvature, and flow over bluff bodies.

The session has six papers that are numerical in nature. The paper by Abrous and Emery on *Turbulent Free Convection in Square Cavities with Mixed Boundary Conditions* is a numerical study using a k-ε closure model for two-dimensional buoyancy-induced flows, including the effects of plumes, infiltration, and horizontal partitions. The paper by Ayyaswamy, Sadhal, and Huang on the *Effect of Internal Circulation on the Transport of a Moving Liquid Drop* is a numerical study in which the governing equations are being solved for the drop and gaseous phases to calculate heat and mass transport during drop motion, and how it is influenced by circulation within the drop. The paper by Asaba, Asako, Nakumura, and Faghri on the *Numerical Solutions of Convection-Diffusion Problems in Irregular Domains. Using a Coordinate Transformation Which Yields a Circular Domain* presents an algorithm for solutions of the governing equations, with two benchmark cases of fully developed laminar flow in a triangular duct and natural convection in a square cavity. The paper by Han and Kuehn entitled *Numerical Simulation of Double Diffusive Natural Convection in a Vertical Rectangular Enclosure* presents computations for heat and mass transfer in a vertical cavity with aspect ratio 4 and aiding and opposed buoyancy. The paper by Bravo and Chen on *Heat Transfer Characteristics of a Finned Heat Exchanger* is a numerical solution for laminar heat transfer and pressure drop associated with flow in a two-dimensional duct with baffles that form fins for increased heat transfer. The paper by Ghassemi and Roux on *Numerical Investigation of Natural Convection Within a Triangular Shaped Enclosure* provides laminar heat transfer results for geometries approximating a residential attic with boundary conditions corresponding to both winter and summer conditions.

The session also has four experimental papers. The paper by Prasad and Koseff entitled *Combined Forced and Natural Convection Heat Transfer in a "Deep" Lid-Driven Cavity Flow* presents liquid crystal flow visualizations and corresponding heat transfer measurements for the bottom-heated cavity. A paper by Maki and Yabe on *Unsteady Characteristics of the Annular Impinging Jet Flow Field and Reverse Stagnation Point Heat Transfer* documents unsteady effects by frequency analysis of pressure fluctuations and correlation of the data with heat transfer measurements, obtained using the mass transfer analogy. A paper by Salce and Simon on *Investigation of the Effect of Flow Swirl on Heat Transfer Inside a Cylindrical Cavity* focuses on heat transfer to the cavity sidewall and endwall using a liquid crystal measurement technique, accompanied by flow visualization to document the flow patterns. The paper by Sato, Hishida, and Maeda on *Turbulent Flow Characteristics in a Rectangular Channel with Repeated Rib Roughness* documents the turbulent velocity field and shear stress field for three rib geometries using laser Doppler velocimetry.

NATURAL AND FORCED CONVECTION

R. S. Figlialo, Clemson University

Convection heat transfer is currently the most active area in heat transfer research. The papers included in this session arose from papers submitted to the General Papers Committee of the ASME Heat Transfer Division. This Committee primarily considers papers which can not be easily included in the various technical sessions sponsored by the Division K committees. Hence, the papers are directed at topics of a general nature and represent research activities over a broad range of circumstances. Both fundamental and applied questions are addressed within this session. The papers indicate the present and future trends in convective heat transfer. The editorial responsibility for the reviews of these papers has been handled by the General Papers Committee members. This Committee consists of V. Carey, M. A. Ebadian, T. Diller, M. Kaviany, K. Vafai, and R. S. Figliola. We thank the authors and the manuscript reviewers for their assistance.

THERMAL CONVECTION WITH VOLUMETRIC HEATING

M. Keyhani, University of Tennessee
N. H. Hussain, San Diego State University

In many transport processes encountered in manufacturing, geophysics, astrophysics, and nuclear reactor safety, to name a few, thermal convection with volumetric heating is involved. It was our aim to present new experimental, analytical and numerical results for problems in which viscous heating and volumetric energy sources are of importance. The papers presented here report on analytical and numerical results related to problems associated with viscous dissipation effects on solidification of a liquid onto a tube wall; the use of rheological fluids in continuous flow electrophoresis with volumetric energy sources and to the existence of multiple solutions in the case of laminar plane poiseuille flow with viscous heating. It is our hope that the readers find these papers to be stimulating and that they share our views that these results are a welcome addition to the heat transfer literature. The session organizers wish to extend special thanks and appreciation to the authors and the reviewers for their contributions to this session.

NATURAL CONVECTION IN ENERGY CONVERSION AND CONSERVATION

N. Lior, University of Pennsylvania

O. A. Plumb, Washington State University

Natural convection plays an important, sometimes dominant, role in energy conversion and conservation. The six papers in this session cover a relatively wide range of such applications: the role of natural convection in boiling heat transfer, the effect of obstructions and baffles on the reduction of natural convection heat transfer in a single zone and on interzonal heat transfer, ways to enhance the chimney effect, and the influence of sudden changes in wall heat flux on natural convection. Four of the papers are based on numerical studies, one on experiments, and one on both.

STABILITY OF CONVECTIVE FLOWS

P. G. Simpkins, AT&T Bell Laboratories

V. Prasad, Columbia University

The papers in this session demonstrate the broad spectrum of interest that currently exists in convective instabilities. This interest is partly due to the need for answers to a variety of engineering applications. At a more fundamental level, however, there is an interest in the insight that convective motions provide in understanding non-linear systems. A blend of theoretical and experimental work is presented here for a variety of topics. The subject matter ranges from channel flows and thermosyphons, to motions in porous media and flows with rotation. Thermocapillarity and doubly-diffusive flows are also discussed in this context. This session was sponsored by the Environmental Heat Transfer Committee (K-19) of the ASME Heat Transfer Division. We should like to thank the authors for their contributions to this meeting and the referees for their help in evaluating the papers.

HEAT TRANSFER IN CONVECTIVE FLOWS

R. K. Shah, Editor

In order to maintain high quality of the papers published by the ASME in the symposium volume or in preprint form, all papers are reviewed by the professionals in the field on a volunteer basis. On behalf of the ASME Heat Transfer Division, Conference Organizers and Session Organizers, we are grateful to the following reviewers for their time and effort to provide thorough and detailed reviews in a timely manner. This list includes reviewers for papers published or rejected for the technical sessions included in this symposium volume.

Abedian, B.
Amano, R. S.
Armaly, B. F.
Arnas, O. A.
Arpaci, V.
Ayyaswamy, P. S.

Beasley, D.
Bejan, A.
Blair, M. F.
Blythe, P. A.
Bradshaw, P.
Burmeister, L. C.

Campo, A.
Catton, I.
Chen, J. L. S.
Chilukuri, R.
Chow, L. C.
Churchill, S. W.

Dhir, V. K.
Diller, T.
Dosajh, S. S.

Ebadian, M. A.
Elghobashi, S.

Faghri, M.
Farouk, B.
Figliola, R. S.
Florschuetz, L. W.
Freitas, C. J.

Gardon, R.
Gebhart, B.
Georgiadis, J. G.

Hansen, M. G.
Han, J. C.
Hickox, C.
Hickox, C. E.
Hodge, S.
Homsey, G. M.
Hussain, N. H.

Imber, M.

Joshi, Y.

Kakac, S.
Karni, J.
Kennedy, K. L.
Kercher, D. M.
Keuhn, T. H.
Keyhani, M.
Kim, I.
Kladias, N.
Korpela, S. A.
Koschmieder, E. L.
Kuehn, T. H.
Kumar, R.

Landis, F.
Lavine, A.
Lavine, A. G.
Liburdy, J. A.
Lior, N.

Majumber, A. S.
Mayle, R. E.
Modi, V. J.

Nansteel, M.
Neitzel, G. P.

Oosthuizen, P.
Ostrach, S.
O'Brien, J. E.

Patera, A. T.
Plumb, O. A.

Raithby, G. D.
Ramadhyani, S.
Ramadyani, S.
Robillard, L.
Rosenhow, W.
Roy, S. K.
Rubesin, M. W.
Rudy, T. M.

Sadhal, S. S.
Schuler, C.
Sen, M.
Sernas, V.
Sherif, S. A.
Siebers, D. L.
Simon, T. W.
Simpkins, P. G.
Smith, R. N.
So, R. M. C.
Strahle, W. C.

Wirtz, R. A.

Yao, L. S.

Zebib, A.

CONTENTS

Plenary Lecture: Application of EPA Computer Models in Buoyant Plume Prediction
  L. R. Davis 1

HEAT TRANSFER IN TURBULENT FLOW
The Effects of Step Changes in the Thermal Boundary Condition on Heat Transfer in the
  Incompressible Flat Plate Turbulent Boundary Layer
  R. P. Taylor, P. H. Love, H. W. Coleman, and M. H. Hosni 9
Experimental Investigation of Turbulent Mixed Convection in Horizontal Tubes With
  Longitudinal Internal Fins
  A. C. Trupp and H. Haine 17
Prediction of Stagnation Point Heat Transfer With Free Stream Turbulence
  L. T. Tran and D. B. Taulbee 27
Turbulent Heat-Transfer Augmentation by Micro-Scale Flow Destabilization of the Viscous
  Sublayer
  H. Kozlu, B. B. Mikic, and A. T. Patera 35
Jet Impingement Heat Transfer From Jet Tubes and Orifices
  M. Gundappa, J. F. Hudson, and T. E. Diller 43
A New Heat Transfer Law for Turbulent Natural Convection Between Heated Horizontal
  Surfaces of Large Extent
  W. K. George 51
Application of a Higher-Order Turbulence Closure Model to Plane Jet
  J. C. Chai and R. S. Amano 63
Fluid Mechanics and Heat Transfer Measurements in Transitional Boundary Layers
  Conditionally Sampled on Intermittency
  J. Kim, T. W. Simon, and M. Kestoras 69
Microscales of Oscillating Turbulent Flows
  V. S. Arpaci 83
Computation of Turbulent Three-Dimensional Mixed Convection Boundary Layers
  B. Afshari and J. H. Ferziger 89
Evaluation of k-ϵ Turbulence Model for Predicting Buoyant Free Round Jets
  D. F. G. Durão, J. C. F. Pereira, and J. M. P. Rocha 99
Application of a Surface Renewal Model to the Prediction of Heat Transfer in an Impinging
  Jet
  T. D. Yuan and J. A. Liburdy 109

HEAT TRANSFER IN RECIRCULATING FLOWS
Turbulent Free Convection in Square Cavities With Mixed Boundary Conditions
  A. Abrous and A. F. Emery 117
Effect of Internal Circulation on the Transport to a Moving Liquid Drop
  P. S. Ayyaswamy, S. S. Sadhal, and L. J. Huang 131
Numerical Solutions of Convection-Diffusion Problems in Irregular Domains, Using a
  Coordinate Transformation Which Yields a Circular Domain
  M. Asaba, Y. Asako, H. Nakamura, and M. Faghri 141
A Numerical Simulation of Double Diffusive Natural Convection in a Vertical Rectangular
  Enclosure
  H. Han and T. H. Kuehn 149
Combined Forced and Natural Convection Heat Transfer in a "Deep" Lid-Driven Cavity Flow
  A. K. Prasad and J. R. Koseff 155
Unsteady Characteristics of the Annular Impinging Jet Flow Field and Reverse Stagnation
  Point Heat Transfer
  H. Maki and A. Yabe 163

Numerical Investigation of Natural Convection Within a Triangular Shaped Enclosure
M. Ghassemi and J. A. Roux 169

Investigation of the Effect of Flow Swirl on Heat Transfer Inside a Cylindrical Cavity
A. Salce and T. W. Simon.. 177

Heat Flow Characteristics of a Finned Heat Exchanger
R. H. Bravo and C. J. Chen 185

Turbulent Flow Characteristics in a Rectangular Channel With Repeated Rib Roughness
H. Sato, K. Hishida, and M. Maeda.. 191

NATURAL AND FORCED CONVECTION

Modified Local Similarity for Natural Convection Along a Nonisothermal Vertical Flat Plate Including Stratification
S. W. Webb 197

The Cavity Width Effect on Immersion Cooling Due to Discrete Flush-Heaters on One Vertical Wall of an Enclosure Cooled From the Top
R. Carmona and M. Keyhani.. 207

Experimental Study of Combined Convective Heat Transfer From Tandem Cylinders in a Horizontal Air Flow
C. Henderson and P. H. Oosthuizen.. 221

Free Convective Flow in an Inclined Square Cavity With a Partially Heated Wall
P. H. Oosthuizen and J. T. Paul... 231

Natural Convection From a Vertical Plate With Step Changes in Surface Heat Flux
S. Lee and M. M. Yovanovich 239

Thermosolutal Natural Convection in a Rectangular Enclosure: Numerical Results
C. Benard, D. Gobin, and J. Thevenin... 249

THERMAL CONVECTION WITH VOLUMETRIC HEATING

Viscous Dissipation Effect on Solidification of a Liquid Onto a Tube Wall
M. T. Ahmadian and L. C. Burmeister.. 255

The Use of Rheological Fluids in Continuous Flow Electrophoresis
I. Kim and T. F. Irvine Jr. 261

Analysis of Multiple Solutions in Plane Poiseuille Flow With Viscous Heating and Temperature Dependent Viscosity
M. Sen and P. Vasseur 267

NATURAL CONVECTION IN ENERGY CONVERSION AND CONSERVATION

Influence of Aperture Height and Width on Interzonal Natural Convection in a Full-Size, Air-Filled Enclosure
C. R. Boardman III, A. Kirkpatrick, and R. Anderson.................................... 273

Natural Convection on a Horizontal Surface with High Gravity
M. E. Ulucakli and H. Merte, Jr. 281

Natural Convection Heat Transfer in a Square Enclosure Containing an Obstruction
J. M. House, C. Beckermann, and T. F. Smith 289

Natural Convection in a Vertical Heated Tube Attached to a Thermally Insulated Chimney of a Different Diameter
Y. Asako, H. Nakamura, and M. Faghri... 299

Transient Natural Convection in a Vertical Cylinder With a Specified Wall Flux
J. Sun and P. H. Oosthuizen... 305

The Effect of Partitions on Natural Convection in Enclosures
P. K.-B. Chao, N. Lior, S. W. Churchill, and H. Ozoe 315

STABILITY OF CONVECTIVE FLOWS

A Linear Stability Analysis of a Mixed Convection Plume: First Order Mixed Convection Effects
R. Krishnamurthy.. 323

Convective Instability in the Thermal Entrance Region of Horizontal Rectangular Channels
   F. C. Chou and J. N. Lin 329
Heat Transfer Augmentation Through Wall Shape Induced Flow Destabilization
   M. Greiner, R.-F. Chen, and R. A. Wirtz 337
Hysteresis Effects in Three-Dimensional Natural Convection in a Tilted Porous Medium
   S. J. Pien and M. Sen 343
Stability of Three Dimensional Flows In a Horizontal Annulus with a Heated Rotating Inner
   Circular Cylinder
   L. Yang and B. Farouk 349
A Numerical Study of the Instability of Double-Diffusive Convection in a Square Enclosure
   with Horizontal Temperature and Concentration Gradients
   R. Krishnan 357
Convective Instabilities in Horizontal Porous Layers Heated From Below: Effects of Grain Size
   and Its Properties
   N. Kladias and V. Prasad 369
The Stability Analysis of Closed-Loop Thermosyphon System
   C. C. Hwang, S. H. Yin, J. T. Teng, and M. J. Tsai 381
Onset of Thermocapillary Convection in Hydromagnetic and Radiating Fluids
   T. T. Lam and S. C. Lee 389
Observations of Secondary Motions in Natural Convection Through Inclined Channels Heated
   From Below
   L. F. A. Azevedo and M. J. Kaskus 399
Influence of Surface Viscosity on Marangoni-Benard Instabilities
   P. Queeckers, J. C. Dupin, and J. C. Legros 405

PLENARY LECTURE

APPLICATION OF EPA COMPUTER MODELS
IN BUOYANT PLUME PREDICTION

L. R. Davis
Department of Mechanical Engineering
Oregon State University
Corvallis, Oregon

ABSTRACT

Computer models suggested by the EPA to calculate near field dilutions for both submerged and surface discharges are discussed. Their basic theory and limitations are presented. The models are applied to several test cases and the results of their predictions compared to experimental and field data. On cases that do not agree with the basic assumptions in the models, suggestions are made as to how approximations to the dilution can be made and which models to use. The results indicate that the models do quite well as long as their basic assumptions are not violated. Some models apply to a wider range of problems than others. Some complicated discharges can be predicted fairly well while others cannot.

NOMENCLATURE

a	transverse diffusion coefficient use in PSY model
B	river width
C_c	plume centerline or shoreline concentration
C_o	effluent concentration
d	average river depth
D	discharge port diameter also transverse diffusion factor used in PSY model
g	gravitational constant
H_o	discharge channel depth
L	spacing between ports
n	Manning's coefficient
P	initial dilution factor ($C_c Q_r / C_o Q_o$) used in PSY model
Q_o	effluent discharge rate
Q_r	river flow rate
U_o	discharge velocity
X	distance downstream or along plume centerline
Y	vertical distance from discharge
$\Delta \rho$	density difference between plume and ambient
ρ_o	discharge density
σ_p	dimensionless distance used in PSY model $(DX)^{1/2}$

INTRODUCTION

The EPA has suggested that specific computer models be used to predict the dilution from various types of outfalls when applying for discharge waivers as outlined in Sections 301h and 301g of the U.S. Clean Water Act (PL 92-500 1977)[1],[2],[3]. These models were selected because of their ease of use, availability and their general agreement with measured data. They are somewhat idealistic and do not describe all situations. Predictions are, however, often required in order to get a discharge permit. As a result, the trend has

been to apply these models to outfalls that push the limits of their applicability. Many users of the models, although highly qualified in their own field, may have little knowledge of how the models work and their limitations.

The object of this paper is to give the background and development of several of the computer models used by the EPA, to discuss the limits of their applicability, and to see how they might possibly be used on selected test cases. On cases where none of the models apply, a discussion is presented as to the alternatives one has in obtaining required predicted dilutions. On some test cases where field or laboratory data are available, the predictions of appropriate models are compared to the data.

The opinions expressed in this paper are of those of the author. They do not reflect procedure or policy of the EPA except when in agreement with published EPA documents.

SUBMERGED PLUME MODELS

The submerged plume models presently used by the EPA are UPLUME, UDKHDEN, UMERGE, ULINE, and UOUTPLM and are described in [3]. They are updated versions of the models found in [2] with the exception of ULINE which was added to the later work. The "U" designation was added to the models when their input was modified so they would all accept the same "Universal input file". None of the models consider boundary effects such as the bottom, shore, or surface. The are all programmed to stop when the plume reaches the surface or trapping level. The trapping level is taken as that level where the plume density reaches the ambient density in a stratified environment. Each will be discussed in detail below.

UPLUME

This model considers a single port discharge issuing at an arbitrary angle into a stagnant, stratified ambient and was developed by Baumgartner, etal[4][5]. The plume is divided into two regions. The zone of flow establishment near the discharge is approximated as a function of the discharge densimetric Froude number using empirical relations and is less than 6 port diameters in length.

In the zone of established flow, it is assumed that flow is steady and incompressible, that the plume is axisymmetric, and that turbulent diffusion can be expressed as a function of an entrainment coefficient. The velocity and concentration profiles within the plume are assumed to be Gaussian in shape. The entrainment of ambient fluid is assumed to be a function of plume size, velocity and entrainment coefficient. With these assumptions, the governing partial differential equations are integrated to yield a system of ordinary differential equations which are solved in steps using a Runge-Kutta integration scheme. The results are the trajectory of the plume centerline, the plume width, centerline velocity and concentration. The concentration is inverted to give the centerline dilution. The average dilution is calculated from the assumed Gaussian profile to be 1.77 times the centerline dilution.

Since the model only considers a single plume it should not be used with diffusers unless the ports are so far apart the plumes have no chance of merging. If used with merging plumes, it will over predict the actual dilution. Since it does not consider current, it is best used to give worst case conditions. In most cases, zero currents are unreasonable and a lowest 10% current is taken as the limiting value. For these cases UPLUME will under predict the desired dilution since current has a pronounced effect on dilution.

UDKHDEN

This is a fully three-dimensional model which can be used for either single or multiple port diffusers with a row of equally spaced ports on one side of the outfall pipe discharging into a moving stratified ambient. It is based on the technical developments of Hirst[6], Davis[7], and Kannberg and Davis[8][9]. It considers variable profiles through the zone of flow establishment and through the merging zone of multiple plumes. Detailed development of the plume is considered through the zone of flow establishment rather than by approximating it in a single step as do most other models. In addition, the gradual changing geometric form of merging multiple plumes is simulated using the method of images instead of a sharply transitioning from multiple, round plumes into a two dimensional slot plume as in some models. When adjacent plumes begin to overlap, the plumes are no longer considered axisymmetric. The distribution of plume properties are superimposed as shown on Figure 1. The entrainment function is modified to account for the interaction of plumes and the reduction in the entrainment surface as the merging process proceeds.

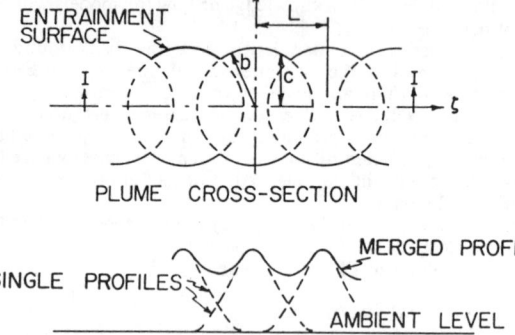

Fig. 1 Superposition method used in UDKHDEN model to simulate merging plumes

Since the model is 3-dimensional, both horizontal and vertical discharge angles are considered. For diffusers, discharge can be plus or minus 45 degrees from the current direction. Vertical angles are allowed from -5 degrees to 135 degrees. For single port discharges, horizontal discharge angles plus or minus 135 degrees from the current direction are allowed.

As a result of being able to consider various horizontal discharge angles and the use of the method of images, this model can also approximate the discharge of a single jet into a shallow moving ambient. Under these conditions, density differences may often be neglected. The image solution can be used to simulate the surface and bottom with the plume in the center by interpreting the port spacing as the water depth and setting the depth to a large number (simulating the lateral extent of the ambient).

The model assumes 3/2 power law approximations to Gaussian profiles, steady flow in the mean, incompressible fluids with density variations included only in the buoyant terms, hydrostatic pressure variations, and no end effects with diffusers. Ambient turbulence is included only in the entrainment function. Entrainment is a function of plume size and shape, centerline excess velocity relative to the current, forced entrainment due the current and buoyancy. The entrainment coefficients within the entrainment function have been tuned to experimental data.[8]

Using the assumed velocity, temperature, and concentration profiles, the governing partial differential equations are integrated to yield a system of ordinary differential equations. If temperature and salinity is input instead of density, densities are calculated using empirical equations obtained from the U.S. Navy Hydrographic Office. The resulting equations are integrated in steps using Hamming's Predictor-Corrector method. Checks on accuracy are made at each integration step. If property changes are greater than a preset value, the integration step is reduced and the calculation is repeated. In some cases, a satisfactory solution cannot be reached. For these cases, the program stops rather than give an answer it is unsure of. This often happens with discharge towards the current, vertical discharges with very high currents, or with discharge Froude numbers approaching unity.

The model follows one single plume, calculating its trajectory, size, velocity, and concentration. If it merges with neighboring plumes, its shape and entrainment are adjusted according to the amount of superposition.

UMERGE

Program UMERGE is a two-dimensional multiple port version of UOUTPLM with minor modifications to the entrainment. They are based on the developments of Winiarski and Frick[10] and Frick[11]. As a result, only UMERGE will be discussed. It considers a row of equally spaced ports on one side of a diffuser pipe with the current direction normal to the diffuser pipe.

The model uses a Lagrangian approach, following a slice of the plume as time proceeds. Entrainment changes the mass of the slice. Buoyancy and entrained fluid change the momentum of the slice. Combining the mass and momentum of the slice yields its new velocity. Stretching of the slice is calculated from the difference in the upstream and downstream faces of the slice. From the mass and thickness of the slice the width is determined. Entrainment is the combination of that forced into the plume by the current and that entrained by fluid shear at the plume-ambient interface. From its velocity and the integration time step, the new slice location if determined. Output includes the plume trajectory, plume size and average dilution.

The model assumes top-hat profiles (average values only), pressures are hydrostatic, and that merging between adjacent plumes does not change the average properties of the plume element but does effect its radius (merging is purely geometric.) Since the model is only two-dimensional, no variations in horizontal discharge angle are allowed. Only discharge in-line with the current is considered although vertical angles from -5 to 90 degrees are allowed.

Because of the simple integration scheme, UMERGE will continue to run with cases where UDKHDEN my stop. The results of these calculations should be used with caution, however.

ULINE

The model ULINE is based on Roberts[12] uniform density flow experiments for slot discharges at various angles relative to the current and is a generalization of Roberts[13] model on dilution in an arbitrarily stratified environment. The model is semi-empirical in that it uses the dilution rate measured in [12] and applies it to the more general case. The average density of the plume is calculated from the entrained fluid. When the average density of the plume reaches the ambient density, trapping is assumed to occur. The output is simply the trapping level and the dilution at that point. If the surface is reached, the dilution at that point is printed out. The advantages of this model are its simplicity and that it will allow diffuser discharge angles plus or minus 90 degrees from the current direction.

SURFACE MODELS

Many discharges are such that the plume floats on the surface of the receiving body of water. An example of this is side channel flow of an buoyant effluent into a river. This type of discharge has several computational difficulties that submerged discharges do not have. The plume is not axisymmetric which requires additional assumptions about its shape. Energy is not conserved due to heat transfer at the surface. Buoyancy causes the central portion of the plume to rise more than the edges. This creates buoyant spreading as well as different entrainment rates in the vertical and horizontal directions. Figure 2 is the cross section of a typical floating surface plume showing the different regions considered. In some cases, the plume may be attached to the bottom, eliminating vertical entrainment all together. In such cases vertical variations in temperature and velocity are often ignored and

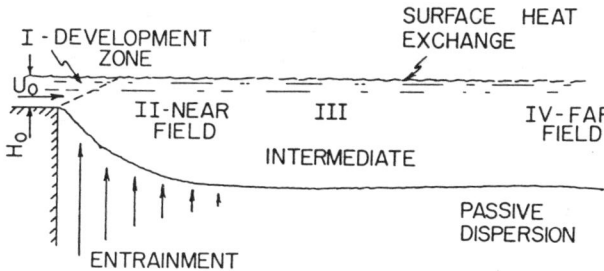

Fig. 2 Cross section of typical surface plume showing different mixing zones

two-dimensional models are constructed.

The model used by the EPA in "Workbook of Thermal Plume Predictions: Surface Discharge"[1] is the PDS model. Other models that have occasionally been used are MOBAN and PSY.

PDS

This model is based on the original work of Prych[14]. It was modified by Shirazi and Davis[1] to improve its spreading characteristics and tuned using a compilation of experimental work found in the literature.

The PDS model is an integral model were Gaussian velocity and concentration profiles are assumed in the vertical and horizontal direction. In the vertical direction, only half of the profile is considered. Two different entrainment models are used. In the vertical direction, entrainment is a function of turbulent shear at the boundary and is reduced as spreading occurs. The horizontal entrainment, however, is increased as spreading forces the plume into the ambient. Vertical and horizontal ambient turbulent entrainment is added to each using eddy diffusion coefficients.

The momentum equation includes both dynamic drag from cross currents and fluid friction at the plume ambient interface. For closure of the equations, the spreading rate is assumed to be a function of the velocity of a local celerity wave as determined by the plume excess density and the plume aspect ratio.

The equations are integrated using the assumed profiles and solved step by step with a Runge-Kutta routine. The zone of flow establishment is approximated using an empirically determined development length. Discharge can be at any horizontal angle relative to the current but the current must be uniform and unstratified. Bottom and shore effects are not allowed, however, a version of PDS exists that uses symmetry to simulate discharge parallel to the current near one shore. The standard PDS can be used for this case by doubling the actual discharge rate and width and only using one half of the predicted plume.

The output from PDS gives the plume centerline trajectory, width, depth, centerline concentrations (temperature) at the surface, and areas within surface concentration contours.

MOBAN

This model was developed under contract by the EPA by Motz and Benedict[15] and considers a two-dimensional jet discharge at the surface. The model is limited to the jet region of the plume since passive diffusion is not considered. The plume is assumed to be vertically mixed and buoyancy effects are ignored. As a result, it is best suited for discharge into shallow rivers where the plume fills the water column but it can also be used with floating jets if the thickness does not vary significantly. Velocity and temperature profiles in the lateral direction are assumed to be Gaussian. Discharge angles relative to the current are variable from 0 to 90 degrees. The discharge velocity is assumed to be at least a factor of four greater than the ambient. Drag in the momentum equation is assumed to be solid body drag with a required drag coefficient. The zone of flow establishment is approximated using empirical expressions for distance and flow direction as functions of initial discharge angle and discharge to ambient velocity ratio.

The differential equations are integrated using the assumed profiles yielding a system of ordinary differential equations. For closure, a drag coefficient is introduced that is a function of the velocity disparity between the jet's centerline and the component of the ambient velocity along the plume axis. Both the entrainment coefficient and drag coefficient are input variables. The resulting equations are integrated step by step using a Runge-Kutta routine. The results yield the jet trajectory, width, velocity, and temperature distribution.

PSY

This model is a vertically mixed, shore attached model that is occasionally used for discharge into shallow rivers where the ambient current is high relative to the discharge velocity. It is based on the theoretical development of Paily and Sayre[16].

The model assumes negligible discharge momentum but includes the effects of both shores and bottom of a river. The standard two-dimensional steady state convection-diffusion equation is transformed into a simple one-dimensional diffusion equation using a space-cumulative discharge coordinate system and a transverse diffusion factor. Its solution is expressed in terms a convolution integral and a Gaussian probability distribution function. When reflections from the banks are taken into account, the maximum shoreline temperature (concentration) from a concentrated source is expressed as:

$$\frac{C}{C_0} = \frac{Q_0}{Q_r} \sqrt{\frac{2}{\pi}} \frac{1}{\sigma_p} \left[1 + 2 \sum_{k=1}^{\infty} \exp\left(-2 \frac{k^2}{\sigma_p^2}\right) \right] \qquad (1)$$

where Q_0 = the discharge rate, Q_r = the river flow rate, $\sigma_p = (Dx)^{1/2}$, D is a transverse diffusion factor, and x is the distance downstream from the source. The factor, D, is a function of Manning's coefficient and the river geometry. The expression used by the authors is:

$$D = \frac{a \, n \, \sqrt{g}}{B^2} \, d^{5/6} \qquad (2)$$

where n is Manning's coefficient, g is the gravitational constant, B is the river width, d is the average river depth and "a" is a dimensionless transverse mixing coefficient. Selecting the appropriate value of "a" is the most difficult part of applying this model. The expression suggested by the authors is:

$$a = 0.4 \left[\frac{B}{R_c}\right]^2 \frac{U^2}{Sgd} \qquad (3)$$

where R_c is the river curvature and S is the slope of the energy gradient. They suggest that "a" can have values ranging from 0.1 to 10. Figure 3 shows the solution of equation (1) for three different initial dilution factors, P. The variable P is defined as the fraction of total river discharge that has been mixed with the effluent at the origin. A value of P=0 is for no premixing.

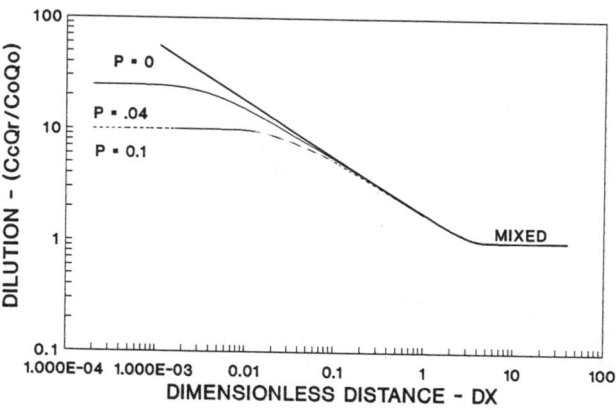

Fig. 3 Paily-Sayer shore attached plume model, dimensionless dilution vs. distance for three values of initial dilution factor, P

MODEL APPLICATIONS

Submerged Discharges

In order to demonstrate the use and limits of the models mentioned above, several different cases will be considered. The first is horizontal, submerged discharge from a diffuser whose ports are spaced 10 discharge diameters apart into a stagnant, unstratified ambient. The densities, port diameter, and discharge velocity are such to give a discharge densimetric Froude number of 11. The Froude number is defined as:

$$F_r = \frac{U_0}{\sqrt{gD\Delta\rho/\rho_0}} \qquad (4)$$

where U_0 is the discharge velocity, g is gravity, D is the port diameter, $\Delta\rho$ is the density difference between the discharge and ambient, and ρ_0 is the discharge density.

Both UMERGE and UDKHDEN apply. Since the current is zero, UPLUME also applies as long as the plumes don't merge. Figures 4, and 5 show the trajectory and average plume concentration as a function of dimensionless distance downstream, X/D as predicted by these three models when compared to data measured by Kannberg[17]. It can be seen that all three models agree very well in both trajectory and concentration up to a distance of about 40 diameters. From that point on, UPLUME starts to over predict dilution as would be expected

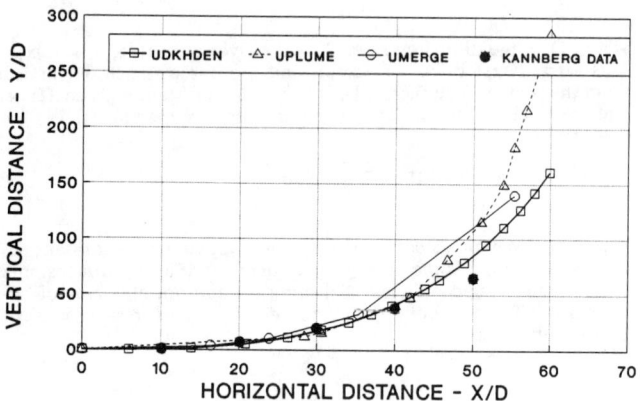

Fig. 4. Trajectory for diffuser with port spacing, L/D = 10, Froude number, F_r = 11, horizontal discharge, and no current. Model predictions and Kannberg[17] data

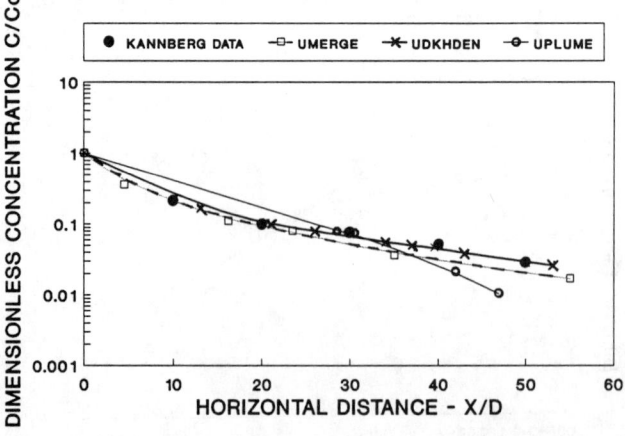

Fig. 5 Plume centerline concentrations for diffuser with port spacing, L/D = 10, Froude number, F_r = 11, horizontal discharge, and no current. Model predictions and Kannberg[17] data

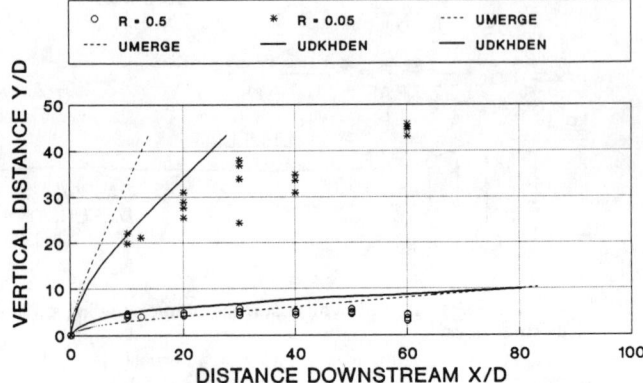

Fig. 6 Trajectory for diffuser with port spacing, L/D = 5, Froude number, F_r = 11, vertical discharge, and current to discharge velocity ratios of 0.5 and 0.05. Model predictions and Kannberg[17] data

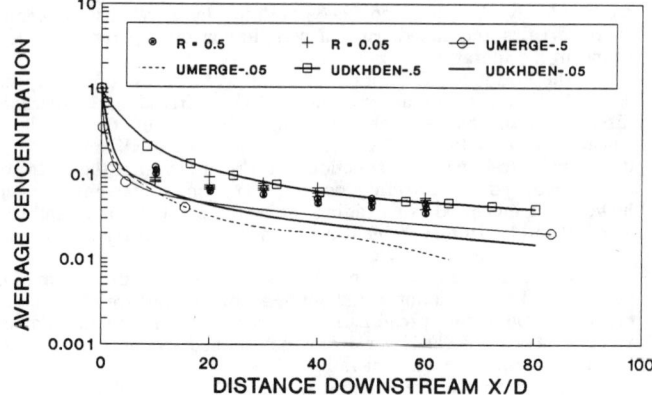

Fig. 7 Centerline concentrations for diffuser with port spacing, L/D = 5, Froude number, F_r = 11, vertical discharge, and current to discharge velocity ratios of 0.5 and 0.05. Model predictions and Kannberg[17] data

when merging starts.

It appears that all three models can be used with confidence for discharge into stagnant ambients. UDKHDEN is slightly better in both trajectory and concentration. If plumes merge, UPLUME should not be used. For this case there appears to be little difference between UMERGE and UDKHDEN.

The next case is vertical discharge from a deep diffuser with ports spaced 5 diameters apart into a flowing, unstratified ambient. Two different ambient velocities are considered. One with an ambient velocity that is 0.5 the discharge velocity, and the other with an ambient velocity that is 0.05 that of the discharge.

Since UPLUME does not consider current, only UMERGE, and UDKHDEN apply. Figures 6 and 7 show the predictions of these models for this case as compared to the data of Kannberg[17]. It can be seen that both models agree reasonably well with the data for trajectory for the higher current case but UDKHDEN seems to do a better job at the lower current case. UDKHDEN tends to under predict the dilution (higher concentrations) for the higher current while UMERGE tends to over predict it by about the same amount. For the lower current case, both models tend to over predict the dilution with UDKHDEN being slightly better.

These runs indicate that both models reasonably predict multiple port discharge into a flowing ambient with UDKHDEN having a slight edge.

One of the more difficult cases to simulate is that of submerged discharge into shallow water with a portion of the plume submerged and a portion of it on the surface. The submerged models do not consider the interaction with the surface and the surface models do not give the high dilution that would occur while the plume is submerged. Attempts to patch two different models have been made by various users with the output from the submerged model used in some way as the input to the surface model.

The problem with this approach is that the transition zone includes very complicated 3-dimensional hydraulics that cannot be simply put aside. In addition, surface models account for a development zone that would not exist if used in conjunction with a submerged plume model. As a result, the predictions are only as good as the users guess on starting conditions for the surface model. One conservative approach to this problem for a single discharge is to ignore the submerged plume and assume a discharge at the surface. This would give an estimate on the lower limit of the dilution. If more accurate answers are needed, physical models may be the answer. Numerical models that divide the system into a finite number of grids or elements have the potential to predict these cases but they require large computers and are difficult to run. Another approach is to use the approximate analyses presented in references(18)(19)(20).

For submerged discharges from a single port where the plume fills the water column, an alternate approach can be made. In such cases, buoyancy can usually be ignored. Both UDKHDEN and UMERGE follow one single plume and account for the blockage that would occur from neighboring plumes if plumes merge. This blocking effect using the method of images can also account for the surface and bottom if the port spacing is interpreted as the water column depth and

the depth is set to a large number, representing the lateral extent of the receiving water as shown on Figure 8.

This method was used to predict the dilution from a single port 18" ocean outfall at Newport, Oregon. The results were compared to a dye study conducted by CH2M-HILL(21). The discharge was 0.175 m³/s of treated sewage into a water column 3.04 m deep. The discharge direction was horizontal and at 90 degrees to a 0.447 m/s current. UDKHDEN was used since it was the only model that would consider horizontal discharge directions other than in line with the current. Multiple ports were simulated with the discharge rate adjusted so each port had 0.175 m³/s and the spacing was set to 3.04 m to simulate the effects of the surface and bottom.

The results are shown on Figure 9. It can be seen that the model does quite well. It under protects the dilution close to the source and slightly over predicts it far downstream. This was partially due to the fact that the discharge was very close to the surf zone causing high mixing near the discharge. Far downstream, coastline effects on dilution begin to appear.

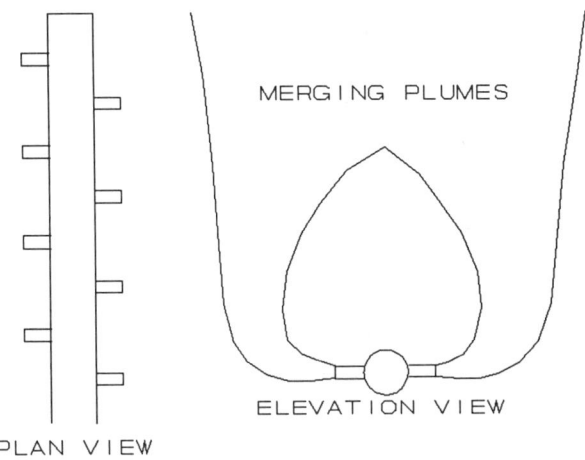

Fig. 10 Sketch of typical diffuser with alternating ports on opposite sides of the diffuser. Also showing merging of plumes from each side

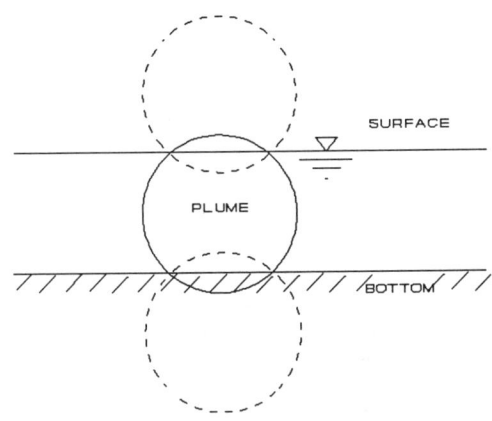

Fig. 8 Image method of simulating shallow discharge in UDKHDEN model

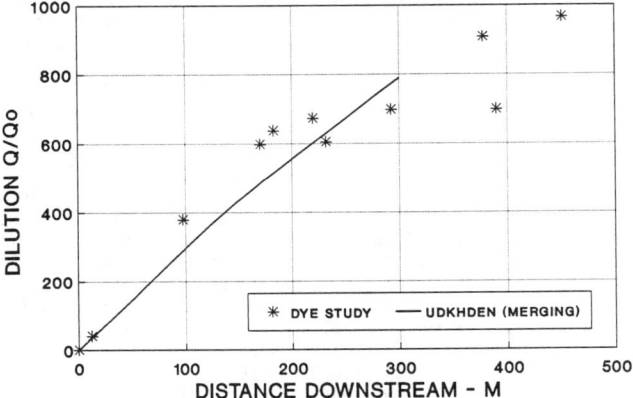

Fig. 9 Dilutions vs distance along plume centerline for single 18 in. outfall normal to current at Newport, Oregon; dye study(21) values and UDKHDEN image solution predictions.

Many diffusers have ports on both sides of the outfall pipe and may be staggered or apposing as shown on Figure 10. Since all of the submerged EPA plume models assume discharge is in the same direction, none of them adequately predict these diffusers. They can be used, at best, to give limits on dilution. The highest dilutions would be when the interaction between ports on opposite sides is ignored and each side is considered independent of the other. The lowest dilution would be when all ports are assumed to be on the same side. The actual dilution would most likely be between these two extremes. The difference between the predictions would decrease as the spacing of the ports increases.

The Encina outfall north of San Diego, California is such a diffuser. It is a straight diffuser located about 2.25 km off shore in 46 meters of water. The actual diffuser is 244 meters long and consists of 88 ports discharging downward at -5.0 degrees from the horizontal on alternating sides of the diffuser. The ports are spaced 3.65 m apart and range in diameter from 6.36 cm at the coastal end to 7.62 cm at the outer end.

Hendricks(22) reports of a in-situ study of this diffuser and the results of the EPA computer models, UPLUME, UMERGE, ULINE, and UDKHDEN. The ambient was stratified and had slow longshore currents that ranged from 9.4 cm/s in the upper layers to 4.8 cm/s near the bottom. Figure 11 presents the results found in this report for three different stations near the center of the plume at the point where the plume was trapped by stratification. The scatter in measurements and predictions is obvious. The models were run, assuming discharge from only one side of the outfall. Since UPLUME does not include the additional dilution caused by the current, it under predicts the dilution at all three stations. ULINE does slightly better than UPLUME, but it also under predicts the dilution. UMERGE and UDKHDEN do reasonably well at station three but over predict the dilution at the

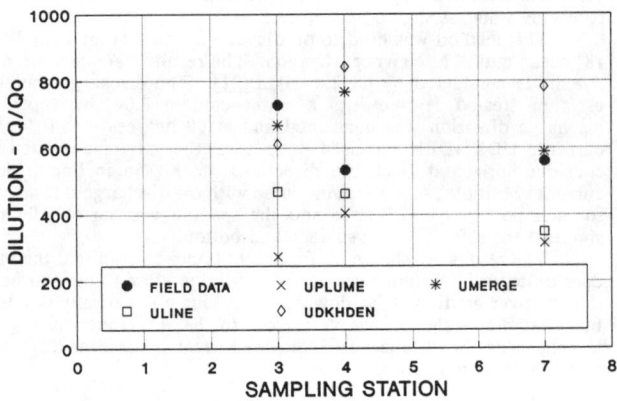

Fig. 11 Dilution results of Encina dye study compared to model predictions(22)

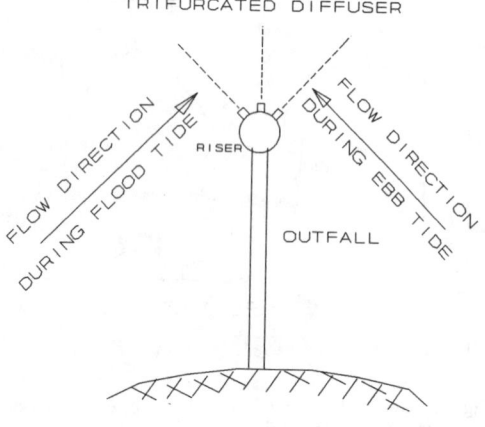

Fig. 12 Plan view sketch of Point Woronzof outfall showing riser and its three discharge ports relative to tidal currents

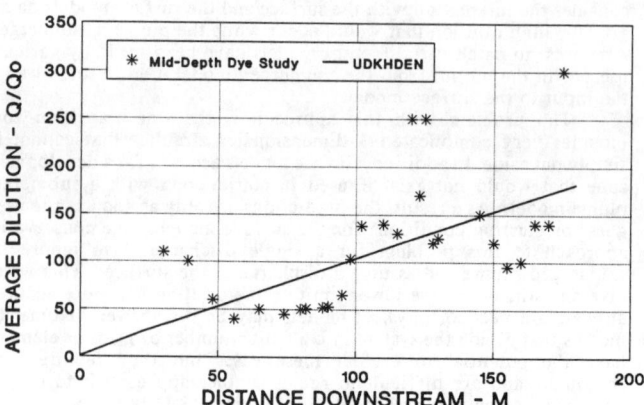

Fig. 13 Dilution vs. distance for Point Woronzof dye study(23) compared to UDKHDEN predictions, depth = 4.6 m, current = 0.4 m/s, discharge = 1.1 m^3/s

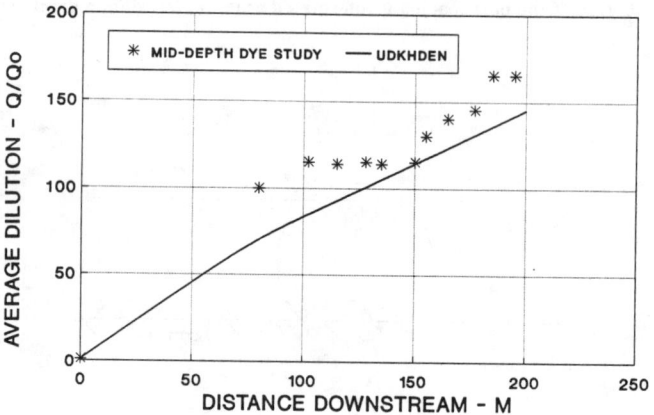

Fig. 14 Dilution vs. distance for Point Woronzof dye study(23) compared to UDKHDEN predictions, depth = 5.28 m, current = 1.28 m/s, discharge = 1.1 m^3/s

other two stations as would be expected with only one side of the diffuser considered by the models.

Another type of diffuser is one that has several nozzles discharging from one single riser such as the Point Woronzof outfall in Alaska. This outfall consists of one riser with three 24 inch ports spaced at 45 degrees to each other as shown on Figure 12. Tidal currents are such that one port is in line with the current, one is at 45 degrees, and the other is at 90 degrees to the current. Depths at the discharge are of the order of 5.0 m. Currents range from 0 to nearly 1.5 m/s. A preliminary analysis indicated that the plumes fill the water column within a few meters of the outfall.

None of the models consider different discharge directions from a single diffuser. One approximation would be to consider each jet independently of each other and to average the results. The image method would have to be used to simulate the plume filling the water column. This was done using UDKHDEN and compared to dye study results obtained by CH2M-HILL(23) for two different tidal conditions at the Point Woronzof diffuser site. The first was for transect 18 with a depth of 4.6 m and a current of 0.4 m/s. The second was for transect 28 with a depth of 5.28 m and a current of 1.28 m/s. The total discharge for both cases was 1.1 m^3/s. The results are shown on Figures 13 and 14. The predictions agree reasonably well with the measurements, falling in the middle of the scattered data for the low current case and slight below the data for the high current case.

One of the more difficult cases to predict using the EPA models is a multiple port diffuser with risers that have multiple "shower head" discharges such as found at on the San Francisco ocean outfall and the North Head, Bondi, and Malabar ocean outfall diffusers near Sydney,

Australia. Figure 15 is a typical riser on one of these diffusers. The heads on the risers are nearly flush with the ocean bottom and have from six to ten discharge ports discharging radially out from the riser. The San Francisco diffuser has 85 risers spaced 36 feet apart on a dog-leg section of the outfall. Each head has eight ports discharging in different directions.

It is obvious that none of the EPA models describe this type of discharge. The best that can be done is to estimate limits on dilution. The worst case would be to ignore the multiple port shower head and assume a multiple port in line diffuser with vertical discharge. The highest dilutions would be obtained by following only one of the ports out of each riser. Another method that would give predictions between the previous two would be to assume the discharges out of each riser are in line on one side of the riser, spaced at some appropriate spacing. This should be closer to the actual but the accuracy would depend on the users ability to select the correct equivalent spacing.

Other diffuser designs that are difficult to predict are where the diffuser pipe is in the shape of a "Y" or "L" or when the ambient current is along the diffuser rather than across. UDKHDEN has the ability to predict dilutions when the current is not at right angles to the diffuser pipe but when the deviation is too great, upstream plumes shield downstream plumes from the ambient current. This happens even when the plumes have not merged. As a result, the horizontal discharge direction in UDKHDEN has been limited to plus or minus 45 degrees from the current direction. This limitation does not apply to single port discharges. ULINE will handle currents parallel to the diffuser pipe but the predictions are quite approximate.

Surface Discharges

The PDS and MOBAN models were compared to laboratory data reported by Stefan etal(24) for discharge at the surface into a stagnant tank. The discharge was from a rectangular channel with a width to

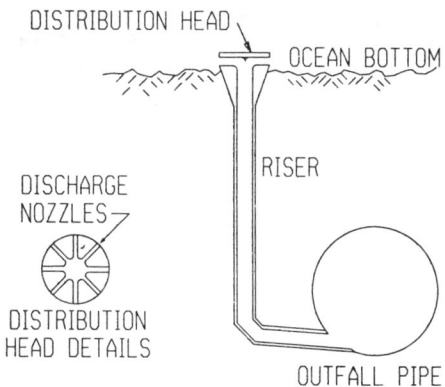

Fig. 15 Typical "shower head" riser and diffuser

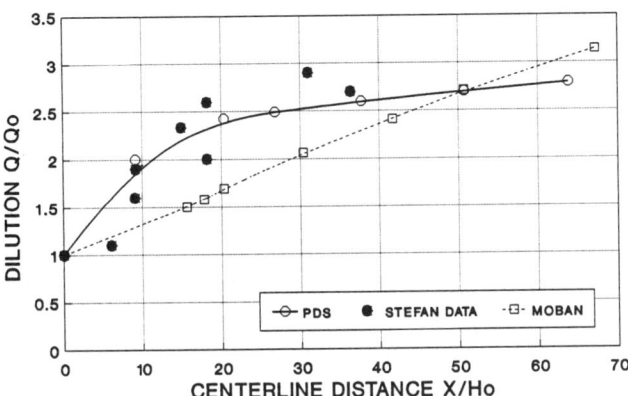

Fig. 16 Comparison of PDS and MOBAN predictions to laboratory data(24)

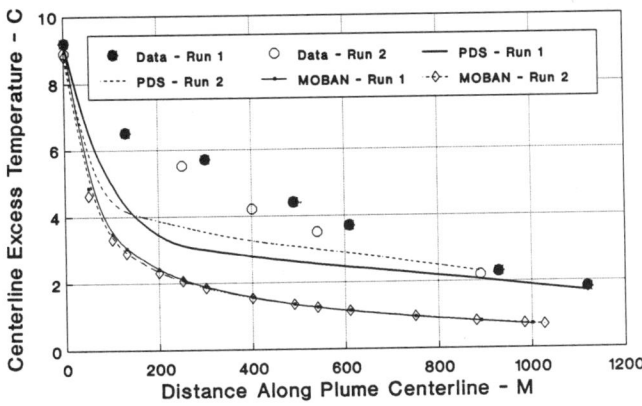

Fig. 17 Point Beach power plant data(25) compared to PDS and MOBAN predictions

depth aspect ratio of 38. The discharge Froude number based on discharge depth was 1.43. The results are shown on Figure 16 which is a plot of average dilution vs. dimensionless distance along the plume centerline. It can be seen that PDS agrees quite well with the data as far as the data goes. The MOBAN model under predicts dilution in the

near field and then gives higher dilutions further out.

These two models were also compared to the field data taken at the Point Beach power plant discharge on Lake Michigan reported by Frigo and Frye(25). Data taken on two different days were used. The first was on May 20, 1971. Only the south unit of the plant was in operation at that time. The discharge was 21.83 m^3/s through a 10.67 m wide by 4.11 m deep channel. The discharge and ambient temperatures were 17.5 and 8.4 C respectively. The current was from the north at about .09 m/s giving a 60 degree discharge angle.

The second case was on July 20, 1971. For this case, flow was from the south. The discharge and ambient temperatures were 21.5 and 12.7 C respectively. The results are shown on Figure 17 as dimensionless centerline excess temperature vs. distance along the plume centerline.

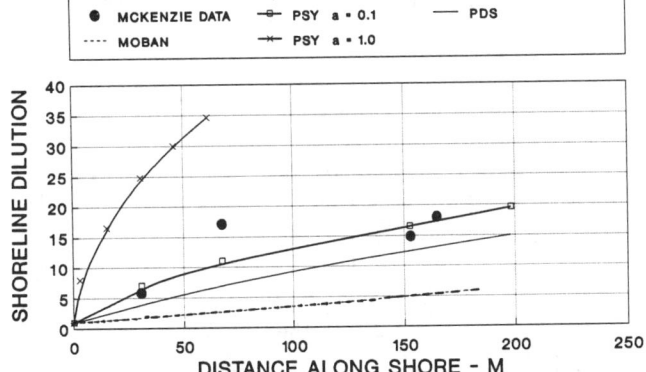

Fig. 18 McKenzie river data(26) compared to PDS, MOBAN and PSY predictions

It can be seen that both models over predict the dilution in the near field. PDS agrees quite well in the far field but MOBAN is still considerably below the data 1000 m from the discharge.

The last case considered is the discharge of 0.68 m^3/s from a paper mill into the McKenzie river near Eugene, Oregon. The average river velocity in the area of interest is about 2.5 m/s. The average depth is 1.52 m. The results of a dye study conducted downstream of the discharge were supplied by Thut(26).

Figure 18 shows the shoreline dilution as a function of distance downstream as determined from the dye study, from PDS, from MOBAN and from PSY with "a" values of 0.1 and 1.0. It can be seen that the predictions of PSY are very sensitive to the transverse diffusion coefficient. In this case a value of 0.1 seems to be good, but a wrong value can give very wrong answers. It is surprising that PDS does as well as it does since it assumes no shore attachment. MOBAN does not do too well in this case.

It appears that PSY is good for fast flowing rivers if the appropriate diffusion coefficient can be selected. PDS is the next best, especially since no input on diffusion coefficient is required.

CONCLUSIONS

The EPA near field computer models appear to be able to predict dilutions with reasonable accuracy as the long outfall being considered agrees with the basic assumptions of the models. Unfortunately, many discharges violate one or more of the assumptions within the models. In these cases, the models can be used to give limits on dilution.

Of the submerged discharge models, UPLUME is the most restrictive, being limited to single port discharge in a stagnant ambient. UDKHDEN is the most versatile. There is no advantage using UDKHDEN or UMERGE for the cases where they both apply but UDKHDEN applies to more cases. The image method of simulating a single port discharge into shallow water gives good results. The only reason for using ULINE is to consider discharges with currents parallel to the diffuser pipe.

For submerged discharges, PDS predicts reasonable values as long as the plume is not attached to the bottom. MOBAN gives the poorest results for the cases considered. PSY can give good predictions in rivers with shore attached plumes but its accuracy depends on the transverse diffusion coefficient that must be supplied by the user.

ACKNOWLEDGMENTS

The author wishes to thank CH2M-HILL, Weyerhaueser Company, and the Souther California Coastal Water Research Project for supplying data on unpublished dye studies.

REFERENCES

1. Shirazi, M.A., and Davis, L.R., "Workbook of Thermal Plume Prediction Volume 2: Surface Discharge," EPA-R2-72-005b, May 1974, NTIS No. PB 235841.

2. Teeter, A.M., and Baumgartner, D.J., "Prediction of Initial Mixing for Municipal Ocean Dischages," EPA CERL-043, Corvallis Environmental Research Laboratory, Corvallis, Oregon, May 1979.

3. Mullenhoff, W.P., Soldat, A.M., Baumgartner, M.D., Davis, L.R., Schuldt, M.D., and Frick, W.E., "Initial Mixing Characteristics of Municipal Ocean Discharges: Vol. I, Procedures and Applications, and Vol. II Computer Programs," US EPA Reports EPA-600/3/85-073a and EPA-600/3/85-073b respectively, November 1985.

4. Baumgartner, D. J., and Trent, D.S., "Ocean Outfall Design: Part I. Literature Review and Theoretical Development," U.S. Department of Interior, Federal Water Quality Administration (NTIS No. PB-203-749), 1970.

5. Baumgartner, D.J., Trent, D.S., and Byram, K.V., "User's Guide and Documentation for Out Fall Plume Model," Working Paper No. 80, U.S. Environmental Protection Agency, Pacific Northwest Water Lab., (NTIS No. PB-204-577/BA), 1971.

6. Hirst, E.A., "Analysis of Buoyant Jets within the Zone of Flow Establishment," Oak Ridge National Laboratory, ORNL-TM-3470 and "Analysis of Round, Turbulent, Buoyant Jets Discharging to Flow Stratified Ambients," ORNL-TM-4586, Oak Ridge, TN, 1971.

7. Davis, L.R., "Analysis of Multiple Cell Mechanical Draft Cooling Towers," EPA-660/3-75-039, U.S. Environmental Protection Agency, Corvallis, Oregon, June 1975.

8. Kannberg, L.D., and Davis, L.R., "An Experimental/Analytical Investigation of Deep Submerged Multiple Buoyant Jets," EPA-600/3-76-101, U.S. Environmental Protection Agency, Corvallis, Or, 1976

9. Kannberg, L.D., and Davis, L.R., "An Analysis of Deep Submerged Multiple-Port Buoyant Discharges," ASME Journal of Heat Transfer, Vol. 99, Series C, No. 4, pp 648-656. Nov. 1977

10. Winiarski, L.D., and Frick, W.E., "Cooling Tower Plume Model," EPA-600/3-76-100, U.S. Environmental Protection Agency, Corvallis, OR, 1976

11. Frick, W.E., "A Theory and User's Guide for the Plume Model MERGE," Technical Memorandum, Tetra Tech, Inc., Corvallis, OR, 1981.

12. Roberts, P.J.W., "Dispersion of Buoyant Waste Discharge from Outfall Diffusers of Finite Length," W.M. Keck Lab. of Hydraulic and Water Resources, Report No. KH-R-35, California Institute of Technology, 1977.

13. Roberts, P.J.W. "A Mathematical Model of Initial Dilution for Deep water Ocean Outfalls," Proceedings of a Specialty Conference on Conservation and Utilization of Water and Energy Resources, San Francisco, CA, American Society of Civil Engineers, Aug, 8-11, 1979.

14. Prych, E.A., "A Warm Water Effluent Analyzed as a Buoyant Surface Jet," Svergis Meterologiska Och Hydrologiska Institut, Serie Hydrologi. Nr 21, Stockholm, 1972.

15. Motz, L.H., and Benedict, B.A., "Heated Surface Jet Discharged into a Flowing Ambient Stream," Department of Environmental and Water Resources Engineering, Vanderbilt University, Nashville, Tenn, Report, NO. 4, August, 1970.

16. Paily, P.P., and Sayre, W. W., "Model for Shore-Attached Thermal Plumes in Rivers," ASCE Journal of the Hydraulics Division, Vol 104, HY5, May 1978, pp 709-723.

17. Kannberg, L.D., "An Experimental/Analytical Investigation of Deep Submerged Multiple Buoyant Jets," Ph.D. thesis, Oregon State University, Corvallis, Oregon, 1977

18. Adams, E.E., "Submerged Multi-port Diffusers in Shallow Water with Current," M.S. thesis, Massachusetts Institute of Technology, Cambridge, Mass., 1972.

19. Jirka, G., and Harleman, D.R.F., "The Mechanics of Submerged Multiport Diffusers for Buoyant Discharges in Shallow Water," Technical Report No. 169, R.M. Parsons Laboratory for Water Resources and Hydrodynamics, Massachusetts Institute of Technology, Cambridge, Mass., 1973.

20. Lee, J.H., Jirka, G.H., and Harleman, D.R.F., "Modeling of Unidirectional Thermal Diffusers in Shallow Water," Technical Report No. 228, R.M. Parsons Laboratory for Water Resources and Hydrodynamics, Massachusetts Institute of Technology, Cambridge, Mass., 1977.

21. Morris, T., Personal communication, CH2M-Hill, Corvallis, Oregon, January 1989.

22. Hendricks, T. J., "In-Situ Measurement of Initial Dilution at an Ocean Outfall," EPA grant report, Southern California Coastal Water Research Project, Long Beach, California, March 1987.

23. Linquist, R., Personal communication, CH2M-HILL, Corvallis, Oregon, July 1988.

24. Stefan, H., Hayakawa, N., and Shiebe, F.R., "Surface Discharge of Heated Water," U.S. EPA Water Pollution Control Research Series, 16130 FSU 12/71, Dec. 1971.

25. Frigo, A. A., and Frye, D. E., "Physical Measurements of Thermal Discharges into Lake Michigan: 1971," Argonne National Laboratory Center for Environmental Studies, report No. ANL/ES-16, October 1972.

26. Thut, R.H., Personal communication, Weyerhaueser Company, Tacoma, Washington, Nov., 1979.

THE EFFECTS OF STEP CHANGES IN THE THERMAL BOUNDARY CONDITION ON HEAT TRANSFER IN THE INCOMPRESSIBLE FLAT PLATE TURBULENT BOUNDARY LAYER

R. P. Taylor, P. H. Love, H. W. Coleman,
and M. H. Hosni
Mechanical and Nuclear Engineering Department
Mississippi State University
Mississippi State, Mississippi

ABSTRACT

Experimental Stanton number data are presented for incompressible turbulent boundary layer flow of air over a smooth flat plate with unheated starting length followed by a heated region with either a constant wall temperature or a constant wall heat flux. For the cases with constant wall temperatures these data extend the available experimental data for this class of problem from the previous maximum x-Reynolds number of 3,500,000 to 10,000,000. For the step heat flux boundary condition, no experimental results exist in the available literature, to the authors' knowledge. The data are compared with the results of the classical solutions of the integral boundary layer equations and with finite difference solutions of the partial differential equations of the boundary layer.

NOMENCLATURE

Roman

A	area of one test plate
C_f	skin friction coefficient = $2\tau_w/u_\infty^2$
c_p	specific heat at constant pressure
c_{pa}	specific heat of dry air
c_{pw}	specific heat of water vapor
g	kernel function; equation (5)
ℓ	Prandtl mixing length
P_{bar}	barometric pressure
Pr_t	turbulent Prandtl number
q_c	conduction heat loss
q_r	radiation heat loss
q''_w	heat flux at the wall
R	recovery factor
Re_x	Reynolds number = $u_\infty x/\nu$
St	local Stanton number based on total temperature
St_t	Stanton number for constant wall temperature
t	temperature
$\overline{t'v'}$	turbulent heat flux factor
T	total temperature = $t + u^2/2c_p$
t_{rail}	temperature of the side rail
t_{wb}	wet bulb temperature
u	x-component of time mean velocity
U	uncertainty in St (95% confidence level)
$\overline{u'v'}$	Reynolds shear stress factor
$(UA)_{eff}$	conductance; $q_c = (UA)_{eff}(t_w - t_{rail})$
v	y-component of time mean velocity
W	electric power to each plate
x	stream-wise direction
y	normal direction
y^+	nondimensional y; $y^+ = y\, u_\infty\sqrt{C_f/2}\,/\nu$

Greek

α	thermal diffusivity
$\beta_r(a,b)$	incomplete beta function; equation (9)
$\Gamma(x)$	gamma function; $\Gamma(x + 1) = x!$
ε	radiative emissivity
ν	kinematic viscosity
ν_t	eddy kinematic viscosity
ξ	dummy integration parameter
ρ	density
τ	shear stress
ϕ	unheated starting length

Subscript

r	recovery
w	values evaluated at the wall
∞	values evaluated in the free stream

INTRODUCTION

This paper presents heat transfer data for the case of incompressible turbulent boundary layer flow of air over a smooth flat plate with an unheated starting length followed by a heated region with either a constant wall temperature or constant wall heat flux. These cases are of interest in their own right as test cases for heat transfer computations. In addition, the case of unheated starting length followed by constant temperature is one of the fundamental problems of convective heat transfer. Under the assumption of incompressible flow with constant fluid properties, the momentum and energy equations become uncoupled, and the energy equation becomes linear. Therefore, the problem of heat transfer in the boundary layer with arbitrary thermal boundary condition becomes amenable to solution by superposition. The simplest boundary condition for which solutions can serve as kernel functions for this superposition is the step wall temperature. The case of unheated starting length followed by constant wall heat flux is of additional interest because it is one of the few cases of nonuniform boundary conditions for which fairly simple closed form solutions can be formulated for the integral boundary layer equations.

To the authors' knowledge the only systematic experimental studies of the step wall temperature boundary condition in turbulent flat plate boundary layers are the limited ($Re_x < 800,000$) work of Scesa (1951) and the definitive work of Reynolds, Kays, and Kline (1958). Reynolds et al. presented 18 experimental runs with streamwise Reynolds numbers, Re_x, ranging from 100,000 to 3,500,000 and Reynolds numbers based on unheated length, Re_ϕ, ranging from 500,000 to 3,000,000. The experiments presented in this paper are an extension of those experiments.

For the step heat flux boundary condition, no experimental results exist in the available literature, to the authors' knowledge. Reynolds, Kays, and Kline presented an integral solution for this case and experimental results for one closely related case of a double pulse heat flux boundary condition. Data are presented in this paper for the step heat flux boundary condition with x-Reynolds numbers ranging up to 10,000,000.

In the following the experimental apparatus and procedures are described, the theories are briefly reviewed, and the results are presented and compared with the theoretical solutions.

EXPERIMENTAL APPARATUS AND PROCEDURES

The experiments were performed in the Turbulent Heat Transfer Test Facility which is shown schematically in Figure 1. A description of the facility and its qualification are presented in Coleman et al. (1988). This facility is a closed loop air tunnel with a free stream velocity range of 6 to 67 m/s. The temperature of the circulating air is controlled with an air to water heat exchanger and a cooling water loop. Following the heat exchanger the air flow is conditioned by a system of honeycomb and screens.

The nominally 2.4 m long by 0.5 m wide by 0.1 m high test section consists of 24 electrically heated flat plates which are abutted together to form a continuous flat surface. Each nickel plated aluminum plate (about 10 mm thick by 0.1 m in the flow direction) is uniformly heated from below by a custom-manufactured, rubber-encased electric heater pad. Design computations showed that with this configuration, a plate can be considered to be at a uniform temperature. The measured average surface roughness is less than 0.5 μm and the step between any two plates is less than 0.0125 mm. The thermal boundary condition is set by controlling the power input to each plate. The heating system is under active computer control and any desired set of thermal boundary conditions can be maintained within the limits of the power supply. For example, the plate temperature can be set and maintained over time at any temperature, ± 0.1°C, between 50°C and 2 degrees above the free stream air temperature which is typically 30°C. To minimize the conduction losses, the side rails which support the plates are heated to approximately the same temperature as the plates. Both the side rails and the plate backs are heavily insulated.

The top wall can be adjusted to maintain a constant free stream velocity. An inclined water manometer with resolution of 0.06 mm is used to measure the pressure gradient during top wall adjustment. Static pressure taps are located in the side wall adjacent to each plate. The pressure tap located at the second plate is used as a reference, and the pressure difference between it and each other tap is minimized. For example, the maximum pressure difference (referenced to the second pressure tap) was 0.30 mm of water for the 42 m/s case.

The boundary layer is tripped at the exit of the 19:1 area ratio nozzle with a 1 mm high X 12 mm wide wooden strip. This trip location is immediately in front of the heated surface.

Measurements in the nozzle exit plane showed the mean velocity to be uniform within about 0.5 percent and the free stream turbulence intensity to be less than 0.6 percent. Measurements 1.1 m downstream of the nozzle exit showed the spanwise variation of momentum thickness to be less than ±5 percent. Profiles of mean temperature and velocity were in good agreement with the usual "laws-of-the-wall." Stanton number data for the baseline uniform temperature cases were in excellent agreement with the data of Reynolds et al. (1958), which is the definitive data set on which the usual Stanton number correlations are based (Kays and Crawford, 1980, for example). The current data fall within the data scatter of this definitive data set.

Stanton Number Determination

The data reduction expression for the experimentally determined Stanton number is

$$St = \frac{W - q_r - q_c}{A \rho c_p u_\infty (t_w - T_\infty)} \qquad (1)$$

The power, W, supplied to each plate heater is measured with a precision wattmeter. The radiation heat loss, q_r, is estimated using a gray body enclosure model where the emissivity of the nickel plated aluminum is estimated as $\varepsilon = 0.11$. The conductive heat loss, q_c, is determined using an experimentally determined plate conductance, $(UA)_{eff}$, which includes both side rail and plate back losses. The conduction losses are minimized by actively heating the side rails. Typical values for q_r/W and q_c/W are 1 percent and 0.5 percent, respectively. The plate area, A, is determined from the length and width dimensions. The density and specific heat are determined from property data for moist air using the measured values of barometric pressure and wet and dry bulb temperatures in the tunnel. The free stream velocity is measured using a pitot probe and specially calibrated precision pressure transducers. The free stream and plate temperatures are measured using specially calibrated thermistors. The free stream total temperature, T_∞, is computed using the measured free stream recovery temperature and a recovery factor for the free stream thermistor probe of $R = 0.86$ (Eckert and Goldstein, 1976). All fluid properties are evaluated at the free stream static temperature.

The experimentally determined Stanton number is then a function of the 13 measured variables and reference parameters

$$St = St(W, \varepsilon, A, t_w, t_r, t_{rail}, (UA)_{eff},$$

$$t_{wb}, P_{bar}, c_{pa}, c_{pw}, u_\infty, R) \qquad (2)$$

The uncertainty in the experimentally determined Stanton number was estimated based on the ANSI/ASME Standard on Measurement Uncertainty (1986). Precision errors associated with all measured variables used to determine the Stanton number were found to be negligible compared to the corresponding bias errors. The bias limits in the measured variables and the input parameters were estimated, and the uncertainty in the Stanton number was computed using the procedures outlined in the standard. Since all of the thermistors were calibrated against the same standard, some elemental contributions to the thermistor bias limits were correlated. This was taken into account in the uncertainty analysis. Values of the computed Stanton number uncertainties are given in Tables 1 - 4. Complete listings of the uncertainties and computations can be found in Coleman et al. (1988) and Love et al. (1988).

The Stanton number ratio, St/St_t, where St_t is the constant wall temperature Stanton number, is used in this paper for comparison of the experimental data and the theoretical solutions. Because the same equipment is used to measure both St and St_t, many of the bias limits involved in the uncertainty calculation of St/St_t are correlated. The effect of the correlated biases is to reduce the uncertainty in the ratio to less than 2 percent for points away from the first heated plate. The first and last heated plates are considered guard heaters. Data taken for these plates are not appropriately corrected for conduction losses.

Plate to plate conduction contributes to the uncertainty of the Stanton numbers in varying amounts. For the constant wall temperature boundary condition, temperatures of neighboring plates are approximately equal so that the effect of plate to plate conduction is negligible. All uncertainties have been calculated using this assumption. For the constant wall heat flux boundary condition, plate temperatures vary along the test section and plate to plate conduction contributes to the uncertainty of the Stanton number measurements. The plates are machined to touch on a thin, 1.5 mm, lip. The plates abut directly with the only thermal resistance being the natural contact resistance. Conduction experiments and uncertainty calculations indicate that an addition of 1 percent to the calculated uncertainty should account for plate to plate conduction except for the first heated plate.

THE THEORIES

The theoretical treatment of heat transfer in the turbulent incompressible flat plate boundary layer flow is mature and well documented (Cebeci and Bradshaw, 1984, and Kays and Crawford, 1980). Here the theory is divided into two subtopics: solutions of the integral boundary layer equations and numerical solutions of the time averaged boundary layer equations.

Integral Analysis

For the integral analysis, we assume constant free stream velocity and temperature and a constant property fluid flow over a flat plate. Here we follow the procedure of Reynolds, Kays and Kline (1958) as presented by Kays and Crawford (1980). The procedure is to use the solution of the integral boundary layer equations for the step wall temperature with an unheated starting length as the kernel function in a superposition integral. Using the 1/7 power law approximation of the velocity and temperature pro-

files, Reynolds, Kays, and Kline established that for an unheated starting length, ϕ, the local Stanton number can be expressed as

$$\frac{St(\phi;x)}{St_t(x)} = \left[1 - \left(\frac{\phi}{x}\right)^{9/10} \right]^{-1/9} \qquad (3)$$

The problem where the wall heat flux is specified can be formulated as

$$t_w - t_\infty = \int_o^x g(\xi;x) q''_w(\xi) d\xi \qquad (4)$$

where for $Z(\xi) = \left[1 - (\xi/x)^{9/10} \right]^{-8/9}$

$$g(\xi;x) = \frac{9/10}{\Gamma(1/9)\Gamma(8/9)\rho u_\infty c_p St_t(x) x} Z(\xi) \qquad (5)$$

or in terms of the local Stanton number

$$\frac{St_t(x)}{St(x)} = \frac{9/10}{\Gamma(1/9)\Gamma(8/9)x \, q''_w(x)} \int_o^x Z(\xi) q''_w(\xi) d\xi \qquad (6)$$

For the case of a step wall heat flux boundary condition equation (6) gives

$$\frac{St_t(x)}{St(x)} = \frac{\beta_r(1/9, \, 10/9)}{\Gamma(1/9)\Gamma(8/9)} \qquad (7)$$

$$r = 1 - (\phi/x)^{9/10} \qquad (8)$$

where Γ is the Gamma function and β_r is the incomplete beta function

$$\beta_r(a,b) = \int_o^r z^{a-1} (1 - z)^{b-1} dz \qquad (9)$$

Numerical Solutions

The numerical solution of the time averaged boundary layer equations for turbulent flat plate flows is by now routine (Cebeci and Bradshaw, 1984). The time averaged boundary layer equations for the flow of a constant property fluid over a flat plate (constant free stream velocity and temperature) are usually reduced to

$$\frac{\partial u}{\partial x} + \frac{\partial v}{\partial y} = 0 \qquad (10)$$

$$u \frac{\partial u}{\partial x} + v \frac{\partial u}{\partial y} = \nu \frac{\partial^2 u}{\partial y^2} - \frac{\partial}{\partial y} \overline{u'v'} \qquad (11)$$

$$u \frac{\partial t}{\partial x} + v \frac{\partial t}{\partial y} = \alpha \frac{\partial^2 t}{\partial y^2} - \frac{\partial}{\partial y} \overline{t'v'}$$

$$+ \frac{1}{c_p} \left(\nu \frac{\partial u}{\partial y} - \overline{u'v'} \right) \left(\frac{\partial u}{\partial y}\right) \qquad (12)$$

Before equations (10) - (12) can be solved, models for the Reynolds stress factor, $\overline{u'v'}$, and the turbulent heat flux factor, $\overline{t'v'}$, must be formulated. For the conditions of interest, the mixing length model and the turbulent Prandtl number concept are adequate. The models are (Kays and Crawford, 1980)

$$- \overline{u'v'} = \nu_t \frac{\partial u}{\partial y} = \ell^2 \left| \frac{\partial u}{\partial y} \right| \frac{\partial u}{\partial y} \qquad (13)$$

$$- \overline{t'v'} = \frac{\nu_t}{Pr_t} \frac{\partial t}{\partial y} \qquad (14)$$

where

$$\ell = 0.4 \, y \, [1 - \exp(- \, y^+/26)] \; ; \; \ell < 0.09\delta$$

$$\ell = 0.09\delta \; ; \; \text{otherwise}$$

$$Pr_t = 0.9$$

The boundary conditions for equations (10) - (12) are

$$x = 0: \quad u = u_\infty, \; t = t_\infty$$

$$y = 0: \quad u = v = 0, \; t = t_w(x) \; (\text{or} \; q'' = q''_w(x)) \qquad (15)$$

$$y \to \infty: \quad u = u_\infty, \; t = t_\infty$$

The momentum and energy equations of the boundary layer are transformed into computational space and solved with a marching implicit finite-difference algorithm. The method is described in detail by Gatlin (1983). For all computations in this paper, the BLACOMP code as verified by Gatlin is used.

RESULTS

For the experimental results and all of the computations, the origin of the momentum boundary layer was taken to be the nozzle exit, and the origin of the thermal boundary layer was taken to be the start of the heated surface. Boundary layer measurements indicate that this is a reasonable approximation for the purpose of this paper. The experimental data and all of the computations were based on constant properties which were evaluated at the free stream static temperature.

Stanton number measurements were made at nominal free stream velocities of 28 m/s and 67 m/s. Three unheated starting length cases were run at each velocity for each boundary condition. The lengths of the unheated regions were chosen so that an appropriate spread in Reynolds numbers, Re_ϕ, was obtained. At u_∞ = 28 m/s, the unheated starting lengths were 0.3 m, 0.7 m, and 1.3 m. At u_∞ = 67 m/s they were 0.5 m, 0.8 m, and 1.3 m. The results of these experiments are presented in Tables 1 - 4. The uncertainty for each Stanton number, U, is presented in percent of the Stanton number value. As indicated in the tables, the end effects of the first heated plate and the last plate are not accounted for in the data reduction. These plates act as guard heaters for the other heated plates. Inspection of the tables reveals that there is some heat leakage into the unheated plates and that the step boundary conditions are not clean. However, both the integral and numerical solutions assume a clean, sharp step in wall temperature or heat flux.

Figure 2 shows a summary of the Stanton number data for a constant wall temperature boundary condition and the step t_w cases for u_∞ = 28 m/s and u_∞ = 67 m/s. The first heated plate is highlighted in each case by plotting its data as a solid symbol. Data from the last plate is not plotted for any case. The figure shows that a step in wall temperature has a large effect on the Stanton number in the heated region near the step. But as the thermal boundary layer develops, the Stanton numbers approach the results for the constant wall temperature boundary

layers. The starting lengths were chosen so that the last case at u_∞ = 28 m/s and the first case at u_∞ = 67 m/s had approximately the same value of Re_ϕ. Based on the data of Reynolds et al., the results of these two cases should coincide when plotted as St versus Re_x. The figure shows that this is true for the present results.

Figure 3 shows a comparison of the results of the step t_w experiments with the integral solution in equation (3), dashed lines, and with the finite difference solutions, solid lines, for u_∞ = 28 m/s. Figure 4 shows the same comparison for u_∞ = 67 m/s. The results are presented in terms of St/St_t for a direct comparison of equation (3). The St_t data were used to normalize the St data, and the finite difference solutions for constant wall temperature, St_t, were used to normalize the finite difference solutions, St. The figures show that both the finite difference and integral solutions are in good agreement with the data in all cases. The integral solutions are consistently low by a small amount in the region of the step in wall temperature. Some disagreement should be expected since the 1/7 power law approximation for the temperature profiles was used in the integral solution. Thus the integral solution is an asymptotic case where the thermal boundary layer has had a chance to develop well. From the comparisons in the figures, it can be concluded that equation (3) is still appropriate for values of $Re_x \to 10^7$.

Reynolds, Kays, and Kline present Stanton number data for step wall temperature boundary conditions for Re_x < 3,500,000. The Reynolds numbers corresponding to the location of the step in wall temperature are approximately equal to those for the present data. The present data and that of Reynolds et al. agree substantially. A good indication of this comparison is the agreement in Figure 3 of the present data and equation (3), which is based on the data of Reynolds et al.

Figure 5 shows a summary of the Stanton number data for a constant heat flux boundary condition and the step q''_w cases for u_∞ = 28 m/s and u_∞ = 67 m/s. The figure shows that as the thermal boundary layer develops, the unheated starting length Stanton numbers approach the results for the constant heat flux boundary condition. The first heated plate is highlighted in each case by plotting its data as a solid symbol. Data from the last plate is not plotted for any case.

Figure 6 shows a comparison of the results of the experiments with the integral solution in equation (7), dashed lines, and with the finite difference solutions, solid lines, for u_∞ = 28 m/s. Figure 7 shows the same comparison for u_∞ = 67 m/s. The results are presented in terms of St/St_t for a direct comparison with equation (7). As before, the St_t data were used to normalize the St data and finite difference St_t solutions were used to normalize the finite difference St solutions. The figures show that the finite difference solutions are in very good agreement with the data in all cases. The integral solutions are also in reasonable agreement with the data, with the maximum difference between the data and the integral solutions being about 10 percent.

SUMMARY

Data are presented from heat transfer experiments in an incompressible turbulent boundary layer flow of air over a smooth flat plate with unheated starting length followed by a heated region with either a constant wall temperature or constant wall heat flux. The step wall temperature data extend the available experimental data for this class of problem from the previous maximum x-Reynolds number of 3,500,000 to 10,000,000. The data are compared with the classical solutions of the integral boundary layer equations and of the finite difference solutions of the partial differential equations of the boundary layer. The comparisons showed reasonable to very good agreement. The integral solution for the step wall temperature case is often used in superposition solutions for arbitrary wall temperature boundary conditions. The results of this work indicate that these solutions are also appropriate at the higher Reynolds numbers.

ACKNOWLEDGMENTS

This work was supported by the U. S. Air Force Office of Scientific Research (Research Grant AFOSR-86-0178). The interest and encouragement of Dr. Jim Wilson and Capt. Hank Helin are gratefully acknowledged. The experimental apparatus was acquired under Grant AFOSR-85-0075.

REFERENCES

_____ (1986), Measurement Uncertainty, ANSI/ASME PTC 19.1-1985, Part 1.

Cebeci, T. and Bradshaw, P. (1984), Physical and Computational Aspects of Convective Heat Transfer, Springer-Verlag, New York.

Coleman, H. W., Hosni, M. H., Taylor, R. P., and Brown, G. B. (1988), "Smooth Wall Qualification of a Turbulent Heat Transfer Test Facility," TFD-88-2, Mechanical and Nuclear Engineering Department, Mississippi State University.

Eckert, R. G. and Goldstein, R. J. (1976), Measurements in Heat Transfer, 2nd edition, McGraw-Hill, New York.

Gatlin, B. (1983), "An Instructional Computer Program for Computing the Steady, Compressible, Turbulent Flow of an Arbitrary Fluid Near a Smooth Wall," MS Thesis, Mechanical and Nuclear Engineering Department, Mississippi State University.

Kays, W. M. and Crawford, M. E. (1980), Convective Heat and Mass Transfer, McGraw-Hill, New York.

Love, P., Taylor, R. P., Coleman, H. W., and Hosni, M. H. (1988), "Effects of Thermal Boundary Condition on Heat Transfer in the Turbulent Incompressible Flat Plate Boundary Layer," TFD-88-3, Mechanical and Nuclear Engineering Department, Mississippi State University.

Reynolds, W. C., Kays, W. M., and Kline, S. J. (1958), "Heat Transfer in the Turbulent Incompressible Boundary Layer, Parts I, II, and III," NASA MEMO 12-1-58W, 12-2-58W, and 12-3-58W.

Scesa, S. (1951), "Experimental Investigation of Convective Heat Transfer to Air from a Flat Plate with a Stepwise Discontinuous Surface Temperature," MS Thesis, University of California.

Table 1. Data for Nominal Free Stream Velocity of 28 m/s, Constant t_w.

P#	x(m)	u_∞ = 27.9 m/s T_∞ = 26.5°C t_w(C)	Re_x X10^{-6}	St X10^3	U%	u_∞ = 28.0 m/s T_∞ = 26.3°C t_w(C)	Re_x X10^{-6}	St X10^3	U%	u_∞ = 28.0 m/s T_∞ = 26.2°C t_w(C)	Re_x X10^{-6}	St X10^3	U%	u_∞ = 27.9 m/s T_∞ = 25.9°C t_w(C)	Re_x X10^{-6}	St X10^3	U%
1	0.05	43.9	0.09	3.70	----*	26.3	....	....	...	26.0	....	....	...	25.8	....	....	...
2	0.15	44.3	0.27	2.75	1.6	27.0	....	....	...	26.2	....	....	...	26.0	....	....	...
3	0.25	43.7	0.45	2.29	1.6	30.3	....	....	...*	26.3	....	....	...	26.0	....	....	...
4	0.36	44.1	0.62	2.25	1.6	44.0	0.63	3.90	----*	26.4	....	....	...	26.2	....	....	...
5	0.46	44.1	0.80	2.06	1.6	43.9	0.81	2.45	1.6	26.6	....	....	...	26.2	....	....	...
6	0.56	44.1	0.98	2.00	1.6	44.1	0.99	2.32	1.6	26.9	....	....	...	26.3	....	....	...
7	0.66	43.8	1.16	1.89	1.6	43.6	1.17	2.15	1.6	28.0	....	....	...*	26.3	....	....	...
8	0.76	43.9	1.33	1.86	1.6	43.7	1.35	2.08	1.6	44.0	1.35	3.27	----	26.3	....	....	...
9	0.86	44.0	1.51	1.82	1.6	43.8	1.53	1.99	1.6	44.1	1.53	2.43	1.6	26.4	....	....	...
10	0.97	44.0	1.69	1.77	1.6	43.9	1.71	1.92	1.6	44.1	1.71	2.25	1.6	26.4	....	....	...
11	1.07	44.0	1.87	1.75	1.7	43.8	1.89	1.87	1.6	44.1	1.89	2.15	1.6	26.6	....	....	...
12	1.17	44.0	2.05	1.70	1.7	43.9	2.07	1.82	1.7	44.2	2.07	2.08	1.6	27.4	....	....	...
13	1.27	44.0	2.22	1.68	1.7	43.8	2.25	1.77	1.7	43.8	2.25	1.93	1.6	30.3	....	....	...*
14	1.37	44.0	2.40	1.66	1.7	43.9	2.43	1.74	1.7	44.1	2.43	1.95	1.6	44.3	2.42	3.41	----*
15	1.47	44.0	2.58	1.65	1.7	43.9	2.61	1.72	1.7	43.9	2.61	1.87	1.6	44.4	2.60	2.27	1.6
16	1.57	44.0	2.76	1.62	1.7	43.8	2.79	1.69	1.7	44.0	2.79	1.85	1.6	44.2	2.78	2.13	1.6
17	1.68	44.0	2.94	1.61	1.7	43.8	2.97	1.68	1.7	44.0	2.97	1.82	1.6	44.2	2.96	2.05	1.6
18	1.78	44.0	3.11	1.61	1.7	43.8	3.15	1.65	1.7	44.0	3.15	1.79	1.6	44.3	3.14	1.99	1.6
19	1.88	44.0	3.29	1.57	1.7	43.7	3.33	1.63	1.7	44.0	3.33	1.75	1.7	44.4	3.32	1.94	1.6
20	1.98	44.0	3.47	1.57	1.7	43.8	3.52	1.62	1.7	43.9	3.51	1.72	1.7	44.3	3.50	1.87	1.7
21	2.08	44.0	3.65	1.57	1.8	43.8	3.70	1.60	1.8	43.9	3.69	1.69	1.7	44.4	3.68	1.84	1.7
22	2.18	44.0	3.83	1.53	1.9	43.7	3.88	1.57	1.9	43.9	3.87	1.66	1.9	44.4	3.86	1.80	1.8
23	2.29	43.9	4.00	1.50	2.2*	43.8	4.06	1.54	2.2	43.9	4.05	1.63	2.1*	44.3	4.04	1.73	2.1*
24	2.39	43.9	4.18	1.55	----	43.8	4.24	1.58	----	43.9	4.23	1.66	----	44.3	4.22	1.77	----

* End effects are not included in the data reduction or the estimates of the uncertainty.

Table 2. Data for Nominal Free Stream Velocity of 67 m/s, Constant t_w.

P#	x(m)	u_∞= 66.9 m/s T_∞= 32.9°C				u_∞= 67.3 m/s T_∞= 32.5°C				u_∞= 67.5 m/s T_∞= 32.3°C				u_∞= 66.9 m/s T_∞= 31.8°C			
		t_w(C)	Re_x $X10^{-6}$	St $X10^3$	U%	t_w(C)	Re_x $X10^{-6}$	St $X10^3$	U%	t_w(C)	Re_x $X10^{-6}$	St $X10^3$	U%	t_w(C)	Re_x $X10^{-6}$	St $X10^3$	U%
1	0.05	44.3	0.21	3.19	---*	32.1	....	....	...	31.7	....	....	...	31.4	....	....	...
2	0.15	44.3	0.64	2.22	2.6	32.4	....	....	...	31.9	....	....	...	31.8	....	....	...
3	0.25	44.1	1.07	1.97	2.6	32.5	....	....	...	32.1	....	....	...	31.9	....	....	...
4	0.36	44.2	1.50	1.84	2.6	32.7	....	....	...	32.1	....	....	...	31.9	....	....	...
5	0.46	44.2	1.93	1.74	2.7	34.0	....	....	...*	32.2	....	....	...	32.0	....	....	...
6	0.56	44.2	2.36	1.66	2.7	43.9	2.33	2.74	---*	32.2	....	....	...	32.0	....	....	...
7	0.66	44.2	2.78	1.63	2.7	44.1	2.76	2.04	2.7	32.3	....	....	...	32.0	....	....	...
8	0.76	44.2	3.21	1.59	2.7	44.1	3.19	1.90	2.7	33.3	....	....	...*	32.1	....	....	...
9	0.86	44.2	3.64	1.55	2.7	44.1	3.61	1.79	2.7	43.9	3.64	2.60	----	32.1	....	....	...
10	0.97	44.2	4.07	1.51	2.7	44.1	4.04	1.73	2.7	44.0	4.05	2.00	2.6	32.1	....	....	...
11	1.07	44.2	4.50	1.50	2.6	44.1	4.46	1.69	2.6	43.9	4.48	1.89	2.6	32.1	....	....	...
12	1.17	44.2	4.93	1.47	2.6	44.1	4.89	1.63	2.6	43.8	4.91	1.78	2.6	32.2	....	....	...
13	1.27	44.2	5.36	1.45	2.6	44.1	5.31	1.60	2.6	43.9	5.34	1.74	2.6	33.3	....	....	...*
14	1.37	44.1	5.78	1.42	2.7	44.1	5.74	1.57	2.6	43.9	5.76	1.69	2.6	44.0	5.77	2.48	----
15	1.47	44.1	6.21	1.44	2.7	44.0	6.16	1.55	2.7	43.8	6.19	1.66	2.6	44.2	6.20	1.91	2.5
16	1.57	44.2	6.64	1.41	2.7	44.0	6.59	1.53	2.7	43.9	6.62	1.63	2.6	44.2	6.63	1.80	2.5
17	1.68	44.1	7.07	1.41	2.7	44.1	7.01	1.52	2.7	43.9	7.04	1.63	2.6	44.2	7.05	1.75	2.5
18	1.78	44.1	7.50	1.41	2.7	44.1	7.44	1.50	2.7	44.0	7.47	1.60	2.6	44.1	7.48	1.69	2.5
19	1.88	44.2	7.93	1.40	2.6	44.1	7.86	1.49	2.6	43.8	7.90	1.56	2.6	44.0	7.91	1.66	2.5
20	1.98	44.2	8.35	1.40	2.6	44.1	8.29	1.47	2.6	43.8	8.33	1.55	2.6	44.1	8.34	1.63	2.5
21	2.08	44.1	8.78	1.36	2.6	44.0	8.72	1.45	2.6	43.8	8.75	1.53	2.6	44.0	8.76	1.59	2.5
22	2.18	44.2	9.21	1.37	2.6	44.1	9.14	1.44	2.6	43.8	9.18	1.51	2.5	44.0	9.19	1.56	2.5
23	2.29	44.2	9.64	1.33	2.6	44.1	9.57	1.40	2.6	43.8	9.61	1.47	2.6	44.1	9.62	1.52	2.4
24	2.39	44.2	10.0	1.36	---	44.0	9.99	1.43	---	43.8	10.04	1.50	---	44.1	10.00	1.54	---

* End effects are not included in the data reduction or the estimates of the uncertainty.

Table 3. Data for Nominal Free Stream Velocity of 28 m/s, Constant q''_w.

P#	x(m)	u_∞= 27.8 m/s T_∞= 26.8°C					u_∞= 27.9 m/s T_∞= 26.5°C					u_∞= 27.8 m/s T_∞= 26.4°C					u_∞= 27.8 m/s T_∞= 26.3°C				
		t_w(C)	Q(w)	Re_x $X10^{-6}$	St $X10^3$	U%	t_w(C)	Q(w)	Re_x $X10^{-6}$	St $X10^3$	U%	t_w(C)	Q(w)	Re_x $X10^{-6}$	St $X10^3$	U%	t_w(C)	Q(w)	Re_x $X10^{-6}$	St $X10^3$	U%
1	0.05	34.3	44.9	0.09	3.95	---*	26.5	....	....	....	...	26.3	....	....	....	...	26.2	....	....	....	...
2	0.15	36.3	44.4	0.26	3.09	4.2	27.0	....	....	....	...	26.6	....	....	....	...	26.4	....	....	....	...
3	0.25	37.7	44.8	0.44	2.71	3.6	28.0	....	....	....	...	26.7	....	....	....	...	26.5	....	....	....	...
4	0.36	38.6	44.4	0.62	2.46	3.3	35.1	45.4	0.62	3.45	---*	26.9	....	....	....	...	26.7	....	....	....	...
5	0.46	39.5	44.6	0.80	2.31	3.0	37.2	45.3	0.80	2.76	3.4	27.0	....	....	....	...	26.7	....	....	....	...
6	0.56	40.1	44.7	0.97	2.21	2.8	38.5	45.3	0.98	2.46	2.9	27.3	....	....	....	...	26.8	....	....	....	...
7	0.66	40.5	44.3	1.15	2.13	2.7	39.2	45.3	1.15	2.33	2.7	27.9	....	....	....	...*	26.9	....	....	....	...
8	0.76	41.1	44.6	1.33	2.05	2.5	40.0	45.4	1.33	2.19	2.5	35.6	44.0	1.33	3.10	----	27.0	....	....	....	...
9	0.86	41.6	44.5	1.50	1.98	2.4	40.5	45.2	1.51	2.11	2.4	37.5	44.2	1.51	2.59	3.6	27.1	....	....	....	...
10	0.97	41.9	44.4	1.68	1.93	2.4	40.9	45.1	1.69	2.03	2.3	38.5	44.2	1.69	2.37	3.2	27.1	....	....	....	...
11	1.07	42.2	44.5	1.86	1.90	2.3	41.3	45.3	1.87	1.99	2.3	39.2	44.1	1.87	2.23	3.0	27.3	....	....	....	...
12	1.17	42.5	44.3	2.03	1.86	2.3	41.7	45.3	2.05	1.94	2.2	39.7	43.7	2.05	2.14	2.9	27.6	....	....	....	...
13	1.27	42.8	44.3	2.21	1.82	2.2	42.1	45.4	2.23	1.90	2.2	40.2	43.9	2.22	2.06	2.7	29.0	....	....	....	...*
14	1.37	43.0	44.3	2.39	1.79	2.2	42.4	45.5	2.40	1.86	2.2	40.6	44.3	2.40	2.01	2.6	35.3	44.7	2.40	3.24	---*
15	1.47	43.2	44.3	2.56	1.78	2.2	42.6	45.3	2.58	1.83	2.2	41.0	43.7	2.58	1.95	2.5	38.0	44.6	2.58	2.49	3.2
16	1.57	43.5	44.2	2.74	1.74	2.2	42.9	45.2	2.76	1.80	2.2	41.4	44.0	2.76	1.91	2.4	39.2	44.4	2.76	2.24	2.8
17	1.68	43.7	44.7	2.92	1.74	2.2	43.1	45.4	2.94	1.78	2.2	41.6	44.0	2.94	1.88	2.4	39.8	44.4	2.94	2.14	2.6
18	1.78	43.8	44.3	3.09	1.71	2.2	43.2	45.1	3.12	1.76	2.2	41.8	43.9	3.12	1.85	2.3	40.3	44.7	3.11	2.08	2.5
19	1.88	44.0	44.3	3.27	1.69	2.2	43.4	45.2	3.30	1.75	2.2	42.0	43.7	3.29	1.82	2.3	40.7	44.4	3.29	2.01	2.4
20	1.98	44.0	44.1	3.45	1.68	2.2	43.5	45.2	3.47	1.73	2.2	42.2	43.8	3.47	1.81	2.3	40.9	44.5	3.47	1.98	2.4
21	2.08	44.2	43.9	3.62	1.66	2.2	43.7	45.4	3.65	1.72	2.2	42.4	44.1	3.65	1.79	2.2	41.2	44.2	3.65	1.94	2.3
22	2.18	44.3	44.1	3.80	1.66	2.2	43.9	45.2	3.83	1.70	2.2	42.5	43.7	3.83	1.76	2.2	41.4	44.1	3.83	1.90	2.3
23	2.29	44.3	44.1	3.98	1.65	2.2*	43.9	45.0	4.01	1.69	2.2*	42.7	43.8	4.01	1.75	2.2*	41.6	44.4	4.00	1.88	2.3*
24	2.39	44.0	43.7	4.15	1.67	---	43.7	45.3	4.19	1.72	---	42.5	44.0	4.19	1.77	---	41.4	44.2	4.17	1.90	---

* End effects are not included in the data reduction or the estimates of the uncertainty.

14

Table 4. Data for Nominal Free Stream Velocity of 67 m/s, Constant q''_w.

P#	x(m)	u_∞= 67.3 m/s T_∞= 33.8°C — t_w(C)	Q(w)	Re_x X10⁻⁶	St X10³	U%	u_∞= 67.4 m/s T_∞= 33.2°C — t_w(C)	Q(w)	Re_x X10⁻⁶	St X10³	U%	u_∞= 67.4 m/s T_∞= 33.2°C — t_w(C)	Q(w)	Re_x X10⁻⁶	St X10³	U%	u_∞= 67.4 m/s T_∞= 33.1°C — t_w(C)	Q(w)	Re_x X10⁻⁶	St X10³	U%
1	0.05	38.5	61.2	0.21	3.65	---*	32.7	....	....	....	...	32.6	....	....	....	...	32.3	....	....	....	...
2	0.15	40.4	61.2	0.63	2.59	5.7	32.9	....	....	....	...	32.8	....	....	....	...	32.7	....	....	....	...
3	0.25	41.4	61.3	1.05	2.25	4.9	33.1	....	....	....	...	33.0	....	....	....	...	32.9	....	....	....	...
4	0.36	42.0	61.1	1.47	2.05	4.4	33.3	....	....	....	...	33.1	....	....	....	...	32.9	....	....	....	...
5	0.46	42.6	61.2	1.89	1.92	4.1	33.8	....	....	....	...	33.1	....	....	....	...	33.0	....	....	....	...
6	0.56	43.1	61.3	2.31	1.83	3.9	39.3	60.4	2.33	2.69	---*	33.2	....	....	....	...	33.0	....	....	....	...
7	0.66	43.3	61.0	2.73	1.78	3.8	40.5	60.3	2.76	2.24	4.7	33.2	....	....	....	...	33.0	....	....	....	...
8	0.76	43.6	61.4	3.15	1.73	3.6	41.2	59.6	3.18	2.07	4.3	33.9	....	....	....	...*	33.1	....	....	....	...
9	0.86	43.9	61.1	3.56	1.68	3.5	41.8	60.1	3.61	1.91	4.0	39.5	60.1	3.61	2.61	---*	33.1	....	....	....	...
10	0.97	44.0	60.5	3.98	1.64	3.4	42.1	60.4	4.03	1.84	3.9	40.7	59.5	4.03	2.15	4.4	33.2	....	....	....	...
11	1.07	44.2	60.8	4.40	1.62	3.4	42.4	60.3	4.46	1.79	3.7	41.3	59.3	4.46	1.98	4.0	33.2	....	....	....	...
12	1.17	44.4	61.1	4.82	1.60	3.1	42.6	60.3	4.88	1.74	3.6	41.8	59.6	4.88	1.89	3.8	33.2	....	....	....	...
13	1.27	44.5	60.7	5.24	1.57	3.2	42.8	59.5	5.31	1.69	3.6	42.1	59.7	5.33	1.82	3.6	34.0	....	....	....	...*
14	1.37	44.7	61.0	5.66	1.55	3.2	43.0	59.6	5.73	1.66	3.5	42.4	59.6	5.73	1.77	3.5	39.3	60.3	5.72	2.65	---*
15	1.47	44.8	60.9	6.08	1.54	3.2	43.2	60.2	6.16	1.65	3.4	42.5	59.2	6.16	1.73	3.5	40.9	59.9	6.15	2.11	4.5
16	1.57	44.9	60.9	6.50	1.52	3.1	43.4	60.4	6.58	1.61	3.3	42.9	60.0	6.58	1.69	3.3	41.7	60.9	6.57	1.94	4.0
17	1.68	45.0	61.3	6.92	1.51	3.1	43.5	60.2	7.01	1.59	3.3	42.9	59.4	7.01	1.66	3.3	42.0	60.8	7.00	1.87	3.9
18	1.78	45.1	61.1	7.34	1.50	3.0	43.6	60.2	7.43	1.58	3.3	43.1	59.4	7.43	1.64	3.3	42.2	60.7	7.42	1.81	3.7
19	1.88	45.2	60.9	7.76	1.48	3.0	43.7	60.0	7.86	1.56	3.2	43.2	59.9	7.86	1.62	3.2	42.5	60.8	7.85	1.76	3.6
20	1.98	45.2	61.0	8.18	1.48	3.0	43.8	60.2	8.28	1.55	3.2	43.3	59.3	8.28	1.60	3.2	42.7	60.5	8.27	1.73	3.6
21	2.08	45.4	61.3	8.60	1.47	3.0	43.9	59.9	8.70	1.53	3.2	43.5	59.8	8.70	1.58	3.1	42.9	61.0	8.69	1.70	3.5
22	2.18	45.4	61.1	9.02	1.46	2.9	44.0	60.1	9.13	1.52	3.1	43.6	59.6	9.13	1.56	3.1	43.0	61.0	9.12	1.67	3.4
23	2.29	45.5	61.3	9.44	1.45	2.9*	44.1	60.4	9.55	1.51	3.1*	43.7	59.8	9.55	1.55	3.1*	43.1	60.4	9.54	1.64	3.4*
24	2.39	45.3	61.6	9.86	1.46	---*	43.9	60.2	9.98	1.53	---*	43.5	59.5	9.98	1.57	---*	43.0	60.6	9.97	1.67	---*

* End effects are not included in the data reduction or the estimates of the uncertainty.

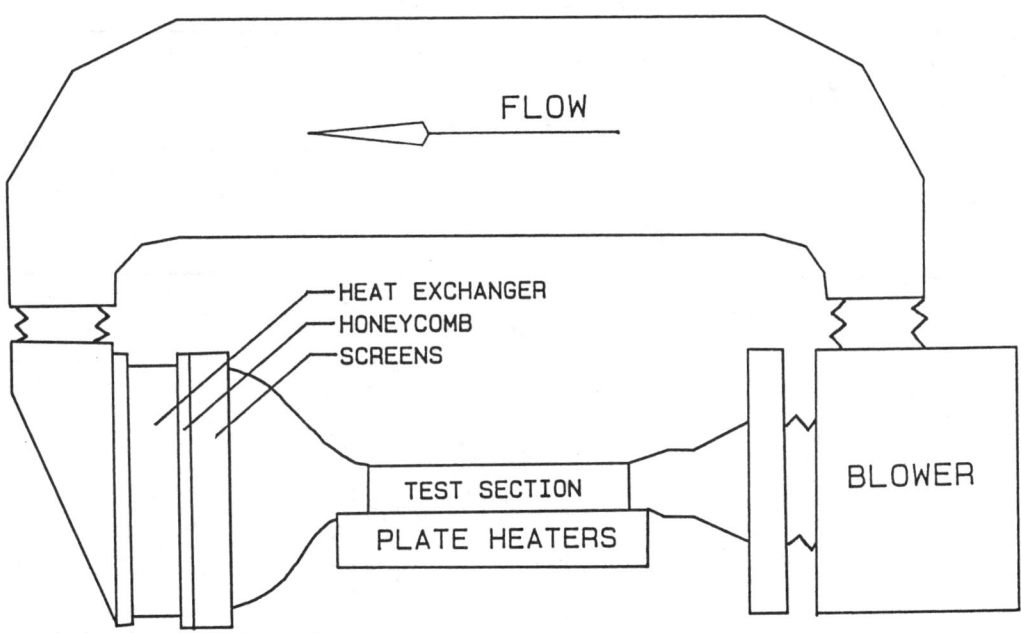

Fig. 1. Schematic of the Turbulent Heat Transfer Test Facility.

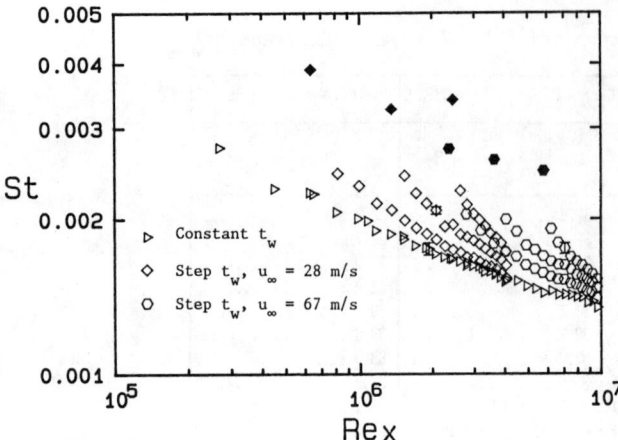

Fig. 2.   Summary of the Stanton Number Data for the
          Constant Wall Temperature and the Step t_w
          Cases. Solid Symbols Indicate the First
          Heated Plate.

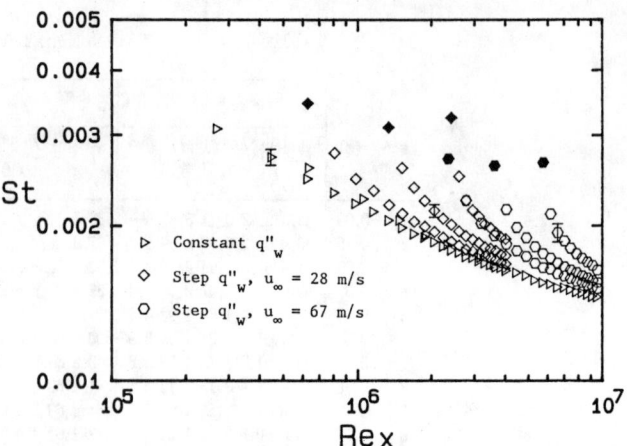

Fig. 5.   Summary of the Stanton Number Data for the
          Constant Heat Flux Boundary Condition and the
          Step q''_w Cases. Solid Symbols Indicate the
          First Heated Plate.

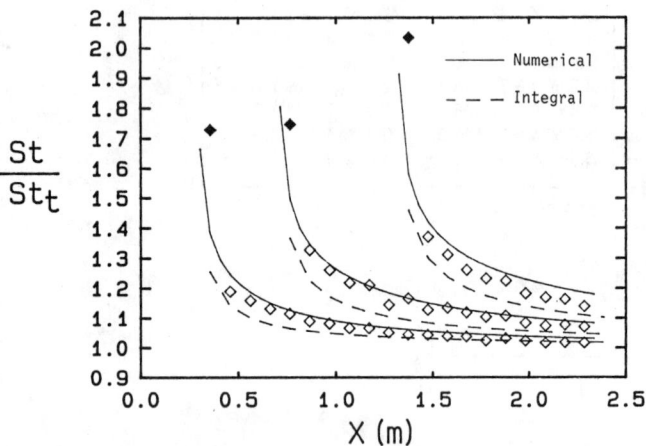

Fig. 3.   Comparison of the Step Wall Temperature Data
          with the Solutions for u_∞ = 28 m/s. Solid
          Symbols Indicate the First Heated Plate.

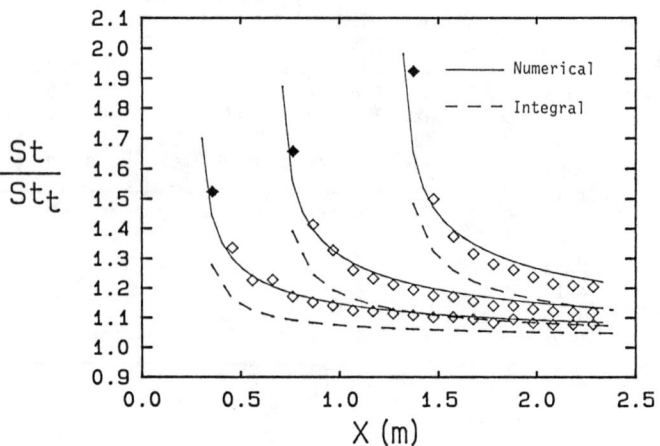

Fig. 6.   Comparison of the Step Wall Heat Flux Data
          with the Solutions for u_∞ = 28 m/s. Solid
          Symbols Indicate the First Heated Plate.

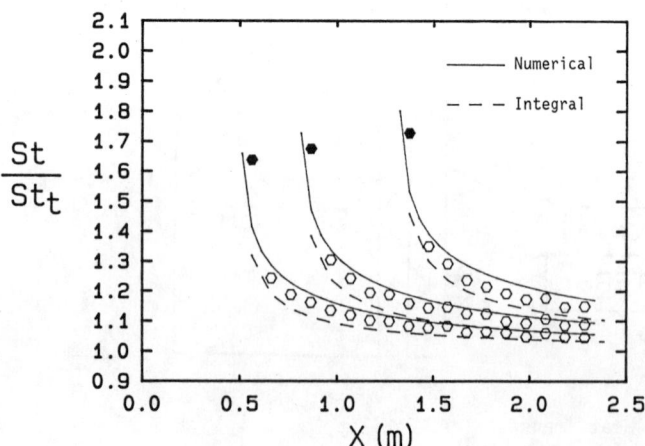

Fig. 4.   Comparison of the Step Wall Temperature Data
          with the Solutions for u_∞ = 67 m/s. Solid
          Symbols Indicate the First Heated Plate.

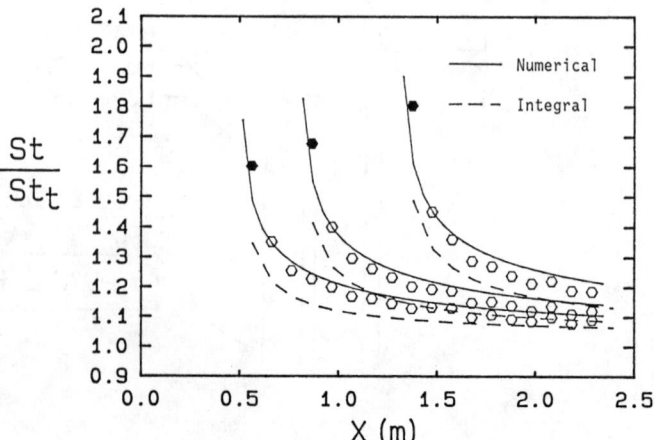

Fig. 7.   Comparison of the Step Wall Heat Flux Data
          with the Solutions for u_∞ = 67 m/s. Solid
          Symbols Indicate the First Heated Plate.

EXPERIMENTAL INVESTIGATION OF TURBULENT MIXED CONVECTION IN HORIZONAL TUBES WITH LONGITUDINAL INTERNAL FINS

A. C. Trupp
Department of Mechanical Engineering
University of Manitoba
Winnipeg, Manitoba, Canada

H. Haine
Chef de Secteur Laminage
Laminoir à Chaud
Annaba, Algeria

ABSTRACT

The heat transfer and pressure drop characteristics were determined for five integral inner finned tubes of various designs for heating water in low Reynolds number turbulent flow. Numerous tests were run at various heating rates (hence varying Grashof numbers) over a range of Reynolds numbers from transition to about 10,000. For the thermal entry regions, novel data were obtained which showed a weak influence of buoyancy effects on the axial distributions of wall temperature and local Nusselt number. The results for the fully developed Nusselt numbers and friction factors were generally in good agreement with the results of previous experimental and theoretical studies.

NOMENCLATURE

A_a = Actual surface area per unit length

A_{fa} = Actual flow area

A_{fc} = Core flow area

A_{fn} = Nominal flow area = $\pi (D_i)^2/4$

A_n = Nominal inside surface area = $\pi D_i L$

b = Average distance between fins (taken as the average of the arc lengths at the root and tip of adjacent fins

c_p = specifc heat of the fluid

D_{ft} = Fin tip diameter

D_h = Hydraulic diameter = $4 A_{fa}/A_a$

D_i = Inside diameter

D_{os} = Outside diameter

f = Fanning friction factor = $\Delta P D_h \rho \ (A_{fa})^2/[2L_p \ (\dot{m})^2]$

g = gravitational acceleration

Gr = Grashof number = $g\beta (\rho)^2 (D_h)^3 (T_w - T_b) / (\mu)^2$

H = Relative fin height = $2l /D_i$

h = Convective heat transfer coefficient

k = Fluid thermal conductivity

L = Heated length

L_p = Distance between pressure taps

l = Fin height

M = Number of fins

$\dot{m}$ = Mass flow rate of fluid

Nu = Nusselt number = $hD_h/k = Q_f D_h/[A_a L(T_w - T_b)k]$

Nu_i = $Q_f D_i/[A_n L(T_w - T_b)k]$

P = Pressure

Pr = Prandtl number = $\mu c_p/k$

Q_e = Rate of electrical heat input

Q_f = Rate of heat gained by the fluid = $\dot{m}c_p(T_{bo} - T_{bi})$

Ra = Rayleigh number = Gr.Pr

Ra^* = Modified Rayleigh number = Ra.Nu

Re_h = Reynolds number = $\dot{m}D_h/\mu A_{fa}$

Re_i = $\dot{m}D_i/\mu A_{fa}$

T_b = Local bulk temperature

T_w = Wall temperature

x = Axial distance from the beginning of heated section

X^+ = Reduced length = $x/(1/2 \ D_h \ Re_h \ Pr)$

α = Helix angle

β = Fluid thermal expansion coefficient

γ = Half-fin angle

μ = fluid viscosity

ρ = Fluid density

Subscripts:

cr = critical (at laminar-turbulent transition);

dia = diabatic; h = based on hydraulic diameter;

i = based on inside diameter; iso = isothermal;

w = wall; x = local value.

INTRODUCTION

Internally finned tubes have gained wide popularity as an augmentation technique due to their relative simplicity. They are now available commercially in various diameters with different fin heights in both straight and spiral configurations of equi-spaced fins. Numerous experimental and theoretical results are now available in the literature regarding heat transfer and pressure drop characteristics of internally finned tubes for both laminar and turbulent flows. These data, together with performance analyses, have clearly shown the effectiveness of internal fins. However, all of these results have dealt basically only with the fully developed region. A few heat transfer results have provided overall-average Nusselt numbers for heated sections which included an entrance length. In addition, for fully developed laminar flow, all analytical treatments with the exception of Mirza and Soliman (1985) have considered only pure forced convection, whereas it is known that free convection effects can be significant. It is necessary for designers to gain some knowledge of friction factor and heat transfer in the entrance regions of internally finned tubes. This is essential particularly when dealing with the short tubes frequently encountered in compact heat exchangers.

The purpose of this paper is simply to report the results of an experimental investigation of the heat transfer characteristics in the thermal entrance region and the fully developed region for horizontal internally finned tubes. Five commercial copper tubes (one spiral) of various diameters, fin numbers and fin heights were tested (together with a smooth tube) under the condition of uniform heat input axially with water as the working fluid. For each test, the flow was fully developed hyrodynamically at the entrance to the heated section. The experimental program encompassed laminar flow, transition and the bottom end of the turbulent flow regime. In addition to the heat transfer results, data were also obtained on both the isothermal and diabatic friction factors. The laminar flow results for combined free and forced convection were recently reported by Rustum and Soliman (1988). The low Reynolds number turbulent flow results are described herein. As shown later, the fully developed turbulent results were found to be consistent with the results of previous investigators. For the thermal entrance region, there are no other results available in the open literature, hence the present results are novel and should be of direct interest to designers.

EXPERIMENTAL FACILITY AND PROCEDURES

General
The research program was designed to experimentally investigate the low Reynolds number turbulent heat transfer and pressure drop characteristics of internally finned tubes. Both the thermal entry regions and the fully developed regions were considered and studied extensively. Data were carefully gathered for each tube and reduced to study the following features: the isothermal and diabatic fully developed friction factors and the critical Reynolds number at which transition to turbulent flow occurred; the axial variation of local Nusselt number at various combinations of Reynolds and Rayleigh numbers; the axial variation of top and bottom wall temperatures; and the variation of fully developed Nusselt number with Reynolds number for each tube. The descriptions of the heat transfer facility and equipment and of the experimental procedures which follow are necessarilly brief, however complete details are available in Haine (1984).

TABLE 1

DETAILED DIMENSIONS OF TEST TUBES

Tube Number	9	10	13	14	20
Tube outside diameter D_{os} - mm	12.7	9.53	9.53	15.9	12.7
Tube inside diameter D_i - mm	10.3	8.00	7.04	13.9	10.4
Fin - tip diameter, D_{ft} - mm	7.75	5.46	4.75	10.9	7.47
Number of fins, M	10	16	10	10	16
Relative Fin Height, H	0.248	0.318	0.325	0.216	0.282
Helix angle, α - degrees	0	0	0	0	2.50
Actual Flow Area, A_{fa} - mm^2	73.6	40.6	29.9	137	68.5
Actual Surface Area, A_a - mm^2/mm	54.0	60.0	35.2	67.3	65.0
Hydraulic Diameter, D_h - mm	5.45	2.71	3.39	8.15	4.21

Heat Transfer Loop
A schematic diagram of the loop is shown in Fig.1. Distilled water from the accumulating tank was pumped at a controlled (by the bypass flow) rate through the test section where energy from the heated tube was absorbed. Using pre-calibrated copper-constantan thermocouples, the upstream bulk temperature was measured immediately before the heated section whereas the downstream bulk temperature was measured at the mixing chamber. The loop water was subsequently cooled in the heat exchangers and later the volumetric flow rate was measured using two variable-area flowmeters prior to return to the accumulating tank.

Test Section
The five finned tubes were supplied by the Forge-Fin Division of Noranda Metal Industries. Each tube had a number assigned to it by the manufacturer and the same numbers were used here. All dimensions for the tubes were provided by the manufacturer and are listed in Table 1. For each finned tube, the longitudinal fins were of similar profile (although fin thicknesses varied somewhat as discussed later) and were uniformly spaced around the inner circumference. This left two major geometric parameters to influence hydraulic and thermal performance, viz the number of fins (M) and the relative fin height (H). Tubes 9, 13 and 14 had the same M but different H, tubes 10 and 13 had almost the same H but different M, whereas tubes 10 and 20 had the same M and fairly similar H but tube 20 had a shallow spiral. A smooth tube of 15.9 mm inside diameter was also tested to provide a base for comparison with the internally finned tubes. Because of the abundant literature available on smooth tubes for comparison, it has been common practise with experiments on finned tubes to first test a smooth tube as a means of verifying the various experimental techniques. The total length of each finned tube test section was 3.2 m of which 2.2 m served as the hydrodynamic entry length. For the smooth tube test section, the total length was 5.65 m and the first 3.4 m was entry length. Fig.2 shows the details of the heated section for each finned tube.

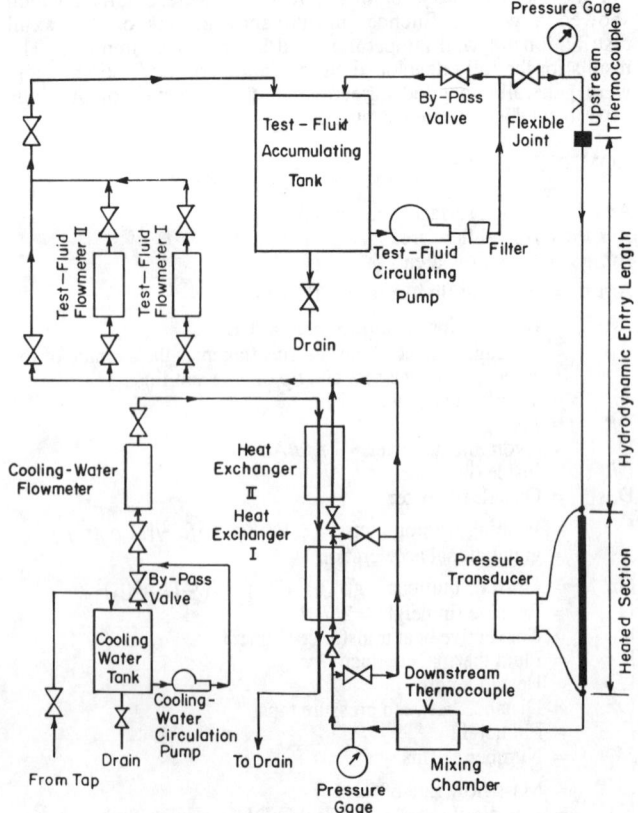

Fig. 1: Schematic diagram of the experimental set-up

Pressure drop measurements across the pressure taps were made using precalibrated pressure transducers. The top and bottom wall temperatures over the test section were measured at 27 axial stations (see Fig.2 for spacing) using copper-constantan thermocouples. Each tube was heated using two insulated electrical heating wires of 0.81 mm diameter and 2.08 ohms/m, tightly wrapped around the tube in parallel. Each tube was completely covered by the wires (except around the thermocouple wells) in order to provide uniform heat input axially. The power supply to the heating wire was controlled by an isolation transformer and a power variac. The input power was measured by a digital wattmeter. In order to minimize external heat losses, each tube (including the thermocouple wells) was covered with a 25 mm layer of fiberglass insulation. For each test, the electrical power input was compared to the energy absorption rate of the cooling water based on water mass flow rate and bulk temperature rise across the test section. Runs with more than 8% heat balance error were rejected. In fact, about 90% of the recorded runs had heat balance errors within 5%.

Procedures and Data Reduction

For each tube, a set of runs without heating was performed at various flow rates to evaluate the isothermal friction factors in the laminar, transition and turbulent regimes. Flow rate increments were kept small enough to detect any change in trend in pressure drop with Reynolds number, so that in particular, an accurate estimate could be obtained of the critical Reynolds number. Heat transfer tests were run on each tube following insitu calibration of the wall thermocouples. For each tube, several sets of runs were conducted varying the flow rate and keeping a constant heat flux, or vice versa. The ranges and limits of flow rates and heat inputs depended on the tubes being tested. The ranges of operating conditions for each finned tube are listed in Table 2. The mean values of Reynolds, Rayleigh and Prandtl numbers were evaluated at the average bulk temperature. The various parameters which were computed from the measured data (friction factor, Nusselt number, etc.) are defined in the nomenclature. For each wall thermocouple station, the local Nusselt number (Nu_x) was based on the local wall and bulk temperatures. The local bulk temperature across the heated section was evaluated by linear interpolation between the inlet and outlet bulk temperatures. The local wall temperature was taken to be the average of the top and bottom thermocouples. The local Reynolds, Rayleigh and Prandtl numbers were evaluated at each station based on the local bulk temperature. For the average fully developed Nusselt number ($\overline{Nu}$), this was taken to be the average of the local Nusselt numbers over the last fourteen stations. Finally, it is noted that the experimental uncertainties were estimated to be as follows: f, ±8%; Re, ±4%; Ra, ±12% and Nu, ±14%. For the latter, the data reported herein were determined assuming a negligible effect due to axial heat conduction within each finned tube wall. Calculations based on a simple conduction model, indicated that conjugate effects were indeed negligible, this being mostly because wall temperature variations were never very non-linear.

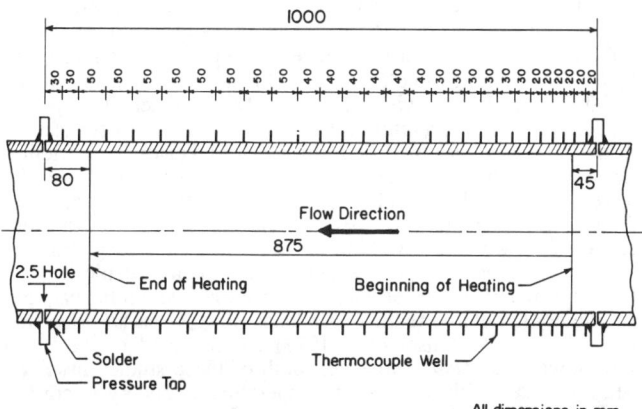

Fig. 2: Details of the heated section for finned tubes

TABLE 2

RANGES OF OPERATING CONDITIONS

Tube Number	Mass Flowrate, ṁ(Kg/Hr)	Input Power, Qe, Watts	Mean Reynolds Number, Re_h	Mean Modified Rayleigh Number, Ra x 10⁻⁶	Mean Prandtl Number, Pr	Number of Test Runs
9	47.3-290.3	740-1470	1640-8900	0.55-1.83	3.7-5.3	18
10	21.7-235.9	240-1725	610-8280	0.011-0.13	3.3-6.4	26
13	24.8-235.2	500-1940	1050-9640	0.093-0.54	4.0-6.5	18
14	101.5-467.5	500-1250	2310-10,800	1.72-5.44	4.3-5.1	14
20	25.1-232.3	1000-1450	740-5130	0.32-0.43	3.7-5.2	7

RESULTS AND DISCUSSION

The experimental results are presented and discussed next. All data were studied carefully and conclusions were drawn where possible with the main concern being the effects of geometry and free convection on the heat transfer and flow characteristics. Comparisons to other investigators were made wherever possible, and correlations were developed where practical.

Smooth Tube

As documented by Haine (1984), the heat transfer results for the smooth tube (Rei up to ~15,000) were generally in good agreement with published circular tube data. For this reason and due to space limitations, no specific results are presented here, but rather only a brief summary is given. Examination of the axial distributions of top and bottom wall temperatures and local Nusselt numbers, revealed the presence of secondary flow and their developing pattern, similar to the results of El-Hawary (1980). The fully developed Nusselt number results were in good agreement with Ede (1961), whereas relative to the Dittus-Boelter correlation, the agreement improved gradually at higher Reynolds number. The effect of free convection was found to decrease with increasing Reynolds number to become negligible at Rei >≈ 6000.

Internally Finned Tubes
a) Friction Factor Results

Fig.3 shows the isothermal pressure drop results with the friction factor and Reynolds number evaluated using the equivalent hydraulic diameter. For each tube, a sharp change in slope occurs to indicate the transition from laminar to turbulent flow. For each tube, transition occurs at a Reh value which is significantly lower than the commonly accepted value of 2100 to 2300 for smooth tubes. The critical Reynolds numbers, on the basis of both Dh and Di, are listed in Table 3. Tube 14 (smallest H) has the highest Recr, whereas tube 13 (largest H) has the lowest Recr. Both of these tubes have 10 fins. In fact, the data presented in Table 3 shows (for fixed fin number, M) that Recr decreases substantially with increasing relative fin height, H. This same trend was noted by Watkinson et al (1975). Regarding the effect of M on Recr (for fixed H), the limited data suggest little influence for H ≈ 0.25, but for higher fin heights (H ≈ 0.32), increasing M increases Recr.

TABLE 3

CRITICAL REYNOLDS NUMBERS BASED ON THE
INSIDE AND HYDRAULIC DIAMETERS

Tube Number	H	M	$(Re_{cr})_i$	$(Re_{cr})_h$
9	0.248	10	1530	810
10	0.318	16	1300	440
13	0.325	10	600	289
14	0.216	10	2130	1249
20	0.282	16	1465	593

From the data shown in Fig.3, friction factors in the laminar flow region were lower than predicted by the theoretical equation for smooth tube (f=16/Re), but no particular trend was noted with respect to the fin height. The effect of the number of fins could not readily be determined for there was insufficient data, but tubes 10 and 20 (both with 16 fins) are by far most detached from the rest. The fitted lines for the various tubes in the laminar region all have similar slopes, which are about the same as that for smooth tubes. Hence the customary dependence of f on Re_h is preserved. This fact has been reported by several previous investigators, e.g. Vasilchenko and Barbaritskaya (1969) and Watkinson et al (1975a).

For the turbulent flow region, as shown in Fig.3, the present low turbulent Reynolds number data deviate considerably from the smooth tube correlation. But the fitted curves for the various tubes show close similarity in slope. This slope suggests a Reynolds exponent of about 0.55 compared to 0.25 to 0.20 for smooth tube correlations beyond $Re \approx 10^4$. A similar slope was also obtained for the present smooth tube for Re < 6000. Furthermore,(and more importantly), a close examination of the friction factor results of previous investigators for $5000 < Re_h < 10^4$, readily shows that higher slopes were obtained in this Re_h interval [see e.g. Hilding and Coogan (1964), Bergles et al (1971), Watkinson et al (1973,1975b) and Vasilchenko and Barbaritskaya (1969)]. In particular, the f vs Re_h plots of Watkinson et al (1973) clearly show higher slopes in this Re_h interval and changes in slope at $Re_h \approx 10^4$ to the familiar Blasius slope. Hence it is believed for the present internally finned tubes that the slopes would approach 0.25 to 0.20 as Re_h approaches approximately 10^4. Direct evidence of this is contained in Fig.4 for tubes 9 and 14; note here that the Carnavos (1980) slope is 0.20.

Fig.4 again shows the isothermal friction factors for each tube, but also now includes the results of pressure drop measurements made while heating. For the diabatic tests, the properties were evaluated at the average of the inlet and outlet temperatures. All of the friction

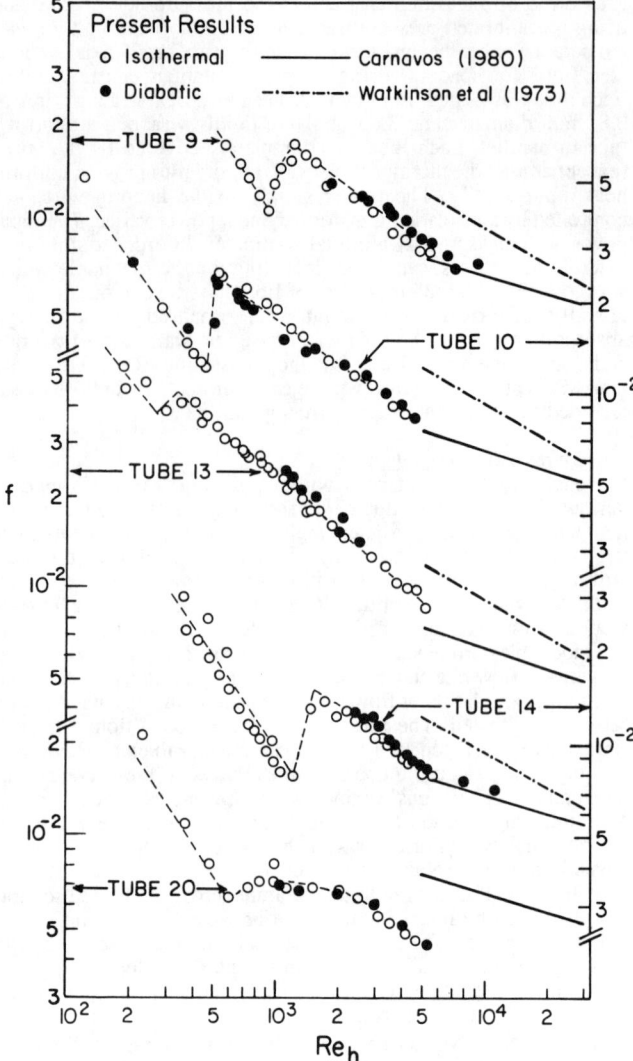

Fig. 4: Isothermal and Diabatic friction factors

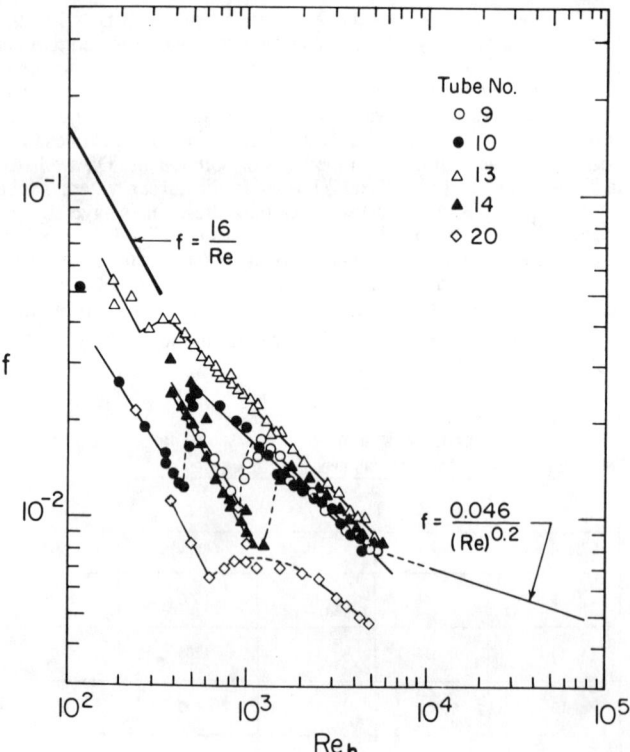

Fig. 3: Isothermal friction factors for internally finned tubes

factors with heating fell within ±10 % of the isothermal values, being generally slightly higher. The diabatic friction factors were expected to be slightly lower than the isothermal ones, due to a drop in the fluid viscosity near the heated surface. The discrepancy is probably due to free convection effects. Most of the present data (with heating) were collected at Ra* numbers in the range 0.11 x 10^5 to 0.54 x 10^7, a fact that suggests the presence of secondary flows strong enough to alter near-wall axial velocity distributions and thereby to override viscosity effect. This argument is supported by the following observations. Data for tubes 9 and 14 correspond to the highest values of Ra*, resulting in higher diabatic friction factors than isothermal ones. Data for tube 10 correspond to the lowest values of Ra* resulting in hardly any differences between diabatic and isothermal factors. Similar effects (but much more pronounced) were observed in the laminar flow results of Rustum (1983). Bergles et al (1971) reported that their diabatic friction factors were slightly lower than the isothermal friction factors, but an attempt to determine the dependence of μ_w/μ_b on f_{dia}/f_{iso} was inconclusive. Watkinson et al (1973) also found f_{dia} to be 5% to 10% lower than f_{iso}. However, both of these studies involved heating water with a constant temperature boundary condition (condensing steam), hence compared to the present uniform heat input boundary condition, free convection effects would be both less and different.

The present friction factor results are compared in Fig.4 to the correlations of Watkinson et al (1973) and Carnavos (1980). The Watkinson correlation for straight fins for isothermal friction factor is given by:

$$f = 0.406 \ (b/Dh)^{0.16}/(Reh)^{0.39}, \quad\quad\quad (1)$$

for $5000 \leq Reh \leq 75,000$ and $0.21 \leq b/Dh \leq 0.49$. The present fins were modelled as trapezoidal in profile and the inter-fin spacing (b) was calculated as the average of the arc lengths at the root and tip of adjacent fins. All (b/Dh) values were within the correlation range. As can be seen in Fig.4, the present results fall substantially below the predictions of equation (1). At Reh=5000, the present isothermal friction factors are about 35% lower for tubes 9, 10 and 14, and about 28% lower for tube 13 (thickest fins). The discrepancies may be due to differences in fin thicknesses and/or surface finish. Alternatively, it may be that equation (1) does not serve well at low Reh. In any case, the agreement with Carnavos (1970) is much better. The Carnavos correlation is given by:

$$f = [0.046/(Reh)^{0.2}] \ (Afa/Afn)^{0.5} \ (cos \ \alpha)^{0.5}, \quad\quad (2)$$

The Reynolds number range for this correlation for diabatic friction factor for both straight and spiral finned tubes was approximately 10^4 to 10^5, however the correlation was extrapolated down to Reh =5000 for comparison purposes. At Reh ≈ 5000, the present diabatic results for tubes 9, 10 and 14 are only slightly higher than equation (2) by about 5% at most, whereas the isothermal result for tube 13 is about 20% high. The worst agreement is for tube 20 (spiral fins) which is about 40% low. This tube was retested, with no change, and the discrepancy remains unexplained.

b) Heat Transfer Results

Wall Temperatures. Two samples (at two different Reh) of the axial variations of walltemperature were selected for each of tubes 9 and 14, and the results are presented in Figs.5 and 6. Each plot shows top and bottom temperatures in an attempt to detect the presence of buoyancy-induced secondary flows. Each sample has the same pattern of a rapid rise in Tw at the start of the heated section, followed by a small dip in the Tw distribution at x/Dh ≈ 30. Since the fluid bulk temperature rises linearly over the heated section, this dip corresponds to a small depression in the wall-to-fluid temperature difference, and hence also a small bump in the Nux distribution. Qualitatively, the Tw behaviours shown in Figs.5 and 6 are similar to our experimental results for the smooth tube where the behaviour was attributed by Haine (1984) to completion of the developing secondary flow. But unlike the smooth tube, the finned tube wall temperatures do not over-shoot, hence the subsequent temperature depressions are much less pronounced than for the smooth tube. Also the difference between top and bottom temperatures (noticeable for the smooth tube) are now negligible. This suggests that the presence of fins tends to minimize the effect of secondary flow. It is also noted that there is no obvious effect of Reynolds number on the two cases for each tube.

Local Nusselt Numbers. Local Nusselt numbers are plotted against reduced length (X+) for tubes 9 and 14 in Figs.7 and 8 respectively. In an attempt to determine the effect of free convection, two plots at different Ra* but similar Reh are included in each figure. Each tube involved M=10 and low H. For tube 9 (Fig.7; H=0.248), the Ra* differences are by a factor of about 2. For tube 14 (Fig.8; H=0.216), the Ra* levels are higher and differ by a factor of about 3. For tube 9 (larger H but smaller Ra* difference), Fig.7 shows no significant difference between the two sets of data. For tube 14, Fig.8 shows higher Nux values at the higher Ra*, hence suggesting the presence of buoyancy effects/secondary flow. However, the differences are no greater than about 5%. In fact, this difference could be attributed to the small differences in Reh and Pr for the two cases. For example, assuming pure forced convection, the local Nusselt number ratio might be estimated as $(2570/2310)^{0.8}$ $(4.31/4.84)^{0.4} = 1.040$, or about a 4% difference. The parallel calculation for tube 9 yields a ratio of 1.002, i.e. essentially unity. But at the same time, this explanation does not necessarily mean that buoyancy effects are negligible. Because of the small Ra* difference, buoyancy effect would not change much, i.e. would remain nearly constant. For both tubes, the general Nux vs X+

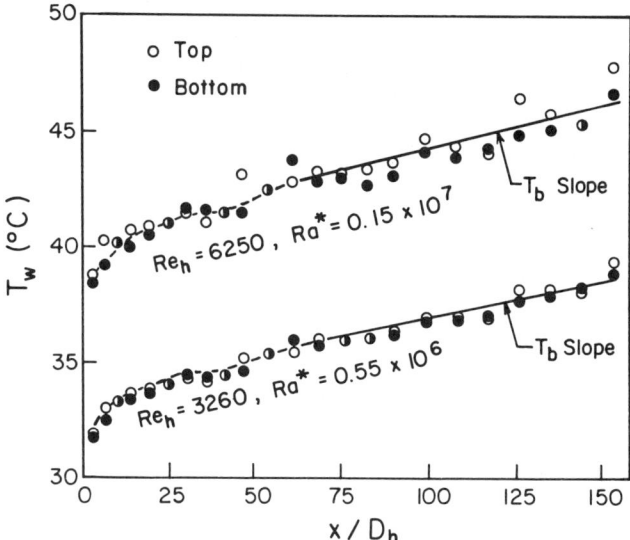

Fig. 5: Axial distributions of wall temperature for Tube No. 9

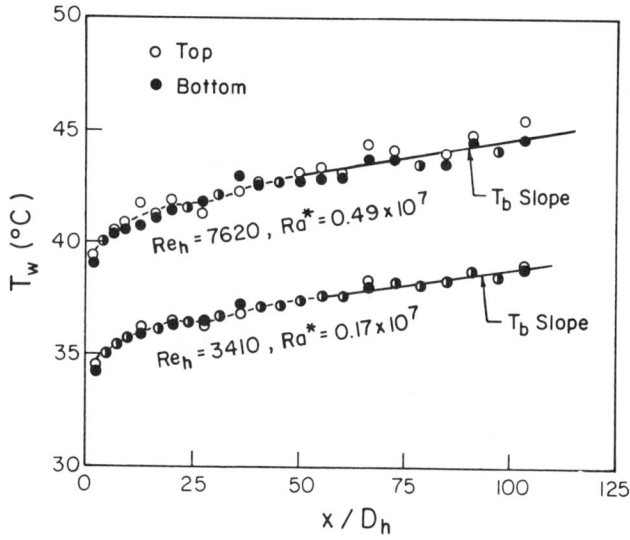

Fig. 6: Axial distributions of wall temperature for Tube No. 14

pattern contains a small bump in Nux in the vicinity of the wall temperature dip. This alone suggests some buoyancy effect, at least for finned tubes having small fin heights. But overall, relative to the smooth tube, buoyancy effects (for comparable Ra*) are probably considerably smaller due to interference by the fins and interactions with turbulence-induced secondary flows of the second kind.

Thermal entry lengths are considered next, and typical data for one tube is shown in Fig.9 which consists of three runs at different Reynolds numbers. These results for tube 9 are presented in the form of a graph of Nux/N̄u vs x/Dh. The scatter of the data, together with the narrow range of Reynolds number (2000-8000), made it difficult to isolate the effect of Reynolds number on the thermal entrance length. Therefore, a single curve was faired through all the data for each tube to facilitate evaluation of the thermal entrance length which was taken to be the axial distance required for the local Nusselt number to first come within 5% of the fully developed Nusselt number.

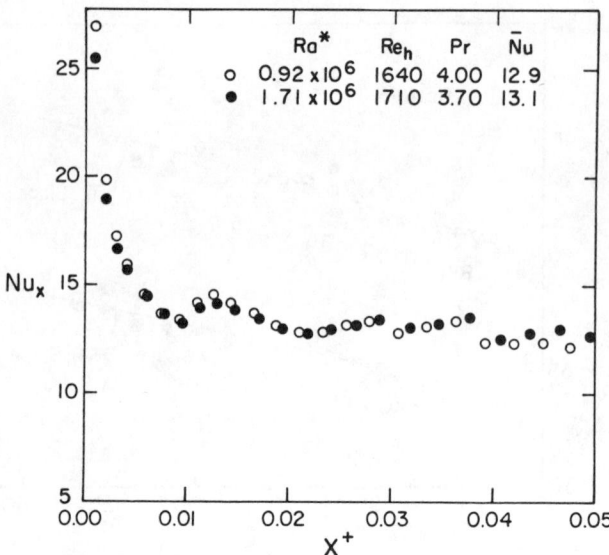

Fig. 7: Effect of Rayleigh number on local Nusselt number for Tube No. 9 for R$eh \approx$ 1680

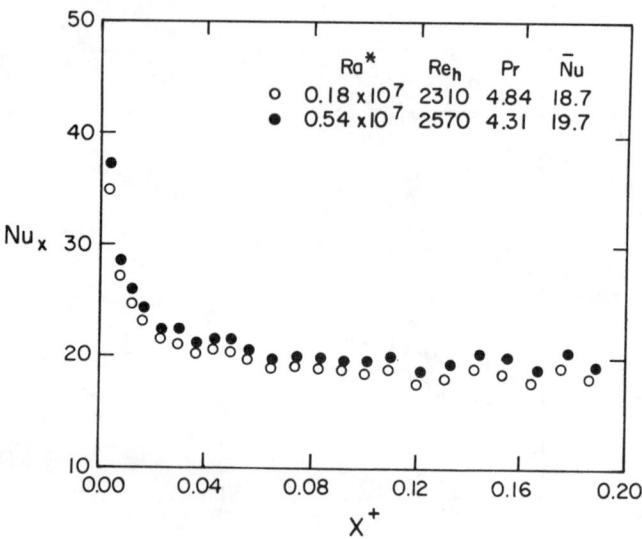

Fig. 8: Effect of Rayleigh number on local Nusselt number for Tube No. 14 for R$eh \approx$ 2440

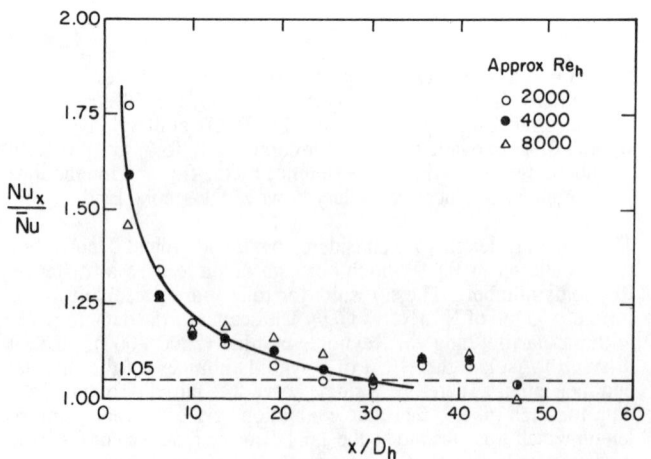

Fig. 9: Thermal entry length for Tube No. 9

TABLE 4

THERMAL ENTRY LENGTH VALUES FOR INTERNALLY FINNED TUBES

Tube Number	H		Pr = 4 - 6	Thermal Entry Length	
		M	Re_h	x/D_h	x/D_i
9	0.248	10	2000-8000	30	15.9
10	0.318	16	2000-8000	41	13.9
13	0.325	10	2000-8000	17	8.2
14	0.216	10	2000-8000	34	19.9
20	0.282	16	2000-5000	13	5.3

The results for thermal entry length are listed in Table 4 in terms of both x/Dh and x/Di.(The reader is reminded that the flow for each tube was fully developed hydrodynamically at the start of the heated section, and these hydraulic entry lengths ranged from about 270 to 800 hydraulic diameters or about 160 to 300 tube inside diameters.) The data in Table 4 show that ≈ tube 10 has the longest thermal entry length in terms of Dh, but it ranks only third longest in terms of Di. Tube 20 (spiral fins) has the shortest entrance length which is not surprising since the induced swirl would promote rapid thermal mixing. Among the tubes with M=10, tube 14 (lowest H) has the longest entrance length while tube 13 (highest H) has the shortest entrance length. This suggests that an increase in H tends to shorten the thermal entry length. This seems plausible since with increasing H, heat from fin tips is progressively injected further into the core flow region. Regarding the effect of M for straight fins, comparing tubes 10 and 13 (H≈0.32) indicates that increasing M increases the thermal entry length. This is not expected if the core region governs thermal development. However, if the inter-fin region governs, the explanation may reside in reductions in velocity and turbulent diffusion as the bay regions become narrower with increasing M. In any event, the present data is far too limited to draw any definite conclusions on either the effect of number of fins or spiralling of fins.

The present results for thermal entry lengths for internally finned tubes are novel; i.e. there are no published data available for comparison. For a smooth tube for the same Reynolds range, thermal entry lengths would be expected to be about 9 to 12 diameters. On the basis of Di, the present values for straight fins average about 14 inside diameters, i.e. only a bit higher than for the smooth tube. But on the basis of Dh, the present entry lengths are substantially greater than for the smooth tube.

Returning again to local Nusselt numbers, the distributions of Nux over the thermal entry regions for the other finned tubes were qualitatively similar to the results for tube 9 shown in Fig.9, however scale details were basically unique for each finned tube. Nonetheless, some quantitative similarities were noted. For example, for tubes 9, 13 and 14 (all with M=10), Nux/N̄u achieved a value of 1.25 at about x/Dh= 8 ± ≈ 1, whereas the average Nux over the thermal entry region was about 120% of N̄u. For tubes 10 and 20 (both with M=16), the average (over the entry length) Nux was noticeably higher, being about 130-135% of N̄u. Although the present results are very limited in scope, it is hoped that they will provide some guidance to designers of compact heat exchangers employing internally finned tubes. A conservative design approach (for similar finned tubes) would be to assume thermal entry lengths to be the same (on a Di basis) as for the smooth tube with average local Nusselt numbers being about 1.20 N̄u. Of course this suggestion is only tentative and would have to be refined when additional data becomes available for a wider assortment of internally finned tubes.

Fully Developed Nusselt Numbers. Fully developed heat transfer data, based on hydraulic diameter and actual surface area, are presented in Fig.10 for the five internally finned tubes. The Nusselt-Prandtl modulus [Nu/Pr$^{0.4}$] is plotted against Reh for each tube in a log-log plot to provide direct comparison to the smooth tube result as given by the Dittus-Boelter equation. Reynolds numbers ranged from 600 to 10,000, but for most data, Reh>1200 for which the flow is definitely turbulent (see Fig.3). Almost all of the data falls below the smooth tube line as is usually the case for internally finned tubes. However, the slopes for the lines drawn through the data for each tube (including the low Reh data for tube 10), were similar but slightly lower than that (0.8) usually encountered for smooth tubes. The data for tubes 10 and 20 (each with M=16) are the most detached from the Dittus-Boelter line. The proximities of the remaining three tubes (each with M=10) to the Dittus-Boelter line is consistent with the predictions of Said and Trupp (1984); viz,for fixed M and Reh, that Nu first decreases with increasing H to reach a minimum at H≈0.3 (for Reh=50,000) before increasing with H. At lower Reynolds number, the minimum could conceivably shift to H≈0.25 as is implied by the ranking seen in Fig.10.

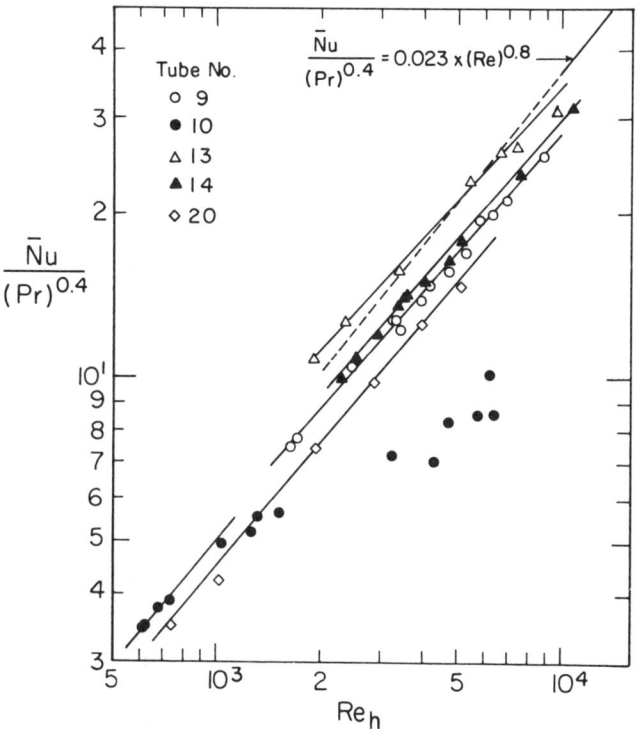

Fig. 10: Fully developed Nusselt numbers for internally finned tubes (based on hydraulic diameter)

The individual results for each finned tube are shown in more detail in Fig.11 where the plot for each tube includes the Carnavos (1980) correlation extrapolated to Reh<10^4. The Carnavos correlation for both straight and spiral fins is given by:

$$\overline{Nu}/(Pr)^{0.4}=0.023\,(Reh)^{0.80}(Afa/Afc)^{0.10}(An/Aa)^{0.50}(\sec\alpha)^3\quad...(3)$$

As can be seen in Fig.11, except for tube 10, the present results are consistent with and generally in good agreement with equation (3). This good agreement includes tube 20; an outcome that is rather surprising in view of the poor agreement with friction factor as noted earlier. For each tube, the experimental data starts a little higher than the Carnavos line and then converges to meet (except tube 10) it at higher Reynolds numbers. This pattern probably reflects some buoyancy effect at the lower Reynolds numbers.

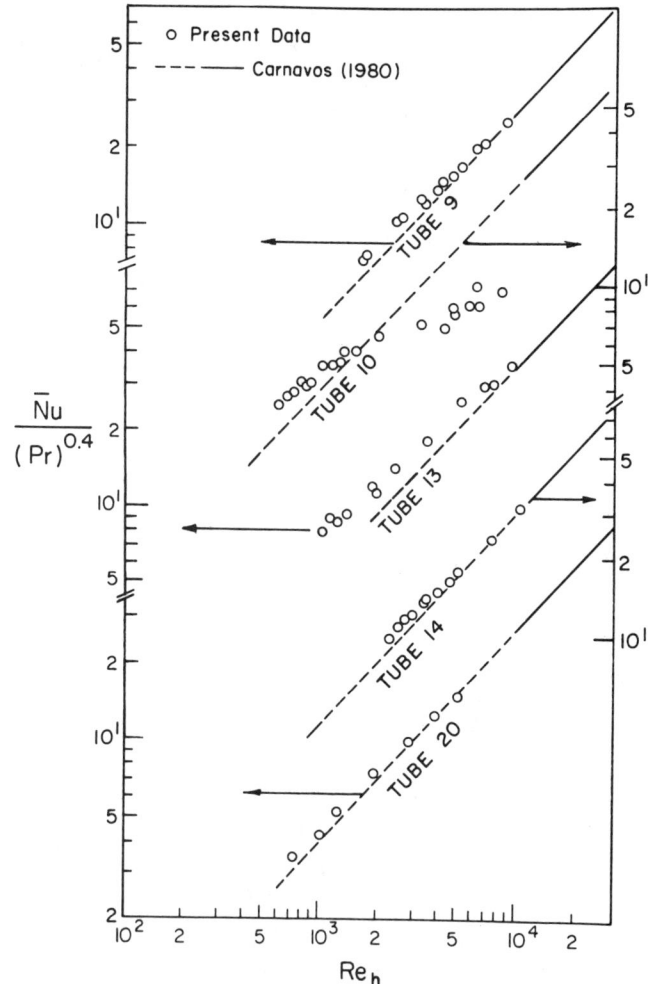

Fig. 11: Fully developed Nusselt numbers

For tube 10 (M=16,H=0.318), as shown in Fig.11, the experimental data crosses the Carnavos line at Reh ≈ 2000 and then falls progressively below the line at higher Reynolds numbers. The data was rechecked carefully and appeared to be valid. The reason for this behaviour is not clear. There were some early difficulties with wall temperature measurements for this tube, but it was thought that the problem was corrected by having reset the thermocouple junctions deeper in the tube wall. Another partial explanation revolves around two peculiarities of tube 10 relative to the other finned tubes. Tube 10 had, by far, the thinnest tube wall (see Table 1). Also, amongst the straight fin tubes, tube 10 had the thinnest fins (see γ values included in Table 5). In addition, all tube 10 tests for Reh<2000 involved input powers of 240 to 990 watts, whereas for Reh>2000, input power levels were in the range of 1000 to 1725 watts. At these higher heat flux levels, because of the relatively thin wall and fins, temperature gradients (both circumferentially in the tube wall and especially radially in the long fins), would be higher than for the other finned tubes which would have the same thermal conductivity but wider conduction cross-sections. But as shown by Haine (1984), fin inefficiencies could only account for a small portion of the discrepancy. Ultimately, since measurement errors on wall temperature (too high) could be neither confirmed nor refuted, the data was retained, however it remains suspect beyond Reh≈2000.

The present fully developed Nusselt numbers were compared, not only to Carnavos (1980), but also to Watkinson et al (1973) and to Said and Trupp (1984). The results of the comparisons are summarized in Table 5. For the Carnavos correlation, equation (3), the agreement varies from excellent to good, as already mentioned. Tubes 9, 14 and 20 averaged within less than 5% difference, whereas tubes 10 and 13 averaged 18.7% and 15.8% difference, respectively. Also, as reflected in Fig.11, apart from tube 10, the agreement for individual data points is best at the higher Reynolds numbers. Considering the facts that the Carnavos correlation is within ±10% and has been extrapolated to $Reh<10^4$, it is considered that the general level of agreement of $\overline{Nu}$ to equation (3) is very good.

In contrast to the good agreement to Carnavos (1980), the present Nu results do not compare well to the correlation of Watkinson et al (1973), which is given by:

$$\overline{Nu}=0.212 \, (Reh)^{0.6} \, (b/Dh)^{0.34} \, (Pr)^{0.33} \, (\mu/\mu w)^{0.14} \, ,.....(4)$$

for the ranges $5000 \le Reh \le 100,000$ and $0.21 \le (b/Dh) \le 0.49$. All of the experimental data for tubes 9, 10 and 14 fall considerably below the predictions of equation 4. The agreement is best for tube 13, as was also the case for the friction factor comparisons discussed earlier. In fact, the agreement with equation (4) for tube 13 is better than for equation (3) (see Table 5), which may be because of its relatively thick fins. Hence the discrepancies for tubes 9, 10 and 14 may be related to fin thickness. On the other hand, Carnavos also experienced poor agreement with equation (4) and noted that the apparatus used by Watkinson et al (1973) did not have the capability for a heat balance and hence the generated data was potentially of a lower accuracy level. Otherwise, it is noted that heated lengths for Watkinson et al (1973) were comparable to those of Carnavos (1980), each being roughly of the order of ten entrance lengths, hence average test-section Nusselt numbers should have been no greater than about 2 to 3% of the fully developed Nusselt numbers. The Said and Trupp (1984) correlation is given by:

$$\overline{Nu}/(Pr)^{0.4}=0.027 \, (Reh)^{0.774} \, (b/Dh)^{0.397} \, (An/Aa)^{-0.168} \, ,......(5)$$

for $0.2 \le H \le 0.8$, $6 \le M \le 14$, $0.31 \le b/Di \le 0.77$, and $25,000 \le Reh \le 150,000$. Before reviewing the comparison to Said and Trupp (1984), it is noted that their correlation is based on theoretical/numerical predictions for pure forced convection for $Reh>25,000$, involving a perfectly conducting, well-defined trapezoidal fin profile having a half-fin thickness angle $\gamma=3°$. In view of this, the agreement reported in Table 5 is remarkable good. The

largest differences are for tube 13 which has the thickest fins (equivalent $\gamma =7.66°$).

Turning next to correlation of the present experimental data, a least-squares-fit was obtained using the correlation form:

$$\overline{Nu}/(Pr)^{0.4}=C1 \, (Reh)^{C2} \, (Afa/Afc)^{C3} \, (An/Aa)^{C4} \, ,.......(6)$$

In view of the good agreement to Carnavos(1980), the geometric parameters in equation (3) were retained for the general form. For the spiral finned tube 20, $(\sec \alpha)^3$ as per equation (3) amounted to only 1.0029 and hence could safely be ignored. The coefficients for best fit of all the data for $600<Reh<10,000$, were C1=0.023, C2=0.787, C3=1.08 and C4=1.11. All the data were fitted by equation (6) to within a standard deviation of 6.5%. Here the standard deviation $\equiv [(1/N) \, \Sigma \, (\overline{Nu}corr - \overline{Nu}exp)^2]^{1/2}$ where N is the number of data points, and is expressed as a percent of the average Nucorr in the Reh range. It is noted that the least-squares-fit for equation (6) produced a Re exponent of 0.787, only slightly below the Carnavos slope of 0.80. The present correlation is not shown in Fig.11 in order to prevent crowding, however it lies close to the Carnavos line and in some cases, is almost coincident. Hence even for equation (6), the present low Reh data approach this correlation from above. Again this reenforces the idea of a lingering buoyancy effect manifested as a changing Re slope towards its asymtotic value of 0.80 at high Re.

The fully developed Nusselt numbers were also computed based on inside diameter and nominal heat transfer area, and these results are presented in Fig.12 together with the smooth tube line as predicted by the Dittus-Boelter equation. The enhancement of heat transfer over the smooth tube ranged considerably depending on Reynolds number, being highest for tubes 10 or 13, and lowest for tube 14. Also obvious is the strong dependence of the heat transfer coefficient on both fin height and fin number. Data for tubes 9, 13 and 14 (all with same number of fins) show an increase of Nusselt number with increasing fin height, while the low Rei data for tubes 10 and 13 (both with similar fin height) show an increase of Nusselt number with increasing fin number. Also comparing the results for tubes 10, 13 and 20 suggest a stronger effect of fin height than fin number; a fact reported by many investigators.

TABLE 5

COMPARISON OF PRESENT $\overline{Nu}$ WITH OTHERS

Ref A = Carnavos (1980), Ref B = Watkinson et al (1973), Ref C = Said & Trupp (1984)

Tube Number	H	M	γ (Deg.)	Ref.[A] % diff.	Ref.[B] % diff.	Ref.[C] % diff.
9	0.248	10	4.84	±4.2	-31.0	+13.2
10	0.318	16	4.05	±18.7	-43.5	±14.5
13	0.325	10	7.66	+15.8	±5.4	+35.2
14	0.216	10	4.54	±4.1	-25.4	+19.5
20	0.282	16	3.42	±2.8	—	—

NOTES:
1) Percentage difference (% diff.) $\equiv \left[\frac{\overline{Nu}_{exp} - \overline{Nu}_{corr}}{\overline{Nu}_{exp}} \right] 100$.

2) Each value listed is the average of the percentage difference magnitudes for all data points over the range of Reynolds number for each tube.

3) The (±) designator simply indicates the experimental data is always above (i.e. higher) the pertinent correlation (+), or always below the pertinent correlation (-), or both above and below (±) in the interval.

4) Each listed γ value is the half-fin angle that would be obtained if the actual fin material cross-sectional area was distributed as M identical trapezoidal fins (as per [C])of height H.

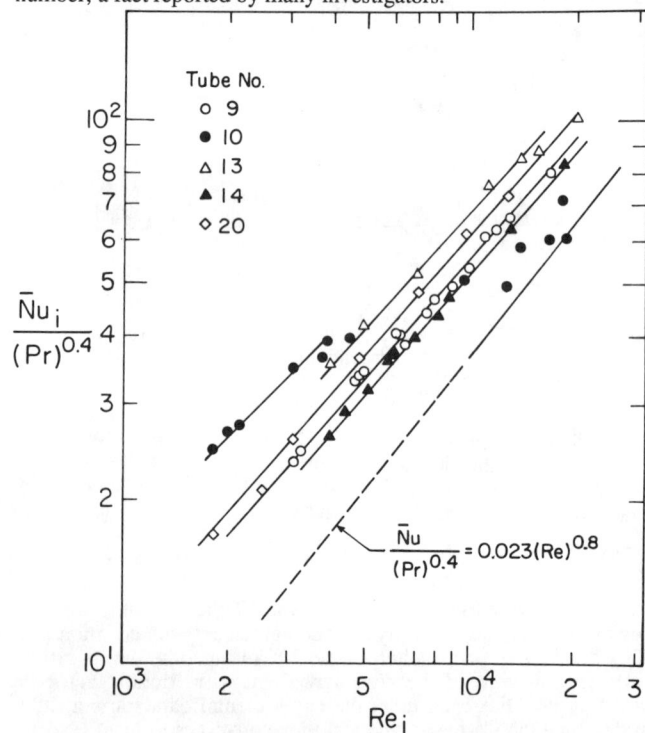

Fig. 12: Fully developed Nusselt numbers for internally finned tubes (based on inside diameter)

CONCLUDING REMARKS

The heat transfer and fluid flow performance were determined for a smooth tube and five integral inner-finned tubes of various designs for heating water in low Reynolds number turbulent flow. Data were presented for both the thermal entry region and the fully developed region. From the present experimental results, the following conclusions can be drawn:

1) Friction factor results for the internally finned tubes were similar to those for the smooth tube, i.e., a high slope of about 0.55 for $Re_h < 6000$. At higher Re_h, the results merged to attain good agreement with Carnavos (1980) except for tube 20. For the finned tubes, the critical Reynolds numbers were found to increase slightly with increasing fin number and to decrease sharply with increasing relative fin height.

2) The wall temperature and local Nusselt number results for internally finned tubes suggested the presence of secondary flows and their developing pattern similar to that of the present smooth tube. However, the effect on the heat transfer characteristics were small, compared to the smooth tube, suggesting that the presence of fins tend to suppress buoyancy effects.

3) The thermal entry lengths for the internally finned tubes decreased with increasing fin height. The effect of number of fins and spiralling of fins on thermal entry length was inconclusive. No particular effect of Re_h was noted in the range covered. The thermal entry lengths, based on the inside diameter, were of the same order of magnitude as those for the smooth tube; but based on the hydraulic diameter, these values were significantly higher.

4) With the exception of tube 10, the fully developed Nusselt numbers, based on hydraulic diameter, were in good agreement with Carnavos (1980) with the level of agreement improving at higher Reynolds numbers. At low Reynolds numbers, the present results were somewhat higher due to the influence of buoyancy effect. On a Re_i, Nu_i basis, at a given Reynolds number, Nu_i increased with both increasing H and M, with fin height having the stronger effect.

REFERENCES

Bergles, A.E., Brown, G.S. and Snider, W.D.,1971,"Heat-Transfer Performance of Internally Finned Tubes", 11th Nat.Heat Transfer Conf., Tulsa, Aug 1971, ASME paper 71-HT-31.

Carnavos, T.C.,1980,"Heat Transfer Performance of Internally Finned Tubes", Heat Transfer Engineering, Vol.1, No.4, pp.32-37.

Ede, A.J.,1961,"The Heat Transfer Coefficient for Flow in a Pipe", Int.J.Heat Mass Transfer, Vol.4,pp.105-110.

El-Hawary, M.A.,1980,"Effect of Combined Free and Forced Convection on the Stability of Flow in a Horizontal Tube", ASME J. of Heat Transfer, Vol.102,pp.273-278.

Haine, H.,1984,"Experimental Investigation of Low Reynolds Number Turbulent Flow and Heat Transfer in the Thermal Entrance Region of Internally Finned Tubes", M.Sc.Thesis, Univ.of Manitoba.

Hilding, W.E. and Coogan,C.H.,1964,"Heat Transfer and Pressure LossMeasurements in Internally Finned Tubes", ASME Sym.on Air-Cooled Heat Exchangers, Cleveland, pp.57-85.

Mirza, S.,and Soliman, H.M.,1985,"The Influence of Internal Fins on Mixed Convection Inside Horizontal Tubes", Int.Comm.Heat Mass Transfer, Vol.12,pp.191-200.

Rustum, I.,1983,"Experimental Investigation of Laminar Heat Transfer in the Thermal Entrance Region of Internally Finned Tubes", M.Sc.Thesis, Univ.of Manitoba.

Rustum, I.,and Soliman, H.M., 1988, "Experimental Investigation of LaminarMixed Convection in Tubes with Longitudinal internal Fins", ASME J. of Heat Transfer, Vol.110,pp.366-372.

Said, M.N.A.,and Trupp, A.C.,1984, "Predictions of Turbulent Flow and Heat Transfer in Internally Finned Tubes", Chem.Eng.Commun.,Vol.31,pp.65-99.

Vasilchenko,Y.A.,and Barbaritskaya, M.S., 1969, "Resistance with Nonisothermal Fluid Flow in Tubes with Longitudinal Fins", Thermal Engrg.,Vol.16,No.1,pp.28-35.

Watkinson, A.P, Miletti, D. L. and Tarassoff, P., 1973, "Turbulent Heat Transfer and Pressure Drop in Internally Finned Tubes", AIChE Sym.Series No.131,Vol.69,pp.94-103.

Watkinson, A. P., Miletti, D. L. and Kubanek, G. R., 1975a, "Heat Transfer and Pressure Drop of Internally Finned Tubes in Laminar Oil Flow", ASME Paper No.75-HT-41.

Watkinson, A. P., Miletti, D. L. and Kubanek, G. R., 1975b, "Heat Transfer andPressure Drop of Internally Finned Tubes in Turbulent Air Flow", ASHRAE Tran.,Vol.81,Part 1,pp.330-349.

ACKNOWLEDGEMENT

This research was supported by the Natural Sciences and Engineering Research Council of Canada.

PREDICTION OF STAGNATION POINT HEAT TRANSFER
WITH FREE STREAM TURBULENCE

L. T. Tran and D. B. Taulbee
State University of New York at Buffalo
Buffalo, New York

Abstract

A method of calculation is presented to predict the effects of free stream turbulence on the stagnation point heat transfer. The mean velocity, temperature, and the turbulence variation along the stagnation streamline are predicted with a Reynolds stress model so as to accurately resolve the turbulent normal stresses which govern the production of turbulence in the stagnation flow. The calculations are performed for various Reynolds numbers, free stream turbulence intensities and length scales. It was found that the surface heat transfer depends on the length scale of the turbulence in the approaching flow in addition to the intensity. Good agreement with experimental data is achieved.

Nomenclature

C_μ	k-ε model constants
f_μ	low-Reynolds number functions
$f' \equiv df/d\eta$	similarity velocity variable
g	similarity temperature variable
H	total enthalpy
k	turbulent kinetic energy
ℓ_0	turbulence scale at the cylinder
ℓ_∞	free stream turbulence scale
$\overline{Nu}$	Nusselt Number
$\overline{p}$	static pressure
$Pr=0.7$	Prandtl number
$Pr_t=0.90$	turbulent Prandtl number
$q^2=\overline{u^2}+\overline{v^2}+\overline{w^2}$	mean squared turbulent velocity
R	cylinder radius
$Re= 2\rho_\infty V_\infty R/\mu_\infty$	Reynolds number
$R_T= k^2/\varepsilon\nu$	turbulent Reynolds number
$Tu= u'_\infty/V_\infty$	free stream turbulence intensity
Tu_0	turbulence intensity at the cylinder
U_i	mean velocity in i-direction
U	mean velocity in x-direction
$u_\tau= \sqrt{\tau_w/\rho}$	friction velocity
V	mean velocity in y-direction
V_∞	mean free stream velocity
x	coordinate parallel to wall
y	coordinate perpendicular to wall
$y^+= u_\tau y/\nu$	wall Reynolds number
$\overline{u_i u_j}$	turbulent Reynolds stresses
$\overline{u^2}, \overline{v^2}, \overline{w^2}$	turbulent normal stresses
$\overline{uv}$	turbulent shear stress
$\overline{u_i H'}$	turbulent heat flux
ε	rate of dissipation of turbulent kinetic energy
μ, μ_t	molecular and turbulent viscosity
η	similarity variable
ρ	fluid density
δ	boundary layer thickness
δ_{ij}	Kronecker delta
σ_ε	k-ε model constant

$\tau_w= (\mu\ \partial U/\partial y)_w$ shear stress at wall

Superscripts

$(\ )^*$	refers to similarity variables
$(\ )'$	fluctuating quantity
$\overline{(\ )}$	time average

Subscripts

$(\ )_w$	refers to wall surface
$(\ )_\infty$	refers to free stream condition

Introduction

Because of its practical importance the stagnation point flow has been the subject for numerous studies. The laminar stagnation point flow solution is presented by Schlichting (1979 ed.) and the stagnation point heat transfer is presented by Squire (1965). The laminar flow theory underpredicted many stagnation point heat transfer experimental data. some of which were presented by Van Der Hegge Zijnen (1958), Kestin et al. (1961), Smith and Kuethe (1966), Kestin and Wood (1971). and Lowery and Vachon(1975). Free stream turbulence was found to be the causes of the discrepancies.

The sensitivity of stagnation point heat transfer to free stream turbulence was explained by the theory put forth by Sutera et al. (1963). and Sutera (1965). They theorized that the vorticity amplification due to stretching of vortex filaments in the stagnation point region is the mechanism responsible for this sensitivity. Sadeh et al. (1970) extended the vorticity amplification theory to include the treatment of the outer flow field. These investigations gave a good understanding of the flow mechanism involved in the augmentation of stagnation point heat transfer, but they did not provide the capability for quantitative predictions.

One approach to predict the turbulent stagnation point flow is turbulence modeling. In the context of modeling turbulence in the free stream is characterized by an intensity and a single length scale which are measures of the energy and the size of the eddies. These quantities change as the flow approaches the stagnation region. In the present study the stagnation point process is analyzed by using the Reynolds stress model. As cited in Taulbee and Tran (1988) there have been other studies of stagnation point heat transfer using turbulence models, particularly two equation models. The advantage of the Reynolds stress model over the two-equation model for this problem was discussed by Taulbee and Tran (1988) and will not be repeated here. The Reynolds stress equations which can be reduced to a set of ordinary differential equations when applied to the stagnation streamline are written to calculate the variation of turbulence quantities along the stagnation streamline. The surface heat transfer is then obtained from the boundary layer solution. The effects of free stream turbulence and Reynolds number on the stagnation point heat transfer are discussed. It is shown that a complete description of the problem requires the specification of a turbulence length scale parameter ℓ/D in addition to the turbulent intensity Tu and the Reynolds number. The calculated results are compared to the experimental data.

Formulation and Solution Method

The stagnation point process can be viewed as made up of two parts. First, in the outer flow, an order of magnitude analysis shows that the turbulence does not significantly affect the mean flow. The turbulent stresses are negligible in the momentum equations and the flow is essentially inviscid. Turbulence in the outer flow is not directly affected by the surface except through interaction with the mean flow. Turbulence, starting in the flow far upstream of the stagnation point, will tend to decay by dissipation, but also will tend to increase by production due to streamwise velocity gradients as the body is approached. Hence, the turbulence is not constant between the far upstream approach flow and a location at the outer edge of the stagnation region boundary layer. In the outer flow the turbulence begins to be significantly changed at several radii upstream by production from the mean flow velocity gradients.

Secondly, the inner flow or near wall region is directly affected by the wall. The flow is affected by viscosity and the turbulent shear stress is important. A rough estimate shows that, for flow over a circular cylinder, this layer at the stagnation point can have a thickness of roughly

$$\delta/R \cong 0.54(Tu\ell_\infty/R)^{1/2} \qquad (1)$$

where Tu is the intensity, ℓ_∞ is the length scale of the far-upstream turbulence, and R is the radius of the cylinder. This relation was obtained from $\delta/R \cong 2.4/Re^{1/2}$ for the laminar stagnation boundary layer but using a turbulent eddy viscosity $\mu_t = c_\mu k^{1/2}\ell_\infty$ with $c_\mu = 0.09$ in place of the laminar viscosity and assuming that the turbulence intensity and length scales do not change in the outer flow. This relation shows that the boundary-layer thickness in the stagnation region can be on the order of 0.01 to 0.1 of the leading edge radius depending on the magnitude of the Tu and ℓ_∞. In this layer the wall directly affects the turbulent diffusion processes.

We consider the stagnation flow on a cylindrical surface as illustrated in Figure 1. To a first approximation the potential flow field near the stagnation point is given by

$$U = 2V_\infty x/R, \qquad V = -2V_\infty y/R \qquad (2)$$

which is reasonably valid for $y/R \lesssim 0.1$. The potential flow velocity field is linear in the region occupied by the boundary-layer region whose thickness is given by equation 1. Hence, similar scal-

ing to that used in the laminar solution can be applied to the turbulent case. Then for the mean flow

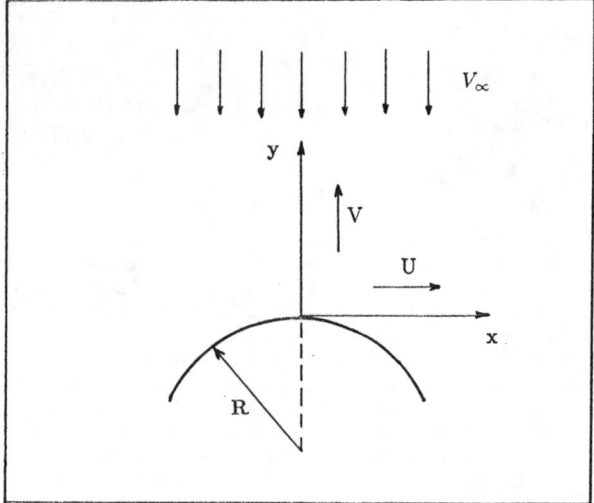

Figure 1   Stagnation Point Nomenclature

$$\eta = Re^{1/2} \int_0^\eta (\bar{\rho}/\rho_\infty)\,dy/R$$

$$U = 2V_\infty(x/R)f', \qquad \bar{\rho}V = -2\rho_\infty V_\infty f/Re^{1/2} \qquad (3)$$

$$\bar{p} = \bar{p}_0 - 2\rho_\infty V_\infty^2[x^2/R^2 + F(\eta)]$$

where $f' = df/d\eta$ and $Re = 2\rho_\infty V_\infty R/\mu_\infty$, and for the turbulence quantities

$$\overline{u^2} = V_\infty^2\,\overline{u^2}^*(\eta)\ , \qquad\qquad \overline{uv} = V_\infty^2(x/R)\,\overline{uv}^*(\eta)$$

$$\overline{v^2} = V_\infty^2\,\overline{v^2}^*(\eta)\ . \qquad\qquad \varepsilon = (V_\infty^3/R)\varepsilon^*(\eta)$$

$$\overline{w^2} = V_\infty^2\,\overline{w^2}^*(\eta) \qquad\qquad\qquad (4)$$

The momentum equation is

$$\bar{\rho}U_j\frac{\partial U_i}{\partial x_j} = -\frac{\partial \bar{p}}{\partial x_i} + \frac{\partial}{\partial x_j}\left[\mu(\frac{\partial U_i}{\partial x_j} + \frac{\partial U_j}{\partial x_i}) - \bar{\rho}\,\overline{u_i u_j}\right] \quad (5)$$

where the mean quantities are taken to be mass averaged to conveniently account for variations in density which are taken to be due only through changes in temperature and not compressibility. Changes in density only occur in the boundary layer where the temperature changes. Substituting equations 3 and 4 into 5 and keeping first order terms gives

$$\left[(\frac{\bar{\rho}}{\rho_\infty}\frac{\mu}{\mu_\infty}\,f'' - \frac{Re^{1/2}}{4}\frac{\bar{\rho}}{\rho_\infty}\,\overline{uv}^*)\right]' + ff'' - (f')^2 + \frac{\rho_\infty}{\bar{\rho}} = 0$$

$$(6)$$

This equation determines the velocity profile in the boundary layer portion of the stagnating flow. The boundary condition at the edge of the boundary layer is $f'(\eta_\infty) = 1.0$ where η_∞ is roughly given by equation 1 in the numerical solution. However, care was taken to make η_∞ sufficiently large so the solution to equation 6 was independent of its choice. In the outer portion of the flow the potential flow solution for flow over a circular cylinder gives

$$f = \frac{Re^{1/2}}{2}\left[1 - \frac{1}{(1+\eta/Re^{1/2})^2}\right] \quad (7)$$

Equation 7 is used outside the point where the solution to equation 6 is f'=0.995.

The thermal field is determined from the energy equation

$$\bar{\rho} U_j \frac{\partial H}{\partial x_j} = \frac{\partial}{\partial x_j}\left[(\frac{\mu}{Pr} + \frac{\mu_t}{Pr_t})\frac{\partial H}{\partial x_j}\right] \quad (8)$$

where $H=h+V^2/2+k$ and the dissipation terms are negligible for relatively low speed stagnation-point flow. Since H significantly changes only in the boundary layer portion of the flow and is mainly influenced by cross-stream transport, a gradient diffusion assumption $\bar{\rho}\,\overline{u_i H'} = (\mu_t/Pr_t)\partial H/\partial x_i$ can be used. For the eddy viscosity we use $\mu_t=\bar{\rho}C_\mu f_\mu k^2/\varepsilon$ where $f_\mu=1-e^{-0.0115y+}$ as given by Chien (1982). The turbulent Prandtl number is taken to be $Pr_t=0.9$. Introducing the similarity variables given by equation 3 gives

$$\left[\frac{\mu}{\mu_\infty}(\frac{1}{Pr} + \frac{\mu_t/\mu}{Pr_t})\,g'\right]' + fg' = 0 \quad (9)$$

where $g=(H-H_w)/(H_\infty-H_w)$. The temperature profile can be obtained from the stagnation enthalpy profile with the velocity profile from the solution of equation 6 and utilizing perfect gas relations. The density profile is obtained from the perfect gas equation of state $\bar{p}=\bar{\rho}RT$ where the pressure is determined from the last of equations 3 and $F(\eta)$ is determined from the y-momentum equation.

To determine the Reynolds stress components the second order turbulence model given by Lumley (1978) was used with the formulation for the wall influence on the pressure-strain correlations given by Shih and Lumley (1986). This equation for the Reynolds stress model when written in component form are quite lengthy and, consequently, are given in the appendix along with the relevant model parameters. The equations are presented in Cartesian form and only those terms which remain after substituting the stagnation point form of the solution given by equations 3 and 4 are given. The very near wall correction f_ε in the destruction term in the dissipation equation A5 was taken from Hanjalic and Launder (1976). This correction was also used by Chien (1982).

To illustrate the final form of the equations we present the turbulent kinetic energy equation even though in our solution we calculate the component energies and find the energy by $k = (\overline{u^2}+\overline{v^2}+\overline{w^2})/2$. The turbulent kinetic energy equation is

$$\bar{\rho}U_j \frac{\partial k}{\partial x_j} = -\frac{\partial}{\partial x_j}(-\mu\frac{\partial k}{\partial x_j} +\bar{\rho}\,\overline{u_j q^2}/2 + \overline{u_j p}) + \bar{\rho}(P-\varepsilon) \quad (10)$$

where the production is $P = -\overline{u_i u_j}\,\partial U_i/\partial x_j$. The pressure diffusion is modeled $\overline{u_j p} = -\bar{\rho}\,\overline{u_j q^2}/5$ as given by Lumley (1978) and retaining first order terms after substituting equations 3 and 4 the turbulent kinetic energy equation for stagnation-point flow becomes

$$\left[\frac{\mu}{\mu_\infty} k^{*\prime} - \frac{3}{5}\frac{\bar{\rho}}{\rho_\infty}\,Re\,\overline{vq^2}/2\right]' + fk^{*\prime} + \frac{\bar{\rho}}{\rho_\infty}(P^*-\varepsilon^*)/2 = 0 \quad (11)$$

and the production is

$$P^* = 2\left[\overline{v^2}^*\,\bar{\rho}\,(f/\bar{\rho})' - \overline{u^2}^*\,f'\right] \quad (12)$$

Note that the production involves the normal stresses $\overline{u^2}$ and $\overline{v^2}$ and for constant density flow flow the production is proportional to their difference. The anisotropy in the turbulence which governs the production arises from the rapid part of the pressure-strain correlation $\overline{p(\partial u_i/\partial x_j + \partial u_j/\partial x_i)}$ which accounts for the distortion of the turbulence by the mean flow. The turbulent interaction part of the pressure strain correlation tends to return the turbulence to isotropy. In the $k-\varepsilon$ model the normal stresses are modeled with a gradient hypothesis which does not properly reflect the nature of the pressure-strain processes. Consequently it is necessary to use a Reynolds stress model to accurately describe the turbulence on the stagnation streamline. The reader is referred to the paper by Taulbee and Tran (1988) and Taulbee et al. (1989) for a detailed discussion of the turbulence in the outer flow. In the present work, the Reynolds stresses are computed from the component equations given in the appendix.

Equations A1-A4, when put into similarity form with equations 3 and 4, determine the Reynolds stress components $\overline{u^2}^*$, $\overline{v^2}^*$, $\overline{w^2}^*$ and $\overline{uv}^*$ along the stagnation streamline. These equations along with equation A5 for ε^* are valid along the entire stagnation streamline although $\overline{uv}^*$ becomes negligibly small in the outer flow. This boundary value problem consisting of a coupled set ordinary differential equations for the Reynolds stresses is solved by finite differences and a block tri-diagonal matrix solver. Zero stress conditions are set at the surface and $\overline{u^2} = \overline{v^2} = \overline{w^2} = 2k_\infty/3$, $\overline{uv} = 0$ in the far free stream at a value of η equivalent to many radii. Similarly, the equation for ε^* is finite differenced and solved by a tridiagonal solver with $\varepsilon = (2\partial k^{1/2}/\partial y)^2$ at the surface as given by Hanjalic and Launder (1976) and $\varepsilon_\infty = k_\infty^{3/2}/\ell_\infty$ where the far free stream kinetic energy $k_\infty=1.5\,(TuV_\infty)^2$ and length scale ℓ_∞ are specified. Equations 6 and 9 are solved for the velocity and enthalpy profiles in the boundary-layer portion of the flow. Equation 7 gives the velocity in the outer portion of the flow and the total enthalpy is essentially constant there. The entire set of equations is iterated until satisfactory convergence is achieved.

Results and Discussions

Stagnation Streamline Turbulence. The results from the Reynolds stress model are shown in Figure 2 for the experiment of Hijikata et al. (1982). The calculation was initiated two radii before the cylinder, the furthest upstream point

29

where data was given. This location is relatively close to the cylinder. However, since this is an initial valued problem, the solution can be started at any point where conditions are known. The turbulence was close to being isotropic at that point. It is seen that the calculated results for the turbulent kinetic energy and the normal Reynolds - stress components are in good agreement with the data which was only presented on the range shown. Equally good agreement was obtained for another case with Tu=0.03. ℓ/D=0.06 for which the data was given by Hijikata et al. (1982). In the experiment by Hijikata et al. the length scale of the free stream turbulence is much smaller than the cylinder radius. Calculations presented by Taulbee and Tran (1988) for the experiment of Britter et al. (1979) show relatively poor agreement with the data. In that experiment the turbulence length scales were on the order of or larger than the cylinder radius. The reason for the relatively poor predictions is that single point closure models really do not apply to the situation where the length scale is on the order of or larger than the cylinder radius. For this situation turbulence eddies approaching the cylinder do not encounter a uniform strain field. Different parts of the same eddy are affected by different strain rates since they are much larger than the local length scale of the mean flow gradient. Single point closure models are based on the assumption that the flow is locally homogeneous which is not the case when the length scale is on the order of the body dimension. Consequently, the free stream turbulence length scale consider in this paper are restricted to be $\ell/D < 0.3$.

As seen in Figure 2 the kinetic energy decreases slightly by dissipation before production begins to significantly increase the energy at about $y/R \approx 0.4$. The energy continues to increase to a maximum value of almost three times the far free stream value at a location of $y/R \approx 0.03$. This position roughly corresponds to the edge of the boundary layer. Inside of this location the wall directly retards the turbulence until the energy approaches zero at the surface. It is seen that in the outer flow the negative longitudinal strain rate attenuates the streamwise energy component $\overline{u^2}$ while the tranverse component $\overline{v^2}$ increases. It is the difference in these two components which governs the growth of turbulent energy since the the production is $P = (\overline{v^2} - \overline{u^2})dU/dx$ for incompressible flow.

Shown in Figure 3 are the boundary layer velocity and temperature profiles for the conditions (Tu= 5.7%, and ℓ/D=0.09) for the case shown in Figure 2. The laminar profiles are shown for comparison. They were obtained from equation 6 and 9 by setting the turbulent shear stress and heat flux to zero. Higher wall friction and heat flux result from steeper gradients in the profiles with turbulence. The relatively small difference between the profiles with and without turbulence suggests that the boundary layer is essentially laminar but is disturbed by the free stream turbulence.

Stagnation Point Heat Tranfer. The nondimensional heat flux can be formulated in terms of the similarity total enthalpy profile $g(\eta)$ as

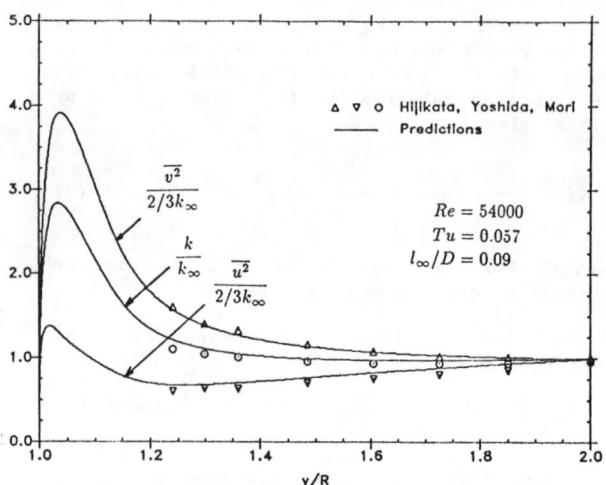

Figure 2   Variation of Turbulence on the Stagnation Streamline

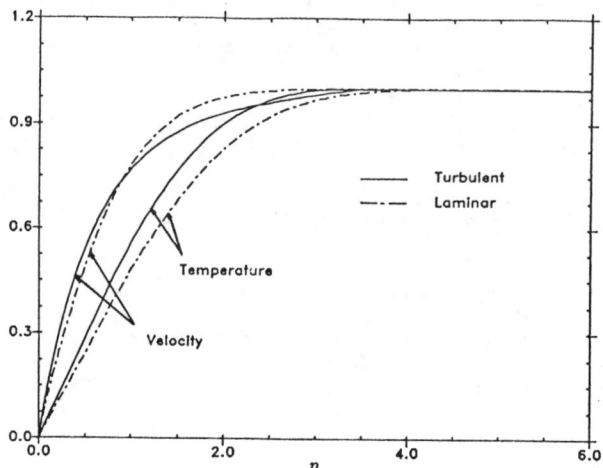

Figure 3   Velocity and Temperature Profiles in the Stagnation Boundary Layer

$$Nu = \frac{q_w D}{k(T_w - T_\infty)} = 2Re^{1/2} \; g'(0) \qquad (13)$$

where q_w is the wall heat flux, Nu is the Nusselt number and k is the thermal conductivity of air. The surface total enthalpy slope $g'(0)$ depends on any parameters in the differential equations and associated boundary conditions. Examination of equations 6.9 and 11 shows there are parameters, μ_t/μ and Pr, in the differential equations and the boundary conditions on the free stream turbulence which are described by the intensity Tu=u'/U_∞ and the length scale ratio ℓ/D. Using the k-ε model formulation for the eddy viscosity $\mu_t = C\mu \; \overline{\rho} \; k^2/\varepsilon = C\mu$ $\overline{\rho} \; k^{1/2} \ell$ since $\varepsilon = k^{3/2}/\ell$ gives

$$\frac{\mu_t}{\mu} \sim \frac{\overline{\rho} k^{1/2} \ell}{\mu} = \frac{\overline{\rho} U_\infty D}{\mu} \; \frac{k^{1/2}}{U_\infty} \; \frac{\ell}{D} \sim Re \; Tu \; \frac{\ell}{D} \qquad (14)$$

Hence, with the turbulence Reynolds number given by this equation, we can write

$$Nu/Re^{1/2} = f(Re \; Tu \; \ell/D, \; Pr, \; Tu, \; \ell/D) \qquad (15)$$

We only make predictions for the experiment in air so we omit the dependence on the Prandtl number. Therefore, in effect, $NuRe^{-1/2} = f(Re, Tu, \ell/D)$.

All experimental investigations reported in the literature have overlooked the dependence of the heat transfer on the turbulence length scale. Of the experimental investigations: Smith and Kuethe (1966), Kestin and Wood (1971) and Lowery and Vachon (1975), only Lowery and Vachon reported turbulence length scale values and they assumed its effect on the heat transfer was negligible.

The effects of the various parameters, and particularly the length-scale ratio, on the heat transfer are shown in Figure 4 which shows predicted Frossling number $Nu/Re^{1/2}$ versus length-scale ratio for various pairs of Reynolds number and turbulence intensity. It is seen that increasing Reynolds number increases the heat transfer. For larger Reynolds number the boundary layer is thinner and the turbulence effects more readily penetrate

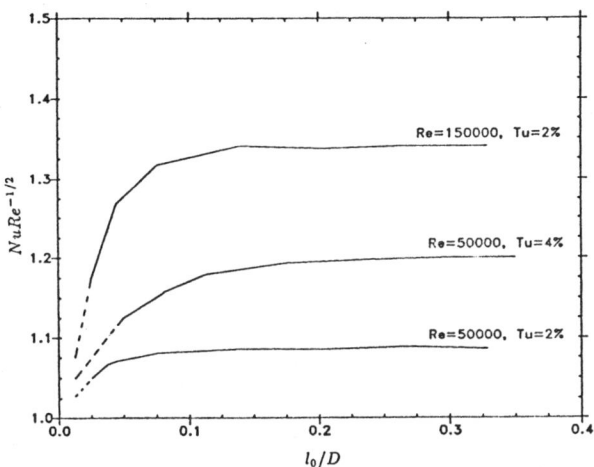

Figure 4   Effect of Turbulence Length Scale on Stagnation Point Heat Transfer

trate to the surface. Heat transfer increases with increasing turbulence as would be expected. Heat transfer also increases with increasing length scale when the length scale is small. When the turbulence scale is very small, much smaller than the boundary layer thickness, viscous action would attenuate the disturbances before they reach the surface. Hence as the length scale goes to zero, the heat transfer approximately assumes the laminar value. As the length scale increases the heat tranfer increases until the length scale becomes somewhat larger than the boundary layer thickness as given by the equation 1. For larger length scales the predicted Frossling number becomes constant and independent of the turbulent length-scale ratio. It should be pointed out that turbulent models do not accurately predict the effects of free stream turbulence on the turbulent boundary layers when the length scale is relatively large. For the flat-plate boundary layer Rodi and Scheuerer (1985) found that the predicted effects of the free stream turbulence continued to increase with increasing length scale whereas data show that the effects become less pronounced. Even so the trends were correctly predicted. Part of the reason why turbulence models do not accurately

predict the action of large-scale of free stream turbulence on the boundary layer is because two turbulences with different length scales are being mixed, whereas, turbulence models are based on a single length scale. The present situation is different because the stagnation point boundary layer does not produce its own turbulence, it only reacts to the external turbulence. In the present situation we would expect the correct trends for length-scale magnitude up to many times the boundary layer thickness.

The length-scale and intensity reference conditions for the results presented in this paper are taken at the cylinder position without it being present in the flow. Reported experimental investigations have also used this location to define the parameter values. These reference conditions are uniquely defined as opposed to arbitrarily selecting a location upstream to specify Tu and ℓ/D. However the solution requires boundary conditions upstream, solutions were initiated 5R ahead of the stagnation point except for the results presented in Figure 2. This location was chosen since there is very little production and only decay of turbulence upstream of this position. Solutions found with a further upstream initial location showed little difference from those initiated at 5R ahead of the stagnation point. To obtain conditions at the upstream position from those at the apparent cylinder location we use the isotropic solution

$$\frac{k}{k_\infty} = \left[1 + (C_2-1)\,\frac{\varepsilon_\infty}{k_\infty}\,\frac{x}{U_\infty}\right]^{\frac{1}{1-C_2}} \tag{16}$$

$$\frac{\varepsilon}{\varepsilon_\infty} = \left[1 + (C_2-1)\,\frac{\varepsilon_\infty}{k_\infty}\,\frac{x}{U_\infty}\right]^{\frac{C_2}{1-C_2}} \tag{17}$$

where the parameter $C_2 = 1.4 + 0.49\exp(-4.25/\sqrt{Re_T})$ as given by Lumley (1978). Given k_0 and ε_0 at the location $x = x_0$ of cylinder, equations 16 and 17 can be solved to determine k_∞ and ε_∞ at $x_0 - x_\infty = 5R$. The length scale ℓ_∞ is determined from $\varepsilon = k^{3/2}/\ell$.

Equations 16 and 17 do not have a solution if the quantity in the squared brackets is negative. This situation corresponds to very small length scales. In reality we should be able to specify a very small length-scale turbulence at the upstream location even though there is no solution to equations 16 and 17. For the purpose of illustrating the trends we simply extrapolated the solution curves for k_∞ and ε_∞ obtained from equations 16 and 17 to lower ℓ_0/D values. The boundary conditions for the solutions given in the dashed-line regions on Figure 4 were obtained in this manner.

As discussed previously, the Frossling number becomes independent of the length-scale ratio ℓ_0/D for values of this parameter greater than roughly 0.1. The heat transfer is then given as $Nu/Re^{1/2} = f(Re, Tu)$. Calculated results for relatively large length-scale ratio are shown on Figure 5. Also, shown are the experimental data of Smith and Kuethe (1966) and Kestin and Wood (1971). Only eight data points given in these papers had length-scale ratio large enough to be included on this plot. Smith and Kuethe (1966) and Kestin and Wood (1971) did not give length scale values, howver, from information given on the spacial variation of the turbulent intensity we were able to deduce length scales from equations 16 and 17. With the exception of the two points noted by arrows on Figure 5, the data agree with predictions for the corresponding Reynolds numbers.

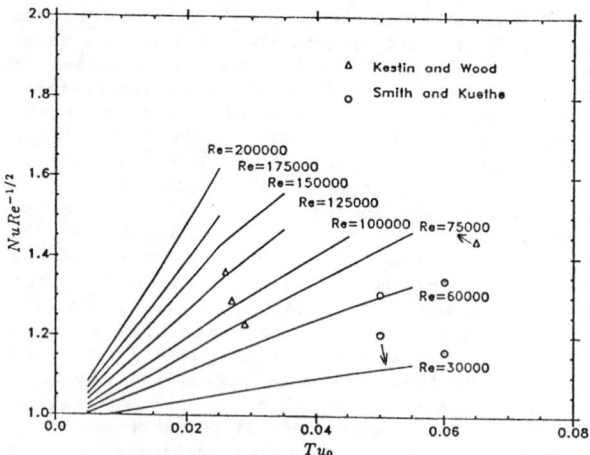

Figure 5   Effect of Turbulence Intensity on Stagnation Point

Heat Transfer for relatively large Length Scale Ratio

For smaller length-scale ratios, a concise plot comparing predictions and experiment could not be made since there are three independent parameters: Re, Tu_0 and ℓ_0/D. There was not too much data given with any one of the parameters held constant. Consequently, predictions were made for many of the experimental cases given in Smith and Kuethe (1966), Kestin and Wood (1971) and Lowery and Vachon (1975). For 42 cases not shown on Figure 5 with smaller length scales, the difference between predicted and measured Frossling numbers were less than ± 5%. Again the comparison is quite acceptable.

Parametric Representation. As illustrated previously, the nondimensional heat transfer depends on the three parameters: Re, Tu and ℓ_0/D. The question arises, can fewer parameters be defined and still correlate the Frossling number? No previous reported work has recognized the role of the length scale so the immediate discussion will be restricted to relatively larger length scales such that $Nu/Re^{1/2}$ is independent of ℓ_0/D.

Smith and Kuethe introduced the parameter $TuRe^{1/2}$ such that $Nu/Re^{1/2}= f(TuRe^{1/2})$. This parameter follows directly from a mixing length assumption ($\mu_t \sim u'_\infty U_\infty y$) for the eddy viscosity in the boundary layer. This formulation ignored the variation in the turbulence quantities outside the boundary layer. However, if one argues that the approaching turbulence is governed by the ratio of production to dissipation, then $P/\varepsilon \rightarrow k(U_\infty/D)/(k^{3/2}/\ell)=(\ell/D)/Tu$. Hence, equation 15 can be written $Nu/Re^{1/2}=f(ReTu\ell/D,Tu/(\ell/D))$. The two parameters appearing in the function could be multiplied together to form a new parameter and

then $Nu/Re^{1/2}=f(ReTu^2,Tu/(\ell/D))$. Then, for large ℓ/D, $Nu/Re^{1/2} \rightarrow f(ReTu^2)=f(TuRe^{1/2})$, thus rationalizing the parameter proposed by Smith and Kuethe.

The predicted results were plotted against $TuRe^{1/2}$. Unfortunately, the plot still showed a significant dependence on Reynolds number similar to the dependence (but not to the same extent) as shown on Figure 5. Hence, $TuRe^{1/2}$ does not seem to be a useful parameter. Smith and Kuethe themselves show that there is an additional Reynolds number variation when they plotted data against $TuRe^{1/2}$. Yet most other investigators have used this parameter to correlate their data. Consequently plots given, for instance by Kestin and Wood (1971) and Lowery and Vachon (1975) of the form $Nu/Re^{1/2}=f(TuRe^{1/2})$, show a good deal of scatter. Of course some of the scatter is due to the effects of length scale which was neglected.

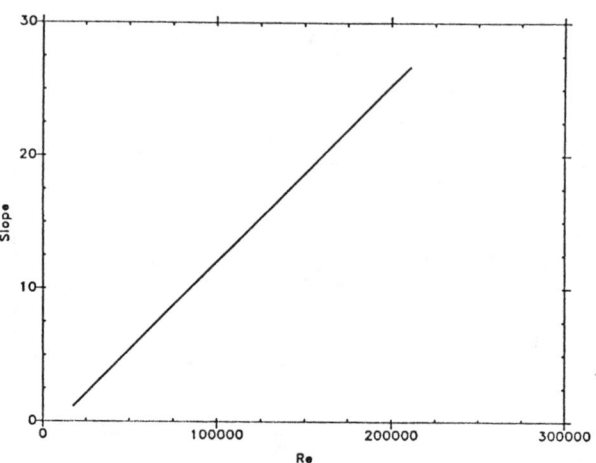

Figure 6   Variation of the Slope of the Lines

shown in Figure 5 versus Reynolds Number

To obtain a correlation formula, it seems reasonable to directly fit the curves on Figure 5. Each line for a given Re is very close to being straight. This suggests a form $Nu/Re^{1/2}=A+B(Tu)$. The average intercept is A=0.965 which is not too different than the Frossling solution $Nu/Re^{1/2}=$ 0.945 for laminar flow. The slope B(Re) is plotted on Figure 6 and turns out to be nearly a straight line B= $-1.547+1.331\times10^{-4}$Re.

In general equation 15 including the length scale ratio specifies the functional form for correlating experimental results and/or predictions. Several unsuccessful attempts have been made to combine the parameters to find a simpler description. It appears that all three parameters: Re, Tu and ℓ_0/D may separately influence the nondimensional heat transfer as indicated by equation 15.

CONCLUSIONS

In general, it is concluded that the model, consisting of an eddy viscosity model for the thermal boundary layer and the Reynolds-stress model of Lumley (1978), without adjusting model constants reasonably predicts the stagnation point process provided the turbulent length scale is somewhat smaller than the body dimension. The effects of Reynolds number, free stream turbulence and length scale on the stagnation point heat transfer are determined in the context of the turbulence model and calculation procedure. The predicted results show good agreement with available experimental data for the 50 cases considered. It was found that the predicted stagnation point heat transfer significantly depends on the length scale of the free stream turbulence, a fact that has been overlooked in experimental investigations. For larger length-scale ratios ($\ell_o/D \gtrsim$ 0.1) the Frossling number $Nu/Re^{1/2}$ becomes independent of length scale. This conclusion is somewhat speculative since there are only eight available data cases for comparing predictions. The widely used parameter $TuRe^{1/2}$ does not correlate the results, the Frossling number still significantly depends on the Reynolds number. In general, it appears that the Frossling number $Nu/Re^{1/2}$ depends individually on the Reynolds number Re, the turbulent intensity Tu, and the length scale ratio ℓ_o/D.

References

Britter, R.E., J.C.R. Hunt and J.C. Mumford, 1979, "The Distortion of Turbulence by a Circular Cylinder," J. Fluid Mech., Vol. 92, pp. 269–301.

Chien, K.Y., 1982, "Predictions of Channel and Boundary Layer Flows with a Low-Reynolds Number Tolerance Model," AIAA Journal, Vol. 20, pp. 33–38.

Hanjalic, K. and B. Launder, 1976, "Contribution Towards a Reynolds-Stress Closure for Low-Reynolds Number Turbulence," J. of Fluid Mechanics, Vol. 74, pp. 593–610.

Hijikata, K. H. Yoshida, and Y. Mori, 1982, "Theoretical and Experimental Study of Turbulence Effects on Heat Transfer Around the Stagnation Point of a Cyliner," Proceedings of the 7th Int. Conference, Munich, West Germany, Sept. 6–10, 1982, Vol. 3, Wash. D.C., Hemisphere Publ. Corp., pp. 165–170.

Kestin, J., P.F. Maeder, and H.H. Sogin, 1961, "The Influence of Turbulence on the Transfer of Heat to Cylinders near the Stagnation Point," ZAMP, Vol. 12, pp. 115–131.

Kestin, J. and R.T. Wood, 1971, "The Influence of Turbulence on Mass Transfer from Cylinders," J. of Heat Transfer, pp. 321–327.

Lowery, G.W. and R.I. Vachon, 1975, "The Effect of Turbulence on Heat Transfer from Heated Cylinders," Int. J. Heat Mass Transfer, Vol. 18, pp. 1229–1242.

Lumley, J.L., 1978, "Computational Modeling of Turbulent Flows," Advance in Applied Mechanics, Vol. 18, pp. 123–176.

Rodi, W. and Scheuerer, G., 1986, "Scrutinizing the k-ε Turbulence Model under Adverse Pressure Gradient Conditions," J. of Fluids Engineering, Vol. 108, pp. 174–179.

Sadeh, W.Z., S.P. Sutera and P.F. Maeder, 1970, "Analysis of Vorticity Amplification in the Flow Approaching a Two Dimensional Stagnation Point," ZAMP, Vol. 21, pp. 699–716.

Schlichting, H., 1979, Boundary layer theory, translated by J. Kestin, McGraw-Hill, New York.

Shih, T-H. and J.L. Lumley, 1986, "Second-order Modeling of Near-wall Turbulence," Phys. Fluids, Vol. 29, pp. 971–975.

Smith, M.C. and A.M. Kuethe, 1966, "Effects of Turbulence on Laminar Skin Friction and Heat Transfer," Physics of Fluids, Vol. 9, No. 12, pp. 2337–2344.

Squire, H., 1965, Modern Developments in Fluid Dynamics, Section 270, Dover, New York.

Sutera, S.P., 1965, "Vorticity Amplification in Stagnation Point Flow and its Effect on Heat Transfer," J. Fluid Mech., Vol. 21, pp. 513–534.

Sutera, S.P., P.F. Maeder and J. Kestin, 1963, "On the Sensitivity of Heat Transfer in the Stagnation Point Boundary Layer to Free Stream Vorticity," J. Fluid Mech., Vol. 16, pp. 497–520.

Taulbee, D.B. and L. Tran, 1988, "Stagnation Streamline Turbulence," AIAA Journal, Vol. 26, pp. 1011–1013.

Taulbee, D.B., L. Tran and M.G. Dunn, 1989, "Stagnation Point and Surface Heat Transfer for a Turbine Stage: Prediction and Comparison with Data," J. of Turbomachinery, Vol. 111, pp. 28–35.

Van Der Hegge Zijnen, B.G., 1958, "Heat Transfer from Horizontal Cylinders to a Turbulent Airflow," Appl. Sci. Res. Vol. 7A, pp. 205–223.

APPENDIX

Reynolds stresses:

$$\bar{\rho}V\frac{\partial \overline{u^2}}{\partial y} = \frac{\partial}{\partial y}\left[(\mu + \frac{5\beta+8}{3\beta(4\beta+10)}\frac{\overline{q^2}}{\varepsilon}\bar{\rho}\overline{v^2})\frac{\partial \overline{u^2}}{\partial y}\right]$$

$$+ \frac{\partial}{\partial y}\left[\frac{\overline{q^2}}{\varepsilon}(\frac{\beta-2}{3\beta(4\beta+10)}\bar{\rho}\overline{v^2}\frac{\partial}{\partial y}(\overline{w^2}+3\overline{v^2})\right.$$

$$\left. + \frac{2(5\beta+8)}{3\beta(4\beta+10)}\bar{\rho}\overline{u^2}\frac{\partial \overline{uv}}{\partial x})\right]$$

$$- 2(4C+1)\bar{\rho}\overline{u^2}\frac{\partial U}{\partial x} + 4C\bar{\rho}\overline{v^2}\frac{\partial V}{\partial y} + 4(C+\frac{1}{10})\bar{\rho}\overline{q^2}\frac{\partial U}{\partial x}$$

$$- \beta\varepsilon\bar{\rho}\frac{\overline{u^2}}{\overline{q^2}} + \frac{\beta-2}{3}\bar{\rho}\varepsilon + C_{2w}(\frac{\overline{u^2}}{\overline{q^2}}-\frac{1}{3})\bar{\rho}\varepsilon F^{1/2}\frac{\ell}{y} \quad (A1)$$

$$\bar{\rho}V\frac{\partial \overline{v^2}}{\partial y} = \frac{\partial}{\partial y}\left[(\mu + \frac{17\beta+20}{5\beta(4\beta+10)}\frac{\overline{q^2}}{\varepsilon}\bar{\rho}\overline{v^2})\frac{\partial \overline{v^2}}{\partial y}\right]$$

33

$$+ \frac{\partial}{\partial y}\left[\left(-\frac{\beta+10}{5\beta(4\beta+10)}\frac{\overline{q^2}}{\varepsilon}\right)\left(\overline{\rho v^2}\frac{\partial \overline{u^2}}{\partial y} + \overline{\rho v^2}\frac{\partial \overline{w^2}}{\partial y}\right.\right.$$

$$\left.\left. + 2\overline{\rho u^2}\frac{\partial \overline{uv}}{\partial x}\right)\right]$$

$$+ 4C\,\overline{\rho u^2}\frac{\partial U}{\partial x} - 2(4C+1)\overline{\rho v^2}\frac{\partial V}{\partial y} + 4(C+\tfrac{1}{10})\bar{\rho}\overline{q^2}\frac{\partial V}{\partial y}$$

$$- \beta\varepsilon\bar{\rho}\frac{\overline{v^2}}{\overline{q^2}} + \frac{\beta-2}{\varepsilon}\bar{\rho}\varepsilon + C_{2w}\left(\frac{\overline{v^2}}{\overline{q^2}} - \frac{1}{3}\right)\bar{\rho}\varepsilon F^{1/2}\frac{\ell}{y} \quad (A2)$$

$$\bar{\rho}V\frac{\partial \overline{w^2}}{\partial y} = \frac{\partial}{\partial y}\left[\left(\mu + \frac{5\beta+8}{3\beta(4\beta+10)}\frac{\overline{q^2}}{\varepsilon}\overline{\rho v^2}\right)\frac{\partial \overline{w^2}}{\partial y}\right]$$

$$+ \frac{\partial}{\partial y}\left[\left(\frac{\beta-2}{3\beta(4\beta+10)}\frac{\overline{q^2}}{\varepsilon}\right)\left(\overline{\rho v^2}\frac{\partial \overline{u^2}}{\partial y} + 3\overline{\rho v^2}\frac{\partial \overline{v^2}}{\partial y}\right.\right.$$

$$\left.\left. + 2\overline{\rho u^2}\frac{\partial \overline{uv}}{\partial x}\right)\right]$$

$$+ 4C\left(\overline{\rho u^2}\frac{\partial U}{\partial x} + \overline{\rho v^2}\frac{\partial V}{\partial y}\right) - \beta\bar{\rho}\varepsilon\frac{\overline{u^2}}{\overline{q^2}} + \frac{\beta-2}{\varepsilon}\bar{\rho}\varepsilon \quad (A3)$$

$$\bar{\rho}U\frac{\partial \overline{uv}}{\partial x} + \bar{\rho}V\frac{\partial \overline{uv}}{\partial y} = \frac{\partial}{\partial y}\left[\left(\mu + \frac{8}{15\beta}\frac{\overline{q^2}}{\varepsilon}\rho\overline{v^2}\right)\frac{\partial \overline{uv}}{\partial y}\right]$$

$$+ \frac{\partial}{\partial y}\left[\left(\frac{1}{3\beta}\frac{\overline{q^2}}{\varepsilon}\rho\overline{uv}\right)\left(\frac{\partial}{\partial y}(\overline{v^2} - \frac{2}{5}\overline{u^2} - \frac{1}{5}\overline{q^2})\right.\right.$$

$$\left.\left. + \frac{8}{5}\frac{\partial \overline{uv}}{\partial x}\right)\right]$$

$$+ \left[-\frac{2C+1}{3}\rho\overline{v^2} - \frac{2(8C+1)}{3}\overline{\rho u^2} + 2(C+\tfrac{1}{10})\bar{\rho}\overline{q^2}\right]\frac{\partial U}{\partial y}$$

$$- \beta\bar{\rho}\varepsilon\frac{\overline{uv}}{\overline{q^2}} + \left[2C_{1w}\bar{\rho}\overline{q^2}\frac{\partial U}{\partial y} + C_{2w}\frac{\overline{uv}}{\overline{q^2}}\bar{\rho}\varepsilon\right]F^{1/2}\frac{\ell}{y} \quad (A4)$$

Dissipation:

$$\bar{\rho}V\frac{\partial \varepsilon}{\partial y} = \frac{\partial}{\partial y}\left[\left(\mu + \frac{9}{5(4\beta+10)}\bar{\rho}\frac{\overline{q^2}}{\varepsilon}(\overline{v^2} + 2\frac{(\overline{v^2})^2}{\overline{q^2}})\right)\frac{\partial \varepsilon}{\partial y}\right]$$

$$+ \frac{\varepsilon}{\overline{q^2}}\psi_1 \text{ Prod} - \bar{\rho}\frac{\varepsilon}{\overline{q^2}}\tilde{\varepsilon}f_\varepsilon\ell_o$$

where

$$\text{Prod} = -\overline{\rho u^2}\frac{\partial U}{\partial x} - \overline{\rho v^2}\frac{\partial V}{\partial y}$$

$$\tilde{\varepsilon} = \varepsilon - 2\left[\frac{\partial(\overline{q^2}/2)}{\partial y}\right]^2$$

$$f_\varepsilon = 1 - \frac{0.4}{1.8}\exp\left(-\left(\frac{R_T}{6}\right)^2\right) \quad (A5)$$

Model Parameters:

$$\beta = 2 + \exp\left(\frac{-7.77}{\sqrt{R_\ell}}\right)\left(\frac{F}{9}\right) \times$$

$$\frac{72}{\sqrt{R_\ell}} + 80.1\,\ell n[1 + 62.4(-\text{II} + 2.3\text{III})]$$

$$F = 1 + 9\text{II} + 27\text{III}$$

$$\text{II} = -\frac{b_{ij}b_{ji}}{2}, \qquad \text{III} = \frac{b_{ij}b_{jk}b_{ki}}{3}$$

$$R_\ell = \frac{(\overline{q^2})^2}{9\varepsilon\nu}, \qquad b_{ij} = \left(\frac{\overline{u_i u_j}}{\overline{q^2}} - \frac{1}{3}\delta_{ij}\right)$$

$$C = -\frac{1}{10}(1 + 0.8\,F)$$

$$C_{1w} = 0.015, \qquad C_{2w} = 0.25, \qquad \psi_1 = 2.4$$

$$\psi_o = \frac{14}{5} + 0.98\left[\exp\frac{-2.83}{\sqrt{R_\ell}}\right]\left[1 - 0.33\,\ell n(1-55\text{II})\right]$$

1989 National Heat Transfer Conference
HTD-Vol. 107, Heat Transfer in Convective Flows

TURBULENT HEAT-TRANSFER AUGMENTATION BY MICRO-SCALE FLOW DESTABILIZATION OF THE VISCOUS SUBLAYER

H. Kozlu, B. B. Mikic, and A. T. Patera
Department of Mechanical Engineering
Massachusetts Institute of Technology
Cambridge, Massachusetts

Abstract

We report here on an experimental study of heat-transfer augmentation in turbulent flow. Enhancement strategies employed in this investigation are based on the near-wall mixing processes induced in the viscous sub-layer through appropriate wall and near-wall streamwise-periodic disturbances. Experiments are performed in a low-turbulence wind-tunnel with a high aspect-ratio rectangular channel having either a) two-dimensional periodic micro-grooves on the wall, or b) two-dimensional micro-cylinders placed in the immediate vicinity of the wall. It is found that the excitation of local instabilities in the viscous sublayer by micro-disturbances induces *favorable* heat-transport augmentation with respect to the smooth-wall case, in that near-analogous momentum and heat-transfer behavior is preserved; a roughly commensurate increase in heat and momentum transport is termed *favorable* in that it leads to a reduction in the pumping power penalty at fixed heat removal rate. The study shows that this favorable performance of micro-cylinder-equipped channel flows is achieved for micro-cylinders placed inside $y^+ \simeq 20$, implying a dependence of the optimal position and size on Reynolds number. For micro-grooved channel-flows favorable augmentation is obtained for a wider range of Reynolds numbers, however optimal enhancement still requires a matching of geometric perturbation with the viscous-sublayer scale.

Nomenclature

a	groove length (Fig. 1a)
b	distance of the micro-cylinders from the wall (Fig. 1b)
c	groove dwell (Fig. 1a)
d	diameter of the micro-cylinders (Fig. 1b)
D_H	channel hydraulic diameter, $4WH/2(W + H)$
e	groove depth (Fig. 1a)
f	friction factor
h	heat-transfer coefficient
H	channel height (Fig. 7)
j	modified Colburn analogy factor, $\frac{f}{2}\frac{Re}{Nu}Pr^{\frac{1}{3}}$
k	thermal conductivity
l	distance between successive micro-cylinders (Fig. 1b)
L	channel length (Fig. 7)
Nu	Nusselt number, hD_H/k
ΔP	pressure drop
Pr	Prandtl number, ν/α
q''	heat flux per unit area
Re	Reynolds number, VD_H/ν
Re_k	Roughness Reynolds number, $u_*e(b)/\nu$
T	temperature
δT	total temperature difference, $T_{w_{out}} - T_{m_{in}}$
u_*	friction velocity, $\sqrt{\tau_w/\rho}$
V	channel-average velocity
W	channel width (Figs. 1a-1b)
x, y, z	Cartesian coordinates
y^+	non-dimensional y coordinate, yu_*/ν
Z_n	set of geometric parameters for channel geometry n

GREEK SYMBOLS

Λ	non-dimensional thermal load, $q''L/k\delta T$
μ	dynamic viscosity
ν	kinematic viscosity
ρ	density
τ	shear stress
Ψ	non-dimensional dissipation per unit width of the channel

1 Introduction

Studies on heat-transfer augmentation are motivated by several important factors: in general terms, from the thermal-hydraulic design point of view, transport enhancement leads to a reduction in momentum-transport penalties at fixed heat removal rate; on a specific level, in certain applications such as aerospace designs and high-power-density electronic equipment, enhancement assures high heat removal rates which are required for functionality and safety of these systems. For these reasons a large number of augmentation techniques (e.g. augmentation hardware modifications, rough surfaces, and flow oscillation) are employed as heat-transfer intensification strategies for laminar and turbulent flows [1].

An important class of heat-transfer augmentation techniques is based on the enhancement of *mixing* processes by generation of *hydrodynamic instabilities* in the region of highest resistance to heat transport. This method has been successfully applied to laminar and turbulent flows in recent studies [2-4]. Intensification of turbulent heat transfer plays a significant role in technological applications since most engineering systems operate under the turbulent-flow conditions. The most important enhancement technique for turbulent flows is based on the augmentation of near-wall mixing processes in the viscous sublayer through appropriate wall and near-wall streamwise-periodic disturbances.

There are a large number of studies on the heat-transfer augmentation employing this type of enhancement procedure. Although they are geometrically different, in all systems the common physical phenomenon is a change in the structure of the viscous sublayer that results in an increase of scalar transport.

- Brouillette et all. [5] studied the thermal-hydraulic behavior of internally grooved tubes. They found that the heat-transfer coefficient is greatly influenced by the groove depth rather than the groove spacing.

- Fortescue and Hall [6] conducted experiments on the longitudinal-finned and transverse-finned fuel elements for the design of the Calder Hall nuclear reactor. Their heat transfer and pressure drop measurements for transverse-finned fuel rods indicate that fins should be closely placed to achieve a better reduction in momentum-transport penalties.

- Dipprey and Sabersky [7] experimentally investigated the heat and momentum transfer in smooth and rough tubes at several Prandtl numbers using water as the working fluid. Their three-dimensional roughness was formed by sand grains. They observed an increase in the heat-transfer coefficient as high as 270% and found that the increase in the heat-transfer coefficient is more than the increase in the friction factor for the Prandtl number of 6.0.

- Zajic [8] reported a study of turbulent heat-transfer augmentation from roughened surfaces (made by a metric profile thread), and extended his study to the case of surface boiling phenomena at high-heat-flux rates.

- Sparrow and Tao [9] performed a study in a flat-channel having streamwise-periodic cylinders attached to the wall. They obtained highly detailed axial distributions of the local mass transfer coefficient using naphthalene sublimation. They also investigated the effect of micro-disturbances on the opposite smooth wall.

- Kawaguchi et all. [10] studied the heat transfer phenomena in a turbulent boundary layer having a cylinder array. They reported the optimum spacing between successive cylinders, and distance of the cylinders from the wall required to achieve the maximum heat-transfer rate.

Work to date on turbulent heat-transfer augmentation has focussed on the enhancement of heat transfer and the associated unavoidable increase in the friction factor with respect to smooth-channel flow. Increases in the heat transfer coefficient as high as 400% were achieved with accompanying changes in the friction factor rising as much as 58 times over the smooth wall case at the same Reynolds number [1]. However, in these studies, effect of destabilization of the viscous sublayer on the thermal-hydraulic behavior of these flows is not adequately explained in terms of governing variables, such as the placement and spacing of hardware modifications as a function of roughness Reynolds number. As a result no general guidelines regarding the optimal heat-transfer design are available.

The aim of the present work is to investigate heat and momentum transfer in turbulent flows under the presence of controlled wall and near-wall disturbances. To gain a better understanding of scalar transport phenomena in turbulent flows, two kinds of streamwise-periodic micro-disturbances are employed: a) wall disturbances (micro-grooves), and b) near-wall disturbances (micro-cylinders). A parametric study for micro-cylinders is conducted by changing the distance between successive micro-cylinders, the diameter of micro-cylinders, and the distance of micro-cylinders from the heated wall. The purpose of the paper is to establish the underlying physical basis for the turbulent transport phenomena which permits choice of the proper enhancement scheme for turbulent flows at a given thermal load.

The outline of the paper is as follows. In Section 2 we present the experimental set-up and geometric characteristics of the micro-disturbances employed. In Section 3 we present and analyze the thermal-hydraulic data for the augmentation schemes of interest. In Section 4 we compare micro-groove and micro-cylinder equipped turbulent-channel flows with respect to minimum-dissipation heat removal. Lastly, in Section 5, conclusions of the study are presented.

2 Experimental Apparatus

We shall consider two heat-transfer augmentation shemes of (a) streamwise-periodic micro-grooves, and (b) streamwise-periodic micro-cylinders in a flat channel as shown schematically in Figures 1a and 1b, respectively. The base geometry, flat-channel flow is obtained using the experimental set-up of Figure 1b with micro-cylinders removed. We denote the base geometry as Z_0 and describe the two augmentation geometries by sets Z_n; $Z_1 = \{e/H, a/H, c/H\}$ for the micro-grooved channels, and $Z_2 = \{d/H, b/H, l/H\}$ for the micro-cylindered channels. Here H is the channel height (see Figures 1a and 1b); e is the micro-groove depth, a the groove length, and c the groove dwell; d is the micro-cylinder diameter, b the distance between the micro-cylinders and the bottom wall, l the distance between successive micro-cylinders in the array.

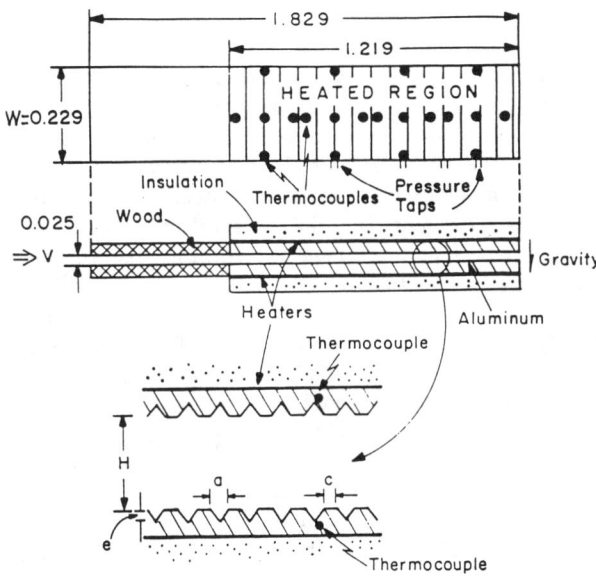

Figure 1a: Details of the test section for the geometry Z_1. The micro-grooves are machined on the aluminum plates to high precision using a shaper. All units are in meters.

We consider one micro-groove geometry (denoted by $Z_1 = \{e/H = 0.025, a/H = 0.035, c/H = 0.015\}$) and seven different micro-cylinder geometries (denoted by Z_2^m, $m = 1, .., 7$) as given in Table 1. The ratio of the width of the channel W, to the channel height H is $W/H = 9.0$ which is considered sufficiently large enough to achieve the two-dimensional flow. The channel is fitted with electrical strip heaters on both channel walls for micro-grooves, and the bottom channel wall for micro-cylinders, to deliver the necessary uniform heat flux q''. These test sections are connected to an open-circuit double-contraction wind tunnel operated in the blower mode.

For the augmentation studies of interest we require a set of Nu, f for each Z_n. Reynolds number, Nusselt number, and friction factor are defined in the conventional way as

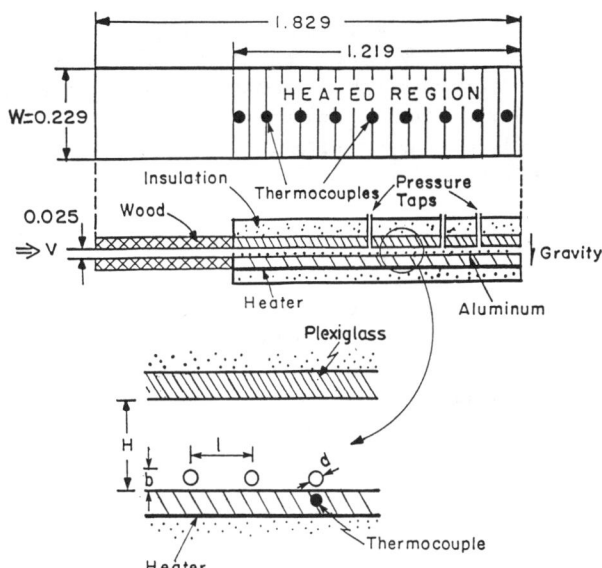

Figure 1b: Details of the test section for the geometry Z_2. For the geometry Z_0 the same test section is used with the micro-cylinders removed. All units are in meters.

Geometry	d/H	b/H	l/H	Micro-cylinder equipped wall	Heated wall
Z_2^1	0.015	0.025	9.33	bottom	bottom
Z_2^2	0.015	0.025	18.66	bottom	bottom
Z_2^3	0.015	0.025	26.00	bottom	bottom
Z_2^4	0.032	0.060	4.13	bottom	bottom
Z_2^5	0.015	0.025	26.00	bottom+top	bottom
Z_2^6	0.015	0.025	26.00	top	bottom
Z_2^7	0.049	0.059	8.65	top	bottom

Table 1: Characteristics of micro-cylinder geometries

$Re = \frac{VD_H}{\nu}$, $Nu = \frac{hD_H}{k}$ and $f = \frac{\Delta P}{L} \frac{D_H}{4} \frac{2}{\rho V^2}$, respectively. Here D_H is the hydraulic diameter for the channel ($D_H = \frac{4WH}{2(W+H)}$ see Figures 1a and 1b), $\frac{\Delta P}{L}$ pressure gradient, ρ the fluid density, V the average velocity; h is the heat-transfer coefficient, k the fluid thermal conductivity. Flow rate is varied to achieve a range of Reynolds numbers. For each Reynolds number the pressure drops and wall and fluid temperatures are measured, thus allowing the evaluation of $Nu(Re, Pr = 0.71; Z_n)$ and $f(Re; Z_n)$. The pressure drop is measured with an MKS Baratron differential pressure transducer, and the flow rate is calculated from traversed pitot-static velocity measurements at inflow. Temperature measurements are made using copper-constantan thermocouples.

Lastly, we note that the flow is allowed to become hydraulically and thermally fully developed in the streamwise direction x before any measurements are taken. All measurements are taken after a distance of roughly $65\,H$ from the inlet of the channel, and roughly $36\,H$ from the beginning of the heated region. The resulting entrance regions are sufficient to obtain hydraulically and thermally fully-developed flat-channel turbulent flow [11].

The entrance length for geometries Z_1 and Z_2^m are much shorter than the flat channel Z_0 as a result of destabilization [9]. In addition to these theoretical considerations of entrance-length effects, it has also been verified directly from measurements that both the time-averaged wall temperature and pressure vary linearly with x, consistent with fully-developed flow.

3   Results and Discussion

We plot in Figures 2 and 3 the $Nu(Re, Pr = 0.71; Z_n)$ and $f(Re; Z_n)$ curves for the flat-channel flow and augmentation schemes of interest. We compare our flat-channel data with existing smooth-channel correlations. As can be seen from Figure 2, the heat-transfer data agree very well (the largest error is about 4%) with the equation given by Kays [11]

$$Nu = \frac{0.152\, Re^{0.9}\, Pr}{0.833\,[2.25\, ln\,(0.114\, Re^{0.9}) + 13.2\, Pr - 5.8]}. \quad (1)$$

The present experimental data are also within the ± 6 percent that is given for the Petukhov-Popov equation for Pr=0.71 [12]. The friction factor data is also compared with the standard relationship [11, 13]

$$f = \frac{0.046}{Re^{0.2}}, \quad (2)$$

and the experimental data all lie within 2 percent of the above equation.

For micro-groove equipped flat-channel flows (Z_1), micro-grooves have negligible effect on transport augmentation for the roughness Reynolds numbers smaller than about 10 as seen in Figures 2 and 3. We define the roughness Reynolds number as $Re_k = u_* e/\nu$, where u_* is the friction velocity $(u_* = \sqrt{\tau_w/\rho}$, where τ_w is the shear stress at the wall) and ν the kinematic viscosity of the fluid. For τ_w we use the value for flat-channel flow at the same Reynolds number; this is a lower bound for

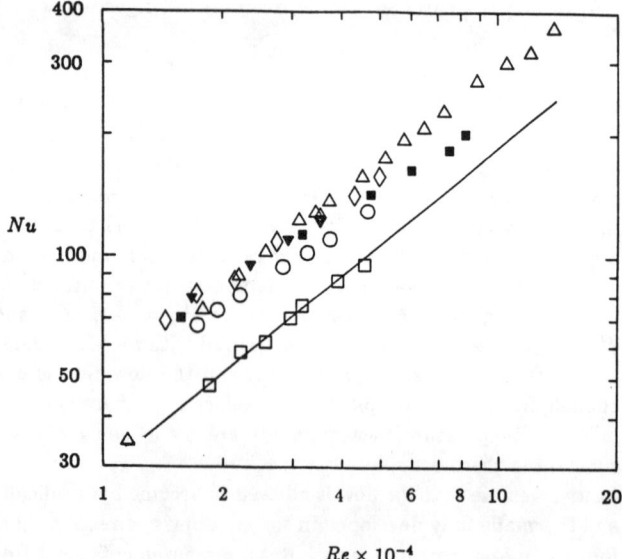

Figure 2: Heat transfer data for $Pr = 0.71$. Z_0 (smooth channel): $\square$, experiment; ———, turbulent correlation [11]. Z_1 (micro-grooves): $\triangle$. Z_2^m, $m = 1, 2, 3, 4$ (micro-cylinders, see table 1): $\diamond$, Z_2^1; $\blacktriangledown$, Z_2^2; $\bigcirc$, Z_2^3; $\blacksquare$, Z_2^4.

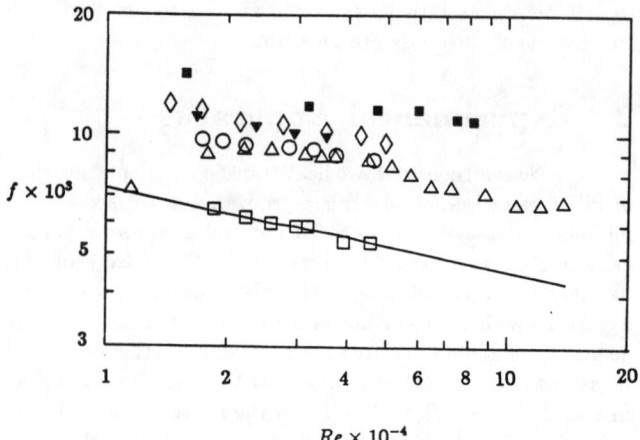

Figure 3: Friction coefficient data. Z_0 (smooth channel) $\square$, experiment; ———, turbulent correlation [11]. Z_1 (micro-grooves): $\triangle$. Z_2^m, $m = 1, 2, 3, 4$ (micro-cylinders, see table 1): $\diamond$, Z_2^1; $\blacktriangledown$, Z_2^2; $\bigcirc$, Z_2^3; $\blacksquare$, Z_2^4.

the roughness Reynolds number and is a reasonable assumption (particularly for micro-cylinders) since we do not measure the wall shear stress directly. When we increase the Reynolds number (or roughness Reynolds number) the effect of micro-grooves become significant on scalar transport, as they geometrically scale with the 'stable' part of the flow, the viscous sublayer. These observations are consistent with past studies on the effect of roughness in the transport processes [14].

We now plot the modified j-factor (defined as $j = \frac{f}{2} \frac{Re}{Nu} Pr^{\frac{1}{3}}$) as a function of roughness Reynolds number, as shown in Figure 4. For flows in which analogous heat and momentum transfer are preserved, the modified Colburn analogy [15] factor has a value of 1. Figure 4 shows that for micro-grooves a roughly equal relative increase in heat and momentum transfer is obtained for $Re_k \leq 80$. As we continue to increase the Reynolds number $(Re_k \geq 80)$, the micro-grooves cease to match the viscous-sublayer scale, and the non-analogous form drag starts dominating in the ensuing 'fully-rough' regime. Thus, the relative increase in pressure drop is larger than the increase in heat transfer; this reconfirms the fact that the roughness Reynolds number is the critical parameter governing the flow, and that there is an optimal placement of micro-disturbances which requires a matching of geometric perturbation with sublayer scale.

For micro-cylinders, we obtain the following results from comparisons of four different data sets $(Z_2^1 - Z_2^4)$. First, as seen in Figures 2 and 3, l/d has a small effect for the values b and d studied until $l/d = 18.66$. It is observed that the results of $l/d = 9.33$ and $l/d = 18.66$ are almost the same, whereas for $l/d = 26.00$ we see a decrease in heat and momentum transfer compared to the previous two case. This is consistent with the expected trend of diminished transport augmentation as $l/d \rightarrow \infty$. As we decrease the distance between the micro-cylinders and increase the diameter of micro-cylinders, an increase in the transport rates is observed.

Referring again to our plot of the modified j-factor as a function of the roughness Reynolds number[1] in Figure 4, in

---

[1]For micro-cylinders we define the roughness Reynolds number based on b, $Re_k = \frac{u_* b}{\nu}$.

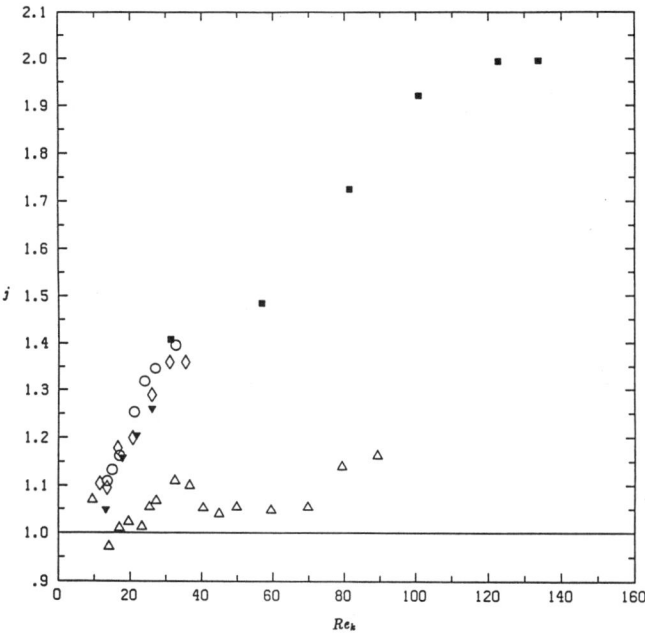

Figure 4: Modified Colburn analogy factor as a function of roughness Reynolds number. Z_1 (micro-grooves): △. Z_2^m, $m = 1, 2, 3, 4$ (micro-cylinders, see table 1): ◇, Z_2^1; ▼, Z_2^2; ○, Z_2^3; ■, Z_2^4.

which we see that all the data show a similar behavior, and that the effectiveness of micro-cylinders (in enhancing heat transfer without generating an unduly high friction loss) is increased when placed inside $y^+ \simeq 20$. This is the reason that Z_2^4 is not favorable compared to other geometric configurations, Z_2^1 and Z_2^2, since the micro-cylinders in the former case were outside the sublayer for the whole range of tested Reynolds numbers, thereby contributing to form drag and non-analogous dissipation.

The internal micro-disturbances cause an increase in scalar transport rates through local effects in the viscous sublayer. The external disturbances from the outer layer are still required to drive the instabilities in the sublayer and hence maintain the turbulent nature of the flow [16]. The response of the viscous sublayer to both these disturbances is manifested through the following flow features. First is the appearance of a destabilized wavy structure of the viscous sublayer through a 'vortex shedding' phenomenon induced by micro-disturbances. Second is the existence of bursting events (perhaps secondary instabilities) whose frequency is controlled by wall-flow parameters. Lastly, there will be a relatively small gross flow displacement (that is, a redistribution of the mean velocity profile caused by the modified near-wall momentum transport characteristics).

The wavy structure of the viscous sublayer due to the 'vortex shedding' phenomenon requires a cylinder-Reynolds number of roughly 40, indicating a dependence of this phenomenon on the micro-cylinder diameter d, and the local velocity V_{lc} (which is a function of l, d and flow Reynolds number). This wavy structure is responsible for enhancing scalar transport rates and also increasing the bursting frequency. Achievement

of favorable heat-transfer augmentation by matching the geometric scale of the augmentation hardware modification to that of the viscous sublayer reconfirms the theory of scale matched destabilization for optimal scalar transport enhancement described in [3]: micro-cylinders placed inside the viscous sublayer achieve the requisite destabilization in the 'laminar-like' wall region while simultaneously controlling the increase of non-analogous drag.

To show the effect of micro-disturbances on the opposite (from the heated) wall, we conducted three different sets of experiments. Thermal-hydraulic data of these tests are presented in Figures 5 and 6. First, for the geometry Z_2^3 we use micro-cylinders at both the heated bottom wall and unheated top wall (geometry Z_2^5). Figure 5 illustrates that there is no change in the heat-transfer coefficient compared to the geometry Z_2^3, indicating that although micro-disturbances effect the flow locally, they have no effect on the core flow, and hence no effect on the opposite wall. To support this conjecture, we have also conducted an experiment with the geometry Z_2^6 in which only the top unheated wall is equipped with micro-cylinders, whereas the bottom heated wall is smooth. As can be seen from Figure 5, the experimental heat-transfer data for Z_2^6 agrees very well with the smooth channel heat-transfer data, strengthening the above explanation. This results are consistent with findings of Sparrow and Tao [9].

When we considerably increased the opposite-wall micro-cylinder diameter (about three times compared to Z_2^5) in geometry Z_2^7, we see no effect at higher Reynolds numbers. However, for low Reynolds numbers an increase in heat transfer up to 10% compared to the smooth-channel flow is observed. This is due to the fact that an increase in micro-cylinder diameter causes an asymmetry in the flow which is responsible for altering the transport characteristics of smooth wall flow for low Reynolds numbers. If we continue to increase the cylinder diameter to the order of channel height, it is clear that transport rates for both walls will be effected by the existence of wavy unsteady secondary flows as studied by Karniadakis, Mikic and Patera [2] for laminar flows; Kozlu, Mikic and Patera [3], and Ichimiya and Yokohama [17] for turbulent flows. This again confirms the scale-matched hypothesis for optimal transport. The extent to which the micro-cylinder high-Reynolds-number and macro-cylinder low-Reynolds-number flows are 'self similar' as regards transport remains to be determined.

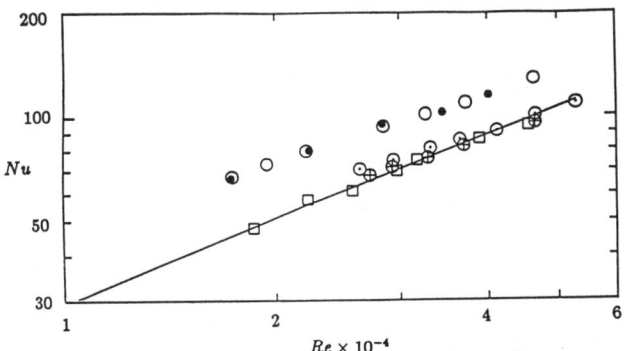

Figure 5: Heat transfer data for $Pr = 0.71$. Z_0 (smooth channel): □, experiment; ——, turbulent correlation [11]. Z_2^m, $m = 3, 5, 6, 7$ (micro-cylinders, see table 1); ○, Z_2^3; ●, Z_2^5; ⊕, Z_2^6; ⊙, Z_2^7.

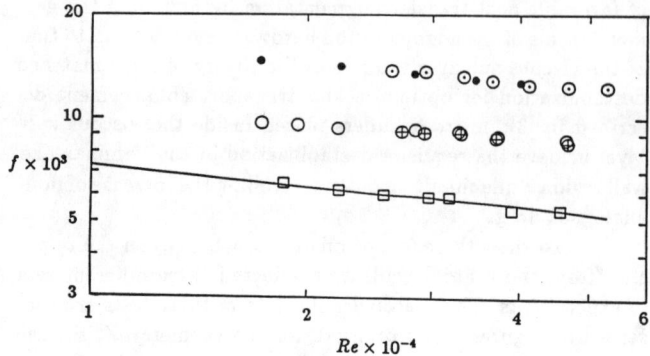

Figure 6: Friction coefficient data. Z_0 (smooth channel): $\square$, experiment; ———, turbulent correlation [11]. Z_2^m, $m = 3, 5, 6, 7$ (micro-cylinders, see table 1): $\bigcirc$, Z_2^3; $\bullet$, Z_2^5; $\oplus$, Z_2^6; $\odot$, Z_2^7.

4 Minimum-Dissipation Heat Removal Considerations

Minimum-dissipation heat removal from a wall to a flowing fluid stream is reported in References [2-3]. We consider the same problem of [3] shown in Figure 7 of incompressible flow in a plane channel of length L and height H, with uniform heat flux q'' imposed on the top wall, and an adiabatic bottom surface. The flow is assumed to be hydraulically and thermally fully developed in x. We also chose the maximum wall temperature δT as our thermal constraint as this is the quantity which typically limits the performance of devices (e.g. computer chips). In terms of these given quantities, a solution of the minimum-dissipation problem is given in [3]. In Figure 8 we plot the non-dimensional minimum dissipation (non-dimensional minimum pumping power $\Psi = \frac{L^2}{\rho \nu^3} \Delta P V H$) versus thermal load ($\Lambda = \frac{q'' L}{k \delta T}$) for flat-channel flows and the augmentation schemes of interest.[2]

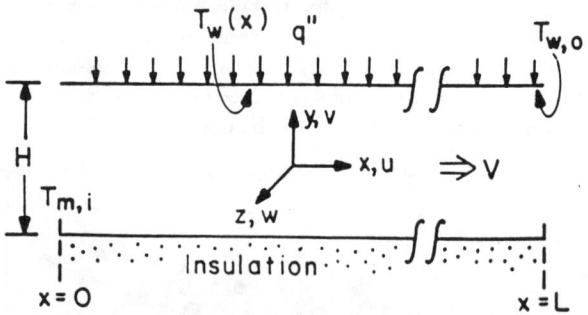

Figure 7: Basic channel geometry for the optimization study considered in [3]. The channel is of length L and height H, with uniform heat flux imposed on the top wall and an adiabatic bottom surface.

---

[2]To obtain one-sided heating/micro-grooved data from experiments we assume $Nu_{1-sided} \simeq Nu_{2-sided}$, and $f_{1-sided} \simeq \frac{1}{2}(f_{2-sided} + f_{smooth})$. We also subtract out the shear stress at the side walls to obtain the friction factor for parallel plates.

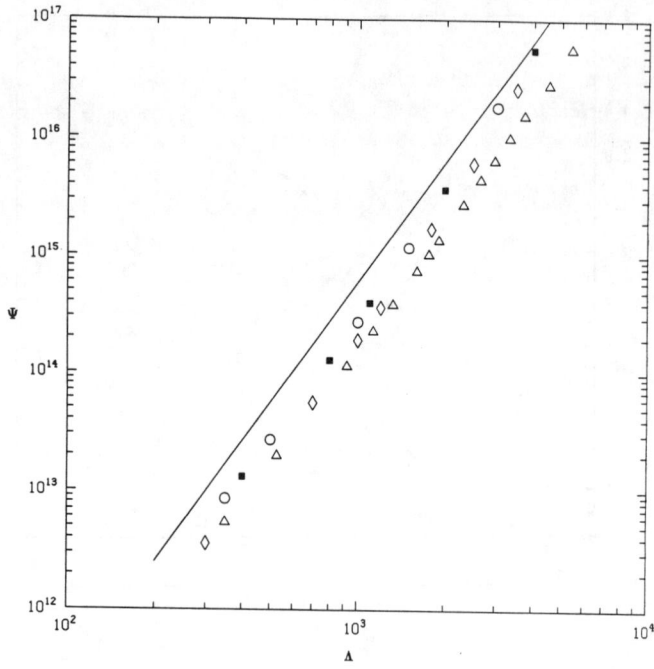

Figure 8: A plot of minimum dissipation as a function of thermal load [3]. The plot is based on the data sets of Figures 2 and 3. Z_0 (smooth channel): ———. Z_2^m, $m = 1, 3, 4$ (micro-cylinders, see table 1): $\diamondsuit$, Z_2^1; $\bigcirc$, Z_2^3; $\blacksquare$, Z_2^4.

Figure 8 illustrates the significant savings in dissipation compared to smooth-channel flows. It is important to note that micro-grooves perform efficiently for a wider range of thermal load than micro-cylinders due to the fact that the matching of geometric perturbation and sublayer scale is less sensitive to Reynolds number. For micro-cylinders placed outside $y^+ \simeq 20$, the relative savings tend to decrease strongly (see Figure 4), implying a significant Reynolds number dependence on position and size of near-wall micro-disturbances.

5 Conclusions

In this experimental study it is shown that turbulent heat-transfer augmentation can be effectively achieved by destabilization of the viscous sublayer by wall and near-wall micro-disturbances. It is found that the optimal placement of micro-disturbances require a matching of geometric perturbation and sublayer scale. The relative performance of micro-disturbances is decreased in the 'fully-rough' regime due to the dominant role of the non-analogous form drag. Significant (almost order-of-magnitude) savings in dissipation are possible through heat-transfer augmentation; this is demonstrated in a sample of minimum-dissipation analysis for heat transfer in a channel.

References

[1] A. E. Bergles, Techniques to augment heat transfer. In *Handbook of Heat Transfer Applications* (Edited by W. M. Rohsenow, J. P. Hartnett and E. N. Ganic), Chap, 3. McGraw-Hill, New York (1986).

[2] G. E. Karniadakis, B. B. Mikic and A. T. Patera, Minimum dissipation transport enhancement by flow destabilization: Reynolds' analogy revisited, *J. Fluid Mech.* **192**, 365 (1988).

[3] H. Kozlu, B. B. Mikic and A. T. Patera, Minimum-dissipation heat removal by scale-matched flow destabilization, *Int. J. Heat Mass Transfer* **31**, 2023 (1988).

[4] B. B. Mikic, H. Kozlu and A. T. Patera, A methodology for optimization of convective cooling systems for electronic devices, submitted to XXth International Symposium of the International Centre for Heat and Mass Transfer, Yugoslavia (1988).

[5] E. C. Brouillette, T. R. Mifflin and J. E. Myers, Heat-transfer and pressure-drop characteristics of internal finned tubes, ASME Paper No: 57-A-47 (1957).

[6] P. Fortescue and W. B. Hall, Heat-transfer experiments on the fuel elements, *J. Brit. Nucl. Energy Conf.* Session 2, 83 (1957).

[7] D. F. Dipprey and R. H. Sabersky, Heat and momentum transfer in smooth and rough tubes at various Prandtl numbers, *Int. J. Heat Mass Transfer* **6**, 329 (1963).

[8] V. Zajic, Some results on research of intensified water cooling by roughened surfaces and surface boiling at high heat flux rates, *ACTA TECHNICA CSAV*, No 5, 602 (1965).

[9] E. M. Sparrow and W. Q. Tao, Enhanced heat transfer in a flat rectangular duct with streamwise-periodic disturbances at one principal wall, *J. Heat Transfer* **105**, 851 (1983).

[10] Y. Kawaguchi, K. Suzuki and T. Sato, Heat transfer promotion with a cylinder array located near the wall, *Int. J. Heat Fluid Flow* **6**, 249 (1985).

[11] W. M. Kays and M. E. Crawford, *Convective Heat and Mass Transfer*. McGraw-Hill, New York (1980).

[12] B. S. Petukhov, Heat transfer and friction in turbulent pipe flow with variable physical properties, *Advances in Heat Transfer*. Academic Press, **6**, 503 (1970).

[13] A. K. M. F. Hussain and W. C. Reynolds, Measurements in fully developed turbulent channel flow, *J. Fluids Engng.* **97**, 568 (1975).

[14] H. Tennekes and J. L. Lumley, *A First Course in Turbulence*. MIT Press, Cambridge, Massachusetts (1972).

[15] A. P. Colburn, A method of correlating forced convection heat transfer data and a comparison with fluid friction, *Trans. Am. Ins. Chem. Eng.* **29**, 174 (1933).

[16] B. B. Mikic, On destabilization of shear flows: concept of admissible system perturbations, *Int. Comm. Heat Mass Transfer* **15**, 799 (1988).

[17] K. Ichimiya and M. Yokoyama, Effects of artificial roughness elements for heat transfer and flow on a smooth heated wall in a parallel plate duct, *Heat Transfer-Japanese Research* **16**, 24 (1987).

JET IMPINGEMENT HEAT TRANSFER FROM JET TUBES AND ORIFICES

M. Gundappa,* J. F. Hudson[†] and T. E. Diller*
* Mechanical Engineering Department
Virginia Polytechnic Institute and State University
Blacksburg, Virginia
[†] Machine Technologies, Incorporated
Martinsville, Virginia

ABSTRACT

A recent trend in the design of large jet-array dryers is to replace orifices with tubes in order to achieve better and more uniform drying. By providing a larger area for the spent air to exhaust, this design reduces the inherent degradation in heat transfer that results from the crossflow of the exhaust. An experimental study was undertaken to quantify the difference in fluid flow and heat transfer from single orifices and jet tubes. Mean and turbulent velocity profiles were measured in the free jet. Local distributions and average heat transfer were measured at the impingement surface. For the same mass flow, the jet tube heat transfer was slightly higher than that from the orifice for most conditions, particularly at the larger spacings. The average heat transfer from the jet tube arrays matched the average transfer from a single jet tube over the comparable area. As a function of open area, the jet tube had high average heat transfer rates for the range of 2 to 3 percent effective open area, with little increase in transfer for higher open areas.

NOMENCLATURE

A	–	area, m^2
C_d	–	discharge coefficient
d	–	diameter, m
d_e	–	effective diameter, m, Eq. 6
E	–	gage output, mV
F	–	gage sensitivity, $mV/(W/m^2)$
h	–	heat transfer coefficient, $W/(m^2\text{-K})$
h_{av}	–	average heat transfer coefficient over the center plate, $W/(m^2\text{-K})$
h_c	–	corrected heat transfer coefficient in convection, $W/(m^2\text{-K})$
h_i	–	heat transfer coefficient based on isothermal gage surface assumption, $W/(m^2\text{-K})$
k	–	thermal conductivity, $W/(m\text{-K})$
K_2	–	thermoelectric constant for copper constantan, 0.042 $mV/^\circ C$
Nu	–	Nusselt number, hd_e/k
OAE	–	effective open area, Eq. 7
P	–	power, W
Q	–	heat transfer, W
r	–	radial coordinate from jet center
Re	–	Reynolds number, $U_e d_e/\nu$
R_h	–	electric resistance of center plate heater, ohms
T	–	temperature, $^\circ C$
Tu	–	turbulence intensity, Eq. 1
u'	–	velocity fluctuation, m/s
U	–	mean velocity, m/s
U_e	–	exit velocity, m/s
U_o	–	centerline velocity, m/s
V	–	rms voltage, V
X	–	amplification factor of amplifier, mV/mV
z	–	axial distance, m
ε	–	surface emissivity
ν	–	kinematic viscosity, m^2/s
ρ	–	air density, kg/m^3
σ	–	Stefan-Boltzmann constant, 5.67×10^{-8} $W/(m^2\text{-K}^4)$

Subscripts

a	–	air
g	–	gage
p	–	plate
T	–	total
∞	–	surroundings

INTRODUCTION

Arrays of impinging air jets are widely used for the heating, cooling or drying of substrates. Applications include internal cooling of gas turbine blades, heat treating glass, and industrial drying of paper, coatings, and textiles. Large quantities of energy are used in these processes. Typically one-half of a textile finishing mill's energy usage, for example, is consumed in drying processes (Beckwith and Beard, 1979). To optimize these processes it is important to develop a detailed understanding of the fluid flow and resulting transfer of impinging jets. The potential rewards are better industrial products, increased production capacity, and large energy savings.

These jet flows are typically highly turbulent and for arrays of jets are very complex because of the presence of fluid recirculation, crossflow, and interaction between jets. In many of the industrial processes it is desired to simultaneously treat large surface areas. Unfortunately, the impingement process is limited by the large crossflows typically generated. Crossflow of the exhaust air from the central regions degrades transfer toward the edges and causes back pressure, which decreases the jet velocities in the central region. These effects not only decrease the overall transfer rate; but, even worse, cause the transfer to be nonuniform. The difference in transfer can easily be 30 percent from center to edge in typical industrial dryers.

One approach to eliminate these crossflow problems, which is coming into wider use, is to replace the standard orifice plates with what are commonly called jet tubes. These are illustrated in Fig. 1. Instead of the jets issuing directly from the plenum through orifices, the air is directed through tubes which extend from the plenum. This allows space for the exhaust to leave without interfering with the jets, while keeping the jet exit close to the transfer surface to maintain high heat transfer rates. Very little work has been reported, however, on the transfer characteristics of turbulent impinging jets which issue from tubes. This paper provides direct, detailed comparisons between jets from orifices and tubes. Both the fluid flow characteristics of the free jets and the impingement heat transfer distributions are quantified. In addition, some of the corresponding average heat transfer results for arrays of jet tubes are presented for comparison.

Background

The velocity fields of single, turbulent, impinging jets have been studied extensively. Several reviews have appeared which detail velocity and temperature distributions and decay characteristics of jets (Abramovitch, 1963; Gauntner et al., 1970; Rajaratnam, 1976). The heat and mass transfer characteristics of impinging jets have also been extensively studied. Because of their greater practical importance, reviews usually focus on transfer from arrays of jets (Livingood and Hrycak, 1973; Martin, 1977; Downs and James, 1987; Mujumdar, 1985).

The usual goal of the many papers on jet heat transfer which have appeared in the literature in the past thirty years has been to obtain a simple empirical correlation for the observed transfer of jet arrays in terms of parameters such as the open area, spacing to the surface, distance between holes, and a variety of Reynolds numbers (e.g., Gardon and Cobonpue, 1963; Huang, 1963; Chance, 1974; Kercher and Tabakoff, 1970; Florschuetz et al., 1981). There is no general agreement among investigators on the appropriate form of the

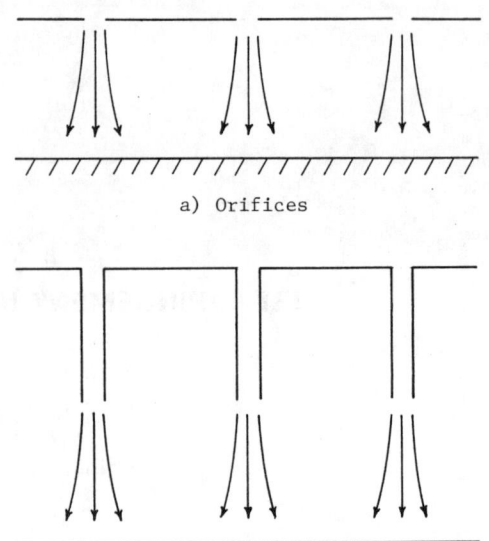

a) Orifices

b) Jet Tubes

Fig. 1 Schematic of Jet Arrays

results relative to the many parameters involved, making it difficult to know how to use the correlations (Livingood and Hrycak, 1973). Much of the problem is due to the different ranges of parameters investigated and the wide error band usually accompanying the measurements (typically ± 15%). Lack of a clear understanding of how the parameters affect the results makes consistent reporting and interpretation difficult.

A more fundamental understanding of the problem can be obtained by studying the details of transfer from single jets and selectively including the additional effects present in jet arrays, e.g., crossflow, jet interaction, and entrainment. Numerous experimental studies have reported the heat or mass transfer resulting from impingement of a single turbulent jet against a flat plate. Attempts have been made to treat the transfer at the stagnation point analytically, although with limited success because of the difficulty in predicting the effect of the free jet turbulence (e.g., Vlachopoulos and Tomich, 1971; Donaldson, et al., 1971; Chia et al., 1977; Hoogendoorn, 1977; Kataoka et al., 1983). Detailed measurements of transfer in the wall jet region have also been made (Donaldson, et al., 1971; Gardon and Cobonpue, 1963; Han and Seely, 1965; Vallis, et al., 1978. All of these experiments have typically been done only with jets from well-designed nozzles, which have discharge coefficients close to unity and low initial turbulence.

Velocity and local heat transfer measurements for single jets from orifices were made by Hollworth and Wilson (1984) and Hollworth and Gero (1985). Obot et al. (1979) measured velocity and heat transfer distributions for jets from tubes of several lengths. Neither set of results included discharge coefficients or comparison to other types of jet openings. The initial velocity and turbulence distributions in the jets are different for these different cases. The effect of these distributions on the resulting heat transfer has not been well documented. For longer tubes, Obot et al. (1979) found higher turbulence and more nonuniform velocity at the tube exit. The measured heat transfer also was generally higher for the longer tubes. This is the same trend as found for laminar jets, which have higher transfer for parabolic velocity profiles than for uniform profiles at the same mass flow rate (Scholtz and Trass, 1970).

44

Several attempts have been made to correlate heat or mass transfer from jet arrays with that measured for single jets. Local measurements, such as those just described, were integrated over an area equivalent to that impinged by an individual jet within an array (Gardon and Cobonpue, 1963; Gardon and Akfirat, 1966; Martin, 1977). The reasoning was that for the same flow conditions, the transfer should be the same for an individual jet as for one within an array. Gardon and Akfirat (1966) did report such a correlation for certain slot nozzles, but were unsuccessful with circular jets. Martin (1977) had significantly lower transfer with jet arrays than for the equivalent single jets. Hollworth and Berry (1978), however, measured the same local heat transfer from individual circular jets with and without the presence of surrounding jets. Striegl and Diller (1984) demonstrated the importance of the temperature of the recirculating air surrounding the jet. They measured the same transfer for single and multiple slot jets only when this thermal entrainment effect was the same.

The present research provides a direct experimental comparison of the flow field and heat transfer between orifices and jet tubes. The diameters and discharge coefficients are matched along with all of the flow and thermal parameters. The heat transfer from a single jet tube is also matched with that of a jet array by eliminating the exhaust crossflow and thermal entrainment effects. This allows prediction of the heat transfer for jet arrays under these conditions for a wide range of open areas.

EXPERIMENTAL APPARATUS AND PROCEDURE

The flow system used in this investigation is shown in Fig. 2. A high-pressure blower provided the air supply. Downstream of the blower is a circular-to-rectangular diffuser and a plenum chamber. The plenum is 36 cm long having a 23 cm x 15 cm cross section. The static pressure in the plenum is measured by a static pressure tap mounted in the wall of the plenum. The pressure tap is connected to a U-tube water manometer. The jet tube plate and the orifice plate used for these measurements were each mounted in turn at the exit of the plenum.

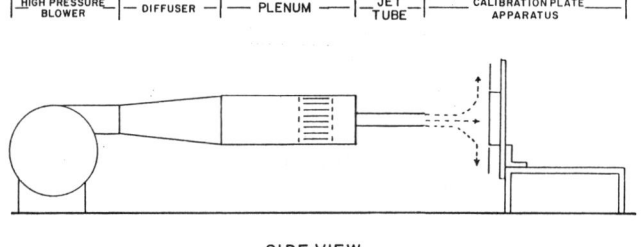

SIDE VIEW

Fig. 2 The Flow System

The jet tube plate consisted of a bank of eleven circular tubes oriented parallel to the flow direction. Each tube is 1.59 cm (5/8 in.) in diameter and 15.2 cm long. Thus, the length to diameter ratio for each tube is about 10. The tubes were spaced to provide an open area of about 4 percent. The tubes were flared at the inlet side and deburred. As described later, the discharge coefficient (C_d) was measured to be 0.84.

The orifice plate consisted of a single orifice 1.59 cm diameter that was drilled and reamed in a 0.32 cm thick steel plate. On the inlet side the orifice hole was countersunk to a depth of 0.16 cm and the hole

was deburred. The discharge coefficient for the orifice was measured to be 0.83.

Velocity Measurements

Mean velocity and turbulence measurements in the free jet were made with a TSI 1210-20 hot film probe which was traversed across the exit of the tube or orifice opening. The hot film was controlled by a TSI IFA-100 hot wire anemometer system. The bridge output from the IFA-100 was linearized using a TSI Model 1072 linearizer. The turbulence intensity referenced to the centerline exit velocity was defined as

$$Tu = \frac{[\overline{(u'^2)}]^{1/2}}{U_e} \qquad (1)$$

This was determined from the true RMS AC and DC components of the linearized signals. The bar indicates time average. In addition to the radial distribution of velocity and turbulence obtained at the exit of the tube and orifice, the axial decay of the mean centerline velocity and the axial variation of the centerline turbulence intensity were also measured.

Heat Transfer Apparatus

The heat transfer measurements were made with the flat plate apparatus used by Borell and Diller (1987), which can be used for both local distributions and average heat transfer. A Gardon heat flux gage (3.2 mm diameter) is mounted in a hole drilled in the center plate, as illustrated in Fig. 3. The gage is positioned so that its sensing element is flush with the top surface of the plate. The plate is made of 1.27 cm (1/2 in.) thick aluminum and measures 15.2 cm (6 in.) on a side. Thirteen type-T thermocouples are embedded in the plate flush with the top surface. They are located three per side, evenly spaced 1.27 cm from the edge, and one near the center. The side perimeter of this plate has balsa wood, 0.48 cm (3/16 in.) thick glued to the edge to reduce side losses. Four aluminum guard plates, 5.1 cm (2 in.) x 15.2 cm (6 in.) x 1.27 cm (1/2 in.) surround the center plate. Each guard plate has a thermocouple embedded directly across from the corresponding one in the center plate. The plates are heated with silicone-rubber wire resistance heaters

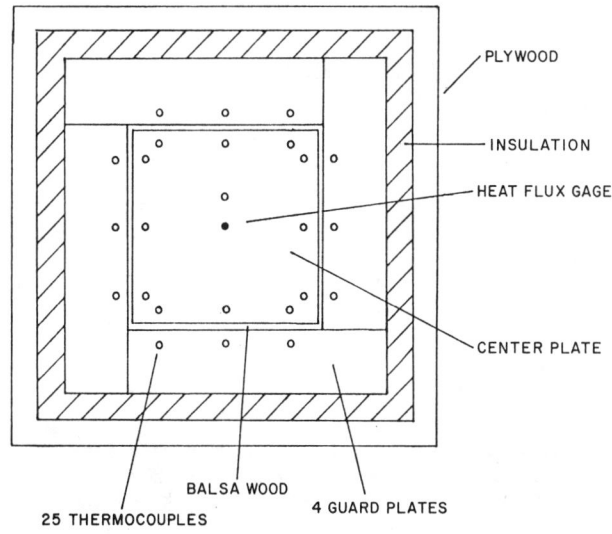

Fig. 3 The Heat Transfer Plate Assembly

that are mounted under each plate. The temperature of each plate is individually controlled to ±0.1°C by a Eurotherm 810 temperature controller and a Eurotherm 831 phase-angle-fired power supply. Behind the plate assembly is 0.6 cm of asbestos mill board for rigidity and 3.8 cm (1-1/2 in.) of Armaflex insulation to reduce back losses. The entire system is placed in a wooden box rigidly mounted on an aluminum frame.

The plate assembly is mounted vertically with the top surface perpendicular to the direction of flow. As illustrated in Fig. 4, the aluminum frame slides along a pair of drill rods mounted horizontally on a large back plate. The box is traversed along these guides by means of a stepping motor that is coupled to a lead screw nut assembly. The stepping motor can be rotated in 200 steps per revolution to provide accurate positioning. The motor is controlled by an IBM PC that was also used for the data acquisition.

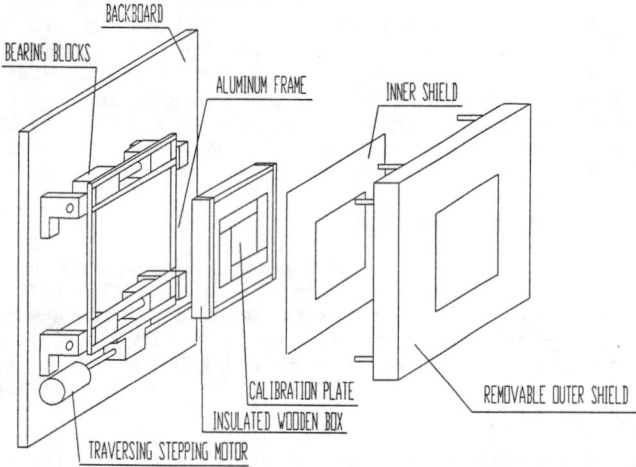

Fig. 4 The Heat Transfer Measurement Apparatus

In front of the measuring plate assembly is a large (66 cm x 66 cm) stationary flow shield. The shield is made of 0.24 cm thick aluminum with a square hole 29.2 cm per side in the center. The hole in the center of the shield exposes the plate assembly to the impinging flow of the jet tube or orifice. The plate assembly is traversed through the flow field while behind the flow shield. Since the large flow shield is stationary and the plate always fills the hole in the shield, the flow remains unchanged during traverses of the flow field.

Heat Transfer Test Procedure and Data Reduction

Both local and average heat transfer measurements were made with the plate assembly. For the average heat transfer measurements, steady-state total power to the center plate was calculated from the measured root-mean-square voltage, $P = V^2/R_h$, where $R_h = 27.2 \ \Omega$ is the electrical resistance of the heater. The resistance was constant to within 0.6 percent over the operating temperature range. The total heat transfer from the top surface of the center plate can be expressed as

$$Q_T = h_{av}A_T(T_p - T_a) + \sigma(\epsilon_p A_p + \epsilon_g A_g)(T_p^4 - T_\infty^4) \qquad (2)$$

The average plate temperature T_p was calculated as the average of the 13 temperatures from the center plate. Q_T represents the input power less back and side losses

$$Q_T = V^2/R_h - Q_{losses} \qquad (3)$$

Losses from the sides of the center plate occur through the back insulation, to the top surface of the balsa wood and to the guard plates. The error introduced by the small difference in temperature between the center plate and guard plates was negligible. The heat loss due to surface convection from the balsa wood was included in the calculations and was always less than 3 percent (VandenBerghe and Diller, 1989). The losses, however, are a function of h_{av}. Therefore, an iterative scheme was used with equations (2) and (3) to determine h_{av} and the losses. Iteration continued until the value of h_{av} converged to within ±0.1 percent. The estimated error for the average heat transfer coefficient is less than 3 percent.

For the local heat transfer measurements, the plate assembly was traversed laterally across the stream in increments of 0.173 cm by means of the stepping motor. This increment is approximately the size of the Gardon gage sensing element. At each position the gage output was averaged and recorded by the IBM PC data acquisition system. The system also recorded the temperature of the thermocouple mounted nearest the gage in the plate and the temperature of the airstream. The temperatures were recorded from a Doric 410A digital thermcouple readout calibrated to an accuracy of 0.1°C. By traversing the plate across the jet, radial distributions of the time-averaged local heat transfer were measured up to a distance of 7.5 jet diameters on either side of the jet stagnation point.

The local heat transfer was calculated from the Gardon gage output as described by Borell and Diller (1987). The radiation heat transfer was subtracted from the total measured transfer to result in a convective heat transfer coefficient, using

$$h_i = \frac{E/FX - \epsilon_g \sigma(T_p^4 - T_\infty^4)}{T_p - T_a} \qquad (4)$$

where E is the measured output voltage. The amplification (X) of the gage output voltage was provided by a Thermogage high-gain DC amplifier connected in series with a second amplifier. In a typical test the combination amplified the signal about 10,000 times. The plate temperature at the gage was taken as the temperature of the thermocouple closest to the gage and the surrounding radiative environment was taken as the ambient temperature.

The gage was calibrated by Borell and Diller (1987) in convection flow using the same apparatus used in this investigation. Their results for this gage gave a gage sensitivity of $F = 2.41 \times 10^{-5}$ mV/W/m^2 with a calibration error of less than 2 percent. Their recommended correction for the nonuniformity of the heat flux across the gage is

$$\frac{h_c}{h_i} = \frac{1}{1 - \frac{0.75 \ Fh_i}{K_2}} \qquad (5)$$

where the Seeback coefficient is $K_2 = 0.042$ mV/°C. The size of the correction was kept small (less than 5 percent for most tests) by using a low sensitivity gage with a very small temperature difference across the gage and therefore small output signal.

The jet diameter used in all of the calculations was an effective diameter that accounted for the discharge coefficient of the tube or orifice,

$$d_e = d\sqrt{C_d} \qquad (6)$$

This allows direct comparison, based on equivalent mass flow rates, between the different orifices, tubes, or nozzles. The Nusselt number and Reynolds number used the effective diameter and were calculated using air properties evaluated at the film temperature. Because the plate was given only about a 30°C overheat, variable property effects were small.

RESULTS

The velocity and heat transfer measurements for a single orifice and jet tube are presented in Figures 5 through 10. While making measurements on a single jet tube the flow issuing out of the other tubes was diverted away from the measuring apparatus. This ensured that the measurements were not influenced by any crossflow or interference effects from the other tubes. A damper on the fan inlet was used to adjust the jet velocity to keep the Reynolds number constant at about 34,000 for all tests performed on both the orifice and the jet tubes.

Velocity measurements

The mean velocity and turbulence measurements are shown in Figs. 5 and 6. Figure 5 shows the velocity and turbulence intensity profiles at the exit of the jet tube and the orifice. The results are plotted as a function of radial distance measured from the jet stagnation point non-dimensionalized by the effective jet diameter (d_e). The pressure difference between the plenum and the atmosphere was 1120 Pa (4.5 in. H_2O) for both the orifice and jet tube. As shown in Fig. 5, the maximum exit velocity is nearly the same for the orifice and the jet tube. This indicates that the pressure drop through the jet tube is negligible (<2 percent). The profile shapes are different, however. The velocity profile appears flat over most of the central portion of the jet for the jet tube. For the orifice the velocity profile assumes a slight concavity in the center accompanied by a small increase in the velocity around the rim. The shape could be caused by the necking of the flow (vena contracta) as it leaves the orifice opening. The coefficient of discharge (C_d) for both the orifice and the jet tube was determined by numerical integration of this velocity profile across the cross section.

Also shown in Fig. 5 is the radial variation of the turbulence intensity defined by Eq. 1. The turbulence measured at the exit of the tube is higher than that measured for the orifice everywhere except at the very edge of the jet. Most of the turbulence is apparently generated by the blower.

Figure 6 compares the axial decay of the centerline velocity and the variation of centerline turbulence intensity between the jet tube and the orifice. Corrsin's (1943) data for a heated jet and the data of Obot et al. (1979) are also included for comparison. The decay in the centerline velocity for the jet tube is gradual, retaining about 80 percent of the exit velocity at $8d_e$. For the orifice the centerline velocity actually increases by about 3% of its initial value due to the vena contracta effect. Unlike the jet tube, the velocity drops more rapidly to about 60% of the exit value at about $8d_e$. The mean velocity data of Corrsin (1943) from a nozzle follows the orifice results, while the data of Obot et al. (1979) from a tube follows the jet tube data, as expected.

The turbulence development is similar for the orifice and the jet tube. The turbulence values of both the orifice and jet tube are higher than the nozzle results of Corrsin (1943) and the tube results of Obot et al. (1979).

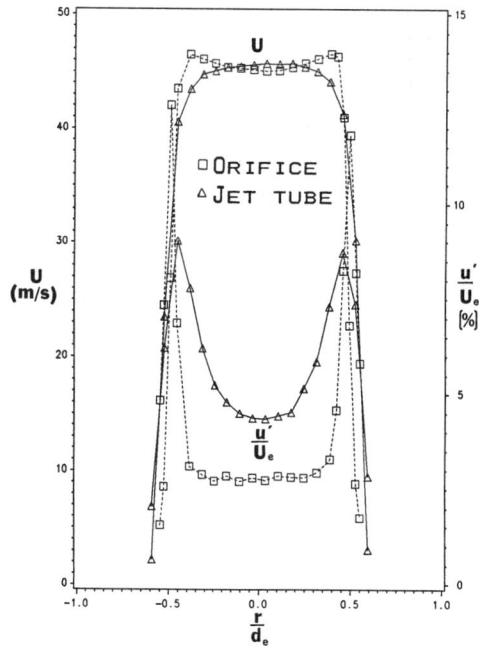

Fig. 5 Velocity and Turbulence Distributions at the Exit

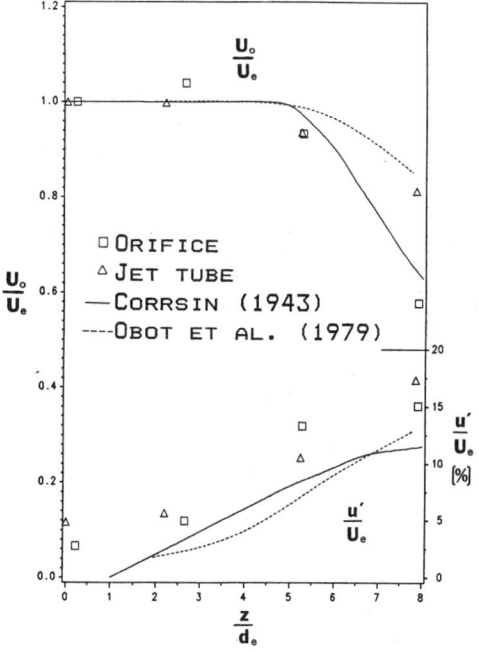

Fig. 6 Axial Decay of Centerline Velocity and Turbulence

Heat transfer measurements

Radial distributions of the time-averaged impingement heat transfer were obtained at axial distances of 3.8 cm, 7.6 cm and 11.4 cm from the exit of the jet tube and the orifice. The results are expressed as the local Nusselt number versus dimensionless radial distance r/d_e in Figs. 7 through 9. The Reynolds number based on effective tube diameter was about 34,000 for all cases. The error in the measurement of the local heat transfer coefficient was estimated to be less than 5 percent.

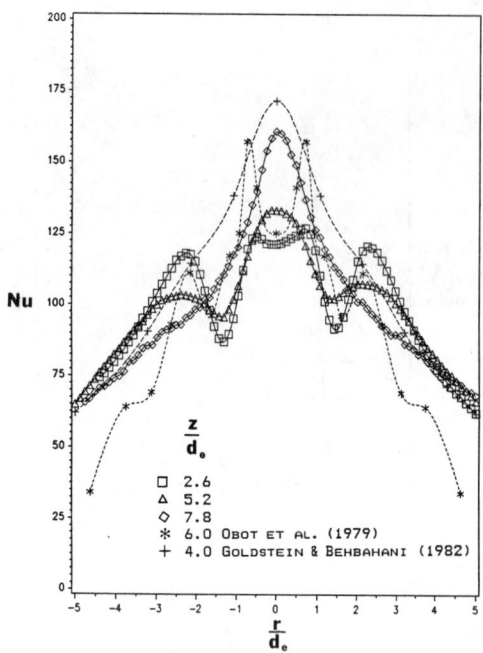

Fig. 7   Radial Distribution of Jet Tube Heat
Transfer

The radial heat transfer distributions for the jet
tube at three axial locations from the tube exit are
plotted in Fig. 7.   At the stagnation point the heat
transfer increases with increasing distance from the
tube exit.   This is due to the increase in free jet
turbulence intensity with axial distance as shown in
Fig. 6.   Gardon and Akfirat (1965) also noticed similar
behaviour in their experiments with two-dimensional
slot jets.   The presence of the characteristic secondary peaks at z/d_e = 2.6 are also similar to the secondary peaks detected by them.   The secondary peaks disappear at z/d_e = 7.8.   At the larger radial distances
in the wall jet (r/d_e > 3.5), the heat transfer appears
to be independent of axial distance.   The tube results
of Obot et al. (1979) at a Reynolds number of 29,000
and Goldstein and Behbahani (1982) at a Reynolds number
of 35,000 match most of the characteristics of the currently reported data.   The heat flux measured by Obot
et al. is lower in the wall jet region and the results
of Behbahani and Goldstein do not show the secondary
peaks.

Figure 8 shows the corresponding radial heat
transfer distributions for the orifice at the same
axial distances.   Like the jet tube, the heat transfer
distribution at z/d_e = 2.6 shows the characteristic
secondary peaks and the heat transfer is independent of
axial distance at the larger radial positions in the
wall jet region.   These measurements are compared with
the mass transfer measurements of Chia et al. (1977)
for a nozzle at a Reynolds number of 34,000.   Their
dimensionless mass transfer coefficients have been
converted to heat transfer coefficients by a standard
mass transfer analogy.

The curves in Figs. 7 and 8 are plotted again in
Fig. 9 to highlight the differences in the heat transfer distribution between the orifice and jet tube at
the same axial location.   In Fig. 9a (z/d_e = 2.6) the
heat transfer for the orifice in the wall jet is
slightly lower than for the jet tube.   At z/d_e = 5.2
(Fig. 9b), the heat transfer for the orifice at the
stagnation point is about 20% higher than the jet tube,

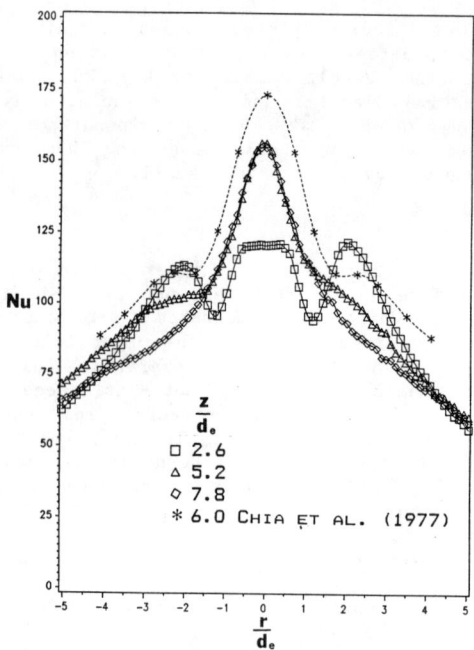

Fig. 8   Radial Distribution of Orifice Heat Transfer

but falls rapidly and is slightly less in the wall
jet.   At z/d_e = 7.8 (Fig. 9c) the local heat transfer
distribution for the orifice is lower than for the jet
tube in both the stagnation and wall jet regions.

The local heat transfer distributions were also
numerically integrated over different areas of the
plate to determine the variation of the average heat
transfer coefficient with effective open area.   The
numerical integration was performed using the trapezoidal rule.   The effective open area ratio (OAE), is
defined as

$$ OAE\ (\%) = \frac{n\ \frac{\pi}{4}\ d_e^2}{A} \times 100 \qquad (7) $$

where A is the area of the surface impinged by the jet
and n is the number of jets in the array.   For a single
jet n = 1 and A is taken as a rectangular area around
the jet tube or orifice.   The dimensions of the rectangle for each open area ratio were proportional to the
spacing of the tubes in the jet array.

The resulting average heat transfer is shown as
solid and dashed lines in Fig. 10 for the three axial
locations used in this investigation.   At z/d_e = 2.6
and 5.2 (Figs 10a and 10b) the average heat transfer
for the jet tube and orifice are nearly the same.
Generally the jet tube appears to perform better at
smaller effective open area ratios (OAE < 4.0 percent).   For higher values of OAE the heat transfer for
the orifice is higher.   One reason for this behavior is
that the average heat transfer variation for the jet
tube is nearly constant for open areas greater than 4.0
percent, while for the orifice the average heat transfer is still increasing with increasing open area.
Conversely, at z/d_e = 7.8 (Fig. 10c), the average heat
transfer for the jet tube is about 8 percent higher
than for the orifice over the entire range shown.   This
is also evidenced in the comparison of the radial distribution (Fig. 9c) between the orifice and the jet
tube.

The average heat transfer measured directly from
the center plate surface is also included in Fig. 10 as
the triangle points.   These measurements were made for
the jet tube for z/d_e = 2.6 and z/d_e = 5.2 at effective

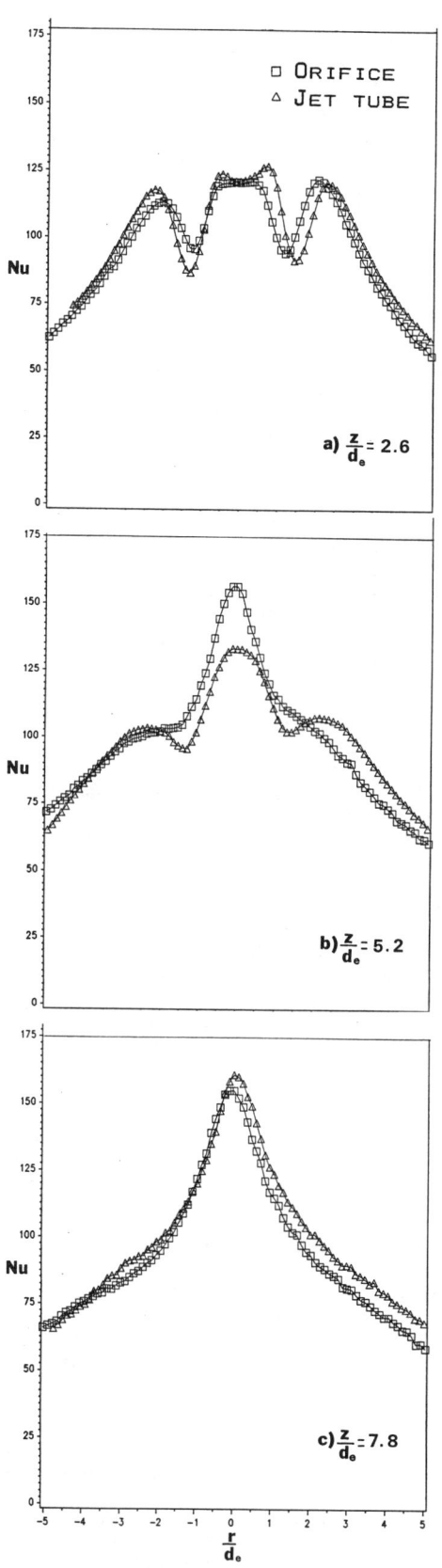

open area ratios of 3.4 and 1.5 percent. The latter open area ratio was obtained by blocking off alternate jet tubes in the tube bank. All four of the average heat transfer measurements for the multiple jet tubes are within five percent of the corresponding integrated average for a single jet tube. This indicates that single jet measurements can be a good predictor of heat transfer for jet arrays when the effects of crossflow and thermal entrainment are negligible.

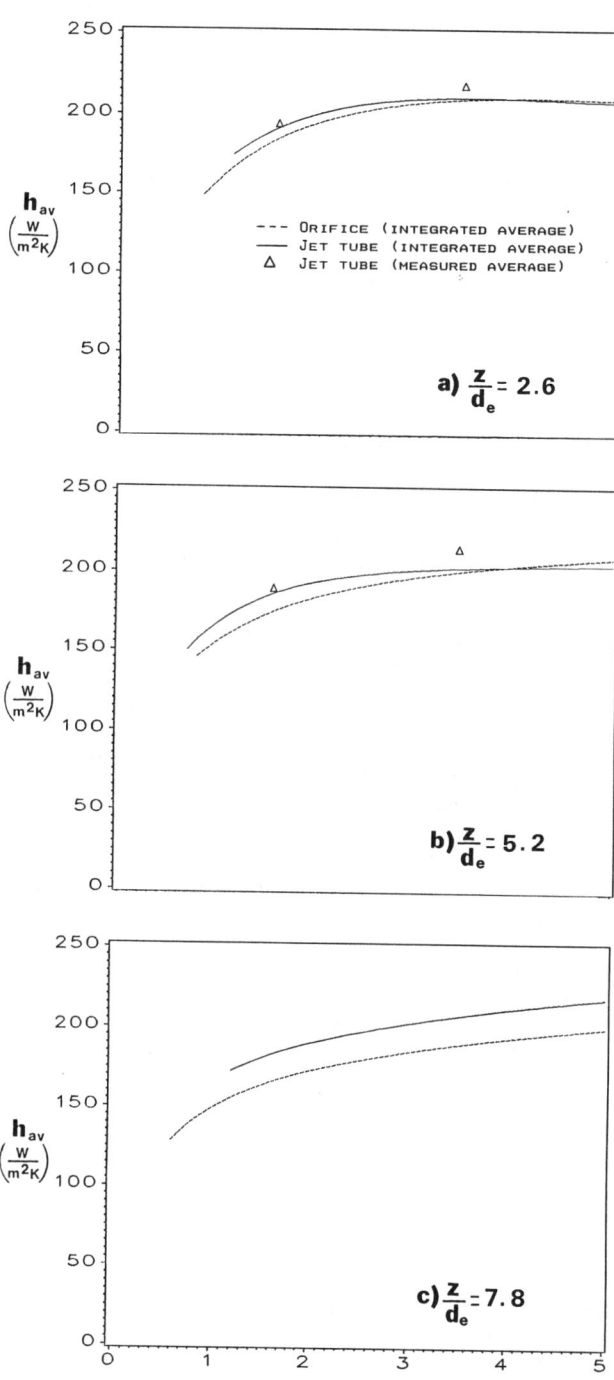

Fig. 9 Comparison of Orifice and Jet Tube Heat Transfer Distribution

Fig. 10 Average Heat Transfer

CONCLUSIONS

The heat transfer and velocity characteristics of impinging jets from jet tubes and orifices were measured. The centerline velocity decay for the jet tube was slower than for the orifice. Correspondingly, the heat transfer was higher for the jet tube at the larger axial distances. Moreover, the average heat transfer at most OAE ratios (OAE < 4 percent) was slightly higher for the jet tube than for the orifice at all axial distances tested. Additional work is continuing to document the advantages of jet tubes for decreasing the effects of exhaust crossflow in large jet arrays.

ACKNOWLEDGMENTS

Support for this work from Machine Technologies, Inc. and the Virginia Center for Innovative Technology is gratefully acknowledged.

REFERENCES

Abramovich, G. N., 1963, **The Theory of Turbulent Jets**, the MIT Press, Cambridge MA.

Beckwith, W. F., and Beard, J. N., Jr., 1979, "A Scheme to Assist in the Evaluation of Tenter Frame Dryer Performance," **ASME J. Engng. for Industry, Trans.**, Vol. 101, pp. 80-84.

Borell, G. J., and Diller, T. E., 1987, "A Convection Calibration Method for Local Heat Flux Gages," **ASME J. of Heat Transfer**, Vol. 109, pp. 83-89.

Chance, J. L., 1974, "Experimental Investigation of Air Impingement Heat Transfer Under an Array of Round Jets," **Tappi**, Vol. 57, No. 6, pp. 108-112.

Chia, C.-J., Giralt, F., and Trass, O., 1977, "Mass Transfer in Axisymmetric Turbulent Impinging Jets," **Ind. Eng. Chem., Fund.**, Vol. 16, pp. 28-35.

Corrsin, S., 1943, "Investigation of Flow in an Axially Symmetrical Heated Jet of Air," NACA WR-94.

Donaldson, C. du P., Snedeker, R. S., and Margolix, D. P., 1971, "A Study of free Jet Impingement, Part 2. Free Jet Turbulent Structures and Impingement Heat Transfer," **J. Fluid Mech.**, Vol. 45, pp. 477-513.

Downs, S. J., and James, E. H., 1987, "Jet Impingement Heat Transfer--A Literature Survey," ASME Paper No. 87-HT-35.

Florschuetz, L. W., Metzger, D. E., and Truman, C. R., 1981, "Jet Array Impingement with Crossflow--Correlation of Streamwise Resolved Flow and Heat Transfer Distributions," NASA CR-3373

Gardon, R., and Cobonpue, J., 1963, "Heat Transfer Between a Flat Plate and Jets of Air Impinging on It," **Int. Developments in Heat Trans.**, Part II, pp. 454-460.

Gardon, R., and Akfirat, J. C., 1965, "The Role of Turbulence in Determining the Heat Transfer Characteristics of Impinging Jets," **Int. J. Heat and Mass Transfer**, Vol. 8, p. 1261-1272.

Gardon, R., and Akfirat, J. C., 1966, "Heat Transfer Characteristics of Impinging Two-Dimensional Air Jets," **ASME J. of Heat Transfer**, Vol. 88, No. 1, pp. 101-108.

Gauntner, J. W., Livingood, J. N. B., and Hrycak, P., 1970, "Survey of Literature on Flow Characteristics of a Single Turbulent Jet Impinging on a Flat Plate," NASA TN D-5652 Technical Note.

Goldstein, R. J., and Behbahani, A. I., 1982, "Impingement of a Circular Jet With and Without Crossflow," **Int. J. Heat Mass Transfer**, Vol. 25, pp. 1377-1382.

Han, S. T., and Seely, T., 1965, "Heat Transfer from a Solid Surface Under Air Impingement," **Tappi**, Vol. 48, pp. 705-708.

Hollworth, B. R., and Berry, R. D., 1978, "Heat Transfer from Arrays of Impinging Jets with Large Jet-to-Jet Spacing," **ASME J. Heat Transfer**, Vol. 100, pp. 352-357.

Hollworth, B. R., and Wilson, S. I., 1984, "Entrainment Effects on Impingement Heat Transfer: Part I--Measurements of Heated Jet Velocity and Temperature Distributions, and Recovery Temperatures on Target Surface," Vol. 106, pp. 797-803.

Hollworth, B. R., and Gero, L. R., 1985, "Entrainment Effects on Impingement Heat Transfer: Part II--Local Heat Transfer Measurements," **ASME J. Heat Transfer**, Vol. 107, pp. 910-915.

Hoogendoorn, C. J., 1977, "The Effect of Turbulence on Heat Transfer at a Stagnation Point," **Int. J. Heat Mass Transfer**, Vol. 20, pp. 1333-1338.

Huang, G. C., 1963, "Investigations of Heat-Transfer Coefficients for Air Flow Through Round Jets Impinging Normal to a Heat-Transfer Surface," **ASME J. Heat Transfer, Trans.**, Vol. 85, pp. 237-245.

Kataoka, K., Kamiyama, Y., Hashimoto, S. and Komai, T., 1983, "Mass Transfer Between a Plane Surface and an Impinging Turbulent Jet: The Influence of Surface-Pressure Fluctuations," **J. of Fluid Mech.** Vol. 119, pp. 91-105.

Kercher, D. M., and Tabakoff, W., 1970, "Heat Transfer by a Square Array of Round Air Jets Impinging Perpendicular to a flat Surface Including the Effect of Spent Air," **ASME J. Eng. Power, Trans.**, Vol. 92, pp. 73-82.

Livingood, J. N. B., and Hrycak, P., 1973, "Impingement Heat Transfer from Turbulent Air Jets to Flat Plates. A Literature Survey," NASA TM X-2778.

Martin, H., 1977, "Heat and Mass Transfer Between Impinging Gas Jets and Solid Surfaces," **Adv. in Heat Transfer**, Vol. 13, pp. 1-60.

Mujumdar, A. S., 1985, "Impingment Drying," in **Handbook of Industrial Drying**, Ed. A. S. Mujumdar, Marcel Dekker, NY, pp. 461-474.

Obot, N. T., Mujumdar, A. S., and Douglas, W. J. M., 1979, "The Effect of Nozzle Geometry on Impingement Heat Transfer under a Round Turbulent Jet," ASME Paper No. 79-WA/HT-53.

Rajaratnam, N., 1976, **Turbulent Jets**, Elsevier, New York.

Scholtz, M. T., and Trass, O., 1970, "Mass Transfer in a Nonuniform Impinging Jet," **AIChE Journal**, Vol. 16, pp. 82-96.

Striegl, S. A., and Diller, T. E., 1984, "The Effect of Entrainment Temperature on Jet Impingement Heat Transfer," **ASME J. of Heat Transfer**, Vol. 106, pp. 27-33.

Vallis, E. A., Patrick, M. A., and Wragg, A. A., 1978, "Radial Distribution of Convective Heat Transfer Coefficient Between an Axisymmetric Turbulent Jet and a Flat Plate Held Normal to the Flow," **Sixth Int. Heat Transfer Conf.**, Vol. 5, pp. 297-303.

VandenBerghe, T. M., and Diller, T. E., 1989, "Analysis and Design of Experimental Systems for Heat Transfer Measurement from Constant-Temperature Surfaces," accepted for **Experimental Thermal and Fluid Sciences**.

Vlachopoulos, J., and Tomich, J. F., 1971, "Heat Transfer from a Turbulent Hot Air Jet Impinging Normally on a Flat Plate," **Can. J. Chem. Eng.**, Vol. 49, pp. 462-466.

A NEW HEAT TRANSFER LAW FOR TURBULENT NATURAL CONVECTION BETWEEN HEATED HORIZONTAL SURFACES OF LARGE EXTENT

W. K. George
Turbulence Research Laboratory
Department of Mechanical and Aerospace Engineering
University at Buffalo/SUNY
Buffalo, New York

ABSTRACT

An analysis is presented of the fully developed turbulent flow between horizontal parallel heated surfaces of large extent (the so-called turbulent Rayleigh problem). The flow is shown to consist of two layers: an inner layer in which viscous and conduction effects are important, and an outer layer in which these effects are negligible. Scaling laws for each region are proposed and inner and outer profiles for the temperature are identified. Matching these profiles in the limit as the ratio of outer to inner scales goes to infinity leads to the identification of a buoyant sublayer in which the mean temperature profile varies as $z^{-1/3}$ with an additive constant. A further matching of the inner and outer representations leads to a heat transfer law which is valid at finite as well as infinite Rayleigh number,

$$Nu = \frac{H_*^{1/4}}{\left[B_{10} H_*^{-1/12} - B_1(Pr) \right]}$$

The two unknowns in this law, one a universal constant and the other a universal function of Prandtl number, are uniquely determined by the mean temperature profile in the buoyant sublayer. The predictions are found to be in excellent agreement with the experimental evidence.

NOMENCLATURE

B_1 Universal function of Prandtl number, eq. (30)

B_{10} Universal constant, eq. (31)

C_p Specific heat at constant pressure

C_H' Coefficient, eq. (41), (42)

C_H Coefficient, eq. (50)

d Separation between surfaces (= 2h)

F_1 Inner temperature profile, eq. (12)

f_1 Inner temperature profile, eq. (20)

F_0 "temperature" flux at wall, eq. (3)

g Gravitational acceleration

H H-number based on half-width and $T_w - T_c$, eq. (44)

h Half-width of separation between surfaces (d/2)

H_* Flux H-number, eq. (14)

K_1 Universal matching constant, eq. (28), (29)

Nu Nusselt number, eq. (37)

P Mean pressure

P_0 Mean reference pressure

Pr Prandtl number, ν/α

P_w Mean pressure at wall (z=0)

q_w Heat flux at wall (z=0)

Ra Rayleigh number based on half width and $T_w - T_c$, eq. (43)

Ra_d Rayleigh number based on width and ΔT_w, $(2)^4 \times Ra$

T Mean temperature

T_c Centerline mean temperature

T_I Inner temperature scale, eq. (10)

T_0 Reference temperature

T_w Wall temperature (z=0)

$\hat{T}$ $(T - T_c)/T_0$

T^+ $(T-T_w)/T_I$

u_o Outer velocity scale, eq. (19)

w Fluctuating vertical velocity

z Vertical coordinate

z^+ z/η_I, inner coordinate

$\tilde{z}$ z/h, outer coordinate

α Thermal diffusivity

β Thermal expansion coefficient

ΔT T_-/T_o

ΔT_w Temperature differences between surfaces $2(T_w-T_c)$

η_I Inner length scale, eq. (11)

ν Kinematic viscosity

ρ Density

ρ_o Reference density

θ Fluctuating temperature

1. INTRODUCTION

The problem of turbulent natural convection above horizontal surfaces has been a subject of considerable interest over the past 35 years. One reason for this popularity lies in the common occurrence of such flows in engineering and natural environments. Other reasons are because such flows represent an important class of flows in which turbulence is produced only by buoyancy, and they can be easily idealized both analytically and experimentally. Examples of these idealized flows include the unsteady and steady boundary layers above a heated horizontal surface of infinite extent, and the steady state flow between horizontal parallel surfaces of infinite extent.

The first theoretical analysis of turbulent natural convection above a heated horizontal surface was put forth by Malkus (1954) in which he argued for the existence of a z^{-2} region in the mean temperature gradient, where z is the distance from the surface. In the same year, Priestly (1954) on dimensional grounds argued instead for the existence of a region in which the temperature gradient varied as $z^{-4/3}$. Long (1976) and Panofsky (1978) used matched asymptotic analyses to support Priestly's hypothesis.

Another aspect of Long's matching analyses (1976) was the derivation of an asymptotic heat transfer law in which Nu ~ $Ra^{1/3}$ with a Prandtl number dependent coefficient. This is consistent with the empirical heat transfer correlation of Globe and Dropkin (1959) who proposed for the flow between parallel surfaces of large extent,

$$Nu = 0.069 \ Pr^{0.074} \ Ra^{1/3}$$

Other investigators (eg. Chu and Goldstein (1973), Fitzjarrald (1976) report values of the Rayleigh number exponent to be somewhat smaller than 1/3, an effect which Long (1976) attributes to the experiments not having reached the asymptotic value.

Adrian et al. (1986) and Fitzjarrald (1976) using their own data argue convincingly for the existence of inner and outer scales, the former proposed by Townsend (1959) and the latter by Deardorff (1970). As to the existence of the power law regions proposed above, Adrian et al. (1986) argue that neither result is particularly satisfactory alone. Instead they provide a qualified endorsement of the meso-layer theory proposed by Chern and Long (1980) in which a z^{-2} layer exists beneath a $z^{-4/3}$ layer in the temperature gradient.

The purpose of this paper is to present a different analysis of the so-called "turbulent Rayleigh problem" using the methodology of George and Capp (1979) for turbulent natural convection next to vertical surfaces. This paper will present only the results for the mean temperature profile and the heat transfer relation. Of special interest is the derivation of the leading correction terms for the heat transfer coefficient at finite Rayleigh number. Some of the features of this analysis have already been proposed, eg. Long 1976. However, it is hoped that this new treatment will provide a more unified approach to the problem of natural convection above horizontal surfaces, and will be particularly useful in understanding the experimental data. The theory is based on an inner-outer scaling approach to the problem, and predicts both the existence of the $z^{-4/3}$ temperature gradient and the asymptotic heat transfer law Nu ~ $Ra^{1/3}$ with a Prandtl number dependent coefficient. An examination of the experimental data from the perspective of this unified theory will be seen to provide strong support for the conclusions.

The remainder of this paper is divided into two parts: Part I will analyze in detail the fully turbulent flow between differentially heated horizontal parallel plates, while Part II will consider the experimental data.

PART I: THE RAYLEIGH PROBLEM

2. The Equations of Motion for Steady Flow

The flow to be analyzed consists of the thermal convection which develops at high Rayleigh number between differentially heated horizontal plates in a gravitational field. It is hypothesized that the flow is fully turbulent and homogeneous in planes perpendicular to the vertical (defined by the gravitational vector). The flow is further assumed to be statistically stationary so that all time derivatives vanish in the averaged equations of motion. The flow and coordinate system is illustrated in Figure 1. The notation utilized throughout will conform as closely as possible to that used by George and Capp (1979).

Since there can be no mean flow in any direction (by hypothesis) the mass conservation equation vanishes identically, and the averaged equations for momentum and temperature to within the Boussinesq approximation reduce to

temperature: $0 = \dfrac{\partial}{\partial z}\left[-\overline{w\theta} + \alpha\,\dfrac{\partial T}{\partial z} \right]$ (1)

z-momentum:
$$0 = \frac{\partial}{\partial z}\left[\overline{-w^2} - \frac{P-P_o}{\rho_o}\right] + g\beta(T-T_o) \tag{2}$$

where α is the thermal diffusivity, g is the gravitational acceleration, and β is the thermal expansion coefficient. The reference values denoted by subscript zero can be taken to represent a hypothetical undisturbed state of uniform density. Note that unlike other boundary layer type flows, the pressure variation across the flow can not be neglected by reference to the momentum equation in other directions. Note also that the absence of the viscous term in equation (2) does not represent a high Reynolds number approximation, but rather is a consequence of the absence of a mean flow. Thus viscosity can affect the averaged motion only by its pressure in the higher moment equations.

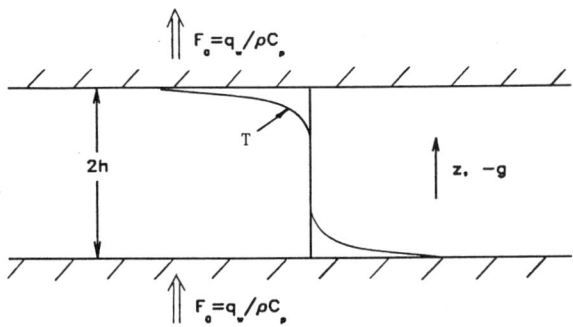

Figure 1. Sketch of Rayleigh flow.

Both equations can be integrated with respect to z to yield

$$-\overline{w\theta} + \alpha\frac{\partial T}{\partial z} = \frac{q_w}{\rho C_p} = -F_o \tag{3}$$

and

$$\left[\overline{w^2} + \frac{P-P_o}{\rho_o}\right] = \int_o^z g\beta\Delta T dz' \tag{4}$$

where q_w is the heat flow at the surface where z=o and must be equal and opposite to that at the other surface for steady flow to exist. Hereafter, for simplicity, the related quantity F_o will be referred to as the heat flux. Equation (3) is of primary importance since it makes it clear that the total heat flux is constant from top to bottom. Thus in the terminology of George and Capp (1979), the entire flow is a constant heat flux "layer". A consequence of this is that the flow is symmetrical about the centerline and only half of it need be considered.

3. The Existence of an Inner Layer

It is instructive to examine the relative order of magnitude of the terms in the governing equations, especially the temperature equations. It is straightforward to show that in the limit as $2F_oh/\alpha\Delta Tw$ (the Nusselt number) increases (it will be shown below that it is in fact the Rayleigh number which governs the development), the conduction term vanishes over most of the flow and the dynamical equations reduce to

$$F_o = -\overline{w\theta} \tag{5}$$

and

$$0 = \frac{\partial}{\partial z}\left[\overline{-w^2} - \frac{P-P_o}{\rho_o}\right] + g\beta\Delta T \tag{6}$$

Therefore, in this high Nusselt (or Rayleigh) number limit, neither the viscosity nor the thermal conductivity directly affect the averaged momentum or temperature equations.

It is clear that equations (5) and (6) can at most describe a region away from the wall since they can no longer satisfy the surface boundary condition

$$F_o = -\alpha\frac{\partial T}{\partial z}\Big|_{z=o} \tag{7}$$

Thus, in like manner to the analysis of George and Capp (1979), equations (5) and (6) must be recognized as governing an outer flow region away from the vicinity of the walls. It follows immediately that there must exist a region near to the walls in which a different set of equations applies which retains both the conduction and the turbulence heat transfer terms. An appropriate scaling length for this wall region must be sought which allows the conduction term to be retained there. If this inner length scale is denoted as η_I, then η_I must be chosen so that the conduction term is at least as important as the turbulence term. Thus in this inner layer the governing equations are exactly equations (3) and (4).

The inner layer can be further subdivided by noting that the kinematic boundary condition ($\underline{u}\cdot\underline{n}=0$ at a fixed surface) causes $\overline{w\theta}$ to vanish identically at z=0. Therefore very close to the wall, the temperature equation reduces to

$$\alpha\frac{\partial T}{\partial z} = -F_o \tag{8}$$

This can be immediately integrated to yield

$$T - T_w = -\frac{F_o}{\alpha}z \tag{9}$$

so that this inner-inner region can be recognized as a layer in which the mean temperature varies linearly. It is appropriate (again following George and Capp (1979)) to refer to this sub-region as the Conductive Sublayer.

An examination of the dynamical equations in the limit as $2F_oh/\alpha\Delta T_w \to \infty$ has led to the recognition that the natural convection flow between parallel horizontal surfaces can be characterized by two layers: an inner layer in which the conduction term allows the surface boundary condition to be met, and an outer layer where conduction effects are negligible. The inner layer has been shown to include a region very close to the wall in which the mean temperature varies linearly, the conductive sublayer. A schematic of the flow is shown in Figure 2 along with and parameters governing each region. Note that the kinematic viscosity has been included for the inner layer, even though it does not directly enter the first-order moment equations but enters only in the higher order equations for the turbulence heat flux and the turbulence kinetic energy.

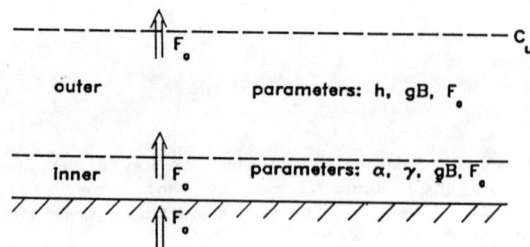

Figure 2. Schematic showing inner and outer layers.

4. ΔT_w versus F_o

In order to establish scaling laws for the inner and outer regions identified above, it is necessary to decide which of the parameters arising in the equations of motion and boundary conditions govern each region. It is easy to see that it is improper to use the temperature difference between the two surfaces, ΔT_w, to scale the temperature variation in either layer since only part of the total temperature drop occurs across each. Thus a proper scaling analysis must include new measures of the temperature scales for the inner and outer regions.

The key to properly scaling the inner and outer layers lies in the recognition that the entire flow is characterized by a constant heat flux (total heat flux independent of height). The inner layer has the surface heat flux, F_o, imposed on it. Because of the constancy of the total heat flux across the inner layer, F_o also provides an inner boundary condition on the outer (or core) flow. Thus F_o is a governing parameter for both the inner and outer regions. The remaining parameters must be obtained from an examination of the governing equations and boundary conditions.

5. The Inner Layer

From the equations of motion for the inner layer (equations 2.3 and 2.4) it is clear that the only parameters arising in the equations are $\alpha, g\beta$, and ν. Thus the inner layer must be governing entirely by the parameters F_o, $g\beta$, α and ν. An inner temperature scale can be defined as

$$T_I = \frac{F_o^{3/4}}{(g\beta\alpha)^{1/4}} \qquad (10)$$

and an inner length scale as

$$\eta_I = \left[\frac{\alpha^3}{g\beta F_o}\right]^{1/4} \qquad (11)$$

Note that these choices are not unique since ν could have been used in place of α. In fact, any combination of α and ν having the correct dimensions could have been used. It is this ambiguity which requires that the Prandtl number dependence be retained in the scaled profiles given below for the inner layer. The particular choices made here were first introduced for this problem by Townsend (1959) and are exactly those used by George and Capp (1979).

The mean temperature profile in inner variables

can now be written as

$$T-T_w = T_I \, F_1(z^+, Pr, H_*) \qquad (12)$$

where z^+ is a dimensionless inner coordinate defined by

$$z^+ = z/\eta_I \qquad (13)$$

$Pr = \nu/\alpha$ is the Prandtl number and H_* is the flux H-number (v. George and Capp (1979) defined as

$$H_* = \frac{g\beta F_o h^4}{\alpha^3} \qquad (14)$$

It is important to note that the temperature has been referenced to its value at the wall in order to avoid the need to directly account for effects from outside the wall region.

The flux H-number takes on additional significance as having a direct relation to the ratio of gap half-width, h, to the inner length scale, η_I. From the definitions of equations (11) and (14) it follows that

$$\frac{h}{\eta_I} = H_*^{1/4} \qquad (15)$$

This is particularly important since h will be identified below as the outer length scale. The inner-outer nature of the problem as $H_* \to \infty$ is again clear.

It is important to note that equation (12) describes not only the temperature variation in the inner region, but across the entire flow, the dependence on H_* reflecting the dependence of the inner flow on the outer at finite values of H_*. In the limit as $H_* \to \infty$, the dependence on this variable must vanish (if the inner variables have been correctly chosen), and equation (12) will describe only the inner temperature profile or a Law of the Wall (Adrian et al.(1986). George and Capp (1979)).

6. The Conductive Sublayer

The linear profile of equation (9) can be rewritten in the inner variables proposed above. The result is

$$\frac{T-T_w}{T_I} = -\frac{z}{\eta_I} \qquad (16)$$

or

$$F_1(z^+) = -z^+ \qquad (17)$$

This particular functional form describes only the conductive sublayer portion of the inner layer where the turbulence heat flux contribution can be neglected. The exact extent of this region will depend on the Prandtl number, since viscous effects control the emergence of $\overline{w\theta}$ as the distance from the wall increases. If the Prandtl number is large, disturbances will be damped to a greater distance from the wall than the distance for which the temperature equation can be expected to be described by the linear term only, As a consequence, the temperature will begin to show a gradual deviation from the linear dropoff between distances characterized by η_I and $Pr^{3/4}\eta_I$, the latter characterizing the extent of the viscous-dominated region. On the other hand if the Prandtl number is low, the linear dropoff in temperature can be expected to continue for distances well beyond the extent of the viscous region.

7. The Outer (or Core) Region

From the equations of motion (equations (5) and (6)), it is clear that the only parameters governing the outer flow are $g\beta$ and F_o, the former occurring directly in the governing equations, and the latter from the boundary condition imposed by the inner layer. Since the inner layer becomes vanishingly small as the H-number increases, the appropriate length scale for the outer flow is the half-distance between the surfaces, h. From these parameters, outer temperature and velocity scales can be uniquely defined as

$$T_o = F_o^{2/3}/(g\beta h)^{1/3} \tag{18}$$

and

$$u_o = (g\beta F_o h)^{1/3} \tag{19}$$

These scales correspond directly to these utilized by George and Capp (1979) for vertical surfaces, and were first introduced into this problem by Deardorff (1970).

The variation of temperature in outer variables can now be written as

$$T-T_c = T_o \, f_1(\tilde{z}, \, H_*, \, Pr^{-3}H_*) \tag{20}$$

where $\tilde{z}$ is a dimensionless outer coordinate defined by

$$\tilde{z} = z/h \tag{21}$$

Here the dependence on H_* and $Pr^{-3}H_*$ reflects the influence of the inner parameters at finite values of H_* and $Pr^{-3}H_*$.

T_c, the centerline temperature, is chosen as a reference temperature to avoid referencing the temperature variation to a value outside the core region. This is necessary to avoid the need to include explicitly the conduction and viscous effects which influence the variation of temperature across the inner layer. (They are, however, included implicitly through H_*.) Thus the term "Deficit Laws" is appropriate to this outer form. Note that similar considerations must also apply to other statistical quantities (which are not considered here). While the earlier analyses have been generally careful about properly referencing the mean temperature, the need to properly reference the other quantities appears to have been overlooked until now.

In the limit as $Pr^{-3}H_*$, $H_* \to \infty$, the function $f_1(\tilde{z}, Pr^{-3}, H_*)$, must become asymptotically independent of both. This is because it represents a profile non-dimensionalized in outer variables and must remain finite in the outer region. Thus in this limit, equation (20) represents a true outer profile, having lost all dependence on viscous and conduction effects.

8. A Matched Layer: The Buoyant Sublayer

It has been noted above that the functional forms of equations (12) and (20) reduce to inner and outer profiles respectively in the limit as $Pr^{-3}H_*, H_* \to \infty$. It is interesting to ask whether there exists a range

of distances for which both the inner and outer scaling laws are valid - in effect, an overlap range. The procedure is exactly equivalent to that followed by George and Capp (1979) and requires matching the inner and outer profiles in the limit of infinite flux H-number. In formal terms: Is there a region of common validity in the limit as $Pr^{-3}H_*$, $H_* \to \infty$ so that

$$\lim_{\substack{z/h \to 0 \\ H_* \to \infty \\ Pr^{-3}H_* \to \infty}} f_1\left[\frac{z}{h}, H_*, Pr^{-3}H_*\right] = \lim_{\substack{z/\eta_I \to \infty \\ H_* \to \infty \\ Pr^{-3}H_* \to \infty}} F_1\left[\frac{z}{\eta_I}, Pr, H_*\right] \tag{22}$$

It is more convenient at this point to match the temperature derivatives given in inner and outer variables as

$$\frac{\partial T}{\partial z} = \frac{T_I}{\eta_I} \, F_1'(z^+, Pr, H_*) \tag{23}$$

and

$$\frac{\partial T}{\partial z} = \frac{T_o}{h} \, f_1'(\tilde{z}, H_*, \, Pr^{-3}H_*) \tag{24}$$

where ' denotes differentiation with respect to the appropriate variable. Equating these in the limit as $Pr^{-3}H_*$, $H_* \to \infty$ yields

$$\frac{T_I}{\eta_I} \, F_1'(z^+, Pr) = \frac{T_o}{h} \, f_1'(\tilde{z}) \tag{25}$$

From equations (10) and (18) it follows that

$$\frac{T_o}{T_I} = \left[\frac{\eta_I}{h}\right]^{1/3} = H_*^{1/12} \tag{26}$$

Using this and multiplying both sides of eq. (25) by $z^{4/3}$ yields

$$(z^+)^{4/3} \, F_1'(z^+, Pr) = \tilde{z}^{4/3} f_1'(\tilde{z}) \tag{27}$$

Since $h/\eta_I \to \infty$ as $H_* \to \infty$, it follows from the definitions of z^+ and $\tilde{z}$ that their ratio becomes undefined in this limit. Thus z^+ and $\tilde{z}$ become independent variables as $H_* \to \infty$. As a consequence, both sides of equation (27) must equal a constant, say $-K_1/3$. Thus,

$$(z^+)^{4/3} \, F_1'(z^+, P_r) = - K_1/3 \tag{28}$$

and

$$(\tilde{z})^{4/3} \, f_1'(\tilde{z}) = - K_1/3 \tag{29}$$

Equations (28) and (29) can be readily integrated to yield

$$F_1(z^+, Pr) = K_1(z^+)^{-1/3} + B_1(Pr) \tag{30}$$

$$f_1(\tilde{z}) = K_1(\tilde{z})^{-1/3} + B_{10} \tag{31}$$

where K_1 and B_{10} are universal constants and $B_1(Pr)$ is a universal function of the Prandtl number.

Equations (31) and (32) can be expressed in physical variables as

$$\text{inner} \qquad \frac{T-T_w}{T_I} = K_1\left[\frac{z}{\eta_I}\right]^{-1/3} + B_1(Pr) \tag{32}$$

and

outer
$$\frac{T-T_c}{T_0} = K_1 \left[\frac{z}{h}\right]^{-1/3} + B_{10} \qquad (33)$$

Equations (32) and (33) thus represent the inner and outer forms of the temperature profile in a matched region between the inner and outer layers. This layer (by analogy with the analysis by George and Capp 1979) will be referred to as the _Buoyant Sublayer_, since it can be shown to be governed by only the distance to the wall and the buoyancy parameter.

The existence of such a region has been previously derived on the following dimensional grounds by Priestley (1954), (see also Turner (1973)). Suppose there exists a region for which η_I and $Pr^{3/4}\eta_I \ll z \ll h$. This can, of course, occur only in the limit of large H_*. Since the region of interest is well-removed from viscous or conductive effects, yet too small to be dependent on gap width, it must be entirely determined by $g\beta$, F_0, and z itself. Thus, on dimensional grounds (considering $g\beta$ to be a single parameter since they always occur in the equations in combination),

$$\frac{dT}{dz} \frac{(g\beta)^{1/3} z^{4/3}}{F_0^{2/3}} = \text{constant} = -\frac{K_1}{3} \qquad (34)$$

This can readily be integrated to yield equations (32) and (33). The origin of the Prandtl number dependence of the constant in equation (32) can be seen to arise from the reference to the wall temperature and the integration over the viscous region.

Although only the temperature profile is being considered here, it is noted that a matching of inner and outer profiles can also be carried out for $\overline{w^z}$, $(P-P_0)/\rho_0$, or any other turbulence quantity, and an appropriate buoyant sublayer form deduced. As for the mean temperature gradient, the functional dependences of $d\overline{w^z}/dz$ and dP/dz could have been deduced on dimensional grounds as noticed by Adrian et al. (1986). It is important to note, however, that because of the need to exclude the dependence on α, ν, and h, dimensional arguments for the buoyant sublayer can only be applied to the _variation_ of T, $\overline{w^z}$ and P within the sublayer, and not to these quantities directly. Thus, it is appropriate to argue that $dT/dz \sim z^{-4/3}$ but not that $T \sim z^{-1/3}$; similarly that $\partial\overline{w^z}/\partial z \sim z^{-1/3}$ but not $\overline{w^z} \sim z^{2/3}$, etc. The matching results above make clear the importance of both the additive constants and the use of appropriate reference values. The failure to recognize this clearly before now is largely responsible for the frustration of experimenters who sought to confirm the existence of the buoyant sublayer by using log-log plots of T and $\overline{w^z}$ versus z (see for example Adrian et al.(1986), Goldstein and Chu (1969)).

9. A Heat Transfer Law

Since both inner and outer forms of the dependent variables in the buoyant sublayer describe the same physical profiles, the actual (unscaled) variables predicted by both must be the same. For example, equations (32) and (33) must yield the same temperature at the same location. Thus

$$T_w + T_I \left\{K_1 \left[\frac{z}{\eta_I}\right]^{-1/3} + B_1(Pr)\right\}$$

$$= T_c + T_0 \left\{K_1 \left[\frac{z}{h}\right]^{-1/3} + B_{10}\right\} \qquad (35)$$

Using the definitions of T_I, η_I and T_0 (equations (10), (11) and (18) it follows after rearranging that

$$\frac{T_w-T_c}{T_I} = \left[\frac{\eta_I}{h}\right]^{1/3} B_{10} - B_1(Pr) \qquad (36)$$

This can be transformed into a more familiar form by using a Nusselt number defined as

$$Nu = \frac{F_0 h}{\alpha(T_w-T_c)} \qquad (37)$$

Note that because of symmetry, the temperature difference between surfaces is

$$\Delta T_w = 2(T_w-T_c) \qquad (38)$$

Thus, since the distance between the plates is 2h, the Nusselt number of equation (38) is just the usual Nusselt number based on temperature difference and gap width.

By using equations (10), (11), (15) and (37), equation (36) can be rewritten as

$$Nu^{-1} H_*^{1/4} = B_{10} H_*^{-1/12} - B_1(Pr) \qquad (39)$$

or

$$Nu = \frac{H_*^{1/4}}{B_{10}H_*^{-1/12}-B_1(Pr)} \qquad (40)$$

Thus the heat transfer law is completely determined to second order by the buoyant sublayer constants of the mean temperature profile. This appears not to have been noticed before now, and provides an important check on the consistency of experimental data as well as on the internal consistency of the theory.

In the limit as $H_* \to \infty$,

$$Nu = C_H' H_*^{1/4} \qquad (41)$$

where

$$C_H' = [-B_1(Pr)]^{-1} \qquad (42)$$

Thus at very large values of H_*, the Nusselt number varies as $H_*^{1/4}$ with a coefficient dependent on Prandtl number. However, for smaller values of as H_* (yet still large enough to ensure the validity of the theory), the $H_*^{-1/12}$ correction term will modify the coefficient. It will be seen later that both $B_1(Pr)$ and B_{10} are negative so that the effect of finite H_* is to _increase_ the Nusselt number above the value which would be predicted by equation (41) alone. Note that the negative value of B_{10} implies that the theory breaks down long before the $H_*^{-1/12}$ term becomes dominant in the denominator; thus an $H_*^{1/3}$ asymptote will not be observed at moderate values of H_*.

It has been customary in the experimental literature to correlate data using the Rayleigh number and the Nusselt number. The Rayleigh number based on h and (T_w-T_c) can be defined as

56

$$Ra = \frac{g\beta(T_w - T_c)h^3}{\alpha\nu} \qquad (43)$$

which can be related to the corresponding H- number by

$$Ra = H \cdot Pr \qquad (44)$$

where

$$H = \frac{g\beta(T_w T_c)h^3}{\alpha^2} \qquad (45)$$

(Note that the usual definitions are based on the temperature difference and distance between the surfaces and are therefore $2^4=16$ times bigger than the H and Ra defined here.)

From the definitions of equations (14), (37), and (45), it follows that

$$H_* = H \cdot Nu \qquad (46)$$

Thus the heat transfer law of equation (40) can be transformed into

$$Nu = \frac{H^{1/4}Nu^{1/4}}{\left[B_{10}H_*^{-1/12}Nu^{-1/12} - B_1(Pr)\right]} \qquad (47)$$

or

$$Nu = \frac{H^{1/3}}{\left[B_{10}H^{-1/12}Nu^{-1/12} - B_1(Pr)\right]^{4/3}} \qquad (48)$$

The transformed heat transfer law of equation (48) is considerably more complicated than equation (40) because of the implicit dependence on the Nusselt number*. Nonetheless, it is easy to see that for very large H- numbers, the Nusselt number varies as the cube root of the H- number and the coefficient is Prandtl number dependent, i.e..

$$Nu = C_H(Pr) H^{1/3} = C_H(Pr)Pr^{1/3}Ra^{1/3} \qquad (49)$$

where

$$C_H(Pr) = \left[\frac{-1}{B(Pr)}\right]^{4/3} \qquad (50)$$

The Prandtl number dependence is not determined by the analysis (to this point, at least) and must be derived from other considerations.

The recognition that $Nu \sim Ra^{1/3}$ with a Prandtl number-dependent coefficient could be derived by an asymptotic matching analysis was apparently first noticed by Long (1975). His arguments were somewhat different, however, and led to two different dependence for smaller values of the Rayleigh number than that proposed here one of which is equivalent to equation (47).. The correction for finite values of H_* given by equation (47) will act to increase the heat transfer coefficient so that the experimentally determined dependence on Rayleigh number will be less than the 1/3 power. This will be discussed further in Part II when experimental data are considered.

---

*This is not a serious limitation in practice because the Nusselt number can first be calculated for given H_* using equation (40), then the value of H assigned using equation (46).

It will be shown in a subsequent publication that a similar matching can be carried for other statistical quantities which enter the governing equations (eg. $\overline{w^z}$, $\overline{\theta^z}$, etc.). The results of matching the appropriate inner and outer buoyant sublayer forms are "laws" which describe the variation of the centerline values of these quantities with H_*. These possibilities do not appear to have been noticed previously.

PART II: EVALUATION OF THEORY USING EXPERIMENTAL DATA

10. Overview of Experiments

There have been a number of laboratory experiments over the past two decades which attempted to simulate the fully developed turbulent Rayleigh problem. There have also been experimental investigations of several related flows including the natural convection between parallel plates with an adiabatic upper surface (Adrian et al. 1986), steady penetrative convection (Townsend 1959), unsteady penetrative convection (Deardorff et al. (1969). In addition there have been numerous investigations of the convective planetary boundary layer (Wyngaard et al. 1971, see also Monin and Yaglom, vol. I 1971). While there is reason to believe that all these flows should obey common scaling laws, there is little reason to believe that the scaled profiles should be the same for all, except perhaps in the inner layer. For now this question will be avoided, and only the Rayleigh experiments will be considered below.

11. The Mean Temperature Profiles: Inner Variables

There have been several experiments carried out which measured the mean temperature distribution at reasonably high Rayleigh numbers (eg. Somerscales and Gazda 1969, Chu and Goldstein 1973, Goldstein and Chu 1969, and Deardorff and Willis 1967a, 1967b). Many of these investigators made a particular effort to evaluate whether or not there was a region in the temperature gradient which varied as $z^{-4/3}$ (or the temperature varied as $z^{-1/3}$). Of these, only Deardorff and Willis (1967b) were partially successful in identifying a limited $z^{-4/3}$ region in the temperature gradient. It is suggested here that the failures were due in part to the relatively low values of the Rayleigh number which limited the extent of the buoyant sublayer, but in larger part to the use of log-log plots in the presence of additive constants, and the failure to understand clearly where the proposed region should lie.

Figure 3 is a plot of the Deardorff and Willis data (1967b) in inner variables. The data were read from enlarged versions of the figures in their paper, and the scatter in the fitted curves is largely due to the errors in this process, especially near the center of the flow. In order to make readily apparent the existence of a region described by equation (32), the data are plotted as $[(T-T_w)/T_I][z/\eta_I]^{1/3}$ versus $[z/\eta_I]^{1/3}$. Thus the buoyant sublayer, should it exist at all, would correspond to a straight line. Note also that since the plot is in inner variables, the data should be expected to deviate at progressively larger values of z^+ as the Rayleigh number of the experiment is increased since the ratio of outer to inner scales increases.

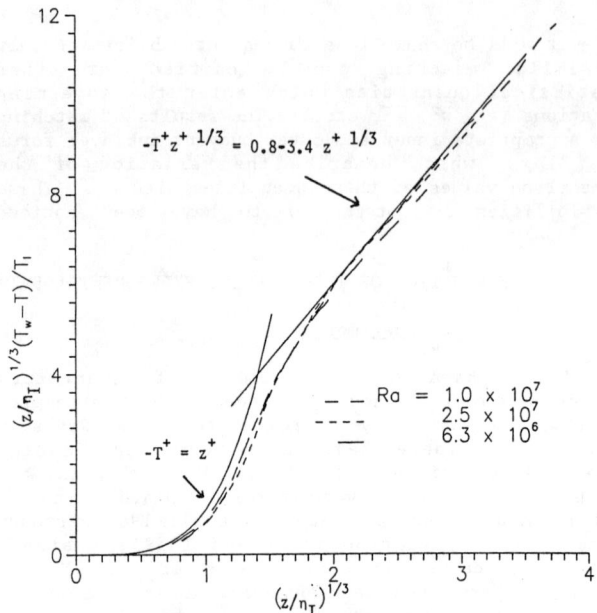

$-T^+ z^{+1/3} = 0.8 - 3.4\, z^{+1/3}$

$-T^+ = z^+$

Figure 3. Mean temperature versus height in inner variables, air data of Deardorff & Willis.

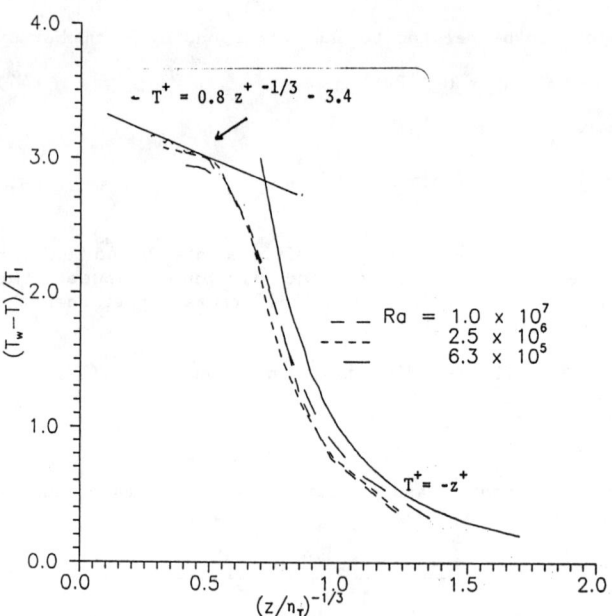

$-T^+ = 0.8\, z^{+-1/3} - 3.4$

$T^+ = -z^+$

Figure 4. Mean temperature versus height in inner variables, air data of Deardorff and Willis.

It is clear from Figure 3 that the data collapse well in inner variables until very large values of z^+ are obtained, and that the value of z^+ for which the curves begin to deviate from each other increases with Ra as expected. The region nearest the wall ($z^+ < 1$) is well-described by the linear dependence of equation (17). Of greater interest is the existence of the $z^{-1/3}$ region as represented by the straight line on the plot. This method of plotting allows determination of both K_1 and $B_1(Pr)$ directly, and also makes clear that the buoyant sublayer (or $z^{-1/3}$ region) is an asymptotic region which will be increasingly apparent with increasing Ra (or H_*).

In Figure 4, the same data are plotted following George and Capp (1979) as $(T-T_w)/T_I$ versus $(z/\eta_I)^{-1/3}$. This method would be of considerable value if profiles from different Prandtl number fluids were being plotted together. According to equation (32), the effect of Prandtl number should be to shift the $z^{-1/3}$ region, but to leave its slope unchanged. This has not been confirmed for this problem since only the air data have been considered to-date.

The constants K_1 and $B_1(Pr)$ of equation (32) have been estimated from a least squares fit to the highest Ra number experiment using data from $8 < z^+ < 27$ ($2 < z^{+1/3} < 3$). The results are

$$K_1 = 0.73 \qquad (51)$$

$$B_1(0.71) = -3.4 \qquad (52)$$

The constant K_1 is related to the C_1 constant of Deardorff and Willis (1967b) by $K_1 = C_1/3$. They found by plotting dT/dz versus z that $C_1 \simeq 2.2$ for the same data which corresponds exactly to the value obtained here. The constant $B_1(0.71)$ is, of course, valid only for air ($Pr = 0.71$) and a different value can be expected for a different fluid. Note that $B(Pr)$ enters directly the heat transfer laws (equations (40) and (50)), and so this value will be considered again below.

12. The Mean Temperature Profile: Outer Variables

The same mean temperature profile data as that used above has been replotted in outer variables in Figure 5. Considerable difficulty was encountered in trying to plot the data in outer variables to make clear the dependence over most of the outer flow. This was because most of the temperature drop occurred in the inner layer (in fact, the linear layer) so that the temperature outside the inner region was very close to the centerline value to which it must be referenced. Thus $(T-T_c)$ was very sensitive to the errors in determining both T and T_c. In view of the lack of alternatives, the data were plotted as $(T-T_c)/T_o$ versus $(z/h)^{-1/3}$. Thus the left ordinate is $(z/h)^{-1/3} = 1.0$ and corresponds to the centerline; increasing values of $(z/h)^{-1/3}$ are progressively closer to the wall. Again the buoyant sublayer (or $z^{-1/3}$ region) should correspond to a straight line.

As expected, the outer scaling works reasonably well for value of $\tilde{z} = z/h$ greater than about 0.1 ($\tilde{z}^{-1/3} < 2$). The deviations at smaller values of $\tilde{z}$ (larger $\tilde{z}^{-1/3}$) are a consequence of the fact that the region nearest the wall collapses only in inner variables. In spite of the scatter and the low values, there does appear to be a straight line region on the plot, especially at the higher Rayleigh numbers. From it the constants K_1 and B_{10} can be determined. Because of the large errors, K_1 was taken at the value determined above, and the value of B_{10} was determined by inspection to be

$$B_{10} \simeq -1.0 \qquad (53)$$

Obviously there is further work needed to refine this estimate. Note that whatever the value is, it should be universal and independent of Prandtl number.

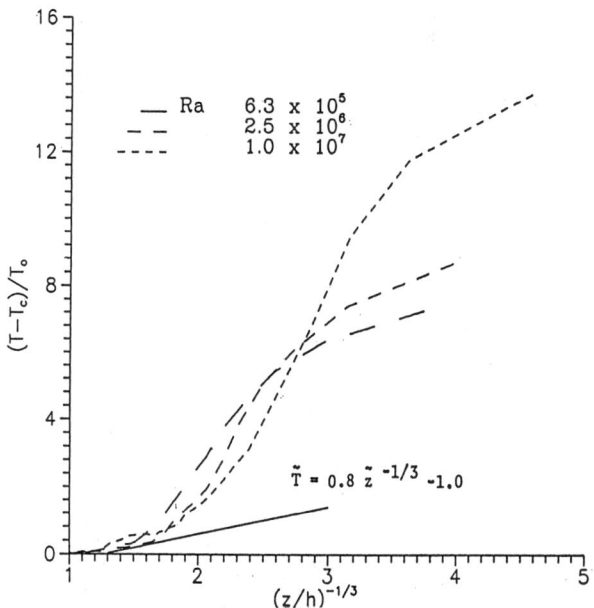

Figure 5. Mean temperature difference from centerline value versus height in outer variables, air data of Deardorff & Willis.

13. The Heat Transfer Law

The relation between Nusselt number and Rayleigh number has been the subject of intense investigation with almost as many proposed laws as investigators. Several of the suggested empirical relations for air are given below

Fitzjarrald (1976) (air, $8 \times 10^4 < Ra < 9 \times 10^9$)

$$Nu = 0.13 \, Ra_d^{0.30} \qquad (54)$$

Goldstein and Chu (1969)(air $3 \times 10^5 < Ra < 10^8$)

$$Nu = 0.123 \, Ra_d^{0.294} \qquad (55)$$

Globe and Dropkin (1959)

$$Nu = 0.0673 \, Ra_d^{1/3} \qquad (56)$$

where Ra_d is 16 times the Ra used earlier. Deardorff and Willis (1967a) note that their experiments in air are not too different from equation (56).

It was shown in Section 9 that the matching of the inner and outer temperature profiles yielded a second order heat transfer law (equation 40) with the coefficients determined only by the matching. Thus for air using $B_1(0.71) = -3.4$ and $B_{10} \simeq -1.0$ in equation (40) yields

$$Nu = \frac{0.294 \, H_*^{1/4}}{\left[1 - 0.294 \, H_*^{-1/12}\right]} \qquad (57)$$

This can be evaluated for given H_*, then the appropriate Rayleigh number calculated from equations (44), (45) and (46). The results are shown in Table I and Figure 6. Also shown are equations (54) and the asymptotic law ($H_* \to \infty$) determined from equation (57) as

$$Nu = 0.294 \, H_*^{1/4} \qquad (58)$$

or

$$Nu = .0690 \, Ra_d^{1/3} \qquad (59)$$

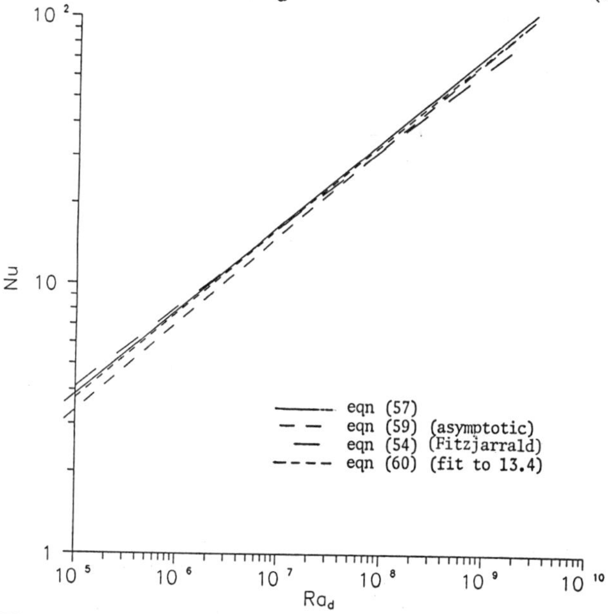

Figure 6. Comparison of theoretical and empirical heat transfer laws.

Table 1

Tabulation of Nusselt Number for Air Using Equations (40) and (41).
$B_1(0.71) = -3.4$, $B_{10} = -1.0$

H_*	Nu	Nuasym	Ra_d
1.00E+04	3.41E+00	2.94E+00	6.62E+04
3.00E+04	4.42E+00	3.87E+00	1.53E+05
1.00E+05	5.89E+00	5.23E+00	3.82E+05
3.00E+05	7.67E+00	6.88E+00	8.81E+05
1.00E+06	1.03E+01	9.30E+00	2.20E+06
3.00E+06	1.34E+01	1.22E+01	5.05E+06
1.00E+07	1.79E+01	1.65E+01	1.26E+07
3.00E+07	2.34E+01	2.18E+01	2.89E+07
1.00E+08	3.14E+01	2.94E+01	7.18E+07
3.00E+08	4.11E+01	3.87E+01	1.65E+08
1.00E+09	5.52E+01	5.23E+01	4.08E+08
3.00E+09	7.23E+01	6.88E+01	9.35E+08
1.00E+10	9.72E+01	9.30E+01	2.32E+09
3.00E+10	1.27E+02	1.22E+02	5.31E+09
1.00E+11	1.72E+02	1.65E+02	1.31E+10
3.00E+11	2.25E+02	2.18E+02	3.00E+10
1.00E+12	3.03E+02	2.94E+02	7.44E+10
3.00E+12	3.98E+02	3.87E+02	1.70E+11
1.00E+13	5.36E+02	5.23E+02	4.20E+11
3.00E+13	7.04E+02	6.88E+02	9.60E+11
1.00E+14	9.49E+02	9.30E+02	2.37E+12
3.00E+14	1.25E+03	1.22E+03	5.42E+12
1.00E+15	1.68E+03	1.65E+03	1.34E+13
3.00E+15	2.21E+03	2.18E+03	3.06E+13
1.00E+16	2.98E+03	2.94E+03	7.56E+13

It is clear from the figures that regardless of the exact value of the constant B_{10}, its effect is to increase the heat transfer above that which would be obtained from the cube root relationship. It is also easy to see that this effect will reduce the effective exponent of the Rayleigh number below 1/3, and that the amount of the reduction will depend on the Rayleigh number range considered. For example, a least squares fit to equation (57) in the range $10^5 < Ra < 10^{10}$ yields

$$Nu = .094 \, Ra_d^{0.32} \qquad (60)$$

The three empirical laws (equations 54, 56 and 60) differ from that given by equation (57) by less than 4% at $Ra = 1 \times 10^7$ and only by 11% at $Ra = 1 \times 10^5$. The agreement could be further improved by selecting a slightly higher value for the outer constant B_{10} (like -2.3) which may be warranted in view of the difficulty in determining it from the outer profile. Thus over the range in which most experiments were performed, the heat transfer law obtained from matching the profiles is in excellent agreement with the earlier empirical correlations. Equation (40) (and 57) must therefore be regarded as an improved heat transfer law valid for at least $Ra > 10^5$.

No effort has been made to determine theoretically or empirically the Prandtl number dependence of the integration constants for the inner buoyant sublayer profiles (eg. B_1 (Pr), etc.). From the correlation of Globe and Dropkin (1959) it can be inferred that

$$B_1(Pr) = 3.06 \, Pr^{-0.305} \qquad (61)$$

It follows that the corresponding heat transfer law should be

$$Nu = \frac{0.327 \, Pr^{0.305} H_*^{1/4}}{\left[1 - .294 \, H_*^{-1/12}\right]} \qquad (62)$$

These must be confirmed by further investigation.

14. Summary and Conclusions

A scaling analysis of fully developed turbulent natural convection between differentially heated surfaces of large extent has led to the identification of inner and outer layers. Only the inner layer is affected by the viscosity and thermal diffusivity. Matching of the mean temperature profiles in the limit as $H_* \to \infty$ and $Pr^{-3}H_* \to \infty$ resulted in the identification of a buoyant sublayer in which the mean temperature varied as $z^{-1/3}$ with an additive constant. Further matching of these profiles yielded a heat transfer law valid at finite values of H_*, as well as in the limit.

The available experimental data for air were analyzed in detail. The data were in excellent agreement with the predictions. Especially gratifying was the success of the temperature profile buoyant sublayer constants in producing a heat transfer law which was able to reproduce the experimental observations over the entire range of Rayleigh numbers measured.

It is likely that the results of the analysis presented here are directly applicable to related natural convection problems. The asymptotic value of the ratio $(T_w-T_c)/T_I$ of 3.4 is very close to its counterpart in both steady and unsteady non-penetrative convection (Adrian et al.(1986)). Also, the temperature profiles obtained by Townsend (1959)

when normalized in inner variables are indistinguishable from those in Figure 3. Thus when a proper accounting is taken of the Prandtl number dependence of the inner layer, it is likely that its characteristics will prove to be flow-independent. It is unlikely that this will be the case for the outer flow, however, (except for the buoyant sublayer part of it) since it will be more directly influenced by the differing boundary conditions away from the surface.

ACKNOWLEDGEMENTS

I first became sensitized to this problem as the result of a presentation at NCAR by Professor A. Libchaber of the University of Chicago during the Prospects for Turbulence Research Symposium in 1987, and immediately suspected the possible relation to my own work with Dr. S. Capp. I am grateful to Dr. J. Herring of NCAR for inviting my participation in the Symposium and to Professor R. Adrian of the University of Illinois for alerting me to the wealth of work on the problem. I should also express my gratitude to United Airlines, whose long flight delays on the return trip made it possible to complete the analysis while the problem was still fresh in my mind without the distractions of home. The diligence of Mrs. Eileen Graber in preparing numerous versions of the manuscript is also deeply appreciated.

REFERENCES

Adrian, R.J., Ferreira, R.T.D.S. and Boberg, T. 1986, "Turbulent Thermal Convection in Wide Horizontal Layers", Exp. in Fluids, 4, 1221-141.

Chern, C.S. and Long, R.R. 1980, "A New Theory of Turbulent Convection Over A Heated Surface" Johns Hopkins Univ., Dept. Earth & Plan. Sci. Rept., Baltimore, Md.

Chu, T.Y. and Goldstein, R.J., 1973, "Turbulent Convection In a Horizontal Layer of Water", J. Fluid Mech., 60, 141-159.

Deardorff, J.W. and Willis, G.E. 1967a, "Investigation of Turbulent Thermal Convection Between Horizontal Plates", J. Fluid Mech., 28, 675-704.

Deardorff, J.W. and Willis, G.E., 1967b, "The Free-Convection Temperature Profile", Q.J. Roy Meteor. Soc., 73, 166-175.

Deardorff, J.W., Willis, G.E. and Lilley, D.K., 1969, "Laboratory Investigation of Non-Steady Penetrative Convection", J. Fluid Mech., 35, 7-31.

Deardorff, J.W., 1970, "Convective Velocity and Temperature Scales for the Unstable Planetary Boundary Layer and the Rayleigh Convection", J. Atmos. Sci., 27, 1211-1213.

Fitzjarrald, D.E., 1976, "An Experimental Study of Turbulent Convection in Air",J. Fluid Mech., 73 693-719.

George, W.K. and Capp, S.P., 1979, "A Theory for Natural Convection Turbulent Boundary Layers Next to Heated Vertical Surfaces", Int. J. Heat & Mass Trans., 22, 813-826.

Globe, S. and Dropkin, D., 1959, "Natural Convection Heat Transfer in Liquids Confined by two Horizontal Plates and Heated From Below", *Heat Trans.*, *81*, 156-65. 219–223.

Goldstein, R.J. and Chu, T.Y., 1969, "Thermal Convection in a Horizontal Layer of Air", *Prog. Heat & Mass Transfer*, *2*, 55-75.

Long, R.R., 1976, "Relation Between Nusselt Number and Rayleigh Number in Turbulent Thermal Convection", *J. Fluid. Mech.*, *73*, 445-451.

Malkus, W.V.R., 1954, "The Heat Transport and Spectrum of Thermal Turbulence", *Proc. Roy. Soc. London*, A225, 195-212.

Monin, A.S. and Yaglom, A.A., 1971, *Statistical Fluid Mechanics*, Vol 1, MIT Press, Cambridge, MD.

Panofsky, H.A., 1978, "Matching in the Convective Planetary Boundary Layer", *J. Atmos. Sci.*, *35*, 272-276.

Priestly, C.H.B., 1954, "A Model for the Simulation of Atmospheric Turbulence", *J. Appl. Meteor.*, *15*, 571-587.

Somerscales, E.F.C. and Gazda, I.W. 1969, "Thermal Convection in High Prandtl Number Liquids at High Rayleigh Number", *Int. J. Heat Mass Trans.*, *12*, 1491-1511.

Townsend, A.A., 1959, "Temperature Fluctuations Over a Heated Horizontal Surface.

Turner, J.S., 1973, *Buoyancy Effects in Fluids. Cambridge Univ. Press*, Cambridge, MA.

Wyngaard, J.C., Cote, O.R. and Izumi, Y., 1971, "Local Free Convection, Similarity, and the Budgets of Shear Stress and Heat Flux, *J. Atmos. Sci.*, *28*, 1171-1182.

1989 National Heat Transfer Conference
HTD-Vol. 107, Heat Transfer in Convective Flows

APPLICATION OF A HIGHER-ORDER TURBULENCE CLOSURE MODEL TO PLANE JET

J. C. Chai and R. S. Amano
Department of Mechanical Engineering
University of Wisconsin-Milwaukee
Milwaukee, Wisconsin

ABSTRACT

This study represents performance tests of a higher-order turbulence closure model for the predictions of turbulent shear flows. Computations of the momentum and temperature fields in the flow domain being considered entail the solutions of the Reynolds-averaged transport equations containing the second-order turbulent fluctuating products.

The computations of the Reynolds stresses are performed with three closure models and $k-\varepsilon$ model (one of the Boussinesq viscosity models). The computations of the second-order temperature velocity products are achieved by closing the pressure-heat flux term. These predictions are compared with three sets of experimental data.

Finally, several advantages have been observed in using the higher-order transport equations for the evaluation of the Reynolds stresses; one of the most important features is that considerable improvement is achieved when higher-order turbulence closures is employed. This is because the Boussinesq viscosity model assumes that turbulence is isotropic, and thus, overpredicts the transverse normal component of the Reynolds stress, $<vv>$.

NOMENCLATURE

C_1, C_2	constants for energy dissipation equation
C_s	constant for diffusion of Reynolds Stresses, $<u_iu_j>$
C_{ϕ_1}, C_{ϕ_2}	constants for pressure-strain correlation
$C_{i\theta,1}, C_{i\theta,2}$	constants for pressure-heat flux correlation
C_μ	constant used in the calculation of turbulent eddy viscosity
D_{ij}	diffusion rate of Reynolds Stresses, $<u_iu_j>$
$D_{i\theta}$	diffusion rate of second-product of velocity-temperature, $<u_i\theta>$
G	generation rate of turbulent kinetic energy
G_{ij}	generation rate of Reynolds Stresses, $<u_iu_j>$
$G_{i\theta}$	generation rate of second-product of velocity-temperature, $<u_i\theta>$
H_{ij}	secondary generation rate of Reynolds Stresses, $<u_iu_j>$
k	turbulent kinetic energy
l	length scale used in the inlet condition for turbulent kinetic energy
P	mean pressure
Pr_ε	Prandtl number in energy dissipation equation
T	mean temperature
T_c	local mean centerline temperature
u	fluctuating velocity in x-direction
U	mean velocity in x-direction
U_c	local mean centerline velocity
v	fluctuating velocity in y-direction
x, y	Cartesian coordinates

Greek Symbols

α	thermal diffusivity
δ_{ij}	Kronecker delta
η	similarity coordinate for velocity in shear layer, $y/y_{\frac{1}{2}U}$
η_θ	similarity coordinate for temperature in shear layer, $y/y_{\frac{1}{2}T}$
ε	dissipation rate of turbulent kinetic energy
ε_{ij}	dissipation rate of Reynolds Stresses, $<u_iu_j>$
$\phi_{ij,1}, \phi_{ij,2}$	pressure-strain correlations for Reynolds Stresses
$\phi_{i\theta}$	pressure-heat flux correlations for second-product of velocity-temperature, $<u_i\theta>$
ρ	density of fluid
ν	kinematic viscosity
ν_t	turbulent kinematic viscosity
μ_t	turbulent molecular viscosity

Subscripts

i, j, k, l	tensor notations
IN	inlet station condition

1 INTRODUCTION

A turbulent shear flow involves high turbulence levels which result in largely augmented heat and momentum transfer. Therefore, mathematical models that can predict complex turbulent shear flows are needed.

In the past decade, research on numerical predictions of turbulent shear flows have been pursued in order to improve the predictions of complex turbulent flow phenomena. As a result, the second-order turbulence closure models have gained increasing attention. In these types of closure models, the second-order turbulent fluctuating products in the turbulent transport of momentum and energy are found directly from their individual transport equation. This type of approach may be contrasted with the conventional two-equation or the $k-\varepsilon$ model which is an isotropic turbulence model that employs the Boussinesq viscosity concept. It has been commonly accepted that the isotropic turbulence model which assumes that turbulence is locally isotropic or second-order turbulence products have equal magnitude in all directions cannot take the non-isotropic effect into account in the evaluation of turbulence levels, whereas this effect is significant in complex turbulent flows. For this reason, the second-order closure model is preferred because turbulence stresses and heat fluxes produced by mean shear interactions can be closed individually taking the non-isotropic effects into account.

In this paper, the second-order closure model for both momentum transport processes and heat transport phenomena are presented. The second-order closure model for momentum transport has been investigated by many researches (Launder et. al., 1975, Dekeyser, 1985, Sini and Dekeyser, 1985, Amano and Goel, 1985, and Amano and Chai, 1988). On the contrary, the applications of the second-order turbulence closure models for the heat transport phenomena have not been considered extensively. This is primarily due to the limitations of both experimental techniques and computational capacity.

The study presented here deals with the computations of heated turbulent plane jet with the temperature being considered as the passive contaminant.

2 MATHEMATICAL MODELS

Computations of the mean velocity and temperature fields in the flow domain entail the solutions of the Reynolds-averaged momentum and energy equations. The Reynolds-stress closure is consolidated by incorporating the equations of k and ε into the Reynolds-stress equations. The Second-order closure for the energy field is achieved by closing the pressure-heat flux correlation.

The transport equations of momentum, turbulent kinetic energy, turbulent energy dissipation and mean temperature can be written in the following form:

$$\frac{\partial}{\partial x_j}(U_j U_i) = -\frac{1}{\rho}\frac{\partial P}{\partial x_i} + \frac{\partial}{\partial x_j}(-<u_i u_j>)$$

$$+\frac{\partial}{\partial x_j}\left[\nu\left(\frac{\partial U_i}{\partial x_j} + \frac{\partial U_j}{\partial x_i}\right)\right] \tag{1}$$

$$\frac{\partial}{\partial x_j}(U_j k) = \frac{\partial}{\partial x_j}\left[(\nu + \nu_t)\frac{\partial k}{\partial x_j}\right] + G - \varepsilon \tag{2}$$

$$\frac{\partial}{\partial x_j}(U_j \varepsilon) = \frac{\partial}{\partial x_j}\left[\left(\nu + \frac{\nu_t}{Pr_\varepsilon}\right)\frac{\partial \varepsilon}{\partial x_j}\right] + \frac{\varepsilon}{k}(C_1 G - C_2\varepsilon) \tag{3}$$

$$\frac{\partial}{\partial x_j}(U_j T) = \frac{\partial}{\partial x_j}\left(\alpha\frac{\partial T}{\partial x_j}\right) - \frac{\partial <u_j\theta>}{\partial x_j} \tag{4}$$

where

$$\nu_t = C_\mu\frac{k^2}{\varepsilon} \tag{5}$$

$$G = -<u_i u_j>\frac{\partial U_i}{\partial x_j} \tag{6}$$

The transport equations for $<u_i u_j>$ can be written as:

$$\frac{\partial}{\partial x_k}(U_k <u_i u_j>) = G_{ij} - \varepsilon_{ij} + \phi_{ij} + D_{ij} \tag{7}$$

where $G_{ij}, \varepsilon_{ij}, \phi_{ij}$, and D_{ij} are the generation, dissipation, pressure-strain correlation and diffusion terms respectively and given as:

$$G_{ij} = -\left(<u_j u_k>\frac{\partial U_i}{\partial x_k} + <u_i u_k>\frac{\partial U_j}{\partial x_k}\right) \tag{8}$$

$$D_{ij} = C_s\frac{\partial}{\partial x_k}\left(\frac{k}{\varepsilon}<u_k u_l>\frac{\partial <u_i u_j>}{\partial x_l}\right) \tag{9}$$

$$\varepsilon_{ij} = \frac{2}{3}\delta_{ij}\varepsilon \tag{10}$$

In closing the Reynolds-stress equations, a proper formulation of the pressure-strain correlation is the focal point for better evaluation of the Reynolds stresses. The authors investigated two pressure-strain correlation models proposed by Naot et. al. (1973), Launder et. al. (1975) and proposed a new pressure-strain correlation for plane jet flows. The first model was developed by removing the isotropic constraint from the double-velocity, two-point correlation tensor. The second model was obtained by approximating the pressure-strain correlation by an arbitrary fourth-order tensor, where the single-point, single-velocity correlations were used. The new model proposed by the authors follows the approach of Launder et. al. (1975) by including the non-isotropic effects of the redistributive actions into account.

Model 1: Naot et. al. (1973)

$$\phi_{ij} = -C_{\phi_2}\left(G_{ij} - \frac{2}{3}\delta_{ij}G\right) + \phi_{ij,1} \tag{11}$$

Model 2: Launder et. al. (1975)

$$\phi_{ij} = -\frac{(C_{\phi_2}+8)}{11}\left(G_{ij} - \frac{2}{3}\delta_{ij}G\right) - \frac{(8C_{\phi_2}-2)}{11}\left(H_{ij} - \frac{2}{3}\delta_{ij}G\right)$$
$$-\frac{(30C_{\phi_2}-2)}{55}k\left(\frac{\partial U_i}{\partial x_j} + \frac{\partial U_j}{\partial x_i}\right) + \phi_{ij,1} \tag{12}$$

Model 3: Amano et. al. (1988)

$$\phi_{ij} = (7C_{\phi_2}-10)G\left(\frac{<u_i u_j>}{k} - \frac{2}{3}\delta_{ij}\right)$$
$$-\frac{2}{5}(2C_{\phi_2}-1)k\left(\frac{\partial U_i}{\partial x_j} + \frac{\partial U_j}{\partial x_i}\right)$$
$$-2(C_{\phi_2}-1)\left(G_{ij} - \frac{2}{3}\delta_{ij}G\right)$$
$$-2(C_{\phi_2}-1)\left(H_{ij} - \frac{2}{3}\delta_{ij}G\right) + \phi_{ij,1} \tag{13}$$

where

$$H_{ij} = -\left(<u_i u_k> \frac{\partial U_k}{\partial x_j} + <u_j u_k> \frac{\partial U_k}{\partial x_i} \right) \quad (14)$$

$$G_{ij} = -\left(<u_j u_k> \frac{\partial U_i}{\partial x_k} + <u_i u_k> \frac{\partial U_j}{\partial x_k} \right) \quad (15)$$

$$\phi_{ij,1} = -C_{\phi_1} \frac{\varepsilon}{k} \left(<u_i u_j> -\frac{2}{3}\delta_{ij}k \right) \quad (16)$$

Following the same procedure, the transport equations of $<u_i \theta>$ can be written as below:

$$\frac{\partial}{\partial x_j}(U_j <u_i \theta>) = G_{i\theta} + D_{i\theta} + \phi_{i\theta} \quad (17)$$

where $G_{i\theta}$, $D_{i\theta}$ and $\phi_{i\theta}$ are the production, diffusion and pressure-heat flux terms of $<u_i\theta>$, respectively. Detailed formulations of these terms can be found in Amano et. al. (1988) and, thus, will not be carried out here. The final form of these terms can be written as follows:

$$G_{i\theta} = -\left(<u_i u_j> \frac{\partial T}{\partial x_j} + <u_j \theta> \frac{\partial U_i}{\partial x_j} \right) \quad (18)$$

$$D_{i\theta} = \frac{\partial}{\partial x_j}[(\alpha + \nu)\frac{\partial <u_i\theta>}{\partial x_j} \\ +0.3\frac{k}{\varepsilon}<u_j u_l>\frac{\partial <u_i\theta>}{\partial x_l}] \quad (19)$$

$$\phi_{i\theta} = -C_{i\theta,1}\frac{\varepsilon}{k}<u_i\theta> +C_{i\theta,2}<u_l\theta>\frac{\partial U_i}{\partial x_l} \quad (20)$$

Table 1 shows the recommended values for the constants used in turbulence modeling.

Table 1: Recommended values for the constants used in turbulence modeling

C_1	C_2	C_s	C_{ϕ_1}	C_{ϕ_2}	$C_{i\theta,1}$	$C_{i\theta,2}$
1.44	1.92	0.25	1.50	1.2 ★	3.2	0.5

★   Model 1: $C_{\phi_2} = 2/3$
Model 2: $C_{\phi_2} = 0.4$

3   NUMERICAL METHOD

3.1   Numerical Procedure

All transport equations are broken into three main parts namely, convection, diffusion, and source/sink terms. These equations are then linearized so that they could be solved by a tridiagonal matrix algorithm through the conventional control-volume discretization approach of Patankar (1980). The staggered grid system is used for the computations of mean velocities. The convection-diffusion string was evaluated through the hybrid differencing schemes noting that the false diffusion rates are almost negligible in jet flow computations. The details of the numerical approach is given in Chai (1988).

The computations are performed by using constant grid meshes of 75 and 150 in transverse direction. As presented in figure 1, comparison of the two computations shows little grid dependence results. Therefore, the 75 transverse grid system is used for all subsequence computations to save time. The computation is terminated when the desired axial location is acquired. The

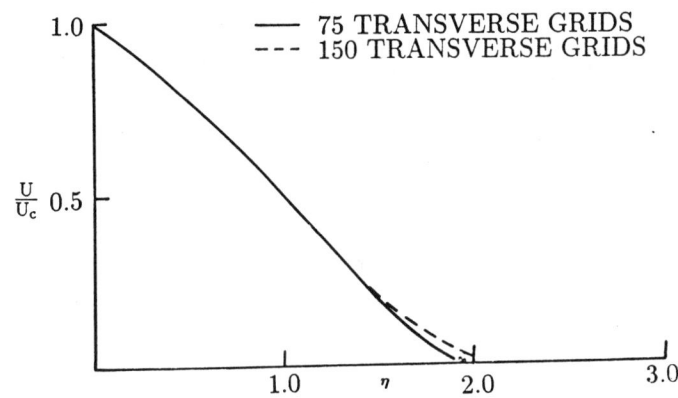

Figure 1: Grid test.

complete process of solving the momentum, turbulent kinetic energy, turbulence energy dissipation, the Reynolds stresses, temperature and second product of velocity-temperature takes about ten seconds of CPU time on a CRAY X-MP computer.

3.2   Boundary Conditions

The computations started in the self-preserved region of a plane jet. At the inflow section, cosine profiles were used for the mean streamwise velocity and mean temperature. The transverse velocity is set to zero. The turbulent kinetic energy and turbulent energy dissipation are prescribed by the following relations:

$$k = C_\mu^{-\frac{1}{2}} l^2 \left(\frac{\partial U}{\partial y} \right)^2$$

$$\varepsilon = C_\mu^{\frac{1}{2}} k \frac{\partial U}{\partial y}$$

The inlet condition for the Reynolds-stresses is given by the Boussinesq viscosity relation. The inlet conditions of the second-product of velocity-temperature are given by assuming that the fluctuation is about ten percent of the mean quantities. These conditions can be written in the following forms:

$$\rho <u_i u_j> = \frac{2}{3}\delta_{ij}k - \mu_t \left(\frac{\partial U_i}{\partial x_j} + \frac{\partial U_j}{\partial x_i} \right)$$

$$<u_i\theta> = i^2 U_{IN}\Delta T_{IN}$$

where i is about 0.1.

Zero gradients and zero values are given for all variables at the symmetry and free boundary, respectively.

4   RESULTS AND DISCUSSION

In general, the jet can be divided into three different zones, namely, the initial region, transition region, and fully-developed region (self-preserved region). In the latter region, the jet appears similar to a flow of fluid from a source of infinitely small thickness which is a point source for an axisymmetric flow and a straight line perpendicular to the plane of flow of the jet in the plane-parallel case. Experimental data of Heskestad (1965),

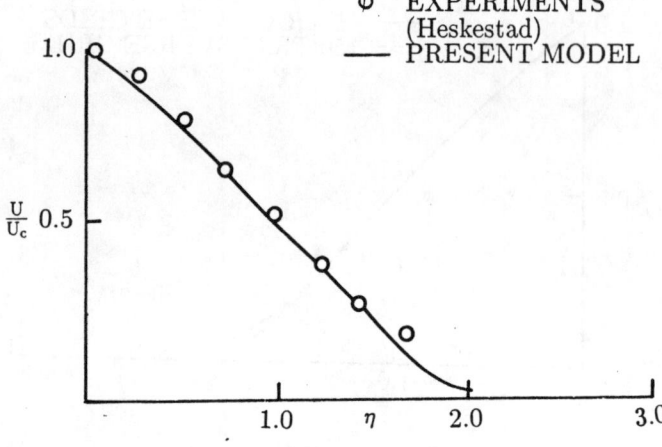

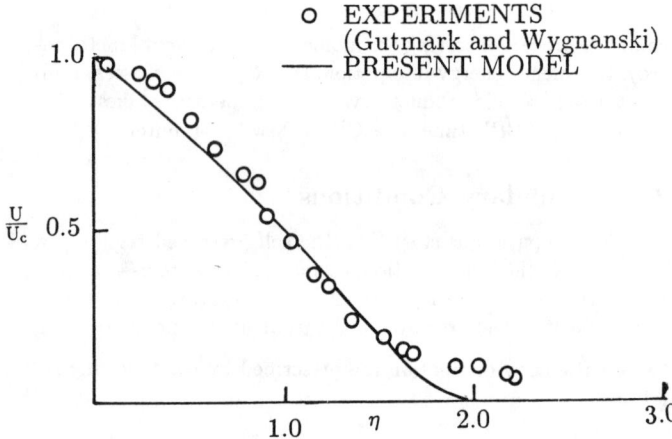

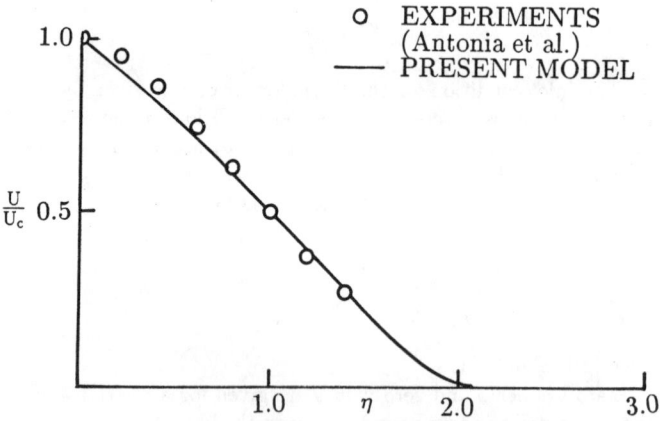

Figure 2: U velocity profile-Comparison with the data of Heskestad (1965), Gutmark and Wygnanski (1976), and Antonia et. al.(1983).

Gutmark and Wygnanski (1976) and Antonia et. al. (1983) were used for comparison with the computational results.

Figure 2 shows the comparisons of the computed velocity profiles in the self-preserved region with the measured velocity profiles of the experimental data mentioned above. The agreement between the computations and the data are very satisfactory in all three cases with about 10% discrepancy.

Figures 3 and 4 show the comparison of computed Reynolds stresses with the experimental data of Heskestad (1965) and Gutmark and Wygnanski (1976), respectively. All three pressure-strain correlations were used for the computations of the Reynolds stress equations. In addition to these, the $k - \varepsilon$ model was also used for comparison. The second-order closure models give relatively good results within 30% discrepancy by taking the non-isotropic effects of turbulence into account. On the contrary, the Boussinesq viscosity model overpredicted normal stresses ($< vv >$) approximately by 100% due to its inability to account for the non-isotropic effects of turbulence.

Figure 5 depicts the computed mean temperature profile with the experimental data of Antonia et. al. (1983). The results compare well with the experimental data, however, the accuracy of the computation is uncertain at the free boundary due to the lack of knowledge of the experimental trend.

Figure 6 compares the computed results of second-product of velocity-temperature, $< u_i \theta >$ with the data of Antonia et. al. (1983). The results for both components compared well with the experiments (within 15% agreement).

5 CONCLUSION

From the above results the following conclusions can be drawn:

1. The present model predicted the mean velocity profiles to within 10% of the measured values.

2. The predictions of the Reynolds stresses are sensitive to the choice of the pressure-strain correlation.

3. The $k - \varepsilon$ model does not predict the Reynolds stresses accurately and thus suggests that the use of higher-order turbulence closure models which can take the non-isotropic effects of turbulence are needed.

6 ACKNOWLEDGEMENTS

This work is currently supported by NASA Marshall Space Flight Center under NAG 8-617.

REFERENCES

Amano, R. S., Chen, J. D., and Chai, J. C., 1988, "Improvement of the Reynolds-Stress Model by a new Pressure-Strain Correlation," ASME Winter Annual Meeting.

Amano, R. S., and Chai, J. C., 1988, "Transport Models of the Turbulent Velocity-Temperature Products for Computations of Recirculating Flows," *Numerical Heat Transfer*, Vol. 14, pp. 75-95.

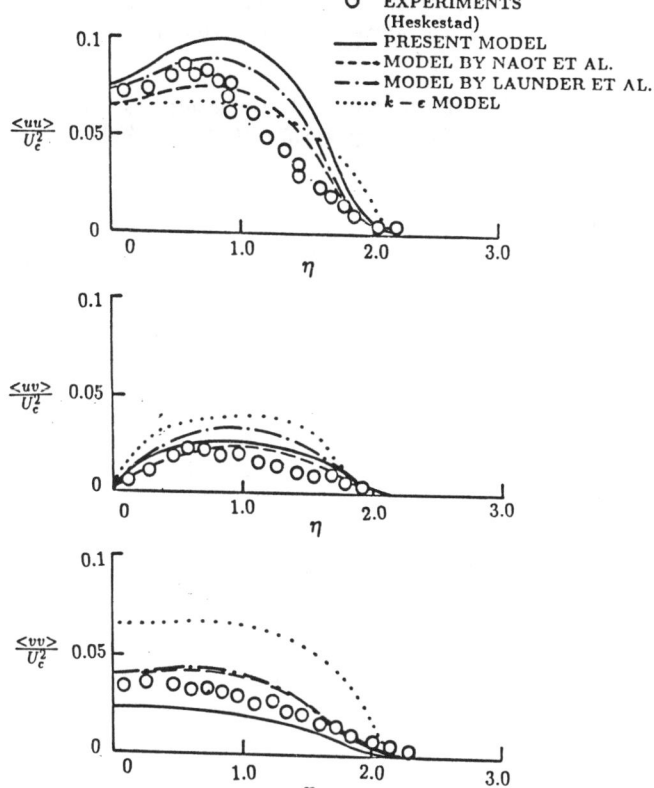

Figure 3: Reynolds stresses profiles - Comparison with the data of Heskestad (1965).

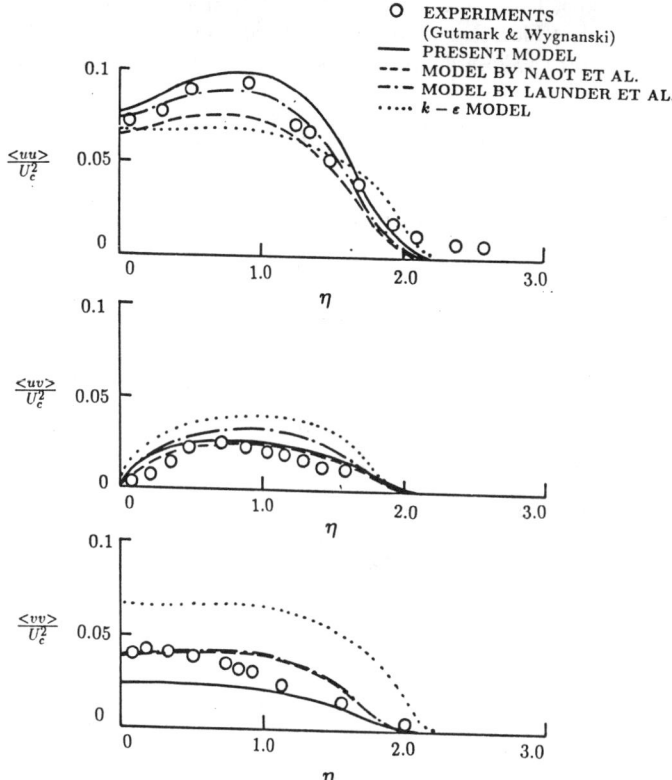

Figure 4: Reynolds stresses profiles - Comparison with the data of Gutmark and Wygnanski (1976).

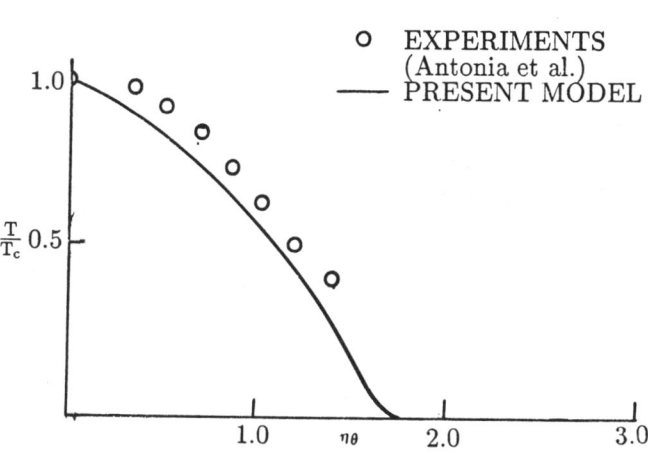

Figure 5: Mean temperature profile - Comparison with the data of Antonia et. al. (1983).

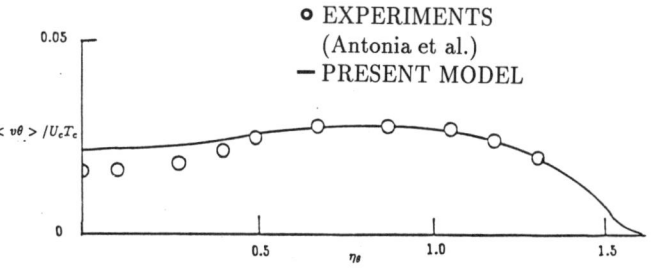

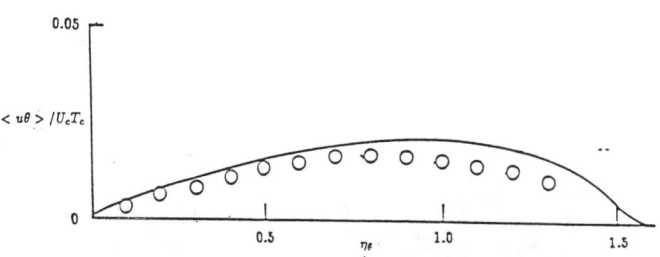

Figure 6: Second-product of velocity-temperature profiles - Comparison with the data of Antonia et. al. (1983).

Amano, R. S., and Goel, P., 1985, "Computations of Turbulent Flow Beyond Backward-Facing Steps Using Reynolds-Stress Closure," *AIAA Journal,* Vol. 23, No. 9, pp. 1356-1361.

Antonia, R. A., Browne, L. W. B., Chambers, A. J., and Rajagopalan, S., 1983, "Budget of the Temperature Variance in a Turbulent Plane Jet," *International Journal of Heat and Mass Transfer,* Vol. 26, No. 1, pp. 41-48.

Chai, J. C., 1988, *Higher Order Turbulence Closure Models and Applications to Backward-Facing Step and Jet Flows,* Master Thesis, University of Wisconsin, Milwaukee, Wisconsin.

Dekeyser, I., 1985, "Numerical Prediction of an asymmetrical Heated Plane Jet with a Second-Moment Turbulence Closure," *International Journal of Heat and Mass Transfer,* Vol. 28, No. 3, pp. 653-662.

Launder, B. E., Reece, G. J., and Rodi, W., 1975, "Progress in the Development of a Reynolds-Stress Turbulence Closure," *Journal of Fluid Mechanics,* Vol. 68, part 3, pp. 537-566.

Gutmark, E., and Wygnanski, I., 1976, "The Planar Turbulent Jet," *Journal of Fluid Mechanics,* Vol. 73, part 3, pp. 563-583.

Heskestad, G., 1965, "Hot-Wire Measurements in a Plane Turbulent Jet," *Journal of Applied Mechanics,* Vol. 32, pp. 721-734.

Naot, D., Shavit, A., and Wolfshtein, M., 1973, "Two-Point Correlation Model and the Redistribution of Reynolds Stress," *The Physics of Fluids,* Vol. 16, No. 6, pp. 738-743.

Patankar, S. V., 1980, *Numerical Heat Transfer and Fluid Flow,* Hemisphere, Washington, D. C.

Sini, J., and Dekeyser, I., 1985, "Numerical Prediction of Heated Turbulent Shear Flows with a Second Moment Turbulence Closure," International Symposium on Refined Flow Modeling and Turbulence Measurements, Iowa.

FLUID MECHANICS AND HEAT TRANSFER MEASUREMENTS IN TRANSITIONAL BOUNDARY LAYERS CONDITIONALLY SAMPLED ON INTERMITTENCY

J. Kim, T. W. Simon, and M. Kestoras
Department of Mechanical Engineering
University of Minnesota
Minneapolis, Minnesota

ABSTRACT

An experimental investigation of the transition process on a flat-plate boundary layer was performed. Mean and turbulence quantities, including turbulent heat flux, were sampled according to the intermittency function. Such sampling allowed segregation of the signal into two types of behavior--laminar-like and turbulent-like. Results show that, during transition, the two forms of boundary layer behavior, identified as laminar-like and turbulent-like, cannot be thought of as separate Blasius and fully-turbulent profiles, respectively. Thus, simple transition models in which the desired quantity is assumed to be an average, weighted on intermittency, of the theoretical laminar and fully turbulent values is expected not to be successful. Deviation of the flow identified as laminar-like from theoretical laminar behavior is shown to be due to recovery after the passage of a turbulent spot, while deviation of the flow identified as turbulent-like from the fully-turbulent values is thought to be due to incomplete establishment of the fully-turbulent power spectral distribution. Measurements were taken for two different levels of free-stream disturbance--0.32% and 1.79%. Turbulent Prandtl numbers for the transitional flow, computed from measured shear stress, turbulent heat flux and mean velocity and temperature profiles, were less than unity.

NOMENCLATURE

C_f	Skin friction coefficient
C_p	Specific heat
C_{pr}	Static pressure coefficient, $(P-P_{ref})/(\rho U_{inf}^2)$
H	Shape factor
Re	Reynolds number
Q	Wall heat flux
St	Stanton number
t	temperature
u	streamwise velocity
v	cross-stream velocity
w	cross-span velocity
x	Streamwise distance
y	Cross-stream distance

Greek

γ	Intermittency
ρ	Density

Superscripts

+	Wall coordinates
'	rms value or fluctuation, depending on context
-	(overbar) time average quantity

Subscripts

w	Wall value
∞	Free-stream value

INTRODUCTION

Despite the attention of many investigators, understanding boundary layer transition remains elusive. The sensitivity of transition to many factors (free-stream acceleration, the level of free-stream disturbance and its characteristics, surface roughness, surface curvature, surface heating, wall suction, compressibility, and unsteadiness, to name a few) makes prediction of transition in complex machines such as gas turbines very difficult. Unlike laminar flow or low-Reynolds number turbulent flows in simple geometries, little can be learned from numerical solutions. The transition process is sufficiently complex that experimental observations must first be made in simple geometries where only a few effects are allowed. Later, as understanding builds, more effects can be added and more realistic geometries can be investigated. The following discusses the latest results of a program that has been underway at the University of Minnesota for the past seven years in which measurements have been made to support a fundamental understanding of transition.

The program began with a study of the effects of free-stream disturbance (Wang, Simon and Buddhavarapu--1985) and streamwise convex-curvature (Wang and Simon--1987), in which Stanton numbers and mean velocity and temperature profiles were measured, as well as the turbulent streamwise normal and shear stresses. Since this study, however, the following important new measurement capabilities have been acquired: 1). direct measurement of the turbulent heat flux and turbulent Prandtl number, 2). conditional sampling on the intermittency function, a bimodal signal that identifies either laminar-like or turbulent-like behavior, 3). spectra, and 4). the three turbulent normal stresses in the free-stream. The purpose of this portion of the experimental program is to document flat-plate transitional boundary layers incorporating the above-mentioned measurement capabilities. The experiments will provide support for the testing and development of transition prediction models. Specifically, the experimental results allow testing the applicability of intermittency-based transition models, the first of which was proposed by Dhawan and Narasimha (1958). It assumes a Blasius-type flow for the laminar portion and a fully-turbulent flow for the turbulent portion. Although many researchers have studied the flat-plate transition process (see Wang and Simon--1987 for a review), only a few have used conditional sampling on intermittency to look at the laminar and turbulent portions of the transitional boundary layer separately. The work of Arnal, Juillen and Michel (1978), Kuan and Wang (1988), Sohn, O'Brien and Reshotko (1989), and Blair (1989) are of special relevance to this study. No researchers, to the authors' knowledge, have directly measured the turbulent heat flux and turbulent Prandtl number in transitional boundary layers, however.

In the tests, a boundary layer is allowed to undergo transition naturally, becoming a fully turbulent boundary layer by the end of the test section. The effects of two levels of free-stream disturbance (0.32%, and 1.79%) were investigated. The measurements consist of the following quantities:

1). The mean and fluctuating components of streamwise and cross-stream velocity and the shear stress. A horizontal hot-wire (TSI Model 1218 Boundary Layer Probe) and a cross-wire (TSI Model 1243 Boundary Layer Probe) were used in the isothermal flows.
2). Mean temperature profiles. The thermocouple probe described by Wang and Simon (1987) was used.
3). Local Stanton number. Thermocouples were embedded in the test wall for this purpose.
4). Intermittency. A special circuit for generating an analog intermittency

function from a hot-wire signal was constructed so that processing based upon the state of the flow (laminar-like or turbulent-like) could proceed as appropriate. Details of this circuit are described below.
5). Profiles of the turbulent heat flux ($\overline{v't'}$) and turbulent Prandtl number were made at streamwise locations where the boundary layer was sufficiently thick using a triple-wire probe developed for this purpose. Details of this probe are described below.

TEST FACILITY DESCRIPTION AND QUALIFICATION

The experiment was conducted in an open-circuit, blown-type wind tunnel. Details of this tunnel are described in Wang and Simon (1987) (Fig. 1). The test channel is rectangular, 68 cm wide, 11.4 cm deep, and 137 cm in length. Higher free-stream turbulence levels were obtained by inserting a coarse grid constructed of 2.5 cm aluminum strips in a square array on 10 cm centers at the entrance of the nozzle.

Mean velocity measurements within the potential core of the flow exiting the nozzle showed a peak-to-peak variation of 0.2% about a nominal velocity of 27 m/s. The free-stream turbulence intensity at the nozzle exit, measured using a cross-wire oriented in two perpendicular directions to get all three velocity components, was 1.79% and 0.32% with

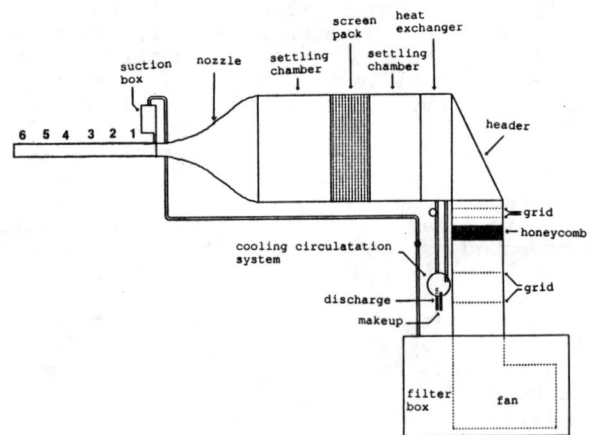

Fig. 1 -- Schematic of test facility.

and without the turbulence generating grid, respectively. Measurements of mean temperature within the flow exiting the nozzle showed a peak-to-peak variation of 0.02°C. The outer wall was adjusted such that the static presure coefficient, C_{pr}, was zero to within 0.015 along the test wall for both the low and high turbulence intensity (TI) cases.

Qualification of the test section with regard to the transition location was initially performed by heating the wall and visualizing transition using a liquid crystal sheet. Transition was assumed to occur at the location where the liquid crystal first changes color as the heat flux is gradually increased. This corresponds to the highest wall temperature, or, since the wall heat flux is essentially uniform, the location of lowest heat transfer coefficient. Various parameters such as the leading edge suction flow rate and the suction slot width were optimized such that transition occurred as far downstream as possible.

This method of determining the transition start location is not entirely reliable, however, due to the destabilizing effect of heating on the boundary layer. It is well known (Schlichting--1979) that the heating of a surface in air causes an inflection in the near-wall velocity profile due to a local increase in viscosity. Thus, transition is expected to occur earlier in a heated boundary layer and the transition length is expected to shorten. The variation of Stanton number along the wall (low TI case) for two wall heat fluxes taken in the test facility is shown on Fig. 2. It is seen that transition occurs over a shorter length with an increase in wall heat flux, as expected. The transition start location is apparently much less influenced by a higher heat flux. A possible explanation of this follows. In the low TI case, it was observed that transition begins in the corner flow region where the test wall meets the tunnel end-walls. This was observed by the liquid crystal technique discussed above. It then propogates on the test-wall toward the centerline with increasing streamwise distance. In the lower TI case, the influence of this encroaching effect had nearly reached the centerline of the test surface at the streamwise location of transition. The transition start location was thus apparently influenced by the corner flow region, reducing

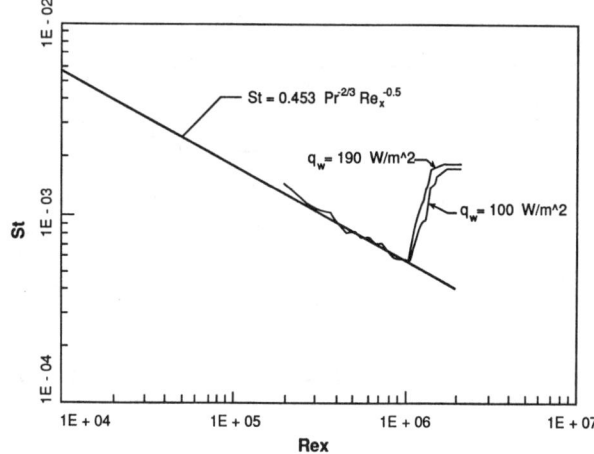

Fig. 2 -- Stanton number vs. Re_x for two wall heat flux values.

the influence of heating. For the remainder of this study, the start of transition is defined as the location where the near-wall intermittency, measured in the unheated flow, reaches 5%.

Except for a small unheated starting length effect (the first five points), Stanton numbers in the laminar boundary layer are in excellent agreement with the accepted correlation for constant wall heat flux (Kays and Crawford--1980). Corrections for streamwise conduction, back heat loss, and

radiation were made in computing St values. The slight dip in Stanton number values below the laminar correlation is caused by a decrease in the wall heat flux, with streamwise distance, below the uniform heat flux values due to increasing radiant heat loss. Radiation exchange tends to make the wall temperature more uniform. A STAN5 (Crawford and Kays-- 1976) simulation, using the measured wall convective heat flux as the boundary condition, yielded Stanton numbers 4% lower than the values given by the constant-wall-heat-flux correlation at the start of transition--in excellent agreement with the above trend in the data.

For the lower-turbulence baseline case, the Reynolds numbers based on displacement and momentum thicknesses at the beginning of transition (unheated flow) were measured to be 1920 and 737, respectively. Transition start Re_x was measured to be 1×10^6. A plot of the transition start Reynolds number based on displacement thickness vs. the free-stream turbulence intensity for the present study is shown on Fig. 3. Transition is seen to occur slightly earlier for the present low TI case than for other researchers due, possibly, to the slight corner flow influence.

An energy balance was performed by integrating the wall heat flux along the centerline of the test wall and comparing this with the increase in energy carried in the boundary layer flow as calculated from the mean velocity and temperature profiles. The closure was within 3%.

Further qualification of the test section and measurement techniques was performed by comparing data measured in the flat wall transitional flow with that of other researchers. Measurements of the mean velocity profiles, shape factor (H), and the

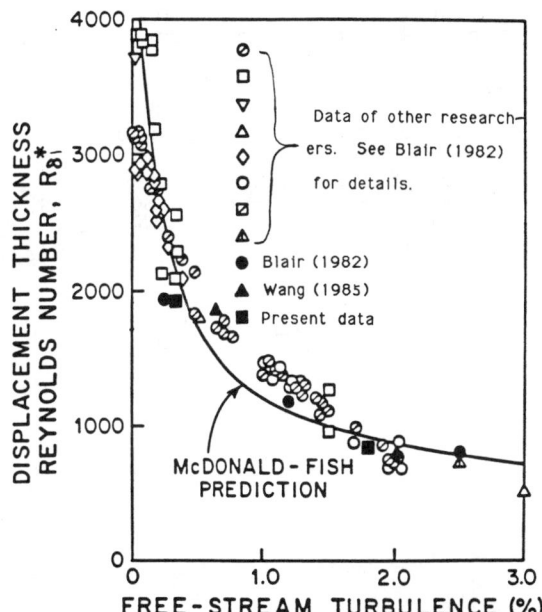

Fig. 3 -- Comparison of transition start location with that of other researchers and with the McDonald-Fish prediction (1973).

intermittency data all were consistent with
other researchers' results (e.g.--Wang, Simon
and Buddhavarapu-1985, Blair-1982, and Kuan and
Wang-1988). For the heat transfer data, it was
decided to work with the lowest wall heat-flux
level which would still give reasonable wall-
to-freestream temperature differences at the
end of transition (the location of smallest
temperature difference). The nominal heat flux
chosen was 160 W/m^2, which yielded a minimum
temperature difference of about 4 $^{\circ}C$ for the
TI=0.32% case. As mentioned earlier, the
transition process was somewhat affected by the
heating even at this low value of wall heat
flux. This unfortunately means that the heated
and unheated data cannot be compared within the
transition zone for the low TI case. For this
reason, the only heat transfer data presented
for the low TI case will be measurements of the
turbulent heat flux and turbulent Prandtl
number (mean temperature profiles are omitted).
Transition for the high TI case (TI=1.79%) was
apparently not affected either by heating (the
intermittency was insensitive to the wall heat
flux) nor by the corner flow. Heat transfer
data and unheated flow data can therefore be
compared to one another in the transition
region for the TI=1.79% case.

INTERMITTENCY CIRCUIT

A circuit for determining when the flow is
laminar-like or turbulent-like was constructed.
This circuit was based on a design by Kim,
Kline and Johnston (1978). The intermittency
circuit declares the flow turbulent if either
the filtered, rectified du/dt or the filtered,
rectified d^2u/dt^2 exceed empirically chosen
thresholds. The output of the circuit is
called the intermittency function. The
intermittency value can be found simply by
time-averaging the intermittency function. The
intermittency function can also be used to
conditionally sample other signals so that data
is processed only when the flow is laminar-like
or only when it is turbulent-like.
    The circuit takes advantage of the much
larger time derivative of the turbulent-like
signal as compared to the time derivative of
the laminar-like signal: the hot-wire-
anemometer signal is processed by a series of
filters, differentiators and rectifiers (Fig.
4). At the level detector, the signal is

compared to an adjustable threshold value. If
it is higher than the threshold, the output
signal of the level detector is high
(turbulent-like). It is low (laminar-like)
otherwise. The analog signal, thus obtained,
can be used to conditionally sample other
quantities, tagging them to either laminar-like
or turbulent-like behavior. The intermittency
measuring unit has two channels which are used
to solve the problem of zero-crossing by using
the "OR" gate to combine the two signals. Its
output is high when either of the two signals
is high and is low only when both inputs are
low. The number of points falsely declared
laminar is thus greatly reduced. A low-pass
filter at the output of the OR gate then
eliminates the few remaining regions falsely
declared turbulent. The threshold values of
the two level detectors are tuned for each
different flow situation. A tuning procedure
that has been found to work well has been
established. An example of the circuit
performance in a transitional boundary layer at
y/δ=0.148 is shown on Fig. 5. It may be seen
that the circuit does a reasonably good job of
discriminating between laminar-like and
turbulent-like flow. Similar behavior was
observed at other locations in the boundary
layer.

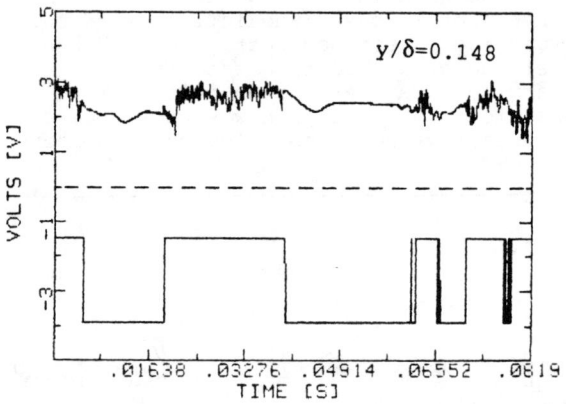

Fig. 5 -- Intermittency circuit performance in
a transitional boundary layer. Position in the
boundary layer is as noted.

TRIPLE-WIRE PROBE FOR TURBULENT HEAT FLUX MEASUREMENTS

A probe to measure $\overline{v't'}$ in turbulent boundary
layers was developed by Kim and Simon (1988)
and used to take measurements in a turbulent
boundary layer over a convex surface, with
recovery from curvature. The probe geometry
was based on a design by Blair and Bennett
(1984). As used in the present study, one wire
is operated in the constant-current mode as a
resistance thermometer to measure the
instantaneous flow temperature, the other two
wires being operated as standard constant-
temperature wires. The method of Hishida and
Nagano (1978) was chosen to compensate for the
low frequency response of the constant-current
cold-wire. In this method, the heat transfer
coefficient over the constant-current cold-wire
is estimated from the parallel hot-wire signal.

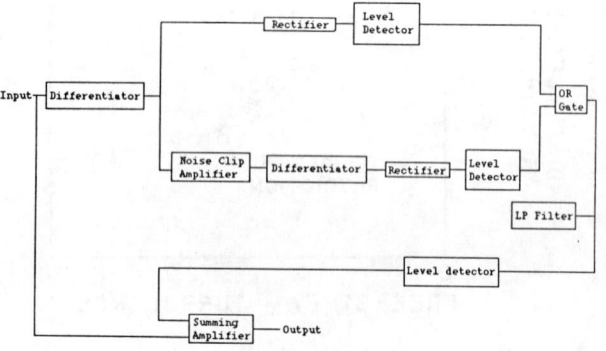

Fig. 4 -- Block diagram of intermittency
measuring unit.

The cold-wire signal and its time derivative are used in the signal processing. The temperature sensing wire was 1.25 μm in diameter.

A circuit built for this purpose consisted of a current source, an amplifier, and a differentiator. A current source of 1 mA drove the cold-wire. The voltage across the wire was amplified 200 times, then sent through a differentiator. The noise of the amplifier was 0.5 μV rms referenced to input, yielding a signal-to-noise ratio of 100; the signal-to-noise ratio at the exit of the differentiator was 30. Careful attention to minimizing the potential for ground loops was essential in obtaining these values. The data taken with the compensated probe in a heated flat-wall turbulent boundary layer was in excellent agreement with the data of Gibson and Verriopoulos (1984).

RESULTS

The Low TI Case -- TI=0.32%

Free-stream Turbulence Intensity and Spectra. A power spectral density (PSD) distribution of the streamwise velocity component using a horizontal wire revealed a spectrum with a relatively strong peak corresponding to 29 Hz. This peak has been traced (using an accelerometer and a vibration analyzer) to a rocking motion of the centrifugal blower on its mounts, resulting in a slight unsteadiness in free-stream velocity. All reasonable effort has been applied to reduce this fan motion. This frequency is not expected to influence the transition process as the minimum critical frequency for amplification of disturbances is estimated from linear stability theory to be 1600 Hz. The spectrum was relatively clean otherwise.

Measurements of the free-stream turbulence intensity vs. streamwise distance using a cross-wire probe revealed that u' was roughly twice the value of either v' or w', with the values remaining constant along the streamwise length of the test section. The low-frequency unsteadiness discussed above is expected to be the source of the anisotropy. It is expected that the isotropy would improve if the unsteadiness at this frequency were filtered out. The average free-stream turbulence intensity calculated from the three fluctuation components (without any filtering) was found to be 0.32%. The free-stream velocity was nominally 26.5 m/s.

Intermittency Profiles. Intermittency profiles obtained at various stations are shown on Fig. 6. The profiles at all stations show the same qualitative behavior, namely a relatively flat value in the near-wall region followed by a decay to zero. This decay is due to both the entrainment of the free-stream flow into the boundary layer (the wake region) and the passage of intermittent turbulent spots. In the following, quantities are discussed in terms of the degree of completion of transition. This is expressed in terms of the intermittency value which corresponds to the intermittency in the near-wall region.

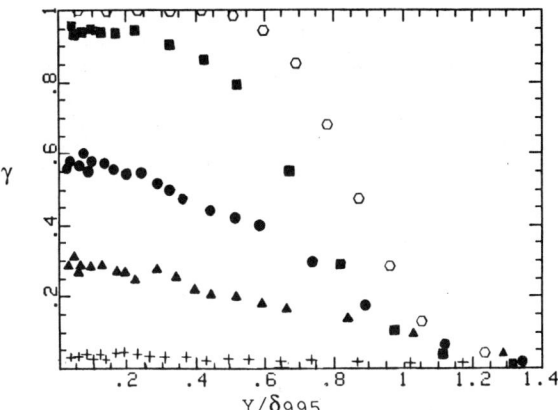

Fig. 6 -- Intermittency profile variation through transition (TI=0.32%). +-station 3, ▲-station 4a, ●-station 4, ■-station 5, ⬡-station 6.

Mean Velocity Profiles. An example of the mean velocity profile sampled within the transitional boundary layer and segregated according to laminar and turbulent behavior using the intermittency function is shown on Fig. 7. The distance away from the wall has been normalized on the boundary layer thickness based on the composite (long-time-average value which includes both laminar-like and turbulent-like portions) flow profile. Two characteristics are immediately apparent. First, the turbulent boundary layer is thicker than the corresponding laminar boundary layer, as expected, due to turbulent transport from bursting and subsequent turbulent spot formation. Second, the turbulent boundary layer profile is flatter than the corresponding laminar profile, resulting in a cross-over between the two. The composite profile is, by definition of the intermittency, a combination of the turbulent and laminar profiles, and must lie between the two.

Plots of the mean velocity, sampled on intermittency and normalized on wall coordinates at various locations within the

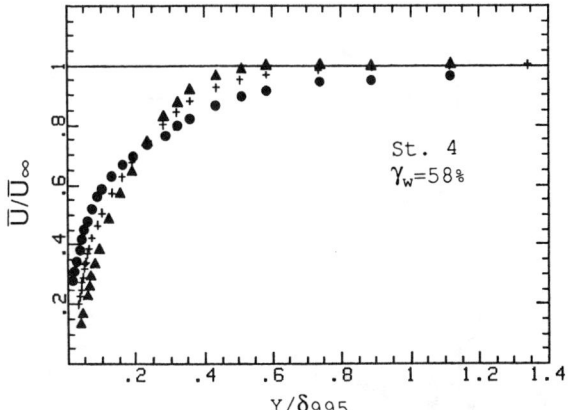

Fig. 7 -- $\bar{u}$ profile sampled on intermittency (TI=0.32%). ▲-laminar, ●-turbulent, +-composite. The profile is normalized on the composite flow profile thickness.

73

transition process, are presented on Fig. 8. The Blasius profile, shown, is the theoretical profile that should exist at the Re_x location where the measurements were made. The composite (long-time average value which includes both laminar-like and turbulent-like portions) flow profiles evolve from the Blasius profile (upstream of the first station shown) to the fully-turbulent log-law profile (the last station shown). The velocity profiles sampled on intermittency, however, do not agree with either the Blasius or log-law profiles in the transition region. The laminar profile increasingly deviates from the Blasius profile as transition proceeds. The turbulent profile has its maximum deviation from the log-law profile early in the transition process.

A plot of the local skin friction, C_f, values deduced from the near-wall velocity gradient (in the laminar-flow case) or by fitting the near-wall data to the log-law relationship (in the turbulent-flow case) is shown on Fig. 9. The skin friction corresponding to laminar flow increasingly deviates from the laminar correlation as

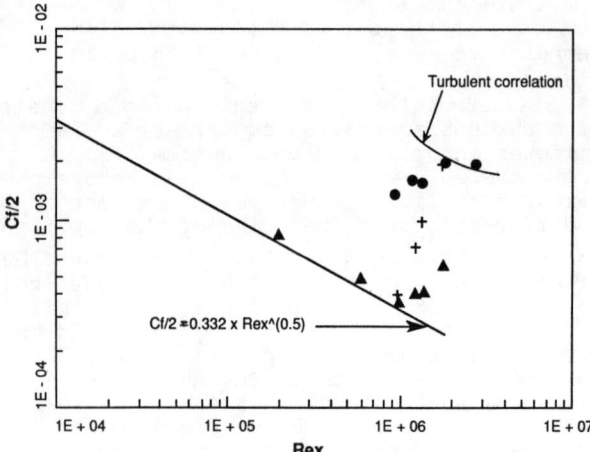

Fig. 9 -- C_f sampled on intermittency (TI=0.32%). ▲-laminar, ●-turbulent, ✛- composite.

transition proceeds. The higher stress at the wall as the transition region is entered is believed to be due to disturbances in the laminar flow regime as a result of the passage of turbulent spots. A near-wall hot-wire voltage trace in the intermittent region (Fig. 10) illustrates this. Although transition from laminar to turbulent flow is quite sharp at the leading interface (end of interval A), the laminar flow signal is slow to relax back to a nominally laminar state (slow relaxation of velocity within interval A). If the intermittency is high, i.e., spots pass frequently, the laminar portion of the boundary layer is continually disturbed, resulting in higher velocities near the wall and, consequently, higher-than-laminar C_f values.

Skin friction coefficients in the turbulent flow, but at the beginning of transition (Fig. 9), are lower than what a fully-turbulent correlation would suggest. This could be due to less than complete establishment of the full turbulence spectrum, i.e., only relatively large eddies are present early in the transition process and turbulence cascading and dissipation are not fully

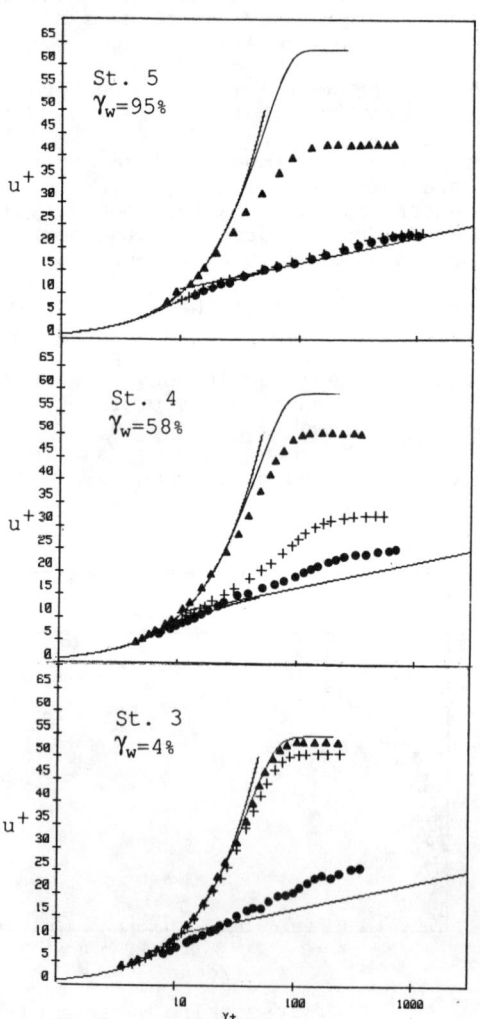

Fig. 8 -- $\overline{u}$ profiles sampled on intermittency in wall coordinates (TI=0.32%). ▲-laminar, ●-turbulent, ✛-composite.

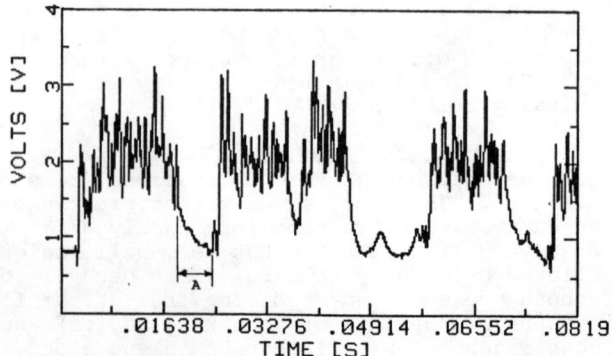

Fig. 10 -- Near-wall hot-wire voltage trace in transition illustrating the different mean velocities between the two regimes and the relaxation of the laminar boundary layer after turbulent spot passage.

74

established. This has not been confirmed, however.

A similar variation was seen for the shape factor, H. As transition proceeded, H for the laminar-like flow increasingly deviated from the laminar value of 2.6 (decreasing to roughly 2.1 toward the end of transition), indicating an increasingly non-Blasius behavior. Similarly, early in transition, the H within the turbulent-like flow deviated substantially from the high-Reynolds-number turbulent value of 1.4. This also illustrates that the laminar-like and turbulent-like flows during transition cannot be thought of as simply Blasius or turbulent flows, respectively. Although this conclusion seems to conflict with the results of Wygnanski, Sokolov, and Friedman (1976) and Blair (1989), who found that the turbulent-zone mean velocity profiles conformed to the standard law-of-the-wall, it is in agreement with the results of Antonia, et al. (1981) who found that unreasonably high values of skin friction were necessary to make the velocity profiles agree with the log-law (as measured along the centerline of the turbulent spot). Similar results were found by Cantwell, Coles and Dimotakis (1978). The results of a flow-visualization study by Gad-El-Hak, Blackwelder, and Riley (1981) also support the present conclusion. They found that the flow in the overhang of the turbulent spot was relatively passive, being cut off from the regeneration mechanism (bursting) in the near-wall region. It should, therefore, not be surprising that in the present study, where turbulent-like flow is accepted for processing regardless of whether it came from the overhang, the wings, or within the core of the turbulent spot, that velocity profiles do not agree with the law-of-the-wall. Furtheremore, there is no guarantee that turbulence within interacting or merging turbulent spots is similar in nature to turbulence within the core of isolated spots.

Velocity fluctuation. The rms values of the streamwise velocity fluctuation at stations 3, 4 and 5 (stations shown on Fig. 1) are seen in Fig. 11. The rms value of the laminar profile at first increases with axial distance, but then reaches a peak value of 8% at station 4, flattening out thereafter. Peaks in the laminar profiles are seen to occur at roughly 30% of the laminar boundary layer thickness for all stations. High rms values of the turbulent profile are seen initially (16% at station 3), indicating a high production of turbulence, but these decay to a peak value of 8% as dissipation in the boundary layer increases. The composite flow profile exhibits quite unexpected behavior. The profile initially follows the laminar profile due to the low intermittency (approx. 29% at station 3), but then jumps to a peak value of 17.5% at station 4, a value larger than the peak in the corresponding turbulent profile. Much of this behavior is due to the change in the mean velocity due to intermittent "switching" of the flow between the laminar-like and the turbulent-like regimes as turbulent spots pass the probe. This was first shown by Arnal, Juillen and Michel (1978). The switching and associated changing of velocity is illustrated

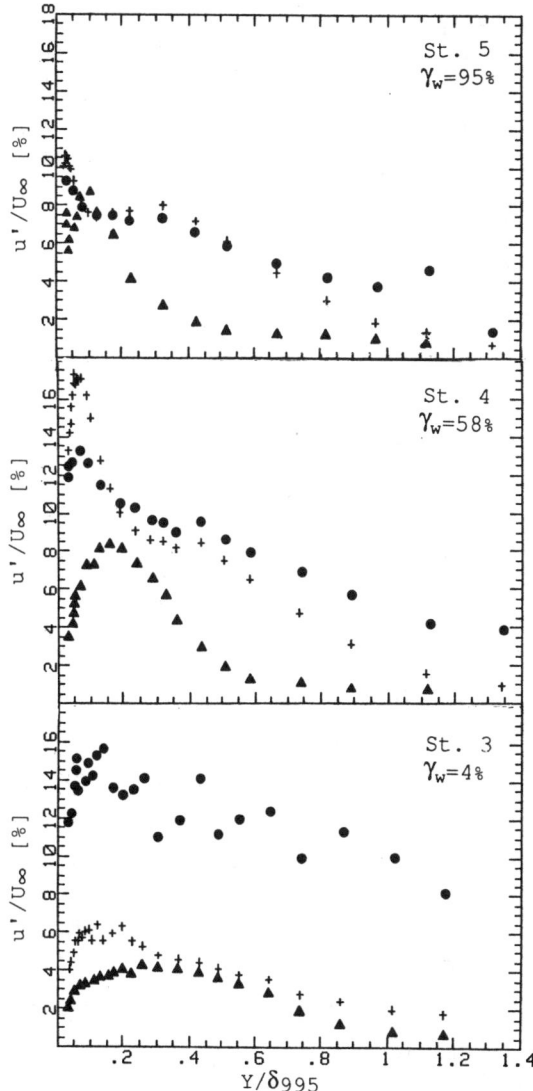

Fig. 11 -- u' sampled on intermittency (TI=0.32%). ▲-laminar, ●-turbulent, ╋-composite.

in the hot-wire voltage trace of Fig. 10. The differences in the mean velocities in the laminar-like and turbulent-like regimes give rise to an rms velocity fluctuation which may be greater than that of either the laminar-like or turbulent-like regime. In fact, the various contributions to u'^2 can be accounted by the expression (Hedley and Keffer-1974):

$$\overline{u'_c{}^2}=\gamma\overline{u'_t{}^2}+(1-\gamma)\overline{u'_l{}^2}+\gamma(1-\gamma)(\overline{u}_t-\overline{u}_l)^2 \qquad (1)$$

where c, t, and l subscripts represent composite, turbulent-like, and laminar-like characteristics, respectively. The level of turbulence as indicated by the composite profile is thus not a good measure of the true turbulent activity (turbulent transport) in the transitional boundary layer.

Shear stress profiles. The variation in the shear stress $\overline{u'v'}$ through transition is

shown on Fig. 12. The laminar contribution to
the shear stress throughout the boundary layer
is seen to be quite small for all stations
except station 5 (this is an artifact of the
small number of samples and the cross-
contamination between the laminar-like and
turbulent-like regimes). A peak in each
profile is seen to move progressively toward
the wall as transition proceeds. The fully-
turbulent profile is reached by station 5.
Although the composite flow profile is between
the turbulent and laminar profiles for all
stations, it also is affected by the
intermittent "switching" from laminar to
turbulent flow. The composite flow profile is,
therefore, also not indicative of the true
turbulent shear stress in the boundary layer.
A more accurate shear stress is found by a
weighted (on intermittency) average of the
laminar-like and turbulent-like contributions.

   Turbulent Heat Flux Measurements. Results
of measurements of the turbulent heat flux

normalized on the wall heat flux and sampled on
intermittency are shown on Fig. 13. A
potential advantage of the present
normalization over the usual normalization
based on the free-stream velocity and the wall-
to-freestream temperature difference is that
the composite flow profile turbulent heat flux
very near the wall should be directly
proportional to the intermittency (if $\overline{v't'}$
sampled on the laminar-like flow is small).
Due to the destabilizing effect of heating on
transition mentioned earlier, the intermittency
values for these profiles taken at stations 3
and 4 are different than those for the unheated
data presented above. This heat flux data
provides insight into the transition process,
though it cannot be directly compared to the
hydrodynamic data.

   It is seen in Fig. 13 that a large
increase in turbulent heat flux above the wall
heat flux value (the normalized quantity
exceeds unity for station 3) occurs within the
turbulent spot. This can occur within the
boundary layer if the eddy diffusion of heat in
the cross-stream direction ($\overline{v't'}$) increases at
the expense of convection of heat in the
streamwise direction ($\overline{u't'}$). The triple-wire
measurements bear this out. The streamwise
heat flux $\overline{u't'}$ was found to decrease almost an
order of magnitude between St. 3 and 4 in the
near-wall region, then remain relatively

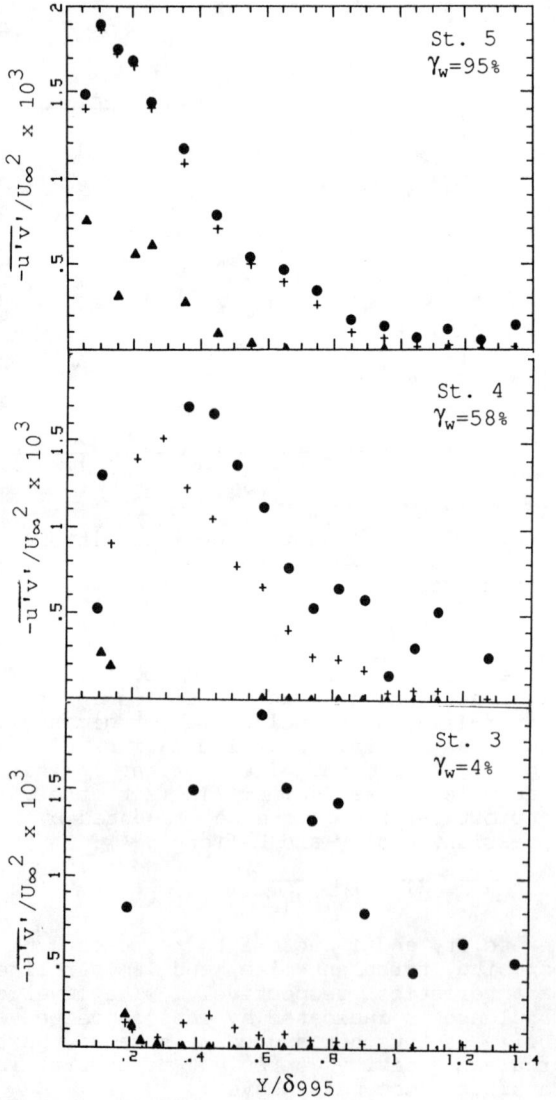

Fig. 12 -- $\overline{u'v'}$ sampled on intermittency
through transition (TI=0.32%). ▲-laminar, ●-
turbulent, +-composite.

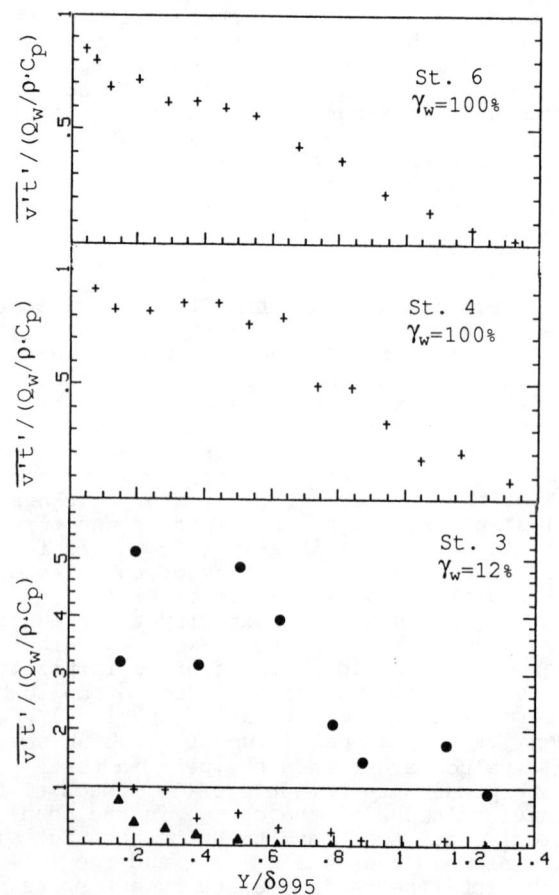

Fig. 13 -- $\overline{v't'}$ sampled on intermittency
(TI=0.32%). ▲- laminar, ●- turbulent, +-
composite.

constant thereafter. Whether the measurements sampled on turbulent-like flow for station 3 approaches unity at the wall is not known. It is possible that the wall transfers more energy to the flow during the passage of a turbulent spot due to the higher heat transfer coefficient than during the times the flow is laminar-like when the heat transfer coefficient is lower. The time-average energy transferred must, of course, be equal to the time-averaged wall heat flux. It is also seen that $\overline{v't'}$ in the laminar-like portion of the boundary layer is not zero. This does not mean that a true turbulent transport of heat is present in this flow, but simply that v' and t' are correlated due to the unsteadiness of the flow. Because $\overline{v't'}$ in the laminar-like regime is not small (station 3 of Fig. 13), the value of $\overline{v't'}$ near the wall for the composite flow profile unfortunately does not seem to be proportional to the near-wall intermittency, as was anticipated.

The variation of turbulent Prandtl number, Pr_t, sampled on intermittency through transition, is shown on Fig. 14. The uncertainty in this data was estimated to be 20%. The data at stations 4 and 6 show Pr_t values consistently close to unity in the inner

half of the boundary layer, as would be expected of fully-turbulent boundary layers. The data in the outer half of the boundary layer is not expected to be reliable due to the very shallow gradients of velocity and temperature. The data for station 3, however, show a drop in Pr_t values (sampled on the intermittency function) substantially below unity in the near-wall region. This implies that the eddy diffusivity of heat increases more rapidly than the eddy diffusivity of momentum. This contradicts the conclusions of other researchers (Blair--1989, Sohn, O'Brien, and Reshotko--1989, and Wang, Simon, and Buddhavarapu--1985). These are the first measurements in which Pr_t was measured directly, however; previous conclusions were inferred from mean profile data. The transition region may be more sensitive to molecular diffusion than is the fully-turbulent boundary layer since the eddies there tend to be relatively large, and the small scale structure (the full turbulence spectrum) has not been established. If so, one would expect a sub-unity turbulent Prandtl number when the molecular Prandtl number is less than one.

The High TI Case -- TI=1.79%

Free-stream Turbulence Intensity and Spectra. The Power Spectral Density (PSD) for this case was similar to that for the low TI case. The peak, corresponding to 27 Hz, was again present, but represented a much smaller fraction of the total turbulence kinetic energy in this case. As in the base case, this peak was caused by a slight rocking of the fan. The frequency was slightly lower in this case, however, since the fan speed was reduced.

The average free-stream turbulence intensity in the test section was measured to be 1.79%, with u' values increasing from 0.85% to 1.00% and v' decreasing from 2.32% to 2.02% between station 1 and station 6. The measured value of w' was 2.0% at station 2. The free-stream velocity was nominally 16.7 m/s.

Stanton number. The Stanton number variation through transition is shown on Fig. 15. The first five points are seen (as in the

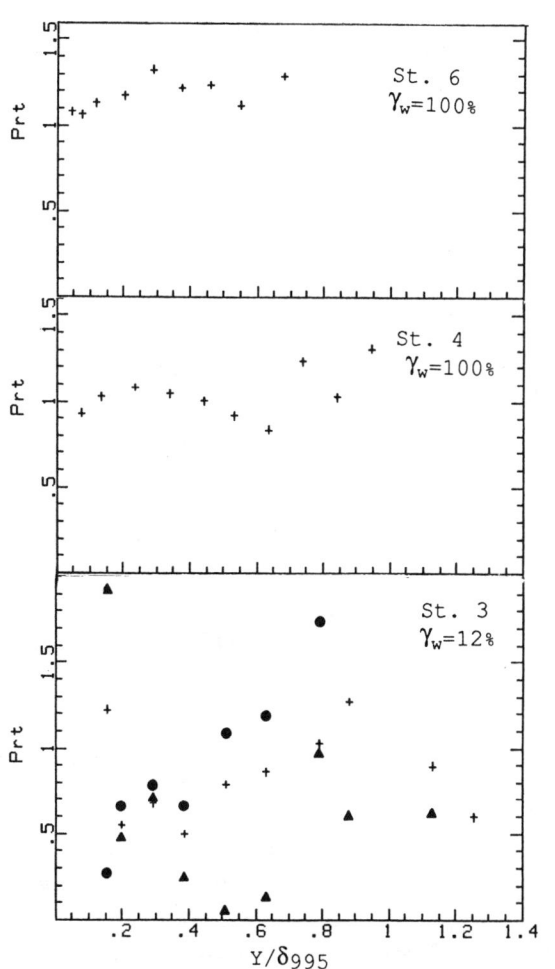

Fig. 14 -- Pr_t sampled on intermittency (TI=0.32%). ▲- laminar, ●- turbulent, +- composite.

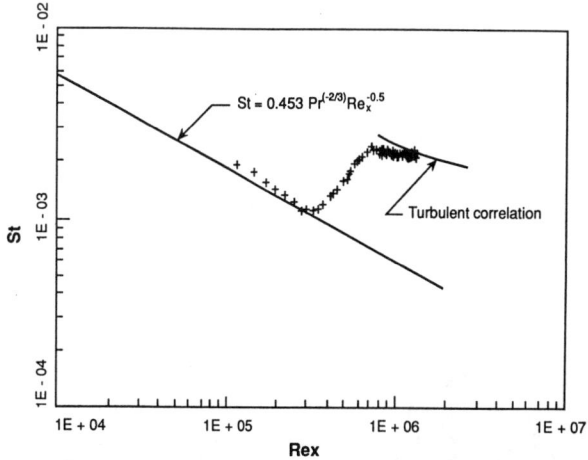

Fig. 15 -- St variation through transition (TI=1.79%).

lower-TI case) to be higher than the laminar correlation due to the unheated starting length effect. The two data points just before transition agree with the correlation. Increasing the free-stream turbulence has a strong effect on transition onset causing transition to move to $Re_x=3\times10^5$, or about one third the value of the low TI case. A comparison of the onset location with other researchers is shown on Fig. 3. The agreement in this case is very good. The Stanton number variation through transition is consistent with the data of Blair (1982).

Mean velocity profiles. Profiles of the mean velocity sampled on intermittency are shown on Fig. 16. The laminar profile is seen to deviate quite strongly from the Blasius profile throughout transition (much more than in the lower TI case), indicating a large perturbation due to increased free-stream turbulence. The turbulent profile, in contrast, agrees with the log-law profile from very early in transition. The above trends are reflected in the skin friction, C_f, values

plotted on Fig. 17. The C_f values in the laminar-like flow deviate strongly from the laminar correlation while the C_f values in the turbulent-like flow are near fully-turbulent values. The C_f values in the turbulent-like flow are not as depressed in the late-transition region as was seen in the lower TI case (Fig. 9--$Re_x=1\times10^6$).

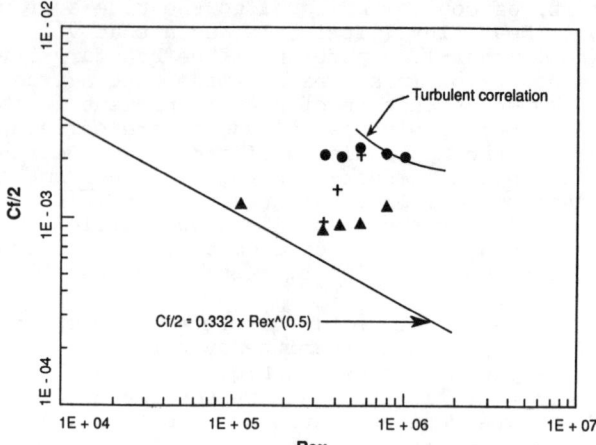

Fig. 17 -- C_f sampled on intermittency (TI=1.79%). ▲-laminar, ●-turbulent, ✛-composite.

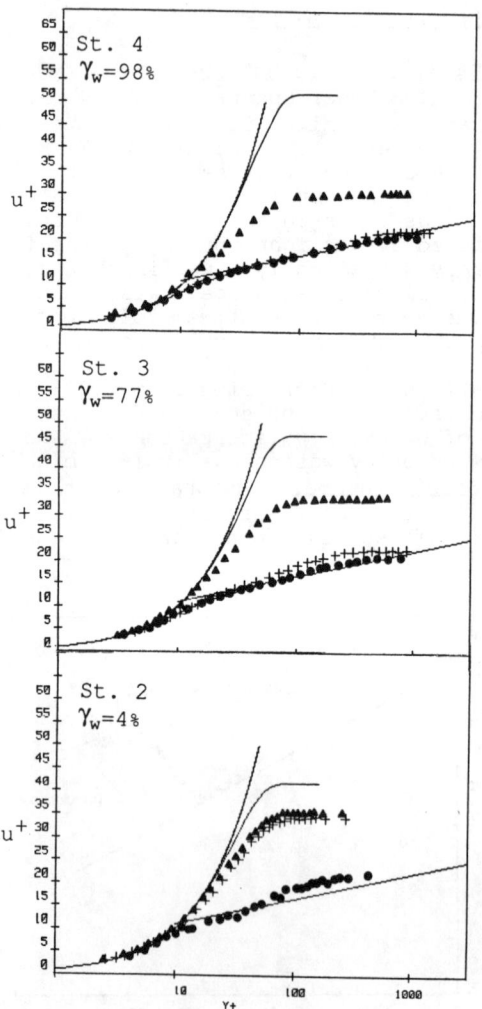

Fig. 16 -- $\overline{u}$ sampled on intermittency in wall coordinates (TI=1.79%). ▲- laminar, ●-turbulent, ✛-composite.

Velocity fluctuation. The streamwise velocity fluctuations (rms) are shown on Fig. 18. The most striking feature of these profiles in comparison with those of the low TI case is the large increase in laminar unsteadiness, which even exceeds some parts of the turbulent profile at station 3. The high values are consistent with the elevated C_f values within the laminar regime. The turbulent profiles evolve as in the lower TI case. The peak values of the turbulence intensity drop more or less monotonically with increasing values of near-wall intermittency for the two cases.

Temperature profiles. The evolution of mean temperature profiles through transition as plotted in t^+ vs. y^+ coordinates is shown on Fig. 19. They smoothly evolve from laminar-like to turbulent-like profiles, as was seen for the mean velocity profiles. The temperature profiles lag the velocity profiles, however, as may be seen by comparing the velocity and temperature profiles at station 3. The temperature profile is still evolving when the velocity profile has assumed a nearly log-law shape. This is consistent with the observations of Blair (1982).

Shear Stress Profiles. The variation in shear stress, $\overline{u'v'}$, profiles through transition is shown on Fig. 20. As in the low TI case, the laminar contribution to the shear stress is small everywhere except very near the wall. With successive downstream positions, passing through transition, the peaks in the turbulent profiles decrease in amplitude while moving toward the wall.

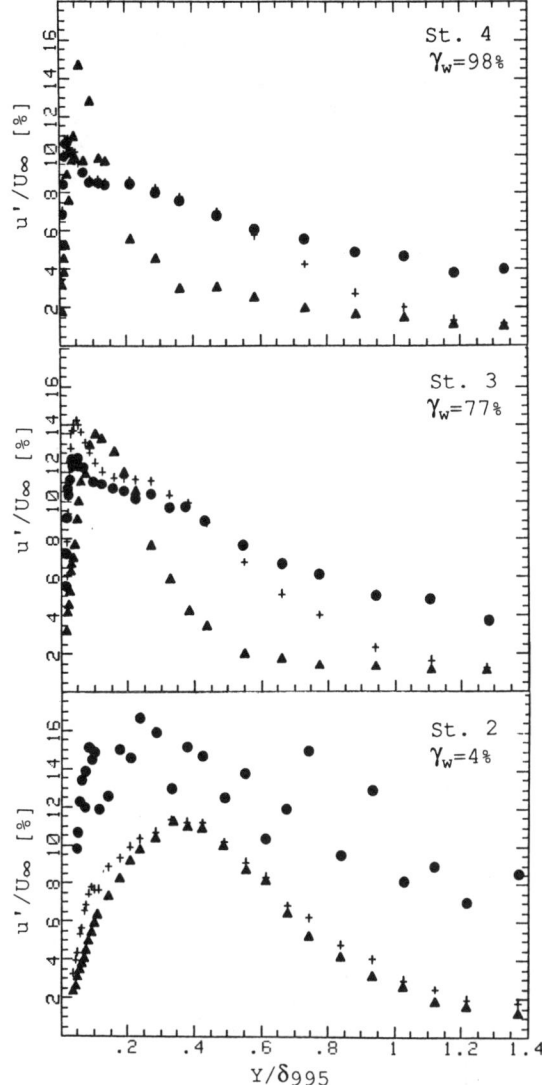

Fig. 18 -- u' sampled on intermittency (TI=1.79%). ▲-laminar, ●-turbulent, +-composite.

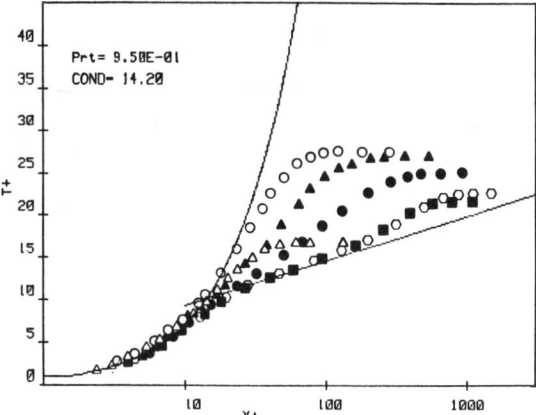

Fig. 19 -- $\overline{T}$ through transition in wall coordinates (TI=1.79%). △-station 1, ○-station 2, ▲-station 3a, ●-station 3, ■-station 4, ⬯-station 5. Station 3a refers to data taken at station 3, but at a reduced free-stream velocity. This was done to obtain a more reasonable intermittency value.

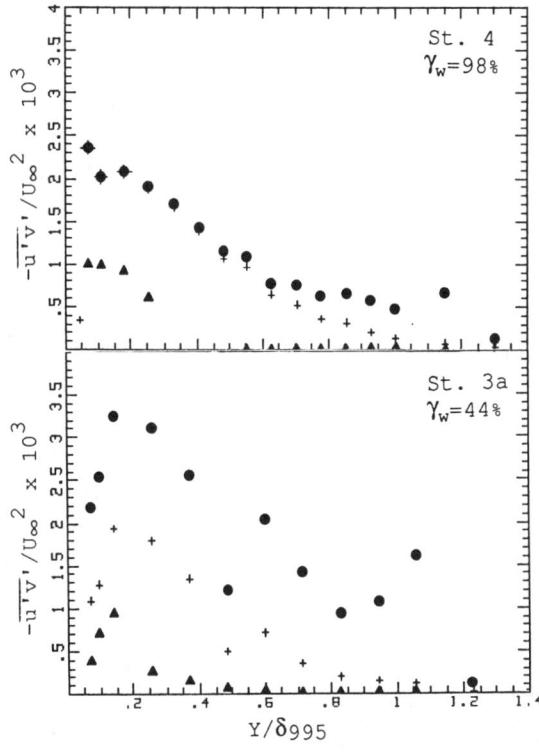

Fig. 20 -- $\overline{u'v'}$ sampled on intermittency through transition (TI=1.79%). ▲-laminar, ●-turbulent, +-composite.

Turbulent Heat Flux. Profiles of the turbulent heat flux, $\overline{v't'}$, are presented on Fig. 21. As in the lower TI case, a strong increase in the turbulent heat flux above the wall heat flux is seen. This peak decays rapidly. The profile achieves a fully-turbulent shape by station 5.

Turbulent Prandtl numbers deduced from the measurements are presented on Fig. 22. The values are all in the vicinity of unity for the fully-turbulent profiles (stations 3 to 5), while dipping below unity in the transitional flow case (station 2), as was seen previously for the low TI case.

CONCLUSIONS

The main conclusions of this study are :
1). The transitional boundary layer cannot be thought of as being a simple composite of a Blasius and a fully-turbulent flow. Transition modelling based on the intermittency function weighting of pure laminar and turbulent flows may be in error.
2). Conditional sampling of turbulence quantities on the intermittency function must be made during transition to obtain

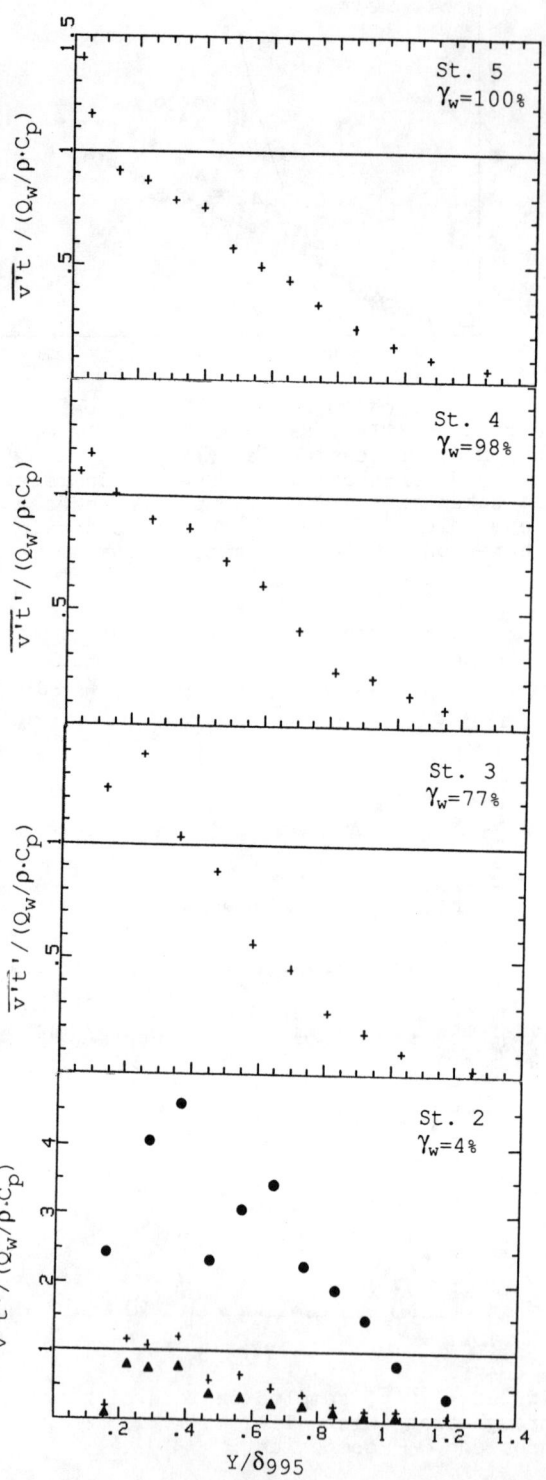

Fig. 21 -- $\overline{v't'}$ sampled on intermittency through transition (TI=1.79%). ▲-laminar, ●-turbulent, +-composite.

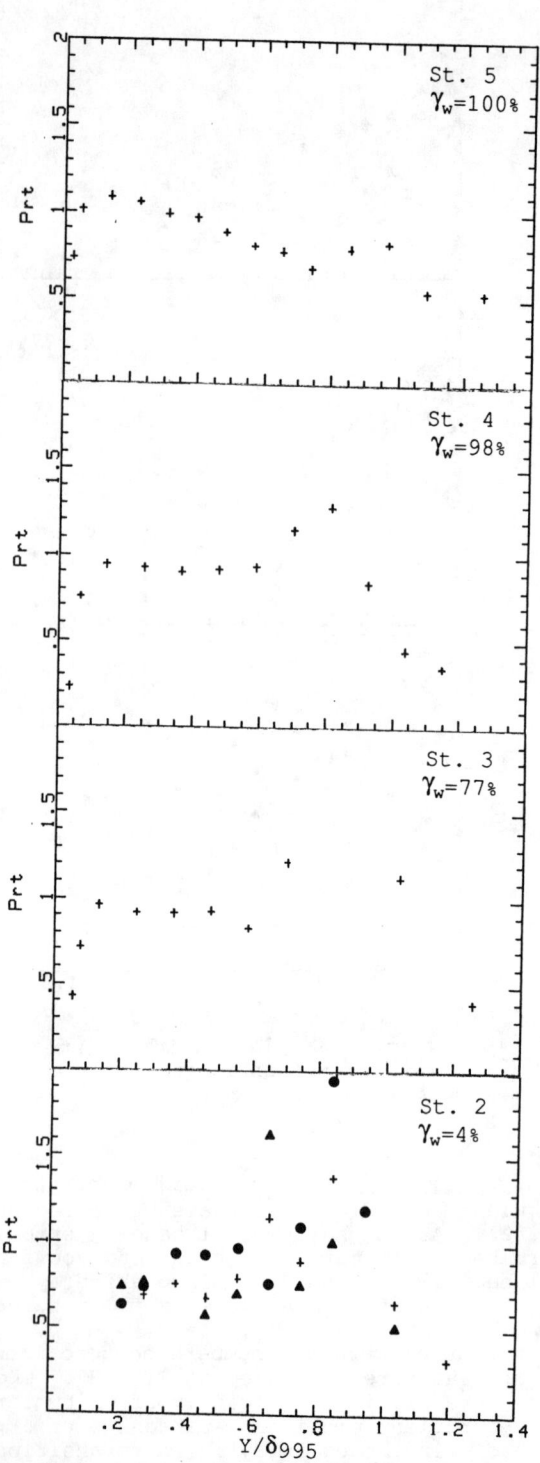

Fig. 22 -- Pr_t sampled on intermittency through transition (TI=1.79%). ▲-laminar, ●-turbulent, +-composite.

an accurate view of the transition process.

3). The turbulent Prandtl numbers in the turbulent core region of the transitional flow are somewhat smaller than unity.

ACKNOWLEDGEMENTS

This study was supported by the NASA Lewis Research Center under grant number NASA/NAG3-881. The grant monitor was Dr. James O'Brien. Further support was provided by the Graduate School of the University of Minnesota and the AMOCO Foundation.

REFERENCES

Antonia, R.A., Chambers, A.J., Sokolov, M. and van Atta, C.W. (1981) "Simultaneous Temperature and Velocity Measurements in the Plane of Symmetry of a Transitional Turbulent Spot", JFM, Vol. 108, pp. 317-343.

Arnal, D., Juillen, J.C., and Michel, R. (1978) "Experimental Analysis and Computation of the Onset and Development of the Boundary Layer Transition", NASA TM-75325.

Blair, M.F. (1982) "Influence of Free-Stream Turbulence on Boundary Layer Transition in Favorable Pressure Gradients", Journal of Engineering for Power, Vol. 104, pp. 743-750.

Blair, M.F. (1989) "Bypass-mode Boundary Layer Transition in Accelerating Flows", submitted for publication in JFM.

Blair, M.F. and Bennett, J.C.(1984) "Hot Wire Measurements of Velocity and Temperature Fluctuations in a Heated Turbulent Boundary Layer", 29th ASME International Gas Turbine Conference, Amsterdam.

Cantwell, B., Coles, D., and Dimotakis, P. (1978) " Structure and Entrainment in the Plane of Symmetry of a Turbulent Spot", JFM, Vol. 87, pp. 641-672.

Crawford, M.E., and Kays, W.M. (1976) "STAN5--A Program for Numerical Computation of Two-Dimensional Internal and External Boundary Layer Flows", NASA CR-2742.

Dhawan, S. and Narasimha, R. (1958) "Some Properties of Boundary Layer Flow During the Transition from Laminar to Turbulent Motion", JFM, Vol. 3, PP 418-436.

Gad-El-Hak, M., Blackwelder, R.F., and Riley, J.J. (1981), "On the Growth of Turbulent Regions in Laminar Boundary Layers", JFM, Vol. 110, pp. 73-95.

Gibson, M.M., and Verriopoules, C.A. (1984) "Turbulent Boundary Layer on a Mildly Curved Convex Surface", Experiments in Fluids II, Springer-Verlag, pp. 73-80.

Hedley, T.B. and Keffer, J.F. (1974) Turbulent/Non-Turbulent Decisions in an Intermittent Flow", JFM, Vol. 64, pp. 625.

Hishida, M., and Nagano, Y. (1978) "Simultaneous Measurements of Velocity and Temperature in Non-isothermal Flows", Journal of Heat Transfer, Vol. 100, pp. 340-345.

Kays, W.M., and Crawford, M.E. (1980) Convective Heat and Mass Transfer, Second Edition, McGraw-Hill.

Kim, J., and Simon, T.W. (1988) "Measurements of the Turbulent Transport of Heat and Momentum in Convexly Curved Boundary Layers: Effects of Curvature, Recovery, and Free-Stream Turbulence", Journal of Turbomachinery, Vol. 110, No. 1, pp. 80-87.

Kim, J., Kline, S.J., and Johnston, J.P. (1978) "Investigation of Separation and Reattachment of a Turbulent Shear Layer Flow over a Backward-Facing Step", Report MD-37, Thermosciences Division, Dept. of Mechanical Engineering, Stanford University.

Kuan, C.L. and Wang, T. (1988) "Some Intermittent Behavior of Transitional Boundary Layers", submitted for publication in Experimental Thermal and Fluid Science.

McDonald, H., and Fish, R.W. (1973) "Practical Calculations of Transitional Boundary Layers", International Journal of Heat and Mass Transfer, Vol. 16, No. 9, pp. 1729-1944.

Schlichting, H. (1979), Boundary Layer Theory, Seventh Edition, McGraw Hill Book Company.

Sohn, K.H., O'Brien, J.E. and Reshotko, E. (1989) "Some Characteristics of Bypass Transition in a Heated Boundary Layer", to be presented at the 7th Symposium on Turbulent Shear Flows, August, 1989, Stanford University.

Wang, T. (1985) " An Experimental Investigation of Curvature and Freestream Turbulence Effects on Heat Transfer and Fluid Mechanics in Transitional Boundary Layer Flows", Ph.D thesis, University of Minnesota.

Wang, T., Simon, T.W., and Buddhavarapu, J. (1985) "Heat Transfer and Fluid Mechanics Measurements in Transitional Boundary Layer Flows", presented at the 30th International Gas Turbine Conference, Houston, TX.

Wygnanski, I., Sokolov, M., and Friedman, D. (1976) "On a Turbulent 'Spot' in a Laminar Boundary Layer", JFM, Vol. 78, pp. 785-819.

1989 National Heat Transfer Conference
HTD-Vol. 107, Heat Transfer in Convective Flows

MICROSCALES OF
OSCILLATING TURBULENT FLOWS

V. S. Arpaci
Department of Mechanical Engineering
and Applied Mechanics
The University of Michigan,
Ann Arbor, Michigan

ABSTRACT

Two microscales for oscillating (with single frequency on the mean) turbulent flows are introduced. The first one is a Taylor scale

$$\lambda \sim \frac{\ell^{1/3} \left(\nu^3/\epsilon_\omega\right)^{1/6}}{\left[1 + \omega \left(\nu/\epsilon_\omega\right)^{1/2}\right]^{1/3}},$$

where ℓ, ν, ϵ_ω, ω denote an integral scale, kinematic viscosity, dissipation and frequency, respectively. In the isotropic limit of this flow,

$$\ell \to \lambda \to \eta ,$$

and the Taylor scale approaches the Kolmogorov scale

$$\eta \sim \frac{\left(\nu^3/\epsilon_\omega\right)^{1/4}}{\left[1 + \omega \left(\nu/\epsilon_\omega\right)^{1/2}\right]^{1/2}}.$$

Two limits of η for $\omega \to 0$ and $\omega \to \infty$ turn out to be the (steady) Kolmogorov scale and the Stokes scale. From the ratio of η and λ,

$$\left(\frac{\eta}{\lambda}\right)^2 \sim \frac{\lambda}{\ell}$$

which agrees with the vortex tube intermittency of turbulence proposed earlier by Tennekes.

These scales are employed in the construction of models for skin friction and heat transfer.

1. INTRODUCTION

The last two decades increased experimental and theoretical attention has been paid to the effect of organized waves on turbulent shear flows. These waves may be externally imposed, say, by a vibrating ribbon near walls of a channel (Hussain and Reynolds 1970, 1972, hereafter called HR, HR2) or may be internally induced, say, in a pulse combustor (Dec and Keller 1989) Also, the dynamical equations which govern oscillating (with single frequency on the mean) turbulent shear flows have been developed (Reynolds and Hussain 1972, hereafter called RH). As expected, these equations require additional information about the wave-induced fluctuations in Reynolds stresses before a close system can be obtained. All closure attempts appear to be statistical, empirical or the usual normal-mode approach of infinitesimal instability theories. The microscales of these flows (a measure for the thickness of sublayer and that for the penetration of diffusion in the core) so far appear to be left untreated and are the motivations of the present study. The approach is intuitive and follows that of Taylor (1935) and Kolmogorov (1941). The same approach have been employed also by Oboukhov (1949) and Corrsin (1951), and by Batchelor (1959) in the development of thermal microscales for small and large Prandtl numbers, respectively.

The study consists of five sections. Following this introduction, Section 2 briefly reviews the dynamics of oscillating turbulent flows, Section 3 develops the appropriate microscales and Section 4 constructs the skin friction and heat transfer models for these flows, and Section 5 concludes the study with some final remarks.

2. DYNAMICS OF UNSTEADY TURBULENCE

The starting point of the present study is the dynamical equations of the unsteady turbulent shear flows which are already available in the literature (see HR and RH). A brief outline of the development leading to these equations is borrowed from the literature and is given below.

Let an instantaneous quantity of turbulence $f(\mathbf{x},t)$ be decomposed as (Fig. 1)

$$f(\mathbf{x},t) = \bar{f}(\mathbf{x},t) + \tilde{f}(\mathbf{x},t) + f'(\mathbf{x},t) \qquad (1)$$

where $\bar{f}$ is the mean (temporal average), $\tilde{f}$ the (statistical) contribution of the organized wave, and f' the turbulent (random) fluctuations. The **temporal average**

$$\bar{f}(\mathbf{x}) = \lim_{T\to\infty} \frac{1}{T} \int_0^T f(\mathbf{x},t)\, dt \ , \qquad (2)$$

the **phase average** is

$$<f(\mathbf{x},t)> = \lim_{N\to\infty} \frac{1}{N} \sum_{n=0}^{N} f(\mathbf{x},t+n\tau) \ , \qquad (3)$$

where τ is the (specified) period of the imposed or induced wave. The phase average is the average over a large ensemble of points having the same phase with respect to the specified wave. Then

$$\tilde{f} = <f> - \bar{f}, \qquad (4)$$

that is, the phase-averaging (ignores the background turbulence and) filters only the coherent oscillations from the instantaneous turbulent motion. The following properties resulting from time and phase averaging,

$$<f'> = 0 \ , \ \ \bar{\tilde{f}} = 0 \ , \ \ \overline{\tilde{f}} = 0 \ ,$$

$$\overline{\bar{f}g} = \bar{f}\,\bar{g} \ , \ \ <\tilde{f}g> = \tilde{f}<g> \ , \ \ <\bar{f}g> = \bar{f}<g> \ , \qquad (5)$$

$$\overline{<f>} = \bar{f} \ , \ \ <\bar{f}> = \bar{f} \ , \ \ \overline{fg'} = <\tilde{f}g'> = 0 \ ,$$

are needed in the development of the dynamic equations.

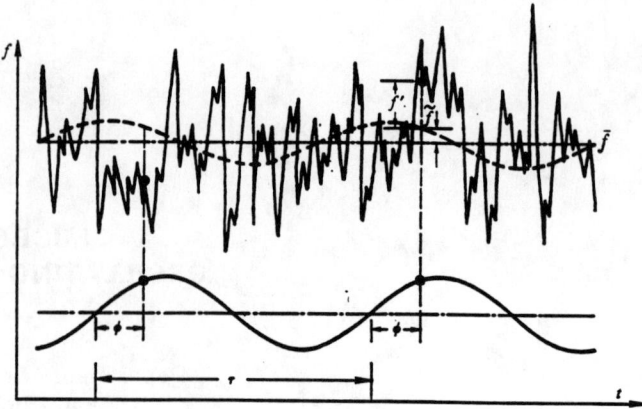

Figure 1. The time and phase averages of a random signal with a weak organized wave. Sketch shows the procedure for obtaining phase average of the signal (upper curve) at phase ϕ of the reference signal (lower curve). Time average $\equiv \bar{f}$, phase average $\equiv <f> = \bar{f}+\tilde{f}$ (from Hussain and Reynolds, 1970).

The continuity and Navier-Stokes equations satisfied by $\bar{u}_i$, $\tilde{u}_i$ and u'_i, and employing some simple closure models, the stability of $\tilde{u}_i$ to infinitesimal disturbances are discussed in RH. The interest of the present study lies in the kinetic energies associated with $\bar{u}_i$, $\tilde{u}_i$ and u'_i, and in the development of microscales from these energies. Let the temporal average of the kinetic energy of instantaneous turbulence be

$$\frac{1}{2}\overline{u_i u_i} = \frac{1}{2}\overline{\bar{u}_i\,\bar{u}_i} + \frac{1}{2}\overline{\tilde{u}_i\tilde{u}_i} + \frac{1}{2}\overline{u'_i u'_i} \qquad (6)$$

or,

$$K_t = K_m + K_o + K \ . \qquad (7)$$

The three components of kinetic energy are obtained by multiplying the momentum equations for $\bar{u}_i$, $\tilde{u}_i$ and u'_i by $\bar{u}_i$, $\tilde{u}_i$ and u'_i, respectively, phase averaging and temporal averaging. The results in terms of

$$u_i = \bar{u}_i + \tilde{u}_i + u'_i \ , \ \ p = \bar{p} + \tilde{p} + p'$$

are

$$\frac{D}{Dt}\left(\frac{1}{2}\overline{\bar{u}_i\,\bar{u}_i}\right) = -\frac{\partial(D_j)_m}{\partial x_j} - \left[-\left(\overline{\tilde{u}_i\tilde{u}_j} + \overline{u'_i u'_j}\right)\right]\frac{\partial\bar{u}_i}{\partial x_j} - 2\nu\,\overline{s_{ij}s_{ij}} \qquad (8)$$

84

$$(D_j)_m = \frac{\overline{p}}{\rho}\,\overline{u}_j + \left(\overline{\tilde{u}_i\tilde{u}_j} + \overline{u_i'u_j'}\right)\overline{u}_i - 2\nu\,\overline{u_i s_{ij}} \qquad (9)$$

$$\frac{D}{Dt}\left(\frac{1}{2}\,\overline{\tilde{u}_i\tilde{u}_i}\right) = -\frac{\partial(D_j)_0}{\partial x_j} - \overline{\left(-<u_i'u_j'>\right)\frac{\partial\overline{\tilde{u}_i}}{\partial x_j}} + \overline{\left(-\overline{\tilde{u}_i\tilde{u}_j}\right)\frac{\partial\overline{u}_i}{\partial x_j}} - 2\nu\,\overline{\tilde{s}_{ij}\tilde{s}_{ij}} \quad (10)$$

$$(D_j)_0 = \frac{\overline{\tilde{p}}}{\rho}\,\tilde{u}_j + \frac{1}{2}\,\overline{\tilde{u}_i\tilde{u}_i\tilde{u}_j} + \overline{\tilde{u}_i<u_i'u_j'>} - 2\nu\,\overline{\tilde{u}_i\tilde{s}_{ij}} \quad (11)$$

$$\frac{D}{Dt}\left(\frac{1}{2}\,\overline{u_i'u_i'}\right) = -\frac{\partial D_j}{\partial x_j} + \overline{\left(-<u_i'u_j'>\right)\frac{\partial\overline{\tilde{u}_i}}{\partial x_j}} - \overline{u_i'u_j'}\frac{\partial\overline{u}_i}{\partial x_j} - 2\nu\overline{s_{ij}'s_{ij}'} \quad (12)$$

$$D_j = \frac{\overline{p'}}{\rho}\,u_j' + \frac{1}{2}\,\overline{u_i'u_i'u_j'} + \tilde{u}_j\frac{1}{2}\,\overline{<u_i'u_j'>} - 2\nu\,\overline{u_i's_{ij}} \quad (13)$$

where

$$\frac{D}{Dt} = \frac{\partial}{\partial t} + \overline{u}_i\frac{\partial}{\partial x_i}. \qquad (14)$$

Now, consider the combined kinetic energy of the coherent oscillations and random fluctuations. Adding equation (10) to (12) yields

$$\frac{D}{Dt}\left(K_0 + K\right) = -\frac{\partial}{\partial x_j}\left[(D_j)_0 + D_j\right] - \left(\overline{\tilde{u}_i\tilde{u}_j} + \overline{u_i'u_j'}\right)\frac{\partial\overline{u}_i}{\partial x_j}$$
$$(15)$$
$$- 2\nu\left(\overline{\tilde{s}_{ij}\tilde{s}_{ij}} + \overline{s_{ij}'s_{ij}'}\right)$$

or,

$$\frac{D}{Dt}(K_0 + K) = -\frac{\partial}{\partial x_j}[(D_j)_0 + D_j] + (\mathcal{P}_0 + \mathcal{P}) - \left(\epsilon_0 + \epsilon\right), (16)$$

again the subscript o indicating the (coherently) oscillating component of energy. Note that the explicit phase averaging disappeared from all terms of equations (15) and (16) except $(D_j)_0 + D_j$, as expected.

For a homogeneous, pure shear flow (all averaged values except $\overline{u}_i$ are independent of position and $\overline{s}_{ij}$ is a constant), equation (16) reduces to

$$\frac{\partial}{\partial t}\left(K_0 + K\right) = \left(\mathcal{P}_0 + \mathcal{P}\right) - \left(\epsilon_0 + \epsilon\right). \qquad (17)$$

The next section develops two microscales from dimensional arguments applied to equation (17).

3. TWO MICROSCALES

Let the rms values of $\tilde{u}_i$ and u_i' be u_0 and u, respectively, and the frequency of oscillations be ω. Then,

$$u_\omega = u_0 + u \qquad (18)$$

and, on dimensional grounds, equation (17) becomes

$$\omega\,u_\omega^2 + u_\omega^2\frac{\overline{u}}{\ell} \sim \nu\frac{u_\omega^2}{\lambda^2} \sim \epsilon_\omega, \qquad (19)$$

where ℓ, λ and ϵ_ω respectively denote an integral scale, the Taylor microscale and the viscous dissipation. For an estimate of spatial production, following Taylor (1935), let

$$\overline{u} \sim u_\omega. \qquad (20)$$

Thus, equation (19) becomes

$$\omega\,u_\omega^2 + \frac{u_\omega^3}{\ell} \sim \nu\frac{u_\omega^2}{\lambda^2} \sim \epsilon_\omega. \qquad (21)$$

Rearrange this result, for a spatially dominated inertial production, as

$$\frac{u_\omega^3}{\ell}\left(1 + \frac{\omega\ell}{u_\omega}\right) \sim \nu\frac{u_\omega^2}{\lambda^2} \sim \epsilon_\omega. \qquad (22)$$

A viscous estimate for u_ω, obtained from the second proportionality of equation (22), is

$$u_\omega \sim \lambda\left(\frac{\epsilon_\omega}{\nu}\right)^{1/2}, \qquad (23)$$

and, an upper bound of u_ω for production is

$$u_\omega \sim \ell\left(\frac{\epsilon_\omega}{\nu}\right)^{1/2}. \qquad (24)$$

Then, the temporal production of equation (22), rearranged in terms of this velocity, becomes

$$\frac{u_\omega^3}{\ell}\left[1 + \omega\left(\frac{\nu}{\epsilon_\omega}\right)^{1/2}\right] \sim \nu\frac{u_\omega^2}{\lambda^2} \sim \epsilon_\omega. \qquad (25)$$

An inertial estimate for u_ω in terms of ϵ_ω is

$$u_\omega \sim \frac{\left(\epsilon_\omega \ell\right)^{1/3}}{\left[1 + \omega \left(\nu/\epsilon_\omega\right)^{1/2}\right]^{1/3}}. \qquad (26)$$

Elimination of u_ω between equations (23) and (26) results in the Taylor microscale,

$$\lambda \sim \frac{\ell^{1/3} \left(\nu^3/\epsilon_\omega\right)^{1/6}}{\left[1 + \omega \left(\nu/\epsilon_\omega\right)^{1/2}\right]^{1/3}} \qquad (27)$$

which explicitly includes the frequency of coherent oscillations. In the isotropic limit,

$$\ell \to \lambda \to \eta , \qquad (28)$$

and the Taylor scale is reduced to the Kolmogorov microscale,

$$\eta \sim \frac{\left(\nu^3/\epsilon_\omega\right)^{1/4}}{\left[1 + \omega \left(\nu/\epsilon_\omega\right)^{1/2}\right]^{1/2}}. \qquad (29)$$

Two limits of equation (29),

$$\lim_{\substack{\omega \to 0 \\ \epsilon_\omega \to \epsilon}} \eta \to \left(\frac{\nu^3}{\epsilon}\right)^{1/4}$$

$$\qquad (30)$$

$$\lim_{\omega \to \infty} \eta \to \left(\frac{\nu}{\omega}\right)^{1/2}$$

are the usual Kolmogorov and Stokes scales, as expected.

From appropriate combinations of equations (27) and (29),

$$\frac{\eta}{\ell} \sim \left(\frac{\lambda}{\ell}\right)^{3/2} \qquad (31)$$

or

$$\left(\frac{\eta}{\lambda}\right)^2 \sim \frac{\lambda}{\ell} \qquad (32)$$

which can be related to the intuitive model introduced by Tennekes (1968) describing the turbulence intermittency by vortex tubes (see also Frisch, Sulem and Nelkin 1979).

The foregoing microscales will be employed in the next section which deals with skin friction and heat transfer in (coherently) oscillating turbulent flows.

4. SKIN FRICTION and HEAT TRANSFER

Let the sublayer of a turbulent flow be characterized by η, and entire dissipation be confined to this layer. Then, the usual definition of skin friction, written in terms of η, gives

$$\left(\frac{1}{2}f\right) \mathrm{Re}_\omega \sim \frac{\ell}{\eta} \qquad (33)$$

or, in terms of equation (31),

$$\left(\frac{1}{2}f\right) \mathrm{Re}_\omega \sim \left(\frac{\ell}{\lambda}\right)^{3/2}, \qquad (34)$$

where

$$\mathrm{Re}_\omega = \frac{u_\omega \ell}{\nu} \qquad (35)$$

is the Reynolds number based on u_ω and ℓ. For the evaluation of ℓ/λ, reconsider equation (22),

$$\frac{u_\omega^3}{\ell}\left(1 + \frac{\omega \ell}{u_\omega}\right) \sim \nu \frac{u_\omega^2}{\lambda^2}$$

which readily gives

$$\frac{\ell}{\lambda} \sim \mathrm{Re}_\omega^{1/2}\left(1 + \frac{\omega \ell}{u_\omega}\right)^{1/2}. \qquad (36)$$

Then, from the combination of equations (34) and (36),

$$\frac{1}{2}f \sim \mathrm{Re}_\omega^{-1/4}\left(1 + \frac{\omega \ell}{u_\omega}\right)^{3/4}. \qquad (37)$$

The limit of this result for $\omega \to 0$

$$\frac{1}{2}f \sim \mathrm{Re}^{-1/4} ,$$

is known to correlate the turbulent data on flat plates.

The heat transfer, in terms of a thermal Kolmogorov scale, is

$$\mathrm{Nu} \sim \frac{\ell}{\eta_\theta}, \qquad (38)$$

Nu being the usual Nusselt number. From the ratio of equations (33) and (38),

$$\frac{Nu}{\left(\frac{1}{2}f\right)Re_\omega} \sim \frac{\eta}{\eta_\theta}, \qquad (39)$$

where

$$\frac{\eta}{\eta_\theta} = f(Pr), \qquad (40)$$

Pr being the usual Prandtl number. When the momentum and thermal energy are similar (term-by-term correspondence in governing equations and boundary conditions)

$$\frac{\eta}{\eta_\theta} = Pr^{1/3} \qquad (41)$$

is known to hold for steady (turbulent as well as laminar) flows. Otherwise, the explicit form of equation (40) needs to be known. However, $Pr \sim 1$ for gases, and

$$\frac{\eta}{\eta_\theta} = \text{Const.}, \qquad (42)$$

and equation (39) becomes

$$Nu \sim Re_\omega^{3/4} \left(1 + \frac{\omega\ell}{u_\omega}\right)^{3/4}, \qquad (43)$$

or, in terms of equation (18),

$$Nu \sim Re^{3/4}\left(1 + \frac{u_0}{u}\right)^{3/4}\left[1 + \frac{\omega\ell}{u\left(1 + u_0/u\right)}\right]^{3/4}, \quad (44)$$

where

$$Re = \frac{u\ell}{\nu}. \qquad (45)$$

Two limits of equation (44) corresponding to Nusselt number dominated by random fluctuations or coherent oscillations are, respectively,

$$Nu \sim Re^{3/4}\left(1 + \frac{u_0}{u}\right)^{3/4}\left(1 + \frac{\omega\ell}{u}\right)^{3/4}, \qquad (46)$$

and

$$Nu \sim Re_0^{3/4}\left(1 + \frac{u}{u_0}\right)^{3/4}\left(1 + \frac{\omega\ell}{u_0}\right)^{3/4}, \qquad (47)$$

where

$$Re_0 = \frac{u_0\ell}{\nu}. \qquad (48)$$

The experimental literature on heat transfer in pulse combustors is being correlated by equations (46) and/or (47), and will be reported later.

5. CONCLUDING REMARKS

Intuitive ideas originated by Taylor and Kolmogorov on the microstructure of turbulence are universal and apply to all turbulent flows. However, the well-known forms of the Taylor and Kolmogorov microscales are for steady (on the mean) isothermal flows. In the present study, following Taylor and Kolmogorov, the microscales appropriate for oscillating (on the mean) turbulent flows have been developed. Also, the skin friction and heat transfer in these flows are given in terms of these scales.

REFERENCES

Batchelor, G. K. 1959, Small-Scale Variation of Convected Quantities Like Temperature in a Turbulent Fluid, *J. Fluid Mech.* **5**, 113.

Corrsin, S. 1951, On the Spectrum of Isotropic Temperature Fluctuations in Isotropic Turbulence, *J. Appl. Phys.* **22**, 469.

Dec, J.E. and **Keller, J.O.** 1989, Heat Transfer Dependence on Frequency, Amplitude and Mean Flow Rate, to appear in *Combust. Flame.*

Frisch, U., Sulem, P. L. and **Nelkin, M.** 1979, A Simple Model of Intermittent Fully-Developed Turbulence *J. Fluid Mech.* **87**, 719.

Hussain, A.K.M.F and **Reynolds, W.C.** 1970, The Mechanics of an Organized Wave in Turbulent Shear Flow **41**, 241.

Hussain, A.K.M.F and **Reynolds, W.C.** 1972, The Mechanics of an Organized Wave in Turbulent Shear Flow. Part 2. Experimental Results **54**, 241.

Kolmogorov, A. N. 1941, Local Structure of Turbulence in Incompressible Viscous Fluid for Very Large Reynolds Numbers, *C. R. Acad. Sci. USSR* **30**, 301.

Oboukhov, A. M. 1949, Structure of the Temperature Field in Turbulent Flows, *Izv. Nauk. SSSR, Geogr. i. Geofiz.* **13**, 58.

Reynolds, W.C. and Hussain, A.K.M.F 1972, The Mechanics of an Organized Wave in Turbulent Shear Flow. Part 3. Theoretical Models and Comparison with Experiments **54**, 263.

Taylor, G. I. 1935, Statistical Theory of Turbulence, *Proc. Roy. Soc. A* **151**, 421.

Tennekes, H. 1968, Simple Model for the Small Scale Structure of Turbulence, *Phys. Fluids* **11**, 669.

1989 National Heat Transfer Conference
HTD-Vol. 107, Heat Transfer in Convective Flows

COMPUTATION OF TURBULENT THREE-DIMENSIONAL MIXED CONVECTION BOUNDARY LAYERS

B. Afshari and J. H. Ferziger
Department of Mechanical Engineering
Stanford University
Stanford, California

Abstract

Orthogonal mixed convection boundary layers combine natural and forced convection. When the temperature difference across the boundary layer, and the free stream velocity are both significant, the flow is highly three-dimensional.

A turbulence model for three-dimensional, orthogonal mixed convection boundary layers on a flat plate with large temperature differences has been developed. This model contains as components: a new mixing length model for turbulent natural convection boundary layers with large temperature differences, a mixing length model for forced convection that allows for variable properties, and a method of combining them. The combination rule is designed so that the model reduces to its two dimensional components in the appropriate limiting cases.

This model has been tested for temperature differences of up to 700 K across the boundary layer and free stream velocities from zero to 7.0 m/s. The predictions match all the experimental data of Siebers (1983) very well.

Nomenclature

A_n	effective sublayer thickness for natural convection
A^+	dimensionless sublayer thickness
c	turbulence model constant see Eq.(28)
c_p	specific heat at constant pressure
c_x	outer layer constant see Eq.(28)
D_n	damping function for the natural convection direction
g	acceleration of gravity
Gr	grashof number
H	stagnation enthalpy
$\hat{H}$	non-dimensional stagnation enthalpy
$H_{w,0}$	stagnation enthalpy at the wall
$H_{\infty,0}$	stagnation enthalpy at the upstream of the plate
H_i	see Eq. (3)
H'	fluctuating component of the enthalpy
j	grid index in the normal direction
J	number of mesh points in the normal direction
k	turbulent kinetic energy
K	the grid stretching parameter constant
l	mixing length

L	a characteristic length
P	turbulence production
Pr	Prandtl number
Pr_t	turbulent Prandtl number
$\dot{q}_w$	heat flux
Re	Reynolds number
St	Stanton number
t	temperature
t_{sl}	sublayer temperature see Eq.(26)
u	component of velocity in x-direction
$\hat{u}$	non-dimensional velocity in x-direction
u_{inner}	inner characteristic velocity for natural convection
u_{outer}	outer characteristic velocity for natural convection
u'	fluctuating component of the horizontal velocity
U	free stream velocity at the upstream edge of the plate
v	component of velocity in y-direction
$\hat{v}$	non-dimensional normal velocity
v'	fluctuating component of the normal velocity
w	component of velocity in z-direction
$\hat{w}$	non-dimensional velocity in z-direction
w'	fluctuating component of the vertical velocity
x	physical coordinate in horizontal direction
x_s	location of start of transition
x_e	location of end of transition
y	physical coordinate in normal direction
y_{umax}	location of the maximum velocity in the natural convection boundary layer
z	physical coordinate in vertical direction

Greek Symbols

α	thermal diffusivity
β	coefficient of thermal expansion
γ_{tr}	transition intermittency factor
δ	scaling function in the normal direction
$\delta_{99.5}$	boundary layer thickness

ϵ	turbulent dissipation
ε	eddy diffusivity
ε_x	eddy diffusivity for forced convection
ε_z	eddy diffusivity for free convection
ε_H	thermal eddy diffusivity
ς	computational coordinate in the vertical direction
η	computational coordinate in the normal direction
θ	nondimensional temperature
κ	von Karman constant, 0.41
μ	dynamic viscosity coefficient
ν	kinematic viscosity coefficient
ν_{sl}	kinematic viscosity at sublayer temperature
ν_t	turbulent viscosity
ξ	computational coordinate in the horizontal direction
ρ	fluid density

1   Introduction

Mixed convection flows exist in many systems in which there is a heated surface in a cross-flow. Mixed convection boundary layers contain a combination of natural and forced convection, and are driven by inertia forces and buoyancy forces across the boundary layer. This study deals with the special case in which the two driving forces are orthogonal to each other and parallel to the surface as shown in Fig.1. If the temperature difference across the boundary layer and the free stream velocity are both large, such flows may be highly three-dimensional, so that neither cross-flow nor gravitational effects can be treated as secondary.

Three-dimensional mixed convection flows exist in a variety of heat transfer systems, for example, external receivers of central solar power plants, nuclear reactor plenums, and silicon wafer coating ovens. A three-dimensional turbulence model and a computer code for simulation of mixed convection boundary layers are developed and applied to the prediction of mixed convection boundary layers with temperature differences of up to 700 K. The process also requires the development of a new turbulence model for high temperature two-dimensional natural convection boundary layers.

An obstacle to the development of turbulence models for mixed convection is the lack of detailed turbulence data for such flows. Consequently, the effects of large temperature differences on turbulence are not well understood. This difficulty, and the requirement of developing a practical engineering code, led to the use of a mixing-length approach to building the turbulence model. Since mixed convection has two distinct driving mechanisms, forced and natural convection, and the flow degenerates to them in the appropriate limits, it is logical to base the three-dimensional model on two-dimensional models for those flows. Examination of existing two-dimensional models and, if necessary, development of new ones for high temperature two-dimensional boundary layers, especially in the natural convection case, is an intermediate objective of this work.

Construction of a robust computer code to compute three-dimensional boundary layers involves finding a suitable numerical scheme. It is equally essential to find a scheme to generate initial and boundary conditions along the upstream and lower boundaries of the flat plate. This is necessary because no velocity or temperature profile data are available at the boundaries of the plate. The issue of initial condition generation is important because calculations cannot be started with a purely laminar profile; initial conditions are needed in transitional and turbulent regions as well as laminar stages of the flow. Also, due to the inherent sensitivity of the Keller box method, the initial conditions have to be accurate; otherwise, there may be oscillations in the solution.

2   Literature Review

The only comprehensive study of three-dimensional mixed convective flows is that of Siebers (1983), who carried out a unique experiment which covered laminar, transitional and turbulent regimes of this flow. Siebers' work provides the target of the present effort. Other literature relevant to mixed convection flows has been thoroughly reviewed in Siebers' report, which should be consulted for additional information.

Siebers' experiment provides the basis for most of our assumptions and physical insight. This experiment consisted of heat transfer and velocity measurements in mixed convection flow over a 3 m by 3 m plate. He reported experimental data for cases with wall temperatures from 320 K to 900 K and free-stream velocities of 0.0 m/sec to 7.0 m/sec.

Siebers pointed out quite a few properties of this flow. Some of the important results pertaining to computation of such flows are:

- There is smooth change in heat transfer mechanism from forced convection to natural convection with increasing Gr/Re^2. The heat transfer rate cannot be predicted with either free or forced convection correlations in the range $0.7 < Gr/Re^2 < 10.0$. Outside of this range, the heat transfer rate could be predicted to within 5% using pure forced or free convection models.

- With the introduction of any free-stream flow in the forced convection direction, forced convection dominates the heat transfer near the vertical leading edge. The flow is nearly independent of height along the leading edge of the plate.

- Wall temperature measurements indicate that with any free-stream flow, the transition to turbulence occurs along a line perpendicular to free stream direction (a vertical line). The location and length of the transition zone are affected by the free-stream velocity, buoyancy effects and temperature difference across the boundary layer.

Young and Yang (1962) published one of the first attempts at calculation of three-dimensional mixed convection flows. They used a perturbation analysis to obtain a solution for a laminar free convection boundary layer over a vertical flat plate with small cross-flow velocity.

Yao and Chen (1979) also using a perturbation analysis to study mixed convection flows over a cylinder concluded that even for small free-stream velocity, the forced convection cannot be treated as a small effect.

Eichhorn and Hassan (1980) made an analytical study of mixed convection flows. They showed that a similarity solution could be obtained if the boundary layer thickness is same at each point on the surface, for the forced and free convection flows taken separately. This restricts the analysis to flows over a wedge with linear temperature variation in free convection direction.

Another directly related work is that of Plumb (1980), who analytically studied the effects of cross-flow on natural convection from vertical heated surfaces. He argues that under certain boundary conditions, similarity solutions can be obtained for three-dimensional laminar boundary layers over a heated vertical wedge in a cross-flow. The boundary conditions which yield such similarity solutions are too restrictive to apply to the general mixed convection problem.

Evans (1981) studied laminar mixed convection experimentally as well as numerically. He used the Keller box numerical scheme and reported oscillations in his results. An explanation of this problem was found during the present work and a modification for its correction is suggested in this report.

Plumb and Evans (1983) presented a numerical study of turbulent mixed convection on a vertical surface in a crossflow. This work, which was the first numerical study of such flows has limitations. No comparison to experimental mixed convection data were made to qualify its results. Also, their mathematical formulation and the assumptions of constant properties and the Boussinesq approximation limits the range of applicability of their results. They used the Cebeci Smith turbulence model which was developed for pure forced convection boundary layers. Our experience showed that this model is inadequate for orthogonal mixed convection. Other than Plumb and Evans' work, nearly all previous works were analytical approaches applicable only to small temperature differences.

3 Mathematical Formulation

The orientation of coordinates and direction of velocity components in formulation of this flow are shown in Fig. 1. To compute the three-dimensional mixed-convection boundary layer flow field over a flat plate, we need to simultaneously solve the equations for x- and z- momentum, the continuity equation, and the energy equation. The standard boundary layer assumptions are the only approximations that are made.

3.1 Computational Coordinates and Transformation

For simplicity and efficiency, the governing equations are nondimensionalized and transformed to new system of coordinates. The y-coordinate, normal to the plate, is normalized by a characteristic length δ, which is formulated so that it follows the boundary layer thickness. The x- and z-coordinates are nondimensionalized with a length, L, which is taken to be the plate length. Thus,

$$\xi = \frac{x}{L}, \qquad \eta = \frac{y}{\delta}, \qquad \varsigma = \frac{z}{L} \qquad (1)$$

An advantage of this transformation is that the mesh structure follows the boundary layer growth very accurately, and little of the free-stream is included in the computational domain.

The hydrodynamic dependent variables are transformed in the following manner:

$$\hat{u} = \frac{u}{U}, \qquad \hat{w} = \frac{w}{U}, \qquad \hat{v} = \frac{vL}{\delta U} \qquad (2)$$

where U is a constant characteristic velocity of the flow.

In the energy equation the stagnation enthalpy is nondimensionalized as follows:

$$\hat{H} = \frac{(H_{w,0} - H)}{H_i}; \qquad H_i = H_{w,0} - H_{\infty,0} \qquad (3)$$

In this definition $H_{w,0}$ is the enthalpy of the fluid at the wall at $x = 0$ and $H_{w,\infty}$ is the enthalpy of the fluid at the free-stream conditions at $x = 0$.

With transformations (1) through (3) the governing equations become the following:
The x-momentum equation is:

$$\frac{\partial \rho \hat{u}^2 U^2}{\partial \xi} - \frac{\eta U^2}{\delta} \frac{\partial \delta}{\partial \xi} \frac{\partial \rho \hat{u}^2}{\partial \eta} + U^2 \frac{\partial \rho \hat{u}\hat{v}}{\partial \eta} + \frac{\partial \rho \hat{u}\hat{w} U^2}{\partial \varsigma} -$$
$$\frac{\eta U^2}{\delta} \frac{\partial \delta}{\partial \varsigma} \frac{\partial \rho \hat{u}\hat{w}}{\partial \eta} = -\left(\frac{\partial u_\infty^2}{\partial \xi}\right) + \frac{LU}{\delta^2} \frac{\partial}{\partial \eta}\left(\mu \frac{\partial \hat{u}}{\partial \eta} - \overline{\rho u'v'}\right) \qquad (4)$$

In this and the following equations all physical properties are assumed to be functions of temperature. In Equation (4) ρ is the fluid density and μ represents the dynamic viscosity. The turbulent shear stress in x-direction is denoted by $\overline{\rho u'v'}$.
The z-momentum equation becomes:

$$\frac{\partial \rho \hat{w}^2 U^2}{\partial \varsigma} - \frac{\eta U^2}{\delta} \frac{\partial \delta}{\partial \varsigma} \frac{\partial \rho \hat{w}^2}{\partial \eta} + U^2 \frac{\partial \rho \hat{w}\hat{v}}{\partial \eta} + \frac{\partial \rho \hat{u}\hat{w} U^2}{\partial \xi} -$$
$$\frac{\eta U^2}{\delta} \frac{\partial \delta}{\partial \xi} \frac{\partial \rho \hat{u}\hat{w}}{\partial \eta} = \frac{LU}{\delta^2} \frac{\partial}{\partial \eta}\left(\mu \frac{\partial \hat{w}}{\partial \eta} - \overline{\rho w'v'}\right) + g(\rho_\infty - \rho) \qquad (5)$$

Here g is acceleration of gravity; the last term in the equation represents the gravitational body force. The term $\overline{\rho w'v'}$ denotes the turbulent shear stress.
The energy equation becomes:

$$\frac{\partial \rho \hat{u}\hat{H}}{\partial \xi} - \frac{\eta}{\delta} \frac{\partial \delta}{\partial \xi} \frac{\partial \rho \hat{u}\hat{H}}{\partial \eta} + \frac{1}{L} \frac{\partial \rho \hat{v}\hat{H}}{\partial \eta} + \frac{\partial \rho \hat{w}\hat{H}}{\partial \varsigma} - \frac{\eta}{\delta} \frac{\partial \delta}{\partial \varsigma} \frac{\partial \rho \hat{w}\hat{H}}{\partial \eta} =$$
$$\frac{1}{U\delta^2} \frac{\partial}{\partial \eta}\left(\frac{\mu}{Pr} \frac{\partial \hat{H}}{\partial \eta}\right) - \frac{1}{U\delta^2} \frac{\partial}{\partial \eta}\left(\overline{\rho H'v'}\right) -$$
$$\frac{U}{\delta^2 H_i} \frac{\partial}{\partial \eta}\left[\mu\left(1 - \frac{1}{Pr}\right) \frac{\partial}{\partial \eta}\left(\frac{u^2 + w^2}{2}\right)\right] \qquad (6)$$

here Pr is the Prandtl number and H is the stagnation enthalpy defined as :

$$H = c_p T + \frac{u^2 + w^2}{2} \qquad (7)$$

where c_p is the specific heat at constant pressure and, like all other fluid properties, is treated as a function of temperature.
Finally, the equation of continuity becomes:

$$\frac{\partial \rho \hat{u} U}{\partial \xi} - \frac{\eta U}{\delta} \frac{\partial \delta}{\partial \xi} \frac{\partial \rho \hat{u}}{\partial \eta} + U \frac{\partial \rho \hat{v}}{\partial \eta} + \frac{\partial \rho \hat{w} U}{\partial \varsigma} - \frac{\eta U}{\delta} \frac{\partial \delta}{\partial \varsigma} \frac{\partial \rho \hat{w}}{\partial \eta} = 0 \qquad (8)$$

Equations (4) through (8) are a complete set of equations describing velocity and temperature fields, provided the turbulent quantities are modeled in terms of mean values. A detailed discussion of the governing equations and their discretization are given by Afshari (1989).

3.2 Special Considerations for Keller's Box Method

The box scheme in its original form described by Keller (1971) permits streamwise oscillations in the velocity profile. These oscillations are small in the outer part of the velocity profiles, but can be very pronounced near the wall, they are not physical and, when the mesh structure is changed, the fluctuations shift with the mesh. This oscillatory behavior is explained by noting that the box scheme averages the equations over the box. When the upstream values are, say, a bit high, the values at the new point will tend to be a bit low to compensate. At the next mesh point, the situation is reversed and the oscillation results.

In order to correct this problem, two measures have been taken. First step is to modify the box scheme and the second step is to improve and generate more compatible initial conditions. The box scheme is modified by solving for the normal component of the velocity v at all eight corners of the box instead of the two corners which are normally used for all the other variables. A complete discussion of modifications to this method are discussed in detail by Afshari (1989).

3.3 Scheme for Generating Initial Conditions

More initial condition data are needed than experiments provide. For mixed convection flows, profiles along both the upstream and lower edges of the plate must be generated. Special care has been taken to render the initial conditions as close to physical conditions as possible. To generate the initial conditions, the governing equations were simplified based on insight from the experimental data and used to generate first guesses at the initial conditions along the edges of the plate. However, these conditions lead to solution oscillations so further improvement is needed. It is achieved by the following procedure.

- First, a new line of grid points is inserted just inside the boundary as shown in Fig. 2.

- Second, the velocity and temperature profiles on this new grid line are calculated.

91

- Third, an average of the variables at the new mesh point and the initial condition is substituted for initial conditions. The idea is that, if the new grid is close enough to the boundary, the profiles at the new point should be very close to the initial condition.

- Fourth, the second and third steps are repeated. The gradients of temperature and velocity at the wall at the new point, are compared to the values at the previous iteration; convergence is assumed when the difference is less than a preset value.

This procedure converges in less than 5 iterations, for most flow conditions, when the fine grid is taken to be $\frac{1}{10}$ of the original grid. After obtaining the converged initial condition, the grid is set to the original size and the main calculation is begun. It also should be noted that a number of other schemes were tried, including a very fine grid to start the calculations. In all other methods the oscilations persisted and the scheme described above was superior to all the other methods.

3.4 Mesh Structure

Keller's box method is readily adapted to constant or variable mesh spacings in all directions. In a boundary layer, the gradients in the normal direction (y-axes) are large near the wall and smaller in the outer region, so it is essential to use a variable grid with more mesh points near the wall. Also, in the streamwise (x) and gravity (z) directions, variable grids are needed to account for boundary layer growth.

3.4.1 Grid Structure in the Normal Direction

The grid generation scheme for the normal direction is that of Cebeci and Smith (1974) as modified by Blottner (1974) for turbulent boundary layers. This method has one adjustable parameter and few limitations. This scheme is

$$\eta = \frac{K^{\tilde{\eta}/\Delta\tilde{\eta}_0} - 1}{K^{1/\Delta\tilde{\eta}_0} - 1} \qquad (9)$$

where η is the computational coordinate, and

$$\tilde{\eta} = (j-1)\,\Delta\tilde{\eta}_0 \qquad (10)$$

where

$$\Delta\tilde{\eta}_0 = \frac{1}{J-1} \qquad (11)$$

K is the stretching parameter and is always greater than one (K=1.14 was suggested by Cebeci and Smith (1974)), j is the mesh index and J is the number of mesh points in the normal direction.

3.4.2 Grid Structure in the Vertical and Horizontal Directions

In the x- and z-directions, the mesh size should be not more than four to five times the local boundary layer thickness. It became apparent from our experience with the box method in mixed convection flows that it is desirable to start the calculations with mesh spacing as small as one boundary layer thickness.

The grid structure used in both directions starts with a small grid size and increases in a geometric progression until the mesh size reaches a user specified constant value. From this point on, mesh size is constant.

4 An Eddy Viscosity Model

The three-dimensional turbulence model for mixed convection flows is based on separate eddy viscosity models for free and forced convection. The proposed model combines the two models to create a model for three-dimensional mixed convection boundary layers. Incorporation of the effect of large temperature differences into those models plays an important role in the success of the proposed three-dimensional model.

An important constraint on any model for mixed convection boundary layers is that it must be able to calculate the limiting cases of two-dimensional natural convection and forced convection. The three-dimensional model suggested here, reduces to the two-dimensional models for free and forced convection. These models have been shown to work for two-dimensional flows.

4.1 Three-Dimensional Model

A method of combining the two-dimensional models is required and a number of formulations, including the methods of Rotta (1977) and Schneider (1979) were tried. The method which was most successful is described below; the results it produces are considerably better than those of its nearest competitor.

According to the well-known k-ε model, the eddy viscosity ν_t is:

$$\nu_t = c \cdot \frac{k^2}{\varepsilon} \qquad (12)$$

where $c = 1.0$ is a constant, k is the turbulent kinetic energy, and ε denotes the turbulent dissipation. In this model, the eddy length scale l is defined by:

$$l = \frac{k^{\frac{3}{2}}}{\varepsilon} \qquad (13)$$

In a equilibrium boundary layer, the rate of production of turbulence P, is equal its rate of dissipation. Combining this notion and Eqs. (13) with (12), one has:

$$\nu_t = c \cdot P^{\frac{1}{3}} l^{\frac{4}{3}} \qquad (14)$$

In a boundary layer or other shear layer, the turbulence production (see Bradshaw et al. (1981)) is:

$$P = -\overline{u'v'}\frac{\partial u}{\partial y} - \overline{w'v'}\frac{\partial w}{\partial y} \qquad (15)$$

In a simple mixing length model, the Reynolds shear stresses in this equation are given by:

$$-\rho\overline{u'v'} = \nu_{tx}\left(\frac{\partial u}{\partial y}\right), \quad -\rho\overline{w'v'} = \nu_{tz}\left(\frac{\partial w}{\partial y}\right) \qquad (16)$$

The models for the eddy viscosities ν_{tx} and ν_{tz} are described below.

Because the ratio of the three components of the turbulence kinetic energy is approximately constant in a thin shear layer, we can use as an approximation:

$$k \approx u'^2 + w'^2 \qquad (17)$$

Inclusion of v'^2 would change constants but not the form of the results.

Since there are no data on the component energies, we approximate Eq. (17) by :

$$k \cong l_x^2\left(\frac{\partial u}{\partial y}\right)^2 + l_z^2\left(\frac{\partial w}{\partial y}\right)^2 \qquad (18)$$

where l_x and l_z are the mixing lengths used in the models for forced and free convection, respectively. Now using this definition of turbulent kinetic energy and Eqs. (13) to (16), we obtain an expression for the eddy length scale l:

$$l = \frac{\left[l_x^2 \left(\frac{\partial u}{\partial y}\right)^2 + l_z^2 \left(\frac{\partial w}{\partial y}\right)^2\right]^{\frac{3}{2}}}{l_x^2 \left|\frac{\partial u}{\partial y}\right| \left(\frac{\partial u}{\partial y}\right)^2 + l_z^2 \left|\frac{\partial w}{\partial y}\right| \left(\frac{\partial w}{\partial y}\right)^2} \tag{19}$$

Finally using this definition of l and Eq. (15) in Eq. (14) we have an eddy viscosity model for three-dimensional boundary layers. This eddy viscosity is used in both momentum equations.

The thermal eddy diffusivity ε_H used in energy equation, is evaluated using the modified Reynolds analogy. Define the turbulent Prandtl number by:

$$Pr_t = \frac{\varepsilon}{\varepsilon_H} \tag{20}$$

where ε_M is the momentum eddy diffusivity $\varepsilon_M = \rho\nu_t$. If Pr_t is taken to be 1.0 we have the exact Reynolds analogy. According to the experiments (Kays and Moffat (1975)), the value of Pr_t changes across the boundary layer, but is close to 0.9. A value of constant 0.9 was used in our calculations.

4.2   Forced Convection Model

The Reynolds stress $(u'v')$ is evaluated using a modified form of the eddy viscosity model for forced convection boundary layers suggested by Cebeci and Smith (1974). This model is described in a concise form by Bradshaw et al. (1981).

The modification needed here is to improve its performance in predicting boundary layers with large temperature differences.

To accomplish this, it is suggested that the viscosity used in estimating the effective sublayer thickness be calculated at the sublayer temperature, where sublayer temperature is defined by Eq.(26) in the following section.

4.3   Natural Convection Model

The natural convection component of the mixed convection model is especially designed for the calculation of boundary layers with large temperature differences.

$$\overline{-w'v'} = \varepsilon_z \left|\frac{\partial w}{\partial y}\right| \tag{21}$$

The eddy viscosity model has two layers, both of the mixing-length type. The inner region model is similar to the forced convection model because the physics of the inner layer is similar in the two cases. The eddy viscosity ε_z has Prandtl's form.

$$\varepsilon_z = l_z^2 \left|\frac{\partial w}{\partial y}\right| \tag{22}$$

In this, or any other model in which the Reynolds stress is proportional to the velocity gradient, Eq.(22) shows that the value of ε_z becomes zero at the point of maximum velocity. There is no evidence to suggest a zero turbulence level anywhere inside the boundary layer. To avoid this problem, at the maximum velocity point the value of the ε_z is calculated from a curve fitted to the values of ε_z at the two neighboring points. This procedure eliminates a sudden, unphysical, drop in the magnitude of eddy-viscosity.

The mixing-length l_z is defined as in the forced convection model:

$$l_z = \kappa y D_n \tag{23}$$

where D_n is a damping factor :

$$D_n = 1 - \exp\left(\frac{-y}{A_n}\right) \tag{24}$$

This definition of damping factor D_n is identical to van Driest damping function except for the definition of A_n.

The definition commonly used for A_n (Bradshaw et al.(1981)) is based on friction velocity u_τ. Although this definition suffices for flows with small temperature differences, it is not sensitive to the inner layer properties and tests revealed that it does not fit the data for natural convection flows with large temperature differences very well. Therefore it is suggested that a better form of A_n for natural convection boundary layers is:

$$A_n = A_n^+ \frac{\nu_{sl}}{u_{inner}} \tag{25}$$

Here ν_{sl} is the viscosity in the sublayer and is evaluated at the temperature t_{sl} at a point half the sublayer thickness from the wall:

$$t_{sl} = \frac{t_0 + t_{A^+}}{2} \tag{26}$$

The characteristic velocity u_{inner} in Eq. (25) is taken from George and Capp's (1979) analysis of turbulent natural convection boundary layers.

$$u_{inner} = \left(\frac{\dot{q}_w}{\rho c_p} g \beta \alpha\right)^{\frac{1}{4}} \tag{27}$$

In this equation $\dot{q}_w$ is the heat flux at the wall, ρ is density, c_p is specific heat at constant pressure, g is acceleration of gravity, β is coefficient of thermal expansion and α is thermal diffusivity. The constant A_n^+ has to be adjusted to account for this change in characteristic velocity. It was found that $A_n^+ = 3.5$ works very well for all cases calculated. Unlike the conventional A_n, the new estimate of the sublayer thickness is sensitive to the temperature in the sublayer.

In the outer layer of the boundary layer, it is assumed that eddy viscosity is independent of distance from the wall. With some simplifying assumptions, this can be demonstrated from experimental data of Cheesewright (1983). In this region, the eddy viscosity takes the form:

$$\varepsilon_z = C_n \delta_{99.5} u_{outer} \tag{28}$$

In this formulation $C_n = 0.13$ is a constant, $\delta_{99.5}$ is the boundary layer thickness, defined as the distance from the wall at which the velocity drops to 0.5% of the maximum velocity and u_{outer} is a characteristic velocity of the outer region. The form of u_{outer} was suggested by George and Capp (1979).

$$u_{outer} = \left(\frac{\dot{q}_w}{\rho c_p} g \beta y_{umax}\right)^{\frac{1}{3}} \tag{29}$$

The characteristic length y_{umax} is the distance from the wall to the point of maximum velocity.

5   Transition Model

According to Siebers, in mixed convection flows over a flat plate, the transition to turbulence shows certain unique characteristics. In three-dimensional mixed convection flow, we might intuitively expect transition to happen in an "L" shaped region on the plate, the vertical part being the result of forced convection transition and the horizontal part, the result of free convection effects. The experiment showed the transition zone to be entirely vertical, that is, perpendicular to the free stream velocity even for the case with the weakest forced convection. This result means that the locations of the start and end of transition zone are only functions of x. Also, although the transition seems to be dominated by forced convection, the start of transition occurs sooner than in pure forced convection flows. As a good correlation for transition was not available, we chose to prescribe the beginning and end of transition by using experimental values.

One way of introducing turbulence into the calculations is by gradually increasing the effective viscosity in the governing equations. We use the intermittency factor γ_t:

$$\nu_{actual} = \gamma_t \cdot \nu_t + \nu \qquad (30)$$

The model used for transition in three-dimensional mixed convection is a modified form of the Abu-Ghannam and Shaw (1980) model, which enables the user to specify the start and end of the transition zone. The intermittency factor given by this model is:

$$\gamma_t = 1.0 - \exp\left(-a\eta_t^b\right) \qquad (31)$$

where a and b are constants chosen to be 5.0 and 2.0, respectively. η_t is given by:

$$\eta_t = \frac{x - x_s}{x_e - x_s} \qquad (32)$$

where x is the distance in the free-stream direction, x_s is the prescribed start of transition zone and x_e is the prescribed end of transition.

6 Results

In this section a few calculations of natural convection boundary layers with large temperature differences, and few sample calculations of three-dimensional mixed convection boundary layers are given.

6.1 Results for Natural Convection

Figure 3 shows a comparison of experimental data of Siebers et al. (1985) to heat transfer predictions of the proposed model and two other models. This is a case of heat transfer from a vertical plate heated to $750\ K$, with the ambient temperature of $298\ K$. The Grashof number at the end of the plate is 2.0×10^{12}. As can be seen, there are considerable differences in the predictions of the existing models and the proposed model. An important difference is that our model correctly predicts the heat transfer coefficient to be independent of streamwise distance in the turbulent region.

Figure 4 shows the prediction of the velocity profiles for a moderate temperature case using the new model and its comparison to the experimental data of Cheesewright (1982); the data are at 2.2 meters up a vertical plate with a constant wall temperature of $380\ K$. The two other predictions are computations using the mixing-length model of Kays used in the boundary layer code STAN5 (Kays and Crawford, (1976)) and the eddy viscosity model of Cebeci and Khattab (1975). These two are chosen as representative models for comparison due to their widespread use and popularity. All three models predict the inner region very well; the new model does better in outer region and at the velocity peak.

6.2 Results for Mixed Convection

The rest of the figures show calculation of three-dimensional mixed convection boundary layers. These results are compared to the experimental results of Siebers et al. (1983), the only available experimental data for high temperature mixed convection flows. The predictions are presented in two parts. In the first part, the heat transfer from the plate is compared to experimental data. In the second part, predictions of velocity and temperature profiles are presented.

The two independent parameters describing mixed convection flow are the free stream velocity and the temperature difference between the wall and the free stream, which are nondimensionalized into the Reynolds and Grashof numbers, where they are defined as follows:

$$Re = \frac{u_\infty L}{\nu}; \qquad Gr = \frac{g\beta(T_w - T_\infty)L^3}{\nu^2} \qquad (33)$$

In these definitions the length scale L is taken to be $3\ m$ to be consistent with Siebers' experimental data. The relative importance of free and forced convection is indicated by the ratio Gr/Re^2.

6.2.1 Heat Transfer

The three-dimensional computer code for mixed convection is written so that either the wall temperature or the heat flux at the wall can be provided as boundary conditions. In the calculations presented here, experimental wall temperature data were used as input to the program. This choice eliminates the need to account for heat transfer due to radiation.

Figure 5 shows a plot of computed heat transfer coefficient as a function of x and z. This figure is for free stream velocity of $4.4\ m/sec$ and wall temperature of $496\ K$, a high velocity, moderate temperature case. The free stream temperature for all the experimental cases is around $300\ K$; the exact values are given on each figure.

In this plot, the upper edge of the graph corresponds to the lower edge of the plate at $z = 0$. Initial conditions are given along the lines $x = 0$ and $z = 0$. There is a laminar region near the leading edge of the plate; it is followed by a transition zone in which the heat transfer coefficient increases. Fully turbulent flow starts near where h reaches its maximum and continues until the end of the plate. Most of the flow is turbulent. There is very little variation in the heat transfer coefficient in the gravity direction (z), consistent with Siebers' experiment. Independence of h from z is not unexpected, since in turbulent natural convection, the heat transfer coefficient becomes independent of distance from the edge. The constancy of h is another indication that the flow is fully turbulent even on the lower part. Note that heat transfer coefficient being independent of z does not mean that the flow is two-dimensional. However, it makes possible comparison of computational results and experimental data using two dimensional plots. In the following figures the heat transfer coefficient is plotted versus x, the distance in the forced convection direction for a constant vertical height of $z = 1.0\ m$ in the free convection direction.

Figures 6 shows a plot of the heat transfer coefficient for a moderate free stream velocity of $4.4\ m/sec$ and wall temperature of $323\ K$. As can be seen from the figure, the predicted heat transfer coefficient is in agreement with the experimental data.

Figure 7 shows four cases with the same free stream velocity of $1.5\ m/sec$ but different wall temperatures. The plots in Figure 7, represent comparisons of predictions to experiment for cases with wall temperatures of $335\ K$, $497\ K$, $690\ K$, $830\ K$, respectively. A point to notice is that although all four cases have the same Reynolds number, the average heat transfer coefficient increases with increasing wall temperature, by almost 50% from the lowest to the highest temperature case. The agreement is very good for all cases, but it seems that for the case with the lowest wall temperature (ID-355) the prediction has a weaker dependency on x than the experimental data. This difference in trends is easier to see if we compare all the plots in this figure with Figure 8, which will be explained next. The point to notice is that the x-dependency in the experimental data seems to disappear for values of Gr/Re^2 of approximately higher than 3.0, where in our predictions h becomes independent of x for Gr/Re^2 of approximately higher than 2.5. The heat transfer coefficient is expected to become independent of x when natural convection is dominant.

Figure 8, shows predictions and experimental data for Siebers' high velocity cases. In this figure, plots of the heat transfer coefficient for the mixed convection flows with free stream velocity $6.2\ m/sec$ and wall temperatures of $333\ K$, $507\ K$, $633\ K$ and $760\ K$ are given. In the two high temperature cases (ID-543 and ID-566), the boundary layer was tripped 0.23 meters downstream of the vertical leading edge therefore, there is no laminar region in these cases. As our computation assumes a laminar region just downstream of the vertical leading edge, a short transition zone beginning at 0.1 meters downstream of the leading edge was included. This explains the difference in the shape of the four curves in Figure 9. Also in the low temperature case (ID-400) the prediction is off for the first data point. In light of fairly good agreement for most of the other first points of different cases, this might be explained by some irregularity in the initial conditions.

An important observation to be made from the plots in Figure 8 is that average turbulent heat transfer coefficient decreases with increasing wall temperature. This is opposite to the trend in Figure 7, which covers the same range of wall temperatures at lower free stream velocity. For free stream velocity 6.2 m/sec, the average heat transfer coefficient drops by about 20.0% as the wall temperature increases by $420K$.

The experiment of Siebers contains further cases. Most of these cases have been computed but are not presented here. The cases presented were chosen to represent the complete range of temperature and free stream velocity to demonstrate that the model predicts all of the essential trends in mixed convection flows. The predictions for the remaining cases are as good as those which have been shown. We conclude that the turbulence model reproduces all of the heat transfer data for mixed convection boundary layers.

6.2.2 Velocity and Temperature Fields

In this section a few sample predictions of velocity and temperature fields are presented and compared to experimental data of Siebers (1983). The velocity field in mixed convection can be highly three dimensional. Figure 9 shows a three-dimensional plot of a typical velocity field in the turbulent region of a mixed convection flow with free stream velocity 3.5 m/sec and wall temperature of 690 K (Case ID-626). As can be seen, the profile is highly three dimensional and the velocity vector rotates considerably in the lower 10% to 15% of the boundary layer.

Figures 10 and 11 show the velocity and temperature fields for a case with free stream velocity of 4.5 m/sec and the wall temperature 830 K. This is the case with highest wall temperature and free stream velocity in Siebers experiment. These two plots are for $x = 0.19\ m$ and at $z = 2.3\ m$. Predictions of profiles compare very well to the experimental data.

Figures 12 and 13 are plots of velocity profiles for a case with free stream velocity of $4.3 m/sec$ and nominally same wall temperature of 620 K. These two plots represent profiles for two locations on the plate at the same x coordinate, $x = 2.7\ m$, but at $z = 0.5\ m$ and $z = 1.8\ m$, respectively. An interesting point is that both the computation and the experimental data show that the maximum of the vertical velocity component is hardly different for these two cases. These two figures represent stations separated more than 1.0 meter in the vertical (natural convection) direction. For pure natural convection and the temperature difference of more that 300 degrees across the boundary layer, there would have been a significant increase in the maximum velocity. The horizontal velocity component profile is also similar at these two locations. From these observations, it can be concluded that flow is almost independent of the distance from the lower edge of the plate (z-direction). This is also demonstrated by the three-dimensional plot of heat transfer rate in Fig. 5.

Our simulations substantiate Siebers' conclusion that the flow state at a given x, z location on the plate is independent of its location on the plate and depends only on the distance that a fluid particle has traveled over the plate, i.e., on the length of the streamline from its first intersection with the plate.

Figures 14 is for a high temperature, low velocity flow condition, one of the limiting cases of the Siebers velocity data set. This profiles are given at $x = 2.7\ m$ and $z = 1.0\ m$. In this case, the w component of the velocity is considerably larger than the previous case and the two velocity components have the same order of magnitude. Predictions for this case, the one with the strongest three-dimensional effects, is very good except that the maximum velocity for the w component seems to be slightly under predicted.

7 Conclusions

In the attempt to construct a model for prediction of three-dimensional mixed convection flows, it became apparent that existing models for two-dimensional high Grashof number natural convection boundary layers are inadequate. Thus, a new mixing length model for two-dimensional natural convection was constructed and validated for Grashof numbers up to 10^{13}.

Existing three-dimensional eddy viscosity turbulence models developed for pressure driven flows do not work well for high temperature mixed convection flows. Thus a new turbulence model had to be constructed for three-dimensional mixed convection. It combines the forced and natural convection turbulence models and, in the appropriate limits, reduces to the models for the corresponding two-dimensional flows. The three-dimensional model was tested against the experimental data of Siebers. The predictions compare very well for the entire range of Siebers experiment.

Choosing correct edge initial conditions was found to be very important for both the smoothness and the quality of the solutions. An iterative scheme for upgrading initial conditions was suggested and successfully implemented. A modification to the Keller box scheme which removes the numerical oscillations inherent in the original method was proposed and implemented.

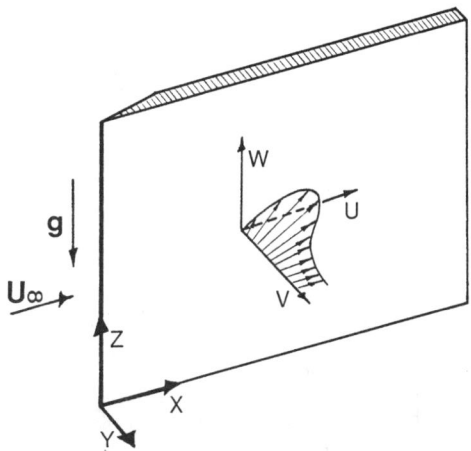

Figure 1   Coordinate System For a Three-Dimensional Mixed Convection Boundary Layer on a Flat Plate.

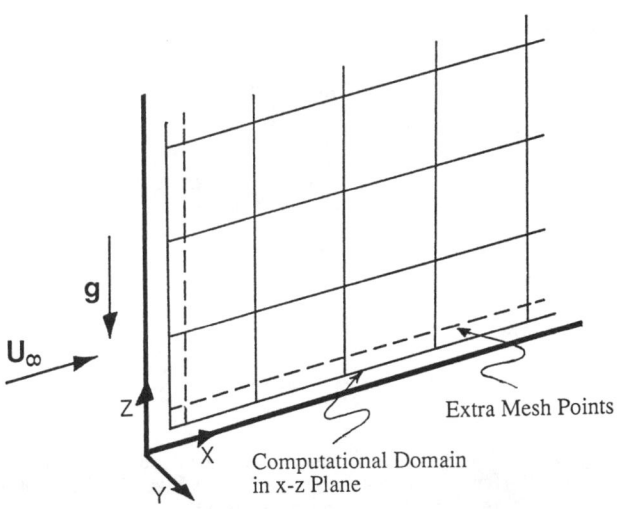

Figure 2   Location of The Extra Grid Points for Generation of Initial Conditions.

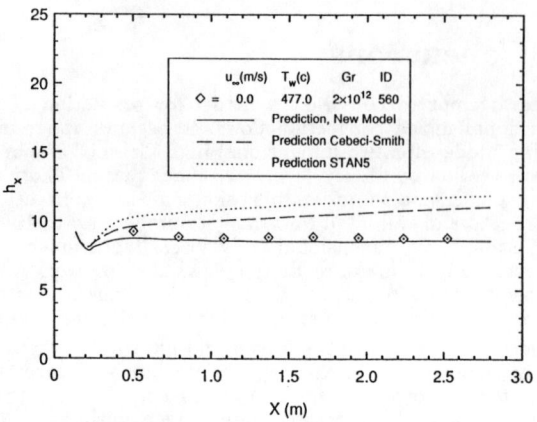

Figure 3 Comparison of Heat Transfer Coefficient Computation with Three Different Turbulence Models For High Temperature Natural Convection Data of Siebers et al.

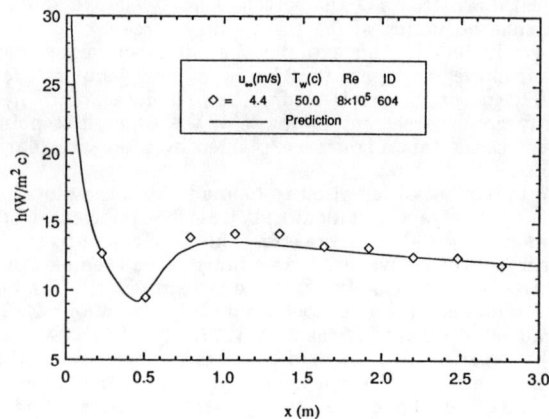

Figure 6 Comparison of Variation of Heat Transfer Coefficient for Forced Convection Over a Flat Plate to Siebers Data.

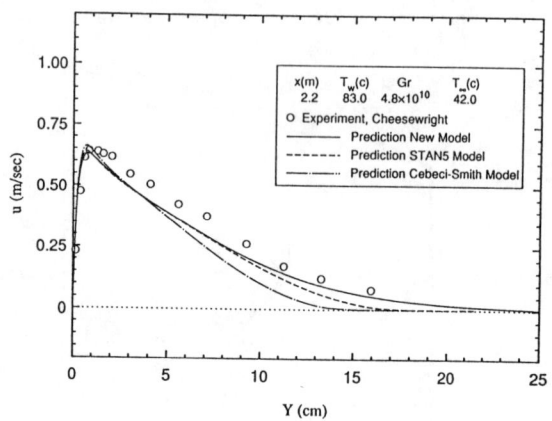

Figure 4 Comparison of Velocity Profile Computation for Moderate Temperature Natural Convection to Data of Cheesewright.

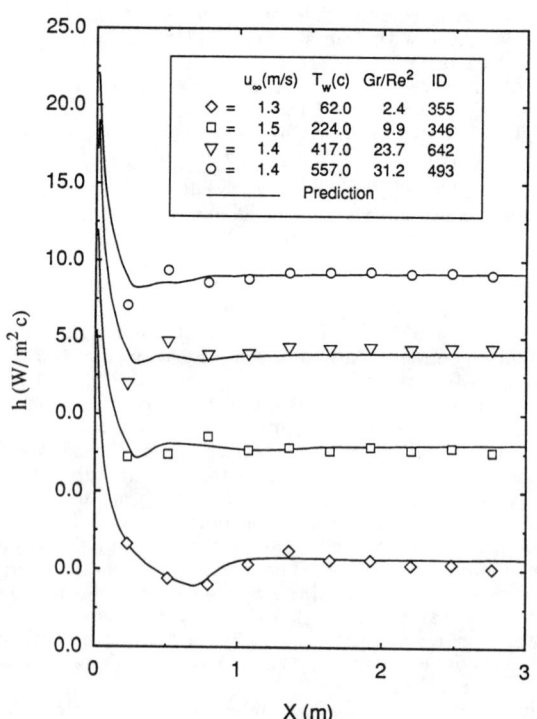

Figure 7 Comparison of Computation of Heat Transfer Coefficients to Siebers Experiment for Four Different Cases of Constant $z = 1.0m$, and Nominally Constant Free-Stream Velocity of 1.5 m/sec.

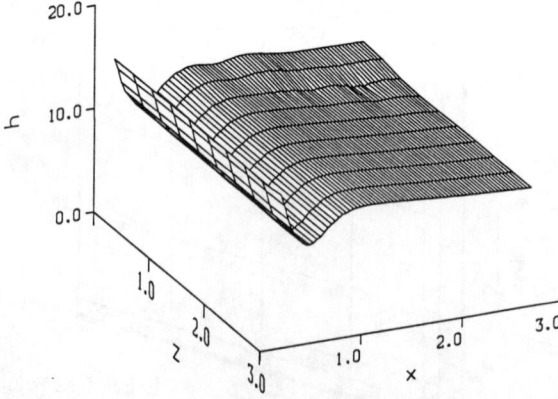

Figure 5 Computational Results for Variation of Heat Transfer Coefficient Over the Entire Plate for $U_\infty = 4.4$ m/sec, $T_w = 496.0K$, (ID-611).

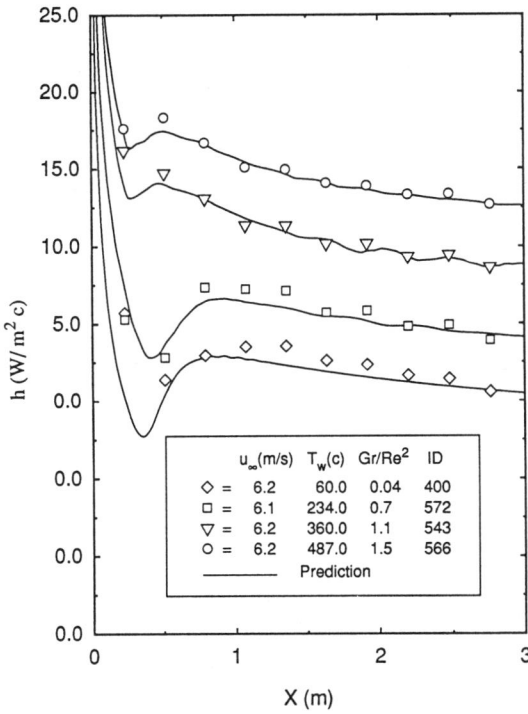

Figure 8   Comparison of Computation of Heat Transfer Coefficients to Siebers Experiment for Four Different
Cases of Constant $z = 1.0m$, and Nominally Constant Free-Stream Velocity of 6.3 m/sec.

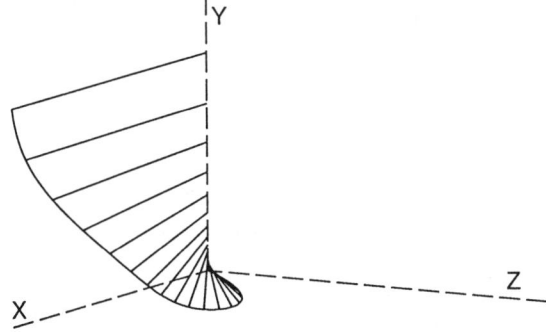

Figure 9   A Computed Three-Dimensional Velocity Profile
for Mixed Convection Flow, $U_\infty = 3.5\ m/sec$,
$T_W = 690.0K$.

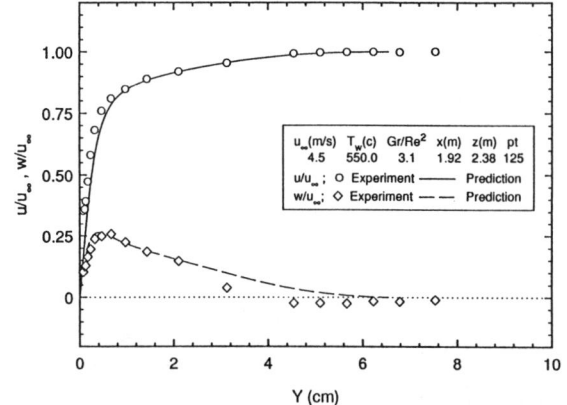

Figure 10   Comparison of Prediction of the Velocity Field to
Data of Siebers for Turbulent Mixed Convection
Flow, ID-648.

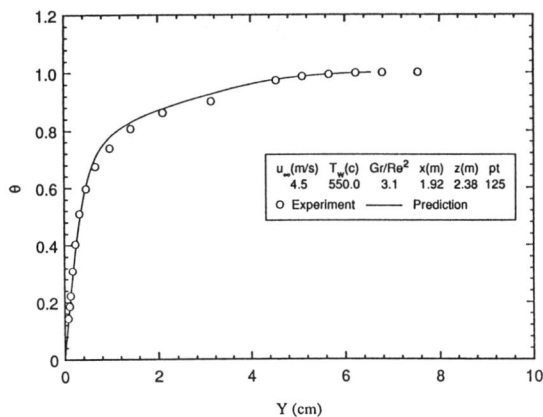

Figure 11   Comparison of Computation of Temperature Field
to Data of Siebers for Turbulent Mixed Convection
Flow, ID-648.

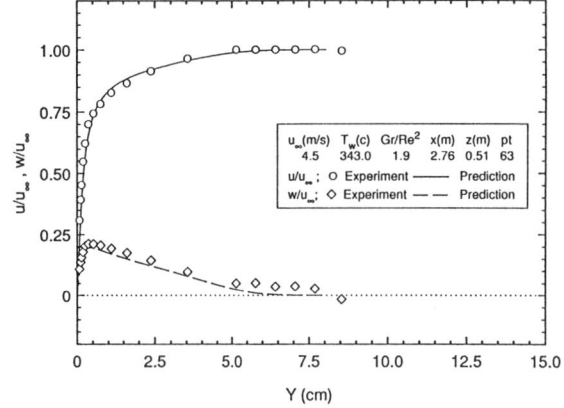

Figure 12   Comparison of Prediction of the Velocity Field to
Data of Siebers for Turbulent Mixed Convection
Flow, at the End of the Plate, ID-620.

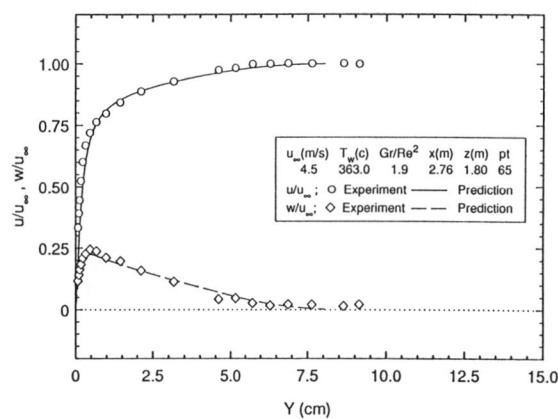

Figure 13   Comparison of Prediction of the Velocity Field to
Data of Siebers for Turbulent Mixed Convection
Flow, at the End of the Plate, ID-620.

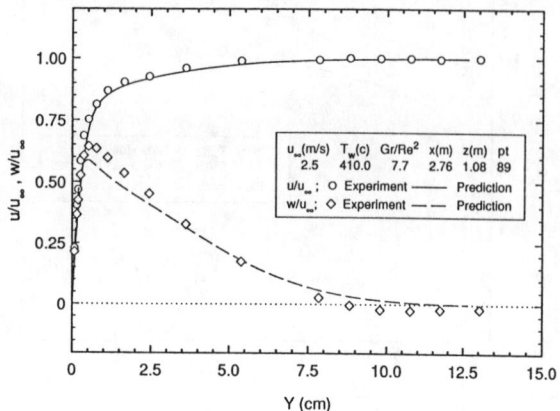

u_∞(m/s)	T_w(c)	Gr/Re^2	x(m)	z(m)	pt
2.5	410.0	7.7	2.76	1.08	89

u/u_∞ ; ○ Experiment —— Prediction
w/u_∞; ◇ Experiment - - - Prediction

Figure 14   Comparison of Prediction of the Velocity Field to Data of Siebers for High Temperature, Low Velocity Turbulent Mixed Convection Flow, ID-631.

References

[1] Abu-Ghannam, B.J. and R. Shaw "Natural Transition of Boundary Layer: The Effects of Turbulence, Pressure Gradient and Flow History," J. Mech. Eng. Sc., Vol.22, No.5, 1980.

[2] Afshari, B. and J.H. Ferziger, "Computation of Orthogonal Mixed-Convection Heat Transfer," Proceedings of ASME-JSME Thermal Engineering Joint Conference, Vol. III, Honolulu, Hawaii, March 20-24, 1983.

[3] Afshari, B., "Computation of Turbulent Three-Dimensional Mixed Convection Boundary Layers ," Ph.D. dissertation, Department of Mechanical Engineering, Stanford University, Stanford, CA, 1989.

[4] Blottner, F.G., "Variable Grid Scheme Applied to Turbulent Boundary Layers," in: *"Computer Methods in Applied Mechanics and Engineering,"* 4, pp.179-194, North-Holland Publishing Co., 1974.

[5] Bradshaw, P.,T. Cebeci and J. Whitelaw, *"Engineering Calculation Methods for Turbulent Flow,"* Academic Press, New York, 1981.

[6] Cebeci, T. and A.M.O. Smith, *"Analysis of Turbulent Boundary Layers,"* Academic Press, N.Y., 1974.

[7] Cebeci, T., and A. Khattab, "Prediction of Turbulent-Free-Convective Heat Transfer From a Vertical Flat Plate." ASME J. of Heat Transfer, Vol. 97, No. 4, p.469, 1975.

[8] Cheesewright, R. and E. Ierokipiotis, "Velocity Measurements In a Turbulent Natural Convection Boundary Layer.", Proceedings of the 7th Int. Heat Transfer Conference, Vol. 2, pp 305-309, Munich, 1982.

[9] Cheesewright, R. and A. Dastbaz, "The Structure of Turbulence in a Natural Convection Boundary Layer," The Fourth Symposium on Turbulent Shear Flows, p. 17.25, Karlsruhe, 1983.

[10] Eichhorn, R. and M.M. Hassan, "Mixed Convection About a Vertical Surface in Cross-Flow: A Similarity Solution.," ASME J. Heat Transfer, vol. 102, 1980.

[11] Evans, G.H. ,"A Numerical Study of Three-Dimensional, Laminar, Mixed Convection, "Ph.D. Thesis, Washington State University, 1981.

[12] George, W.K. Jr. and S.P. Capp,"A Theory for Natural Convection Turbulent Boundary Layers Next to Heated Vertical Surfaces.," Int. J. Heat Mass Transfer, vol. 22, pp. 813-826, 1979.

[13] Kays,W.M. and R.J. Moffat, "The Behavior of Transpired Turbulent Boundary Layers," *Studies in Convection*, Vol.1: *Theory, Measurment and Application*, Academic Press, London, 1975.

[14] Keller, H.B. and T. Cebeci, "Accurate Numerical Methods for Boundary Layer Flows," Lecture Notes in Physics, Proc. of Second Int. Conf. on Numerical Methods in Fluid Dynamics, vol.8, Springer-Verlger, 1971.

[15] Plumb, O.A., "The Effect of Crossflow on Natural Convection from Vertical heated Surfaces," ASME Publication 80-HT-71, 1980.

[16] Plumb, O.A. and G.H. Evans, "Turbulent Mixed Convection From a Vertical Heated Surface in a Crossflow," Proceedings of ASME-JSME Thermal Engineering Joint Conference, Vol. III, pp. 47-53, Honolulu, Hawaii, March 20-24, 1983.

[17] Purtell, L.P., P.S. Klebanoff and F.T. Buckley, "Turbulent Boundary Layers at Low Reynolds Numbers," Phys. Fluids, vol. 24, No. 5, p. 802, 1981.

[18] Rotta, J.C.,"A Family of Turbulence Models for 3-D Boundary Layers," Proc. of First Turbulent Shear Flow Conference, Editors: F.Durst, B.E.Launder, F.W. Schmidt, J.H. Whitelaw, University Park, Pennsylvania, 1977.

[19] Schneider, G.R.,"Calculation of the Turbulent Boundary Layer on an Infinite Swept Wing Using a Three-Dimensional Mixing Length Model," European Space Agency Technical Translation, ESA-TT-534, 1979.

[20] Siebers, D., "Experimental Mixed Convection Heat Transfer From a Large, Vertical Surface in a Horizontal Flow," Ph.D. dissertation, Department of Mechanical Engineering, Stanford University, Stanford, CA, 1983.

[21] Siebers, D., R.F. Moffat and R.J. Schwind "Experimental, Variable Properties Natural Convection From a Large, Vertical, Flat Surface," ASME J. Heat Transfer, Vol. 107, p. 124, February 1985.

[22] Yao, L.S. and F.M. Chen,"Analysis of Convective Heat Loss From the Receiver of Solar Power Plants," ASME Publication 79-WA/HT-36.

[23] Young, R.J. and K.T. Yang,"Effect of Small Crossflow and Surface Temperature variation on Laminar Free Convection Along a Vertical Plate," J. of Appl. Mech., Paper 62-WA-81, 1962.

1989 National Heat Transfer Conference
HTD-Vol. 107, Heat Transfer in Convective Flows

EVALUATION OF k-ε TURBULENCE MODEL FOR PREDICTING BUOYANT FREE ROUND JETS

D. F. G. Durão, J. C. F. Pereira and J. M. P. Rocha
Mechanical Engineering Department
Instituto Superior Técnico
Lisboa, Portugal

ABSTRACT

This paper describes the application of a k-ε-$<t^2>$ bouyancy-extended turbulence model to the computation of a buoyant flow that arises from the discharge of an axismmetric jet into a still uniform ambient. Improved forms of eddy-viscosity/eddy-diffusivity concepts were tested and compared with standard formulae. In addition, a transport equation for the dissipation rate of scalar-variance was incorporated into the model. The performance of the model was analysed by means of a detailed comparison between experimental data. An elliptic calculation procedure was used to solve the governing equations. Computation results show that the flow mean behavior is accurately reproduced. Some important discrepancies remain in the turbulance description, mainly in the axial transport which is strongly underpredicted.

NOMENCLATURE

b   jet integral length-scale
C's   model constants
F   densimetric Froude number
e   exponential base (2.718)
g   gravitational constant
k   turbulent kinetic energy
L_M $= M_0^{3/4}/F_0^{1/2}$
M   momentum flux
P   pressure mean-value, production of turbulent kinetic energy
P_t production of temperature-variance
$R_{1/2U}$ distance at which velocity attains one-half of maximum
r   radial coordinate
u   vertical velocity fluctuation
u_i velocity fluctuation in i-direction
U_i velocity mean-value in i-direction
$<u_it>$ turbulent heat flux in i-direction
$<ut>$ axial turbulent heat flux
v   radial velocity fluctuation
$<vt>$ radial turbulent heat flux
T   temperature mean-value
t   temperature fluctuation

$<t^2>$ temperature-variance
x_i spatial coordinate in i-direction
x   axial coordinate

Greek Symbols

Λ   anisotropic model parameter (defined in the text)
β   coefficient of volumetric expansion
δ_{ij} Kronecker delta
Γ   molecular diffusivity
κ_e isotropic eddy-diffusivity (defined in the text)
κ_{eA} anisotropic eddy-diffusivity (defined in the text)
ε   dissipation rate of turbulent kinetic energy
ε_t dissipation rate of temperature-variance
μ   molecular viscosity
ν_t kinematic turbulent viscosity
Π_{ij} anisotropic eddy-diffusivity term (defined in the text)
σ_k Prandtl number for k
σ_ε Prandtl number for ε
σ_t turbulent Prandtl number
τ's   anisotropic model constants

Special Symbols

<>   mean value

Subscripts

a   ambient conditions
i,j   spatial coordinate indexes, jet regime
m   maximum value
0   source conditions
t   turbulent quantity
U   axial velocity
T   temperature
p   plume regime

1. Introduction

The study of turbulent buoyant jets is of interest for a wide range of engineering applications with special relevance to atmospheric pollutant dispersion as the rise and spreading of discharged fluid from smoke stacks, cooling towers and fires. The ability to predict the turbulent transport of scalar quantities such as heat is of great importance for the design of such discharge devices and for the study and control of environmental flow phenomena.

Different models have been proposed for predicting turbulent flows with scalar dispersion [1,2]. Despite the development of second-order closure models, the k-ε model of turbulence still remains a valuable engineering tool, suitable to be applied to a variety of complex flow geometries. Nevertheless, one of the major k-ε's shortcomings is the isotropic eddy-viscosity hypothesis, which in many flow situations cannot be held to represent the Reynolds-stress tensor [3]. In addition, in scalar dispersion flow computations the diagonal form of the diffusivity tensor was shown to be not appropriate to other than isotropic turbulence [4], being inadequate for general modelling of complex flows [2].

The applicability and relevance of the eddy-viscosity concept is discussed by Pope [3] and Speziale [5]. These authors have presented improved eddy-viscosity representations of turbulent momentum fluxes, in which the whole mean velocity-gradient tensor affects the Reynolds-stresses. In a different approach, Yoshizawa [6] using Direct Interaction Approximation (DIA), together with a scale-parameter expansion method has formulated a two scale expansion of Navier-Stokes equations in which the interaction between scales becomes explicit (TSDIA). It was derived the asymptotic expansion form of the terms that must be modelled in one-point closures, such as the diffusive transport of turbulent kinetic energy k and its dissipation rate ε, and with special interest for the present work, anisotropic representations for eddy-viscosity. The new constitutive relations for $<u_i u_j>$ were applied to channel and duct flows [7], leading to the capture of the experimentally observed anisotropy of turbulent momentum fluxes.

It has been shown [4] that the diffusivity coefficient must be in its general form an asymmetric second-rank tensor depending on the turbulence time-scales and mean-field gradients [8]. Several numerical studies have used Daly and Harlow's [9] proposal for the diffusivity tensor. Recently, Yoshizawa [10] presented a new formulation for this quantity in which was included the dependency of mechanical and scalar turbulent time-scales and mean-field gradients.

For scalar dispersion flows, an obstacle to the prediction of turbulent scalar-variance (like temperature fluctuations) remains in the evaluation of the dissipation terms in the respective transport equation. An almost generalized approach is to assume that the mechanical and scalar time-scale ratio, defined as R = 1/2 $(<t^2>/\varepsilon_t)/(k/\varepsilon)$, is a known constant [2], usually taken equal to 0.5. In order to discard the constancy in the proportionality coefficient of scalar and mechanical time-scales, Launder *et al.* [11] suggest to introduce a dependency of R on a scalar-flux parameter. Nevertheless, this formulation cannot account for different production mechanisms of $<t^2>$ and k which may occur in many practical flows. Several computational studies of atmospheric shear layers have employed modelled versions of an ε_t transport equation [12, 13]. Newman *et al.* [14] devised a more general form for the equation which was extended later by Elghobashi and Launder [15]. All the subsequent reported tests were performed aiming to calibrate the model constants in a variety of simple shear flows.

The objective of the present paper is to compare the performance of the k-ε-$<t^2>$ turbulence model in the prediction of the mean-field and turbulence development of a vertical buoyant jet issuing into a still ambient. In a number of practical applications there is a considerable interest in studying the flow-source near region, where

initial conditions and geometry dictate the flow evolution, as well as the far-field where the behavior of the flow can display markedly different features by means of buoyancy increasing influence. Experiments of Corrsin and Uberoi [16] were chosed to test the models' performance in the near exit, non-buoyant region, while recent measurements in buoyancy-driven flow zone of Shabbir and George [17] were used to compare with predictions of flow's far-field region. The standard model comprises transport equations for the turbulent kinetic energy, its dissipation rate and temperature-variance. An extension is provided by the inclusion of a transport equation for ε_t. The eddy-viscosity/eddy-diffusivity's improved expressions proposed in [6, 10] are included and compared with standard formulae (see [1]).

The next section of the paper provides a brief description of the physical and mathematical model. This is followed by the presentation of the numerical model in section 3. Comparison between predictions and referred experimental data is given in section 4 and the conclusions and final remarks are provided in the last section.

2. The Physical and Mathematical Model

2.1 Mean flow equations

The instantaneous velocity field in a turbulent flow is described by the Navier-Stokes equations. In its time-averaged form, obtained by Reynolds decomposition of instantaneous quantities in the equation into the sum of mean and fluctuating parts and averaging, the result reads, for a statistically stationary flow:

Continuity

$$\frac{\partial}{\partial x_j}\left(\rho\, U_j\right) = 0, \tag{1}$$

Momentum

$$\frac{\partial}{\partial x_j}\left(\rho\, U_j\, U_i\right) = -\frac{\partial P}{\partial x_i} + \frac{\partial}{\partial x_j}\left(\mu\left(\frac{\partial U_i}{\partial x_j} + \frac{\partial U_j}{\partial x_i}\right)\right)$$
$$- \frac{\partial}{\partial x_j}\left(\rho <u_i u_j>\right) + \left(\rho - \rho_a\right)g_i. \tag{2}$$

Similary, an energy equation (that stands also for some scalar property) can be written as:

energy

$$\frac{\partial}{\partial x_j}\left(\rho U_j T\right) = \Gamma\frac{\partial T}{\partial x_j \partial x_j} - \frac{\partial}{\partial x_j}\left(\rho <u_j t>\right). \tag{3}$$

It is assumed that only small variations in density occur, due to temperature differences, and the Boussinesq approximation prevails.

2.2 The turbulence models

The closure of the above system of equations requires the introduction of additional expressions for the unknown turbulent quantities, the Reynolds-stress $<u_i u_j>$ second-rank tensor and the turbulent heat flux $<u_i t>$ vector. The constitutive relation for the turbulent momentum fluxes is given, in the k-ε modelling framework, by:

$$-<u_i u_j> = \nu_t\left(\frac{\partial U_i}{\partial x_j} + \frac{\partial U_j}{\partial x_i}\right) - \frac{2}{3}\delta_{ij}k + \Pi_{ij}. \tag{4}$$

The first two terms in the RHS of (4) are the usual isotropic eddy-viscosity relation, while the rightmost term represents the induced anisotropy effects of mean-velocity strain rates on the Reynolds stresses proposed by Yoshizawa [6] and given by:

$$\Pi_{ij} = \frac{1}{3}\left(\sum_{m=1}^{3}\tau_m S_{mij}\delta_{ij} + \Lambda_{ij}\right), \tag{5}$$

$$\tau_m = C_{\tau m} T_M^2, \tag{6}$$

$$S_{1ij} = \frac{\partial U_i}{\partial x_k}\frac{\partial U_j}{\partial x_k}, \tag{7}$$

$$S_{2ij} = \frac{1}{2}\left(\frac{\partial U_i}{\partial x_k}\frac{\partial U_k}{\partial x_j} + \frac{\partial U_j}{\partial x_k}\frac{\partial U_k}{\partial x_j}\right), \tag{8}$$

$$S_{3ij} = \frac{\partial U_k}{\partial x_i}\frac{\partial U_k}{\partial x_j}, \tag{9}$$

$$\Lambda_{ij} = -\sum_{m=1}^{3}\tau_m S_{mij}, \tag{10}$$

where $T_M = k/\varepsilon$ represents a characteristic time-scale of mechanical turbulent field. From Yoshizawa´s analysis, it was only retained the expansion resulting terms until first order of the scale parameter (see [6] for details).

The gradient transport hypothesis is based on the assumption that turbulent heat flux, $<u_i t>$ and the mean gradient are proportional,

$$-<u_i t> = \kappa_{ij}\frac{\partial T}{\partial x_j}, \tag{11}$$

where κ_{ij} represents the diffusive coefficient tensor. In the standard k-ε model

$$\kappa_{ij} = \kappa\,\delta_{ij}, \tag{12}$$

is diagonal constant, being the scalar diffusivity coefficient $\kappa_t = \nu_t/\sigma_t$. Algebraic parametric functions for σ_t and c_μ derived from simplification of turbulent fluxes transport equations were proposed by Hossain and Rodi [18], introducing in this way some dependency of the turbulent Prandtl number σ_t and c_μ constant calculation on mean-fields strains as

$$C_\mu = \frac{2}{3}(1 - C_c)\,C_c\left(1 + \frac{1}{3}T_M\,\beta\,g\,\frac{\partial T/\partial r}{\partial U/\partial r}\right), \tag{13}$$

$$\sigma_t = 3\,C_c\left(1 + \frac{1}{3}T_M\,\beta\,g\,\frac{\partial T/\partial r}{\partial U/\partial r}\right), \tag{14}$$

where Cc is a constant, which value is given in table 1.

Daly and Harlow [9] have suggested a non-diagonal form for κ_{ij} as

$$\kappa_{ij} = C_t T_M <u_i u_j>. \tag{15}$$

A more elaborate proposal is suggested by Yoshizawa [10] in which the diffusive tensor is given by the sum of two contributions as

$$\kappa_{ij} = \kappa'_e\,\delta_{ij} + (\kappa_{cA})_{ij}, \tag{16}$$

the first term in RHS of (16) represents the isotropic contribution to eddy-diffusivity,

$$\kappa'_e = C_k T_S^2\,\varepsilon, \tag{17}$$

where C_k is a model constant and $T_S = <t^2>/\varepsilon_t$ is the scalar time-scale. The second term is, in its general form, an asymmetric second-rank tensor, which includes the effects of mean-strains as

$$(\kappa_{eA})_{ij} = -T_S^3\,\varepsilon\left(C_{\kappa A}(\frac{\partial U_i}{\partial x_{kj}} + \frac{\partial U_{kj}}{\partial x_i}) \right.$$
$$\left. + C'_{\kappa A}(\frac{\partial U_i}{\partial x_{kj}} + \frac{\partial U_{kj}}{\partial x_i})\right), \tag{18}$$

where $C_{\kappa A}$ and $C'_{\kappa A}$ are model constants, given by table 2.

Together with the usual buoyancy extended equations for turbulent kinetic energy and its dissipation rate (see, e. g. [19]), the present model embodies an equation for the transport of temperature-variance in the form

$$\frac{\partial}{\partial x_j}\left(\rho U_j <t^2>\right) = \frac{\partial}{\partial x_j}\left(c_t T_M k\frac{\partial}{\partial x_j}<t^2>\right) + 2\rho P_t - 2\rho\varepsilon_t. \tag{18}$$

The main difficulty in closing equation (18) is the evaluation of dissipation rate of scalar-variance. A common approach of nearly all previous computational studies (e.g. [18, 19, 20]) is from definition of mechanical and scalar time-scale ratio R to obtain ε_t in terms of the other quantities, assuming R as a known constant. Launder et al. [11] suggest to discard the constancy of R by means of

$$R = (2.5 + 2\Pi_t)^{-1}, \tag{20}$$

where $\Pi_t = <u_i t>^2/k <t^2>$ is the invariant of the heat flux ´anisotropy´ tensor. A more general approach is yielded by considering a transport equation for ε_t given in its modelled form by [14]:

$$\frac{\partial}{\partial x_j}\left(\rho U_j\varepsilon_t\right) = \frac{\partial}{\partial x_j}\left(c_{\varepsilon t}\rho\,T_M\,k\frac{\partial\varepsilon_t}{\partial x_j}\right) + c_{\varepsilon t1}\rho\frac{\varepsilon_t}{<t^2>}P_t + c_{\varepsilon t2}\rho\frac{\varepsilon_t}{k}P$$
$$- C_{\varepsilon t3}\rho\frac{\varepsilon\varepsilon_t}{k} - C_{\varepsilon t4}\rho\frac{\varepsilon_t^2}{<t^2>}, \tag{21}$$

and the model constants $c_{\varepsilon t1}\rho$, $c_{\varepsilon t2}\rho$, $C_{\varepsilon t3}\rho$ and $C_{\varepsilon t4}\rho$ are given in table 3.

3. Numerical Solution Procedure

The numerical procedure is based on a finite-volume discretization of the governing equations, employing a staggered grid for the mean-velocities relatively to scalar properties. Computations have been made with a modified version of an elliptic code TEACH [21], specially adapted for free flows, and in particular for free jets. An elliptic solution is obtained in a comparatively small extension of the computational domain (called box), with four and a half jet diameters in the axial direction. The output of the solution at each box is the inlet condition for the next box. Pressure was prescribed far from the jet boundary. A zero

101

second-derivative was used at the outlet of each box as the inlet condition. With this procedure, a free flow can be accurately calculated (there may be as many boxes as desired). When compared with the usual parabolic procedure, one has the advantage of solving the full elliptic form of the equations to find the strong coupling between vertical heat flux $<ut>$ and $<t^2>$ and most importantly, the second-moment correlations are accurately computed [22].

A numerical grid containing 20×16 nodes is used for each box. A total number of 10 boxes are calculated in order to obtain predictions up to $X/D=45$. The numerical accuracy of the results is about 8% on a 40×32 grid, as concluded from test runs with grids ranging from 20×16 to 64×64. Since the numerical errors are essentially the same for all the tested models, the conclusions of the present comparison are not clouded by numerical errors. Convergence was controlled by monitoring the absolute residual sum of momentum and continuity equations, normalized by the inlet mass and momentum flux. The iterative sequence was stopped after the residual sum to fall below 1%.

4. Results and Discussion

In this section, comparisons between the predicted and measured behavior of a hot and a buoyant axisymmetic jet are presented and discussed. The jet's predictions may not contain any new information. Nevertheless, they are included as a mean of validation of the present models in their extension to buoyant flows. The comparisons were made using the data of Corrsin and Uberoi [16] corresponding to source mean-temperature excess of 170° C, at the axial location $X/D = 15$. The predicted development of the buoyant jet in the plume regime is compared with Shabbir and George's [17] data. These experimenters have measured mean and turbulence quantities at axial locations above the dimensionless distance $x/L_M > 5$, which insures that the jet had already attained the plume regime (see [17]).

4.1. Mean-field predictions

The mean-field velocity and temperature computed and measured profiles for the jet and plume regimes are plotted in figures 1, 2 (a, b). The results of $k\varepsilon 1$, $k\varepsilon 4$ and $k\varepsilon 5$ models are almost identical and close follow the experimental profiles, except in the vincinity of plume's boundary, where the mean-temperature excess is overpredicted. The same discrepancy was observed in previous studies ([19, 23]), though not completely explained. In the present computations, the use of an elliptic procedure has probably amplified this model's feature, as also found in [23]. The flatter comparatively to velocity temperature predicted distribution is experimentally confirmed [16, 25], and reflects the preferential transport of heat over momentum; the same evidence is given by the measured [16] value of turbulent Prandtl number, approximately equal to 0.7 for the round jet. In the plume regime, the predicted and measured behavior indicates similar developments for the mechanical and thermal mean-fields, which implies a turbulent Prandtl number of about 1.

The standard model $k\varepsilon 1$ predicts a jet's spreading rate of 0.131, while Corrsin and Uberoi didnot reported the value of the spreading rate for $T_0 - T_a = 170^\circ$ C experiment. The measured value for mean-temperature excess equal to 15° C is $dR_{1/2u}/dx = 0.10$, while for $T_0 - T_a = 300^\circ$ C Corrsin and Uberoi reported a value of 0.133.

Some integral parameters of interest in both jets and plume regimes where computed and compared with reported experimental values. The mean-velocity and temperature length-scale, defined by the point of the

profile where the value $1/e$ times the maximum radial value, is predicted as $(b_u/x)_j = 0.117$, $(b_T/x)_j = 0.144$ for the jet and $(b_u/x)_p = 0.132$, $(b_T/x)_p = 0.132$ in the plume.

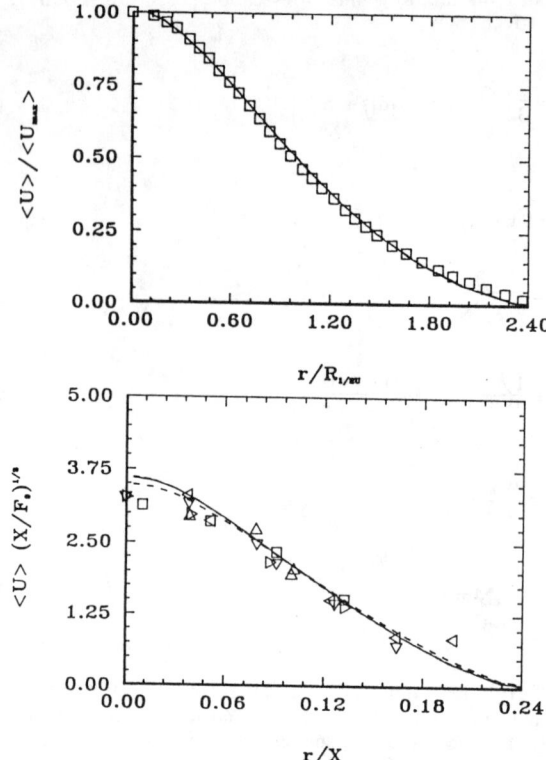

Figure 1. Mean axial velocity profiles (a) Heated jet; Symbols, experiment of Corrsin and Uberoi [16], □ $x/D = 15$; (b) Buoyant jet; Symbols, experiment of Shabbir and George [17], △ $X/D = 10$, ▷ $X/D = 14$, ▽ $X/D = 19$, ◁ $X/D = 21$, □ $X/D = 25$; Present predictions: ___ $k\varepsilon 1$, --- $k\varepsilon 4$, - - - $k\varepsilon 5$

Dimensional arguments imply that these parameters are constant in both jet and plume regime, which is confirmed experimentally [24, 25]. Shabbir and George [17] didnot present explicitly values for these quantities, although it can be inferred from the good agreement observed between measured and computed profiles that these scales have been accurately computed. Papanicolaou and List [24] reported $(b_u/x)_j = 0.108$, $(b_T/x)_j = 0.139$ for the jet regime and $(b_u/x)_p = 0.105$, $(b_T/x)_p = 0.110$ for the plume; these values are consistent with and close to present predictions. The observed wider temperature length scale of the jet relatively to plume flow may be attributed to a non-linear growth of the buoyant jet [24, 25].

Another related integral length-scale is the local jet-width parameter $Cp = (2\pi)^{1/2}(b_U/x)$. Dimensional reasoning leads to Cp to be constant in jets and plumes; the predicted values, $(Cp)_j = 0.27$ for the jet and $(Cp)_p = 0.26$ in the plume-like flow, are in close agreement with experimental value in [24], who have presented $Cp = 0.27$.

The Richardson number Ri in a buoyant jet provides an estimative of the local degree of plume-like behavior exhibited by the flow. Dimensional analysis shows that, for plume regime, there is an asymptotic limiting value for Ri of about 0.70, while for jets, Ri is proportional to axial distance from the source. The predicted asymptotic Ri is equal to 0.66, which is near to 0.716 measured by [24].

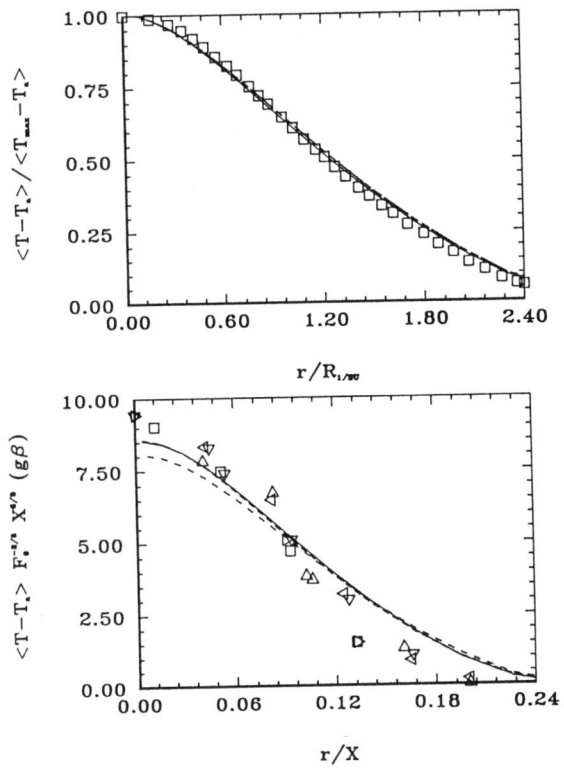

Figure 2. Mean temperature profiles (a) Heated jet; Symbols, experiment of Corrsin and Uberoi [16], □ x/D = 15; (b) Buoyant jet; Symbols, experiment of Shabbir and George [17], △ X/D = 10, ▷ X/D = 14, ▽ X/D = 19, ◁ X/D = 21, □ X/D = 25; Present predictions: ___ kε1, --- kε4, - - - kε5

4.2. Turbulence properties Predictions

The normalized axial radial Reynolds stress profiles are shown in figures 3 (a, b) and 4 (a,b) for the jet and the plume, respectively. The standard model kε1 is compared with kε4 and kε5. All the predicted profiles are similar. In the jet, the predicted axial and radial normal stress maximum is 0.27 and 0.25 at the axis, respectively, while the measured values are 0.17; Papanicolaou and List [24] reported 0.25 for axial stress and Chevray and Tutu [27] 0.23 and 0.19. In the plume region, the [17] reported turbulence intensities (shown in figures in a different scale) are 0.32 and 0.26, considerably greater than in the non-buoyant flow. This increase in turbulence intensities and in the anisotropy level is a result of additional buoyancy production, preferentially in axial direction. All the models failed to capture this important flow feature. Nevertheless, it is obvious that eddy-viscosity form (4) has no means to accomodate any direct buoyancy effects on turbulence. Nevertheless, an interesting features of both jets and plumes stress profiles is that they are similar in shape, taking in account the different mechanisms responsibly by turbulence production which govern each flow regime.

Figures 5 (a,b) show the predicted temperature-variance profiles in both jet and plume zones. The radial development of the quantity is found considerably different in each case. In the jet, the models compared (kε1, kε3, kε5 and kε6) exhibit markedly different behaviors. While model kε5 shows considerably discrepancies from the measured values, all the remaining models follows acceptably the experimental values; the model kε6, which

embodies a transport equation for ε_t, produces a more particular profile shape, thougth not very far from the experimental one. The kε1 predicted centerline value is 0.20

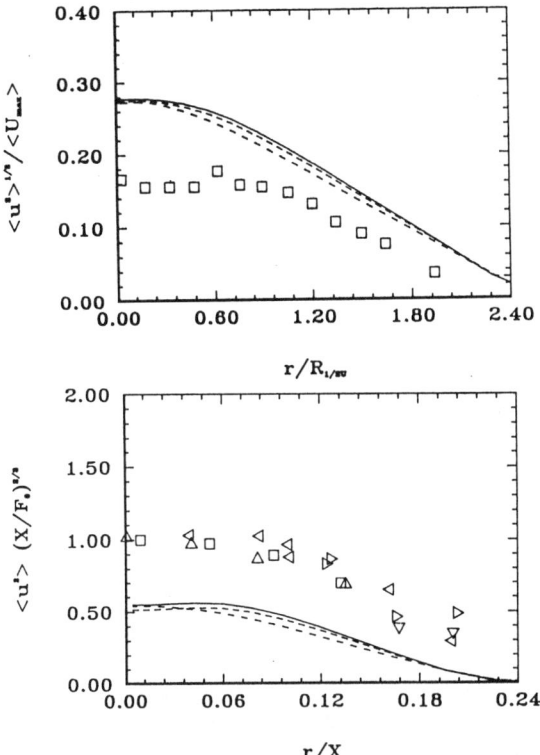

Figure 3. Turbulent axial normal Reynolds stress profiles (a) Heated jet; Symbols, experiment of Corrsin and Uberoi [16], □ x/D = 15; (b) Buoyant jet; Symbols, experiment of Shabbir and George [17], △ X/D = 10, ▷ X/D = 14, ▽ X/D = 19, ◁ X/D = 21, □ X/D = 25; Present predictions: ___ kε1, --- kε4, - - - kε5

with a peak of 0.21 at r/x = 0.08, which is consistent with the measurements and also with [24], who reported a value of 0.25 at r/x = 0.07. Figure 5 (b) evidenciates again that buoyancy work is an important agency contributing to augment turbulence intensity; kε6 predicted temperature-variance centerline level is 0.34, closer to measurements. Comparison between models kε5 and kε6 confirm that the transport equation for ε_t has improved considerably the predicted levels of $<t^2>$, as depicted in figure 5 (b). The usual hypothesis R = 0.5 (model kε1) and also its functional form (model kε3) show also a good agreement with measurements. The predicted levels of $<t^2>$ with kε2 (not shown here) were found too hight to have any physical meaning. As found also by [24, 25, 26], the peak observed in the jet tends to disapear in a plume-like flow $<t^2>$-profile, indicating that in the plume zone the controlling production process is buoyancy rather than shear. This may be a strong ground to the inadequacy of a constant value for R in general buoyant flow modelling.

The computed shear stress profiles are plotted together with measurements in figures 6 (a,b), for both flow regimes. The predicted maximum, 0.023 occurs at about r/x = 0.08, while Corrsin and Uberoi found a smaller value of 0.011; Papanicolaou and List [24] reported a peak of 0.015 at r/x = 0.09 in both jet and plume regions. The predicted maximum shear in the plume is 0.28 at 0.08, which is close experiments.

Figures 7 (a, b) show the axial heat flux turbulent correlation development. In the jet (figure 8,a), models kε1,

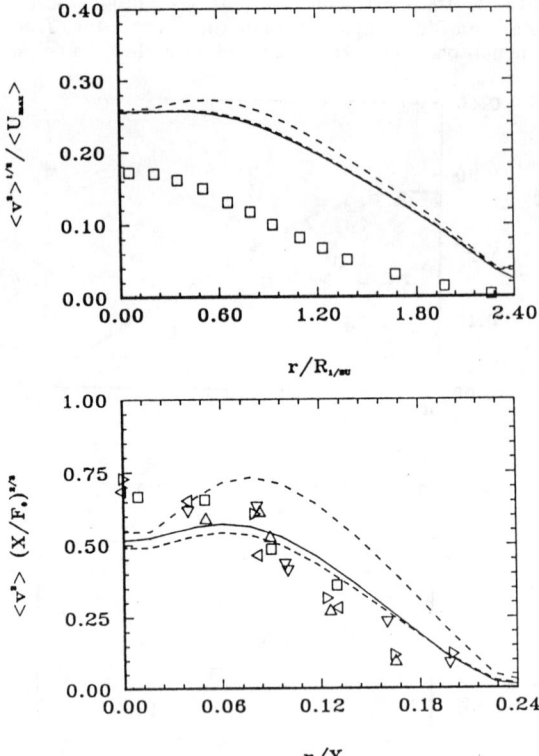

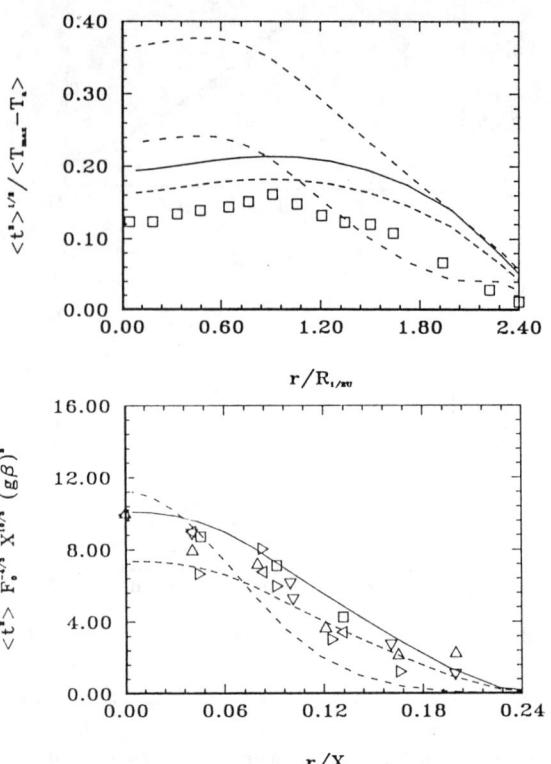

Figure 4. Turbulent radial normal Reynolds stress profiles (a) Heated jet; Symbols, experiment of Corrsin and Uberoi [16], □ x/D = 15; (b) Buoyant jet; Symbols, experiment of Shabbir and George [17], △ X/D = 10, ▷ X/D = 14, ▽ X/D = 19, ◁ X/D = 21, □ X/D = 25; Present predictions: __ kε1, --- kε4, - - kε5

kε2, kε4 and kε6 display different profiles. Though kε1 profile follows closely the experimental values, a more careful inspection may bring to light an important defect of diagonal representation for diffusive tensor: in the outer region of the jet, some negative signed values are obtained which is not physically plausible in this flow [16]. The other models produce almost identical values for $<ut>$, although considerably higher than those found experimentally. However, accordingly with [28], Corrsin and Uberoi's measured heat flux levels would be too low; Papanicolaou and List [24] reported a maximum value of 0.020 and Chevray and Tutu [27] found 0.030 at $r/R_{1/2} = 0.70$. The predicted profile shapes kε2, kε4, kε6 are qualitatively consistent with measurements [24, 25]. In contrast, plume axial turbulent heat transport is drastically underpredicted. The same reasoning presented above is used to explain this result: none of the tested forms of GTM includes any representation of the physical mechanism which contributes to amplify the turbulent axial heat flux levels.

The radial turbulent heat flux is plotted in figure 8 (a, b), predicted by models kε1, kε2, kε4 and kε6. Because the new formulation (16) for GTM is stronger linked with scalar fields, the correct evaluation of $<t^2>$ and ε_t provides an accurate prediction of the radial development of the heat flux correlation; otherwise, and due to these strong dependence on scalar time-scale, the model (kε5) yields to unphysical levels for $<vt>$. The remaining models kε1 and kε4 produce profiles which are in poorer agreement with experiments while kε2 overpredicts the measured values. In the plume zone, kε6 produces lower values than the other models, although accurs of prediction may be not clear due the scatter in the data. The kε6 predicted peak is $<vt>/(<T_{max}$-

$t_a><U_{max}>) = 0.018$ at r/x = 0.09, while kε1 gives 0.020 and kε4 cleary overpredicteds the experimental data. These results may be compared with [24], who reported peak values ranging from 0.020-0.025 at r/x = 0.10.

Figure 5. Temperature-variance profiles (a) Heated jet; Symbols, experiment of Corrsin and Uberoi [16], □ x/D = 15; Present predictions: __ kε1, --- kε3, - - - kε5, - - - kε6 (b) Buoyant jet; Symbols, experiment of Shabbir and George [17], △ X/D = 10, ▷ X/D = 14, ▽ X/D = 19, ◁ X/D = 21, □ X/D = 25; Present predictions: __ kε1, --- kε3, - - - kε6

An important hypothesis of integral methods in the analysis of jets and plumes is to neglect the turbulent contribution to axial transport. In the jet, the predicted ratio between advective and turbulent flux values of 0.16 and 0.05 for momentum and enthalpy, respectively; Papanicolaou and List [24] presented 0.15 and 0.07-0.12 for the same quatities. In the plume regime, experimental values reported by these authors are 0.15 and 0.15-0.20, while prediction display 0.10 and 0.03. These predicted lower values for the enthalpy fluxes may be understood under the previous discussion concerning turbulent transport prediction.

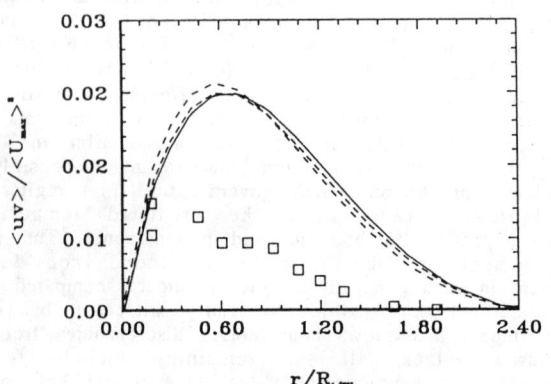

104

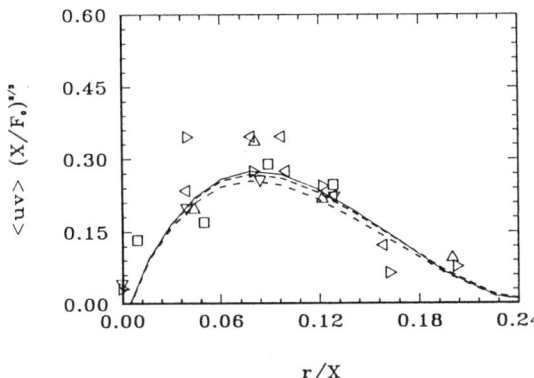

Figure 6. Turbulent shear stress profiles (a) Heated jet; Symbols, experiment of Corrsin and Uberoi [16], □ x/D = 15; (b) Buoyant jet; Symbols, experiment of Shabbir and George [17], △ X/D = 10, ▷ X/D = 14, ▽ X/D = 19, ◁ X/D = 21, □ X/D = 25; Present predictions: ___ kε1, --- kε4, - - - kε5

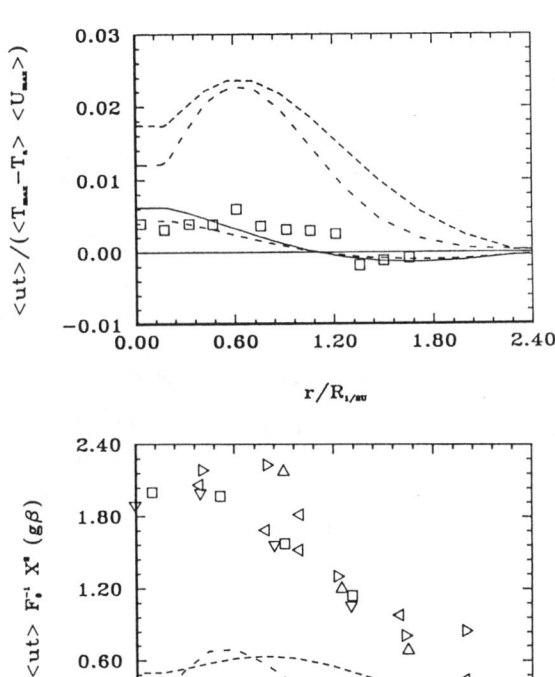

Figure 7. Turbulent axial heat flux profiles (a) Heated jet; Symbols, experiment of Corrsin and Uberoi [16], □ x/D = 15; (○) Buoyant jet; Symbols, experiment of Shabbir and George [17], △ X/D = 10, ▷ X/D = 14, ▽ X/D = 19, ◁ X/D = 21, □ X/D = 25; Present predictions: ___ kε1, --- kε2, - - - kε4, - - - kε6

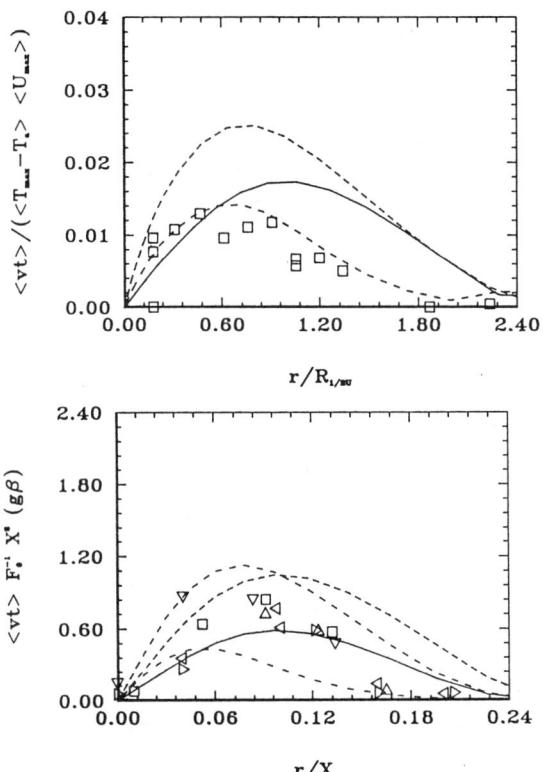

Figure 8. Turbulent radial heat profiles (a) Heated jet; Symbols, experiment of Corrsin and Uberoi [16], □ x/D = 15; Present predictions: ___ kε1, --- kε4, - - - kε6 (b) Buoyant jet; Symbols, experiment of Shabbir and George [17], △ X/D = 10, ▷ X/D = 14, ▽ X/D = 19, ◁ X/D = 21, □ X/D = 25; Present predictions: ___ kε1, --- kε2, - - - kε4, - - - kε6

5. Conclusions and Final Remarks

In the present study, a buoyancy-extended k-ε-$<t^2>$ turbulence model was applied to the computation of mean and turbulence properties of axisymmetric jets, both in jet-like and plume-like flow regimes. Improved eddy-viscosity/eddy-diffusivity representations for the turbulent fluxes were compared in their performance with standard formulae. In addition, it was incorporated in the model a transport equation for the dissipation rate of temperature-variance. An elliptic calculation procedure was employed to solve the flow governing equations.

The experimentally observed mean and turbulence characteristics were, in general, reasonably captured by some of the tested models. The velocity and temperature mean-field was accurately computed and calculated jet´s integral parameters showed little discrepancies compared with reported experimental values. Earlier studies that have employed simple algebraic turbulence models yielded to a similar accurate reproduction of mean characteristics of flow (see e.g. [29]). It is well known that the boussinesq hypothesis works well in a free thin shear-layer type flows. It is the present case, and despite the existence of buoyancy effects on the flow, it may be inferred that as far as the mean flow is concerned, the eddy-viscosity/eddy-diffusivity model (or even simpler mixing-length models) can yield to acceptable predictions. However, the turbulent field cannot held to be represented by such models and predictions evidenciate serious defects in its representation. A major difficulty concerns the buoyancy-generated anisotropy in second order turbulent correlations. The tested models showed to underpredict strongly the axial turbulent transport. None of the models can account for the additional turbulent energy supply by buoyancy forces. This suggests that an important improvement would be to add to expressions (4) and (11) a term proportional to buoyancy production $g_i<t^2>/T$. This additional directional source of turbulent kinetic energy is crucial to the correct representation of Reynolds stress and heat flux in a bouyancy-driven flow. Although this defect may induce an important misleading in the turbulence description, the overall flow representation provided by the $k-\varepsilon-<t^2>$ model may be considered acceptable to most engineering purposes.

ACKNOWLGEMENTS

This work was supported by Direcção Geral da Qualidade e Ambiente (DGQA), under the grant 25/87. The authors are grateful to CTAMFUTL-INIC and to Ms. Rosa Maria Ribeiro for typing the manuscript.

TABLES

Table 1.- Model Constants used in the Standard $k-\varepsilon-<t^2>$ model

C_μ	C_D	$C_{\varepsilon 1}$	$C_{\varepsilon 2}$	$C_{\varepsilon 3}$	σ_k	σ_ε	σ_t	C_t	C_g	C_c
0.09	1.0	1.44	1.92	1.44	1.0	1.3	0.71	0.13	0.35	0.205

Table 2.- Anisotropic $<u_i u_j>$ and $<u_i t>$ representations [6, 10]

$C_{\tau 1}$	$C_{\tau 2}$	$C_{\tau 3}$	C_K	C_{KA}	C'_{KA}
0.0288	4.78×10^3	0.024	0.2	0.1	0.08

Table 3.- ε_t-Equation Constants

$C_{\varepsilon t 1}$	$C_{\varepsilon t 2}$	$C_{\varepsilon t 3}$	$C_{\varepsilon t 4}$	$C_{\varepsilon t}$
1.7	1.6	1.7	1.8	0.11

Table 4.- Summary of $k-\varepsilon$ model Modifications

Model Specification	Scales	$<u_i u_j>$	$<u_j t>$	Constants Modification
$k\varepsilon 1$	$k,\varepsilon,<t^2>$	(4)	(12)	---
$k\varepsilon 2$	$k,\varepsilon,<t^2>$	(4)	(15)	---
$k\varepsilon 3$	$k,\varepsilon,<t^2>$	(4)	(12)	C_μ,σ_t (13),(14)
$k\varepsilon 4$	$k,\varepsilon,<t^2>$	(4)	(12)	---
$k\varepsilon 5$	$k,\varepsilon,<t^2>$	(4),(5)	(16)	---
$k\varepsilon 6$	$k,\varepsilon,<t^2>,\varepsilon_t$	(4),(5)	(16)	---

REFERENCES

1. Rodi, W., "Examples of Turbulence Models for Incompressible Flows". AIAA Journal, Vol. 20, Nº7, 1981, pp. 872-879

2. Launder, B. E.,"Heat and Mass Transport in Turbulence". Topics in Applied Physics, Ed. P. Bradshaw, Vol 12, 1976

3. Pope, S., "A more general effective-viscosity hypothesis" J. Fluid Mech. 72, 1975, pp. 331-340

4. Corrsin, S., "Limitations of Gradient Transport Models in Random Walks and in Turbulence". Turbulent Diffusion in Environmental Pollution, Eds. F.N. Frankiel and R.E. Munn, Academic Press, Vol 18A, 1974, pp. 25-59

5. Speziale, C. G. "On Non Linear k-l and k-ε Model of Turbulence". J. Fluid Mech. 178, 1987, pp. 459-475

6. Yoshizawa, A., "Statistical Analysis of the Deviation of the Reynolds Stress from its Eddy-Viscosity Representation". Phys. Fluids, Vol. 27, (6), 1984, pp. 1377-1387

7. Nisizima, S. and Yoshizawa, A., "Turbulent Channel and Couette Flows Using an Anisotropic k-ε Model". AIAA Journal, Vol. 25, Nº3, 1987, pp. 414-420

8. Tavoularis and Corrsin

9. Daly, B. J. and Harlow, F. H., "Transport Equations in Turbulence". Phys. Fluids, Vol 13, 1970, pp. 2634-2649

10. Yoshizawa, A., "Statistical Modelling of Passive-Scalar Diffusion in Turbulent Shear Flows". J. Fluid Mech. 195, 1988, pp. 541-555

11. Launder, B. E., Haroutonian, V. and Ince, N., "Turbulent Time Scale Ratios in Plumes". UMIST 3rd Bienniel. Coll. on Computational Fluid Dynamics, University of Manchester, 1988-A

12. Donaldson, C. du P., Sullivan, R. D. and Rosenbaum, H. "A Theoretical Study of the Generation of Atmospheric Clear Air Turbulence". AIAA Journal, Vol. 10, 1972, pp. 162-170

13. Meroney, R. N., "An Algebraic Stress Model for Stratified Turbulent Shear Flows". Computer and Fluids, Vol. 4, 1976, pp.93-107

14. Newman, G. R., Launder, B. E. and Lumley, J. L., "Modelling the Behavior of Homogeneous Scalar Turbulence". J. Fluid Mech., 111, 1981, pp. 217-232

15. Elghobashi, S. E. and Launder, B. E., "Turbulent Time Scales and The Dissipation Rate of Temperature Variance in the Thermal Mixing Layer". Phys. Fluids, Vol. 16, (9), 1983, pp. 2415-2419

16. Corrsin, S. and Uberoi, M. S.(1949), "Further Experiments On The Flow and Heat Transfer in a Heated Turbulent Air Jet". NACA Tech. Note 1865

17. Shabbir, A. and George, W. K., "Energy Balance Measurements in an Axisymmetric Turbulent Buoyant Plume in a Neutral Environment". Sixth Symposium on Turbulent Shear Flows, Toulouse, 1987

18. Hossain, M. S. and Rodi, W., "A Turbulence Model for Buoyant Flows and its Application to Vertical Buoyant Jets". Turbulent Jets and Plumes, Ed. W. Rodi, Pergamon Press, 1982

19. Chen, C. J. and Rodi, W. "A mathematical model for stratified turbulence flows and its application to buoyant jets". 16th Congress IAHR, 1975

20. Gibson, M. M. and Launder, B. E., "On the Calculation of Horizontal Non-Equilibrium Turbulent Shear Flows Under Gravitational Influence", ASME J. of Heat Transfer, 1976, pp. 81-87

21. Gosman, A. D. and Pun, W. M., "Calculation of Recirculating Flows", Rep. N⁰ HTS/74/2, Dept. of Mech. Engn., Imperial College, London, 1974

22. Launder, B. E., Haroutonian, V. and Ince, N., "Assessment of Thin-Shear Flow Approximation in Jets". UMIST 3rd Bienniel. Coll. on Computational Fluid Dynamics, University of Manchester, 1988-B

23. Sini, J. F. and Dekeyser, I. "Numerical prediction of turbulent plane jets and forced plumes by use of the k-ε model of turbulence", Int. J. Heat Mass Transfer, Vol. 30 N⁰ 9, 1987, pp. 1787-1801

24. Papanicolaou, P. N. and List, E. J., "Investigations of round vertical turbulent buoyant jets", J. Fluid Mech. Vol. 195, 1988, pp. 341-391

25. Papanicolaou, P. N. and List, E. J., "Statistical and spectral properties of tracer concentration in round buoyant jets" Int. J. Heat and Mass Transfer, Vol. 30, 1987, pp. 2059-2071

26. Kotsovinos, N. E., "Temperature measurements ina turbulent round plume" Int. J. Heat Mass Transfer, Vol. 28, N⁰4, 1985, pp. 771-777

27. Chevray, R. and Tutu, N. K., "Intermittency and Preferencial transport of heat in a round jet", J. Fluid Mech, 88, pp. 133-160, 1978

28. Chen, J. C. and Rodi, W. "Vertical turbulent buoyant jets, a review of experimental data", Pergamon Press, 1980

29. Madni, I. K., Pletcher, R. H., "Prediction of Turbulent Forced Plumes Issuing Vertically Into Stratified or Uniform Ambients" ASME J. Heat Transfer, Feb. 1977, pp. 99-104

1989 National Heat Transfer Conference
HTD-Vol. 107, Heat Transfer in Convective Flows

APPLICATION OF A SURFACE RENEWAL MODEL TO THE PREDICTION OF HEAT TRANSFER IN AN IMPINGING JET

T. D. Yuan and J. A. Liburdy
Thermal Fluids Laboratory
Department of Mechanical Engineering
Clemson University
Clemson, South Carolina

ABSTRACT

The concept of surface renewal is applied to impinging heated jets to predict the local heat transfer coefficients. The modeling concept is based on a zonal approach which identifies an outer flow and an impingement region. The goal is to accurately determine the local variation of the surface heat transfer coefficient near the impingement region. The outer region flow is determined by the use of the standard k-ε model. The impingement region is modeled using the surface renewal concept. In the application of the surface renewal it is found that the solution is sensitive to the matching and initial conditions imposed. We show very good agreement with experimental data when the jet centerline conditions are used to scale the initial velocity and temperature renewal process. Also, we show that the temperature renewal is best modeled when the frequency associated with the thermal energy is more than twice as large as the frequency associated with the momentum transport. Overall, the use of the surface renewal model in the impingement region is shown to be an alternative approach in the prediction of the surface interactions.

NOMENCLATURE

C_f Wall friction factor
d   Width of the slot
f   Frequency
g   Acceleration of gravity
Gr   Grashof number, $g \beta (T_j - T_w) d^3/\nu^2$
H   Slot height from the surface
k   Thermal conductivity, turbulence kinetic energy
Nu   Nusselt number, h d/k
p   Pressure
P   Nondimensional pressure, $(p - p\infty)/\rho v_j^2$
Pr   Prandtl number, ν/α
Re   Reynolds number, $v_j d/\nu$
Ri   Richardson number, Gr/Re^2
s   Mean turbulent dominant frequency

S   Nondimensional frequency used in the surface renewal model, $= sd/v_j$
S_t Nondimensional dominant frequency used for thermal transport
St   Strouhal number, defined in equation (11)
S_ψ Source term
t Temperature fluctuation
T   Thermodynamic mean temperature
u   Velocity component or velocity fluctuation in x-direction
U   Nondimensional mean velocity component in x-direction, u/v_j
v   Velocity component or velocity fluctuation in y-direction
V   Nondimensional mean velocity component in y-direction, v/v_j
x   Coordinate along the impinging surface from the stagnation point
X   Nondimensional x, x/d
y   Coordinate normal to the surface
Y   Nondimensional y, y/d

Greek Symbols

α Thermal diffusivity
β Thermal expansion coefficient
Γ Diffusion coefficient
ε Turbulence kinetic energy dissipation rate
θ Nondimensional temperature, $(T-T_w)/(T_j-T_w)$
λ Time variable during the surface renewal process
μ Dynamic viscosity
ν Kinematic viscosity
ρ Density
φ Statistical age distribution
ψ Generalized variable

Subscript

i   Index
l   Inrush condition
j   Slot exit

k   Turbulence kinetic energy
o,s Stagnation point
t   Thermal
w   Impinging surface
x   Longitudinal direction
∞   Ambient
ε   Turbulence kinetic energy dissipation rate

INTRODUCTION

In the treatment of turbulent flows near a wall, the application of the standard k-ε model, which is normally used for regions of very high Reynolds number, is not satisfactory. Usually a wall function or a low Reynolds number modification are used in the near wall regions. The application of such models is often based on the assumption of local equilibrium between turbulent production and dissipation. In the near impingement region of an impinging jet, the existence of strong curvature of streamlines and large velocity gradients tends to reduce the shear stress in this region and causes anisotropy. Therefore, a different approach for treating the effects of the near wall turbulence is needed to study the thermal and flow fields.

Childs and Nixon [1985] indicate a need for a higher order turbulence model to better predict the influence of curvature as well as the near impingement anisotropy turbulence. The disagreement between the predicted and the experimental heat transfer coefficients, as documented by Looney and Walsh [1984], can be attributed to the complex structure in the near impingement region. This warrants a better turbulence model which is consistent with some of the physics of the flow field that has been found in experimental studies.

Recently there has been significant insight into some aspects of the surface renewal mechanism in complex flow geometries using flow visualization techniques. The physical behavior in the near wall region seems to be characterized by the interaction between large eddies and the surface. It seems logical to pursue the surface interaction mechanism through a turbulence model based on the surface renewal concept.

Kestin and Maeder [1957] postulated the presence of a mechanism causing the amplification of vorticity fluctuations to be responsible for the observed enhancement of heat transfer in the neighborhood of the stagnation point. Experimental work to study the enhanced heat transfer in the impingement region was carried out by the Kataoka and Mizushita [1984]. They found, for a Prandtl number of 10^3, that the local enhancement of Nusselt numbers was attributed to the penetration of nonuniform turbulence of the free-stream across the laminar boundary layer and the subsequent transition from a laminar to a turbulent boundary layer. For their experimental conditions the local Nusselt number was a maximum at a plate distance of 6 times the nozzle diameter. Yokobori et al. [1978], following the vorticity amplification theory of Kestin and Maeder [1957], investigated the production mechanism of large-scale eddies using flow visualization. They found that large-scale longitudinal vortex-like structures are formed discretely in space and time in this region. They concluded that these structures are predominant in the passive transport in the impingement region. In

several other experimental results the maximum heat transfer coefficient has been found to occur at H/d between seven and eight, where H is the plate distance from the nozzle exit and d is the size of the nozzle opening. Most previous explanations of this have been based on a two-dimensional flow consideration that there is a maximum interaction between the centerline turbulence intensity, which increases with increasing H, and the mean arrival velocity, which decreases with H. Aside from this explanation, Yokobori et al. [1978] reasoned this result from the scaling of the three-dimensional eddy structure based on nozzle size and exit velocity. The dominant frequency of the large eddies which exists at the end of the potential core was found by Browand and Laufer [1975] to be independent of the Reynolds number.

Gutmark et al. [1978] studied far field impinging turbulent jets, H/d = 100. Menon [1989] studied jet impingement for H/d=8 and 32. They found that the effects of the wall occurred at approximately 0.2H from the surface. They also found a neutral frequency which distinguishes those eddies being amplified from those being dissipated by viscous effects near the surface. A similar result was observed by Pelfrey and Liburdy [1986] in the impingement region of an offset jet. The neutral frequency was found to scale with nozzle size, impinging plate separation distance and exit velocity.

The concept of a surface renewal model was initially developed for flow over flat plates. The notion is that eddies intermittently move from the turbulent core into the wall region while exchanging momentum and energy during the life of the eddies. This was proposed by Danckwerts [1951]. In a theoretical study of the rate of gas absorption in a liquid flow by Higbie [1935] it was concluded that the conventional concept of a viscous sublayer as a stagnant film was not valid in many situations when the contact time of absorption is less than the penetration time. Rather, the turbulence penetrates into the surface continually and is replaced with high speed renewed fluid during the absorption period. Einstein and Li [1954] also observed the unsteady nature of the viscous sublayer and proposed a mathematical model based on a cyclic growth and decay phenomenon. They concluded that with the quasi-steady sublayer concept it is impossible to explain the periodic creation of turbulence in this region. The scale of the sublayer thickness was estimated, in accordance with experimental data, to be on the order of the local wall unit. The inrush velocity was chosen iteratively so that the predicted velocity profile matched experimentally measured values. However, their assumption of a constant velocity in the turbulent flow adjacent to the sublayer is artificial and, according to the authors, was introduced only for mathematical convenience.

In this paper we present an application of the surface renewal model to an impinging jet. In so doing we incorporate some characteristics of the renewal process based on previous experimental results. Agreement with experiments is very good. Further detailed experimental verification is required to determine the validity of the proposed procedure. It is necessary to determine if a relatively simple approach to surface interaction can be useful for a broad range of conditions. However, if valid, this model will prove to be easy to use and computationally efficient.

MATHEMATICAL MODEL

In the surface renewal model it is necessary to describe the existence of multiple eddies existing within the domain of interest. Each large scale eddy is assumed to have a different age at any given time. At an instant, many large scale eddies reside in the region, all with different ages. The life of each eddy within the impingement region is characterized by the flow inrush at early times and flow ejection at later times. In an impinging jet flow the large scale energy continuing eddies are assumed to impinge with a characteristic frequency. The determination of this frequency is discussed later.

Based on the formulation by Thomas [1982] we treat the surface renewal mechanism as an unsteady transport process. The instantaneous equations governing the transport of mass, momentum and heat associated with the life of the large eddies within the impingement region describe the renewal process. During the inrush and ejection periods, when large eddies reside in the near wall region, the mass, momentum and energy equations are:

$$\partial u_i / \partial x_i = 0 \qquad (1)$$

$$\partial u_i / \partial \lambda + u_j \partial u_i / \partial x_j = - \partial p / \rho \partial x_i + \nu \, \partial / \partial x_j (\partial u_i / \partial x_j) \qquad (2)$$

$$\partial T / \partial \lambda + u_j \partial T / \partial x_j = \alpha \, \partial / \partial x_j (\partial T / \partial x_j) \qquad (3)$$

Where λ represents time during the life of an eddy and the analysis is limited to nonbuoyant, incompressible, constant property flow.

The flow field is transformed using a statistical age distribution. The ensemble average of a generalized variable ψ, which represents a velocity component or temperature, is related to its instantaneous distribution by assuming a probability function, ψ. The probability function is prescribed an age distribution based on a characteristic mean turbulent burst frequency, s, and information regarding the boundary and initial conditions. We choose $\psi = s \exp(-\lambda s)$, suggested by Danckerts [1951] and used by Thomas [1982] for boundary layer flow. Assuming the ergodic conditions that time and ensemble averages are identical for a stationary process, ψ represents the time average distribution for steady, turbulent flow. It is shown by Danckerts [1951] that the results are more sensitive to the choice of s rather than the form of ψ.

After the statistical transformation of the flow field for ψ, the Navier Stokes equations can be reformulated by averaging and put into the form:

$$\overline{\partial u_i} / \partial x_i = 0 \qquad (4)$$

$$s(\overline{u_i} - U_{1i}) + \overline{u_j \partial u_i / \partial x_j} = - \overline{\partial p} / \rho \partial x_i + \nu \, \partial^2 \overline{u_i} / \partial x_j^2 \qquad (5)$$

$$s(\overline{T} - T_1) + \overline{u_j \partial T / \partial x_j} = \alpha \, \partial^2 \overline{T} / \partial x_j^2 \qquad (6)$$

where U_{1i} and T_1 are defined as the velocity and temperature fields at the instant of the inrush process and the barred terms represent the ensemble average.

The governing equations require further explanation. The term $s(\overline{u_i} - U_{1i})$ represents the inrush-ejection contribution of the eddy transport mechanism. Similarly, the term $s(\overline{T} - T_1)$ is the thermal transport and is assumee to have a similar age distribution as the velocity. The terms $\overline{u_j \partial u_i / \partial x_j}$ and $\overline{u_j \partial T / \partial x_j}$ represent the unsteady convective interaction between eddies which contribute to the eddy transport mechanism, see Thomas [1982]. The other terms represent the conventional terms in the Navier-Stokes equations. The assumption which allows the convection terms to be written as $\overline{u_j \partial u_i / \partial x_j}$ and $\overline{u_j \partial T / \partial x_j}$ requires that the contribution of the convective interaction to the establishment of the mean flow field be small relative to the effects of the inrush process. This can be satisfied by assuming that the momentum influx by large eddies during inrush tends to be balanced by the momentum efflux during ejection near the impingement region.

To complete the formulation of the problem for the surface renewal model, initial and boundary conditions are required. The above equations are only applied near the impingement region. The initial condition is taken as the flow field at the instant of the inrush process which transfers momentum and heat with velocity U_{1i} and temperature T_1 at $\theta = 0$. Specification of these conditions is discussed later. The boundary conditions for this two dimensional elliptic problem require all four boundaries to be specified. At the wall no slip conditions are specified and along the center line of the jet symmetry is assumed. The upper boundary conditions account for the interaction between large scale structures near the wall. The specifications for this boundary are based on a zonal model concept which matches the solution from a $k-\varepsilon$ model prediction away from the wall to the surface renewal model domain of interest. The downstream boundary conditions are taken to be zero gradients along the flow direction for u, v and T. Based on previous numerical experiments, this latter condition was found to have little or no effect on the solution provided the boundary was far enough downstream to be in the wall jet region.

Earlier studies using the surface renewal model for turbulent boundary layer flows suggested that the initial condition, U_{1i}, be the free stream velocity. This was used successfully in the prediction of the skin friction coefficient by Thomas [1982]. From a physical viewpoint it is not necessarily a justified assumption. In the impinging jet flow, at the instant of inrush, it is assumed that the incoming flow is as if the wall were not present. In the following instant the effects of the wall become evident and the ejection of fluid from the wall region follows. Based on this we set the initial conditions for the impingement region as equal to the mean jet conditions at a distance sufficiently far from the surface. The condition at the upper boundary uses the calculated velocity just outside of the surface renewal region based on predictions using a $k-\varepsilon$ model. In this sense the surface renewal

111

application becomes a zonal model linked to the k-ε model predictions of the flow-field from the nozzle exit up to impingement region.

With the above conditions and assumptions the two dimensional governing equations are given below in nondimensional form. The scaling parameters are v_j, the velocity at the jet exit, d, the width of the jet exit and $(T_j - T_w)$, the temperature difference between the jet exit and the wall.

$$\partial U/\partial X + \partial V/\partial Y = 0 \tag{7}$$

$$S(U-U_1) + U\partial U/\partial X + V\partial U/\partial Y = -\partial P/\partial X +$$
$$1/Re(\partial^2 U/\partial X^2 + \partial^2 U/\partial Y^2) \tag{8}$$

$$S(V-V_1) + U\partial V/\partial X + V\partial V/\partial Y = -\partial P/\partial Y +$$
$$1/Re(\partial^2 V/\partial X^2 + \partial^2 V/\partial Y^2) \tag{9}$$

$$S_t(\theta-\theta_1) + U\partial\theta/\partial X + V\partial\theta/\partial Y = 1/Re\,Pr\,(\partial^2\theta/\partial X^2 + \partial^2\theta/\partial Y^2) \tag{10}$$

where the nondimensional parameters Re, Gr, Pr, θ and S are defined in the Nomenclature.

The selection of the computation domain for the surface renewal model should be determined by the region in which the presence of the wall influences the flow characteristics. From the data of Gutmark et al. [1978], for H/d = 100 and Menon [1989], for H/d = 8 and 32, the mean flow and turbulent intensity is affected by the wall within a region 0.2H from the wall. The domain size of the surface renewal model is based on this criteria. The solution procedure is as follows. Given a flow condition which includes the exit Reynolds number, exit temperature, and surface position, H/d, the problem is first solved using a k-ε model with a standard wall function near the surface. After a converged solution is achieved, the surface renewal model is applied to the impingement region of interest. This region extends 0.2H away from the wall and along the wall to a point where the results no longer change due to extension of the domain in the wall direction. For example, for a total domain of 36d by 16d (the longer side along the impinging surface direction) applied to the flow conditions of Re = 11000 and H/d = 16, the domain for the surface renewal model is 11.5d by 3.2d.

The frequency, s, may be specified based on inferences from existing experimental data. However, it must be cautioned that there is no definitive evidence as to the universal nature of the results at this time. The neutral frequency found by Gutmark et al. [1978] was identified from the power spectra of different components of the turbulent intensity along the jet centerline. They nondimensionalized this frequency using local velocity and length scales based on the corresponding values of centerline velocity and jet width the flow would have at the surface if the surface were not present. This can be expressed as a Strouhal number,

$$St = f\,H/v_j(H/d)^{1/2} \tag{11}$$

The Strouhal number associated with the neutral frequency was found to be 5.6. The value of St for the most amplified frequency was approximately 2.5. The study by Menon [1989] found the neutral frequency

to have a value of St = 6.2 for H/d = 32. This is discussed further in the results.

NUMERICAL PROCEDURE

The governing equations were solved using the control-volume approach of Pantankar [1980] applied to both the initial k-ε formulation and the surface renewal model, equations (7, 8, 9, and 10). The equations can be cast in a similar form as:

$$\partial(\rho u_i\psi)/\partial x_i = \partial(\Gamma\partial\psi/\partial x_i)\partial x_i + S_\psi \tag{12}$$

where ψ is a generalized variable, Γ is a diffusion coefficient, and S_ψ is the source term. The quantities Γ and S_ψ are specific for a particular variable ψ. For the surface renewal formulation the variables are listed in Table I.

The computational domain and boundaries are shown in Figure 1. The specification of the initial conditions and boundary conditions at the interface between the near wall region and free stream region are based on the numerically predicted results using the k-ε model. The boundaries to be specified are identified in Figure 1. Since the staggered grid is used throughout the numerical domain, linear interpolation was used to specify the boundary

Table 1. The Transport Equation for Surface Renewal Model

Equation	ψ	Γ	S_ψ
x-Momentum	U	1/Re	$-\partial p/\partial x - Sd/U_j(U - U_1)$
y-Momentum	V	1/Re	$-\partial p/\partial y - Sd/U_j(V - V_1)$
Energy	θ	1/(Re Pr)	$-Sd/U_j(\theta - \theta_1)$

conditions as needed. The boundary conditions at II and IV are specified based on symmetry, no slip at the surface and an isothermal surface. At boundary III, the velocity and temperature derivatives in the X direction are assumed to be negligible and were tested using successively larger grids until no change in the results were detected. A summary of the boundary conditions for the dimensionless velocity components and temperatures are given in Table II. Grid independent solutions were obtained by varying the grid from 18 x 18 to 25 x 30 nonuniform grid. The closest node to the surface was 0.005d or y^+ of approximately 0.4, where y^+ is a nondimensional distance based on the local friction velocity and viscosity.

In the application of the initial conditions, or the so-called inrush contribution, two approaches are presented in this paper. One approach uses the interface boundary conditions determined by the k-ε model as the inrush contribution terms, U_1, V_1, and θ_1. In this way the entire upper boundary represents an initial condition for inrush. This results in a variation of the inrush condition across the renewal region which depends on the local mean conditions. The other approach introduced in this study uses the local velocity and temperature at the jet centerline on the outer edge of the near wall region as the initial contribution. Physically, the latter specification represents the presence of a dominate momentum and thermal energy characterized by the centerline value which controls the development of the renewal process.

Table II. Boundary Conditions (boundaries are indicated in Figure 1)

Boundary I: Upper Boundary.

   U and V are based on k-ε model solution.

   $\theta = \theta_{k-\varepsilon}$ (y = 0.2 H from the plate)

Boundary II: Axis of Symmetry.

   $U = 0$, $\partial V / \partial X = 0$

   $\partial \theta / \partial X = 0$

Boundary III: Outflow Boundary

   $\partial U / \partial X = 0$, $\partial V / \partial X = 0$

   $\partial \theta / \partial X = 0$

Boundary IV: Impingement Wall.

   $U = 0$, $V = 0$

   $\theta = 0$

RESULTS

The surface interaction coefficients results, skin friction and heat transfer, predicted by the surface renewal model are compared with available experimental data in the literature. The range of flow conditions considered is H/d = 16 and Re = 11000, 16500 and 22000. The value of Ri in all cases was less than 10^{-4} indicative of nonbuoyant flows. The skin friction coefficients have been measured by various researchers. Here, the data of Beltas and Rajaratnam [1973] provide the necessary range of flow conditions to compare the skin friction prediction. The heat transfer coefficients are compared with the experimental data of Gardon and Akfirat [1966] who provide a wide range of flow conditions.

Skin Friction Prediction

The surface renewal model was tested with two different initial inrush conditions mentioned previously. The first approach based on the k-ε distribution of the mean velocity is first tested for Re = 11000 and H/d = 16. Using a range of frequencies, S, as the input to the momentum equation

the local skin friction coefficients distributions are compared with experimental results in Figure 2. According to the data of Beltas and Rajaratnam [1973], the local skin friction coefficient is dependent on the Reynolds numbers and H/d. The results from their measurements are shown as the shaded band in the figure. The calculations fail to predict the location of the peak value of C_f for a range of values of S. Comparison is also made for Re = 7000 and H/d = 16 in Figure 3. The results for both values of Re illustrate that the profiles shift towards the center of impingement and deviate significantly from the experimental results. It appears that the distribution of the renewal process is not modeled appropriately. Efforts were made to improve the results by adjustment of the domain size and grid specification with no satisfactory results. However, the inaccuracy of the prediction can be improved by redefining the initial inrush conditions. If it is assumed that the inrush is dominated by the jet maximum velocity upon entering then impingement region, then the initial condition is to be specified based on the local centerline velocity at 0.2H from the surface. Using this approach the local skin friction coefficients are presented for Re = 7000, 11000, 16500 and 22000 for H/d = 16 in Figure 4 where S = 0.0875. This value of S resulted in the most agreeable distribution with the experimental data of Beltos and Rajaratnam [1973]. The results are in good agreement in terms of both the magnitude and distribution near the impingement region. The value of S, 0.0875, is the nondimensional frequency based on a Strouhal number of 5.6 and H/d = 16. Therefore, the choice of a frequency near the large scale amplification region of the spectrum as determined by Gutmark et al. [1978], gives satisfactory results. Note that the specification of S, based on a given value of the Strouhal number, is independent of Reynolds number and only dependent on H/d.

Heat Transfer Prediction

The preliminary assumption in this model, as stated earlier, is that the surface renewal mechanism dominates the heat transfer process as well as the momentum transfer process. However, there is no evidence that they both occur at the same frequency, this would result in different values of S used in the energy equation (identified as S_t) and the momentum equation (identified as S). Because of the success of using the centerline velocity scale for the initial conditions in the skin friction coefficients, we continue to use this same approach in the heat transfer predictions. That is, the nondimensional centerline temperature at the edge of the renewal domain defines the initial condition of the thermal renewal process.

The prediction of the local surface heat transfer coefficient is presented in Figure 5 for a range of S_t, with S = 0.0875, for Re = 11000 and H/d = 16. The results indicate that the value of S_t must be significantly larger than that of S in order to match the experimental results of Gardon and Akfirat [1966]. The value of S_t which gives the closest predictions is 0.22. The heat transfer coefficients near the stagnation point are slightly lower than the experimental data but the results, in general, are significantly superior to the prediction using the k-ε model which uses the wall function given by Launder and Spalding [1974].

The surface renewal model was further tested with S = 0.0875 and S_t = 0.22 for different flow conditions. The results for Re = 11000 and 22000 are compared to the experimental data of Gardon and Akfirat [1966] shown in Figure 6. The results are acceptable, except at the centerline, for Re = 22000 where the prediction is about 10% lower than the experimental data. This can be improved if the value of S_t used for Re = 22000 is increased slightly higher than 0.22. The specification of the mean frequency of renewal for heat transfer greater than that for the momentum transfer requires further experimental exploration. A larger value of S_t compared to S implies a greater enhancement of surface transport of heat compared to momentum by surface renewal.

CONCLUSIONS:

We have demonstrated the use of a surface renewal model coupled with a k-ε model to predict the local heat transfer in an impinging jet. The method is easy to implement and provides good agreement with experimental data for the range of conditions studied. The general application of surface renewal for surface interactions requires further insight into the physical process. Experiments need to specify the spectral characteristics of momentum and heat transport. Particular attention needs to be paid to the distinction between momentum and heat transport processes.

REFERENCES

Beltaos, S. and Rajaratnam, N., 1973, "Plane Turbulent Impinging Jet," Journal of Hydraulic Research, Vol. 11, No. 1, pp. 28-59.

Browand, F. K. and Laufer, J., 1975, "The Role of Large Scale Structure in Initial Development of Circular Jet," Proceedings of Fourth Biennial Symposium on Turbulence in Liquids, Univ. of Missouri-Rolla, pp. 333-344.

Childs, R. E., and Nixon, D., 1985, "Simulation of Impinging Turbulent Jets," AIAA-85-0047, AIAA 23rd Aerospace Science Meeting, Jan.

Danckwerts, P. V., 1951, "Significance of Liquid-film Coefficients in Gas Absorption," I&CE, v43, pp. 1460-1467.

Einstein, H. A. and Li, H., 1954, "The Viscous Sublayer Along a Smooth Boundary," Proc. ASCE J. Engr. Mech. Div., v82, pp. 1-27.

Gardon, R. and Akfirat, J. C., 1966, "Heat Transfer Characteristics of Impinging Two-Dimensional Air Jets," Journal of Heat Transfer, v88, n1, pp. 101-108.

Gutmark, E., Wolfshtein, M. and Wygnanski, J., 1978, "The Plane Turbulent Impinging Jet," Journal of Fluid Mechanics, v88, pt. 4, pp. 737-756.

Higbie, R., 1935, "The Rate of Absorption of a Pure Gas into a Still Liquid During Short Periods of Exposure." Trans. AIChE, v31, pp. 365-390.

Kataoka, K. and Mizushita, T., 1984, "Local Enhancement of the Rate of Heat Transfer in an Impinging Round Jet by Free Turbulence," Fifth International Heat Transfer Conference, Tokyo, paper FC-8-3, Sept.

Kataoka, K., Suguro, M., Megawa, H., Maruo, K. and Mihata, I., 1987, "The Effect of Surface Renewal Due to Large Scale Eddies on Jet Impingment Heat Transfer," Int. J. Heat Mass Transfer, v5, pp. 305-310.

Kestin, J. and Maeder, P. F., 1957, "Influence of Turbulence on Transfer of Heat from Cylinders," NACA Tech. Note No. 4018.

Launder, B. E. and Spalding, D. B., 1974, "The Numerical Computation of Turbulent Flow," Comp. Methods in Applied Mechanics of Engineering, v3, pp. 269-289.

Looney, M. K. and Walsh, J. J., 1984, "Mean Flow and Turbulent Characteristics of Free and Impinging Jet Flows," Journal of Fluid Mechanics, v147, pp. 397-429.

Menon, R., 1989, "Impingement Characteristics of Heated Plane Turbulent Jets," Ph.D. Dissertation, Clemson University, Clemson, SC.

Pantankar, S. V., 1980, Numerical Heat Transfer and Fluid Flow, Hemisphere, Washington, DC.

Pelfrey, J. R. R. and Liburdy, J. A., 1986, "Effect of Curvature on the Turbulence of a Two-Dimensional Jet," Experiments in Fluids, v4, pp. 143-149.

Thomas, L. C., 1982, "A Turbulent Burst Model on Wall Turbulence for Two Dimensional Turbulent Boundary Layer Flow," Int. J. Heat Mass Transfer, v25, pp. 1127-1136.

Yuan, T. D., 1988, "The Study of Buoyancy Effects on Laminar and Turbulent Plane Impinging Jets," Ph.D. Dissertation, Clemson University, Clemson, SC.

Yokobori, S., Kasagi, N., Hirate, M. and Nishiwaki, N., 1978, "Role of Large Scale Eddy Structure on Enhancement of Heat Transfer in Stagnation Region of two Dimensional, Submerged, Impinging Jet," Proceedings of Sixth International Heat Transfer Conference, v5, pp. 305-310.

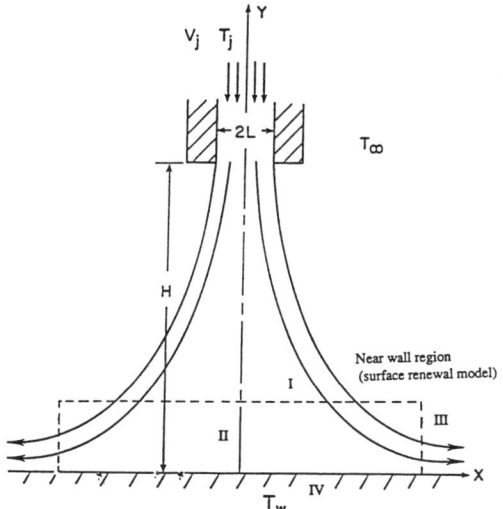

Figure 1. Flow configuration and coordinate system.

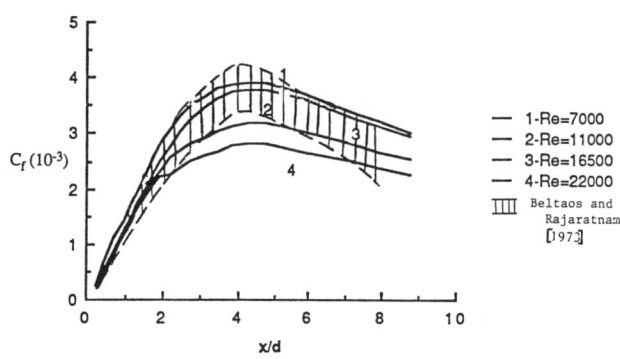

Figure 4. Predicted skin friction coefficient for S = 0.0875 using the centerline velocity for the initial condition; H/d = 16.

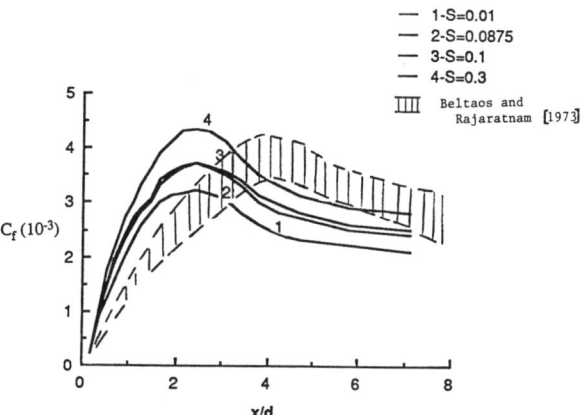

Figure 2. Predicted skin friction coefficient for different values of the renewal frequency using the distributed initial condition; Re = 11000, H/d = 16.

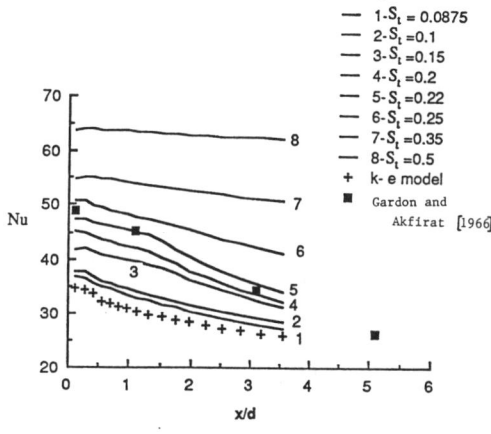

Figure 5. Predicted local Nusselt number distribution for S = 0.0875 and various values of S_t; Re = 11000, H/d = 16.

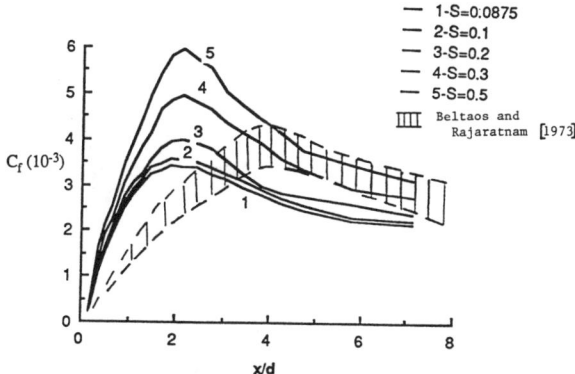

Figure 3. Predicted skin friction coefficient for different values of the renewal frequency using the distributed initial condition; Re = 7000, H/d = 16.

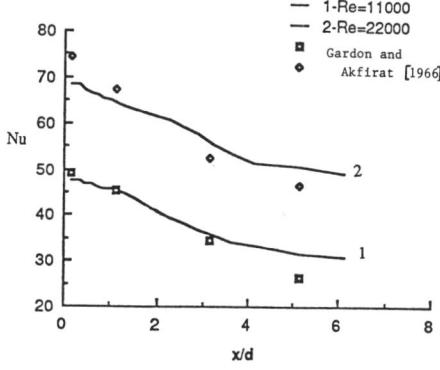

Figure 6. Predicted local Nusselt number distribution for S = 0.0875 and S_t = 0.22 for two values of Re.

115

TURBULENT FREE CONVECTION IN SQUARE CAVITIES
WITH MIXED BOUNDARY CONDITIONS

A. Abrous and A. F. Emery
Department of Mechanical Engineering
University of Washington
Seattle, Washington

ABSTRACT

In this paper we describe a numerical study performed on the laminar and turbulent two-dimensional air flow in simple spaces using the $k - \epsilon$ model of turbulence. With this numerical study, we achieved solutions for two-dimensional buoyancy-induced flows inside square cavities characterized by Rayleigh numbers up to (10^{12}). The physical model used is a 2-D model of a square cavity with the vertical walls maintained at different, but constant, temperatures and the horizontal walls either adiabatic (*Adiabatic Case*) or perfectly conductive (*Conductive Case*). The numerical results are compared with existing data from other numerical and experimental results; the agreement is very good. Using these results, the effects of the nature of the horizontal boundary condition upon the free convection flow patterns, turbulence fields, and the local surface heat transfer coefficients are discussed. Various turbulence model constants are tested to determine their effects on the results. A variety of applications of the numerical technique are presented by examining several configurations which simulate the effect of plumes, infiltration and horizontal partitions in order to determine their effects on the basic flow pattern in occupied spaces. The results indicate that the *Base Flow* can easily be upset by small perturbations and that once the flow has been disturbed, further increases in perturbations do not elicit stronger responses.

NOMENCLATURE

| English Symbols |

C_μ
C_1
C_2 Empirical constants in the turbulence model
C_3

C_p Specific heat at constant pressure $[J/Kg\ K]$
g Acceleration of gravity $[m/s^2]$
Gr Grashof number $(Gr = g\beta W^3 (T_h - T_c)/\nu^2)$
h Convective heat transfer coefficient $[W/m^2\ K]$
k Kinetic energy of turbulence $[m^2/s^2]$

L Height of cavity $[m]$
l Characteristic length of turbulence $[m]$
l_m Prandtl mixing length $[m]$
Nu_H Nusselt number along the hot wall
Nu_L Nusselt number along the lower wall
P Pressure $[Pa]$
p' Pressure fluctuation $[Pa]$
Pr Prandtl number $(Pr = \mu C_p / \lambda)$
q Heat flux $[W/m^2]$
R Resultant Velocity $[m/s]$
Ra Rayleigh number $(Ra = Pr\ Gr)$
R_t Turbulence Reynolds number $(R_t = k^{1/2} l / \nu)$
s Size of heated surface $[m]$
t Time $[s]$
T Temperature $[K]$
T' Temperature fluctuation $[K]$
T_c Temperature of the cold wall $[K]$
T_h Temperature of the hot wall $[K]$
T_M Mean temperature $(T_M = (T_h + T_c)/2)$ $[K]$
U Time-averaged velocity in the X-direction $[m/s]$
U_{maxh} Maximum U velocity at vertical midplane $[m/s]$
V Time-averaged velocity in the Y-direction $[m/s]$
V_{maxh} Maximum V velocity at horizontal midplane $[m/s]$
X Horizontal coordinate of the grid $[m]$
Y Vertical coordinate of the grid $[m]$

| Greek Symbols |

λ Thermal conductivity $[W/m\ K]$
β Coefficient of thermal expansion $[K^{-1}]$
ϵ Dissipation rate of turbulence kinetic energy $[m^2/s^2]$
μ Molecular viscosity $[Kg/m/s]$
μ_t Turbulent viscosity $[Kg/m/s]$
ν Kinematic viscosity $[m^2/s]$
ρ Density $[Kg/m^3]$
κ Von Karman's constant (0.42)
$\sigma_k, \sigma_\epsilon$ Prandtl-Schmidt numbers for k and ϵ
σ_T Turbulent Prandtl number

INTRODUCTION

The determination of buoyancy flow characteristics and the rate of heat transfer in rectangular enclosures has been the topic of much research in the past two decades. The physical mechanisms characterizing the flow are frequently encountered, particularly in double pane windows, solar collectors, double wall insulation, nuclear reactor insulation, ovens and rooms. It is therefore of practical importance to be able to predict the flow structure and heat transfer processes associated with this problem. Catton [1] and Ostrach [2] give detailed reviews of the work done in this area.

A problem considered very extensively as the classical research problem of internal natural convection flows is the buoyancy-driven natural convection in differentially heated upright cavities with adiabatic ends as indicated by the large number of studies compiled in [3]. Unfortunately, most of them deal only with laminar flow calculations. In fact, in many real physical situations, e.g. a typical room where the Rayleigh number is in the order of 10^{12}, the flow is turbulent and therefore such limited algorithms cannot be used to understand the processes.

The stimulus behind the present study is to develop a numerical technique which can simulate flows in rooms subjected to complex boundary conditions such as infiltration, plumes, partitions etc... However, such flows are not only three dimensional in nature but are also turbulent. Moreover, the data concerning these configurations are very scarce. Therefore, as a first step, we decided to develop a two dimensional turbulent model which we can validate with available numerical and experimental data using simple cavities before applying it to study more complex configurations.

In this paper, we present the *Adiabatic* and the *Conductive* results and compare them in order to examine the effect of the nature of the horizontal boundary condition on the flow pattern and turbulence fields. Moreover, we conduct a sensitivity study on the turbulence constants involved with the low turbulence Reynolds number formulation of the $k - \epsilon$ model and we describe their effects. Finally, as an application, we will subject the cavity to different types of boundary conditions and describe the resulting effects on the flow pattern, the turbulence field, and the heat transfer characteristics.

PHYSICAL MODEL

The physical model used is a 2-D square cavity and the working fluid is air. The vertical walls are held at different, but constant, temperatures and the floor and ceiling are either adiabatic, Figure 1a, or perfectly conductive, Figure 1b. The fluid in the immediate neighborhood of the hot wall will become warmer as a result of conduction. This temperature increase causes a decrease in the density of the fluid layers resulting in an upward motion. At the cold wall, the density of the adjacent fluid will increase and a downward motion is formed. In this way, a circulation of the fluid exists in the cavity. Laminar and turbulent flow regimes may simultaneously exist depending on the Rayleigh number.

MATHEMATICAL MODEL

Mathematically, the problem of buoyancy-driven flow is described by a set of coupled non-linear partial differential equations in which the buoyancy dictates the flow patterns. All the thermophysical properties of the fluid are evaluated at the mean cavity temperature and assumed to be constant except for the density which, under the Boussinesq approximation, is assumed to be constant everywhere except in the buoyancy term. The time averaged momentum and energy equations include some turbulent correlations known as the Reynolds stresses and turbulent fluxes which must be approximated. In this study, we use the well known two-equation model of turbulence ($k - \epsilon$).

In this model, the Reynolds stresses are related to the mean velocity gradients as:

$$-\overline{\rho u_i u_j} = \mu_t \left(\frac{\partial U_i}{\partial x_j} + \frac{\partial U_j}{\partial x_i} \right) - \frac{2}{3} \rho k \delta_{ij} \quad (1)$$

The proportionality parameter μ_t is the turbulent viscosity given by:

$$\mu_t = C_\mu \rho k^2 / \epsilon \quad (2)$$

The turbulent heat fluxes are handled using Reynolds analogy [4,5] which is based on the similarity between the transport of heat and momentum of the form:

$$-\overline{\rho u_i T'} = \frac{\mu_t}{\sigma_T} \frac{\partial T}{\partial x_i} \quad (3)$$

Using the non-dimensional variables adopted in [9], namely

$$T^* = \frac{T - T_M}{T_h - T_c} \qquad x^* = \frac{x}{L} \qquad y^* = \frac{y}{L}$$

$$U^* = \frac{U}{\frac{\nu}{L} Ra^{1/4}} \qquad V^* = \frac{V}{\frac{\nu}{L} Ra^{1/4}} \qquad k^* = \frac{k}{\frac{\nu^2}{L^2} Ra^{1/2}} \quad (4)$$

$$\epsilon^* = \frac{\epsilon}{\frac{\nu^3}{L^4} Ra^{3/4}} \qquad \mu_t^* = \frac{\mu_t}{\mu} \qquad P^* = \frac{P}{\frac{\rho \nu^2}{L^2} Ra^{1/2}}$$

$$t^* = \frac{t}{\frac{L^2}{\nu} Ra^{-1/4}}$$

we obtain the following closed set of equations describing a 2-D buoyancy driven flow *(omitting the stars)*:

1. Continuity

$$\frac{\partial U}{\partial x} + \frac{\partial V}{\partial y} = 0 \quad (5.a)$$

2. $X-$ Momentum Equation

$$\frac{\partial U}{\partial t} + \frac{\partial (UU)}{\partial x} + \frac{\partial (UV)}{\partial y} = -\frac{\partial P}{\partial x} + Ra^{-1/4} \frac{\partial}{\partial x} \left[(1 + \mu_t) \frac{\partial U}{\partial x} \right]$$

$$+ Ra^{-1/4} \frac{\partial}{\partial y} \left[(1 + \mu_t) \frac{\partial U}{\partial y} \right] - \frac{2}{3} \frac{\partial k}{\partial x} \quad (5.b)$$

118

3. $Y-$ Momentum Equation

$$\frac{\partial(V)}{\partial t} + \frac{\partial(UV)}{\partial x} + \frac{\partial(VV)}{\partial y} = -\frac{\partial P}{\partial y} + Ra^{-1/4}\frac{\partial}{\partial x}\left[(1+\mu_t)\frac{\partial V}{\partial x}\right]$$

$$+Ra^{-1/4}\frac{\partial}{\partial y}\left[(1+\mu_t)\frac{\partial V}{\partial y}\right] + \frac{Ra^{1/2}}{Pr}T - \frac{2}{3}\frac{1}{A}\frac{\partial k}{\partial y} \qquad (5.c)$$

4. Energy Equation

$$\frac{\partial T}{\partial t} + \frac{\partial(UT)}{\partial x} + \frac{\partial(VT)}{\partial y} = Ra^{-1/4}\frac{\partial}{\partial x}\left[(\frac{1}{Pr}+\frac{\mu_t}{\sigma_T})\frac{\partial T}{\partial x}\right]$$

$$+Ra^{-1/4}\frac{\partial}{\partial y}\left[(\frac{1}{Pr}+\frac{\mu_t}{\sigma_T})\frac{\partial T}{\partial y}\right] \qquad (5.d)$$

5. $k-$ Equation

$$\frac{\partial k}{\partial t} + \frac{\partial(Uk)}{\partial x} + \frac{\partial(Vk)}{\partial y} = Ra^{-1/4}\frac{\partial}{\partial x}\left[(1+\frac{\mu_t}{\sigma_k})\frac{\partial k}{\partial x}\right]$$

$$+Ra^{-1/4}\frac{\partial}{\partial y}\left[(1+\frac{\mu_t}{\sigma_k})\frac{\partial k}{\partial y}\right] - \epsilon - \frac{Ra^{1/4}}{Pr}\frac{\mu_t}{\sigma_T}\frac{\partial T}{\partial y}$$

$$+\mu_t Ra^{-1/4}\left[2(\frac{\partial U}{\partial x})^2 + 2(\frac{\partial V}{\partial y})^2 + (\frac{\partial U}{\partial y}+\frac{\partial V}{\partial x})^2\right] \qquad (5.e)$$

6. ϵ Equation

$$\frac{\partial\epsilon}{\partial t} + \frac{\partial(U\epsilon)}{\partial x} + \frac{\partial(V\epsilon)}{\partial y} = Ra^{-1/4}\frac{\partial}{\partial x}\left[(1+\frac{\mu_t}{\sigma_\epsilon})\frac{\partial\epsilon}{\partial x}\right]$$

$$+Ra^{-1/4}\frac{\partial}{\partial y}\left[(1+\frac{\mu_t}{\sigma_\epsilon})\frac{\partial\epsilon}{\partial y}\right] - C_2\frac{\epsilon^2}{k} - C_3\frac{Ra^{1/4}}{Pr}\frac{\mu_t}{\sigma_T}\frac{\epsilon}{k}\frac{\partial T}{\partial y}$$

$$+C_1\mu_t\frac{\epsilon}{k}Ra^{-1/4}\left[2(\frac{\partial U}{\partial x})^2 + 2(\frac{\partial V}{\partial y})^2 + (\frac{\partial U}{\partial y}+\frac{\partial V}{\partial x})^2\right] \qquad (5.f)$$

The values of the constants contained in this model are usually the ones recommended by Launder and Spalding [6], who report

$$\begin{array}{lll} C_\mu = 0.09 & C_1 = 1.44 & C_2 = 1.92 \\ \sigma_k = 1. & \sigma_\epsilon = 1.3 & \sigma_T = 1. \end{array} \qquad (6)$$

for turbulent flow remote from walls. The above constants are empirical and have been optimized for higher speed flows. The constant C_3 varies linearly between the value suitable for a vertical boundary layer, $C_3 = C_1$, and the one suitable for a horizontal boundary layer, depending on the local velocity vector orientation. In this study, C_μ and C_2 are made functions of R_t, the turbulence Reynolds number, in order to account for low Reynolds number effects on the flow in the regions close to the solid walls where the local turbulence Reynolds number is so small that viscous effects dominate turbulent effects [4].

$$C_\mu = 0.09\, exp[-3.4/(1+R_t/50)^2] \qquad (7.a)$$

$$C_2 = 1.92\,[1 - 0.3\,exp(-R_t^2)] \qquad (7.b)$$

The boundary conditions characterizing the flow are:
1. The U- and V-velocities are both zero at the walls.
2. Either the temperature or the temperature gradient is specified at the walls.
3. The kinetic energy of turbulence is zero at the walls.
4. The boundary condition for ϵ requires special attention. The procedure usually employed is to take the first mesh point near enough to the wall to fall in the region where the transport of turbulence is negligible and to assume that there is a local equilibrium of production and dissipation. By assuming that the Prandtl mixing length l_m varies linearly in the boundary layer with the distance from the wall such that if d is the distance from the wall to the first mesh line, then the value of ϵ at the first mesh line would be expressed as:

$$\epsilon_1 = (C_\mu^{3/4}\,k_1^{3/2})/(\kappa\,d) \qquad (8)$$

where k_1 is the kinetic energy of turbulence at the first mesh-line.

In order to set $\epsilon = 0$ at the solid walls, some extra terms need to be added to the k and ϵ equations. However, when this was implemented, it caused the results for k and ϵ to converge to zero for all grid points.

MESH GENERATION

As the Rayleigh number increases, relatively thin boundary layers are formed adjacent to the solid walls and the resolution in these regions becomes inadequate if a uniform grid is used. Thus, an analytical function of the form x^n is used to generate a stretched mesh such that more nodes are placed next to the solid boundaries. For both the x and y directions, the functions used to generate the grid are :

$$(2x/L)^n/2\ for\ (x \leq L/2)$$

$$(-(2-2x/L)^n +2)/2\ for\ (x \geq L/2) \qquad (9)$$

For $n = 1$, a uniform grid is generated. Increasing the value of n will generate a fine grid next to the boundaries so that the boundary layers could be adequately resolved. We use $n = 1$ for $10^3 \leq Ra \leq 10^4$, $n = 1.5$ for $10^5 \leq Ra \leq 10^6$, $n = 2$ for $Ra = 10^7$, $n = 3$ for $10^8 \leq Ra \leq 10^{11}$ and finally, $n = 3.5$ for $Ra = 10^{12}$. With these values, at least six nodes are in the boundary layer. In test cases in which the values of n were increased further, no effects on the results were observed.

NUSSELT NUMBER CALCULATION

The heat transfer at the hot wall is

$$q = -\lambda(\frac{\partial T}{\partial x})_{x=0} = h(T_h - T_c) \qquad (10)$$

With the gradient $\partial T/\partial x$ evaluated in terms of $(T_h - T_c)$ as the reference temperature and W as the reference length, the Nusselt number along the hot and lower walls are respectively given by:

$$Nu_H(y) = -(\frac{\partial T^*}{\partial x^*})_{x^*=0} \qquad Nu_L(x) = -(\frac{\partial T^*}{\partial y^*})_{y^*=0} \qquad (12)$$

The temperature gradients are evaluated by using the wall temperature and the closest internal node then the average Nusselt numbers are calculated by numerical integration using the trapezoidal rule.

NUMERICAL SOLUTION

The difference equations are solved using a control volume finite difference technique based on the SIMPLE algorithm [7]. The general algorithm is as follows:

1. Estimate the dynamic pressure and the values of U, V, T, k and ϵ.
2. Solve for the V velocity using the estimated values of the other variables. The line iteration is used by sweeping in a horizontal direction from bottom to top.
3. Solve for the U velocity the same way as for V.
4. Solve for the pressure correction p' by requiring continuity of mass in each control volume. This involves the solution of a Poisson equation for p'.
5. Correct the pressure and velocities.
6. Solve for the temperature and kinetic energy of turbulence with the new velocities. An iterative method similar to that used for U and V is used.
7. Compute ϵ at the first mesh-lines to be used as a boundary condition for ϵ.
8. Solve for ϵ for the remaining nodes inside the cavity.
9. Compute the turbulent viscosity everywhere.
10. Regard the new values as improved estimates. Return to step 2 and repeat until convergence is obtained.

The criterion for stopping the program is when the maximum relative change of a variable is less than 0.001% for laminar calculations and 0.005% for turbulent calculations. An average of 200 iterations for laminar flow and 400 iterations for turbulent flow were used. Computations for higher Rayleigh numbers are started from the results of the next lower Rayleigh number. The results are independent of the initial conditions. The equations are written in their transient forms and the use of relaxation factors is equivalent to the use of smaller time steps.

The turbulent model has been implemented in the program in such a way that the user can choose wether to trigger the model or not. For each Rayleigh number, the calculations have been performed with and without the use of the model. For those flows which we term laminar, the use of the turbulence model didn't produce any noticeable changes in the mean flow variables when compared to results obtained without the model.

BASE CASE RESULTS

Qualitative results for both the *Adiabatic Case* and the *Conductive Case* are similar. At $Ra = 10^3$, the resulting velocity pattern is that of a single vortex, centered at the center of the cavity and the isotherms are nearly parallel to the heated walls, indicating that most of the heat transfer throughout the fluid is by heat conduction. As the Rayleigh number increases $(Ra = 10^4)$ the velocity pattern becomes distorted into an elliptical shape and the effect of convection is more pronounced. Temperature gradients are now more severe near the vertical walls, but diminish in the centre. This behavior continues to $Ra = 10^5$; but the velocity pattern shows the appearance of two secondary vortices near the central region of the cavity. Increasing Ra to 10^6 causes the secondary vortices to move towards the walls and

to be moved towards the corners. A third vortex appears in the centre of the section. As the Rayleigh number is further increased, the secondary vortices generated in the central core region are convected further towards the corners and closer to the differentially heated walls. The boundary layers on the heated walls become increasingly thin. The velocity maximum moves closer to the wall as the Rayleigh number increases. The vertical velocity maximum is close to the centre of the hot wall for all Rayleigh numbers, while the maximum horizontal component regresses towards the corners as the Rayleigh number increases. The maximum turbulent viscosity occurs in the lower left corner of the cavity. In the centre of the cavity, the turbulent viscosity is negligible compared with the actual molecular viscosity but its value becomes important as the Rayleigh number increases. The turbulent kinetic energy and the dissipation rate are seen to be important in the vertical boundary layers. Figures 2 and 3 show typical velocity patterns and contours for T, U, V, k, μ_t at $Ra = 10^8$ for both cases.

TABLE 1

Comparison of Nu, U and V
Adiabatic Case

$Ra = 10^3$	Nu	Nu_{max}	y	Nu_{min}	y	U_{maxh}	y	V_{maxh}	x
De Vahl Davis	1.116	1.494	0.097	0.697	1.00	3.649	0.813	3.696	0.179
Jones I	1.117	1.507	0.100	0.692	1.00	3.651	0.792	3.697	0.179
Portier	1.143	1.525	0.089	0.685	1.00	3.660	0.814	3.690	0.178
Hans	1.128	1.493	0.099	0.728	0.99	3.590	0.834	3.690	0.176
Wan Hassan	1.138	1.559	0.110	0.678	1.00	3.633	0.798	3.686	0.184
Present Work	1.124	1.530	0.079	0.685	0.974	3.665	0.816	3.712	0.184
$Ra = 10^4$	Nu	Nu_{max}	y	Nu_{min}	y	U_{maxh}	y	V_{maxh}	x
De Vahl Davis	2.243	3.507	0.147	0.589	1.00	16.20	0.823	19.64	0.119
Jones I	2.242	3.550	0.146	0.586	1.00	16.23	0.823	19.63	0.120
Portier	2.368	3.657	0.141	0.578	1.00	16.19	0.823	19.53	0.119
Hans	2.265	3.437	0.173	0.680	1.00	16.10	0.827	19.55	0.106
Wan Hassan	2.253	3.564	0.156	0.599	1.00	16.05	0.790	19.46	0.151
Present Work	2.349	3.814	0.132	0.576	0.974	15.73	0.816	19.41	0.132
$Ra = 10^5$	Nu	Nu_{max}	y	Nu_{min}	y	U_{maxh}	y	V_{maxh}	x
De Vahl Davis	4.517	7.605	0.083	0.739	1.00	34.81	0.855	68.68	0.066
Jones I	4.523	7.816	0.088	0.734	1.00	35.36	0.855	68.65	0.067
Portier	4.869	8.386	0.075	0.705	1.00	35.16	0.855	68.14	0.067
Hans	4.538	7.568	0.087	0.808	1.00	34.18	0.846	68.26	0.068
Wan Hassan	4.429	7.686	0.073	0.775	1.00	34.41	0.927	67.00	0.068
Present Work	4.584	8.039	0.069	0.795	0.991	34.41	0.836	69.32	0.068
$Ra = 10^6$	Nu	Nu_{max}	y	Nu_{min}	y	U_{maxh}	y	V_{maxh}	x
De Vahl Davis	8.797	18.64	0.039	1.065	1.00	64.96	0.850	221.6	0.038
Jones I	8.783	16.89	0.058	1.011	1.00	69.07	0.850	219.5	0.040
Portier	9.924	20.76	0.042	0.832	1.00	65.46	0.858	217.2	0.040
Hans	8.871	17.76	0.033	1.011	1.00	65.59	0.857	219.7	0.033
Wan Hassan	8.648	17.86	0.026	1.253	1.00	69.17	0.886	222.5	0.041
Present Work	9.052	19.97	0.033	1.417	0.991	69.01	0.887	222.7	0.033

Note: $U^* = U/(\nu/\alpha)$ and $V^* = V/(\nu/\alpha)$

Table 1 shows a comparison of the *Adiabatic Case* results with the benchmark solution [3] and other comparable solutions. It is apparent that there is a good agreement between them over the range of Rayleigh numbers from 10^3 to 10^6 given our modest grid size (21x21). For higher Rayleigh numbers, 10^7 to 10^{12}, similar results are presented in Table 2. Table 3 lists a comparison of the average Nusselt numbers for high Rayleigh numbers with numerical and experimental correlations. The Nusselt numbers we predict are higher than the ones given by Wan Hassan [9] but they are in good agreement with ones published by Markatos [11] who used a much finer grid size and the experimental correlation of Jakob [12]. It is to be noted that the experimental data for high Rayleigh numbers are usually obtained from correlations that are extrapolated to the turbulent regime. For $Ra \geq 10^{10}$, we found that the local Nusselt numbers and thus the average Nusselt numbers, are very sensitive to the type of temperature fit used to evaluate the temperature gradients. Part of this sensitivity is attributed to the lack of sufficient meshing inside the boundary layer. Table 4 shows the results obtained for the *Conductive Case*. Table 5 shows a comparison of the vertical and horizontal Nusselt numbers and maximum vertical and horizontal velocities in the cavity with the data of Fraikin and Portier [13]. The average Nusselt numbers at the hot wall are within a 10% agreement with those predicted in [13]. The maximum horizontal and vertical velocities agree within 7% for the whole range of Grashof numbers. Turbulent quantities contours agree very well qualitatively but direct quantitative comparisons are not possible since Fraikin and Portier did not disclose how they accounted for the effects of the low turbulence number phenomena.

TABLE 2

Data of Turbulent Calculations
at High Rayleigh Numbers
Adiabatic Case

Ra	10^7	10^8	10^9	10^{10}	10^{11}	10^{12}
U_{max}	306.	1007.	2234.	6583.	17654.	36773.
V_{max}	703.	2100.	5921.	16053.	50157.	120852.
U_{maxh}	130.	544.	896.	2494.	5573.	9414.
V_{maxh}	675.	2036.	5806.	15935.	49292.	117124.
Nu	16.18	35.63	67.26	160.4	297.9	672.8
$Nu_{1/2}$	14.29	32.84	62.25	146.65	277.60	619.80
Nu_{max}	40.76	73.82	174.16	364.25	906.20	1656.5
Nu_{min}	1.92	5.21	12.56	11.73	47.37	103.40
k_{max}	1.07	79.71	257.	891.	1545.	4030.
ϵ_{max}	19.25	1.22E4	2.87E5	2.44E6	2.48E7	1.91E8
$\mu_{t\,max}$	1.54E-2	7.8	16.9	38.0	90.7	203.0

TABLE 3

Comparison of Average Nusselt Numbers
at High Rayleigh Numbers
Adiabatic Horizontal Walls

Ra	10^7	10^8	10^9	10^{10}	10^{11}	10^{12}
Ozoe et Al (E)	18.8	31.7	53.3	146.1	-	-
Wan Hassan (N)	15.65	32.46	61.45	129.0	238.8	432.6
Markatos (N) $Nu = 0.082Ra^{.329}$	16.47	35.14	74.96	159.88	341.05	727.48
Jakob (E) $Nu = 0.072Ra^{1/3}(\frac{L}{D})^{-1/9}$	15.53	33.21	71.51	153.93	331.38	713.40
Catton (E) $Nu = .18[\frac{RaPr}{.24+Pr}]^{.29}$	17.74	34.59	67.46	131.53	256.47	500.08
Emery and Chu (N) $Nu = 0.05Ra^{1/3}$				108.	232.	500.
Present results	16.18	35.63	67.26	160.4	297.9	672.8

E = Experimental   N = Numerical

TABLE 4

Results for the Conductive Case

Ra	10^3	10^4	10^5	10^6	10^7	10^8
U_{max}	3.82	20.20	59.50	191.96	590.87	1772.15
V_{max}	3.84	22.96	82.06	255.44	788.34	2423.80
U_{maxh}	3.81	20.19	54.35	154.62	444.05	1322.14
V_{maxh}	3.84	22.95	81.74	254.83	781.90	2403.20
Nu_H	1.05	1.81	3.43	7.73	16.36	35.76
Nu_L	0.34	1.27	2.17	4.47	10.11	21.46
$Nu_{1/2}$	1.08	2.19	4.17	8.32	16.71	34.76
Nu_{max}	1.23	2.66	5.09	12.79	27.28	65.74
Nu_{min}	0.87	0.857	0.987	1.041	1.270	1.103
k_{max}	-	-	-	18.18	51.01	137.83
ϵ_{max}	-	-	-	168.18	3,281.	37,679.
$\mu_{t\,max}$	-	-	-	4.58	11.23	25.76

TABLE 5

Comparison with Data of Fraikin and Portier

Gr	$6\ 10^6$ FP	$6\ 10^6$ PR	10^7 FP	10^7 PR	$5\ 10^7$ FP	$5\ 10^7$ PR	10^8 FP	10^8 PR
Nu_H	12.20	12.46	14.23	14.67	22.38	24.77	26.10	29.49
Nu_L	6.64	7.30	7.53	8.50	10.64	13.40	11.71	15.20
V_{max}	495.	535.	612.	679.	1461.	1376.	2107.	1948.
U_{max}	383.	390.	492.	504.	986.	1052.	1458.	1383.

FP = Data of Fraikin and Portier [13]
PR = Present Results

EFFECTS OF THE HORIZONTAL BOUNDARY CONDITION

In this section, we examine the floor and ceiling boundary conditions on the two dimensional natural convection temperature, flow profiles, heat transfer between the vertical walls and turbulence structure inside the cavity. Table 6 illustrates quantitatively the effects of the horizontal boundary condition. The results for the *Adiabatic Case* and the *Conductive Case* are presented simultaneously for the Rayleigh range $10^3 \leq Ra \leq 10^8$.

We first notice that the maximum horizontal and vertical velocities inside the cavity are consistently higher for the *Conductive Case*. Whereas the maximum vertical velocity inside the cavity is higher by 15%, the maximum horizontal velocity almost doubles for $Ra = 10^7$. It appears that the imposed horizontal boundary condition has a greater effect on the horizontal flow and a lesser effect on the vertical flow.

For $10^3 \leq Ra \leq 10^6$, corresponding to the laminar range of the *Adiabatic Case*, the predicted average Nusselt number at the hot wall is lower for the *Conductive Case*. However, for higher Rayleigh numbers, the average Nusselt numbers are slightly higher, about 5%. For $Ra = 10^8$, the average Nusselt numbers agree within 3%.

Of interest in the *Conductive Case* as opposed to the *Adiabatic Case*, is the onset of turbulence. In the conductive case, it starts at a lower Rayleigh number. This is due to the effect of the imposed horizontal boundary condition on the temperature distribution inside the cavity. In fact, as shown by the isotherms, as the Rayleigh number increases, some *unstable stratification* occurs next to the horizontal boundaries thereby constituting a source of production of kinetic energy whereas in the adiabatic case, the adiabatic condition causes a zero temperature gradient at the horizontal walls. This explains why turbulence is enhanced for the *Conductive Base Case*.

TABLE 6

Illustrating the Effect of the Nature of the
Horizontal Boundary Condition

Ra	10^4 AC	10^4 CC	10^5 AC	10^5 CC	10^6 AC	10^6 CC	10^7 AC	10^7 CC	10^8 AC	10^8 CC
U_{max}	15.74	20.20	42.73	59.50	113.31	191.96	308.48	590.87	1007.00	1772.15
V_{max}	19.41	22.96	69.46	82.06	222.78	255.44	705.28	788.34	2099.67	2423.80
U_{maxh}	15.73	20.19	34.41	54.35	69.00	154.62	136.14	444.05	544.16	1322.14
V_{maxh}	19.41	22.95	69.32	81.74	222.68	254.83	691.49	781.90	2035.75	2403.20
Nu_H	2.35	1.82	4.58	3.43	9.05	7.88	16.15	17.42	35.63	36.60
Nu_L	0.	1.27	0.	2.17	0.	4.47	0.	10.12	0.	21.46
$Nu_{1/2}$	2.44	2.19	4.57	4.17	8.20	8.32	14.29	16.71	32.84	34.76
Nu_{max}	3.81	2.66	8.04	5.09	19.97	12.75	41.26	27.28	73.82	68.30
Nu_{min}	0.58	0.86	0.79	0.99	1.42	1.04	1.90	1.27	5.21	1.10
k_{max}	-	-	-	-	-	18.18	-	51.01	79.71	137.82
ϵ_{max}	-	-	-	-	-	168	-	3,281	12,247	37,679
$\mu_{t max}$	-	-	-	-	-	4.58	-	11.23	7.78	25.77

AC = Adiabatic Case   CC = Conductive Case

DERIVED $Nu - Ra$ CORRELATIONS

The following correlations for average Nusselt numbers are derived from the present predictions by *least-square linear regression*. For the turbulent regimes two forms are given. The first one is obtained by correlating the average Nusselt number with Ra and the second one with $Ra^{1/3}$. The second alternative is offered since many experimental correlations force the 1/3 power for the turbulent regime.

Adiabatic Case

Laminar $10^3 \leq Ra \leq 10^6$

$$Nu = 0.145\, Ra^{0.299} \qquad (12.a)$$

Turbulent $10^7 \leq Ra \leq 10^{12}$

$$Nu = 0.092\, Ra^{0.321} \; or \; Nu = 0.072\, Ra^{1/3} \qquad (12.b)$$

Conductive Case

Laminar $10^3 \leq Ra \leq 10^5$

$$Nu = 0.175\, Ra^{0.257} \qquad (13.a)$$

Turbulent $10^5 \leq Ra \leq 10^8$

$$Nu = 0.078\, Ra^{0.332} \; or \; Nu = 0.077\, Ra^{1/3} \qquad (13.b)$$

SENSITIVITY STUDY ON THE TURBULENCE CONSTANTS

A sensitivity study was carried out in order to determine the effect of the various model constants on the calculated results. The constants studied are the three constants C_1, C_2, C_3 in the ϵ equation because they determine the balance between the influence of shear and buoyancy on the turbulent length scale. The constant C_μ is also considered to see how the low-Reynolds number formulation affects it and its consequences for the mean flow variables. We show in Table 7 the influence of the constants on the following parameters: mean Nusselt number along the hot wall, mean Nusselt number along the lower wall, maximum vertical velocity, maximum horizontal velocity, maximum turbu-

lent viscosity, maximum turbulent kinetic energy and maximum dissipation rate. The reference values are obtained for the Conductive Case at $Gr = 10^7$ using the constants in equation 8 along with the low turbulence Reynolds formulation for the constants C_μ and C_2 as shown in equations 9.

A variation of C_1 by $\mp 20\%$ has a negligible effect on V_{max} and U_{max} but has a moderate effect of the order of $\pm 16\%$ on the average Nusselt numbers at the hot wall. Its effects on the turbulent quantities are large. The maximum turbulent viscosity changes by $+30\%$ and -27% respectively. In general, *an increase in C_1 leads to a decrease in the turbulent quantities whereas its decrease causes them to increase.*

A variation of C_2 by $\mp 20\%$ also has negligible effects on V_{max} and U_{max} and a moderate effect on the average Nusselt numbers, as indicated in Table 7. The effects on the turbulent quantities is generally large, leading in some cases to significant changes in the maximum turbulent viscosity. Note the reverse effect the constant C_2 has on the maximum turbulent quantities. *Unlike C_1, an increase in C_2 will produce an increase in the maximum turbulent quantities and vice a versa.*

The constant C_3 is assigned a value of 1.44 for vertical flows and 0.4 for horizontal flows and in between them, it varies linearly as a function of the angle of inclination of the velocity vector from the horizontal. We tested this constant by fixing its value to 0.4 over the whole flow domain and the results obtained are not much different from the reference results as indicated in Table 7. However, using a value of 1.44 instead produced large effects on the average Nusselt numbers and maximum velocities and almost eliminated turbulence.

TABLE 7

Results of Sensitivity Study of
the Turbulence Constants, $Gr = 10^7$
(Conductive Case)

Constants Tested	Nu_H	Nu_L	V_{max}	U_{max}	$\mu_{t max}$	k_{max}	ϵ_{max}
Reference Values	15.52	9.11	18.29	13.57	10.18	44.99	2320.63
$C_1 = 1.15\ (-20\%)$	18.31	10.24	17.62	13.01	13.36	53.62	2805.23
	M	M	N	N	L	M	L
$C_1 = 1.73\ (+20\%)$	13.09	7.46	18.83	12.94	7.41	35.50	1286.72
	M	M	N	N	L	L	L
$C_2 = 1.54\ (-20\%)$	12.50	6.15	18.28	11.77	4.49	23.31	807.85
	M	L	N	M	L	L	L
$C_2 = 2.30\ (+20\%)$	17.34	9.98	17.80	12.99	13.01	51.85	2679.42
	M	S	N	N	L	M	M
$C_3 = 0.40\ (-43\%)$	15.76	9.11	18.43	13.94	10.62	47.20	2385.80
	N	N	N	N	N	N	N
$C_3 = 1.44\ (+106\%)$	12.08	5.55	16.75	10.55	0.93	14.94	380.64
	L	L	S	L	L	L	L
$C_\mu = constant$ $C_2 = f(R_t)$	14.09	7.69	17.74	12.27	5.50	27.25	2981.99
	S	M	N	S	L	L	L
$C_\mu = constant$ $C_2 = constant = 1.44$	14.09	7.69	17.74	12.27	5.50	27.25	2981.86
	S	M	N	S	L	L	L
$\sigma_T = 0.70\ (-30\%)$	17.87	10.71	19.87	15.20	10.66	55.16	2990.67
	M	M	S	M	N	L	L
$\sigma_T = 1.30\ (+30\%)$	14.21	8.13	17.31	12.33	9.60	39.02	1955.34
	S	S	S	S	S	L	M

$L \equiv$ LARGE or Change is $\geq 20\%$
$M \equiv$ MODERATE or Change is 10% to 20%
$S \equiv$ SMALL or Change is 5% to 10%
$N \equiv$ NEGLIGIBLE or Change is $\leq 5\%$

A calculation is performed by ignoring the low turbulence Reynolds number formulation of constant C_μ and fixing its value to 0.09 everywhere in the flow field. Although small effects are produced on the velocities and average Nusselt numbers, the maximum turbulent viscosity decreased by 50%. However, a similar test done for the constant C_2 did not produce any effect on any of the variables.

In the $k - \epsilon$ model, we use a constant σ_T which is very important to obtain a good description of the turbulent heat fluxes. Therefore, we tested the constant σ_T as well. The result of varying σ_T by $\mp 30\%$ is shown in Table 7. We see that its effects are reasonable for most of the variables except for the dissipation rate.

A choice of the values of the different constants could be made if experimental data were available. We need, above all, measurements of turbulent quantities as we have seen the influence of the constants is most critical on them. This conclusion is in agreement with the study of Fraikin and Portier who also found that the constants have a great influence on the turbulent quantities.

APPLICATIONS

A major problem in the design of passive solar homes and in the general evaluation of occupant comfort is the interaction of the free convection due to heat loss from cold windows and the convection caused by the absorption of solar radiation on the floor and walls. The augmentation or reduction in surface heat transfer coefficients due to these interactions can substantially alter occupant comfort and can change the required heating system loads by as much as 50 % [15]. When drafts (due to window openings or leakage through internal partitions or under doors) interact with the normal internal free convection flow, the entire flow field may be significantly distorted [16]. Since comfort is dominated by the local air temperature and velocity [17], the resulting complex flow patterns often create multiple zones of circulation and discomfort. There are only a few papers [16,18] which discuss the effects of infiltration, either spatially diffuse or concentrated in a jet (as is common with most drafts) or of local cold or hot spots (as found at windows or wall electrical heaters) or of partitions (as found below cold windows). Since such effects can completely alter the usual flow patterns by reversing flow direction, creating multiple cells, or increasing or decreasing local heat transfer coefficients, it is important to be able to predict the resultant effects.

As we mentioned previously, such flows are not only three dimensional in nature but they are also turbulent. In this study, we apply a validated numerical algorithm developed for two dimensional turbulent free convection in cavities in an attempt to understand the physical mechanisms associated with cavities subjected to more complex boundary conditions. Although the results may not be representative of the actual complex three dimensional flow under such conditions, they certainly help in understanding the effects of perturbations on the basic two dimensional flow.

In this section, we present the results obtained for several configurations which simulate the effects of plumes, infiltration and horizontal partitions on the basic flow pattern in two dimensional cavities. The physical model is a square enclosure with the hot wall and the cold wall maintained at constant temperatures and a Rayleigh number of 10^7. This case will be taken as a reference case for comparison.

EFFECTS OF PLUMES *

With the results of configuration 4a taken as a reference case, we consider two variations of the physical model. First, The two vertical walls are at the same temperature, and a plume source ** of width s is centered at the floor. In this case, the flow pattern will be dictated by the plume only. Secondly, we establish a differential temperature of $2C$ between the vertical walls in order to simulate the effect on the base circulation flow of the sun entering through a window and heating a central portion of the floor creating a plume, Figure 4b.

Results and Discussion

(a) *The case of plume only*

In this case, the vertical walls are kept at the same temperature, 21C, and a plume source of 30C, occupying 53% of the floor area, is centered on the floor. The hot spot of 30C, centered on the floor creates a vertical flow of hot air which splits at the ceiling, thereby dividing the cell into two distinct regions, Figure 5a. After the flow splits at the ceiling, it moves along the ceiling towards the vertical walls and then descends along the vertical walls. The fluid then moves horizontally towards the plume which entrains it towards the ceiling, thereby closing the loop. The problem is symmetrical and the flow thus obtained is also symmetrical. *The most interesting feature to note is that the flow is now turbulent.* In fact, a very high temperature gradient exists in the vicinity of the hot spot, thereby favoring turbulent kinetic energy generation by buoyancy. The flow is subjected to no resistance and consequently, the turbulent kinetic energy is transported by the vertical flow and spread throughout the cavity. From the contour representations of the kinetic energy of turbulence and the turbulent viscosity throughout the cavity, Figures 5c and 5d, it is seen that although most of the cavity flow is turbulent; the turbulence kinetic energy is especially important where the plume originates and where it splits.

(b) *The case of plume with base flow*

In this case, the base flow competes with the plume in controlling the flow pattern. The hot wall is maintained at 21C while the cold wall is maintained at 19C. Three different strengths of the plumes are considered by examining a central plume source with temperatures of 24C, 30C and 36C. For each of these temperatures, computations are carried out for central plume sources occupying respectively, 13%, 35%, 53% and 68% of the floor area. In this way, both the effects of plume temperature and size will be observed.

Figure 6 illustrates the effect of increasing the plume temperature for the case of $s/L = 0.35$, in terms of velocity patterns and isotherms. At a low plume source temperature, 24C, the velocity pattern indicates that the flow structure was not significantly affected. In fact, the plume which rises at the hot spot is almost completely anihilated by the stong base flow. A small vortex at the lower left corner and another in the top left corner of the cavity, although weak, are the primary indication of the start of the plume effect. At a higher plume source temperature, 30C, the effect is stronger and characterized by an enhancement of the two vortices which become more noticeable. At the high-

---

A plume refers to the convection created by an isothermally heated portion of the floor and/or vertical side walls.

A plume source refers to the actual heated spot which creates the plume.

est plume source temperature, 36C, the velocity pattern shows a vertical flow originating at the hot spot which splits at the ceiling, thereby creating two distinct cells in the cavity. The asymmetry of the flow is caused by the base flow which still creates a counterclockwise flow therby pushing slightly the vertical flow to the left. The flow is now totally disturbed and dictated by the plume effect. Further increases in the plume source temperature did not affect the flow pattern significantly.

Table 8 summarizes the quantitative results obtained. In general, they indicate that both the size and the temperature of the plume source contribute to affecting the base flow. Increasing the size of the plume source or its temperature both affect the flow by increasing the circulation and enhancing turbulence.

TABLE 8

Characteristics of a Room with a Central Plume
$Ra = 10^7$

	Base Case	$T_{plume} = 24C$	$T_{plume} = 30C$	$T_{plume} = 36C$
$s/L = 0.13$				
R_{max}	18.40	24.26	30.81	33.68
Nu_1	17.38	4.16	-12.45	-34.04
Nu_3	17.38	25.42	41.98	63.26
Nu_5	0.	169.54	492.06	895.60
$\mu_{t\,max}$		9	11	19
$s/L = 0.35$				
R_{max}	18.40	28.29	34.18	45.21
Nu_1	17.38	-4.90	-43.73	-78.64
Nu_3	17.38	34.01	70.87	133.38
Nu_5	0.	123.24	388.30	748.45
$\mu_{t\,max}$		12	23	36
$s/L = 0.53$				
R_{max}	18.40	29.65	36.12	49.28
Nu_1	17.28	-10.93	-65.55	-116.89
Nu_3	17.28	39.31	91.41	163.67
Nu_5	0.	107.34	349.87	675.25
$\mu_{t\,max}$		13	31	44
$s/L = 0.68$				
R_{max}	18.40	30.33	37.20	51.65
Nu_1	17.38	-16.17	-83.13	-145.85
Nu_3	17.38	43.61	103.68	188.13
Nu_5	0.	99.61	333.12	650.10
$\mu_{t\,max}$		15	36	49

1 = Hot wall surface
3 = Cold wall surface
5 = Heated spot surface

EFFECTS OF INFILTRATION *

With the configuration shown in Figure 4a taken as the reference case, we consider the following situations

1. Air is injected at the midheight of the hot wall and allowed to exit at the midheight of the cold wall. This configuration is studied to gain some understanding of the infiltration velocities needed to create multiple cells and to simulate the effects of horizontal drafts and cross ventilation.

Infiltration refers to the injection and exit of air through the vertical side walls, e.g., drafts, window openings, or leakage through internal partitions or under doors

2. Air is injected at the upper edge of the hot wall and is allowed to exit at the lower edge of the cold wall. This case simulates the conditions in a living space connected to a Trombe wall, one of the more common passive solar architectural constructions.

3. Air is injected at the lower edge of the hot wall and is allowed to exit at the upper edge of the cold wall. This simulates the flow conditions in the solar energy absorbing portion of a Trombe wall and spaces with floor heating vents.

4. Air is injected at the lower part of the hot wall and is allowed to exit at the top of the same wall. This configuration simulates the flow conditions of many exterior facing offices in which a cold draft, from an adjoining corridor, sweeps under the door and exhausts through a highly placed vent into the corridor.

In all of the cases with infiltration, the infiltrated air enters with a temperature equal to the mean temperature of the room and $Ra = 10^7$. For each case, three infiltration velocities will be considered in order to examine the effects of increased injection.

The injection velocities which we refer to subsequently are normalized such that each velocity unit represents 2.34×10^{-3} m/s and they are chosen as a function of the maximum non-dimensional velocity of the Base Case ($15.57 \equiv 0.037m/s$). However, since the injection velocity by itself does not describe adequately the situation, for each normalized injection velocity, we associate its corresponding normalized mass flow rate defined as the average mass flow rate that occurs through the half midplane of the reference Base Case, $m_{ref} = 0.04$. All the quantitative comparisons are made with respect to the Base Case results.

Results and Discussion

(a) _The case of midheight infiltration_

At low injection velocity (0.03 m/s), of the order of the maximum velocity characterizing the reference case, $m_{in} = 25\,m_{ref}$, the flow rising from the lower part of the hot wall is slightly turned and a zone of relatively high horizontal velocity is created just above the injection site. The average temperature at the lower part of the hot wall increases by 14% causing a 15% decrease in the correponding Nusselt number. In addition, the average temperature at the upper part of the cold wall decreases by 6% causing a 6% decrease to its associated Nusselt number. The average velocity at the upper part of the hot wall decreases by 18%. In fact, the injected air blocked the flow rising from below and this disturbance caused all the boundary layer velocities to decrease as shown in Table 9. Note, however, that the velocities at the hot wall are the most affected. For a higher injection velocity (0.046 m/s), or $m_{in} = 39\,m_{ref}$, a local reverse flow cell is almost completely formed directly above the injection site. The flow in the lower part of the space retains its original direction, Figure 7a. For this increased jet velocity, the trends noted above are continued consistently. The average temperature at the lower part of the hot wall now decreases by 18% resulting in an 18% decrease in the corresponding Nusselt number and the average velocities at the walls decrease especially at the hot wall. Note that the average velocities on the floor and ceiling are slightly affected. With an increase in the jet velocity to 0.06 m/s such that, $m_{in} = 51\,m_{ref}$, some of the trends noted above are continued, but the effects are much smaller, as shown

124

TABLE 9

Characteristics of a Room with Midplane Injection
$Ra = 10^7$ and $m_{ref} = 0.04$
subscripts refer to surfaces depicted in Figure 4c

	$U_{in} = 0.$ $m_{in} = 0.$	$U_{in} = 13.$ $m_{in} = 1.02$	$U_{in} = 19.5$ $m_{in} = 1.53$	$U_{in} = 26.$ $m_{in} = 2.04$
T_1	0.251	0.287	0.295	0.301
T_2	-0.442	-0.402	-0.398	-0.387
T_3	0.442	0.429	0.430	0.432
T_4	-0.251	-0.266	-0.276	-0.287
R_1	11.09	9.16	8.91	8.80
R_2	7.65	7.75	7.19	7.24
R_3	7.65	4.72	4.10	3.94
R_4	11.09	10.58	10.35	10.14
T_{up}	0.319	0.327	0.320	0.311
T_{dw}	-0.319	-0.296	-0.287	-0.272
R_{up}	1.522	1.442	1.459	1.504
R_{dw}	1.522	1.529	1.528	1.625
R_{max}	15.57	17.16	20.01	25.96
Nu_1	29.19	24.97	24.04	23.30
Nu_2	6.80	11.45	11.90	13.18
Nu_3	6.80	8.33	8.18	7.95
Nu_4	29.19	27.44	26.28	24.95
Nu_{draft}	0.	5.59	5.96	6.88

TABLE 10

Characteristics of a Room with Injection
from the Upper Edge of the Hot Wall
1=Hot wall and 2=Cold wall
$Ra = 10^7$ and $m_{ref} = 0.04$

	$U_{in} = 0.$ $m_{in} = 0.$	$U_{in}=60.$ $m_{in}=1.05$	$U_{in}=90.$ $m_{in}=1.575$	$U_{in}=120.$ $m_{in}=2.10$
T_1	0.341	0.409	0.418	0.408
T_2	-0.341	-0.335	-0.329	-0.313
R_1	10.160	8.522	8.862	9.834
R_2	10.160	16.238	18.512	21.935
R_{max}	15.57	59.19	89.28	119.18
T_{up}	0.319	0.027	0.000	-0.025
T_{dw}	-0.319	0.012	0.081	0.068
R_{up}	1.522	20.766	38.593	60.356
R_{dw}	1.522	5.045	8.738	11.763
Nu_1	18.63	10.60	9.65	10.79
Nu_2	18.63	19.36	20.03	21.91
Nu_{draft}	0.	8.76	10.38	11.12

TABLE 11

Characteristics of a Room with Injection
from the lower Edge of the Hot Wall
$Ra = 10^7$ and $m_{ref} = 0.04$

	$U_{in} = 0.$ $m_{in} = 0.$	$U_{in}=60.$ $m_{in}=1.05$	$U_{in}=90.$ $m_{in} = 1.575$	$U_{in}=120.$ $m_{in}=2.10$
T_1	0.362	0.378	0.380	0.383
T_2	-0.362	-0.380	-0.389	-0.353
R_1	10.07	9.780	9.801	9.570
R_2	10.07	8.284	9.017	9.155
T_{up}	0.319	0.210	0.222	0.227
T_{dw}	-0.319	-0.205	-0.143	-0.046
R_{up}	1.522	8.031	11.990	13.952
R_{dw}	1.522	17.868	35.841	58.789
R_{max}	15.57	59.12	89.25	119.21
Nu_1	16.12	14.30	14.01	13.70
Nu_2	16.12	14.07	12.97	17.22
Nu_{draft}	0.	0.23	1.04	3.52

in Table 9. For instance, increasing the jet velocities didn't produce stronger effects for the average temperatures at the top of the hot wall and the ceiling. In fact, none of the variables presented in the Tables varied proportionately to the jet velocity. It thus appears that once the boundary layer of the hot wall has been affected, further increases in infiltration velocity are less effective in altering the flow.

(b) *The case of Infiltration from the Upper Left Edge*

For a low injection velocity (0.14 m/s), in the order of 4 times the maximum velocity characterizing the reference case, or $m_{in} = 26.25\ m_{ref}$, the flow is accelerated at the top and right sides of the cavity, Figure 7b. The average temperature at the ceiling decreases substantially to close to the temperature of the injected air but increases at the floor. The average temperature at the hot wall increases by 20% causing a 43% decrease in its associated Nusselt number. The average velocity at the cold wall increases by 60% but only a 16% decrease is noted at the hot wall. Note also in Table 10 that the velocities at the ceiling, cold wall and floor have substantially increased. However, the effect on the velocities at the hot wall are less dramatic. At a higher injection velocity (0.21m/s), or $m_{in} = 52.5\ m_{ref}$, the flow is accelerated everywhere in the cavity. The velocities increase everywhere except at the hot wall where a 10% decrease is observed. We also note that this increased jet velocity didn't alter further the average temperature at the hot wall and thus the average Nusselt number was unchanged. Increasing the injection velocity to 0.28m/s such that $m_{in} = 52.5\ m_{ref}$, further accelerated the flow inside the cavity. As shown in Table 10, this increased jet velocity didn't produce stronger effects on the Nusselt number and average velocity at the hot wall. It appears that the effect of this type of infiltration is restricted to the ceiling, the cold wall and the floor.

(c) *The Case of Infiltration from the Lower Edge of the Hot Wall*

At low injected velocity (0.14 m/s), about 4 times the maximum velocity characterizing the base case, or $m_{in} = 26.25\ m_{ref}$, the injected air counters the flow sweeping across the floor and turns it in the upward direction, away from the hot wall. In contrast to the case of infiltration from the upper edge of the hot wall, injection at the bottom forces a substantial realignment of the flow. A strong recirculation cell is formed at the injection site. The average temperature at the floor decreases by 36% due to the relatively low temperature of the injected air. However, the average temperature at the vertical walls changed only slightly, less than 5%. The average Nusselt number decreased by 12% at both walls, Table 11. For a higher injection velocity (0.21 m/s), or $m_{in} = 52.5\ m_{ref}$, the jet infiltrates further across the floor, Figure 7c. The average velocities at the floor and ceiling increase. Again, in this case, the increased jet velocity didn't alter further the average velocities at the vertical walls and the Nusselt numbers. Increasing the injection velocity to 0.28 m/s) such that $m_{in} = 52.5\ m_{ref}$, causes the infiltration to become so forceful that it controls the whole flow inside the cavity. A large vortex dominates the central portion of the cavity and two smaller vortices appear at the top corners, Figure 7d. With this increased jet velocity, the heat transfer at the hot wall isn't altered significantly.

TABLE 12

Characteristics of a Room with Injection
from the lower Edge of the Hot Wall
and Exit at the Top of the Same Wall
$Ra = 10^7$ and $m_{ref} = 0.04$

	$U_{in} = 0.$ $m_{in} = 0.$	$U_{in} = 60.$ $m_{in} = 1.05$	$U_{in} = 90.$ $m_{in} = 1.575$	$U_{in} = 120.$ $m_{in} = 2.10$
T_1	0.356	0.351	0.349	0.336
T_2	-0.348	-0.424	-0.428	-0.387
R_1	10.632	13.387	13.522	14.721
R_2	9.652	7.388	7.490	7.029
T_{up}	0.319	-0.043	-0.105	-0.123
T_{dw}	-0.319	-0.208	-0.143	-0.048
R_{up}	1.522	4.569	7.325	10.144
R_{dw}	1.522	17.986	35.842	58.958
R_{max}	15.57	59.11	89.25	119.22
Nu_1	16.92	17.49	17.64	19.23
Nu_2	17.78	8.85	8.43	13.18
Nu_{draft}	0.	8.64	9.21	6.05

(d) *The Case of Infiltration from the Lower Edge of the Hot Wall and Exit at the Top of the Same Wall*

At low injection velocity (0.14 m/s), about 4 times the maximum velocity characterizing the reference case, or $m_{in} = 26.25 \, m_{ref}$, the injected air diverts the flow sweeping across the floor and a local vortex is formed right at the injection site. We observe a 26% increase in the average velocity at the hot wall and a 23% decrease at the cold wall. This was to be expected since part of the air exits at the top of the left vertical wall. Due to the relatively low temperature of the injected air, we note a substantial decrease in the average temperature of the ceiling. The average temperature at the hot walll decreases by 1% causing a 3% increase in its associated heat transfer. The average temperature at the cold wall drops by 22% causing a 50% decrease in the corresponding heat transfer coefficient, Table 12. For higher injection velocity (0.21 m/s), or $m_{in} = 53.5 \, m_{ref}$, the injected air sweeps further across the floor, Figure 7e. However, the average temperature, velocity and Nusselt number at the hot wall didn't respond strongly to this increased infiltration velocity. By increasing the injection velocity to 0.28 m/s such that $m_{in} = 53.5 \, m_{ref}$, the effect of the base flow is almost completely anihilated. The injected air sweeps along the whole floor, encounters the right vertical wall, rises and encounters the descending fluid at the cold wall and is directed towards the center of the cavity. Overall, a strong vortex is created inside the cavity. Most of the cavity is characterized by high velocities, Figure 7f. The average temperature at the cold wall decreases by 11% leading to a 26% decrease in the corresponding average Nusselt number. Note also in this case the insignificant effect on the characteristics of the hot wall.

INTERACTION OF PLUMES AND INFILTRATION

In this application, we have three effects which will compete in trying to control the flow, namely, the *base flow*, the *flow created by the plume* and the *flow created by the infiltrating air*. The physical model used is similar to the one depicted in Figure 4b with a central plume source occupying 53% of the floor area and with air injected at the mid-height of the hot wall and allowed to exit at the mid-height of the cold wall. The enclosure is such that the hot wall is maintained at 21C and the cold wall at 19C and

$Ra = 10^7$. However, in this case, *instead of changing the inlet velocity of infiltrating air, we will change its temperature*. This will complement the previous application in understanding the effect of inlet temperature on the flow pattern. Since infiltration is involved in this case, the reference mass flow rate we use will be the average mass flow rate through the half midplane of the corresponding reference case, $m_{ref} = 0.034$. The magnitude of the injection velocity is 0.047 m/s.

Results and Discussion

Figure 8a shows the flow pattern when no injection exists. We inject air at a temperature of 20C and with a velocity of magnitude equal to 75% of the maximum velocity characterising the reference case, or $m_{in} = 74 \, m_{ref}$. The resulting flow pattern shown in Figure 8b shows that the infiltration has tremendously changed the flow structure inside the cavity. In fact, the injected air descends immediately due to its relatively low tempearture and sweeps strongly across the floor, thereby, anihilating the usual vertical flow effect of the plume. Most of cavity is characterized by higher velocities. Figure 8c shows the flow pattern resulting from the same injection as previously but at a temperature of 25C. This flow pattern is very different from the previous one. In this case, the injected air is hotter and therefore the down draught is less dramatic. The ascending air due to the plume is directed towards the exit opening and results in the creation of two distinct vortices above the injection site and below the exit site. Figure 8d shows the flow pattern resulting from the same injection as previously but at a temperature of 28C. The same trends as above are noted with the difference that the vortex above the injection site is smaller. The injected air is hotter than previously and therefore descends less.

EFFECTS OF HORIZONTAL PARTITIONS

For this last application, the Rayleigh number chosen is 10^4. The presence of any solid partition inside the cavity will necessitate a fine grid distribution in its neighborhood to describe adequately the gradients. Since our grid size is only 21x21, we chose a low Rayleigh number to be able to use a uniform grid spacing and carry out the calculations by anticipating no loss in generality.

Several diverse situations are considered by examining a room, $Ra = 10^4$, with the following boundary conditions

1. The room is closed, Figure 4e. A heater is placed at the bottom of the right wall and a cold window occupies half of the right vertical wall. *All the remaining walls are adiabatic.* This case simulates the winter flow conditions in a room.

2. The same configuration as above but a horizontal partition is added at the midheight of the cold wall, Figure 4f. This a very common configuration, more realistic than the previous one. We study this case to determine the effect of such partitions on the flow conditions inside the room.

3. The same conditions as above with air entering the room at the bottom of the door and exiting at the top of the cold window.

Results and Discussion

1. The Reference case

Figures 6a and 6b show the resulting velocity pattern and isotherms. The cold air descending at the cold window strongly counters the rising air at the heater to form a jet. The effect of the heater is characterized by a local vortex in its neighborhood. Note from the temperature contours in Figure 9b that most of the cavity is cold.

2. The Reference Case with Horizontal Partition

Figures 9c and 9d show the resulting velocity pattern and isotherms for the case of a small partition, $s/L = 0.20$. We observe a significant change in the flow structure inside the cavity. In this case, the flow rising at the heater is stronger. It appears that the horizontal partition somewhat weakened the descending flow at the cold window thereby allowing the warm air from the heater to rise further. The jet thus created divides the cavity flow into two important vortices moving in different directions. The vortex associated with the heater dominates the cavity, thereby isolating the flow of cold air to the window location. Note from Figure 9d that most of the cavity is now warmer. The partition size is further increased, $s/L = 0.30$, and Figures 9e and 9f show the corresponding velocity pattern and isotherms. No effect is noticed on the flow structure. However the temperature contours indicate a further domination of the cavity temperature by the heater temperature. Except for the region next to the cold window, most of the room is warm.

3. The Reference Case and Infiltration at Bottom of Door and Exit at Top of Window

Air is injected at the bottom of the door and allowed to exit at the top of the cold window and the partition size is such that $s/W = 0.10$. The injection velocity is about twice the maximum velocity of the reference case, $m_{in} = 1.43\ m_{ref}$. The temperature of the injected air is set to the mean temperature T_M. Figures 10e and 10f show the corresponding velocity pattern and isotherms. The injected cold air sweeps across the floor and ascends along the heater. Originally, the flow in the lower portion of the cavity was characterized by a strong vortex moving counterclokwise. The location of infiltration simply accelerates the flow. When the configuration is subjected to infiltration at a higher temperature T_h, Figures 10 show the correponding velocity and isotherms. The injected air rises almost immediately and the vortex next to the heater is unaffected. Most of the cavity is characterized by high velocities. Note that the high jet velocities at the partition location can cause uncomfortable feelings to the neck of a person sitting closeby.

The perturbation in this case is the partition. We have observed that a small horizontal partition placed directly below the cold window caused a dramatic change in the flow pattern inside the cavity. However, further increases in its size didn't produce stronger effects as far as the flow pattern was concerned.

CONCLUSIONS

With the numerical technique presented here, predictions have been obtained for buoyancy-induced flows inside a cavity for a wide range of Rayleigh numbers, $10^3 \leq Ra \leq Ra^{12}$, with a modest grid size 21x21. Two types of horizontal boundary conditions were examined, namely, adiabatic (*Adiabatic Case*) and perfectly conductive (*Conductive Case*). Through numerical experimentations with different boundary conditions, we conclude the following:

1. Good predictions for a wide range of Rayleigh numbers have been obtained for both the *Base Case* and the *Conductive Base Case*.

2. Useful correlations for the average Nusselt numbers have been established for both cases.

3. Numerical predictions agree very well with published numerical and experimental data.

4. Changing horizontal boundary condition from adiabatic to perfectly conductive produces large effects on horizontal velocity and small effects on the average Nusselt number. More importantly, turbulence is enhanced.

5. The low turbulence Reynolds number formulation requires only a readjustment of the constant C_μ. Using a constant value of C_μ everywhere in the flow field will cause a serious underprediction of turbulent quantities.

6. Changing the turbulent constants produces large effects on the turbulent quantities. Generally, the effects are reasonable as far as the velocities and the average Nusselt numbers are concerned. An increase in C_1 leads to a decrease in the turbulent quantities whereas its decrease causes them to increase. The variation of the constant C_2 produces opposite effects.

7. Isolated hot spots on the floor of the cavity in some cases overcome the base flow generated by internal convection. Plumes, when strong, not only alter flow configurations but are a high source of turbulence. The data indicated that both the plume temperature and its size are important in affecting the flow structure of the base flow.

8. In the case of infiltration, the base flow is easily upset by slight amounts of infiltration but further increases in the infiltration velocities do not elicit stronger responses. This conclusion is in agreement with the common observation that slight drafts can be easily felt by occupants, but stronger drafts do not produce a stronger response. Part of this is due to the nature of human comfort, being related to both skin temperature and moisture diffusion, but these results clearly indicate that very small infiltrations can cause significant effects in terms of fluid velocities, direction and formation of recirculating cells.

9. In the event of infiltration, the temperature of the injected air is just as important as the injected velocity. When cold air is injected, it tends to descend quickly and when it's warm, it rises quickly, thereby affecting the flow pattern differently.

10. While a partition causes changes in the flow pattern of the cavity, further increases in its size didn't produce stronger responses. A small partition placed directly below a cold window and above a wall heater blocks the strong downdraft along the window and allows the warm air from the heater to rise further and enhance the mixing throughout the cavity. We also observed that the region next to the window is characterized by high jet velocities and this explains the discomfort often felt when sitting there.

11. A robust numerical scheme has been developed which takes into account all the above special boundary conditions. The predicted results are independent of the initial conditions.

REFERENCES

1. Catton I., "Natural Convection in Enclosures", *6th International Heat transfer Conference*, Toronto, 1978, Vol. 6, 1979, pp. 13-43

2. Ostrach S., "Natural Convection in Enclosures", *Adv. Heat Transfer*, Vol. 8, 1972, pp. 161-227

3. De Vahl Davis G., and Jones I. P., "Natural Convection in a Square Cavity: a Comparison Exercise", *International Journal for Numerical Methods in Fluids*, Vol. 3, pp. 227-248, 1983

4. Rodi W., "Influence of Buoyancy and Rotation on Equations for the Turbulent Length Scale", 2nd Symposium on turbulent shear flows, *Journal of Fluid Engineering*, Vol 97, No. 3, pp 386-389, 1975

5. Rodi W., "A Note on the Empirical Constant in the Kolmogorov-Prandtl Eddy-Viscosity Expression", *Journal of Fluids Engineering*, Vol. 97,No. 3, pp 386-389, 1975

6. Launder B. E. and Spalding D. B., "Numerical Computation of Turbulent Flows", *Computer Methods in Applied Mechanics and Engineering*, Vol. 3, pp 269-289, North-Holland, 1974

7. Patankar S. V. and Spalding D. B., "A Calculation Procedure for Heat, Mass and Momentum Transfer in Three Dimensional Parabolic Flows", *International Journal of Heat and Mass Transfer*, Vol. 15, pp. 1787-1806, 1972

8. De Vahl Davis G., "Natural Convection of Air in a Square Cavity: a Benchmark Solution", *International Journal for Numerical Methods in Fluids*, Vol 3, pp 249-264, 1983

9. Wan Hassan M., "The Natural Convection in a Cavity at High RAyleigh Numbers", *PhD Thesis*, Department of Mechanical Engineering, University of Washington, 1986.

10. Ozoe H. et Al, "Numerical Calculations of Laminar and Turbulent Convection in Water in Rectangular Channels and Cooled Isothermally on Opposing Vertical Walls", *International Journal of Heat and Mass Transfer*, Vol. 28, No 1, 1985, pp. 125-137.

11. Markatos N. C. and Pericleous C. A., "Laminar and Turbulent Natural Convection in an Enclosed Cavity", *HTD* Vol 26, pp. 59-68, 1983.

12. Jakob M., *Heat Transfer*, Vol 1, John Wiley & Sons, New York, 1949.

13. Fraikin M. P., Portier J. J. and Fraikin C. J., "Application of a k-ε Model to an Enclosed Buoyancy Driven Recirculating Flow", *Chem Eng Commun*, V 13 n 4-6, pp. 289-314, 1982

14. Emery A. F., and Chu N. C., "Heat Transfer Across Vertical Layers", *Journal of Heat Transfer*,Transactions of the ASME, Series C, Vol 85, pp. 110-114, 1965

15. Emery, A.F., Kippenhan, C.J., Heerwagen, D.R., and Varey, G.B., "The Simulation of Building Heat Transfer for Passive Solar Systems", *Energy and Buildings*, vol 3, pp. 287-294, 1981

16. Euser, H., Hoogendoorn, C.J., and Van Ooijen, H., "Airflow in a Room Induced by Natural Convection Streams", *Energy Conservation in Heating, Cooling, and Ventilating Buildings*, Ed. Hoogendoorn and Afgan, Hemisphere Publ., pp. 259-270, 1978

17. Fanger P.O., *Thermal Comfort*, McGraw Hill Book Co., 1970

18. Abrous, A.,"A Numerical Study of the Turbulent Free Convection in Rectangular Cavities", *Ph.D Thesis*, University of Washington, 1988

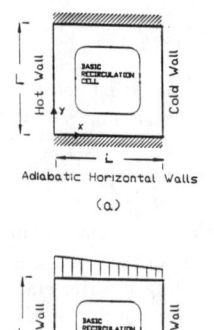

Figure 1. Geometry and Coordinate System

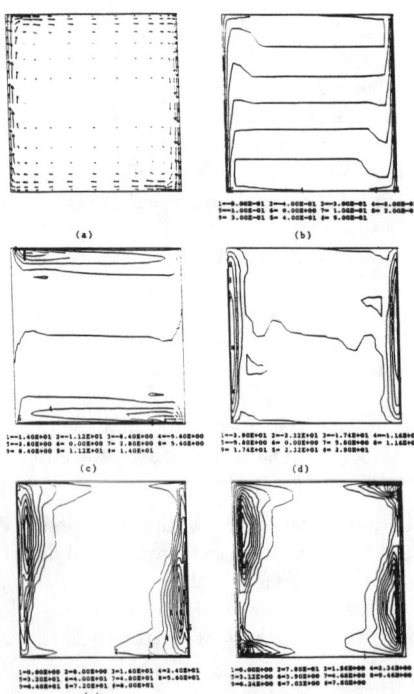

Figure 2. Velocity Pattern and T, U, V, k, μ_t Contours
$Ra = 10^8$, *Adiabatic Case*

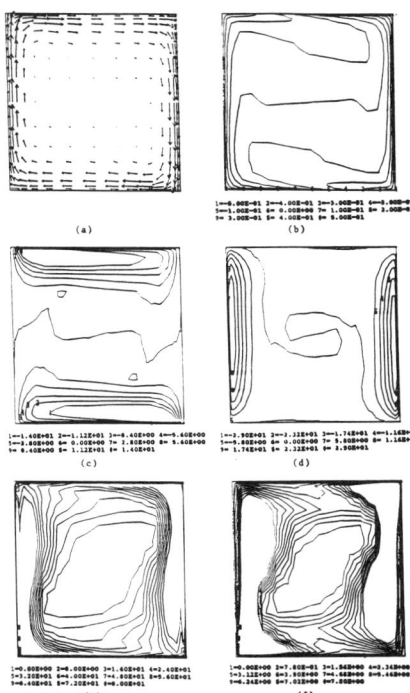

Figure 3. Velocity Pattern and T, U, V, k, μ_t Contours
$Ra = 10^8$, *Conductive Case*

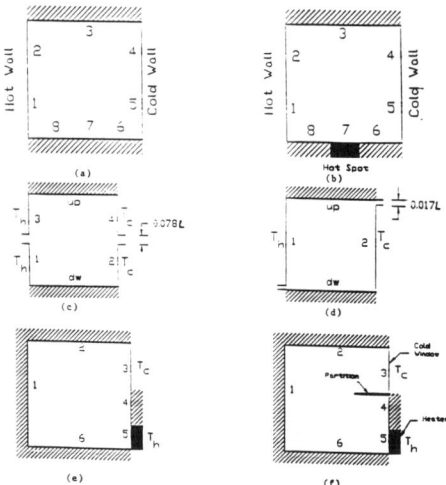

Figure 4. Illustration of the Different Configurations Studied

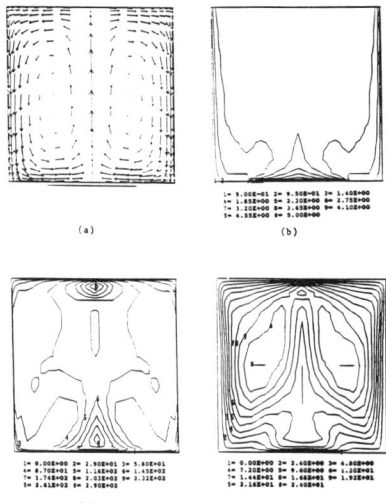

Figure 5. Velocity Pattern and T, k, μ_t Contours for the Case
of Central Plume, $\Delta T = 0C$, $T_{plume} = 30C$

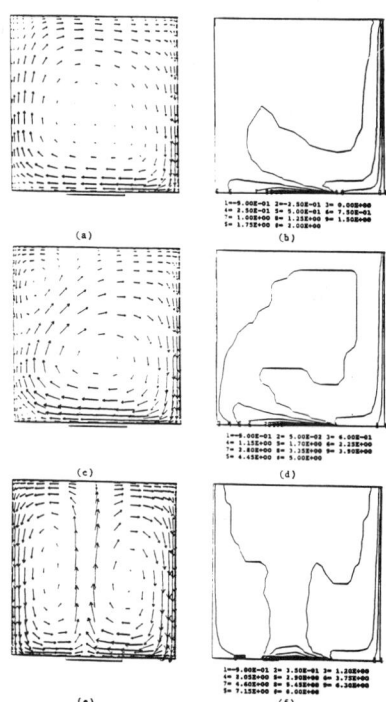

Figure 6. Velocity Patterns and Isotherms the **Case of Central
Plume** $T_{plume} = 24C(a, b), 30C(c, d), 36C(e, f)$ and
$s/L = 0.35$

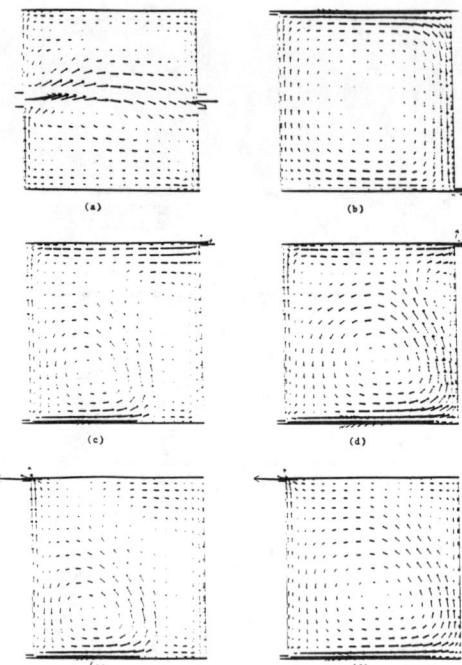

Figure 7. Velocity Patterns For Different Injection Sites Midheight Injection (a), Injection at Top of Hot Wall (b), Injection at Bottom of Hot Wall (c,d), Injection at Bottom of Hot Wall and Exit at Top of Hot Wall (e,f)

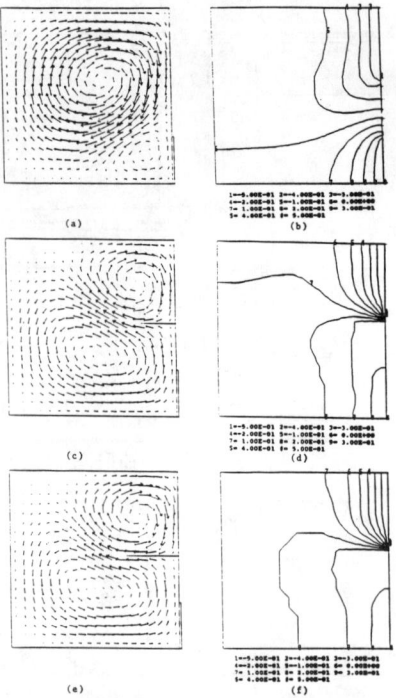

Figure 9. Velocity Pattern and Isotherms for the Reference Case with no Partition (a,b), with a Partition (s/W=.20) (c,d) and with a Partition (s/L=.30) (e,f), $Ra = 10^4$

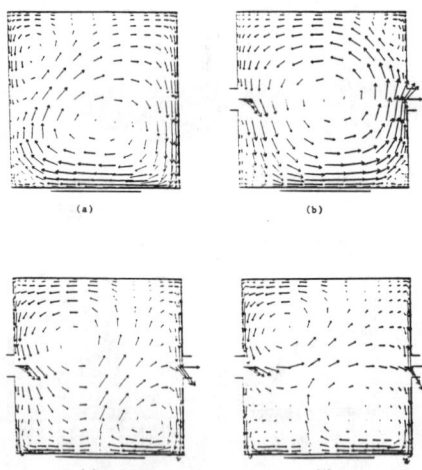

Figure 8. Velocity Patterns for the Case of Central Plume with Midheight Infiltration at Different Temperatures $\Delta T = 2C$, $T_{plume} = 30C$, No Injection (a), $T_{in} = 20C(b), 25C(c), 28C(d)$ and $s/L = 0.53$

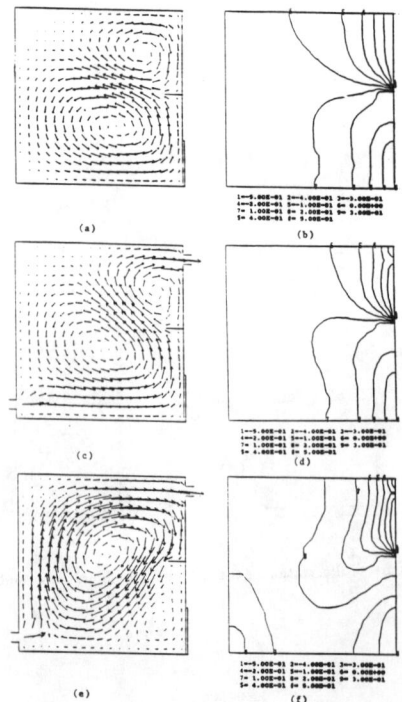

Figure 10. Velocity Pattern and Isotherms in a Cavity with a Partition and No Injection (a,b), with Injection of Cold air (c,d), with Injection of Warm Air (e,f)

EFFECT OF INTERNAL CIRCULATION ON THE TRANSPORT TO A MOVING LIQUID DROP

P. S. Ayyaswamy,* S. S. Sadhal, and L. J. Huang***
* Department of Mechanical Engineering and Applied Mechanics
University of Pennsylvania
Philadelphia, Pennsylvania
** Department of Mechanical Engineering
University of Southern California
Los Angeles, California

ABSTRACT

The existence of internal circulation in fluid droplets has been demonstrated by a number of investigators. The internal circulation, where present, provides an important mechanism for the heat and/or mass transfer associated with a moving liquid drop. In this paper we discuss the different roles played by this mechanism in the context of a moving drop experiencing condensation (inward radial field) or evaporation (outward radial field). It is noted that the strength of the internal vortex is not only a highly useful parameter in the modelling of transport to moving drops but is also necessary in the actual evaluation of the extent of transport. We provide analytical expressions for estimating the vortex strength in circumstances where the drop translation is in the low Reynolds Number regime while the radial field has inertia. It is found that with increasing radially outward velocity the strength of the vortex decreases. On the other hand, for an inward normal velocity the strength increases. With a slowly moving evaporating drop, a sufficiently large outward radial velocity may completely inhibit internal circulation, and interestingly at very high values may cause the sense of circulation to reverse. For a large sized drop (intermediate Reynolds number regime) experiencing condensation, the presence of the radial field inhibits the appearance of the secondary internal vortex motion in the drop. Convective mixing plays an important role during the immediate transient following the introduction of the drop into a condensing field. With increasing time, the center temperature approaches the surface value, and the vortex streamlines tend to become isothermal loops. Subsequent heat transfer occurs primarily in directions normal to vortex lines, with diffusion as the dominant mechanism. It is argued that as a result of internal circulation, beyond about one percent of the total condensation period, the surface temperature of a moving drop (Re = O(100)) experiencing condensation may be regarded as uniform.

NOMENCLATURE

$a_1(\theta)$	enhancement due to translation	C_D	total drag coefficient
A_o	evaporation velocity at drop surface in the absence of translation	c_p	specific heat
		D	drop diameter
		D_{12}	binary diffusion coefficient
A_{oo}	radial Reynolds Number $(= A_o R/\nu)$	E_o	Eötvös number $(= g\,\Delta\rho\,D^2/\sigma)$
		F	total drag

F_m	evaporation drag
F_P	pressure drag
F_μ	viscous drag
$\mathbf{g}$	gravity
$\mathbf{i}$	unit vector in x-direction
$\mathbf{j}$	unit vector in y-direction
k	thermal conductivity
Le	Lewis number ($= \alpha/D_{12}$)
m	non-condensable mass fraction
p_∞	far-stream pressure
Pe	Peclet number ($= U_\infty D/\alpha$ or $U_\infty D/D_{12}$)
q	heat flux
r	radial coordiante
R	instantaneous radius of the drop
R_o	initial radius of the drop
$\dot{R}$	dimensionless rate of change of drop-radius (scaled by U_∞)
R_∞	radius of the outer boundary in the numerical scheme
Re	Reynolds number ($= U_\infty D/\nu$)
Sc	Schmidt number ($= \nu/D_{12}$)
t	time
T	temperature
T_b	instantaneous bulk temperature of the drop
T_s	surface temperature of the drop
$\mathbf{u}$	velocity
$\hat{u}_c$	dimensionless condensation velocity
u_c	dimensionless condensation velocity at the drop surface scaled by $D_{12}/2R$
U_0	initial velocity of drop
U_∞	far-stream translational velocity
u_r, u_θ	velocity components
w	normalized mass fraction
W	condensation parameter ($= 1 - m_\infty/m_s$)
x	x-coordinate
y	y-coordinate
z	transformed radial coordinate ($= \ell n\ r$)

Greek Symbols

α	thermal diffusivity
ζ	vorticity
θ	polar angle
λ	latent heat of condensation
μ	dynamic viscosity
ν	kinematic viscosity
ρ	density
σ	surface tension
ϕ	azimuthal angle
ψ	stream function

Subscripts

b	bulk
c	condensation
f	friction
ℓ	liquid-phase
m	mass transfer
s	drop-surface
t	thermal, total
0	at initial time; stagnant drop
∞	far-stream

Superscripts

–	average
*	dimensionless quantity

INTRODUCTION

The presence of internal circulation in fluid droplets has been demonstrated by a number of investigators. See, for example, Garner and Haycock (1959), Spells (1952), Winnikow and Chao (1966). In particular, Winnikow and Chao have experimentally demonstrated that in a highly purified system, a moving droplet invariably exhibits internal circulation. Spells' study shows that circulation patterns agreed fairly well with the theoretical predictions of Hadamard-Rybczynski, although the vortices did not quite exhibit the fore and aft symmetry. Linton and Sutherland (1954) postulate that droplets should be circulating irrespective of their size, although they remark that this could be very difficult to prove experimentally, as perfectly pure fluids would then be required. As pointed out by Kintner (1963), for clean surfaces circulation should occur in any drop regardless of size. According to Sandry (1973), a strong internal circulation pattern has been noted during the droplet formulation and separation from a nozzle.

The internal circulation, where present, provides an important mechanism for the heat and/or mass transfer associated with a moving liquid drop [Chung, Ayyaswamy and Sadhal (1984); Sundararajan and Ayyaswamy (1984); Gogos, Sadhal, Ayyaswamy and Sundararajan (1988)]. In this paper, we discuss the different roles played by internal circulation in the context of a moving drop experiencing condensation (inward radial field) or evaporation (outward radial field). We have examined both a slowly translating drop (Re < 1), and a drop translating in the intermediate Reynolds Number regime (Re =O(100)).

The analysis calls for the simultaneous solution of the flow fields in the drop and the gaseous phases. The magnitude of the radial velocity is allowed to be very large, but the drop motion is restricted to slow translation. We analyze the problem by regular perturbation method. The solution has been developed by considering a uniform radial flow with the translatory motion introduced as a perturbation. The intermediate Reynolds Number problem is studied by a hybrid numerical scheme. We present expressions for the various flow and transport quantities, and evaluate their magnitudes to delineate the role played by the circulation inside the drop. The internal vortex is not only a highly useful parameter in the modelling of transport to moving drops but is also necessary in the actual evaluation of the extent of transport. We provide expressions for estimating the vortex strength. It is found that with increasing radially outward velocity the strength of the vortex decreases. This is because of the reduced vorticity at the surface. On the other hand, for an inward normal velocity the strength increases due to the increased shear stress. With a slowly moving evaporating drop, a sufficiently large outward radial velocity may completely inhibit internal circulation and at very high values may cause the sense of circulation to reverse. This is due to the non-uniformity of the radial field, which, with its maximum at the front of the drop, provides a shear stress to oppose the usual circulation. With increasing radial velocity, the vorticity resulting from translation is convected away, while the shear stress due to the non-uniformity in the normal velocity persists. Consequently, the internal circulation may weaken to a point where the latter force dominates and causes a reversal. For a large sized drop (intermediate Reynolds number regime) experiencing condensation, it is noted that the presence of the radial field inhibits the appearance of the secondary internal vortex motion in the drop. The primary vortex shifts towards the drop equatorial plane as the condensation level is increased. At high levels of condensation, the wake volume at the rear of the drop is reduced. The reduction in wake volume leads to greater fore-and-aft symmetry in the shear stress profile at the drop surface; there is a corresponding reduction in the asymmetry in the internal circulation. Independently of the shear stress profile at the surface, at high circulation Reynolds number, the circulation pattern resembles a Hill's vortex owing to the advection of vorticity. Convective mixing plays an important role during the immediate transient following the introduction of the drop into a condensing field. The much colder fluid from the drop center is brought to the surface and a high condensation rate is achieved. With increasing time, the center temperature approaches the surface value, and the vortex streamlines tend to become isothermal loops. Subsequent heat transfer occurs primarily in directions normal to vortex lines, with diffusion as the dominant mechanism. It is argued that as a result of internal circulation, beyond about one percent of the total condensation period, the surface temperature of a moving drop experiencing condensation may be regarded as uniform.

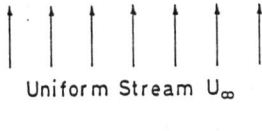

Uniform Stream U_∞

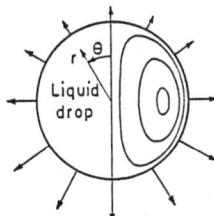

Non-uniform Radial Velocity $A_0 + a_1(\theta)$

Fig. 1. Schematic of the flow problem.

ANALYSIS
a) Low Translational Reynolds Number
Consider a spherical, liquid droplet of radius R with a uniform radial velocity A_0 at the outer surface. The A_0 would correspond, for example, to either condensation or evaporation velocity in a quiescent situation. The droplet translates at a velocity U_∞ in a gaseous medium. In addition to A_0, the radial field is enhanced by an amount $a_1(\theta)$ due to translational effects (Figure 1). The flow is axially symmetric with velocity $\mathbf{u}(r,\theta)$ having two components (u_r, u_θ). The velocity fields in both the liquid and the gas phases are taken to be quasisteady. The quasisteady aspect is justified in view of the large ratio of the liquid to the gas phase densities [Sadhal and Ayyaswamy, 1983]. The transient effects due to size changes are negligible, and because of the large density ratio, the normal velocity on the liquid side of the surface is negligible.

The equations governing the transport are:

continuity: $\nabla \cdot \mathbf{u} = 0$ (1a)

$\nabla \cdot \mathbf{u}_\ell = 0$ (1b)

momentum: $\rho \mathbf{u} \cdot \nabla \mathbf{u} + \nabla p = \mu \nabla^2 \mathbf{u}$ (2a)

$\rho_\ell \mathbf{u}_\ell \cdot \nabla \mathbf{u}_\ell + \nabla p_\ell = \mu_\ell \nabla^2 \mathbf{u}_\ell$ (2b)

energy: $\mathbf{u} \cdot \nabla T = \alpha \nabla^2 T$ (3a)

$\partial T_\ell / \partial t + \mathbf{u}_\ell \cdot \nabla T_\ell = \alpha_\ell \nabla^2 T_\ell$ (3b)

mass: $\mathbf{u} \cdot \nabla m = D_{12} \nabla^2 m$ (4)

The boundary conditions are:

(i) uniform stream at infinity:

$u_r = U_\infty \cos \theta, \ u_\theta = -U_\infty \sin \theta$

$T = T_\infty, \ m = m_\infty$ (5a)

(ii) normal velocity:

$u_r |_{r=R} = A_0 + a_1(\theta)$

$u_{\ell r} |_{r=R} = 0;$ (5b)

(iii) continuity of tangential velocity:

$u_\theta |_{r=R} = u_{\ell \theta} |_{r=R};$ (5c)

(iv) continuity of shear stress:

$$\mu\left[r\frac{\partial}{\partial r}\left(\frac{u_\theta}{r}\right) + \frac{1}{r}\frac{\partial u_r}{\partial\theta}\right]_{r=R}$$

$$= \mu_\ell\left[r\frac{\partial}{\partial r}\left(\frac{u_{\ell\theta}}{r}\right) + \frac{1}{r}\frac{\partial u_{\ell r}}{\partial\theta}\right]_{r=R} \quad (5d)$$

(v) Axi-symmetric conditions at $\theta=0$ and $\theta=\pi$:

$$u_\theta = \partial u_r/\partial\theta = \partial p/\partial\theta = 0$$

$$\partial T/\partial\theta = \partial m/\partial\theta = 0$$

$$u_{\ell\theta} = \partial u_{\ell r}/\partial\theta = \partial p_\ell/\partial\theta = 0 \quad (5e)$$

(vi) continuity of mass flux across the interface:

$$\rho(u_r - \dot{R}) = \rho_\ell(u_{\ell r} - \dot{R}) \quad (5f)$$

(vii) impermeability condition:

$$u_r\, m_s = D_{12}\, \partial m/\partial r\,|_{r=R} \quad (5g)$$

(viii) continuity of heat flux across the interface:

$$k\partial T/\partial r|_{r=R} - \rho u_r\lambda|_{r=R} = k_\ell\partial T_\ell/\partial r|_{r=R}\,(5h)$$

(ix) continuity of temperature:

$$T = T_s = T_\ell \quad (5i)$$

(x) mass fraction at interface:

$$m = m_s \quad (5j)$$

where, m_s is evaluated by using Clapeyron equation with the assumption of local thermodynamic equilibrium.

The translation-induced normal velocity $a_1(\theta)$ in (5b) is treated as an arbitrary parameter which may be, in any given circumstance, determined by the prevailing thermodynamics of system. However, it is noted that in many situations involving interfacial transport, it has the same characteristics as the translational field [Chung, Ayyaswamy, and Sadhal, 1984]. Hence a velocity variation of the type

$$a_1(\theta) = a_{01} + a_{11}\cos\theta \quad (6)$$

will be examined. In analyzing the actual evaporation of a moving liquid droplet, the quantities A_0, a_{01}, and a_{11} would be directly related to the fuel mass fraction at the drop surface [Gogos et al. (1986)]. On the other hand for a condensing droplet, the same quantities are related to the vapor mass fraction at the interface as shown in the paper by Chung et al. (1984). The enhancement due to translation is more pronounced at the front stagnation point ($\theta = \pi$, Figure 1), and the maximum value of $|a_1(\theta)|$ occurs there. The sign of a_{11} is thus opposite to that of A_0 and a_{01}.

The leading-order velocity field is taken to be purely radial flow with a velocity A_0 at the surface. In the absence of translation, the velocity $\mathbf{u}_0$ is

$$\mathbf{u}_0 = A_0\frac{R^2}{r^2}\,\hat{r} \quad (7)$$

To account for translation, corrections $\mathbf{u}'$ and $\mathbf{u}_\ell'$ are implemented, viz

$$\mathbf{u} = \mathbf{u}_0 + \mathbf{u}' \quad (8)$$

$$\mathbf{u}_\ell = \mathbf{u}_{\ell0} + \mathbf{u}_\ell' \quad (9)$$

where $\mathbf{u}_{\ell0}$ is equal to zero. The corresponding pressures are

$$p = p_0 + p' \quad (10)$$

$$p_\ell = p_{\ell0} + p_\ell' \quad (11)$$

where $p_{\ell0}$ is a constant.

The equations are rendered dimensionless by introducing

$$\mathbf{u}^* = \mathbf{u}R/\nu,\quad \mathbf{u}_\ell^* = \mathbf{u}_\ell R/\nu_\ell,\quad p^* = pR^2/\mu\nu$$

$$p_\ell^* = p_\ell R^2/\mu_\ell\nu_\ell,\quad \mathbf{u}_0^* = \mathbf{u}_0/A_0$$

$$\mathbf{u}'^* = \mathbf{u}'/U_\infty,\quad \mathbf{u}_\ell'^* = \mathbf{u}_\ell'/U_\infty$$

$$A_{01} + A_{11}\cos\theta = (a_{01} + a_{11}\cos\theta)/U_\infty$$

$$r^* = r/R,\quad \epsilon = U_\infty R/\nu,\quad p_0^* = p_0/(A_0\mu/R)$$

$$p'^* = p'/(U_\infty\mu/R),\quad p_\ell'^* = p_\ell'/(U_\infty\mu_\ell/R)$$

$$T^* = (T-T_\infty)/(T_0-T_\infty),\quad w^* = m-m_\infty$$

and

$$\nabla^* = R\nabla$$

where ϵ is the translational Reynolds number. Thus

$$\mathbf{u}^* = A_{00}\mathbf{u}_0^* + \epsilon\mathbf{u}'^* \quad (12)$$

$$\mathbf{u}_\ell^* = Re_\ell\,\mathbf{u}_\ell'^* \quad (13)$$

$$p^* = A_{00}p_0^* + \epsilon p'^* \quad (14)$$

and

$$p_\ell^* = A_{\ell00}p_\ell^*{}_0 + Re_\ell\,p_\ell'^* \quad (15)$$

A perturbation scheme

$$\mathbf{u}' = \mathbf{u}_1 + \epsilon\mathbf{u}_2 + \cdots,$$

$$p' = p_1 + \epsilon p_2 + \cdots,$$

is introduced where $\mathbf{u}'$ and p' are dimensionless variables with the asterisks dropped. In view of (12) and (14),

$$\mathbf{u} = A_{00}\,\mathbf{u}_0 + \epsilon\mathbf{u}_1 + \epsilon^2\mathbf{u}_2 + \cdots, \qquad (16a)$$

$$p = A_{00}\,p_0 + \epsilon p_1 + \epsilon^2 p_2 + \cdots, \qquad (16b)$$

where $\mathbf{u}$ and p also are dimensionless. However, $\mathbf{u}_\ell{}'$ is not perturbed, and instead an exact solution of the liquid-side momentum equation is obtained. The temperature and mass fraction quantities are expanded in a similar fashion.

By substituting (16) into dimensionless governing equations and equating powers in ϵ, equations appropriate to Orders ϵ^0 and ϵ^1 have been developed and solved [Sadhal and Ayyaswamy, 1983]. To facilitate a discussion of the results of this study, stream functions ψ and ψ_ℓ for the gas and the liquid phases are introduced. A detailed analysis provides the following results:

The dimensionless stream function for the gas phase is given by

$$\psi = -A_{00}\bar{\mu} + \epsilon[-A_{01}\bar{\mu} + \tfrac{1}{2}f(r)(1-\bar{\mu}^2)] + 0(\epsilon^2) \qquad (17)$$

where $\bar{\mu} = \cos\theta$, and

$$f(r) = r^2 + \frac{B}{r} + C\left[G - \frac{1}{5}\left(\frac{r}{A_{00}}\right)^4 + \frac{1}{6}\left(\frac{r}{A_{00}}\right)^2\right] \qquad (18)$$

For the liquid phase the stream function is given by

$$\psi_\ell = \frac{1}{2}B_\ell(r^4 - r^2)(1-\mu^{-2}) \qquad (19)$$

which is Hill's spherical vortex with strength $Re_\ell\,B_\ell$. The constants B, C and B_ℓ in Equations (17-19) have been determined by satisfying the boundary conditions. These are:

$$B = \left[(1-A_{11})[1-\tfrac{1}{5}(3+2\phi_\mu)-(1+A_{00}-\tfrac{1}{3}\phi_\mu A_{00}^2)e^{-A_{00}}]\right.$$
$$\left. +2A_{11}(-\tfrac{1}{5}+\tfrac{1}{6}A_{00}^2)\right]/\Delta \qquad (20)$$

$$C = \left[-\Gamma A_{00}^4\right]/\Delta \qquad (21)$$

$$G = \frac{A_{00}}{r}\int_{1/A_{00}}^{r/A_{00}}(\xi^4+\xi^3)e^{-1/\xi}d\xi \qquad (22)$$

$$B_\ell = \phi_\mu\left[(1-A_{11})[1-\tfrac{1}{3}A_{00}^2-(1+A_{00}+\tfrac{1}{6}A_{00}^2)e^{-A_{00}}]\right.$$
$$\left. +\tfrac{1}{3}A_{00}^2(1-e^{-A_{00}})\right]/\Delta \qquad (22)$$

where

$$\phi_\mu = \mu/\mu_\ell$$

$$\Delta = -1+\tfrac{1}{6}(3+2\phi_\mu)\,A_{00}^2+(1+A_{00}-\tfrac{1}{3}\phi_\mu A_{00}^2)e^{-A_{00}}$$

$$\Gamma = (3+2\phi_\mu)-A_{11}(1+2\phi_\mu)$$

The dimensionless velocities are:

$$u_r = A_{00}\frac{1}{r^2} + \epsilon\left[\frac{A_{01}}{r^2} + \left[1 + \frac{B}{r^3}\right.\right.$$
$$\left.\left. + \frac{C}{r^2}\left(G - \frac{1}{5}\left(\frac{r}{A_{00}}\right)^4 + \frac{1}{6}\left(\frac{r}{A_{00}}\right)^2\right)\right]\cos\theta\right] \qquad (23)$$

$$u_\theta = -\frac{1}{2}\epsilon\left[2 - \frac{B}{r^3} + \frac{C}{r^2}\left[-G\right.\right.$$
$$+ \left(\left(\frac{r}{A_{00}}\right)^4 + \left(\frac{r}{A_{00}}\right)^3\right)e^{-A_{00}/r}$$
$$\left.\left. - \frac{4}{5}\left(\frac{r}{A_{00}}\right)^4 + \frac{1}{3}\left(\frac{r}{A_{00}}\right)^2\right]\right]\sin\theta \qquad (24)$$

$$u_{\ell r} = Re_\ell\,B_\ell(r^2-1)\cos\theta \qquad (25)$$

and,

$$u_{\ell\theta} = -Re_\ell\,B_\ell(2r^2-1)\sin\theta. \qquad (26)$$

With regard to the drag forces experienced by an evaporating droplet, it is noted that the drag forces consist of contributions from the viscous stresses, the pressure and the momentum flux at the interface. The viscous drag in dimensionless form is

$$F_\mu^* = \frac{F_\mu}{6\pi\mu U\,R_\infty}$$

$$= \frac{1}{3\epsilon}\int_0^\pi[\sigma_{rr}\cos\theta-\sigma_{r\theta}\sin\theta]_{r=1}\sin\theta\,d\theta \qquad (27)$$

where σ_{rr} and $\sigma_{r\theta}$ are the viscous stresses. In terms of the dimensionless velocities,

$$F_\mu^* = -\frac{2}{9}\Gamma\left[(2+2A_{00}+A_{00}^2)e^{-A_{00}}-2\right]/\Delta \qquad (28)$$

The dimensionless pressure drag is

$$F_p^* = \frac{1}{3\epsilon}\int_0^\pi(-p_1|_{r=1}\cos\theta)\sin\theta\,d\theta \qquad (29)$$

$$= \frac{2}{9}\left[A_{00}A_{11}+\Gamma[(2+A_{00})e^{-A_{00}}+A_{00}-2]/\Delta\right] \qquad (30)$$

135

The force $F_m^* = F_m/6\pi\mu U_\infty R$ due to the momentum flux at the interface is given by

$$F_m^* = \frac{1}{3\epsilon} \int_0^\pi -[u_r u_r \cos\theta - u_r u_\theta \sin\theta] \sin\theta \; d\theta \quad (31)$$
$$\hspace{10cm} r=1$$

$$= \frac{2}{9} \left[-A_{00}(3+A_{11}) + \right.$$

$$\left. + \left(\Gamma[(A_{00}+A_{00}^2)e^{-A_{00}} - A_{00} + \frac{1}{2}A_{00}^3] \right)/\Delta \right] \quad (32)$$

The total drag is the sum $F^* = F_\mu^* + F_p^* + F_m^*$ given by

$$F^* = -\frac{2}{9} \left[3A_{00} + \frac{C}{2A_{00}} \right]$$

$$= \frac{2}{9}\left[-3A_{00} + \left(\frac{1}{2} [(3+2\phi_\mu) - A_{11}(1+2\phi_\mu)]A_{00}^3 \right)/\Delta \right]$$
$$\hspace{12cm} (33)$$

The corresponding instantaneous drag coefficient C_D is defined by

$$C_D = \frac{\text{Total drag force}}{0.5\rho_\infty[U_\infty(t)]^2 \; \pi R^2} \quad (34)$$

b) Intermediate Reynolds Number

The hydrodynamics and heat/mass transport associated with condensation on a moving droplet for the intermediate Reynolds number range of droplet motion ($Re = 0(100)$) have been investigated earlier [Sundararajan and Ayyaswamy (1984), and Huang and Ayyaswamy (1987)]. The quasi-steady assumption has been invoked in the first paper while the second paper addresses the fully transient formulation. The presence of noncondensable in the gaseous environment has also been taken into account. The flow-solutions and transport rates to the droplet have been obtained in terms of two non-dimensional parameters (one, Re, representing the flow conditions, and the other, W, representing the thermodynamic conditions). The governing differential equations for the gaseous-phase have been solved by a hybrid finite-difference scheme [Sundararajan and Ayyaswamy (1985)]. The drop interior is solved by the Crank-Nicolson procedure.

Condensation causes a radially inward flow towards the drop surface. The non-zero mass flux at the interface alters the translational flow field and modifies the drag on the drop. Also, the radial flow leads to a build-up of the non-condensable concentration near the drop-surface above that in the free stream. The accumulation results in a mass-transfer resistance and a consequent reduction in the transport rates.

In the present paper, we shall concentrate on the effects of recirculation alone. For $Re_g = 0(10^2)$ flow separates on the rear of the drop. A recirculating wake is formed and the radial flow due to condensation reduces the wake size. We observe this feature by numerically solving the governing equations subject to the appropriate initial and boundary conditions. The governing equations are similar to those of part (a), and will not be repeated here for the sake of brevity.

Briefly, in the numerical procedure, the governing equations and boundary conditions are transformed in terms of the dimensionless stream function ψ (scaled with $U_\infty R^2$) and vorticity ζ (scaled with U_∞/R). In spherical coordinates the stream function and vorticity are

$$\mathbf{u} = \nabla x \left[-\frac{\psi}{r\sin\theta} \; \mathbf{e}_\phi \right], \quad \zeta \; \mathbf{e}_\phi = \nabla x \mathbf{u} \quad (35)$$

and

$$\mathbf{u}_\ell = \nabla x \left[-\frac{\psi_\ell}{r\sin\theta} \; \mathbf{e}_\phi \right], \quad \zeta_\ell \; \mathbf{e}_\phi = \nabla x \mathbf{u}_\ell \quad (36)$$

Due to the elliptic nature of the governing equations, the far-stream boundary conditions, and the location of their specification, significantly influence the numerical solutions. The far-stream conditions are specified on a large but finite spherical surface of radius R_∞, and the value of R_∞ has been judiciously chosen as a compromise between the computational effort and the desired accuracy for the solutions. The solution domain is divided into a grid with a variable step-size. A fine spacing is employed near the drop where the gradients are steep. A coarse spacing seems to be adequate in the far-stream where the gradients are weak. An exponential grid spacing is generated by making a transformation $r = e^z$, and considering equal spacing in z. A constant angular step-size is used for the θ coordinate. To accurately predict the transport to the drop, a central difference scheme (CDS) with a second order accuracy is employed near the drop. But CDS has not been used throughout the solution domain for the sake of computational economy. An upwind difference scheme (UDS) is used far away from the drop. The non-linear, algebraic difference equations are solved iteratively, starting from suitable guess solutions. A successive over-relaxation procedure is used to accelerate the convergence. Computations are carried out until the changes in the predicted transport quantities are less than 10^{-7} (absolute error) or less than 0.1% (relative error), between successive iterations.

A Crank-Nicolson procedure is used to evaluate the transient heat-up of the drop interior. The spatial derivatives are central differenced. The difference equations are arranged in a tri-diagonal matrix form, and a computationally inexpensive tri-diagonal matrix solver algorithm has been employed. From known initial conditions, computations are carried out until the drop temperature approaches that of the far-stream. The liquid-phase numerical solution procedure is facilitated by noting that :

$$u_{\ell r} \approx -0.1(r^2-1)\cos \theta, \qquad (37)$$

and,

$$u_{\ell \theta} \approx -0.1(2r^2-1) \sin \theta . \qquad (38)$$

The factor 0.1 approximately accounts for the effect of liquid viscosity on the strength of Hill's spherical vortex solution in the intermediate Reynolds number range of drop motion [Prakash and Sirignano (1980), Chung and Ayyaswamy (1981)].

RESULTS AND DISCUSSION
a). Low Translational Reynolds Number: Evaporating Droplet.

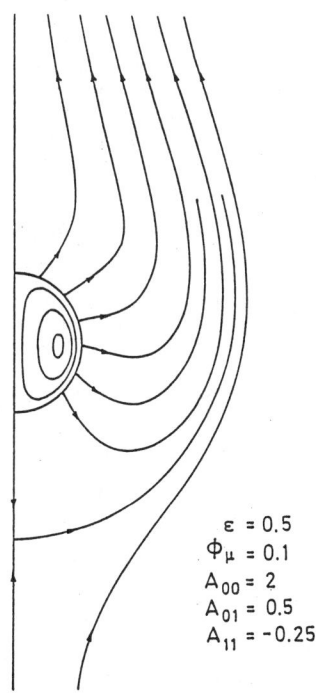

$\epsilon = 0.5$
$\phi_\mu = 0.1$
$A_{00} = 2$
$A_{01} = 0.5$
$A_{11} = -0.25$

Fig. 2. Flow streamlines for an evaporating drop.

The streamlines of the flow corresponding to $A_{00} = + 2$, $A_{01} = 0.5$, $A_{11} = -0.25$, $\phi_\mu = 0.1$ and $\epsilon = 0.5$ are shown in Figure 2. Near the front of the drop, the radial flow and the uniform stream oppose each other, and as a result a stagnation point is formed there. In Figure 3, the drag coefficient based on the instantaneous

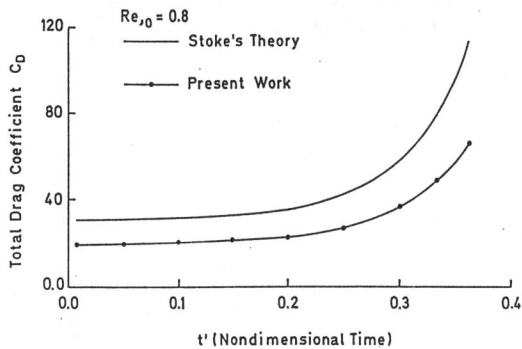

Fig. 3. Total drag coefficient for Re = 0.8 (evaporating n-hexane drop).

droplet Reynolds number is plotted for Re = 0.8. The Stokes' drag coefficient 24/Re is also shown based on the instantaneous value of Re. The reduction in the drag is due to the convection of the vorticity away from the droplet surface in the presence of the radial flow field and the reduction in the pressure drop from the front to the rear stagnation points. With regard to internal circulation, the results indicate that for increasing evaporation velocity the strength of the vortex decreases. This is because of the reduced vorticity at the surface. An expression for this strength is given by (22), and the numerical values are plotted in Figure 4. A remarkable feature of a slowly translating evaporating droplet is that with a sufficiently large radial velocity the internal circulation may vanish. A further increase in A_{00} may reverse the circulation. This very interesting result is due to the non-uniformity of the radial field, which, with its maximum at the front of the droplet, provides a shear stress to oppose the usual circulation. With increasing radial velocity, the vorticity resulting from translation is convected away, while the shear stress due to the non-uniformity in the normal velocity persists. Consequently the internal circulation weakens to a point where the latter force may dominate and cause a reversal.

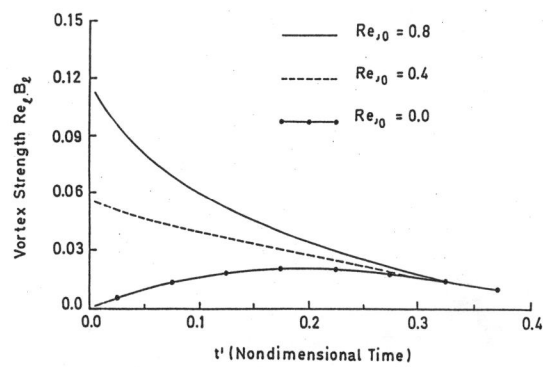

Fig. 4. Temporal variation of the vortex strength (evaporating drop).

b). Low Translational Reynolds Number: Condensing Droplet.

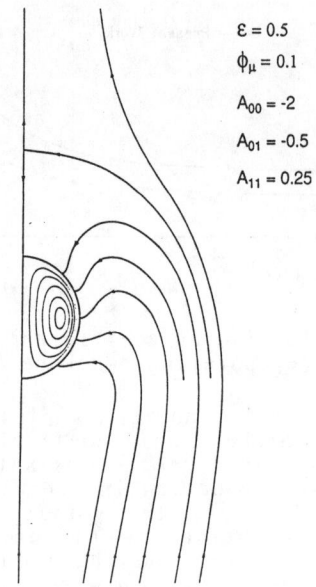

$\epsilon = 0.5$

$\phi_\mu = 0.1$

$A_{00} = -2$

$A_{01} = -0.5$

$A_{11} = 0.25$

Fig. 5. Flow streamlines for a condensing drop.

The streamlines of the flow corresponding to $A_{00} = -2$, $A_{01} = -0.5$, $A_{11} = 0.25$, $\phi_\mu = 0.1$ and $\epsilon = 0.5$ are shown in Figure 5. For the inward flow ($A_{00} < 0$) a stagnation point is formed near the rear of the drop. In this case, some of the streamlines from the uniform stream end on the surface of the drop. In figure 6, the drag force $F/6\pi\mu U_\infty R$ is shown as a function of increasing radial velocity. The drag increases monotonically. The vorticity is convected towards the drop surface, and hence the viscous drag increases. The variation of the strength of internal vortex with radial velocity is shown in figure 7. The strength of the vortex increases with increasing inward normal velocity. This is because of the increased vorticity at the surface.

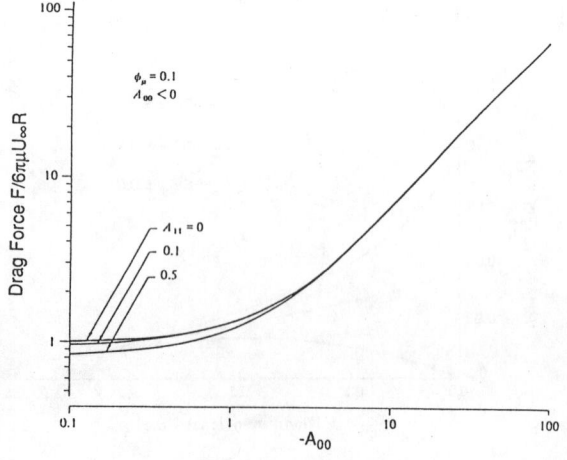

$\phi_\mu = 0.1$

$A_{00} < 0$

$A_{11} = 0$

0.1

0.5

Fig. 6. Total drag force as a function of condensation velocity.

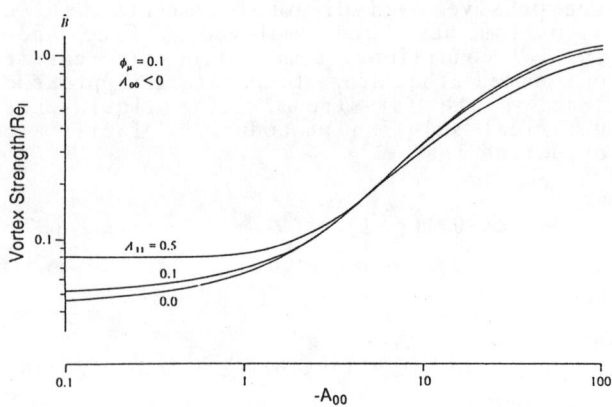

$\phi_\mu = 0.1$

$A_{00} < 0$

$A_{11} = 0.5$

0.1

0.0

Fig. 7. Variation of the strength of internal vortex with condensation velocity.

c). Intermediate Reynolds Number: Condensing Droplet.

The changes in the flow structure inside and outside the drop due to condensation on a drop translating at Re = 300 are shown in Figure 8. The flow patterns on the left and right halves of the axis (hatched) correspond to a noncondensing (W = 0) and a typical condensing situation (W = 0.7), respectively, at $Re_g = 300$. For W = 0, a detached recirculatory wake is present in the rear of the drop in the gaseous phase. Within this wake, fluid particles recirculate. For the interior of the drop, a primary liquid vortex generated by the positive shear stress in the front portion of the drop and a secondary vortex generated by the negative shear stress in the rear, are noted. The secondary vortex strength is

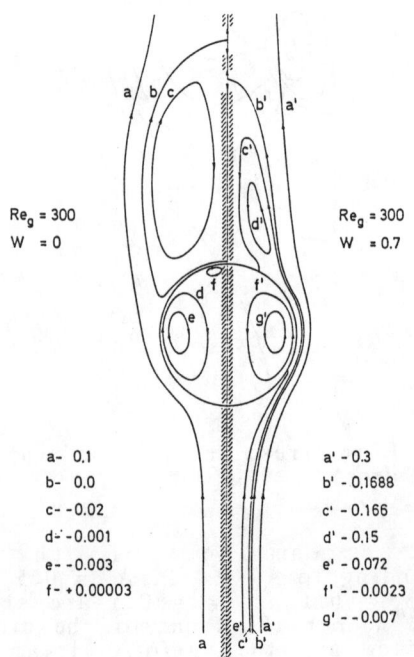

$Re_g = 300$
W = 0

$Re_g = 300$
W = 0.7

a- 0.1
b- 0.0
c- -0.02
d- -0.001
e- -0.003
f- +0.00003

a'- 0.3
b'- 0.1688
c'- 0.166
d'- 0.15
e'- 0.072
f'- -0.0023
g'- -0.007

Fig. 8. Effect of condensation on flow pattern.

very small compared to that of the primary and the sense of circulation in the secondary vortex is opposite to that of the primary vortex. In the presence of condensation, the wake-length and volume are reduced. A dividing stream surface (b') exists, inside which fluid particles either condense or recirculate. Inside the drop, the strength of the primary vortex is higher. This is due to the increased shear stress at the interface caused by the transport of momentum to the drop. The internal secondary vortex does not exist at high rates of condensation because the wake volume is severly reduced by condensation. As W is increased, the primary vortex center shifts towards the drop equatorial plane and there is a reduction in the asymmetry in the circulation. Independantly of the shear stress profile, at high Re_ℓ, the circulation pattern resembles a Hill's vortex owing to the increased advection of vorticity.

In figure 9, the effects of internal circulation on heat transfer and average condensation velocity at the drop surface are displayed. The dimensionless quantities $\bar{q}$ and $\bar{u}_c$ are plotted as functions of time for a vertically falling drop introduced with an initial velocity, $U_0 = 10$ m/s. The far stream thermodynamic conditions are taken to be prescribed. Following Huang and Ayyaswamy (1987), we define the average heat transfer by

$$\bar{q} = \frac{1}{2} \int_0^\pi q_s \sin\theta \, d\theta \qquad (39)$$

with,

$$\bar{q} = \frac{\bar{u}_c}{2 \, Le} \qquad (40)$$

where,

$$\frac{\bar{u}_c}{u_{c,0}} = 1 + 0.261 \, Re_g^{1/2} Sc^{1/3} (1-W)^{-1/5} \qquad (41)$$

also for $W < 1$, and

$$u_{c,0} = 2 \ln (1-W) \qquad (42)$$

for $W < 1$.

In the immediate transient period following the introduction, both $\bar{q}$ and $\bar{u}_c$ are very high because of the larger thermal driving force. With increasing time, both the surface and bulk temperatures increase. The internal circulation in the drop is vigorous and provides an efficient mechanism for heat transfer. The thermal driving force keeps getting weaker. Eventually the drop thermally equilibrates with the outside and the condensation ceases. Owing to the smaller heat capacity, the smaller drops equilibrate with the outside sooner than the

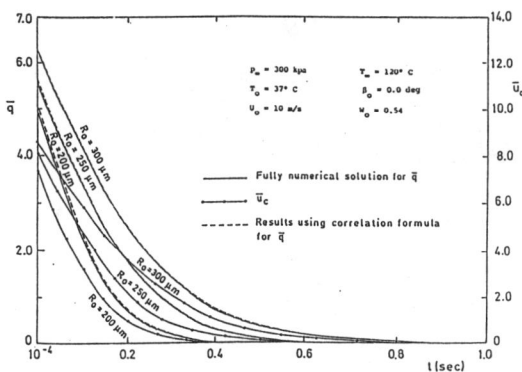

Fig. 9. Effects of internal circulation on heat transfer and average condensation velocity.

larger drops. The figure provides the time histories for $\bar{u}_c$ and $\bar{q}$ as derived both from the complete numerical solution of both flow and transport equations and from employing the correlation (40) in the numerical computations. The predicted results agree very well. The correlation correctly accounts for the effects of internal circulation.

d). Effect of Internal Circulation on the Surface Temperature of a Condensing Droplet at Re = O(100).

The surface temperature of a moving drop experiencing condensation in a vapor-gas atmosphere will never be uniform until the drop thermally equilibrates with the outside (condensation ceases). However, the parameter of interest to us is the ratio Ω of the angular variation in surface temperature $\hat{T}_s$ to the average surface temperature $\bar{T}_s$. When Ω is small the assumption of a uniform surface temperature is reasonable.

During the initial transient period following drop introduction, the high condensation heat flux will raise the surface temperature rapidly. A moving fluid element near the drop surface will be heated continuously along its path and $\hat{T}_s$ will be comparable to $\bar{T}_s$ ($\Omega = O(1)$). The time of this period is $\hat{O}(R/U_s)$, or the time for few internal circulation cycles. With an increased number of cycles, $\bar{T}_s$ increases, wheras $\hat{T}_s$ actually decreases. By about ten circulation cycles (roughly one percent of the total condensation period for $Pe_\ell = O(10^3)$), Ω drops to a value $O(10^{-1})$, and the uniform - T_s assumption is justified for later periods. As the drop approaches the thermal equilibrium with the outside, $\Omega \to 0$ and T_s becomes truly uniform.

ACKNOWLEDGEMENT
The numerical calculations were made by using the Pittsburgh Supercomputer under NSF Grant ECS-0000000/8515068. The authors are grateful for PSC services and to NSF.

REFERENCES

1. P. S. Ayyaswamy, "Combustion Dynamics of Moving Droplets," Chapter 20 in <u>Encyclopedia of Environmental Control Technology, vol. 1 : Thermal Treatment of Hazardous Wastes,</u> Ed: P.N. Cheremisinoff, pp. 479-532, Gulf Publishing Co., Houston, TX, (1989).

2. J. N. Chung and P. S. Ayyaswamy "Laminar Condensation Heat and Mass Transfer to a Moving Drop", <u>AIChE Journal,</u> <u>27</u>, No. 3, 372-377 (1981).

3. J. N. Chung, P. S. Ayyaswamy, and S. S. Sadhal, "Laminar Condensation on a Moving Drop. Part I: Singular Perturbation Technique," <u>J. Fluid Mech.,</u> <u>139</u>, 105-130 (1984).

4. J. N. Chung, P. S. Ayyaswamy, and S. S. Sadhal, "Laminar Condensation on a Moving Drop. Part 2: Numerical Solutions," <u>J. Fluid Mech.</u>, <u>139</u>, 131-144 (1984).

5. F.H. Garner and P.J. Haycock, " Circulation in Liquid Drops," <u>Proc. Roy. Soc.(London)</u>,A252,457 (1959).

6. G. Gogos, S. S. Sadhal, P. S. Ayyaswamy, and T. Sundararajan, "Thin-flame Theory for the Combustion of a Moving Liquid Drop: Effects Due to Variable Density," <u>J. Fluid Mech.</u>, <u>171</u>, 121-144 (1986).

7. L. J. Huang and P. S. Ayyaswamy, "Heat and Mass Transfer Associated With a Spray Drop Experiencing Condensation: A Fully Transient Analysis", <u>Int. J. Heat Mass Transfer,</u> <u>30</u>, No. 5, 881-891 (1987).

8. R.C. Kintner," Drag Phenomena Affecting Liquid Extraction," <u>Advances in Chemical Engineering</u>, vol. 4, Academic Press, New York and London (1963).

9. M. Linton and K.L. Sutherland," Dynamic Surface Forces , Drop Circulation and Liquid-Liquid Mass Transfer," <u>Proc.2nd Int'l. Congress Surface Activity</u>, Butterworths Sci.Publ.Co.(London),1957.

10. S. Prakash and W. A. Sirignano, "Liquid Fuel Droplet Heating with Internal Circulation," <u>Int. J. Heat Mass Transfer</u>, <u>23</u>, 885-895 (1978).

11. S. Prakash and W. A. Sirignano, "Theory of Convective Droplet Vaporization with Unsteady Heat Transfer in the Circulating Liquid Phase," <u>Int. J. Heat Mass Transfer</u>, <u>23</u>, 253-268 (1980).

12. S. S. Sadhal and P. S. Ayyaswamy, "Flow Past a Liquid Drop with a Large Non-Uniform Radial Velocity," <u>J. Fluid Mech.</u>, <u>133</u>, 65-81 (1983).

13. T.D. Sandry, " Drop Shapes and Internal Flow Patterns by Numerical Solution of the Navier-Stokes Equations," Doctoral Dissertation, Iowa State University (1973).

14. K.E. Spells,"Circulation Patterns in Liquid Drops," <u>Proc. Roy. Soc.</u>, BGS,541(1952).

15. T. Sundararajan and P. S. Ayyaswamy, "Hydrodynamics and Heat Transfer Associated with Condensation on a Moving Drop: Solutions for Intermediate Reynolds Numbers", <u>J. Fluid Mech.</u>, <u>149</u>, 33-58 (1984).

16. T. Sundararajan and P. S. Ayyaswamy, "Heat and Mass Transfer Associated with Condensation on a Moving Drop: Solutions for Intermediate Reynolds Numbers by a Boundary Layer Formulation", <u>J. Heat Transfer</u>, <u>Trans. ASME</u>, <u>107</u>, No. 2, 409-416 (1985).

17. T. Sundararajan and P. S. Ayyaswamy, "Numerical Evaluation of Heat and Mass Transfer to a Moving Liquid Drop Experiencing Condensation", <u>Numerical Heat Transfer</u>, <u>8</u>, No. 6, 689-706 (1985).

18. S. Winnikow and B. T. Chao, "Droplet Motion in Purified Systems," <u>Phys. Fluids</u>, <u>9</u>, 50-61 (1966).

NUMERICAL SOLUTIONS OF CONVECTION-DIFFUSION PROBLEMS IN IRREGULAR DOMAINS, USING A COORDINATE TRANSFORMATION WHICH YIELDS A CIRCULAR DOMAIN

M. Asaba, Y. Asako, and H. Nakamura
Department of Mechanical Engineering
Tokyo Metropolitan University
Tokyo, Japan

M. Faghri
Department of Mechanical Engineering
and Applied Mechanics
University of Rhode Island
Kingston, Rhode Island

ABSTRACT

A coordinate transformation methodology has been developed for convection-diffusion problems with an arbitrary solution domain. An algebraic coordinate transformation is used which maps the solution domain onto a circle. The transformed conservation equations are discretized by a control-volume finite difference technique. Sample computations are performed for fully developed flow and heat transfer in a polygonal duct, and for natural convection in a square cavity. The results are compared with the available values.

NOMENCLATURE

a	:	thermal diffusivity
b	:	source term, equation (26) for V, equation (27) for U
c_p	:	specific heat
D_h	:	equivalent hydraulic diameter
$\vec{e}_R, \vec{e}_\theta$	:	unit vectors in R, θ direction
$\vec{e}_\eta, \vec{e}_\xi$	:	unit vector in η, ξ direction
k	:	thermal conductivity
L	:	characteristic length
Nu	:	Nusselt number
n	:	refers to the number of sides of the polygonal duct
$\vec{n}$	:	unit vector along normal to surface
Pr	:	Prandtl number
q	:	heat flux
R, r	:	radial coordinate : dimensionless and dimensional
Ra	:	Rayleigh number [$= g\beta L^3(t_h-t_c)/a\nu$]
S	:	control surface for main control volume
T, t	:	temperature : dimensionless and dimensional
t_b	:	bulk temperature
t_c	:	cold wall temperature
t_h	:	hot wall temperature
Δt	:	temperature difference
t_{ref}	:	reference temperature
U, u	:	velocity component in θ direction: dimensionless and dimensional
V, v	:	velocity component in r direction: dimensionless and dimensional
$\vec{V}$	:	velocity vector
U_ξ, V_η	:	dimensionless velocity component in ξ and η directions
$\bar{w}$	:	mean axial velocity
Z, z	:	axial coordinate : dimensionless and dimensional
α	:	geometric function [$= 1 + \beta^2$]
β	:	geometric function [$=(1/\delta)(\partial\delta/\partial\xi)$] and also volume expansion coefficient
Γ	:	diffusion coefficient
γ	:	diffusion term [$= -(1/\eta)(\partial\phi/\partial\xi)$]
$\Delta(\theta)$	:	dimensionless radius of solution domain
$\delta(\theta)$	:	radius of solution domain
δ'	:	first derivative of δ
δ''	:	second derivative of δ
δ'''	:	third derivative of δ
η	:	transformed coordinates [$\eta = LR/\delta$]
θ	:	peripheral coordinate
Λ	:	pseudodiffusion term [$= \beta(\partial\phi/\partial\xi)$]
ν	:	kinematic viscosity
ξ	:	transformed coordinate [$\xi = \theta$]
ρ	:	density
σ	:	bulk temperature gradient parameter
Ψ	:	pseudodiffusion term [$= \beta(\partial\phi/\partial\eta)$]
Ω	:	diffusion term [$= -\alpha\eta (\partial\phi/\partial\eta)$]

INTRODUCTION

One way to solve convection-diffusion problems for fluid flow in and around complex geometries is the utilization of boundary-fitted coordinates of Saitoh (1978) in conjunction with a finite volume technique. This method has been widely used in melting and solidification problems [e.g. Sparrow et. al. (1977)].

Faghri et. al. (1984) proposed an algebraic coordinate transformation which transforms an irregular solution domain onto a rectangle. The solution domains considered were limited to the geometries bounded by one arbitrary curved side and three straight sides. An extension of the previous work by Faghri et al. (1984) for more general domains which will be mapped onto a circle, is given here. Sample computations will be performed for fully developed flow and heat transfer in polygonal ducts and for the natural convection in a square cavity.

FORMULATION

Description of the Problem

A schematic view of the type of physical domain being considered is shown in Fig. 1. As seen there, the origin of the coordinate is located in the domain and the distance from this origin to the boundary is a function of θ which is denoted by $\delta(\theta)$.

Conservation Equations

The governing equations to be considered are the continuity, momentum, and energy equations. Constant thermophysical properties are assumed and viscous dissipation and compression work are omitted in the energy equation. The following non-dimensional variables are used:

$$R=r/L, \quad U=u/(a/L), \quad V=v/(a/L)$$
$$P=p/\rho(a/L)^2, \quad T=(t-t_{ref})/\Delta t \qquad (1)$$

where L, t_{ref} and Δt are reference quantities. Then, upon the introduction of the non-dimensional variables, the governing equations take the form:

Continuity Equation:

$$R(\partial V/\partial R) + V + (\partial U/\partial \theta) = 0 \qquad (2)$$

Momentum Equations:

$$V(\partial V/\partial R) + (U/R)(\partial V/\partial \theta) - U^2/R = -(\partial P/\partial R)$$
$$+ Pr[\nabla^2 V - V/R^2 + (2/R^2)(\partial U/\partial \theta)] \qquad (3)$$

$$V(\partial U/\partial R) + (U/R)(\partial U/\partial \theta) - VU/R = -(1/R)(\partial P/\partial \theta)$$
$$+ Pr[\nabla^2 U - U/R^2 + (2/R^2)(\partial V/\partial \theta)] \qquad (4)$$

Energy Equation:

$$V(\partial T/\partial R) + (U/R)(\partial T/\partial \theta) = \nabla^2 T \qquad (5)$$

where $\nabla^2 =(\partial^2/\partial R^2) + (1/R)(\partial^2/\partial R^2) + (1/R^2)(\partial^2/\partial \theta^2)$

The boundary conditions will be specified for the illustrative example discussed later.

Coordinate Transformation

The coordinate transformation methodology in Cartesian coordinates is well documented in an earlier paper by Faghri et. al. (1984). In this paper, the polar coordinates r and θ are transformed into η and ξ coordinates such that the physical domain bounded by an arbitrary boundary maps onto the circle by the relations

$$\eta = R/\Delta(\theta), \quad \xi = \theta \quad \text{where} \quad \Delta(\theta) = \delta(\theta)/L \qquad (6)$$

In terms of the new coordinates, the solution domain is defined by $0 < \eta < 1$, and $0 < \xi < 2\pi$.

Attention will be focused on a new velocity component. Equations (3) and (4) are the components of the vector momentum equation in R and θ directions, respectively. Equation (3) contains the terms that multiply unit vector $\vec{e}_R$, while equation (4) contains the terms that multiply unit vector $\vec{e}_\theta$ in the vector momentum equation. The relation between the orthogonal unit vectors $\vec{e}_R$, $\vec{e}_\theta$ and the non-orthogonal unit vectors $\vec{e}_\eta$, $\vec{e}_\xi$ as illustrated in Fig. 2, will be first obtained. Since lines of constant ξ coincides with the θ coordinates lines, $\vec{e}_\eta = \vec{e}_R$. To determine $\vec{e}_\xi$, consideration is first given to the unit vector $\vec{n}$ that is perpendicular to a line of constant η. The gradient of η is normal to a line of constant η such that

$$\vec{n} = \nabla\eta/|\nabla\eta| = (\vec{e}_R - \beta\vec{e}_\theta)/\alpha^{(1/2)} \qquad (7)$$

where $\beta = (\partial\delta/\partial\xi)/\delta$, $\alpha = 1 + \beta^2$

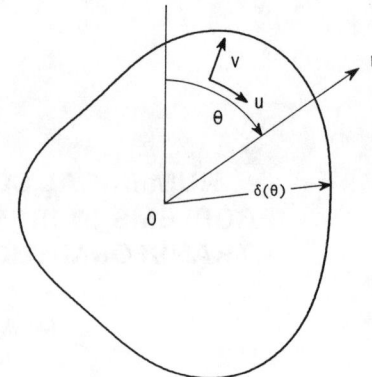

Fig. 1 Illustration of the class of problems under investigation

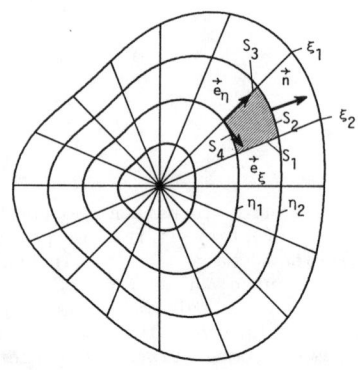

Fig. 2 Constant η and ξ in the physical domain

The unit vector $\vec{e}_\xi$ is normal to $\vec{n}$, hence

$$\vec{e}_\xi = (\beta\vec{e}_R + \vec{e}_\theta)/\alpha^{(1/2)} \qquad (8)$$

The following inverse relations can be obtained directly from $\vec{e}_\eta = \vec{e}_R$ and from equation (8),

$$\vec{e}_R = \vec{e}_\eta, \quad \vec{e}_\theta = \alpha^{(1/2)}\vec{e}_\xi - \beta\vec{e}_\eta \qquad (9)$$

Then any velocity vector $V=(U_\xi, V_\eta)$ can be expressed as

$$\vec{V} = V_\eta\vec{e}_\eta + U_\xi\vec{e}_\xi \quad \text{or} \quad \vec{V} = V\vec{e}_R + U\vec{e}_\theta$$

so that

$$V_\eta = V - \beta U, \quad U_\xi = U\alpha^{(1/2)} \qquad (10)$$

A new momentum equation, in which V_η is the primary dependent variable, is introduced by multiplying $-\beta$ by the U momentum equation and by adding it to V momentum equation to obtain

$$\vec{V}\cdot(\nabla V_\eta + U\nabla\beta) - U^2/R - \beta(UV_\eta/R) - \beta^2 U^2/R =$$
$$- [(\partial P/\partial R) - (\beta/R)(\partial P/\partial \theta)] + Pr[\nabla^2 V_\eta + U\nabla^2\beta + 2\nabla U\nabla\beta]$$
$$+ Pr[-V_\eta/R^2 - (2/R^2)(\partial U/\partial \theta)] - 2Pr\beta[(1/R^2)(\partial V_\eta/\partial \theta)$$
$$+ (U/R^2)(\partial\beta/\partial \theta) + (\beta/R^2)(\partial U/\partial \theta)] \qquad (11)$$

142

This equation will be solved instead of the V momentum equation and the velocity component V in the diffusion terms of equation (4) will be replaced by the new velocity component V_η. Then, equation (4) can be rewritten as

$$\vec{V} \cdot \vec{\nabla} U + V_\eta U/R + \beta U^2/R = -(1/R)(\partial P/\partial \theta) + Pr\nabla^2 U$$
$$+ Pr[-(U/R^2) + (2/R^2)(\partial V_\eta/\partial \theta) + (2\beta/R^2)(\partial U/\partial \theta)$$
$$+ (2U/R^2)(\partial \beta/\partial \theta)] \qquad (12)$$

Integral Forms

The momentum equations (11), (12) and the energy equation (5) are integrated over a control volume in physical space bounded by lines of constant η and constant ξ. Such a control volume is illustrated in Fig. 2. Using the divergence theorem, equation (11) becomes

$$\int_S (\vec{V} \cdot \vec{n}) V_\eta ds - Pr \int_S (\vec{n} \cdot \vec{\nabla} V_\eta) ds = \int_v [-(\partial P/\partial R)$$
$$+ (\beta/R)(\partial P/\partial \theta)] dv - \int_v \vec{V} \cdot (U \vec{\nabla} \beta) dv + Pr \int_v (U \nabla^2 \beta + 2 \nabla U \nabla \beta) dv$$
$$+ \int_v [U^2/R + \beta(U/R)V_\eta + (\beta U)^2/R] dv - Pr \int_v [(V_\eta/R^2)$$
$$+ (2/R^2)(\partial U/\partial \theta)] dv - 2Pr\beta \int_v [(1/R^2)(\partial U/\partial \theta)$$
$$+ (\beta/R^2)(\partial U/\partial \theta) + (U/R^2)(\partial \beta/\partial \theta)] dv \qquad (13)$$

and equation (12) becomes

$$\int_S (\vec{V} \cdot \vec{n}) U ds - Pr \int_{Sn} \vec{\nabla} U ds = -\int_v (1/R)(\partial P/\partial \theta) dv - \int_v [(UV_\eta/R)$$
$$+ (\beta U^2/R)] dv + Pr \int_v [(-U/R^2 + (2/R^2)(\partial V_\eta/\partial \theta)$$
$$+ (2\beta/R^2)(\partial U/\partial \theta) + (2U/R^2)(\partial \beta/\partial \theta)] dv \qquad (14)$$

and equations (5) and (2) become

$$\int_S (\vec{V} \cdot \vec{n}) T ds = \int_{Sn} \vec{\nabla} T ds \qquad (15)$$
$$\int_v (\vec{\nabla} \cdot \vec{V}) dv = \int_S (\vec{V} \cdot \vec{n}) ds = 0 \qquad (16)$$

where v and s represent the dimensionless volume and surface of the control volume, respectively. For the evaluation of the surface integrals, expressions are needed for the surface element ds, the gradient operator $\vec{\nabla}$ and the unit vector $\vec{n}$. To derive these quantities, it is first necessary to consider a formal coordinate transformation from R and θ to η and ξ, where η and ξ have been defined by equation (6). The transformation is

$$(\partial/\partial R)_\theta = (1/\Delta)(\partial/\partial \eta)_\xi$$
$$(\partial/\partial \theta)_R = -\eta\beta(\partial/\partial \eta)_\xi + (\partial/\partial \xi)_\eta \qquad (17)$$

To facilitate the evaluation of the surface integrals that appear in equations (13) to (16), reference may be made to the shaded control volume of Fig. 2. As suggested there, the surface integral may be subdivided into a sum of four surface integrals over the segments, S_1, S_2, S_3, and S_4. For surface 1, $\vec{n} = \vec{e}_\theta$, $\vec{V} \cdot \vec{n} = U$; also $\xi = $ constant along S_1. Therefore, an element of surface area can be expressed as

$$ds = dR = \Delta d\eta \qquad (18)$$

For surface 2,

$$\vec{n} = (\vec{e}_R - \beta\vec{e}_\theta)/\alpha^{(1/2)}, \quad \vec{V} \cdot \vec{n} = V_\eta/\alpha^{(1/2)} \qquad (19)$$

and

$$ds = [dR^2 + (Rd\theta)^2]^{(1/2)} = R(\beta^2 + 1)^{(1/2)} d\xi$$
$$= \alpha^{(1/2)} R d\xi \qquad (20)$$

For surface S_3 and S_4, ds is identical to those for S_1 and S_2, with the exception that the outward normal $\vec{n}$ has the opposite sign. The volume integrals appearing on the right-hand side of equations (11) and (12) can now be evaluated. The volume element dv can be written as

$$dv = RdRd\theta = \Delta^2 \eta d\eta d\xi \qquad (21)$$

The standard form of the $\vec{\nabla}$ operator in cylindrical coordinates is

$$\vec{\nabla} = (\partial/\partial R)\vec{e}_R + (\partial/\partial \theta)\vec{e}_\theta \qquad (22)$$

which, after substitution for $\partial/\partial R$ and $\partial/\partial \theta$ from equation (17), becomes

$$\vec{\nabla} = (1/\Delta)(\partial/\partial \eta)\vec{e}_R + (1/\eta\Delta)[(\partial/\partial \xi) - \eta\beta(\partial/\partial \eta)]\vec{e}_\theta \quad (23)$$

The evaluation of the all surface integrals and the volume integrals in equations (13) to (16) can be performed by using equations (17) to (23) in the same manner as described by Faghri et. al. (1984). Because of the similarities among equations (13) to (15), it is possible to rewrite them in a compact form by introducing the abbreviations

$$\int_1 [U\phi\Delta + \Gamma(\gamma + \Psi)] d\eta - \int_3 [U\phi\Delta + \Gamma(\gamma + \Psi)] d\eta$$
$$+ \int_2 [V_\eta\phi\Delta\eta + \Gamma(\Omega + \Lambda)] d\xi - \int_4 [V_\eta\phi\Delta\eta + \Gamma(\Omega + \Lambda)] d\xi = b \qquad (24)$$

where ϕ stands for U, V_η, or T and Γ is the diffusion coefficient which is equal to Pr for the momentum equations and equal to one for the energy equation. The variables Ω, γ, Λ, and Ψ are abbreviations which are defined as:

$$\Omega = -\alpha\eta \, (\partial\phi/\partial\eta), \quad \gamma = -(1/\eta)(\partial\phi/\partial\xi),$$
$$\Lambda = \beta(\partial\phi/\partial\xi), \quad \Psi = \beta(\partial\phi/\partial\eta) \qquad (25)$$

The Λ and Ψ are the pseudodiffusion terms.

Discretization

Equation (24), is discretized by the widely known control volume scheme of Patankar (1981). The pseudodiffusion terms in equation (24) are included in the source term. The staggered grids are used in which the grid points for temperature are located at the center of the main control volumes and the grid points for the velocity components are located on the sides of the main control-volume. The power-law finite difference approximation of Patankar (1981) is selected for the discretization of the differential equation. The discretized equations are solved by using a line-by-line method. The pressure and the velocity are linked by the SIMPLE algorithm of Patankar (1980). It is well known that the discretization of the differential equation causes false diffusion depending on the Peclet number. A detailed discussion of false diffusion has been given in a book by Patankar (1980).

SAMPLE COMPUTATIONS

Heat Transfer in Fully Developed Region of Polygonal Duct

The first example problem deals with the fully developed laminar flow and heat transfer for regular polygonal ducts with constant wall temperature. A summary of the literature on this problem has been brought together by Shah and London (1978). The cross section of the duct is shown in Fig. 3. The thermal boundary condition is assumed to be uniform temperature both axially and peripherally. This is the T boundary condition of Shah and London (1978). As seen in this figure, the origin of the coordinate is located at the center of the cross section, and the distance from the origin to the walls is denoted by $\delta(\theta)$. The geometry of the duct is specified by length L and parameter n. The shaded area in the figure is the solution domain. Because of symmetry, it is sufficient to confine the solution domain to only (1/n)th of the cross section. The expression of $\delta(\theta)$ is as follows:

$$\delta(\theta) = 0.5L/[\tan(\pi/n)\cos(C\pi/n + \theta)] \qquad (26)$$

where C is a constant and changes with θ.

$$
\begin{aligned}
0 < \theta < \pi/n &\quad : C = 0 \\
\pi/n < \theta < 2\pi/n &\quad : C = -2
\end{aligned}
\qquad (27)
$$

The governing equations to be considered here are the momentum, and energy equations. Constant thermophysical properties are assumed. Using the following non-dimensional variables

$$R = r/D_h, \ Z = z/(D_h/Re), \ W = w/\bar{w},$$
$$P = p/\rho\bar{w}^2, \ T = (t-t_w)/(t_b-t_w) \qquad (28)$$

where z is the axial direction of the duct, and D_h is the hydraulic diameter, 4(Area)/(Perimeter), $\bar{w}$ is the average velocity in z direction, Re is the Reynolds number, (wD_h/ν), and the t_b is the bulk temperature, the governing equation takes the following form:

$$0 = -(\partial P/\partial Z) + \nabla^2 W \qquad (29)$$
$$0 = \nabla^2 T - PrWT\sigma \qquad (30)$$

where $\sigma = (dt_b/dZ)/(t_b-t_w) \qquad (31)$

which is the parameter arising from the assumptions of the T boundary condition. This value is determined as part of the solution process from the identity of $\int TW \, dA = \int W \, dA$. It is obvious from the equations (29) and (30) that the problem to be considered reduces to the diffusion problem.

The friction factor f is defined as

$$f = -(2D_h)(dp/dz)/\rho\bar{w}^2 = 2(dP/dZ)/Re \qquad (32)$$

In the fully developed region, the friction factor becomes inversely proportional to the Reynolds number, and $(f \ Re)_{f.d.}$ becomes independent of z and equal to $2(dP/dZ)$. The heat flux from the walls can be expressed as

$$q = c_p \ \dot{m} \ (dt_b/dz)/(nL) = c_p \ \bar{w} \ (D_h/4)(dt_b/dz) \qquad (33)$$

where the $\dot{m}$ is the mass flow rate, n refers to the number of sides of polygonal duct, and nL represents the perimeter. The Nusselt number based on the hydraulic diameter can be expressed as

$$Nu = qD_h/[k(t_w-t_b)] = -Pr \ \sigma/4 \qquad (34)$$

The computation were performed for the triangular, rectangular, pentagonal and hexagonal ducts. To investigate the grid size effect on the friction factor and the Nusselt number, the computations were performed with various number of grid points. The computed friction factor and Nusselt number are listed in Tables 1 and 2, respectively. Shih (1967) obtained the friction factor and the Nusselt number values for the fully developed region of regular polygonal ducts. Shih's results (1967) are also listed in these tables. The error indicates the deviation from the values of Shih (1967). As seen from these tables, the percentage error

decreases with increasing number of grid points. This tendency is accentuated for pentagonal and hexagonal ducts.

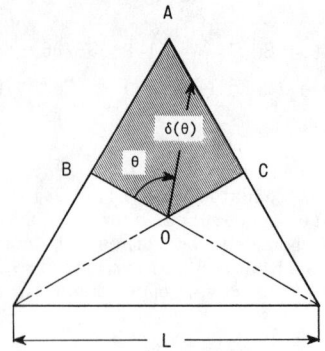

Fig. 3   Schematic diagram of cross section of triangular duct

Table 1   Fully developed value of $(f \ Re)_{f.d.}$

n	grid points	$(f \ Re)_{f.d.}$	error
3	48 X 36	54.845	2.90%
	78 X 36	54.317	1.91%
	178 X 36	53.782	0.90%
	Shih (1967)	53.3	
4	48 X 36	57.653	1.31%
	78 X 36	57.372	0.82%
	148 X 36	57.145	0.42%
	Shih (1967)	56.910	
5	48 X 36	59.385	0.74%
	78 X 36	59.216	0.45%
	158 X 36	59.072	0.21%
	Shih (1967)	58.948	
6	48 X 36	60.508	0.48%
	78 X 36	60.390	0.24%
	158 X 36	60.290	0.12%
	Shih (1967)	60.216	

Table 2   Fully developed Nusselt number of T boundary

n	grid points	Nu	error
3	48 X 36	2.611	5.72%
	78 X 36	2.570	4.04%
	178 X 36	2.530	2.43%
	Shih (1967)	2.47	
4	48 X 36	3.033	1.91%
	78 X 36	3.013	1.23%
	148 X 36	2.996	0.66%
	Shih (1967)	2.976	
5	48 X 36	3.243	0.46%
	78 X 36	3.231	0.09%
	158 X 36	3.221	-0.20%
	Shih (1967)	3.228	
6	48 X 36	3.363	0.40%
	78 X 36	3.355	0.16%
	158 X 36	3.349	-0.04%
	Shih (1967)	3.35	

<u>Natural Convection in a Square Cavity</u>

The second example problem deals with the two-dimensional natural convection in a square cavity of side L with differentially heated vertical walls as shown in Fig. 4(a). One side wall is heated at temperature t_h, and the opposite wall is cooled at temperature t_c. The top and bottom walls are thermally insulated. The velocity components are zero on the boundaries. This problem is well known bench mark numerical solution given by de Vahl Davis (1982). As seen from the figure, the origin of the coordinate is located at the center of the cavity. The distance from the origin to the walls, $\delta(\theta)$, is given by

$$\delta(\theta) = 0.5L/\sin(C\pi - \theta) \qquad (35)$$

where C is a constant and changes with θ.

$$
\begin{aligned}
0 < \theta < \pi/4 &\quad : C = 0.5 \\
\pi/4 < \theta < (3/4)\pi &\quad : C = 1.0 \\
(3/4)\pi < \theta < (5/4)\pi &\quad : C = 1.5 \\
(5/4)\pi < \theta < (7/4)\pi &\quad : C = 2.0 \\
(7/4)\pi < \theta < 2\pi &\quad : C = 2.5
\end{aligned}
\qquad (36)
$$

and its derivatives can be easily obtained.

Using the Bussinesq approximation, the governing equations of this problem are equations (2) to (5), with the extra source terms which are derived from the buoyancy force. These terms are $(RaPrT \sin\theta)$ and $(RaPrT \cos\theta)$ which are added to equations (3) and (4), respectively.

Computations were performed for Ra = 10^3, 10^4 and 10^5 with grid points of (40 peripheral)x(12 radial) and for Pr = 0.71. These grid points are distributed uniformly. A typical main control volume distribution is illustrated in Fig. 4(b). The average temperature $t_m = (t_h + t_c)/2$ is selected for the characteristic temperature t_{ref}. The temperature difference between the two walls, (t_h-t_c), is selected as the characteristic temperature difference, Δt. So that the nondimensional temperature of the hot wall, T_h, is 0.5, and of cold wall, T_c is -0.5. The average heat flux from the hot wall, q, can be expressed as

$$q = -(k/L)\int_0^L (dt/dx)dy = -(k/L)(t_h-t_c)\int_{5\pi/4}^{7\pi/4} \alpha(\partial T/\partial \eta)d\xi \qquad (37)$$

The average Nusselt number based on the characteristic length L is defined as

$$Nu = qL/[k(t_h-t_c)] = \int_{5\pi/4}^{7\pi/4} \alpha(\partial T/\partial \eta)d\xi \qquad (38)$$

Supplementary runs based on the orthogonal x, y coordinates were also performed for the comparison. The grid points were (22x22) which were distributed uniformly.

The representative streamline and isotherm maps obtained from coordinate transformation solutions for Ra = 10^4 are shown in Fig. 5(a) and 5(b). The results of the computation based on the orthogonal x, y coordinates are also shown in Fig. 5(c) and 5(d) for comparison. The deviation of streamlines near the corners can be seen from these figures.

The representative horizontal velocity profile on the vertical plane of x/L = 0.5 is shown in Fig. 6 for Ra=10^4. This vertical plane is located at the mid point between the hot and the cold walls. The representative vertical velocity profile on the horizontal plane of y/L = 0.5 is also shown in the same figure. The solid lines

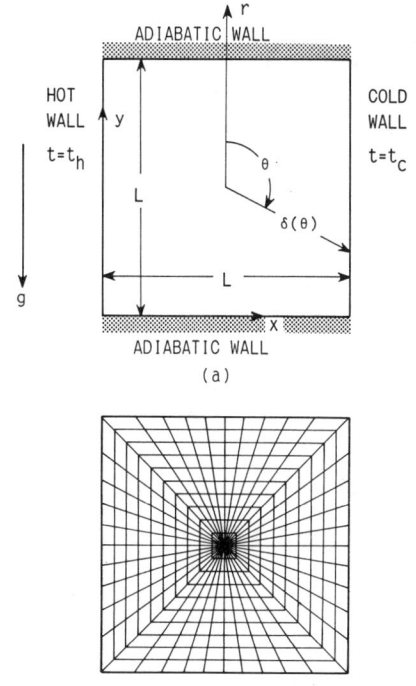

Fig. 4(a) Schematic diagram of square cavity
(b) typical control volume distribution

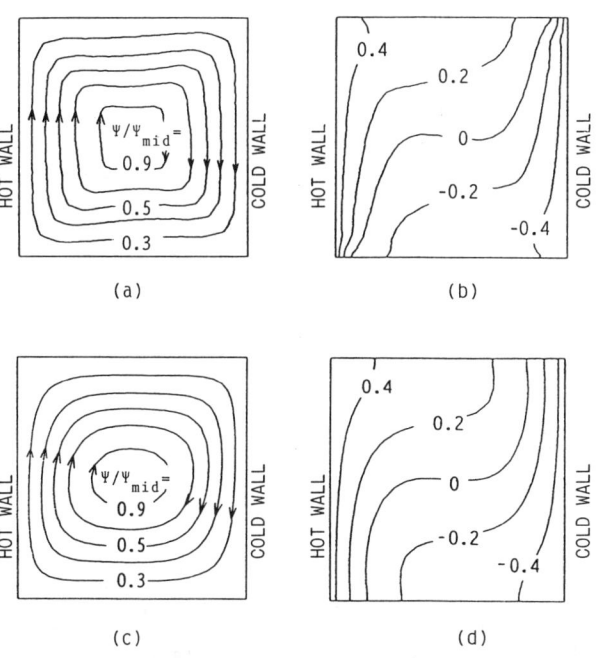

Fig. 5 Streamlines and isotherms obtained from
(a),(b), the coordinate transformation (40x22)
and (c), (d), x-y coordinate computation (22x22)

145

indicate the results of the computation based on the coordinate transformation methodology, and the dashed lines indicate the results of the computation based on the x, y orthogonal coordinate. As seen from the figure, the solid lines and the dashed lines almost coincide with each other.

The representative temperature profile on the horizontal plane of $y/L = 0.5$ is shown in Fig. 7 for $Ra=10^4$. The solid line indicates the results based on the coordinate transformation methodology, and the dashed line indicates the results based on the x, y orthogonal coordinate. As seen from the figure, the solid lines and the dashed lines almost coincide with each other.

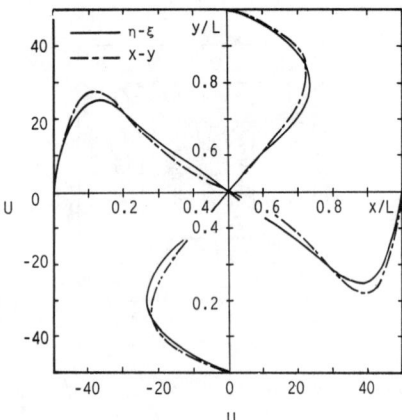

Fig. 6 Comparisons of velocity distributions

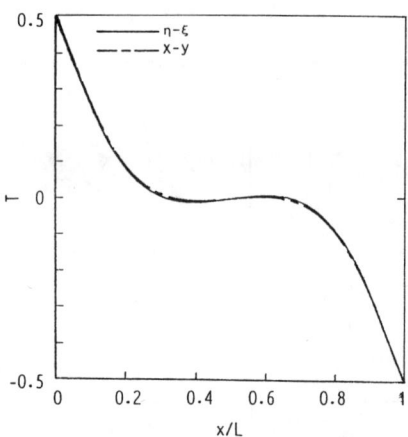

Fig. 7 Comparisons of temperature distribution at $y/L=0.5$

Nusselt number results are listed in Table 3. Results of the supplementary runs based on x-y coordinates are also listed in the table. de Vahl Davis (1982) performed two computations with fine and coarse meshes and calculated the extrapolated solution from these two original solutions. de Vahl Davis' extrapolated values are also listed in the same table.

To investigate the grid size effect on the Nusselt number, supplementary computations were performed for $Ra = 10^4$ with fine meshes. Number of grid points are

(40-peripheral x 22-radial) and (40x27). The Nusselt number results are listed in Table 4. The Nusselt number found by the present methodology with fine mesh approaches de Vahl Davis' extrapolated value.

Equations (13) and (14) involve the first, second and third derivatives of $\Delta(\theta)$. In many works, in which the boundary-fitted coordinate transformation methodology was adopted, the derivatives of $\Delta(\theta)$ were often neglected in the computation [e.g. Sparrow et. al. (1986)]. Therefore, to investigate the effect of the omitted derivatives, supplementary computations were performed with the following assumptions:

(a) $\partial\Delta/\partial\xi =0,\ \partial^2\Delta/\partial\xi^2 =0,\ \partial^3\Delta/\partial\xi^3 =0$
(b) $\partial^2\Delta/\partial\xi^2 =0,\ \partial^3\Delta/\partial\xi^3 =0$
(c) $\partial^3\Delta/\partial\xi^3 =0$.

The number of grid points was (40x22) and the Rayleigh number was $Ra = 10^4$. The Nusselt number results are listed in Table 5. It is noteworthy that the effect of the omitted derivatives on the Nusselt number were small.

Table 3   Nusselt number values

	Rayleigh number		
	10^3	10^4	10^5
$\eta - \xi$ coordinates	1.219	2.398	4.900
$x - y$ coordinates	1.123	2.300	4.902
de Vahl Davis (1982)	1.118	2.238	5.505

Table 4   Grid size effect on the Nusselt number

grids points	Nu
40 X 12	2.398
40 X 22	2.371
40 X 27	2.365
de Vahl Davis (1982)	2.238

Table 5 Effect of omission of the derivatives on Nusselt number

omitted derivatives	Nu
(a) $\Delta''' = \Delta'' = \Delta' = 0$	2.247
(b) $\Delta''' = \Delta'' = 0$	2.414
(c) $\Delta''' = 0$	2.351
no-omission	2.398
de Vahl Davis (1982)	2.238

CONCLUDING REMARKS

The development of a coordinate transformation methodology for convection-diffusion problems which maps the arbitrary domain onto a circle is presented. Sample computations were performed for the heat transfer in a fully developed region of a polygonal duct and for the natural convection in a two dimensional square cavity. Comparisons were made with the available values.

REFERENCES

de Vahl Davis, G., 1982, "Natural Convection of Air in a Square Cavity: A Bench Mark Numerical Solution", Report 1982/FMT/2, Univ. N.S.W.,School of Mech. and Indust. Eng.

Faghri, M., Sparrow, E.M. and Prata, A.T., 1984, "Finite-Difference Solution of Convection-Diffusion Problems in Irregular Domains, Using a nonorthogonal coordinate Transformation", <u>Numerical Heat Transfer</u>, Vol. 7, pp.183-209.

Patankar, S.V., 1980, Numerical Heat Transfer and Fluid Flow, Mcgraw-Hill.

Patankar, S.V., 1981, "A Calculation Procedure for Two-Dimensional Elliptic Situation," <u>Numerical Heat Transfer</u>, Vol. 4, pp.409-425.

Saitoh, T., 1978, "Numerical Method for Multi-Dimensional Freezing Problem Arbitrary Domains", <u>ASME J. Heat Transfer</u>, Vol.100, pp. 294-299.

Shah, R.K., and London, A.L., 1978, Laminar Flow Forced Convection in Ducts, Academic Press, Inc., New York.

Shih, S.F., 1967, "Laminar Flow in Axisymmetric Conduits by a Rational Approach", <u>Can. J. Chemical Engineering</u>, Vol.45, pp. 285-294.

Sparrow, E.M., Patankar, S.V., and Ramadhyani, S., 1977, "Analysis of Melting in the Presence of Natural Convection in the Melt Region", <u>ASME Journal of Heat Transfer</u>, Vol.99, pp. 520-526.

Sparrow, E.M., and Ohkubo, Y., 1986, "Numerical Analysis of Two-Dimensional Transient Freezing Including Solid-Phase and Tube-Wall Conduction and Liquid-Phase Natural Convection ", <u>Numerical Heat Transfer</u>, Vol.9, pp.59-77.

1989 National Heat Transfer Conference
HTD-Vol. 107, Heat Transfer in Convective Flows

A NUMERICAL SIMULATION OF DOUBLE DIFFUSIVE NATURAL CONVECTION IN A VERTICAL RECTANGULAR ENCLOSURE

H. Han and T. H. Kuehn
Department of Mechanical Engineering
University of Minnesota
Minneapolis, Minnesota

ABSTRACT

A numerical study has been performed on double diffusive natural convection fluid flow in a vertical rectangular cavity when the temperature and concentration gradients are imposed in the horizontal direction. A finite difference algorithm has been adopted to solve the non-linear momentum equations coupled with the energy and concentration equations. Different flow structure regimes have been obtained as a function of the Grashof number ratio for aiding and opposing buoyancy conditions. Double diffusive multicell flow structures observed in the experiments by authors have been simulated numerically. Details of the multicell flow characteristics as well as the overall heat and mass transfer characteristics in the cavity are presented.

NOMENCLATURE

C	Concentration
ΔC	Concentration difference between plates
c	Dimensionless concentration ($= \frac{(C-C_{low})}{(C_{high}-C_{low})}$)
D	Diffusivity (m^2/sec)
Gr_T	Thermal Grashof number ($g\,\beta_T\,\Delta T\,H^3\,/\nu^2$)
Gr_M	Solutal Grashof number ($g\,\beta_M\,\Delta C\,H^3\,/\nu^2$)
h	Heat transfer coefficient (W/m^2 °C)
h_m	Mass transfer coefficient (m/sec)
g	Gravitational constant
H	Cavity height
k	Thermal conductivity (W/m°C)
L	Cavity width
Le	Lewis number ($= \frac{Sc}{Pr}$)
L1	Number of grids in the x direction
M1	Number of grids in the y direction
N	Grashof number ratio ($= \frac{Gr_M}{Gr_T}$)
Nu	Nusselt number (h H/k)
P	Pressure
Pr	Prandtl number (ν/α)
Sc	Schmidt number (ν/D)
Sh	Sherwood number (h_m H/D)
T	Temperature
ΔT	Temperature difference between plates
t	Dimensionless temperature ($= \frac{(T-T_{cold})}{(T_{hot}-T_{cold})}$)
U	Horizontal velocity component
V	Vertical velocity component
X	Horizontal coordinate
Y	Vertical coordinate

Greek

α	Thermal diffusivity
β_T	Volumetric expansion coefficient due to temperature gradient
β_M	Volumetric expansion coefficient due to concentration gradient
ν	Kinematic viscosity
ρ	Density

INTRODUCTION

This paper considers the natural convection induced by simultaneous heat and mass transfer in a rectangular enclosure. Uniform temperature and concentration boundary conditions are prescribed along the vertical walls, while the horizontal walls are insulated for both heat and mass transfer. The governing parameters are found to be thermal Grashof number, solutal Grashof number, Prandtl number, and Schmidt number for a given aspect ratio. The main objectives of this investigation are to characterize the double diffusive fluid flow structure and to determine the heat and mass transfer characteristics in the cavity.

The term double diffusive convection was first used by oceanographers who have been working on the ocean curiosities. Their interest lies mostly on stability problems in stratified salt solutions. Some of the double diffusive convection phenomena, such as salt fingers and sharp diffusive interfaces, have been verified in laboratory experiments (Turner, 1979). The double diffusion theory has been developed and applied to other fields such as geology, astrophysics, and many engineering fields. Some of the engineering applications include energy storage in solar ponds, storage of liquified natural gas, manufacturing crystals, and solidification of metals. The unique double diffusion characteristics, however, have not been observed when the Lewis number is near unity, which includes simultaneous heat and moisture transfer in air.

There are only a few studies on natural convection induced by simultaneous heat and mass transfer in an enclosed cavity, while there have been numerous studies concerning the natural convection induced by a single diffusive component in an enclosure. Hu and El-Wakil (1974) were the first to conduct a simultaneous heat and mass transfer experiment in an enclosure using air-water and air-N heptane systems. An electrochemical experiment was performed by Kamotani et al. for an application to crystal growth processes. Layered stratified flow structures were reported when both temperature and concentration gradients were imposed in shallow enclosures for a large Lewis number fluid.

A flow visualization study conducted by the present authors exhibited photographs of the transient multicell flow structure in an enclosure of aspect ratio, H/L, 1 and 4. Three cell flow structures were observed for both aiding and opposing buoyancy conditions. Details of experimental apparatus and procedures can be found in Han and Kuehn (1988).

In this study, it is attempted to simulate the double diffusive three cell flow structures in a cavity of aspect ratio 4, numerically. The parameters used in the experiments are extended to a wider range. Comprehensive information of velocity, temperature, and concentration distributions in a cavity are presented in comparison with the experimental results. The Prandtl number and the Schmidt number are set at experimental values of 8 and 2000, respectively. Thermal Grashof numbers vary from -10^5 to 10^5 for a solutal Grashof number fixed at 10^5 or 3×10^6.

MATHEMATICAL FORMULATION

A schematic diagram of the present problem is presented in Fig. 1. Depending on the directions of the buoyancy forces, the problem can be in either an aiding or opposing buoyancy condition. The solutal boundary layer is considered to be thinner than the thermal boundary layer since the solutal diffusivity is much smaller than the thermal diffusivity.

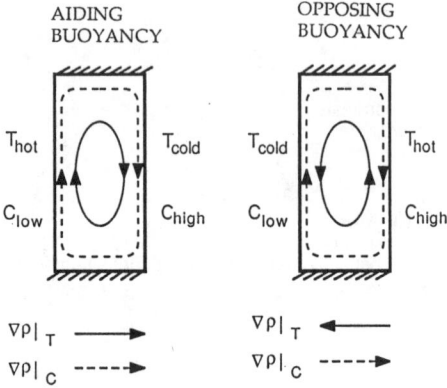

Fig. 1 Schematic drawing of the present problem when Le>>1

The flow in the cavity is considered to be two-dimensional and steady, and follows classical Boussinesq approximation. The properties of the fluid are assumed to be constant except for the density in the buoyancy terms. The fluid is assumed to be incompressible and Newtonian in behavior with negligible viscous dissipation. The heat flux by concentration gradients (thermal diffusion or Soret effect) and the mass flux by temperature gradients (diffusion thermo or Dufour effect) are neglected. By employing the above assumptions into the conservation equations of mass, momentum, energy, and species, a set of governing equations can be obtained as

$$\frac{\partial u}{\partial x} + \frac{\partial v}{\partial y} = 0 \tag{1}$$

$$u\frac{\partial u}{\partial x} + v\frac{\partial u}{\partial y} = -\frac{1}{\rho}\frac{\partial p}{\partial x} + v\left(\frac{\partial^2 u}{\partial x^2} + \frac{\partial^2 u}{\partial y^2}\right) \tag{2}$$

$$u\frac{\partial v}{\partial x} + v\frac{\partial v}{\partial y} = -\frac{1}{\rho}\frac{\partial p}{\partial y} + v\left(\frac{\partial^2 v}{\partial x^2} + \frac{\partial^2 v}{\partial y^2}\right) + (Gr_T t + Gr_M c) \tag{3}$$

$$u\frac{\partial t}{\partial x} + v\frac{\partial t}{\partial y} = \frac{1}{Pr}\left(\frac{\partial^2 t}{\partial x^2} + \frac{\partial^2 t}{\partial y^2}\right) \tag{4}$$

$$u\frac{\partial c}{\partial x} + v\frac{\partial c}{\partial y} = \frac{1}{Sc}\left(\frac{\partial^2 c}{\partial x^2} + \frac{\partial^2 c}{\partial y^2}\right) \tag{5}$$

The above equations have been made dimensionless using the following parameters.

$$x = \frac{X}{H}, \quad y = \frac{Y}{H}, \quad u = \frac{U}{v/H}, \quad v = \frac{V}{v/H}, \quad p = \frac{P}{\rho v^2/H^2}$$

$$t = \frac{T - T_{cold}}{T_{hot} - T_{cold}}, \quad c = \frac{C - C_{low}}{C_{high} - C_{low}} \tag{6}$$

The dimensionless boundary conditions for the physical system considered in the present study are

$$t = 1, \quad c = 0 \quad \text{(aiding)}$$
$$\qquad\quad c = 1 \quad \text{(opposing)} \quad \text{at } x = 0$$

$$t = 0, \quad c = 1 \quad \text{(aiding)}$$
$$\qquad\quad c = 0 \quad \text{(opposing)} \quad \text{at } x = \frac{L}{H}$$

$$\frac{\partial t}{\partial y} = \frac{\partial c}{\partial y} = 0 \quad \text{at } y = 0 \text{ and } y = 1$$

$$u = v = 0 \quad \text{at } x = 0, \frac{L}{H} \text{ and at } y = 0, 1 \tag{7}$$

For aiding buoyancy conditions, the vertical hot surface is maintained at a low concentration, while the cold surface is maintained at a high concentration. For opposing buoyancy conditions, this is reversed. In actual programming, a buoyancy force acting in the reverse direction is implemented by changing the sign of the Grashof number instead of changing the values of the boundary conditions. A buoyancy force acting in the clockwise direction is considered to be positive and that in the counterclockwise direction is considered to be negative.

SOLUTION PROCEDURE

Finite difference equations are derived by integrating the governing differential equations over an elementary control volume. A power law scheme is adopted for the convection-diffusion formulation. The coupling between the non-linear algebraic equations is handled using the SIMPLER algorithm. The velocity values are defined on staggered grids. The other dependent variables are located at the nodal points of the main grid. The discretized equations obtained are solved iteratively, using a line-by-line application of the Thomas algorithm. The non-linear coefficients are substituted successively with updated values. Underrelaxation is required to ensure the convergence of the iterative procedure. A block correction scheme is incorporated to accelerate the convergence rate (Patankar, 1980).

Non-uniform grid spacing is used in the x-direction. Grid spacing is minimum near a vertical wall and is increased exponentially away from the wall according to the following expression.

$$XU(I) = \left(\frac{I - 2}{\frac{L1}{2} - 2}\right)^{XPOWER} \tag{8}$$

where XU(I) are the locations of control volume faces, L1 is the number of grids in the x-direction. The value of XPOWER determines the non-uniformity of the grid spacing. Uniform mesh spacing is used in the y-direction in order to resolve the possible multicell flow structure. Since the location of the cell interfaces are not known a priori, fine meshes are required everywhere in the vertical direction. Typical numbers of grids used are 22x82 for $Gr_M = 10^5$, and 34x130 for $Gr_M = 3 \times 10^6$ in the x and y directions, respectively.

The solution is considered to be fully converged when the maximum value of the mass source and the changes of the dependent variables from iteration to iteration are smaller than a prescribed value, i.e.10^{-5}. Due to the small solutal diffusivity of the fluid, the convergence rate is very slow, especially for solutal buoyancy dominant flows. A typical number of iterations to obtain a fully converged solution is approximately 1000 for simple flows, and more than 5000 for complex flows. Most of the runs have been made on a Cray-2 with 4 processors at the University of

150

Minnesota Supercomputer Institute. The CPU time for calculating 1000 iterations for a 34x130 grid is approximately 10 minute.

Nusselt and Sherwood numbers are calculated after convergence is attained. The local Nusselt number at the wall is defined as

$$Nu_W(y) = \frac{hH}{k} = -\frac{\partial t}{\partial x}\bigg|_{wall} \qquad (9)$$

The non-dimensional temperature gradients are calculated by approximating the linear temperature profile between grids. The local Nusselt numbers are integrated to obtain an overall average Nusselt number. An overall Nusselt number in the middle of the cavity could be calculated by adding a convection term to the diffusion term. Theoretically, it should be identical with the Nusselt number at the wall. The difference between them is an indication of the convergence and the correctness of the finite difference approximation.

Similar equations can be applied in calculating the Sherwood numbers. The local Sherwood number at a wall is expressed as

$$Sh_W(y) = \frac{h_m H}{D} = -\frac{\partial c}{\partial x}\bigg|_{wall} \qquad (10)$$

The average Sherwood numbers are obtained by integrating the local Sherwood numbers in the y coordinate direction as well.

RESULTS

Test of Grid Dependence

In order to validate the present numerical code, and to report the error range of the present results, numerical aspects are presented first, which include the effects of the grid size on the overall Nusselt numbers for pure heat transfer and a comparison with available benchmark results.

Pure heat transfer results have been obtained for $Gr=10^5$ and $Pr=2000$ in a cavity of H/L=4 using various numbers of grids. Fig. 2a shows the overall Nusselt number on a wall as a function of ΔX_{min}. The number of x direction grids, L1, varies from 22 to 82 with XPOWER=2. ΔX_{min} is the minimum grid spacing next to the vertical wall, by which the temperature gradients on the wall are calculated from the linear approximation of the temperature profiles. The extrapolation of the curves to $\Delta X_{min}=0$ provides the Nusselt number at zero grid spacing in the x direction. The extrapolated value of the Nusselt number is within 1 % of the Nusselt number obtained at L1=22.

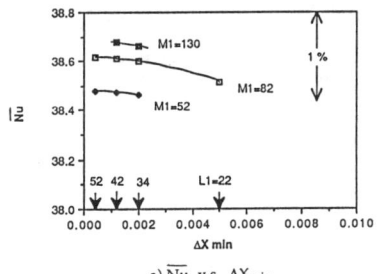

a) $\overline{Nu}$ v.s. ΔX_{min}

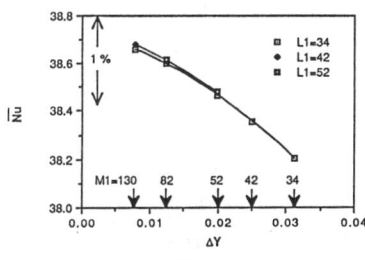

b) $\overline{Nu}$ v.s. ΔY

Fig. 2 Effect of grid spacing on average Nusselt number results for pure heat transfer: $Gr=10^5$, Pr=2000, and H/L=4

Fig. 2b shows the Nusselt number variations depending on the grid spacing in the y-direction, ΔY. The number of grid points in the y-direction, M1, is varied from 34 to 130. The effect of the grid spacing in the y-direction is found to be significant. It indicates that the number of grids in the y-direction should be larger than 52 for the present range of parameters to maintain the discretization error within 1 %.

Solutions obtained for $Ra=10^5$, Pr=0.71 in a square cavity with non-uniform grid spacings in both directions (POWER=1.5) are compared with the benchmark results of de Vahl Davis. A converged solution at one grid spacing was used as the starting condition for the next finer grid spacing with the intermediate points obtained by interpolation. The extrapolated average Nusselt number using a second order polynomial is 4.519, which matches the benchmark results up to 4 significant digits. The Nusselt number obtained using 22x22 grid points is within 1 % of the extrapolated value.

Effect of Buoyancy Ratio on Flow Structures and Transfer Rates

With the solutal Grashof number fixed at 10^5, various thermal Grashof numbers are superimposed either in the opposite direction or in the same direction. Streamline plots for different opposing buoyancy conditions are shown in Fig. 3 as a function of buoyancy ratio. For $|Gr_T|<10^4$, the solutal buoyancy force is dominant. The overall structure is unicellular motion rotating in the clockwise direction, which is similar to pure mass transfer convection. The streamlines of pure solutal convection, $N=\infty$, are shown in Fig. 3a. For $|Gr_T|>5x10^4$, the thermal buoyancy force is dominant over the solutal buoyancy force. The overall flow structure is again a unicell motion rotating in the counterclockwise direction driven by the thermal buoyancy. Compared to pure thermal convection, however, there are separation points near the bottom of the cold wall (left wall) and near the top of the hot wall. In the recirculation zones, the flow recirculates very slowly in the opposite direction driven by the solutal buoyancy. The streamlines of pure thermal convection, $Gr_M=0$, are shown in the Fig. 3e for reference. The boundary layer is much thicker than that of pure solutal convection, since the more-diffusive species (heat) is dominant.

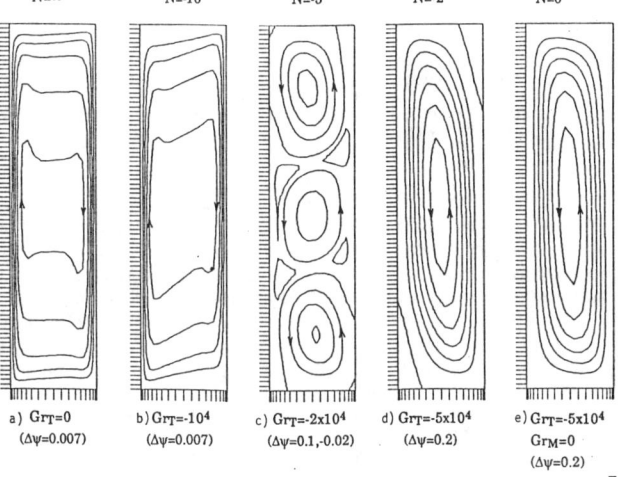

| $N=\infty$ | $N=-10$ | $N=-5$ | $N=-2$ | $N=0$ |

a) $Gr_T=0$ b) $Gr_T=-10^4$ c) $Gr_T=-2x10^4$ d) $Gr_T=-5x10^4$ e) $Gr_T=-5x10^4$
($\Delta\psi=0.007$) ($\Delta\psi=0.007$) ($\Delta\psi=0.1,-0.02$) ($\Delta\psi=0.2$) $Gr_M=0$
($\Delta\psi=0.2$)

Fig. 3 Streamline plots for opposing buoyancy conditions: $Gr_M=10^5$, Pr=8, Sc=2000 and H/L=4

For a range of Grashof number ratios between -10 and -2, thermal and solutal forces have the same order of magnitude. It could have been expected to have a transitional buoyancy ratio at which the flow direction is reversed. It is very interesting to have a three-cell flow structure generated for N=-5 ($Gr_T=2x10^5$). Streamlines of the three cell structure are shown in Fig. 3c. Each cell rotates counterclockwise driven by the thermal buoyancy force. There are high shear regions between cells. The thermal buoyancy forces from the side walls overcome the shear stress between cells, and rotate the fluid in each cell in the same direction. Small idling cells rotating clockwise can be observed between cells and walls. Details of the multicell flow structure will be included in the next section.

The dependence of the flow structure on the grid spacing has been checked for N=-5. The center cell could not be obtained with M1 less than 62. With large grid spacings in the y-direction, the thin shear layers between cells could not be resolved, and false diffusion is a serious concern. The distinct three cell structure was obtained with grids finer than 22x82. The convergence becomes extremely slow for very fine grids. Furthermore, it is recommended that the average grid size in both coordinate directions should be of the same order of magnitude to ensure convergence. Hence the grid 22x82 is used for the present range of Grashof numbers.

Fig. 4 shows streamline plots for aiding buoyancy conditions. Fig. 4a is a streamline plot at $Gr_T=10^3$, and shows a unicell flow structure rotating clockwise. For the mildly aiding condition with a small Gr_T superimposed, the results are similar to pure solutal convection as was shown in Fig. 3a. Fig. 4e shows streamlines at $Gr_M=1.5x10^4$. Even though the magnitude of Gr_T is much smaller than Gr_M (N=6.67), the thermal buoyancy is dominant over the solutal one. The overall flow is unicellular motion similar to pure thermal convection as shown in Fig. 3e, but rotating in the opposite direction.

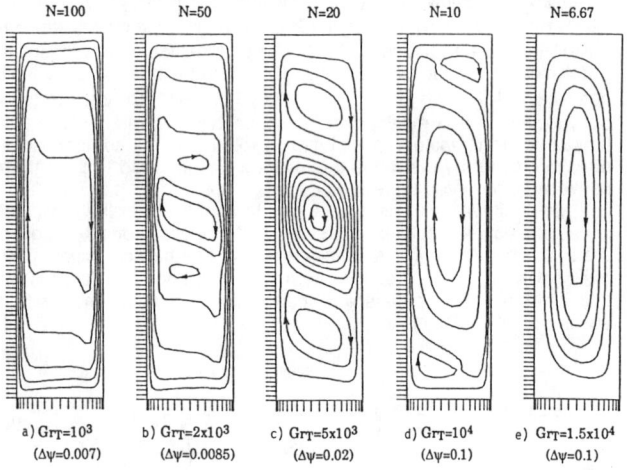

N=100   N=50   N=20   N=10   N=6.67

a) $Gr_T=10^3$ b) $Gr_T=2x10^3$ c) $Gr_T=5x10^3$ d) $Gr_T=10^4$ e) $Gr_T=1.5x10^4$
($\Delta\psi=0.007$)   ($\Delta\psi=0.0085$)   ($\Delta\psi=0.02$)   ($\Delta\psi=0.1$)   ($\Delta\psi=0.1$)

Fig. 4 Streamline plots for aiding buoyancy conditions: $Gr_M=10^5$, Pr=8, Sc=2000 and H/L=4

For 10<N<50, three cell flow structures are obtained even for aiding buoyancy conditions. As the superimposed Gr_T increases, a three cell structure begins at $Gr_T=2x10^3$ as shown in Fig. 4b. Inside the solutal dominant convection pattern, two small inside cells driven by weak thermal convection appear. The three cell structure is clearly seen in Fig. 4c. Each cell rotates in the same direction (clockwise). As Gr_T is further increased to 10^4, the top and bottom cells are pushed to the top and bottom of the cavity by the vigorous center cell motion driven by the thermal buoyancy force (Fig. 4d). The range of N for the three cell flow structure in the aiding buoyancy conditions is found to be larger compared to the opposing buoyancy conditions. The exact range of buoyancy ratios for the three cell flow structure is yet to be determined.

The overall Nusselt numbers are shown in Fig. 5 as a function of Gr_T. The results for both aiding and opposing flow conditions are superimposed with the pure heat transfer results shown with a dotted line. As the thermal Grashof number increases, the Nusselt numbers of both aiding and opposing conditions converge to the dotted line. In the other limit ($Gr_T=0$), the Nusselt numbers approach the value of pure conduction, which is 4. The overall Nusselt numbers for negative Gr_T are reduced compared to the pure heat transfer results in the intermediate range of buoyancy ratios. It is interesting to note that the Nusselt numbers are also decreased slightly even for positive Gr_T in the intermediate range of buoyancy ratios compared to the pure heat transfer case. This is due to the multicell flow structure explained above. However, the effect of the superimposed Gr_M on the overall Nusselt number appears to be very weak in the present range of parameters.

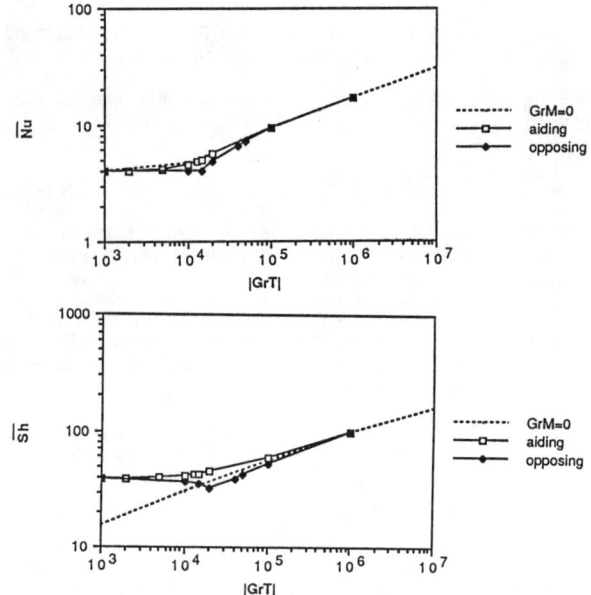

Fig. 5 Overall Nusselt numbers and Sherwood numbers as a function of superimposed Gr_T: $Gr_M=10^5$, Pr=8, Sc=2000 and H/L=4

In Fig. 5b, the overall Sherwood numbers are presented as a function of thermal Grashof number for aiding and opposing buoyancy conditions. The dotted line here indicates accompanied mass transfer rates due to a purely thermal-driven convection flow (not pure mass transfer convection). For large magnitudes of the thermal Grashof number, the mass transfer rates converge to this line regardless of the aiding or opposing buoyancy conditions. As the thermal Grashof number decreases to zero, the Sherwood numbers of both aiding and opposing conditions approach an asymptotic value, which is the mass transfer rate at $Gr_M=10^5$ driven by pure solutal convection. Over the entire range of Gr_T, the Sherwood number for the opposing buoyancy condition is observed to be smaller than that for the aiding buoyancy condition.

Details of the Multicell Flow Structure

Converged solutions have been obtained for a higher range of Grashof numbers ($Gr_M=3x10^6$), which is the same range as previous experiments (Han and Kuehn, 1988). For the larger Grashof numbers, the boundary layers are expected to be thinner than before, and finer grid spacings are required. A grid of 34x130 is used with XPOWER=2 and with uniform ΔY. Fig. 6 shows the velocity vectors, streamlines, isotherms and iso-concentration contours of the converged solution for the opposing buoyancy condition: $Gr_T=-4x10^5$, $Gr_M=3x10^6$. The three cell flow structure is more distinct for this higher Grashof number range, which is well within the boundary layer regime. The temperature field is stably stratified in each cell, and the concentration is nearly uniform in each cell except near the walls and cell interfaces, as shown in Figs. 6c and 6d. The concentration of the center cell is at the initial bulk concentration, and the concentration differences of the the top and bottom cells are symmetric with respect to the center cell.

The two-dimensional plots include the complete information of the flow field, but cross-sectional profiles of the velocity can help to understand the flow field more clearly. The horizontal velocity along the vertical centerline and the vertical velocity along the horizontal centerline are shown in Figs. 7a and 7b, respectively. The horizontal velocity profile shows a zig-zag pattern which can be expected from the multicellular streamline plot as shown in Fig. 6b. The cross-sectional vertical velocity profile exhibits bidirectional flow in the boundary layers near the walls. The fluid in the thin concentration boundary layer is flowing in the opposite direction to the fluid motion in the relatively thick thermal boundary layer. However, the direction of the fluid motion in the concentration boundary layer does not appear to affect the overall fluid motion in the core region significantly. There are high shear regions near the concentration boundary layer edges due to the bidirectional flows. This type of velocity profile has not been observed in single diffusive convection flows.

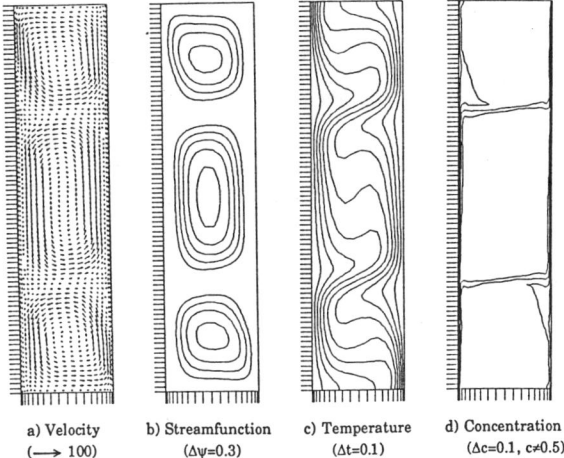

a) Velocity
(→ 100)

b) Streamfunction
(Δψ=0.3)

c) Temperature
(Δt=0.1)

d) Concentration
(Δc=0.1, c≠0.5)

Fig. 6 Two-dimensional plots of velocity vectors, streamlines, isotherms and iso-concentration contours for opposing buoyancy: Gr_T=-4.0x10^5, Gr_M=3.0x10^6, Pr=8, Sc=2000 and H/L=4

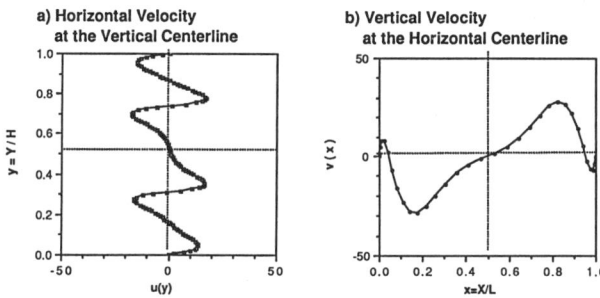

a) Horizontal Velocity at the Vertical Centerline

b) Vertical Velocity at the Horizontal Centerline

Fig. 7 Cross-sectional velocity profiles a) u(y) at x=0.5, b) v(x) at y=0.5 for opposing buoyancy condition: Gr_T=-4.0x10^5, Gr_M=3.0x10^6, Pr=8, Sc=2000 and H/L=4

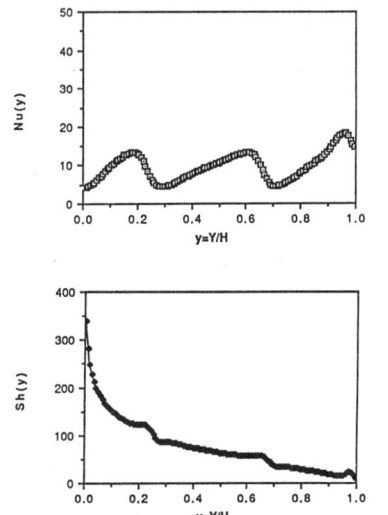

Fig. 8 Local Nusselt number and local Sherwood number distributions along the vertical cold cathode (x=0.0) for opposing buoyancy condition: Gr_T=-4.0x10^5, Gr_M=3.0x10^6, Pr=8, Sc=2000 and H/L=4

The local Nusselt number distribution along the vertical cold cathode plate, where concentration is zero, is shown in Fig. 8a. The Nusselt number shows the local maximums and minimums along the vertical plate. The maximums occur where the fluid flows toward the plate, and the minimums occur where the fluid flows away from the plate. Fig. 8b shows the local Sherwood number distribution along the plate. The Sherwood number distribution appears to be nearly monotonically decreasing with respect to the vertical coordinate. The mass transfer rate appears to be less affected by the presence of the double diffusive convection flows in the core than the heat transfer rate does, since the concentration gradients are mostly confined within the thin concentration boundary layers.

A solution with aiding buoyancy condition has also been obtained for Gr_T=-3x10^5, Gr_M=-3x10^6, Pr=8, Sc=2000, and H/L=4. The same grid system used for the opposing buoyancy solution was utilized. A streamline plot is shown in Fig. 9a with a Schlieren photograph. It is interesting to note that the basic overall flow structure is similar to the previous opposing buoyancy condition, even though the details are not the same. The cross sectional vertical velocity profile does not show a bidirectional behavior, but it is similar to the pure thermal and/or pure solutal convection. The present numerical solutions confirm that each cell behaves separately, and the thermal buoyancy force determines the direction of fluid rotation in each cell.

The vertical temperature and concentration profiles are similar to those of the opposing buoyancy case. However, the concentration difference between cells is smaller than that in the opposing condition. The calculated concentration distribution is compared with a still photograph taken through the cavity in Fig. 9b. The cupric ion has blue color, and the darkness of the color determines the concentration of cupric ion in the solution.

a) Calculated streamlines and Schlieren photograph

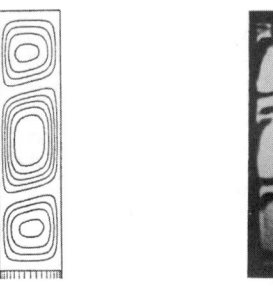

b) Concentration distributions

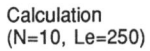

Calculation
(N=10, Le=250)

Experiment
(N=8.5, Le=280)

Gr_T=-3.0x10^5, Pr=8.0
Gr_M=-3.0x10^6, Sc=2000

Gr_T=-3.8x10^5, Pr=8.5
Gr_M=-3.2x10^6, Sc=2400

Fig. 9 Comparison of the numerical results with the experiments a) Calculated streamlines and Schlieren photograph b) Calculated and experimental concentration distributions. (Fluid rotates counter-clockwise in each cell.)

The Nusselt number distribution along the hot cathode shows similar local maximums and minimums as the opposing buoyancy condition. The Sherwood number distribution along the cathode is also similar to that of opposing buoyancy. The Sherwood number is maximum at the bottom of the cathode, and decreases with respect to y. Even though the local heat and mass transfer rates show similar trends, the overall heat and mass transfer rates are increased compared to the above opposing buoyancy condition.

The discrepancy in simulating the double diffusive convection experiments may be due to the assumption of constant properties, especially the viscosity and the diffusion coefficients. It may also be due to the non-uniform concentration boundary conditions on electrode surfaces. The assumption of uniform concentration may not be appropriate for an electrochemical experiment in a stratified fluid in an enclosure.

CONCLUSIONS

Double diffusive natural convection flows have been investigated for various conditions in a vertical rectangular cavity with an aspect ratio of 4. The flow regimes can be categorized as thermally dominant, solutally dominant, and equally important. The thermally dominant and solutally dominant convection results in a unicell flow structure as in pure thermal and pure solutal convection. In the intermediate range of Grashof number ratio, a three cell flow structure is obtained regardless of the direction of the buoyancy forces for a high Lewis number. The minimum Lewis number resulting in the multicell flow structure is yet to be determined.

The present numerical solutions agree with the previous experimental results qualitatively. The characteristics of the multicell flow structure are given in detail. The flow direction in the thin concentration boundary layer does not appear to affect the overall flow structure significantly. The temperature and concentration profiles in the cavity are unique to double diffusive natural convection. The peculiar Nusselt number and Sherwood number distributions along the vertical walls are also due to the double diffusive multicell flow structure.

In order to plot multicell flow regimes on a parametric map, more calculations are required. Transient modeling of the problem will provide a more realistic picture of the multicell flow structure formations.

ACKNOWLEDGMENTS

This work is partially supported by a grant from the Minnesota Supercomputer Institute of the University of Minnesota.

REFERENCES

Bottemanne, F. A., 1971, "Theoretical Solution of Simultaneous Heat and Mass Transfer by Free Convection about a Vertical Flat Plate," *Appl. Sci. Res.*, Vol. 25, pp. 137-149.

Churchill, S. W., Chao, P., and Ozoe, H., 1981, "Extrapolation of Finite-Difference Calculations of Laminar Natural Convection in Enclosures to Zero Grid Size," *Numerical Heat Transfer*, Vol. 4, pp. 39-51.

de Vahl Davis, G., 1983, "Natural Convection of Air in a Square Cavity: A Bench Mark Numerical Solution," *Int. J. for Numerical Methods in Fluids*, Vol. 3, pp. 249-264.

Han, H., 1988, "Double Diffusive Natural Convection in a Vertical Rectangular Enclosure," Ph.D. Thesis, University of Minnesota.

Han, H. and Kuehn, T. H., 1988, "Flow Visualization of Double Diffusive Natural Convection in a Vertical Rectangular Enclosure," *Proc. 25th National Heat Transfer Conf.*, Vol. 3, pp. 409-414.

Hu, C. Y. and El-Wakil, M. M., 1974, "Simultaneous Heat and Mass Transfer in a Rectangular Cavity," *Proc. 5th Int. Heat Transfer Conf.*, Vol. 5, pp. 24-28.

Kamotani, Y., Wang, L. W., Ostrach, S., and Jiang, H. D., 1985, "Experimental Study of Natural Convection in Shallow Enclosures with Horizontal Temperature and Concentration Gradients," *Int. J. Heat Mass Transfer*, Vol. 28, pp. 165-173.

Nilson, R. H., 1985, "Countercurrent Convection in a Double-Diffusive Boundary Layer," *J. Fluid Mech.*, Vol. 160, pp. 181-210.

Ostrach, S., 1983, "Fluid Mechanics in Crystal Growth - The 1982 Freeman Scholar Lecture," *J. Fluids Eng.*, Vol. 105, pp. 5-20.

Patankar, S. V., 1980, '*Numerical Heat Transfer and Fluid Flow*,' Hemisphere Pub. Co.

Thorpe, S. A., Hutt, P. K., and Soulby, R., 1969, "The Effects of Horizontal Gradients on Thermohaline Convection," *J. Fluid Mech.*, Vol. 38, Pt. 2, pp. 375-400.

Turner, J. S., 1979, '*Buoyancy Effects in Fluids*,' Cambridge University Press, London.

Wang, L. W., 1983, "Experimental Study of Natural Convection in a Shallow Horizontal Cavity with Different End Temperatures and Concentrations," Ph.D. Thesis, Case Western Reserve University.

1989 National Heat Transfer Conference
HTD-Vol. 107, Heat Transfer in Convective Flows

COMBINED FORCED AND NATURAL CONVECTION HEAT TRANSFER IN A "DEEP" LID-DRIVEN CAVITY FLOW

A. K. Prasad and J. R. Koseff
Environmental Fluid Mechanics Laboratory
Stanford University
Stanford, California

ABSTRACT

In this study, we describe the combined forced and natural convection (also known as mixed-convection) heat transfer process within a recirculating flow in an insulated lid-driven cavity of rectangular cross-section (150 mm × 450 mm), and depth varying between 150 mm and 600 mm. The forced convection is induced by a moving lid which shears the surface layer of the fluid in the cavity, thereby setting up a recirculating flow, while the natural convection flow is induced by heating the lower boundary and cooling the upper one. By appropriately varying the lid-speed, the vertical temperature differential, and the depth, we obtained Gr/Re^2 ratios for these flows from 0.01 to 1000. Flow visualization using liquid crystals, and heat flux measurements at specific locations over the lower boundary provided an insight into the nature of the heat transfer process under different flow and temperature conditions. The mean heat flux values over the entire lower boundary were analyzed to produce Nusselt number and Stanton number correlations which should be useful for design applications.

NOMENCLATURE

B	cavity width
c_p	specific heat of water
D	cavity depth
Gr	Grashof number based on cavity depth
h	heat transfer coefficient
k	thermal conductivity of water
L	cavity length or span
Nu	Nusselt number based on cavity depth
q	heat flux (W/m²)
Re	Reynolds number based on lid-velocity and cavity width
St	Stanton number based on lid-velocity
U_B	lid-velocity
ΔT	temperature difference between lower (hotter) and upper (colder) cavity boundaries
ν	kinematic viscosity of water
ρ	density of water

ABBREVIATIONS

DAR	Depthwise Aspect Ratio (D/B)
DSE	Downstream Secondary Eddy
DSW	Downstream Side Wall
SAR	Spanwise Aspect Ratio (L/B)
TGL	Taylor-Görtler-Like
TLC	Thermochromic Liquid Crystals
USW	Usptream Side Wall

INTRODUCTION

Heat transfer in flows in which the influence of forced convection and natural convection are of comparable magnitude (commonly referred to as "mixed-convection" flows) occur frequently in engineering situations. Gebhart [1] separates mixed convection processes into external flows and internal flows, and provides a review of some of the more common geometries. Mixed convection flows may be further sub-divided into those where the inertia force is parallel to the buoyancy force, and those where the inertia force is perpendicular to the buoyancy force. Lloyd and Sparrow [2] developed a similarity solution for a mixed convection flow on a vertical surface for a range of Prandtl numbers. In this study, the forced convection flow aided the free convection component. Siebers [3] carried out experiments with a heated vertical wall in the presence of a cross stream, with a view towards predicting the behavior of a central receiving station in a solar power plant. Kays and Crawford [4] list a number of studies of mixed convection over horizontal plates. For instance, Wang et al [5] conducted heat transfer measurements in a water channel flow that was uniformly heated from below, and compared their results with existing correlations. Similarly, Imura et al [6] focused on the transition from the laminar to the turbulent regime, by measuring the heat flux on a horizontal heated plate. At about the same time, a numerical investigation of a mixed convection boundary layer flow on a horizontal surface was performed by Chen et al [7].

All the studies mentioned in the preceding paragraph pertain to external flows. However, internal, or confined flows are also of great interest to designers of alternate energy systems, such as solar ponds and solar storage devices. For example, Muñoz and Zangrando [8] describe an experimental investigation of mixing in a double-diffusive fluid layer. They studied the effect of surface shear as well as bottom heating on entrainment, and developed correlations for the same. Cha and Jaluria [9] performed a numerical investigation of a recirculating mixed convection flow in an stratified energy storage system, involving withdrawal of hot fluid, and discharge of cold fluid into the storage chamber. For the range of governing parameters studied, their results indicate that the recirculating flow is strongly affected by buoyancy, as well as the geometry. In addition, the cooling of electronic components on a printed circuit board, and the heat transfer from rectangular cut-outs on the surface of a heat-exchanger in the presence of a oncoming stream could also be modeled as internal or recirculating flows, and provide a special incentive for our study.

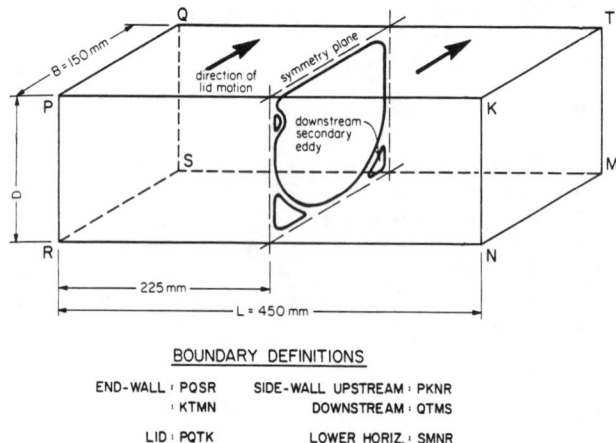

BOUNDARY DEFINITIONS

END-WALL : PQSR   SIDE-WALL UPSTREAM : PKNR
        : KTMN   DOWNSTREAM : QTMS
      LID : PQTK   LOWER HORIZ. : SMNR

Figure 1   Definitions for the lid-driven cavity flow

The configuration described in the present investigation consists of an internal flow within a rectangular cavity which is driven by two mechanisms. First, by shearing the surface layer by means of a lid moving at a constant velocity, we induce a forced convection flow within the cavity (see Figure 1). The lid-speed can be increased continuously to produce Reynolds numbers (Re, based on the lid-speed, U_B, and the cavity width, B) up to 12,000. This recirculating motion is extremely complex and three-dimensional in nature and has been described in detail by Koseff and Street [10] and Prasad and Koseff [11]. Second, by heating the lower boundary of the cavity, and cooling the upper boundary, we can create a natural convection flow, as described by Rhee et al [12]. The temperature differential between the lower and upper boundaries, and the depth of the cavity D, can be varied to obtain Grashof numbers for these flows (Gr, based on the cavity depth) between 10^7 and 5×10^9. The simultaneous use of both forcing mechanisms enables us to create flows in which the parameter, Gr/Re^2, varies between 0.01 and 1000. By varying the dimensions of the cavity, it is also possible to create flows of widely differing geometric attributes, yet possessing the same Gr/Re^2 ratio.

Flow visualization using Thermochromic Liquid Crystals (TLC) provides a simultaneous view of the temperature and velocity fields, thereby identifying the relevant flow structures, and their influence on the heat transfer mechanisms in the cavity. This

was supplemented by quantitative data, in the form of heat flux measurements at specific locations over the lower boundary, using microfoil heat flux sensors. We are currently examining the heat-flux data from the individual heat-flux meters (such as of time-traces and power spectra) in order to isolate the influence of the three-dimensional flow features that are unique to lid-driven cavity flows. The heat flux data in this paper, however, are averaged over the entire lower boundary of the cavity (hereafter referred to as the cavity floor).

The ratio Gr/Re^2 has been traditionally used to represent the relative magnitudes of forced- and natural-convection in a mixed-convection flow. In our configuration, we have produced the same value of the ratio, but by using different combinations of ΔT, D and U_B. By examining the heat-flux data from these different cases, we wish to determine if the ratio Gr/Re^2 is sufficient to characterize this class of mixed-convection flows, or if it needs to be modified by factors such as the Depthwise-Aspect-Ratio (DAR).

FACILITY AND INSTRUMENTATION

The Lid-Driven Cavity Facility

The lid-driven cavity facility, constructed from 12.5 mm thick Plexiglas, consists of two attached "shoe-boxes," as shown in Figure 2. The lower of the two boxes is the main area of interest. It is rectangular in cross-section, with streamwise width (B) of 150 mm, spanwise length (L) of 450 mm, and depth (D) varying between 150 mm and 600 mm. The upper box houses the lid, and the driving system for the lid, which consists of a variable speed motor connected by a chain drive to one of a pair of rollers. The "lid" is a 0.08 mm thick copper belt which is mounted on and driven by the two rollers. During operation, the belt-and-drive assembly in Figure 2 is lowered down into the upper box such

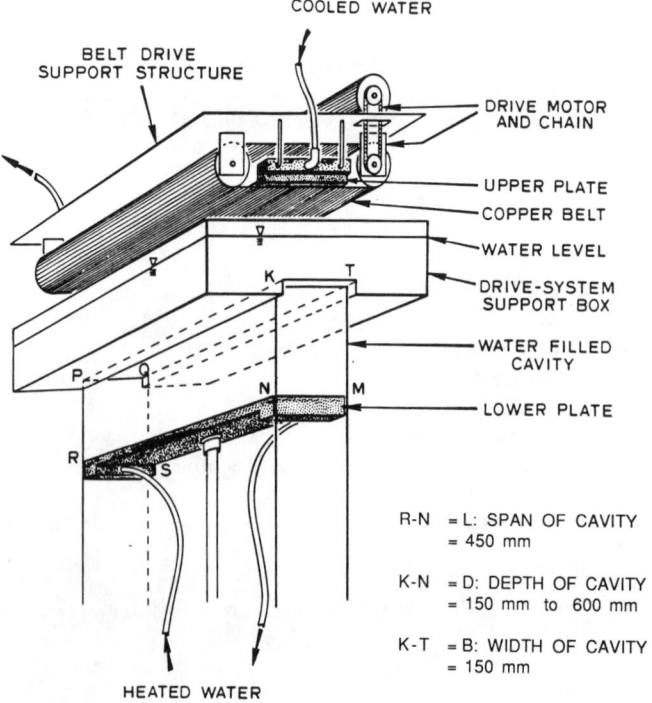

R-N = L: SPAN OF CAVITY
      = 450 mm

K-N = D: DEPTH OF CAVITY
      = 150 mm to 600 mm

K-T = B: WIDTH OF CAVITY
      = 150 mm

Figure 2   The lid-driven cavity facility (from Koseff and Street [10])

that the belt just touches the upper edges of the working area. The belt speed can be varied continuously to produce Reynolds numbers up to 15,000. At a particular setting, the belt speed is constant to within ±0.25%.

The lower boundary of the cavity is a heat exchanger plate. In addition, the copper belt is in continuous contact with a similar plate. By circulating water of the appropriate temperature through each plate, it is possible to generate a natural convection flow within the cavity. The maximum vertical temperature differential that can be attained is $10 \pm 0.05°$ C. The four vertical walls of the cavity are adiabatic. This is ensured in practice by adding styrofoam insulation to the side-walls and by maintaining the mean fluid temperature as close to the laboratory ambient temperature as possible.

Finally, the cavity-depth (D) can be varied from 150 mm to 600 mm by lowering the bottom heat exchanger plate. In this study, heat flux measurements were performed at four cavity-depths: 150, 300, 450 and 600 mm. These correspond to Depthwise-Aspect-Ratios (DAR = D/B) of 1:1, 2:1, 3:1, and 4:1 respectively. For each case, three temperature differentials were studied: 1, 4 and 8 °C. Furthermore, for each combination of DAR and temperature differential, nine belt-speeds were used providing Re's ranging from 0 to 12,000. This resulted is a total of 108 different cases.

Flow Visualization Technique

Thermochromic liquid-crystals possess the ability to change color with temperature, and therefore, can be used to visualize the temperature field in mixed-convection flows. The liquid crystals are encapsulated in micro-spheres which are advected by the flow, and thereby enable visualization of the flow field as well. For the present study we chose a chiral nematic type of TLC manufactured by Hallcrest. This particular sample is available in the the microencapsulated form with 50-100 μm sized capsules. The active range for this TLC starts at 23 °C and a has bandwidth of 2 °C. A concentration of 0.2 gm/ltr is adequate to visualize the flow satisfactorily.

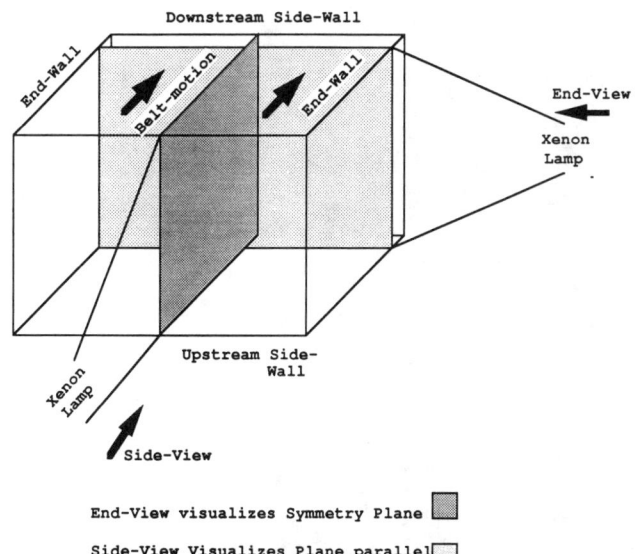

Figure 3   Illuminated planes for flow visualization

Specific planes of the cavity were illuminated as shown in Figure 3. The "end-view" refers to the view of the symmetry plane through the end-wall, while the "side-view" offers a view of a plane parallel to the DSW through the Upstream Side Wall (USW). The illumination for both views was provided by an 800 W ILC Technology Xenon lamp in conjuction with a "slit-and-lens" arrangement, which focused the light into a thin (3 mm) sheet. Kodacolor 400 ASA film, and an exposure time of 3 seconds with an f-stop of 5.6, produced the best results.

Visualization photographs (color) were obtained for $\Delta T = 8$ °C, at Re's of 3200 and 7500, for all four DAR's. In this paper, however, we have included only two representative prints, due to the fact that much of the temperature field information is lost when the color photographs are reduced to a black and white format.

Heat Flux Measurements Technique

The heat flux meters used for the present study were Micro-Foil Heat Flow Sensors manufactured by the RdF Corporation. These heat flux meters are in the form of a thin, rectangular strip or foil and are mounted flush on the surface at which the heat flux is to be measured. A good thermal contact between the sensor and the surface can be achieved with double-sided adhesive tape. The heat flux meters used in this study measured 12 mm × 30 mm, with a foil-thickness of 0.11 mm.

The foil consists of an electrical insulator, in which a number (typically twenty) of copper-constantan thermocouple junctions are embedded. The junctions are connected in a single series circuit, and are arranged such that alternate junctions are on opposite surfaces of the foil. Therefore, the voltage generated at the leads of the heat flux meter is directly proportional to the difference in the temperatures of the two surfaces, which in turn, depends on the heat flux through the foil. The manufacturer-supplied thermal resistance of the sensors used in our experiments was 5.3×10^{-4} K/(W/m²) with a response time of 50 Hz.

In our measurement configuration, the heat flux meters were submerged in water for prolonged periods of time. Consequently, the manufacturer-supplied calibration constants could not be relied upon for high accuracy. Therefore, we devised an *in-situ* calibration technique, and re-calibration was performed at frequent intervals. On the basis of the slight drift in the calibration constants, we estimate that the uncertainty in the heat flux measurements is limited to ±5 %.

A total of eight heat flux meters were attached to the lower boundary of the cavity as shown in Figure 4. This particular arrangement was chosen, in order to distribute the measurement locations uniformly over the lower boundary of the cavity, and thus obtain an accurate estimate of the average, overall heat transfer from the lower boundary. The voltage output of the meters (which typically, is about 6 μvolts for every 100 W/m² of heat flux) was then amplified by a factor of 40,000 using an eight-channel amplifier, before digitization. An HP2100 data acquisition system was used to sample the heat flux output at a rate of 100 Hz, for a period of 11.92 minutes. The heat flux from each meter was then analyzed on an HP1000 system, providing time-traces and spectra, and the mean heat flux over the entire lower surface was determined by taking the average of all eight heat flux meters. (However, as mentioned previously, this paper focuses specifically on the averaged heat-flux values.)

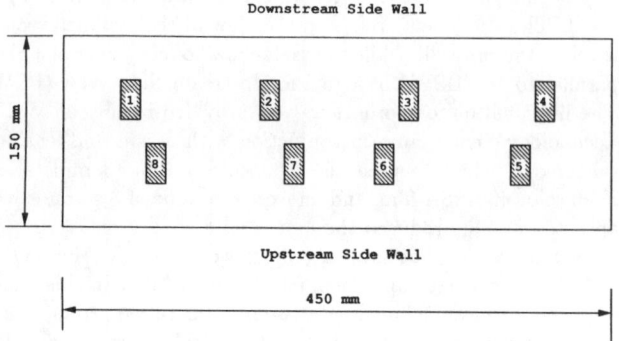

Figure 4   Location of heat flux meters over the cavity floor

OBSERVATIONS

Flow Visualization Results

Visualization of two perpendicular planes of the cavity provided valuable insight into the nature of the three-dimensional flow structures in the lid-driven cavity flow. At the outset, the heat exchangers were set to produce the required temperature differential and allowed to run for a period of 24 hours. The two light sources and cameras were then positioned as shown in Figure 3. Next, the belt was started and set at the desired speed and the flow eventually reached a fully-developed state. Subsequently, a measured quantity of TLC's was introduced slowly through a small port in the Downstream Side Wall (DSW), and distributed uniformly by the mixed-convection flow.

Figures 5(a) and (b) form an end-view/side-view pair for a DAR of 2:1, ΔT of 8 °C, and Re of 3200. Figure 5(a) is an end-view picture of the symmetry plane. The belt moves from left to right at the top of the picture. The fluid adjacent to the belt is cold, as it is cooled by the upper heat exchanger plate. This cold fluid (which is red in color) is stripped off at the corner formed by

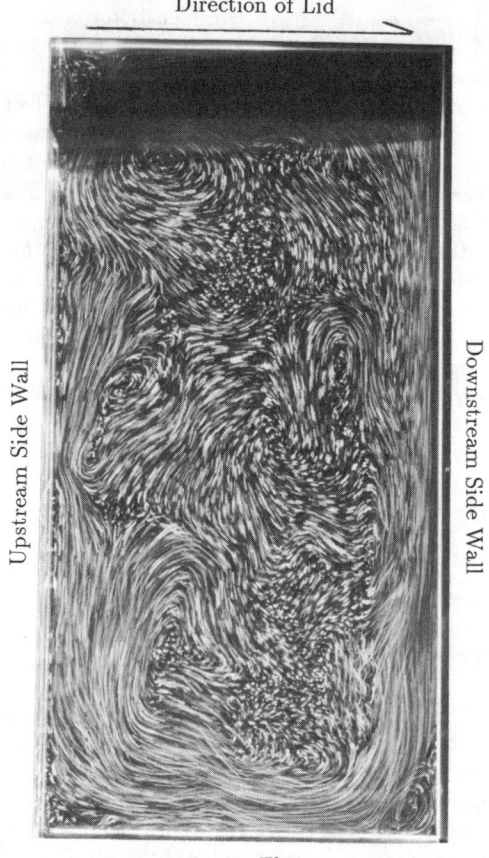

Figure 5(a)   Flow visualization with liquid crystals: Symmetry Plane, DAR 2:1, Re 3200, ΔT 8 °C, (End-View)

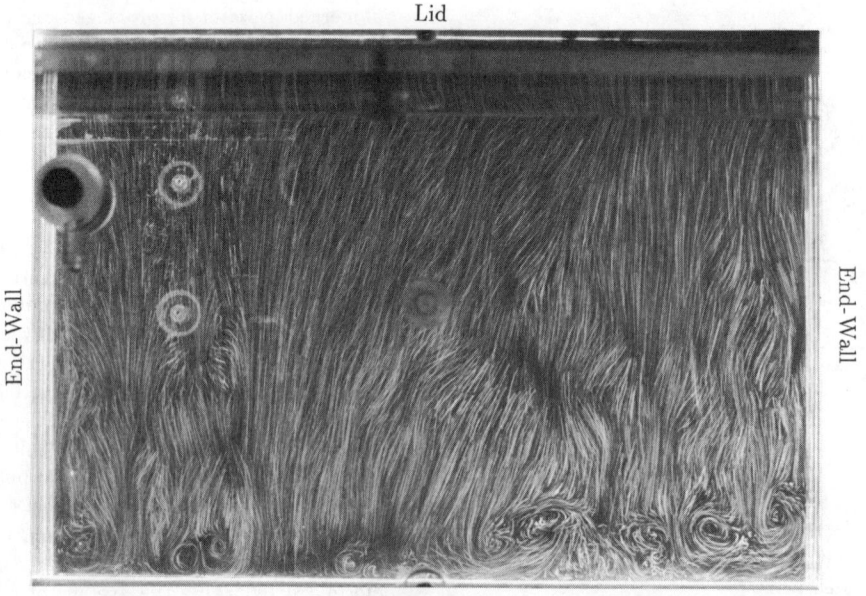

Figure 5(b)   Flow visualization with liquid crystals: Plane parallel to DSW and 20 mm from it, DAR 2:1, Re 3200, ΔT 8 °C, (Side-View)

the lid and the DSW, and plunges down along the DSW in the form of a wall-jet. The inertia imparted to this fluid by the lid is aided by its larger density (the red fluid is colder, and therefore, denser), and the wall-jet continues all the way to the floor of the cavity. At the floor, the fluid absorbs heat provided by the lower heat exchanger plate, and the fluid becomes warmer and changes in color to a dark blue (the hot end of the color spectrum). Finally, the warmed-up fluid ascends the Upstream Side Wall (USW) in the form of a buoyant plume and penetrates into the core of the recirculating fluid.

Figure 5(b) is the corresponding side-view picture that was exposed four seconds after Figure 5(a), in a plane parallel to the DSW, and 22 mm from it. It is possible to pick out counter-rotating pairs of longitudinal vortices along the floor, separated at regular intervals in the spanwise direction. These are the Taylor-Görtler-Like (TGL) vortices that originate over the concave separation surface of the Downstream Secondary Eddy (DSE) (see Figure 1). The mechanism that generates these vortices involves the centrifugal instability resulting from the flow over a concave surface, and have been extensively described in [10] and [11] for isothermal flows at a DAR of 1:1. Rhee et al [12] discovered that TGL vortices also form at a DAR of 1:1 in a mixed convection flow ($\Delta T = 4$ °C, $Re = 3200$); however, the size of the TGL vortices are diminished when compared with that of the isothermal case.

Heat Flux Results

Heat flux measurements were performed in the following manner. The calibrated heat flux meters were attached on the cavity floor at the locations indicated in Figure 4, and the lower plate was moved to the appropriate depth. As in the case of flow visualization, the heat exchangers were run for a period of 24 hours at the required temperatures before data acquisition commenced. The first case studied during an experiment was the natural convection case (lid is stationary, $Re = 0$). Subsequently, the belt was started and the belt-speed was set to produce the next Re. Data was acquired only after allowing sufficient time for the new flow conditions to reach a fully-developed state (about one hour). This process was repeated every hour, until the highest Reynolds number had been studied.

Figures 6(a), (b), (c) and (d) show the mean heat flux over the cavity floor (averaged over all eight heat flux meters for a period of 12 minutes) for DAR's 1:1, 2:1, 3:1 and 4:1 respectively. Each plot contains heat flux data for the three temperature differentials, as a function of the Reynolds number. The corresponding uncertainties ($\pm$ 5%) are also indicated. These plots show that the heat flux is strongly influenced by the temperature differential, and at a given ΔT, it usually increases with the Reynolds number. This behavior is seen for all DAR's. We see that the DAR seems to have a rather weak influence on the heat flux. However, for the DAR's of 3:1 and 4:1, we see that at higher Reynolds numbers (> 8000), the heat flux begins to level off: an increase in Re does not cause an increase in the heat flux. In fact, for the $\Delta T = 1$ °C case at these DAR's, the heat flux actually decreases with Re, for $Re > 5000$.

The dependence of the heat flux on the temperature differential can be removed by dividing the heat flux by ΔT. This yields the the heat transfer coefficient h:

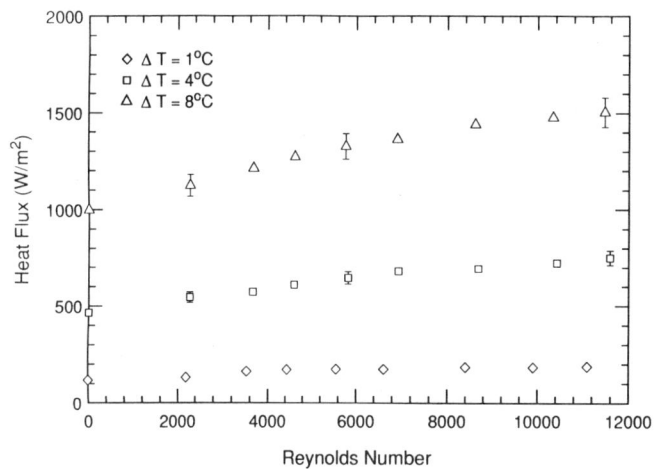

6(a)   DAR 1:1

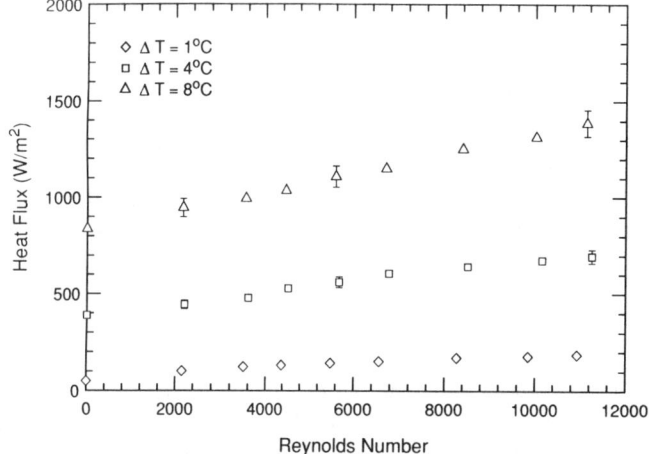

6(b)   DAR 2:1

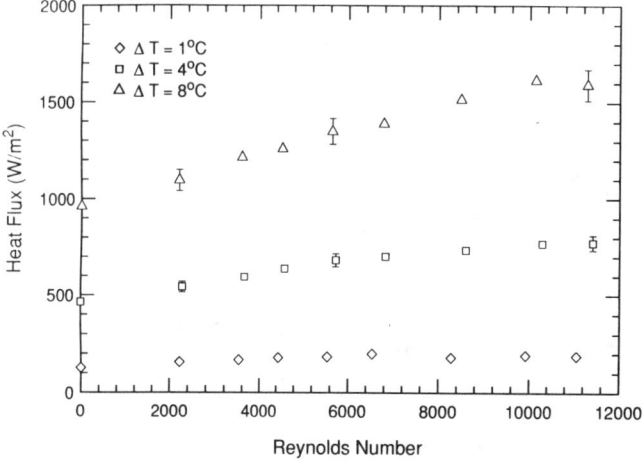

6(c)   DAR 3:1

$$h = \frac{q}{\Delta T} \qquad (1)$$

159

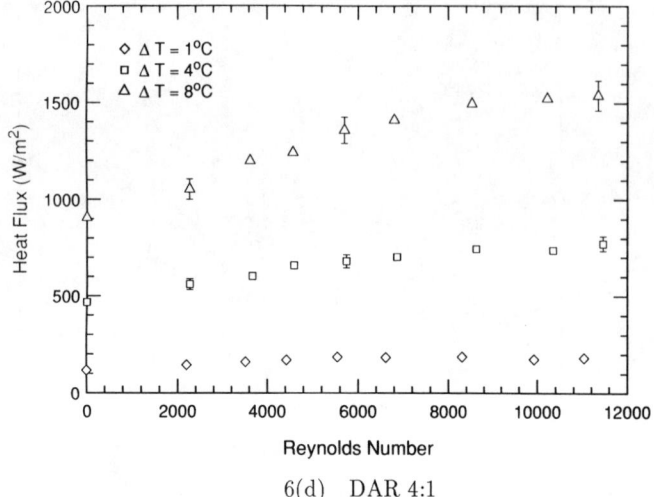

6(d)   DAR 4:1

Figure 6   Variation of Heat Flux with Reynolds number
and the temperature differential

Figure 7 shows the variation of the heat transfer coefficient as
a function of Reynolds number. This plot contains the data for
the three ΔT's and four DAR's. The good collapse of the data-
points suggests that the heat transfer processes occuring along
the cavity floor are fairly uniform for all the cases studied. A
least-squares fit of the data provided the following correlation:

$$h = 23.6 Re^{0.22} \qquad (2)$$

which is also indicated in Figure 7. In this correlation, the units
of h are W/m^2 °C. The amount of uncertainty in the calculated
value of the heat transfer coefficient is somewhat larger than that
in the corresponding heat flux. This additional uncertainty arises
due to the error in the measurement in the ΔT (± 0.05 °C), and
we expect the relative error to larger for a smaller ΔT. Therefore,
the uncertainty in the heat transfer coefficient is approximately
± 10 % for the 1 °C case, ± 7 % for the 4 °C, and ± 6 % for the
8 °C case. These uncertainties are indicated in Figure 7.

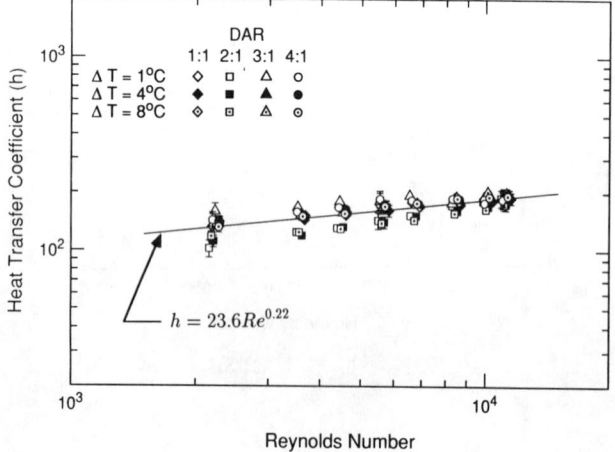

Figure 7   Variation of the heat transfer coefficient with
Reynolds number and temperature differential
for all DAR's

The heat flux data can also be presented in the form of the
Nusselt number, Nu, which is defined as:

$$Nu = \frac{q}{\Delta T} \frac{D}{k} \qquad (3)$$

The resulting Nusselt numbers for all four DAR's were plotted
on a single graph, against the ratio Gr/Re^2 (Figure 8). As a result
of this non-dimensionalization, the Nusselt number distribution
indicates a large spread. However, for a given DAR (especially
DAR = 1:1) the data appear to collapse quite well. It is obvious
from Equations 1 and 3, that the uncertainty in the value of Nu
is the same as that in the corresponding h.

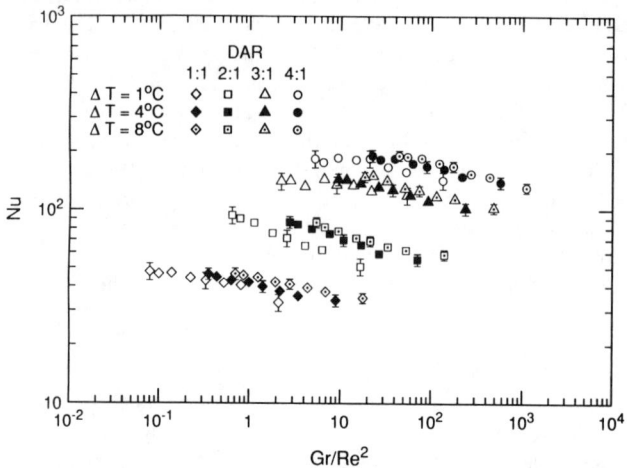

Figure 8   Variation of the Nusselt number with Gr/Re^2
for all DAR's

From Figure 8, it is seen that as the DAR is increased, the corresponding Nusselt numbers also increased at roughly the same
rate. Therefore, we decided to incorporate the DAR as an extra variable and calculated the coefficients for the the following
correlation, which is valid for the data from all four DAR's.

$$Nu = 7.96 Re^{0.18} \left[\frac{Gr}{Re^2} \right]^{-0.02} \left[\frac{D}{B} \right]^{1.1} \qquad (4)$$

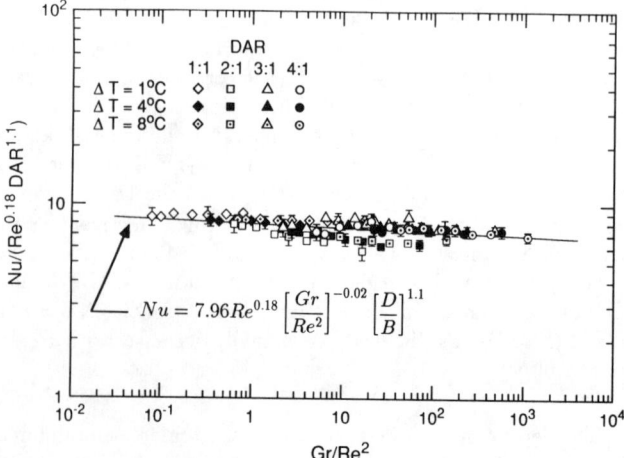

Figure 9   Variation of the Nusselt number as a function
of Reynolds number, DAR, and Gr/Re^2

The corresponding least-squares-fit is shown in Figure 9. The coefficient for the DAR is close to unity, which supports our estimate that the Nusselt number increases almost linearly with DAR. Furthermore, the coefficient of Gr/Re^2 is close to zero, which implies that the heat transfer process in this particular flow is Grashof number independent.

DISCUSSION

The Heat Transfer Coefficient

The good collapse of the data in Figure 7 is important because it indicates that the mixed convection flow in the cavity is relatively insensitive to the Grashof number for the range $10^7 \leq Gr \leq 5 \times 10^9$. A clue as to why this is so may be obtained from the flow visualization pictures. Figure 5(a) shows that the flow along the walls of the cavity is well organised and dominated by inertia. For instance, the region adjacent to the DSW is always occupied by a cold, plunging wall-jet, while the fluid adjacent to the USW is a warm, buoyant plume. In addition, the lid and the floor have boundary layer flows attached to them. In contrast, the internal, core region of the recirculating fluid is dominated by the more chaotic, natural convection flow structures.

This pattern of a well-organised boundary layer flow with a choatic internal flow is exactly duplicated at DAR's of 1:1, 3:1, and 4:1. Therefore, since the flow dynamics over the cavity floor are similar for all DAR's, we expect the magnitudes of the heat transfer coefficient to exhibit a similar behavior over a wide range of DAR's and ΔT's.

As the Reynolds number is increased, the inertia effects begin to grow inwards from the walls. Increasing regions of the flow are dominated by the more organized, inertia-driven flow, and the random motions in the core are suppressed. This is reflected by a rise in the heat transfer rates at the lower boundary: as the boundary layer over it becomes more vigorous it is able to transport heat more efficiently.

The Nusselt Number

The ratio Gr/Re^2 is used traditionally to indicate the relative strengths of the two modes of convection in a mixed-convection environment. Gebhart [1] deduced from a simple manipulation of the Navier-Stokes equations that the buoyancy effects become noticeable when Gr/Re^2 approaches unity. For $Gr/Re^2 \ll 1$, the forced convection component controls the heat transfer processes, while for $Gr/Re^2 \gg 1$, buoyancy effects predominate. Based on this principle, Siebers [3] defined the mixed convection limits for heat transfer from a vertical heated plate in a horizontal stream as $0.7 \leq Gr/Re^2 \leq 10$.

Unfortunately, the ratio Gr/Re^2 cannot be applied universally to every flow configuration. Researchers have often modified this parameter in order to interpret their data. For instance, Wang [5] et al. delineated the forced, mixed and free convection regions with the parameter $Gr/Re^{3/2}$. They determined the mixed convection regime to apply for $10 \leq Gr/Re^{3/2} \leq 400$. Similarly, Imura [6] found that transition from laminar forced convection to turbulent convection occured in the range of $100 \leq Gr/Re^{3/2} \leq 300$. Chen et al [7] report that for $Gr/Re^{5/2} > 0.05$, buoyancy effects became significant.

In keeping with the traditional method of presenting heat transfer results we have also determined a correlation for the Nusselt number as a function of Gr/Re^2. However, as Figure 9 indicates, we found it necessary to modify Gr/Re^2 with the factor (D/B), which is linked to the fact that the heat transfer coefficient is independent of the Grashof number. The correlation in Equation (4) also shows a Reynolds number dependence of 0.18. This is a fairly weak dependence when compared to the $Re^{1/2}$ seen, for instance, in the Pohlhausen relationship for laminar flow over an isothermal, flat plate. In fact, the exponent of 1/2 on the Reynolds number has been applied to laminar flows in other configurations as well. The lower exponent obtained in this study suggests that the flow is apparently tending towards turbulent behavior. However, our visualization studies confirm that the lid-driven cavity flow in not strictly turbulent even, at the highest Re of 12,000. We, therefore, suspect that the confined, recirculating nature of the flow provides a turbulent-like coefficient in the heat transfer correlations.

The Stanton Number

From the discussion in the preceding paragraphs, it is apparent, that the heat transfer data can be adequately represented by the Nusselt number, only when a modifying factor of D/B is introduced into the correlation in Equation 4. In fact, if we neglect the term involving Gr/Re^2 (on account of its very small exponent) and approximate the exponent of D/B as unity, then Equation 4 can be rewritten in a form very similar to Equation 2:

$$h \sim C Re^{0.18} \tag{5}$$

where C is a constant ($C = 8k/B$ W/m^{2}K). This exercise reveals that the heat transfer coefficient h, constitutes the physically correct way to represent the heat transfer within the lid-driven cavity. The Nusselt number is not as well-suited for this purpose because it introduces a length scale D, that does not seem to be relevant to this problem (as we have already explained, the heat transfer is Grashof number independent).

However, Equation 2 relates a dimensional quantity h, to a non-dimensional number Re. This drawback can be circumvented by employing the Stanton number, which does not require a length scale:

$$St = \frac{h}{\rho c_p U_B} \tag{6}$$

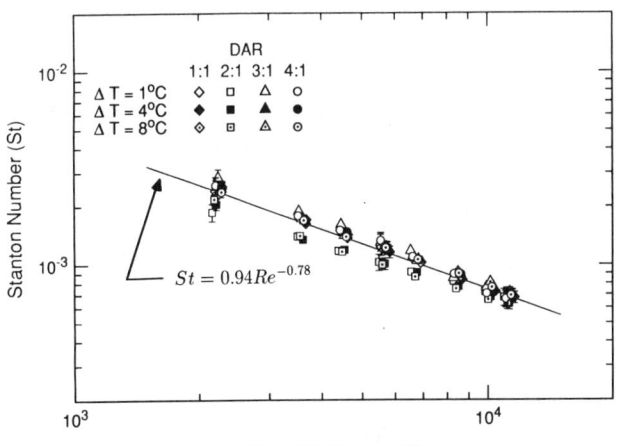

Figure 10   Variation for the Stanton number with Reynolds number and temperature differential for all DAR's

The Stanton number is a measure of the heat flux normal to the surface, realtive to the heat flux parallel to the surface. Figure 10 displays the Stanton number variation with Re for all the DAR's and ΔT's studied in this investigation. The least-squares-fit shown in Figure 10 is given by:

$$St = 0.94Re^{-0.78} \tag{7}$$

This expression now correlates one non-dimensional parameter St, with another, Re. The negative exponent on Re indicates that as the lid-speed is increased, the Stanton number decreases, because the rate at which heat is convected in a parallel direction to the surface increases much faster than heat transfer coefficient h. The uncertainty in St is essentially equal to the uncertainty in the corresponding h on account of the low error in U_B; this uncertainty is also indicated in Figure 10.

CONCLUSIONS

Flow visualization and heat heat flux measurements have been conducted in a bottom-heated lid-driven cavity flow. The range of Reynolds numbers studied were from 0 to 12,000. The vertical temperature differential, as well as the cavity depth were varied to produce Grashof numbers from 10^7 to 5×10^9. Our results indicate that:

1. The heat transfer coefficient (obtained by dividing the averaged heat flux over the entire floor by the temperature differential) is insensitive to the Grashof number for the range mentioned above. Flow visualization studies (for DAR's 1:1, 2:1, 3:1 and 4:1) indicate that this is because the nature of the flow adjacent to the cavity floor remains unchanged with DAR and ΔT.

2. The Nusselt number correlation as a function of Gr/Re^2 requires a modifying factor of $(D/B)^{1.1}$. Furthermore, as seen in Equation 4 there is essentially no functional dependence of Nu on Gr/Re^2. The low exponent on the Reynolds number when compared with laminar flow correlations, suggests that the flow under consideration is turbulent in nature. However, because our visualization confirms that the flow is not really turbulent, we believe that the reason for the low exponent is probably linked to the confined, recirculating nature of the flow.

We are presently examining the heat flux data from the individual heat flux meters in the form of time-traces and spectra to provide a quantitative link between the hydrodynamics and heat transfer.

ACKNOWLEDGEMENTS

This work has been supported by the Department of Energy under Grant DE-FG03-84ER13240. The authors appreciate the valuable contributions of Profs. R.L. Street and R.J. Moffat, and Mr. Keith Hollingsworth in interpreting the experimental results.

REFERENCES

[1] B. Gebhart, *Heat Transfer*, 2nd ed, McGraw-Hill Book Co., 1971.

[2] J.R. Lloyd and E.M. Sparrow, "Combined Forced and Free Convection Flow on Vertical Surfaces," *Int. J. Heat and Mass Transfer*, Vol. 13, pp. 434-438, 1970.

[3] D.L. Siebers, "Experimental Mixed Convection Heat Transfer from a Large Vertical Surface in a Horizontal Flow," Ph.D. Diss., Dept. of Mech. Eng., Stanford University, Stanford, California, March 1983.

[4] W.M. Kays and M.E. Crawford, *Convective Heat and Mass Transfer*, 2nd ed, McGraw-Hill Book Co. 1980.

[5] G.S. Wang, F.P. Incropera and R. Viskanta, "Mixed Convection Heat Transfer in a Horizontal Open-Channel Flow with Uniform Bottom Heat Flux," *J. Heat Transfer*, Vol. 105, pp. 817-822, Nov. 1983.

[6] H. Imura, R.R. Gilpin and K.C. Cheng, "An Experimental Investigation of Heat Transfer and Buoyancy Induced Transition from Laminar Forced Convection to Turbulent Free Convection over a Horizontal Isothermally Heated Plate," *J. Heat Transfer*, Vol. 100, pp. 429-434, Aug. 1978.

[7] T.S. Chen, E.M. Sparrow, and A. Mucoglu, "Mixed Convection in a Boundary Layer Flow on a Horizontal Plate," *J. Heat Transfer*, Vol. 99, pp. 66-71, 1977.

[8] D. Muñoz and F. Zangrando, "Mixing in a Double-Diffusive, Partially Stratified Fluid," SERI/TR-252-2942, Solar Energy Research Institute, Golden, Colorado, Dec. 1986.

[9] C.K. Cha and Y. Jaluria, "Recirculating Mixed Convection Flow for Energy Extraction," *Int. J. Heat Mass Transfer*, Vol. 27, No. 10, pp. 1801-1812, 1984.

[10] J.R. Koseff and R.L. Street, "The Lid-Driven Cavity Flow: A Synthesis of Qualitative and Quantitative Observations," *J. Fluids Eng.*, Vol. 106, pp. 390-398, Dec. 1984.

[11] A.K. Prasad, C-Y Perng and J.R. Koseff, "Some Observations on the Influence of Longitudinal Vortices in a Lid-Driven Cavity Flow," in *A Collection of Technical Papers, Part 1*, presented at the AIAA/ASME/SIAM/APS 1st National Fluid Dynamics Congress, Cincinnati, Ohio, pp. 288-295, July 25-28, 1988.

[12] H.S. Rhee, J.R. Koseff and R.L. Street, "Visualization of Natural and Mixed Convection Flows in a Cavity," *Proc. of Intl. Symp. on Refined Flow Modeling and Turbulence Measurements*, Univ. of Iowa, Iowa, pp. I11 1-9, Sept. 1985.

1989 National Heat Transfer Conference
HTD-Vol. 107, Heat Transfer in Convective Flows

UNSTEADY CHARACTERISTICS OF THE ANNULAR
IMPINGING JET FLOW FIELD
AND REVERSE STAGNATION POINT HEAT TRANSFER

H. Maki
Science University of Tokyo
Noda, Chiba, Japan

A. Yabe
Mechanical Engineering Laboratory
Ministry of International Trade and Industry
Ibaraki, Japan

ABSTRACT

This paper deals with the unsteady behavior of the recirculating flow formed by the annular impinging jet. Large scale fluctuations, cross-spectra, and cross-correlation functions of pressure are obtained by varying a distance H between the annular nozzle and the impinged flat plate and a velocity issuing from the annular nozzle. Three critical distances H_{c1}, H_{c2}, and H_{c3} are defined from unsteady behavior of the recirculating flow. Heat transfer for the reverse stagnation point on the impinged flat plate in the region of $H_{c3}<H<H_{c2}$, where large scale fluctuation does not exist and of $H_{c2}<H<H_{c1}$ where large scale fluctuation exists, are experimentally studied to make clear that the large scale pressure fluctuations have influence about 20% increase on the reverse stagnation heat transfer.

NOMENCLATURE

b : Slot width of the annular nozzle = $(D-d)/2$
d : Outer diameter of the inner circular tube
D : Inner diameter of the outer circular tube
f : Frequency
f_n (n=A,B,C and D) : Frequencies of the large scale pressure fluctuations
f_n^* : Dimensionless fequencies $=f_n H/\overline{U}_\theta$
l_n (n=1,2 and 3) : Lengths of initial, developed, and impinging regions of the annular impinging jet
H : Distance between the annular nozzle and the impinged flat plate
H_{c1} : First critical distance
H_{c2} : Second critical distance
H_{c3} : Third critical distance, $H_{c3}<H_{c2}<H_{c1}$
Nu : Nusselt number = $\alpha b/\lambda$
$\overline{p}$: Mean pressure
p : Fluctuating pressure
q : Heat flux
r : Radialwise coordinate
r_θ : Radius of the inner circular tube = $d/2$
R : Cross-correlation function
R_e : Reynolds number = $\overline{u}_\theta b/\nu$
R_{eR} : Reverse Reynolds number = $(\overline{u}_{H/2} b/\nu)$
S : Precision index
t : Time

T : Time period or temperature
$\overline{u}_\theta$: Mean velocity at the annular nozzle exit
$\overline{u}_{H/2}$: Reverse velocity at a location of r=0 and z=H/2
V : Voltage
z : Axialwise coordinate
α : Heat transfer coefficients
θ : Phase lag
λ : Thermal conductivity
ν : Kinematic viscosity
ρ : Density
τ : Time lag

Subscripts
n : Annular nozzle side
i : Impinged flat plate side

INTRODUCTION

One remarkable phenomenon generated in an annular impinging jet is the formation of a reverse stagnation point at the center of the annular impinging jet on an obstacle or a flat plate. The reverse stagnation heat transfer has come to have various applications, such as the augmentation of convective and boiling heat transfer by utilizing an electro-hydrodynamical(EHD) liquid jet (YABE and MAKI,1988), which is the jet flow ejected through the ring electrode in the direction coming from the plate. This reverse stagnation flow is caused by applying high electric voltage between the ring electrode and the plate electrode. Furthermore, the flow characteristics of the reverse stagnation point would be important from the viewpoint of analyzing the basic aspects of the turbulent flow. However, the heat transfer experiments on the reverse stagnation flow and the unsteady characteristics of the flow field have not been adequately reported so far.

An annular jet was characterized by forming a recirculation zone just downstream of an annular nozzle. When impinging the annular jet on a flat plate, behavior of the recirculation zone governs a flow field. Denoting H by a distance between the annular nozzle and the flat plate, it has been reported by the authors (MAKI and YABE 1989) that there were three critical distances, i.e., H_{c1}, H_{c2}, and H_{c3}, depending on measured pressure distributions on the flat plate and velocity profiles in the flow field (MAKI et al., 1980) Nusselt number on the annular stagnation line for $H<H_{c1}$ and the Nusselt number on the reverse stagnation point for $H_{c2}<H<H_{c1}$ were also obtained from naphthalene sublimation experiments by using the analogy between the heat transfer and the mass transfer. However, the measured quantities in those reports were limited to mean quantities, though it is considered that unsteady characteristics for the recirculation zone would be also necessary to make clear the details of the flow field and heat transfer. Therefore, in this paper the unsteady characteristics of the recirculation zone have been made clear experimentally by measuring the pressure fluctuations in air and also in water for various Reynolds number within the range of 1.2×10^3 and 3.6×10^4. Consequently, the critical distances already defined from the measured mean quantities coincided exactly with those obtained by the unsteady behavior of the recirculating flow. Furthermore, heat transfer at a reverse stagnation point formed on the flat plate in the regions of $H_{c3}<H<H_{c2}$ and $H_{c2}<H<H_{c1}$ are experimentally studied to make clear that the large scale pressure fluctuations have influence about 20% increase on the reverse stagnation point heat transfer.

UNSTEADY CHARACTERISTICS

Experimental Apparatus And Procedures

Water and air were used as working fluids. A schematic drawing of an experimental apparatus arrangement for water is shown in Fig.1 as a typical example. Water is drawn from an overhead tank to a cylindrical surge tank (270mm--inner diameter and 330mm--length) through a valve and a float-area-type flow meter and issues from the annular nozzle made of acrylic plastic (60mm--outer diameter of the inner circular tube, 68mm--inner diameter of the outer circular tube and 200mm--length). The water then impinges on a flat plate placed at right angles to an axisymmetrical line of the annular nozzle. The surge tank, the annular nozzle, and the impinged wall are immersed in water in an overflow tank, (1800mm--length, 900mm--width and 800mm--depth) as shown in Fig.1. The height of the water head in the overflow tank is stabilized by overflow.

A variation of H was performed by a screw attached to a supporting rail of the surge tank with an accuracy of ± 0.05mm. The impinged wall (flat pate) which traverses the r-direction was performed by a specially manufactured apparatus with an accuracy of ± 0.02mm. The accuracy of the float-area-type flow meter used was ± 2% of the Indicated Scale. Water temperature was measured by installing a mercury thermometer of 0.1 ℃ accuracy at the inlet of the tank.

In order to measure unsteady pressure on the impinged wall and on a disk installed at the downstream end of the inner tube, two pressure transducers made of semi-conductor were used, as shown in Fig.2(a). Outputs from these pressure transducers were led to an electric computer through DC-amplifiers and low-pass filters separately and then an A/D converter. A block diagram of them is also shown in Fig.2(a).

Cross-spectra and cross-correlation functions of the pressures were obtained between the pressure on the impinged wall and the pressure on the plate containing the annular nozzle. The experiments were conducted by varying H and a mean velocity $\bar{u_\theta}$ issuing from the annular nozzle. The mean velocity $\bar{u_\theta}$ at the exit of the nozzle were choiced to make the flow entirely

turbulent. Cutting frequency of the low-pass filter used was always set on half of the sampling frequencies. The sampling frequencies were selected from 10Hz to 200Hz according to the object of the experiments and the number of sampling was 512 points or 1024 points. The "ensemble average" was performed eight times. Moreover, flow visualization by dyes was also used, as shown in Fig.2(a) and estimated four kinds of flow patterns were sketched in the Fig.2(b).

Experimental Results And Discussions

The presentation of experimental results is organized so that each of the distinct regions of different pressure fluctuating behavior, which are formed according to the change of the distance H, is concentrated on. These regions are easily identified on the cross-spectra presented, since they are delineated by sharp changes for the variation of the distance H.

New definition for the second critical distance H_{c2}

Figures 3 and 4 show the cross-spectra (a) and the cross-correlation functions (b) for $H/H_{c1}=0.39$ and 0.65, respectively, as typical examples. The obtained value of H_{c1} for the annular nozzle used was 77mm from the preliminary experiment, by measuring mean pressure

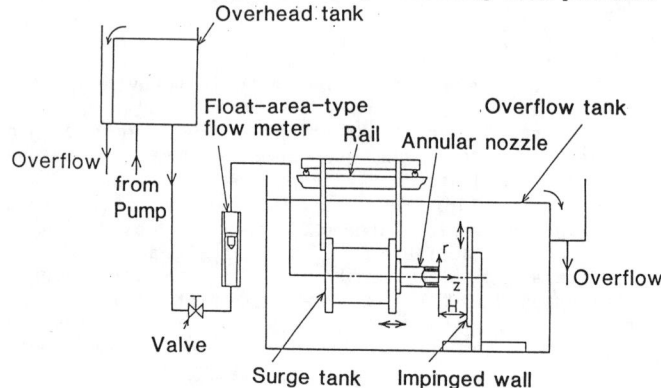

Fig.1 Experimental apparatus arrangement

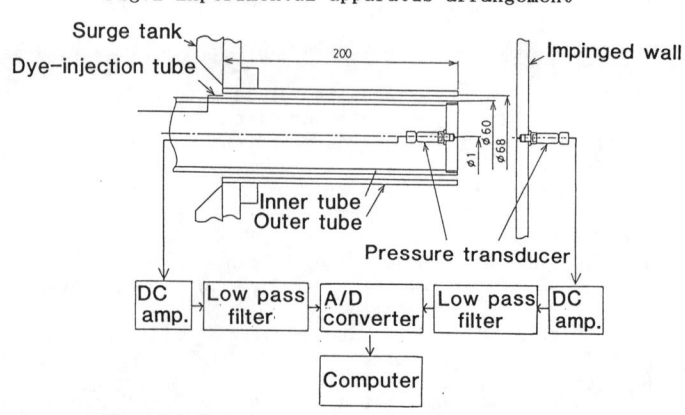

Fig.2(a) Detailed drawing of instrumentation

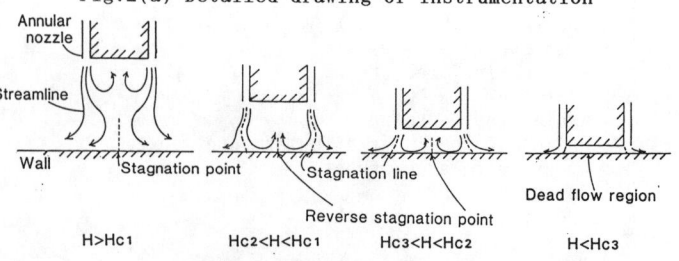

Fig.2(b) Flow patterns

distributions on the impinged wall (flat plate). The first critical distance H_{c1} is a boundary which took the forms of the annular impinging jet flow field and the usual round jet stagnation flow as shown in Fig.2(b). H_{c1} is a function of both d and b, and is a little affected by U_0 (MAKI et al.,1980). Moreover, the origin of the time of τ starts from the annular nozzle side.

As seen in Fig.3(a), there are no distinguishable large scale pressure fluctuations over the wide range of frequencies and the distribution of the cross-spectrum is considered to be characteristic of simple turbulent pressure fluctuation. On the other hand, three distinguishable peaks of pressure for cross-spectrum are recognized in Fig.4(a) and the frequency values are entered in the same figure. The frequency of roughly 1.8Hz coincided with that of the vibration of the annular jet in the r-direction, which was visualized by the dye method.

Calculating the ratio of those frequencies, a value of roughly 1:2:3 was obtained. It is considered that those distinguishable peaks consist of a fundamental fluctuation and its higher harmonics, judging from the ratio of the integer. As seen in Fig.4(b), the curve for R peaks at $\tau=-120$ms. This implies that the phase of pressure fluctuation on the impinged wall is advanced as compared with that on the disk.

In order to study the influence of a variation of the distance H, H was decreased by means of the screw. The large scale pressure fluctuations abruptly vanished when H was reached at roughly 38mm ($H/H_{c1}=0.49$) and nearly the same curves shown in Fig.3 appeared. A boundary distance of this variation of pressure fluctuations is used newly and additionally as the difinition for the second critical distance

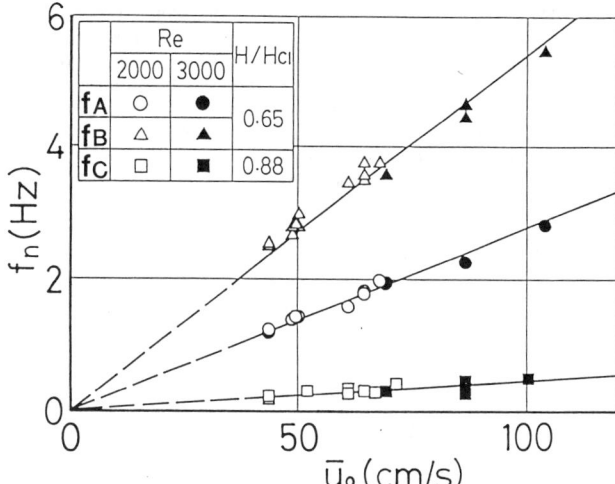

Fig.6   Relations between $\bar{u}_0$ and f_n

H_{c2}. It was already reported by the authors from the experiments of mean pressure distributions (MAKI and YABE, 1989) that the location of the peak in the mean pressure distribution on the impinged flat plate abruptly changed from inside to outside of the slot at H_{c2}, according to a decrease of H. Therefore, the physical meaning of H_{c2} is as follows.

Generally speaking, the flow field of the impinging jet is divided into three regions, i.e., the initial, the developed, and the impinging regions. When a length of the initial region is denoted by l_1, that of the developed region by l_2 and that of the impinging region by l_3, it is always $H = l_1 + l_2 + l_3$. When decreasing H, only the length l_2 is decreased, and when $l_2 = 0$, i.e., $H = l_1 + l_3$, large scale pressure fluctuations disappear due to interference in the initial region and in the impinging region and the length $l_1 + l_3$ is equivalent to H_{c2}.

When decreasing H more, no recirculating flow exists, i.e., the so-called dead flow region. Its boundary is defined by the third critical distance H_{c3} and its value was roughly 13mm in this experimental apparatus ($H/H_{c1}=0.17$).

Frequencies of the large scale pressure fluctuations
When increasing H in the region of $H_{c2}<H<H_{c1}$, the large scale pressure fluctuations shown in Fig.4 were abruptly changed. Figure 5 shows the cross-spectrum and the cross-correlation function at $H/H_{c1} = 0.88$. The peak of 0.315Hz shown in Fig.5(a) is new and the peak of 1.41Hz is the same fundamental fluctuation shown in Fig.4(a). As seen in Fig.5(b), the curve is reversed and the delay time τ is 250ms as compared with the curve shown in Fig.4(b). Denoting, now, the fundamental fluctuation by f_A, the second harmonic of it by f_B, and the newly appeared fluctuation by f_C, they will be accurately represented by f_n (n=A,B,and C).

From Figures 4 and 5 and the flow visualization experiments, the self-exciting fundamental pressure fluctuation f_A of the recirculating flow field was supposed to be yielded by a feedback mechanism where the pressure fluctuation f_C near the impinged wall was carried back to the upstream of the flow field by the reverse stagnation flow and where the pressure on the annular nozzle p_n influenced the pressure on the impinged flat plate p_i through the annular jet also. By increasing H/H_{c1} in the region $H/H_{c1}<1$, newly appeared self-exciting fluctuation became larger and the other fundamental fluctuations became smaller. Figure 6 shows the relationship between f_n and $\bar{u}_0$

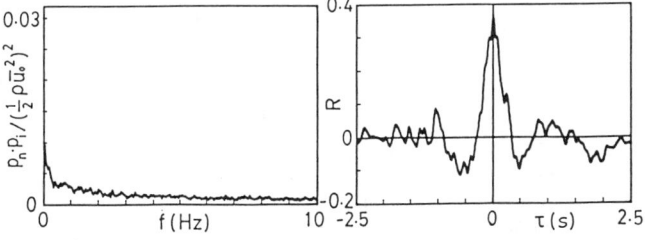

(a) Cross-spectrum   (b) Cross-correlation function
Fig.3 $H/H_{c1}=0.39$ (Re=2000, $\bar{u}_0=64.4$cm/s, $r/r_0=0$)

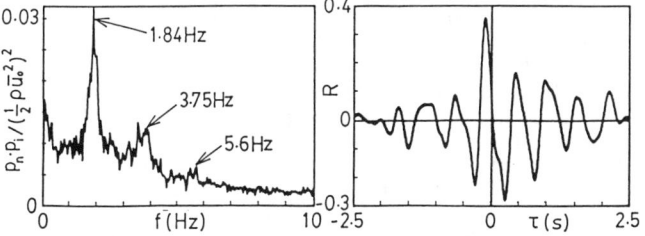

(a) Cross-spectrum   (b) Cross-correlation function

Fig.4 $H/H_{c1}=0.65$ (Re=2000, $\bar{u}_0=64.4$cm/s, $r/r_0=0$)

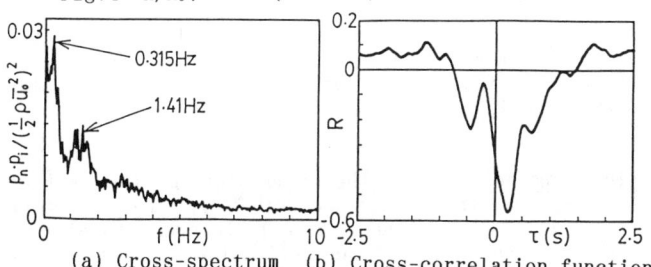

(a) Cross-spectrum   (b) Cross-correlation function
Fig.5 $H/H_{c1}=0.88$ (Re=2000, $\bar{u}_0=64.4$cm/s, $r/r_0=0$)

obtained by maintaining the constant value of Re, varying ν, i.e., the temperature of the water varied. It becomes clear that f_n is a liner function of $\bar{u}_0$ alone. In order to examine the relationship shown in Fig.6 over a wide range of $\bar{u}_0$, air was used as a working fluid instead of water. Figure 7 shows the experimental results together with those shown in Fig.6. As seen from Figs.6 and 7, all of the frequencies of the large scale pressure fluctuations which exist in the annular impinging jet flow field are determined by the reaching time alone from the annular nozzle to the impinged wall. Figure 8 shows differences of phase angles of the cross-spectra for f_A and f_C. It is clear in Fig.8 that they are unchanged by $\bar{u}_0$.

Figure 9 shows a cross-spectrum and a cross-correlation function for $H/H_{c1} = 1.17$. It is clear in Fig.9 that there is one kind of large scale pressure fluctuation in the region $H/H_{c1} > 1$ and it is denoted by f_D. The region $H/H_{c1} > 1$ means there is no annular impinging jet flow field as shown in Fig.2(b), and the frequencies of f_C and f_D coincided with the flow visualization experiments.

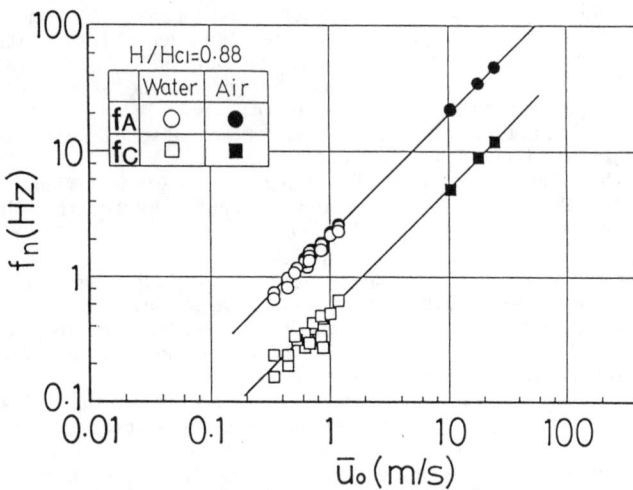

Fig.7   Relations between $\bar{u}_0$ and f_n at $H/H_{c1} = 0.88$

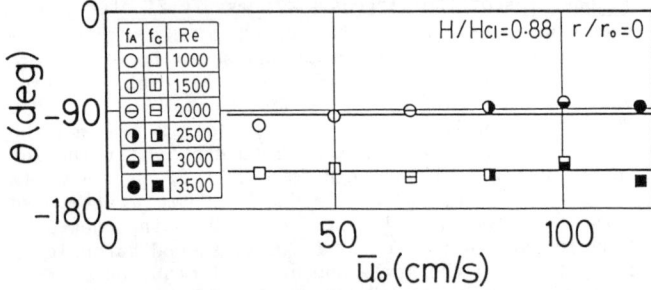

Fig.8   Relations between $\bar{u}_0$ and θ at $H/H_{c1} = 0.88$ and $r/r_0 = 0$

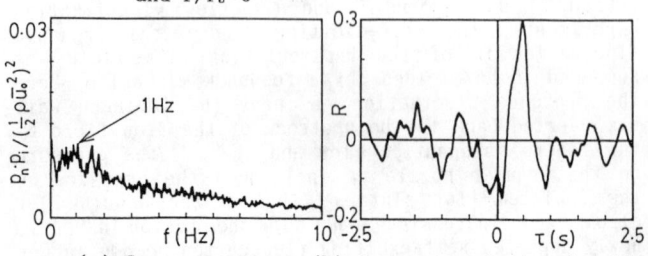

(a) Cross-spectrum   (b) Cross-correlation function

Fig.9 $H/H_{c1} = 1.17$ (Re=2000, $\bar{u}_0$=64.4cm/s, r/r_0=0)

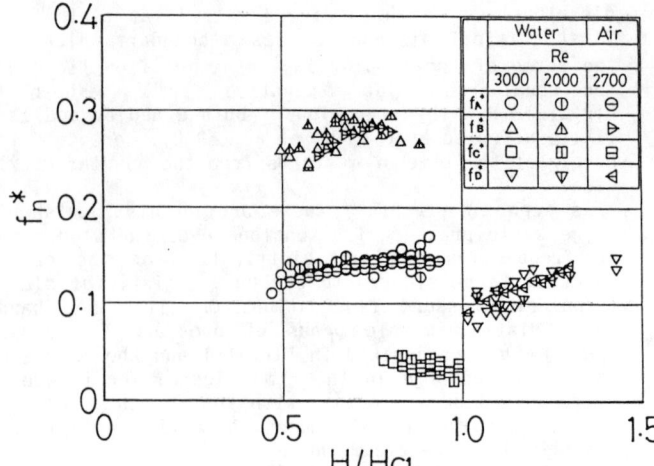

Fig.10   Relations between H/H_{c1} and f_n^*

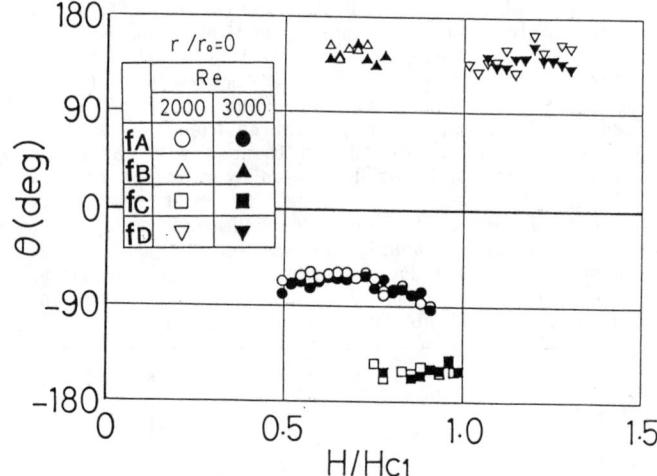

Fig.11   Relations between H/H_{c1} and θ

Influences of H on f_n In order to study the influences of H on F_n (n=A,B,C and D in this case), the same experiments described in the preceding section were performed for the wide range of H. The experimental values for f_n were arranged by dimensionless form, adopting $\bar{u}_0/H$ as a characteristic time, and denoting them by $f_n^* = f_n H/\bar{u}_0$.

Figures 10 and 11 show the relationships of f_n^* and θ vs. H/H_{c1}, respectively. As seen from Fig.10, the dimensionless frequencies f_A^* and f_C^* of the basic fluctuations existing in the annular impinging jet flow field are slightly varied by varying H/H_{c1}. They are, however, considered to be generally unchanged. The reason for scattering f_B^* is based on a little difficulty to recognize the accurate value of f_B, since it is the second harmonic of f_A^*. Judging from roughly constant values of θ shown in Fig.11, vibrating patterns of the respective fluctuations are considered to be unchanged by varying H. f_D is maintained to be constant (about 1Hz) by changing the value of H for the range of $H/H_{c1}>1$, since the size of the reverse flow region just downstream of the annular nozzle would not be changed by the value of H. Therefore, f_D^* is not constant in Fig.10.

Vibrating patterns of the recirculating flow. All of the experimental values described in the preceding sections were obtained at r = 0. In order to clarify vibrating patterns of the recirculating flow in the range of $H_{c2}<H<H_{c1}$, distributions of θ in the r-direction were measured at $H/H_{c1} = 0.65$ for f_A and

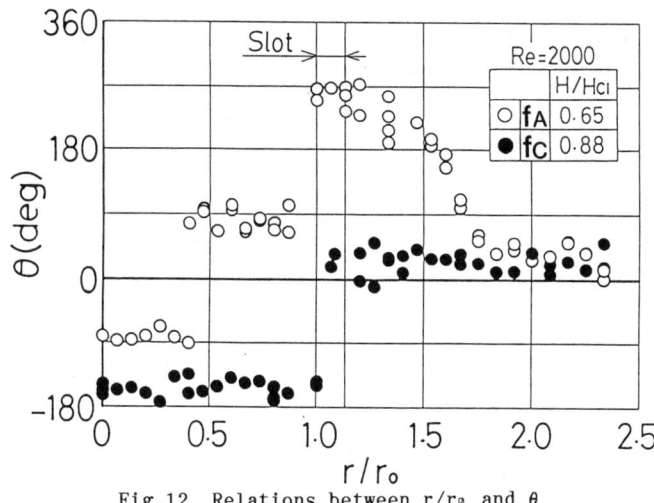

Fig.12 Relations between r/r_0 and θ

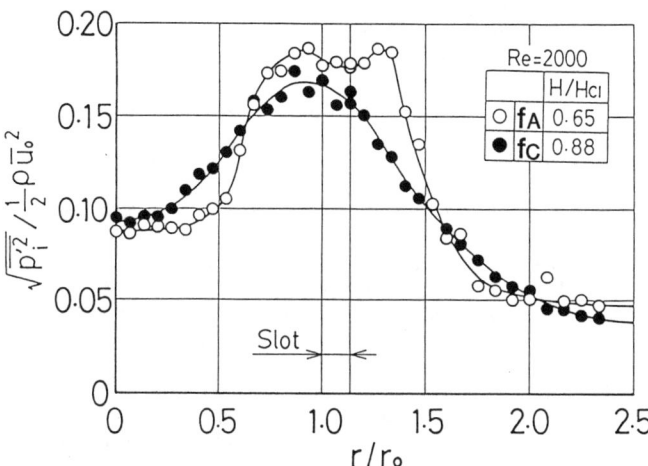

Fig.13 Large scale fluctuating pressure distributions on the impinged flat plate

0.88 for f_C, as typical examples.

Figure 12 shows the experimental results. The reason for the discrepancy of θ being roughly 180 degrees for f_A in the vicinity of $r/r_0 = 1$ is due to the vibration of the annular impinging jet in the direction of r. Moreover, the existence of a reverse stagnation region is suggested, judging from the discrepancy of θ being roughly 180 degrees for f_A at $r/r_0 = 0.4$. Furthermore, the extent of the influence for fluctuating f_A is roughly $r/r_0 = 1.8$, because of a rapid decay of θ in the region of $1.5 < r/r_0 < 1.8$, which means that the influenced region of the pressure fluctuation corresponding to f_A would be flown mainly by the impinging jet to the wall jet region.

Figure 13 shows the large scale fluctuating pressure distributions on the impinged wall for f_A and f_C. With the aid of the flow visualization, the appearance of two peaks on the curve for f_A was cleared to be caused by the vibration of the annular jet in the the region where r/r_0 is nearly equal to 1 and also z/H is about 1 corresponding to the neighborhood of the impinged wall. On the other hand, the vibration of the annular jet for f_C was limited to the middle region of the annular impinging jet between the annular nozzle and the impinged wall. In this case, it is seen in Fig.12 that the reverse stagnation region formed in the case of f_A was not yielded, judging from the fact that the discrepancy of θ at roughly

$r/r_0 = 0.4$ has not been appeared. The discrepancy of θ for f_C at roughly $r/r_0 = 1$ (shown in Fig.12) is due to the vibration of the middle region of the annular impinging jet in the direction of r.

REVERSE STAGNATION POINT HEAT TRANSFER

As described in the preceding chapters, the flow pattern of the annular impinging jet must be devided by using H_{c2} and H_{c3}. Since the reverse stagnation heat transfer for the region of $H_{c2} < H < H_{c1}$ has been experimentally reported by using the naphthalene sublimation (MAKI and YABE, 1989), in this chapter the reverse stagnation heat transfer for the regions of $H_{c3} < H < H_{c2}$ and $H_{c2} < H < H_{c1}$ has been researched by making the heat transfer experiments..

Experimental Apparatus And Procedures

Air was used as a working fluid for the heat transfer experiments. Another annular nozzle was employed. Its outer diameter of the inner circular tube was 163mm and the inner diameter of the outer circular tube was 181mm. The critical distances of the annular nozzle were roughly H_{c1} = 205mm, H_{c2} = 92mm, and H_{c3} = 41mm in the preliminary experiments.

Air from a 37kW Roots blower was issued from the annular nozzle through an orifice flow meter, a flow rate adjusting valve and a surge tank. Consequently, experiments up to Re=6.13×10⁴ were practical to perform. In order to measure surface temperature accurately, thin C-C thermocouples with a 50μm thick starching pattern with plastic paste of 0.2mm thickness (as shown in Fig.14(a)) was employed. The purpose of the pattern of the thermocouple was to avoid heat loss through the thermocouple itself, by reducing the temperature gradient along it.

The impinged wall was made of Pyrex glass with a thickness of 3mm and a diameter of 50mm. The thermal conductivity of the impinged wall was measured with the preliminary experiment and (0.08 ± 3%) W/mK was decided on. The Pyrex glass was assembled as shown in Fig.14(b).The underside of the Pyrex glass was heated by hot water and a vessel of hot water was insulated by foamed styrol as shown in Fig.14(b). The outputs of the thermocouples were measured by means of a potentiometer with an accuracy of 0.1μV and the temperature of air issuing from the annular nozzles was measured at the annular nozzle exit by means of a mercury thermometer with an accuracy of 0.1℃. The reverse velocity $\bar{u}_{H/2}$ at r = 0 and z = H/2 was measured by means of a hot-wire anemometer.

Experimental Results And Discussions

Figure 15 presents experimental data on Nu_R vs. $Re_R \cdot r_0/H$. It should be noted that the data was clearly separated into two groups according to the regions of $H_{c3} < H < H_{c2}$ and $H_{c2} < H < H_{c1}$.

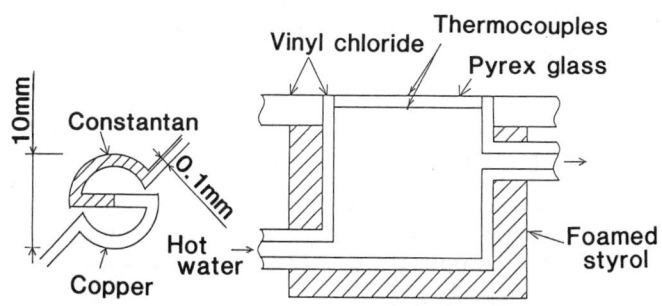

(a) Pattern of thermocouple (b) Experimental apparatus

Fig.14 Apparatus for heat transfer experiments

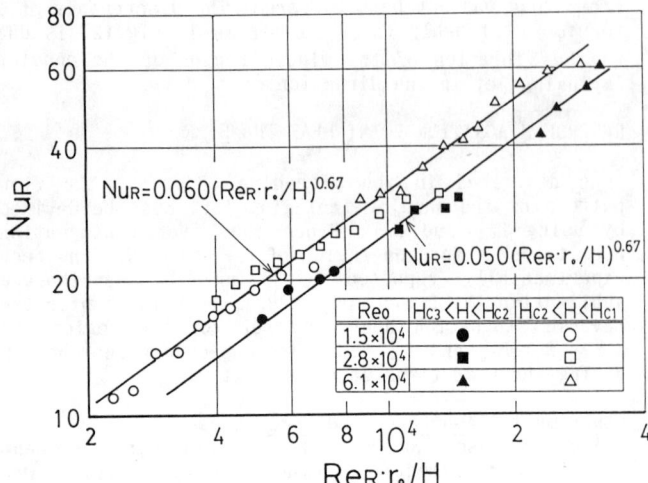

Fig.15 Heat transfer characteristics for reverse
stagnation point

The empirical formulae drawn in Fig.15 by the
solidlines are

$Nu_R = 0.060(Re_R \cdot r_0/H)^{0.67}$ for $H_{c2} < H < H_{c1}$
$Nu_R = 0.050(Re_R \cdot r_0/H)^{0.67}$ for $H_{c3} < H < H_{c2}$ (1)
within an accuracy of ± 12%

As seen in Fig.15, it is experimentally concluded
that the large scale pressure fluctuations, which were
appeared in the region of $H_{c2} < H < H_{c1}$, have influence
about 20% increase on the reverse stagnation point
heat transfer.

Error Analysis

Concerning the possible error sources for the
measurement of unsteady characteristics, the indicated
accuracy of the instruments could only be used for
various instruments such as the DC-amplifier, the
low-pass filter, the analog-to-digital converter, and
the electric computer employed.

Therefore, in this section the uncertainty of the
heat transfer experiments was focused and analyzed.
Uncertainty of surface tempereature measurement by
means of a thermocouple is generally considered to
consist of calibration before and during measurements.
The calibration of the thermocouples employed in the
experiments was carefully performed by the usual oil-
bath method with a temperature range from 5℃ to 60℃.

Uncertainty for an accurate-type mercury thermometer
becomes ± 0.05℃ because its minimum scale was 0.1 ℃.
Thermoelectromotive force from the thermocouple was
measured by means of a potentiometer supplied by the
manufacturer having a minimum scale of 0.1µV and an
accuracy of ±(0.02%+0.05µV). Maximum uncertainty at
60℃ was obtained from calibration curves described
below as ± 0.12℃. The calibration curves formulated
by the method of least squares were T = 0.1314+27.14V-
0.3018V² for the thermocouple of the upper side of the
Pyrex glass and T = -0.2219+28.5V-0.8563V² for that of
the lower side of it. The maximum deviation at 60 ℃
between the experimental values and those formulae was
considered to be 0.472℃.

The precision index S (ASME,1984) for the
calibration process is the root-sum-square of all the
elemental precision indices. It was obtained as
$S = \pm[0.05^2 + 0.12^2 + 0.472^2]^{1/2} = \pm 0.490(℃)$ (2)
Several measurement errors which were considered were
a conduction error through thermocouple, a setting
error of a junction of the thermocouple, and

temperature fluctuation of the flow from the annular
nozzle. Among them, the conduction error was
considered to be negligible, judging from the setting
pattern shown in Fig.14(a).

Heat flux q was calculated by the following equation
$q = \lambda \cdot \Delta T/\Delta l = \alpha \cdot \Delta T$ (3)
where ΔT denotes the temperature difference between
the flow and the surface of the wall and Δl the
distance between the thermocouples buried in the
plastic plate as described in the preceding sections.

The accuracy of λ was ± 3% and the setting error for
the thermocouples was calculated as ± 4.7% and the
temperature fluctuation of the flow was measured as ±
0.8% at 30℃. Referring to Eq.(2), the precision index
± 0.490℃ at 60℃ corresponded to 0.82%.

The precision index in this case was calculated as ±
5.68%.

CONCLUSIONS

The unsteady behavior and heat transfer of the
recirculating flow formed by the annular impinging jet
has been clarified experimentally and the following
conclusions have been obtained.
(1) The unsteady behavior of the annular impinging jet
was classified into three regions by using the
differences of the cross-spectra and the cross-
correlation functions of pressure, depending on the
distance H between the annular nozzle and the impinged
flat plate. It was ascertained that obtained three
critical distances denoted by H_{c1}, H_{c2}, and H_{c3}
coincided entirely with those determined by the
measured mean quantities. Furthermore, it was
cleared that the large scale pressure fluctuations existed in
the region of $H_{c2} < H < H_{c1}$, and that any particular
frequency has not been recognized for the turbulent
pressure fluctuations in the region of $H_{c3} < H < H_{c2}$.
(2) Heat transfer characteristics for the reverse
stagnation point formed on the impinged flat plate
were divided into two groups according to the regions
of $H_{c2} < H < H_{c1}$ and $H_{c3} < H < H_{c2}$. The large scale
pressure fluctuations, which were appeared in the
region of $H_{c2} < H < H_{c1}$, had influence about 20% increase
on the reverse stagnation point heat transfer.

ACKNOWLEDGEMENTS

These experiments were partly performed by Messrs.
Mitsuru KURIHARA and Taro OKUTSU as part of their
undergraduate theses at Science University of Tokyo
and the authors are grateful to them for their
cooperation.

REFERENCES

ANSI / ASME, MFC-2M-1983, Measurement
Uncertainty for Fluid Flow in Closed Conduits, ASME,
Aug. 1984.
MAKI,H. and YABE,A., Heat Transfer by the Annular
Impinging Jet, Int. J. Experimental Heat Transfer,
to be published.
MAKI,H., AIDA, E., and AKIMOTO,K., Fundamental
Studies on the Annular Impinging Jet, Trans. of the
JSME, Vol.46, No.410B, Oct. 1980, pp. 1959 - 1967.
(in Japanese)
YABE,A. and MAKI,H., Augmentation of Convective and
Boiling Heat Transfer by applying an Electro-
hydrodynamical Liquid Jet, Int. J. Heat Mass Transfer,
Vol.31, No.2, 1988, pp.407-417.

NUMERICAL INVESTIGATION OF NATURAL CONVECTION
WITHIN A
TRIANGULAR SHAPED ENCLOSURE

M. Ghassemi and J. A. Roux
University of Mississippi
University, Mississippi

ABSTRACT

A fundamental study of natural convection heat
transfer within a triangular shaped enclosure was
considered. A numerical method based on Patankar's
control volume approach was employed to solve the
governing non-linear partial differential equations.
For model verification the numerical results were
compared to existing experimental data for a triangular
enclosure. Two sets of boundary conditions have been
considered: summer conditions (hot sloped wall and
cold horizontal wall) and winter conditions (cold
sloped wall and warm horizontal wall). In this study,
steady-state, two-dimensional solutions were obtained
for a range of aspect ratios (Ar) varying from 0.2 to
0.9; the Grashof number (Gr_H) ranged from 10^4 to 10^6.
The heat transfer results are presented in terms of
temperature, heat fluxes, and Nusselt number as a
function of Ar and Gr_H (Pr was held constant at 0.7).

NOMENCLATURE

Ar, Aspect ratio, H/D
A, Area
D, Width of triangular enclosure (Fig. 1)
g, Acceleration of gravity
Gr_H, Grashof number, $g\beta(\theta_H - \theta_C)H^3/\nu^2$
h, Heat transfer coefficient, $q/(\theta_H - \theta_C)$
H, Height of triangular enclosure (Fig. 1)
k, Thermal conductivity
L, Hypotenuse of Fig. 1 $(0 \leq X_s \leq L)$
Nu_x, Local Nusselt number, hX/k [Eqs.(13) and (16)]
$\overline{Nu}$, Average Nusselt number
P, Pressure
Pr, Prandtl number ν/α
Ra, Rayleigh number, $Gr_H Pr$
q, Heat flux
S_ϕ, General source term
T, Dimensionless temperature, $(\theta - \theta_C)/(\theta_H - \theta_C)$

θ, Temperature
θ_C, Temperature, cold wall
θ_H, Temperature, hot wall
θ_1, Angle in enclosure (Fig. 1)
U, Velocity component in X direction
u, Dimensionless velocity component in x direction
V, Velocity component in Y direction
v, Dimensionless velocity component in y direction
X, Vertical coordinate, (Fig. 1)
X_S, Coordinate along the sloped wall, (Fig. 1)
x, Dimensionless X coordinate
Y, Horizontal coordinate, (Fig. 1)
y, Dimensionless Y coordinate
N, Normal component of sloped wall, (Eq. 17)
n, Dimensionless normal component of sloped wall, N/H
α, Thermal diffusivity
β, Volumetric coefficient of thermal expansion
Γ, General diffusion coefficient
ρ, Density
ν, Kinematic viscosity
ϕ, General dependent variable

INTRODUCTION

Natural convection in an enclosure is as varied
as the geometry and orientation of the enclosure.
According to Bejan [1] it can be divided into two
separate classes: enclosures heated from the sides, and
enclosures heated from below. By comparison enclosures
heated from the side represent a much more applied
subfield of convective heat transfer. Many
investigations have been performed both theoretically
and experimentally for a wide range of Grashof numbers
and aspect ratios. Their important applications are
solar collectors, double pane windows, double wall
insulation, etc. A triangular enclosure heated from
below corresponds to the functioning of a thermal
insulation oriented horizontally; for example, this is
an important application for heat transfer through a
triangular shaped attic space (where different
temperatures along the sloped wall and horizontal wall
represent winter and summer conditions). Previously if
one was interested in predicting heat transfer rates
due to natural convection in these geometries, it was

sometimes necessary to rely on data provided by titled enclosures [2-3].

The phenomenon of steady-state laminar natural convection in a triangular enclosure was experimentally studied by Flack, Konopnicki, and Rooke [4]. The geometry presented was a triangular shaped enclosure with a solar collector (hot) on one side and a cold condition on the other side with an adiabatic bottom wall. The aspect ratio (Ar) was varied between 0.29 and 0.87, and the Grashof number (Gr_H) was varied between 2.9×10^6 and 9.0×10^6.

Flack and Witt [5] presented experimental velocity measurements throughout a two-dimensional triangular enclosure which consisted of two isothermal sloped walls (one sloped was wall hot and the other cold), insulated bottom, and with glass end plates; the results were compared to inclined isothermal plate data. Flack [6] conducted an experimental study of heat transfer in a triangular shaped enclosure in order to represent a residential attic during winter or summer conditions respectively; the same geometries studied in [4] were also examined in [6]. This experiment was performed for Rayleigh numbers ranging from 7.0×10^4 to 10^6. When the upper walls were heated (summer condition) the flow field was stable and maintained a laminar flow profile in all cases. Secondly, when the horizontal wall was heated (winter condition), the flow eventually became turbulent due to increases in the Grashof number (large temperature difference). Since the range of Rayleigh numbers in the above study were too low to be applicable to a real attic space, Poulikakos and Bejan [7] decided to experimentally study the triangular shaped attic configuration for the high Rayleigh number regime (up to $Ra = 10^9$).

Poulikakos and Bejan [7] conducted a natural convection heat transfer experiment in a relatively shallow triangular enclosure. This experiment simulated winter conditions where the temperature measurements and flow visualization showed that the attic flow field consists of nearly an isothermal core surrounded by a boundary layer type counter flow. Poulikakos and Bejan [8,9] reported an analytical study of natural convection in a triangular shaped enclosure filled with a fluid saturated porous medium for winter conditions. The study [8] was carried out in two separate parts: the shallow attic limit and the boundary layer region. For the shallow attic limit (considering small angle limit), a perturbation solution was considered based on asymptotic expansions. In this region the circulation pattern was influenced strongly by the shape of the sloped wall (roof): however, in the boundary layer regime a solution based on Ostrach's Oseen linearization technique for the flow field and heat transfer was developed. This boundary layer regime was established along the sloped wall (roof) as well as a core flow moving slowly along the horizontal wall (ceiling). In Ref. [9] a theoretical and numerical analysis of the enclosure proposed in [8] was performed to predict the transient and steady-state flow and temperature fields. First, the transient behavior of the triangular enclosure was conducted based on a scaling analysis to interpret the later numerical results. Second, a numerical solution based on a finite difference method (using stream function) for a mesh of 41 grid points was conducted in the Rayleigh number domain of $Ra = 10^2$ and $Ra = 10^3$ with a constant aspect ratio of 0.2. The effect of Rayleigh number at $Ra = 10^2$ resulted in conduction heat transfer across the triangular enclosure, and for $Ra = 10^3$ the heat transfer departed from conduction and here convection played a significant role in the heat

transfer across the porous wedge. The numerical solution confirmed the results obtained in the theoretical analysis.

Poulikakos and Bejan [10] presented a three part study for triangular shaped attics for winter conditions. First, a theoretical evaluation of the flow field and temperature field was conducted via an asymptotic analysis. It was indicated that the net heat transfer was dominated by pure conduction. Second, the transient behavior of the triangular enclosure was examined by sudden cooling of the sloped wall. Finally, in order to verify the results developed theoretically, a transient numerical simulation based on a finite difference method (using stream and vorticity functions) covering values of Ar (0.2, 0.4, 1.0), Gr_H (10, 10^2, 10^3), and Pr (0.72, 6.0) was performed. For a 41 point grid mesh, the theoretical and numerical results deviated slightly (only about 3 percent).

As the above literature indicates, most of the studies for triangular enclosures were experimentally simulating winter conditions for low ranges of Raleigh number and for shallow enclosures (small Ar). The numerical study of Bejan [10] was performed for Rayleigh numbers (up to $Ra = 10^3$) and the shallow enclosure filled with a porous medium (low aspect ratios) considered winter conditions only. Thus the objective of this present paper was to numerically investigate the phenomenon of natural convection heat transfer inside the triangular enclosure experimentally studied by Flack [6]. Therefore, a numerical technique based on Patankar's method [11] was used to compute the velocity field and temperature field. The Grashof number ranged from 10^4 to 10^6 and the aspect ratio ranged between 0.2 to 0.9. Two sets of boundary conditions were investigated: summer (hot sloped walls and cold horizontal wall) and winter (cold sloped walls and hot horizontal wall). The effect of Gr_H and Ar variation on Nusselt number were thoroughly investigated; also the peak heating location within the enclosure was determined.

STATEMENT OF PROBLEM

The physical problem is shown in Fig. 1. Due to symmetry, this work includes only one half of the geometry about the plane of symmetry. The sloped and horizontal walls are considered to have uniform temperatures (i.e. the hot sloped wall (θ_H) and the cold horizontal wall (θ_c) correspond to summer conditions and vice versa for winter conditions). The vertical boundary is treated physically as a plane of symmetry. The convective motion generated by the buoyancy forces was considered as two dimensional.

The following sets of boundary conditions are considered: summer conditions and winter conditions. For summer conditions the horizontal boundary temperature was cooled (θ_c) while the sloped boundary temperature was hot (θ_H). For winter conditions the horizontal boundary (high temperature) was hot while the sloped boundary (low temperature) was cooled. In all of the above conditions, the vertical plane (Fig. 1) was considered as insulated due to symmetry.

After examination of different aspect ratios (Ar between 0.2 and 0.9) it was clear that as the angle θ_1 between the two isothermal walls increased (for Ar = 0.289, 0.5, 0.865; θ_1 = 60, 45, 30) that the height of the triangular enclosure increased and hence the enclosure aspect ratio increased. Increasing (θ_1) should yield lower convective heat transfer and a higher percentage of conduction heat transfer. Since the convective motion is primarily in the vertical direction, it is likely that the height of a triangular enclosure also has an influence on heat transfer.

One purpose of this study was the computation of the heat transfer along the sloped and horizontal walls. The heat transfer results have a direct implication to energy conservation in residential attic situations.

ANALYSIS

In formulating the governing equations which describe the physical phenomenon shown in Fig.1, the following assumptions were made: 1) the flow is laminar, 2) the motion is two-dimensional and steady-state, 3) the flow is incompressible, 4) the flow is Boussinesq, 5) the fluid properties are considered to be constant (ν, k, α) and evaluated at the reference temperature θ_C (from Holman [12]).

Governing Equations and Boundary Conditions

The governing equations for two-dimensional laminar natural convection in a triangular shaped enclosure are as follows

continuity

$$\frac{\partial U}{\partial X} + \frac{\partial V}{\partial Y} = 0 \tag{1}$$

momentum

X-momentum

$$U \frac{\partial U}{\partial X} + V \frac{\partial U}{\partial Y} = - \frac{1}{\rho} \frac{\partial P}{\partial X} + \nu \left[\frac{\partial^2 U}{\partial X^2} + \frac{\partial^2 U}{\partial Y^2} \right] + g\beta(\theta - \theta_c) \tag{2}$$

Y-momentum

$$U \frac{\partial V}{\partial X} + V \frac{\partial V}{\partial Y} = - \frac{1}{\rho} \frac{\partial P}{\partial Y} + \nu \left[\frac{\partial^2 V}{\partial X^2} + \frac{\partial^2 V}{\partial Y^2} \right] \tag{3}$$

and energy

$$U \frac{\partial \theta}{\partial X} + V \frac{\partial \theta}{\partial Y} = \alpha \left[\frac{\partial^2 \theta}{\partial X^2} + \frac{\partial^2 \theta}{\partial Y^2} \right] \tag{4}$$

The following boundary conditions must be satisfied by the above equations

$$\begin{array}{llll}
U = V = 0 & \theta = \theta_C & \text{at } X = H & 0 < Y < D \quad \text{(5a)} \\
V = 0 & \partial\theta/\partial Y = 0 & \text{at } Y = D & 0 < X < H \quad \text{(5b)} \\
V = 0 & \partial U/\partial Y = 0 & \text{at } Y = D & 0 < X < H \quad \text{(5c)} \\
U = V = 0 & \theta = \theta_H & \text{at } X = F(Y) & Y = G(X) \quad \text{(5d)}
\end{array}$$

where $F(Y) = HY/D$ and $G(X) = DX/H$. To transform the governing equations into dimensionless form let

$x=X/H$; $y=Y/H$; $u=U/(\nu/H)$; $v=V/(\nu/H)$; $T=(\theta-\theta_C)/(\theta_H-\theta_C)$; $p=P/\rho g(\theta_H-\theta_C)\beta H$ (6)

Substituting Eq. (6) into Eqs. (1)-(4) and simplifying yields in dimensionless form

continuity

$$\frac{\partial u}{\partial x} + \frac{\partial v}{\partial y} = 0 \tag{7}$$

momentum

x-momentum

$$\frac{1}{Gr_H} \frac{\partial}{\partial x} \left[uu - \frac{\partial u}{\partial x} \right] + \frac{1}{Gr_H} \frac{\partial}{\partial y} \left[uv - \frac{\partial u}{\partial y} \right] = - \frac{\partial p}{\partial x} - T \tag{8}$$

y-momentum

$$\frac{1}{Gr_H} \frac{\partial}{\partial x} \left[uv - \frac{\partial v}{\partial x} \right] + \frac{1}{Gr_H} \frac{\partial}{\partial y} \left[vv - \frac{\partial v}{\partial y} \right] = - \frac{\partial p}{\partial y} \tag{9}$$

and energy

$$\frac{\partial}{\partial x} \left[uT - \frac{1}{Pr} \frac{\partial T}{\partial x} \right] + \frac{\partial}{\partial y} \left[vT - \frac{1}{Pr} \frac{\partial T}{\partial y} \right] = 0 \tag{10}$$

The boundary conditions in dimensionless form become

$$\begin{array}{llll}
u = v = 0 & T = 1 & \text{at } x = Ar & 0 < y < 1 \quad \text{(11a)} \\
v = 0 & \partial T/\partial y = 0 & \text{at } y = 1 & 0 < x < Ar \quad \text{(11b)} \\
v = 0 & \partial u/\partial y = 0 & \text{at } y = 1 & 0 < x < Ar \quad \text{(11c)} \\
u = v = 0 & T = 1 & \text{at } x = f(y) & y = g(x) \quad \text{(11d)}
\end{array}$$

Outline of the Solution

Equations (7)-(10) can all be written in the general form proposed by Patankar [11] as

$$C \frac{\partial}{\partial x} \left[u\phi - \Gamma \frac{\partial \phi}{\partial x} \right] + C \frac{\partial}{\partial y} \left[v\phi - \Gamma \frac{\partial \phi}{\partial y} \right] = S_\phi \tag{12}$$

To solve systems of equations in the form of Eq. (12), Patankar's [11] SIMPLE finite difference scheme was applied. In this method a "staggered grid" concept was employed. The finite difference approximation was obtained by using the so-called power law scheme [11].

Heat Transfer Equations

The parameters of primary importance are the heat transfer and Nusselt number. The local Nusselt number along the horizontal wall is given by

$$Nu_X = \frac{\partial\theta/\partial Y \,)_{Y=H} \; H}{\theta_H - \theta_C} \tag{13}$$

According to Ref. [13] the heat flux along the sloped wall is given by

$$q = |\,q\,| = \sqrt{(-k \frac{\partial\theta}{\partial X})^2 + (-k \frac{\partial\theta}{\partial Y})^2} = k \frac{\partial\theta}{\partial N} \tag{14}$$

or

$$q = h (\theta_H - \theta_C) \tag{15}$$

and thus from Eqs. (14) and (15), the local Nusselt number along the sloped wall can be expressed as

$$Nu_{X_S} = \frac{(\partial\theta/\partial N) \; H}{\theta_H - \theta_C} \tag{16}$$

where X_S is the distance along the sloped wall (Fig. 1)

and where

$$\frac{\partial \theta}{\partial N} = \sqrt{(\frac{\partial \theta}{\partial X})^2 + (\frac{\partial \theta}{\partial Y})^2} \tag{17}$$

In terms of dimensionless variables, Eqs. (13) and (16) can be written as

$$Nu_X = \frac{\partial T}{\partial y}\Big|_{y=1} \tag{18}$$

and

$$Nu_{X_s} = \frac{\partial T}{\partial n} \tag{19}$$

Up to 231 grid points were considered within the triangular enclosure. The result of varying grid size was not significant (for example, the average Nusselt number results for 231 grid points varied only slightly (0.1%) as compared to 45 grid points), however the computational time (CPU) was increased by a factor proportional to the total number of grid points. Since increasing the number of grid points increased the computational time (CPU) and did not influence the results significantly, 45 grid points were used within the triangular enclosure to numerically model the natural convection heat transfer. For $Gr_H = 6.25 \times 10^6$ and 45 grid points, the computational time (CPU) on a AMDAHL 470/V8 for Ar = 0.289, 0.50, 0.865 (θ_1=60, 45, 30) was equal to 0.29, 0.42, 0.54 seconds respectively. Numerical convergence was considered to have been achieved when the absolute value of two successive interations on average Nusselt number were less than 0.0001. The sloped wall was approximated by a simple stair-step model [11].

NUMERICAL RESULTS AND DISCUSSION

The present numerical results were compared with the experimental results from Ref. [6] for verification purposes. After verification, the computer model was employed to predict results for the physical problem shown in Fig. 1. To verify the present methodology, the verification was divided into two parts: rectangular geometry and triangular geometry. First, in this study a rectangular geometry was selected and the results were validated by comparison with experimental and numerical results from Refs. [14,15]; the numerical results obtained agreed well with the experimental data [14,15].

Comparison with Triangular Geometry Data

The numerical model for the triangular enclosure was verified by comparing the numerical results with the experimental results of Ref. [6]. Figures 2 and 3 show the flow pattern inside the triangular geometry for the hot and cold sloped wall cases. In Fig. 2 (hot sloped wall) the flow near the sloped wall moves toward the top of the enclosure due to upward bouyancy forces. The flow then moves downward to the bottom of the enclosure. In the bottom corner, where the hot and cold walls meet, the velocity decreases due to the closeness of the walls and hence the higher viscous forces. The highest velocity occurs near the insulated (vertical) plane as shown in Fig. 2. In Fig. 3 (cold sloped wall) the flow moves downward along the sloped wall due to gravity pulling the cooler (higher density) fluid downward. Again, the velocity in the bottom corner (where the hot and cold walls meet) decreases due to the higher viscous forces; the velocity is highest near the vertical plane of symmetry as shown in Fig. 3.

Figures 4 and 5 depict the non-dimensional centerline temperature profile along the vertical plane of symmetry shown in Fig. 1. In Fig. 4, where the sloped wall is hot, the temperature profile reveals a drop near the cold wall and a montonically increasing temperature as it approaches the hot wall. In Fig. 5 (cold sloped wall), moving away from the apex (top of enclosure) along the centerline yields an increasing temperature which was in very good agreement with the experimental data. From X/H of 0.2 to 0.8 there appears to be an essentially isothermal region which in general again agrees with the experimental data. Here the numerical results agree well with the experimental data of Ref. [6] as shown in Figs. 4 and 5.

Now it is convenient to define a term X_S/L which is the non-dimensional distance along the sloped wall. Figures 6 and 7 illustrate the behavior of the heat transfer along the sloped and horizontal walls as a function of X_s and horizontal distance (Y) for summer and winter conditions. Figure 6 shows the heat transfer for the hot and cold sloped wall cases utilizing different Gr_H where the hot sloped wall heat transfer is compared with experimental data from Ref. [6]; no experimental heat transfer data were available along the cold sloped wall. As shown, moving away from the apex along the hot or cold sloped walls yields an increased heat flux. This increase in heat transfer is due to the decrease in distance between the two isothermal walls (sloped and horizontal walls). Figure 7 shows the heat transfer along the hot and cold horizontal walls using different Gr_H. The results for the hot horizontal wall cases were compared with the experimental data of Ref. [6]; no experimental data were available for comparison for a cold horizontal wall. Moving along the cold or hot horizontal wall shows the heat transfer rate decreases as the distance nears the center of the triangular enclosure. The heat transfer reaches a maximum at the bottom corner (where the two isothermal walls meet). Again, the agreement between the above numerical results and the experimental data of Ref. [6] was excellent as seen in Fig. 7.

Figures 8 and 9 indicate the dependence of $\overline{Nu}$ on Gr_H. Figure 8 shows the dependence of $\overline{Nu}$ on Gr_H for both the hot and cold sloped wall cases and for various θ_1 values. For the cold sloped wall, the results obtained agree with the experimental data of Ref. [6] very well. Results obtained for the hot sloped wall and a given Gr_H where θ_1=30 also agree with the experimental data quite well. However, for a given Gr_H, as the distance between the two walls increases (increasing θ_1) the results obtained differ from the experimental data by as much as 20 percent. In Fig. 8, for a given Gr_H as the angle between the hot and cold wall increases (distance between two walls increase), the $\overline{Nu}$ decreases for the cold sloped wall case; the $\overline{Nu}$ appears to increase for the hot sloped wall case. This illustrates that $\overline{Nu}$ is strongly dependent on Ar. Figure 9 shows the dependence of the $\overline{Nu}$ on Gr_H for both hot and cold horizontal wall cases. As anticipated, $\overline{Nu}$ for the cold sloped wall (winter condition) is higher than for the hot sloped wall (summer condition) for all θ_1 values. This is due to greater bouyancy forces for the cold sloped wall condition. As θ_1 increases the contribution of bouyancy forces decreases (increasing aspect ratio) so less heattransfer by convection and hence $\overline{Nu}$ decreases. For Fig. 9 no experimental data were available for comparison.

The results from Figs. 6 and 7 show that the peak heat transfer occurs near the bottom corner (between the two isothermal walls) for both summer (hot sloped wall and cold horizontal wall) and winter (cold sloped wall and hot horizontal wall) cases. The heat transfer results have an application to energy conservation for a residential attic insulation. These results imply that a greater thickness of insulation is needed near the bottom corner of the horizontal surface since the heat transfer in this area is higher for both summer and winter conditions. By contouring the thickness of the insulation an energy saving could be realized.

SUMMARY

A numerical investigation of natural convection inside a triangular shaped enclosure using Patankar's [11] SIMPLE method was performed. First, natural convection inside of a rectangular enclosure for different Gr_H (1×10^3 - 1×10^6) was examined. The results obtained were in excellent agreement with available data, Refs. [14,15]. Second, natural convection inside a triangular shaped enclosure for Gr_H as high as 10^6 and variation of θ_1 ($30 < \theta_1 < 65$) was investigated. The above conditions were studied for both summer and winter situations. Due to symmetry, this study addressed only one half of the attic. The results obtained for one half are a mirror image of the other half. Temperature field, velocity field, local Nusselt number and average Nusselt number results were presented.

The results for $\overline{Nu}$ indicate a montonic relationship between θ_1 and $\overline{Nu}$; specifically, as θ_1 increases the average Nusselt number decreases for the cold sloped wall case and increases for the hot sloped wall case. Furthermore, the heat transfer increases near the bottom corner (where the hot and cold sloped wall meet). The maximum heat transfer occurs at this corner during both the summer and winter conditions. For energy conservation purposes, a thicker insulation is recommended in this corner.

REFERENCES

1. Bejan, A., Convective Heat Transfer , New York: John Wiley and Son, Inc., 1984.
2. Ozoe, H., Sayam, H., and Churchill, S. W., " Natural Convection in an Inclined Rectangular Channel at Various Aspect Ratios and Angles --- Experimental Measurements, " International Journal of Heat and Mass Transfer , Vol. 18, No. 12, Dec., 1975, pp. 1425-1431.
3. Kierkus, W. T., " An Analysis of Laminar Free Convection Flow and Heat Transfer About an Inclined Isothermal Plate, " International Journal of Heat Transfer , Vol. 11, No. 2, Feb. 1968, pp. 241-253.
4. Flack, R. D., Konopnicki, T. T., and Rooke, J. H., " The Measurement of Natural Convective Heat Transfer in Triangular Enclosures ", ASME Journal of Heat Transfer , Vol. 101, No. 4, Nov. 1979, pp. 648-654.
5. Flack, R. D. and Witt, C. L., " Velocity Measurements in Two Natural Convection Air Flows Using a Laser Velocimeter ", ASME Journal of Heat Transfer , Vol. 101, No. 2, May 1979, pp. 256-260.
6. Flack, R. D., " The Experimental Measurement of Natural Convection Heat Transfer in Enclosures Heated or Cooled from Below ", ASME Journal of Heat transfer , Vol. 102, Nov. 1980, pp. 770-772.
7. Poulikakos, D., and Bejan, A., " Natural Convection Experiment in a Triangular Enclosure ", Trans. ASME Journal of Heat Transfer, Vol. 105, Aug. 1983, pp. 652-655.
8. Poulikakos, D., and Bejan, A., " Natural Convection in an Attic Shaped Space Filled With Porous Material ", Trans. ASME Journal of Heat Transfer, Vol. 104, May 1982, pp. 241-247.
9. Poulikakos, D., and Bejan, A., " Numerical Study of Transient High Rayleigh Number Convection in An Attic-Shaped Porous Layer ", Trans. ASME Journal of Heat Transfer, Vol. 105, Aug. 1983, pp. 476-484.
10. Poulikakos, D., and Bejan, A., " The Fluid Dynamic of An Attic Space " , Journal of Fluid Mechanics , Vol. 131, April 1982, pp. 251-269.
11. Patankar, S. V., Numerical Heat Transfer and Fluid Flow, New York : McGraw-Hill, Inc., 1980.
12. Holman, J. P., Heat Transfer, New York : McGraw-Hill, Inc., 1981.
13. Arpaci, V. S., Conduction Heat Transfer, Massachusetts : Addison-Wesley, Inc., 1966.
14. Schinkel, W. M. M., Linthorst, S. J. M., and Hoogendoorn, C. J., " The Stratification in Natural Convection in Vertical Enclosures, " Journal of Heat Transfer, Vol. 12, 1954, pp. 209-233.
15. Shohadaee. S. A. A., " The Experimental Measurement of Natural Convection Heat Transfer in Enclosures Heated or Cooled from Below ", Proceeding of 4th International Conferences on Applied Numerical Modeling, Dec. 1984, pp. 551-558.

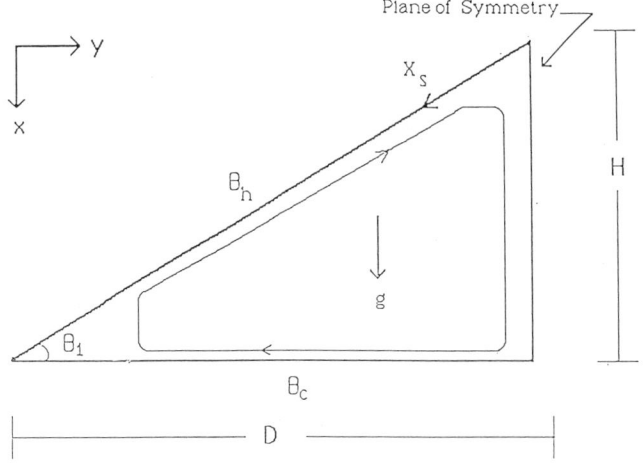

Fig. 1   Schematic illustration of triangular enclosure

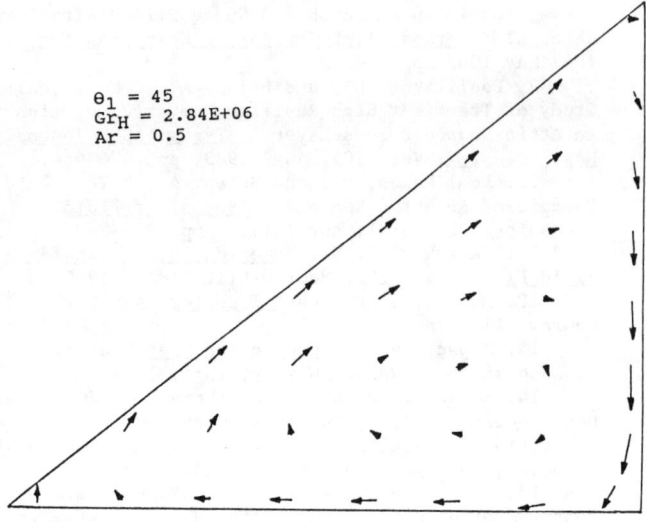

$\theta_1 = 45$
$Gr_H = 2.84E+06$
$Ar = 0.5$

Fig. 2   Velocity field illustration for hot sloped
wall (H=.0762 m D=.1524 m)

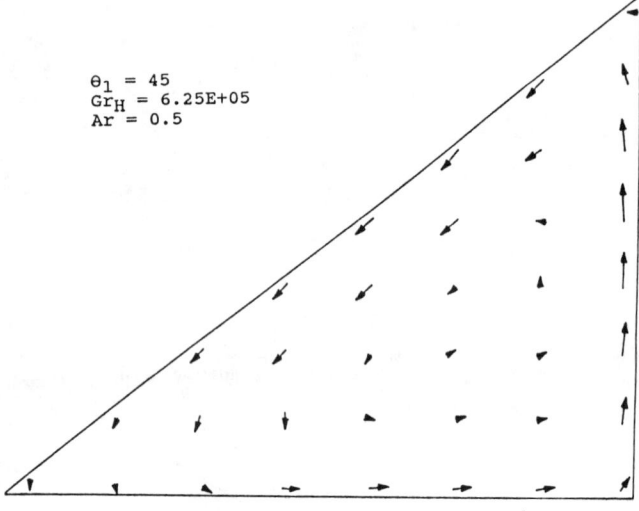

$\theta_1 = 45$
$Gr_H = 6.25E+05$
$Ar = 0.5$

Fig. 3   Velocity field illustration for cold sloped
wall (H=.0762 m D=.1524 m)

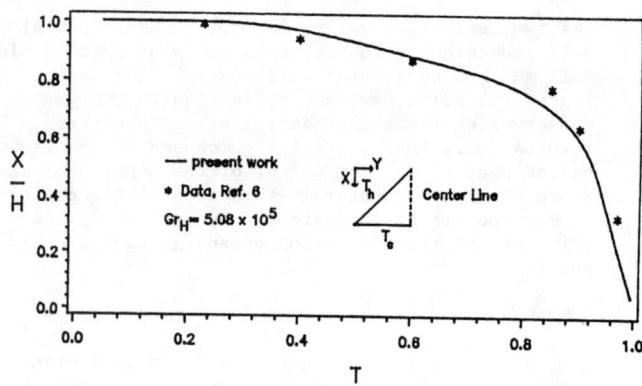

Fig. 4   Non-dimensionalized centerline temperature
profile for hot slope wall

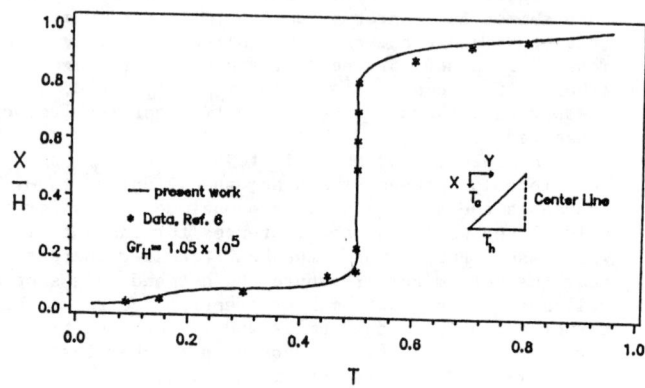

Fig. 5   Non-dimensionalized centerline temperature
profile for cold sloped wall

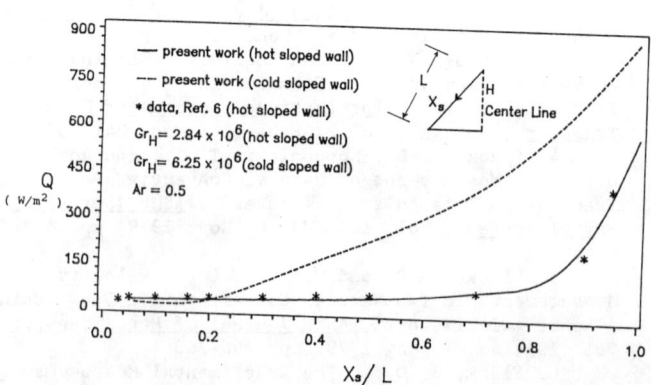

Fig. 6   Local heat flux distribution along the hot and
cold sloped wall

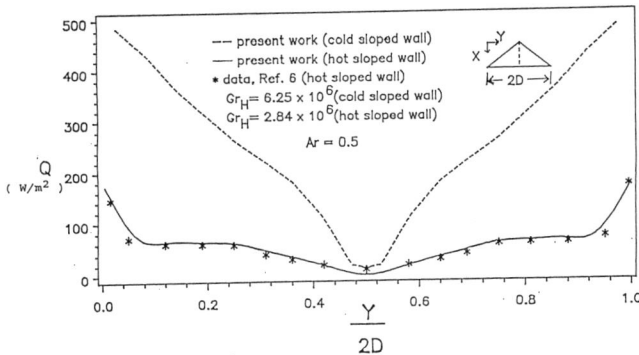

Fig. 7   Local heat flux distribution along Y for hot
          and cold sloped wall

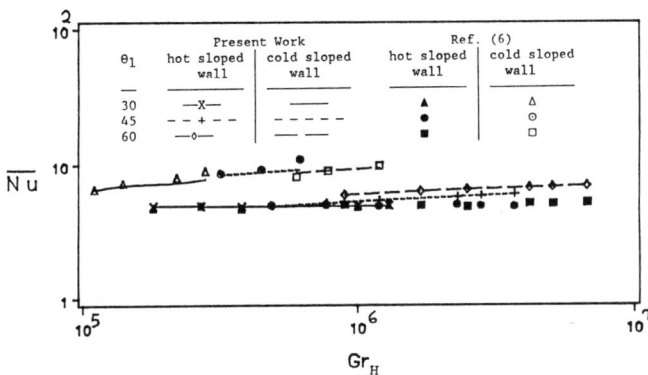

Fig. 8   Average Nusselt number along the sloped wall
          as a function of Gr_H

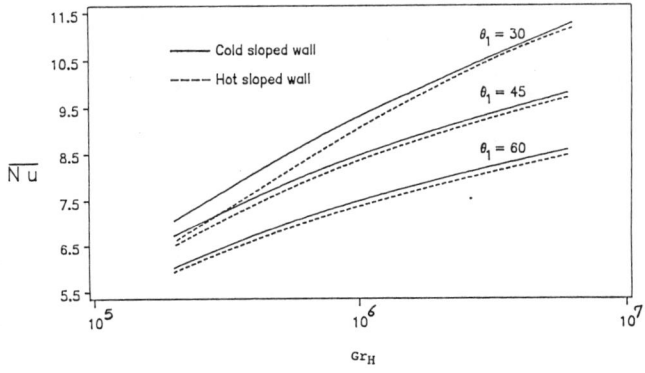

Fig. 9   Average Nusselt number along the horizontal
          wall as a function of Gr_H

INVESTIGATION OF THE EFFECT OF FLOW SWIRL ON HEAT TRANSFER INSIDE A CYLINDRICAL CAVITY

A. Salce
FluiDyne Engineering Corp.
Minneapolis, Minnesota

T. W. Simon
University of Minnesota
Minneapolis, Minnesota

ABSTRACT

Experiments were conducted to determine local heat transfer coefficients on the inside surfaces of a cylindrical cavity that is cooled by a swirling air flow. Temperature sensitive liquid crystals were used as temperature sensors.

Five blowing (cooling) modes were tested: three with swirl numbers of 0.36, 0.84, and 1.73, a fourth with no swirling component (axial flow) and a fifth which is similar to the fourth but has the flow direction reversed.

Flow visualization and static pressure measurements were performed to improve understanding of the situation. The smoke-wire technique was successfully used to picture the flow patterns.

Plots of local Nu number along the cavity surfaces were obtained for the five blowing modes and for three different Reynolds numbers.

The swirling cases have similar flow fields to one another. Higher heat transfer rates are found near the cavity top and lower rates are found near the cavity bottom (just the opposite of the non-swirling cases). A tornado-like structure was observed in the swirling cases on the cavity bottom. The tornado structure is stronger and more violent as the degree of swirl and the Reynolds number are increased.

It is interesting that for the two non-swirling cases (cases four and five) the Nu number curves are of similar shape, though the flow direction is reversed.

NOMENCLATURE

C_p = heat capacity, J/(kg K)
d = swirl generator diameter, m
D = cavity diameter, m
h = heat transfer coeficient = $q/(T_w-T_i)$, W/(m^2 K)
k = thermal conductivity, W/(m K)
L = cavity heigth, m
$\dot{m}$ = mass flow rate, kg/s
Nu = Nusselt number , hd/k
Pr = Prandtl , $\mu C_p/k$
q = heat flux, W/m^2
Re_D = Reynolds number = $4\dot{m}/\pi D\mu$
S ' = Swirling number, W/U
T = temperature, K

U = axial velocity component, m/s
W = tangential velocity component, m/s
β = swirl generator flow angle
μ = viscosity , Ns/ m^2
ρ = density, kg/m^3

Subscripts

i = inlet
LC = liquid crystal
w = wall

INTRODUCTION

It has been found that the heat transfer inside circular ducts can be enhanced by using swirl flow. This method of heat transfer can be specially useful where high heat transfer rates are needed. This is the case of cooling of internal cavities. Applications may be the curing of mold linings, carrying of moisture above grain bins, moldings of cast metal components, instrumentation inserted into high temperature regions which must be internally cooled and "dead-end" heat exchanger tubes, just to name a few. Swirl is also used for augmenting combustion or purging of cylinders in internal combustion engines

It is clear that this heat transfer situation is not peculiar to the above examples; that the results can be applied anywhere that the internal walls of a closed-end geometry must be cooled and the introduction of swirl is a reasonable option.

There are many reports in the literature of the effects on flow swirl inside pipes or ducts. They all agree that there is an increase on the heat transfer rate between the fluid and the duct wall due to swirl. Khalil [1982] reported increases of local heat transfer coefficients in the inlet vicinity of a turbulent pipe flow by using a vane swirlier. Enhancements of 40 per cent with respect to axial flow at the same Reynolds number based upon axial velocity were observed. The designer is faced with a trade-off between a higher heat transfer rate and a larger pressure drop across the swirl generator.

It is also well known that free swirling jets are applied in the combustion and plasma fields. The increase of mixing by the use of swirl results in a more efficient combustion and plasma formation. Beer and Chigier [1972] found that, for combustion applications, one of the most significant and useful phenomena of swirling jet flows is the recirculation bubble generated centrally for high swirl numbers (S>0.6). The recirculation bubble plays an important role in flame stabilization by producing a hot flow of recirculated combustion products and a reduced velocity region where flame speed and flow velocities are matched. Flame lengths and distances from the burner at which the flame is stabilized are shortened significantly. The size and shape of the reverse flow region depend on the degree of swirl.

PRESENT STUDY, VARIABLES AND SIMILARITY

In the present study, air enters and leaves the cavity through concentric openings located at the top of the cavity. Heat transfer measurements are performed on the circular walls of the cavity as well as on the bottom surface using a liquid crystal technique. The cooling fluid is air. Flow visualization studies are conducted to improve understanding of the situation using the smoke-wire technique.

The geometry studied is a cylindrical cavity with a height to diameter ratio of 1.70 (see Fig 1).The two independent variables of the experimental program are: the mass flow rate of cooling air and the swirl degree of the flow entering the cavity.

Three flow rates are used during the experiments: 9.3, 14.0 and 19.5 gr/s of air corresponding to Reynolds numbers based upon cylinder diameter, of 2540, 3750 and 5270, respectively.

Four swirl strengths were tested, each one had a different flow angle, ß, at the exit of the swirl generator. The four angles tested are ß=0°, 20°, 40° and 60°, corresponding to swirl numbers, S', of 0.0, 0.36, 0.89 and 1.73, respectively. This swirl number is defined by:

$$S' = \frac{W}{U} \qquad (1)$$

where U is the axial velocity component and W is the tangential velocity component of the flow, both at the exit plane of the swirl generator (which is the entry plane of the cavity).

In the fifth case, axial (no tangential component) flow enters the cavity through the center (which was the exhaust tube for the cases above). This, then, is the S'=0.0 case but with the flow direction reversed. Herein, this case is called the "Inverted Flow" case.

Fig. 1 is a sketch of the basic geometry and the parameters and variables considered important for this problem. The parameters are the fluid properties (ρ, C_p, μ, k) and geometric dimensions (d, L, D). The variables are the flow characteristics (U, W, $\dot{m}$) and the heat transfer coefficient, h (the dependant variable to the problem).

Dimensional analysis is used to find the important dimensionless groups. They are the following:

$$\frac{hD}{k}, \quad \frac{4\dot{m}}{\pi D \mu}, \quad \frac{\mu C_p}{k}$$

which are the Nusselt, Reynolds and Prandtl numbers, respectively. With the dimensionless, length, L/D, and entry tube diameter, d/D, the dimensionless parameter list is complete.

The task of the present study is to find a relation of the form:

$$Nu = f(Re, S')$$

for a chosen Pr, L/D and d/D.

A small cavity model of D=0.146 m was built during the development stage of the experimental program and used to test the similarity parameters of the experiment by comparison with the final model (D=0.260 m). The two models are geometrically similar, having the same L/D and d/D ratios. Similarity is confirmed by the following test: Heat transfer data are taken on the small and large models while matching Reynolds number. Figs. 2 and 3 show the results. One can observe that the Nusselt number distributions are nearly the same, an indication of a similar flow field and that Reynolds number is appropriate for characterizing this flow situation.

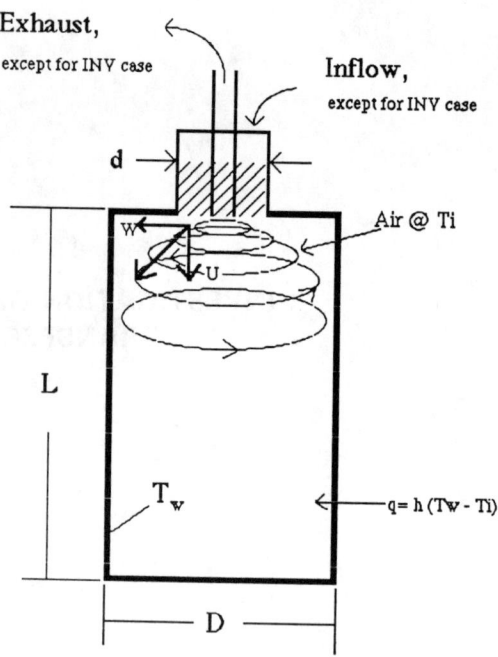

Fig. 1 The test section geometry.

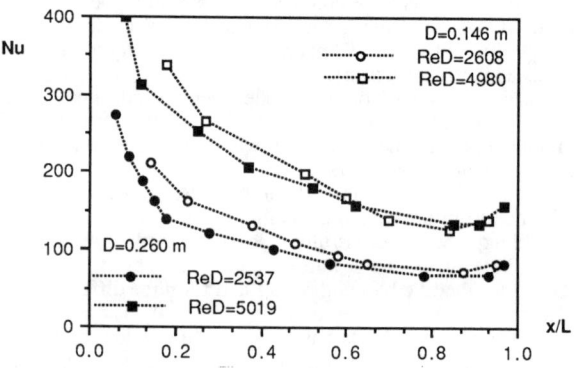

Fig. 2 Results of the similarity study: Nu Vs. wall position for two cavities, matching Reynolds number. The inverted flow case.

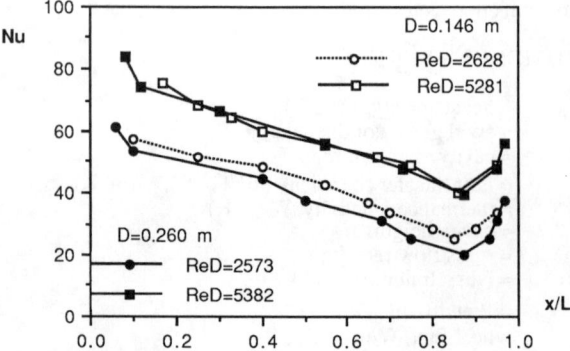

Fig. 3 Results of the similarity study: Nu Vs. wall position for two cavities, matching Reynolds number. S'=0.0 case.

TECHNIQUES USED

Heat Transfer

Local heat transfer coefficients are obtained using the liquid crystal technique. This technique has been successfully used on other convective heat transfer applications (Simonich and Moffat [1984] and Hippensteele, et al.[1985]). Cholesteric liquid crystals are substances that exhibit different colors at different temperatures by selectively reflecting only certain wavelengths of the light with which they are illuminated, i.e. cholesteric liquid crystals change colors with temperature.

The liquid crystal film is sandwiched between a transparent wall (acrylic or glass) and an electric heater element, as shown in Fig. 4. On the outside of the test wall, water at a fixed temperature which is the mapping temperature, T_{LC}, is circulated. The mapping temperature is a temperature for which the liquid crystal color is known (the color chosen for doing the temperature mapping). For our experiment, T_{LC} is 41.8°C and the mapping color is red. In the absence of cooling air inside the cavity and power to the heater and with water circulation outside the liquid crystal, the liquid crystal displays the same color, red, over the entire test section.

When power is applied and cooling air is blown, the liquid crystal shows different colors at different locations across the test section, blues and violets for the higher temperature areas, red, yellow and black for the lower temperature areas.

The areas displaying the mapping color are the areas of interest. At these positions, energy balances are easily done to find the heat flux. The areas of colors different from the mapping color are of no interest for this step of the study because there is heat exchange at these positions with the circulating water.

For the areas of interest, the temperature outside the test section wall is the same as the temperature on the inside of the wall, T_w (See Fig. 4). This is so because the liquid crystal and the circulating water on the outside are at the mapping color temperature. Thus, no heat conduction is allowed across the wall and all the heat generated in the heater is convectively removed by the inside cooling air.

$$h = \frac{q}{(T_w - T_i)} \qquad (2)$$

Where q is the uniform heat flux from the heater and T_i is the air temperature at the inlet of the swirl generator. All the terms needed to calculate the heat transfer coefficient are known at these positions where the liquid crystal displays the mapping color. Changing the electric power to the heater will move the mapping color on the test section to other locations. In doing so, Q is the only term in equation (2) to change. The heat transfer coefficient, h, is then calculated for the areas to which the mapping color has moved.

The entire heat transfer coefficient distribution can be calculated by continuing until the entire surface has, at one heat flux or another, been at the mapping temperature. One of the main advantages of this technique is that it can be used in applications where a large variation in heat transfer coefficient is present.

Flow Visualization

Flow visualization studies were conducted using the smoke-wire technique.

Originally developed by Ruspet and Moore (discussed by Goldstein [1983])in the early 1950's, this technique has been successfully used in many different flows especially in wind tunnel studies of bluff bodies. Mueller and Batill [1980] and Nagib [1977] are among several researchers that have used the technique on studies of complex flows.

The smoke-wire technique utilizes the evaporation of a thin oil film from a tiny metal wire immersed in the flow field. Evaporation is by means of electric heating. Surface tension forms small beads on the wire. When the wire is heated, a smoke filament (streakline) is produced downstream of each of the oil beads. Precise control over the time of heating and the voltage applied to the wire is essential to generate good smoke for visualization and to prevent the wire from burning out.

An electric control box was designed and built to provide the desired synchronization between the smoke generation and the triggering of the recording instruments. This control box also provides control over the time of heating and the magnitude of the power applied to the wire.

Pressure Measurements

Static pressure measurements are also taken along the cylinder walls and the bottom of the cavity. Pressure taps were machined through the surface at various places and measurements were taken with respect to a reference point situated mid-height on the cylinder wall.

APPARATUS

The apparatus discussion is divided into two parts; the heat transfer and the flow visualization facilities.

Heat Transfer Facility

The heat transfer facility consists of three sub-systems: The constant temperature water supply system, the air supply system and the test section, see Fig. 5. The constant temperature water supply system was designed to provide up to 20 li/min of water at a fixed temperature (nominally 41.8 °C) as required by the liquid crystal technique. The air supply system provides the air (the working fluid) which cools the inside of the cavity. The test section is a model of the cavity in which the heat transfer coefficient measurements are performed. Thermocouple junctions, type T (copper-constantan), were used to record the relevant temperatures of the experiment.

Flow Visualization Facility

The flow visualization facility consists of three sub-systems: the same air supply system used in the heat transfer experiments, the smoke-wire electric control box and the flow visualization model (see

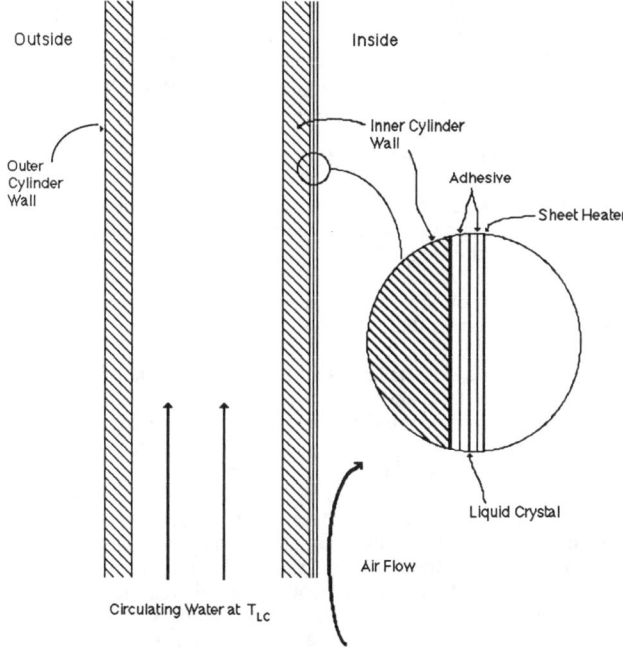

Fig. 4 Heated wall construction.

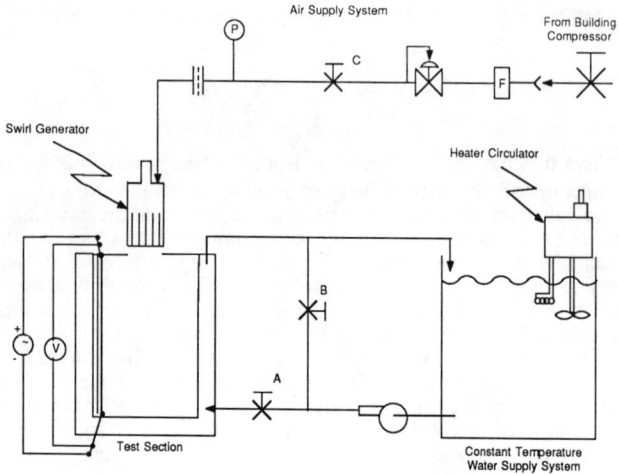

Fig. 5   Schematic of the heat transfer facility.

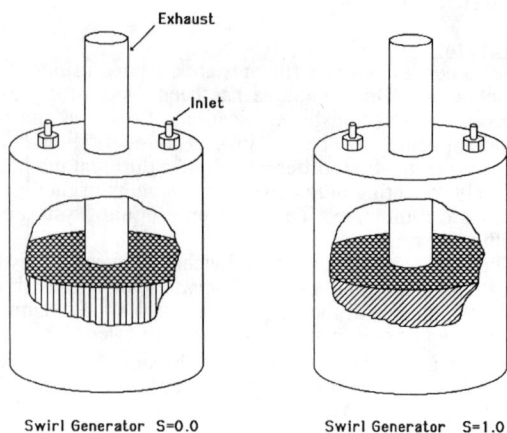

Fig. 7   Swirl generators of ß=0° and ß=45°.

Fig. 6). The flow visualization model has drilled holes for installing the smoke wire and for pressure taps; it was constructed as a separate piece because of the required holes. The visualization model is geometrically identical to the heat transfer model, in fact some of its components were interchangeable. The major differences are the absence of the external water circulation system, which is not needed for the flow visualization studies, and the use of transparent walls as required for viewing inside the cavity.

The pressure signal is carried via Tygon® tubing to a variable-reluctance pressure transducer. The transducer's D.C. voltage signal from the signal conditioner is proportional to the pressure. This voltage was then displayed on a 8044A A-D Fluke voltmeter.

Swirl Generator

Commonly used swirl generators such as twisted tape, tangential-plus-axial entry, tangential entry or tangential vane types don't provide a uniform and known flow distribution. This makes it difficult to characterize the exit flow and detailed flow measurements for every flow condition must be made.

In order to avoid this, a swirl generator with a uniform and known velocity distribution at its exit plane for all flow rates is used on this experiment. A sketch of this swirl generator is shown in Fig. 7. It is made of two concentric PVC circular tubes. The space between them contains pieces of Tygon tubing, 4.8 mm (3/16") ID and 6.4 mm (1/4") OD, glued together with RTV silicone rubber forming something similar to a twisted honeycomb.

By aligning all the Tygon tubing to the same angle, the tangential-to-axial velocity ratio becomes the same at all positions on the exit plane and the the swirl number, S', is well defined. The exit flow is, thus, uniform across the exit plane of the generator in both axial and tangential velocity.

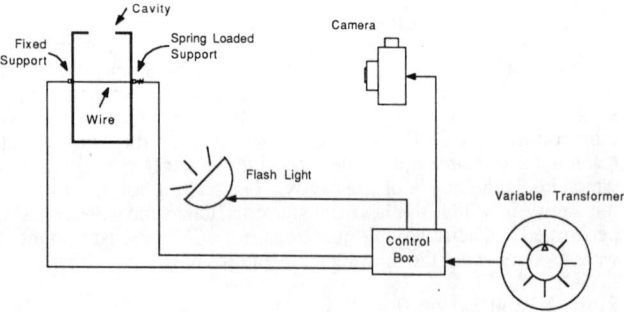

Fig. 6   Flow visualization layout and control schematic.

A perforated plate (not shown) was placed in the space between the top plane (where the air enters) and the plane of entrance to the Tygon tubing pack. This plate produces a uniform velocity distribution into the tubing. The center tube is the exit path for the air. The swirl generator is mounted on the top of the cavity, as shown in Fig. 1, such that its exit plane is aligned even with the cavity top plane.

Four swirl generators were built, each one for a specific swirl strength. They have different inclination angles, ß of 0°, 20°,40° and 60°, creating swirl strengths of S'=0.0, 0.36, 0.89 and 1.73 respectively. During the experiments, a swirl generator is positioned in either the heat transfer or the flow visualization facility as shown in Fig. 1.

Smoke-Wire

Fig. 6 shows the smoke-wire arrangement used for the flow visualization studies of this experiment. The system consists of a variable transformer, a control box, wire supports, wire, SLR camera and flash.

Two supports were used to hold the wire in place and serve as electrical contacts. The supports were placed directly opposite one another along a diameter of the cavity model. One was fixed while the other was able to move in the radial direction to remove slack due to the elongation of the wire produced by heating.

EXPERIMENTAL PROCEDURE

Heat transfer measurements start with a 3 hour warm up of the water in the storage tank of the circulator system and all the components in the water loop. The heater circulator is set at a temperature slightly above the mapping temperature, allowing for a temperature drop on the piping system (pump, hoses, etc.). During this warm-up process there is no water circulating through the test section.

Once the water reaches its steady-state temperature, the model is placed into the loop and water is allowed to circulate through it. An additional 20 or 30 minutes is needed for the model to reach the mapping temperature. At this moment all the areas covered with liquid crystal show the color red. The appropriate swirl generator is placed on top of the test section and the air flow rate is set at its desired point on the rotameter. The cooling effect of the air turns the liquid crystal areas black. Just enough electric power is then applied to the heater, to see the first appearance of a red area on the liquid crystal. After 1 or 2 minutes, the isothermal red areas at the mapping temperature are well established. Measurements of the inlet air temperature, the inlet and outlet water temperatures, voltage applied to the heater and the position of the red isotherm are recorded. All this information is used to record the regions of a single value of convective heat transfer coefficient.

Without changing any setting in the air and water supply systems, a slightly higher voltage is applied to the heater. The isothermal red areas then move to new positions. A time of 1 or 2 minutes is allowed for the isotherms to settle to new positions on the test surface. Again, temperature, voltage, and position of the isotherm are recorded for the regions of a new convective heat transfer coefficient.

The process is repeated several times, moving the isotherm across the entire test surface and generating plots of Nu vs. position on the test surface, for given values Re_D and S'.

Different Re_D values are achieved by adjusting the air mass flow rate. Different S' values are obtained by using different swirl generators.

An uncertainty analysis based on the 95% confidence interval was performed on the Nusselt and Reynolds numbers data. The Nusselt number uncertainty was found to be 3.9% and the Reynolds number uncertainty 3.6%.

RESULTS

The heat transfer data is presented along with the best estimate of the corresponding flow fields (deduced from the pressure measurements and smoke-wire results). Some pictures of the flow visualization and the pressure data are also presented and explained.

Though the flow has three non-zero velocity components, there was no evidence in the data of a departure from an axisymmetric situation.

For simplicity for the reader, a convention in presenting the data has been adopted.

Heat Transfer Results

The heat transfer data is presented in terms of a local Nusselt number, Nu, defined as:

$$Nu = \frac{hD}{k}$$

where h is the local heat transfer coefficient, D is the cavity diameter, and k is the thermal conductivity of the air based upon the air entry temperature which is slightly cooler than the air exit temperature. The local heat transfer coefficient is calculated as:

$$h = \frac{q}{T_w - T_i}$$

where q is the heater total heat per unit area, T_w is the wall temperature obtained from the liquid crystal mapping temperature and corrected by the temperature gradient across the heater electrical insulation, and T_i is the inlet air temperature.

The data is presented as plots of Nu number distribution along the surface of the cavity. The ordinate is the Nu number and the abscissa can be x/L or x/D. The dimensionless height, x/L, is zero at the bottom of the cavity and one at the top. The dimensionless diameter, x/D, is zero at one side of the bottom surface and one at the opposite side of the diameter, shown in Figs. 8, 9 and 10.

INVerted Case (no swirl - reverse flow). The air enters the cavity by the exhaust tube; no swirl component is present (see Fig. 8).

The high velocity inlet stream travels downward through the center of the cavity, the flow impinges upon the bottom surface then changes direction, returning upward by the outer region near the cylindrical wall.

Since the inlet and outlet streams are very close to one another and the two streams move in opposite directions, a region of high shear is established near the cavity top. In this high-shear region, some of the incoming air is drawn out by the exit stream without having a chance to cool the cavity walls.

The Nu number is high near the center of the bottom surface because of the unheated, high-velocity jet impinging on this region.

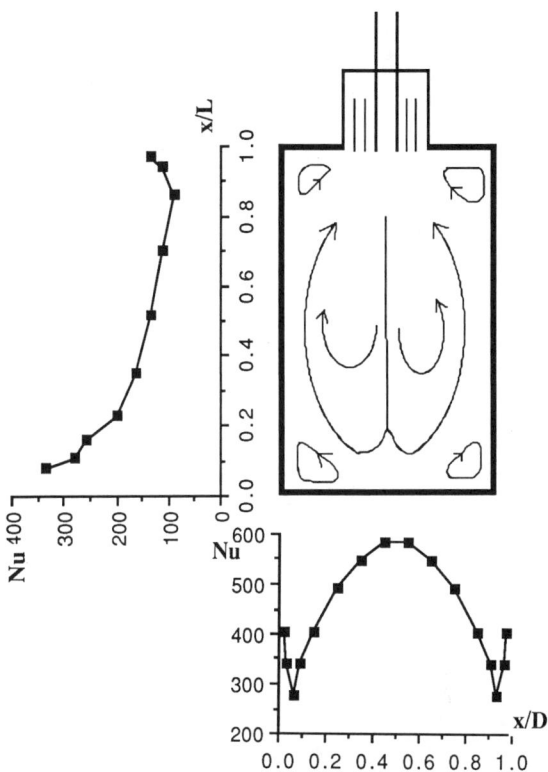

Fig.8 Nusselt number distributions on the side and bottom of the cavity and the flow pattern sketch for the INVerted case Re_D=3750.

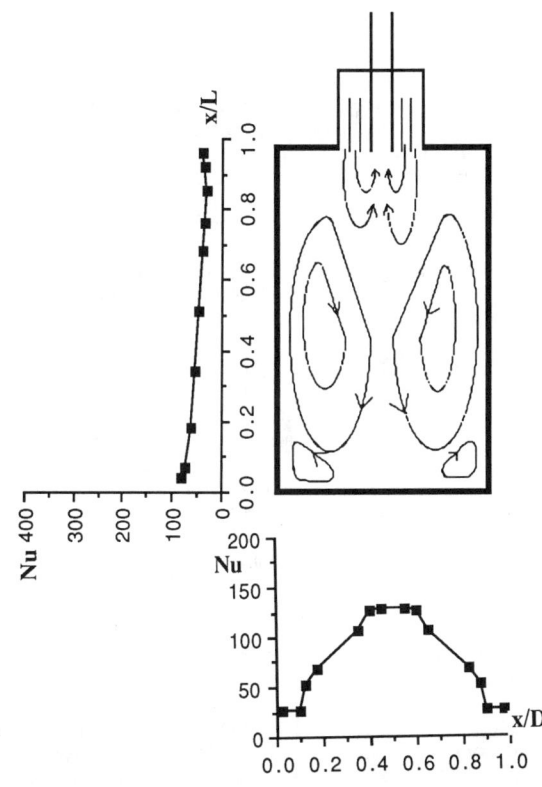

Fig.9 Nusselt number distributions on the side and bottom of the cavity and the flow pattern sketch for the S'=0.0 case, Re_D=3750.

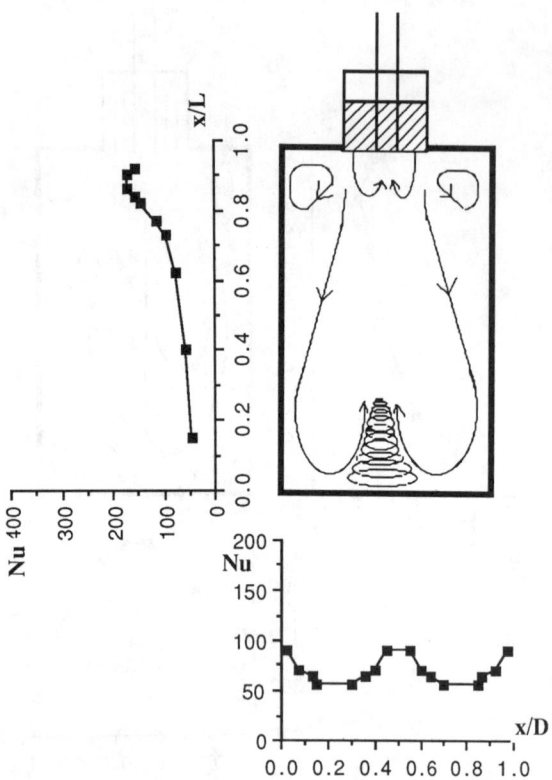

Fig.10 Nusselt number distributions on the side and bottom of the cavity and the flow pattern sketch for the S'=0.36 case, Re$_D$=3750.

As the jet spreads radially outward, the velocity decreases and the Nu number drops. Separation of the flow occurs and a toroidal recirculation zone is created in the lower corners of the cavity. This recirculation zone is responsible for the high Nu numbers in this area (x/D=0.05, x/D=0.95, and x/L=0.05).

The air continues its upward movement, washing the cylindrical walls. But, as the fluid remains in contact with the wall its temperature rises creating a decreasing Nu as x/L increases.

As the flow approaches the top surface, it separates again from the wall creating another toroidal recirculation zone in the top corner. This top recirculation zone is responsible for the slight increase in the Nu number.

The S'=0.0 case (no swirl - forward flow). Again, there is no swirl component in this case. The air enters the cavity through an annular region and exits by the central exhaust tube, Fig. 9. The inlet and outlet streams are located close to one another and, since they have flows of opposite direction, a strong shear region is formed. This shear region is again responsible for making a large portion of the incoming air reverse and leave the cavity without cooling the heated surfaces. The result of this is that much of the potential cooling effect of the fresh air is lost and the Nu numbers are low.

The cavity is cooled by a very large recirculating flow that behaves similar to that of the INV case (see Figs. 8 and 11). As a result of this, the Nu number distributions are nearly the same in shape, thought the INV case has much larger values of Nu.

The S'=0.36 case (mild swirl). This time the tangential component of the velocity imposed on the flow by the swirl generator causes the inlet stream to spread radially outward soon after entering the cavity (Fig. 10). This outward movement allows more cool air to wash the cavity walls. As a result, the Nu numbers in the upper portions of the cavity are significantly higher than those for S'=0.0.

The flow travels down near the wall and returns along the cavity centerline. An interesting phenomenon was observed on the bottom surface of the cavity. The flow reaches the bottom traveling near the cylindrical wall, turns and travels radially inward to the center where it turns again to move upward. As it approaches the center, the flow accelerates to satisfy continuity. Simultaneously, the flow also swirls more intensely to conserve angular momentum. A tornado like structure of high tangential velocity is thus formed in the lower part of the cavity near the centerline (see Figs. 10 and 12).

The highest Nu values are found on the cavity side walls near the top of the cavity (x/L=0.9). Attachment to the side-wall and the upper toroidal recirculation zone are responsible for this.

The S'=0.84 case (moderate swirl). The flow field in this case is very similar to that of the mild swirl but with all the swirling effects augmented.

The higher angular velocity causes the flow to reattach to the cylindrical wall nearer the cavity top than for the S'=0.36 case, compressing the upper recirculation zone.

In general, the Nu number profiles are also alike, they are shifted upward because of the higher near-wall velocity gradients produced by the higher swirl number (Figs. 13-15). Also, since the incoming flow spreads at a wider angle, there is less cool inlet flow entrained into the exiting stream.

The tornado structure formed on the bottom surface is stronger and larger (Figs. 13-15).

The S'=1.73 case (strong swirl) Flow field and Nu number profiles are very similar to those for the moderate swirl case (Figs. 13-15). The curves are raised a bit, however. The tornado structure is stronger, more violent, and is visible along the entire axis of the cavity. In the region of x/L=0.9, the Nu number is higher for the S'=0.84 than for the S'=1.73 case. The reason for this apparent contradiction is unknown.

Plots of Nu number vs. x/L or x/D have been constructed to compare the five different blowing modes (INV, S'=0.0, S'=0.36, S'=0.84, and S'=1.73) at the same Reynolds number or mass flow rate, see Figs. 13, 14 and 15.

Flow Visualization

Fig. 11 shows the large-scale recirculation zone that is created on the S'=0.0 case. Note that the flow in the center is traveling downward and the flow near the walls is traveling upward.

Fig. 12 shows the tornado-like structure that is formed in the swirling cases S'=0.36, S'=0.84 and S'=1.73. The tornado shape is similar for these three cases except that the tangential velocity is higher as S' increases.

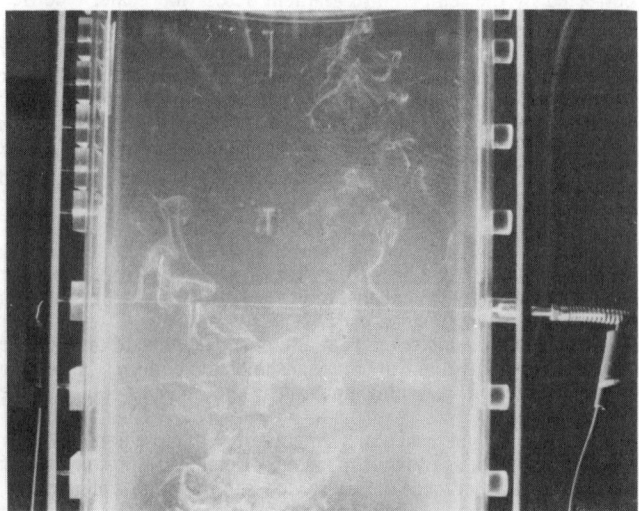

Fig. 11 Smoke-Wire photography. S'=0.0

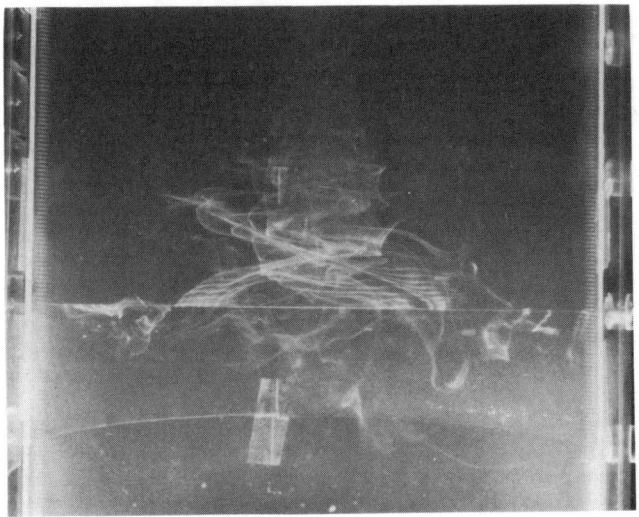

Fig. 12 Smoke-Wire photography. S'=0.36

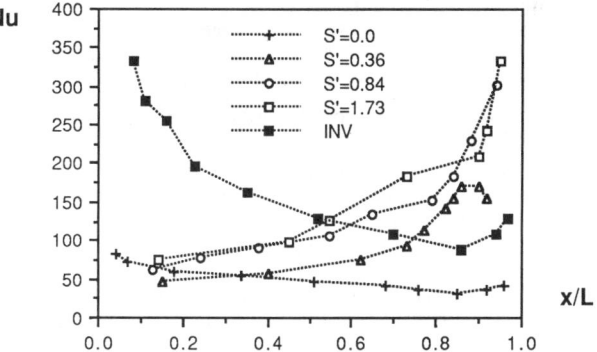

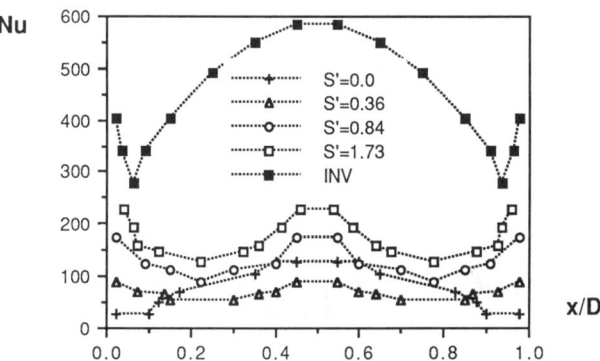

Fig. 14 Nusselt number distributions for various entry conditions at Re_D= 3750.

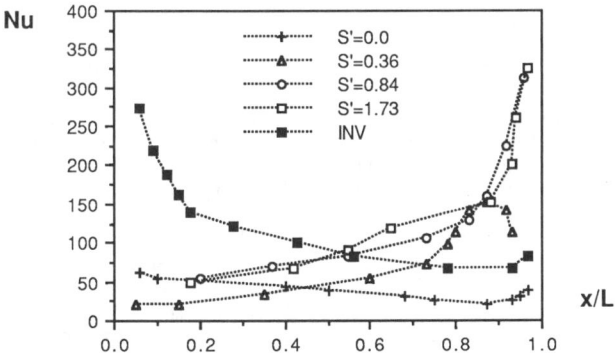

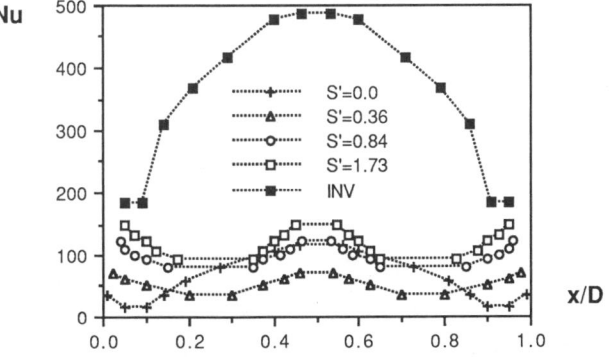

Fig. 13 Nusselt number distributions for various entry conditions at Re_D= 2540.

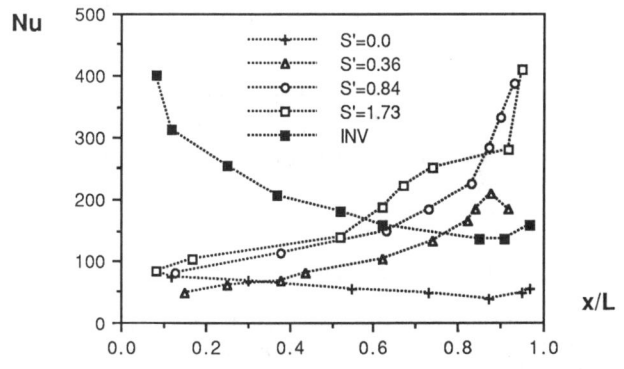

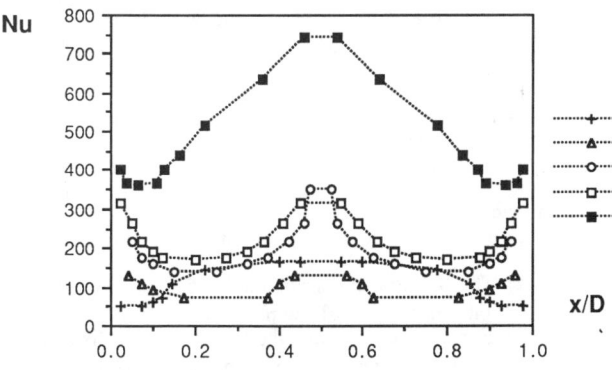

Fig. 15 Nusselt number distributions for various entry conditions at Re_D= 5270.

Pressure Measurements

Static pressure measurements were taken along the cylinder wall and bottom surfaces. All pressures are with reference to the static pressure on the side-wall at mid-height, P_r.

Only those of the bottom surface are to be shown. It was found that no significant pressure difference $(P-P_r)$ exists on the cylindrical walls for any of the five blowing modes at any Re_D number. The absolute values of P and P_r increased as Re_D increased, but the difference remained nearly zero. This result was not expected, especially in the recirculation zones and near the reattachment point near the top of the cavity.

The only place where pressure differences were detected was on the bottom surface. Negative values of the pressure (relative to the cylinder wall pressure) were detected on the bottom near the centerline of the cavity (Fig. 16). It is presumed that high velocities of the tornado structure are responsible for this. The magnitude of the negative pressure is a measure of the local velocity and, thus, the swirl strength. Two opposing factors affect the swirl strength in this area: Re_D and the length of the cavity. As Re_D increases, the swirl is stronger and more influential on the bottom wall. When the cavity is longer the swirl strength is dissipated more and it is less influential on the cavity bottom.

P-Pr (cm H2O)

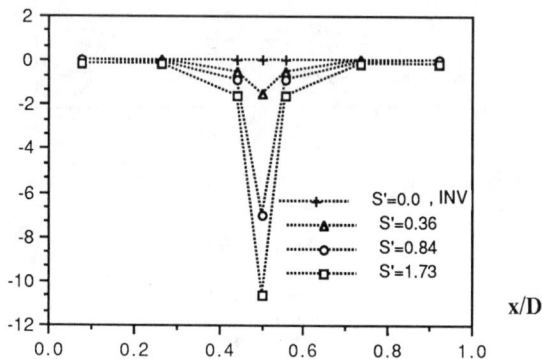

Fig. 16 Pressure distribution on the cavity bottom. ReD=3750.

CONCLUSIONS AND COMMENTS

The liquid crystal technique used in this study is well suited for the present problem and capable of measuring over a wide range of Nusselt numbers (18 to 1800).

Experiments preformed at the same Reynolds number on two geometrically similar (same L/D and d/D) cavities, but of different size (D=0.146 m and D=0.260 m) and with different mass flow rates demonstrated that Reynolds number, Re_D, is an appropriate parameter for characterizing the flow.

Flow visualization studies and pressure measurements are qualitatively consistent with the heat transfer data and are shown to be very useful in understanding the flow.

Though the flow has three non-zero velocity components, there was no evidence in the data collected of a departure from an axisymmetric situation.

From the data collected, the following conclusions can be drawn:

1) The Nu number distribution is a function of Re_D number, the degree of swirl and the flow direction.
2) For the geometry and the Re_D range of this experiment, the flow pattern is a function only of the blowing mode. The flow pattern does not change with Re_D.
3) The Nusselt number increases as the Re_D increases.

4) The two non-swirling cases, one with central exhaust (S'=0.0) and the other with central entry (INV), produce Nu number distributions of similar shape, though they have different flow directions. The S'=0.0 case has much lower values of Nu because the cavity is washed by a recirculation zone driven by the entry flow, whereas in the INV case, the cavity is directly washed by the entry flow.
5) The swirling cases (S' of 0.36, 0.84, and 1.73) have similar flow fields to one another. Higher heat transfer rates are found near the cavity top due to swirl and lower rates are found near the cavity bottom. This is the opposite to that of the non-swirling cases. A tornado-like structure on the cavity bottom increases in strength as S' or Re_D is increased.

If one is looking for uniformity in cooling, the non-swirling (S'=0.0) case is better. However, if the intention is to cool the region near the cavity top, swirl is clearly beneficial. If the intention is to have effective cooling of the bottom, the INVerted flow case is the best

ACKNOWLEDGEMENTS

The authors would like to thank the AMOCO Foundation for partial support and AT&T for an equipment grant in support of this project.

REFERENCES

Beer, J. M. and Chigier, N. A., 1972,Combustion Aerodynamics, Applied Science Publishers LTD.

Goldstein, R.J.,1983, Fluid Mechanics Measurements. Hemisphere Publishing Corp.

Hippensteele, S. A., Russell, L. M., and Torres, F. J., 1985, "Local Heat-Transfer Measurements on a Large Scale-Model Turbine Blade Airfoil Using a Composite of a Heater Element and Liquid Crystals." J. of Eng. for Gas Turb. and Power., Vol. 107, pp. 953-960.

Khalil, E. E., 1982, "Heat Transfer to Turbulent Pipe Flows with Swirl and Following a Sudden Enlargement." Proc. of the Seventh Int. Heat Transfer Conf. , FC10.

Mueller, T. J. and Batill, S. M.,1980, "Experimental Studies of the Laminar Separation Bubble on a Two-Dimensional Airfoil at Low Reynolds Numbers." AIAA Paper 80-1440.

Nagib, H. M., 1977, "Visualization of Turbulent and Complex Flows Using Controlled Sheets of Smoke Streaklines." Proc. of the Int. Symp. on Flow Visualization, Tokyo, pp. 181-186.

Simonich, J. C. and Moffat, R. J., 1984, "Liquid Crystal Visualization of Surface Heat Transfer on a Concavely Curved Turbulent Boundary Layer." J. of Eng. for Gas Turb. and Power, Vol. 106, pp. 619-627.

HEAT FLOW CHARACTERISTICS OF
A FINNED HEAT EXCHANGER

R. H. Bravo and C. J. Chen
Department of Mechanical Engineering
Iowa Institute of Hydraulic Research
The University of Iowa
Iowa City, Iowa

ABSTRACT

In this study, fluid flow and heat transfer in a two dimensional finned heat exchanger is analyzed by the Finite Analytic Numerical Method. The passage of the heat exchanger is formed by plates and staggered fins maintained at constant temperature. The fluid is considered to be incompressible and the flow laminar. Flow and heat transfer from the inlet of the heat exchanger to the outlet are simulated by solving Navier-Stokes and energy equations. Results are presented for three basic configurations of heat exchangers, namely heat exchangers with (1) no fins, (2) a pair of staggered fins and (3) two pairs of staggered fins. For each configuration with staggered fins, three different fin heights are considered. Computations are made for Reynolds numbers from 100 to 500, and Prandtl numbers 0.7 and 4.0. The results are presented in the form of velocity vector fields, isotherms, velocity component profiles and local and average Nusselt numbers. The characteristics of the heat transfer and pressure drop in different configurations and different fin height are analyzed.

INTRODUCTION

Finned heat exchangers are used extensively in industry [1,2,3]. The objective of these devices is to transfer heat at an optimum rate considering the energy required to pump the fluid, the cost of their manufacturing, and the limits in their size. Although many of these devices work in the range of turbulent flow, a well designed low Reynolds number flow heat exchanger will yield high heat transfer rates with a reasonable low pressure drop. The number of applications for laminar flow heat exchangers is on the increase [3], especially in the area of compact heat exchangers.

The first stage to improve overall heat transfer coefficients is to use an extended surface. This surface extension increases the total area of heat transfer and, if it is done correctly, increases the efficiency of this process. The efficiency is enhanced when the growing boundary layers are disrupted and when a cross-stream mixing is promoted. Usually this is accomplished by introducing finite plates into the flow and staggering them in the channel to cause local changes in the mean flow direction and magnitude. The operating price paid for the enhancement of heat transfer is an increased pressure drop. However, if the fins are improperly designed, arranged, or their size is not adequately selected, they can reduce the efficiency of heat transfer on some surfaces, and consequently, reduce the efficiency of the exchanger and greatly increase the cost of operation.

A numerical simulation of different designs of heat exchangers allow engineers to gain insight into the flow and temperature fields, enabling them to select an optimal design for these devices. The objective of this paper is to analyze the flow and heat transfer in heat exchangers with staggered fins of different heights in three basic configurations, as shown in Figure 1. Configuration 1 is selected as a reference for comparison purposes with configurations 2 and 3. The inlet is located at the lower left side with the outlet at the upper right side. Both inlet and outlet have a width L. In configurations 2 and 3, the staggered fins are uniformly spaced. Three fin sizes, F, equal to 0.5L, 1.0L, and 1.5L are considered in the study. For a given flow rate, an increase in the fin height will certainly result in an increase of the required driving pressure, and usually, in an increase in the heat transfer rate. On the other hand, shorter fins and large separation between fins will reduce the residence time of fluid in the exchanger and, consequently, such an arrangement of fins in the heat exchanger will not significantly increase the heat transfer. In this study, the effects of different fin heights in configurations 2 and 3 on the heat transfer rate and pressure drop are analyzed for various Reynolds and Prandtl numbers.

A general discussion of flow in baffled heat exchangers is presented in the text by Knudsen and Katz [4]. Previous flow visualization for the effects of blaffled geometry has been conducted by Gunter, Sennstrom, and Kopp [5]. A related problem was solved by Kelkar and Patankar [6]. In their work, instead of the entire heat exchanger, the fluid flow and temperature in one period of staggered fin was analyzed. This situation represents a long heat exchanger where the flow can be considered mainly periodic. In the present study, heat and flow characteristics from inlet to outlet of a short heat exchanger are analyzed. This is particularly useful for compact heat exchangers where the flow in the exchanger could never become fully periodic. Berner, Durst and McEligot [7] found that for a Reynolds number of 600, and with fin heights close to 1.6L, the flow must pass over three to five baffles before it appears periodic. For shorter baffles this number can increase to seven or eight.

In the present work, the flow is considered laminar, and Reynolds numbers up to 500 are used. This assumption is acceptable according to the evidence provided by Berner, et.al. [7], which indicates that the flow is laminar and free of vortex shedding for Reynolds number of 600.

MATHEMATICAL FORMULATION

Governing Equations

In this study, the fluid is considered to be viscous, incompressible, with constant properties and the flow is laminar and two dimensional without gravitational effects. If the viscous dissipation effect is neglected in the energy equation, the general governing equations for predicting two dimensional incompressible, viscous flow and heat transfer in dimensionless form are

$$\frac{\partial u}{\partial x} + \frac{\partial v}{\partial y} = 0 \tag{1}$$

$$\frac{\partial u}{\partial t} + u\frac{\partial u}{\partial x} + v\frac{\partial u}{\partial y} = -\frac{\partial p}{\partial x} + \frac{1}{Re}\left[\frac{\partial^2 u}{\partial x^2} + \frac{\partial^2 u}{\partial y^2}\right] \tag{2}$$

$$\frac{\partial v}{\partial t} + u\frac{\partial v}{\partial x} + v\frac{\partial v}{\partial y} = -\frac{\partial p}{\partial y} + \frac{1}{Re}\left[\frac{\partial^2 v}{\partial x^2} + \frac{\partial^2 v}{\partial y^2}\right] \tag{3}$$

$$\frac{\partial \theta}{\partial t} + u\frac{\partial \theta}{\partial x} + v\frac{\partial \theta}{\partial y} = \frac{1}{Pe}\left[\frac{\partial^2 \theta}{\partial x^2} + \frac{\partial^2 \theta}{\partial y^2}\right] \tag{4}$$

The quantities u, v, p, θ, x, and y are, respectively, the dimensionless variables for X, Y, velocity components, pressure, temperature and space coordinates. They are made dimensionless by U_0, U_0, ρU_0^2, $(T-T_0)/(T_w-T_0)$, L, and L respectively, where L is the inlet and outlet width of the heat exchanger. U_0 is the maximum velocity of the inlet or reference velocity, ρ is the density of fluid, T_0 is the temperature of incoming fluid, T_w is the constant temperature of the walls and baffles, Pe is the Peclet number or RePr and Re is the Reynolds number or $U_0 L/\nu$.

In order to represent the heat transfer rate on the heated thin blocks, the local Nusselt number is defined by $Nu = hL/k$ with the local heat transfer coefficient $h = q/(T_w-T_0)$. The quantity q is the rate of local heat transfer per unit area dQ/dA. In this form the local Nusselt number adopts the form $Nu = \partial\theta/\partial v|_{wall}$. Where n is the dimensionless distance in the outward normal direction to the wall. The local Nusselt number, Nu, is thus proportional to the local heat transfer rate per unit surface area. The average Nusselt Number Nu_{av} is defined by $Nu_{av} = \int_A NudA/\int_A dA)$. Then the total heat transfer Q in the heat exchanger is $Q = h_{av} A(T_w - T_0) = \backslash I(A_,,qdA) = \frac{kA(T_w - T_0)}{L} Nu_{av}$, where A is the total surface area of the blocks and k the conductivity of the fluid.

Initial and Boundary Conditions

Three kinds of boundary conditions must be specified around the domain of computation shown in Figure 1. They are inlet, outlet and wall boundary conditions. At the inlet a parabolic profile for velocity and a uniform temperature are assumed. This is an approximate entry condition for a heat exchanger with a long upstream duct and the fluid has a uniform temperature. In the analysis, the maximum dimensionless velocity U/U_0 at the entrance is 1.0 and the dimensionless temperature θ equal to 0.0.

At the solid surface, the no slip condition is imposed for the velocity components u and v. Since a constant temperature on the wall is assumed, the dimensionless temperature at the wall is $\theta = 1.0$. The constant temperature on the surface of the fins is similar to assuming highly conducting fins. At the outlet boundary, continuous flow conditions are applied where gradients of u, v, p and θ in the flow direction are taken to be zero. This outlet condition is approximately equivalent to the conditions when the fluid exits from the heat exchanger into a large plenum. The initial conditions for u and v are taken to be zero or at rest while the initial fluid temperature is set to be zero everywhere or equal to the entry fluid temperature.

COMPUTATIONAL DETAILS

The governing equations, Eqs. (1) to (4), are solved numerically. To discretize the momentum and energy governing equations, Eqs. (2), (3) and (4), the FAM (Finite Analytic Method) [8], [9] was used. The central idea of the Finite Analytic Method is to obtain the algebraic representation of the governing equations (2), (3) and (4) by an analytic solution of the locally linearized governing equations in a small element. Using this analytic solution, the nodal points in the element are related to each other by an algebraic equation. Usually a central node is related to its eight neighboring nodes. The main advantage of this analytic solution is the proper representation of the convective and diffusive terms of the governing equations even in complicated flow patterns. A brief summary of the Finite Analytic Solution is given in the Appendix.

The continuity equation is used to solve for pressure in staggered grid coordinates. The SIMPLER procedure of Patankar [10] was used in the calculation. The system of algebraic equations is solved by the line by line iteration method. The time marching computation was stopped when a difference of the dependent variables between two time steps $\Delta t= 0.3$ was less than 0.0001. The flow and heat transfer is then considered to be steady state. All computations were performed on an 80 by 80 grid on an Apollo DN4000. A computation for Re=150, F = 1.0L, Pr = 4.0 was also done on a grid of 40x40. A change on the average Nusselt number of less than 5% between the two grid sizes was observed. As this change is small, the accuracy of the solution for a grid of 80x80 was considered satisfactory.

The effect of the Reynolds number on the flow field was studied for Re equals 100, 150, 300 and 500. Three fin heights, namely, F=0.5L, 1.0L and 1.5L were analyzed. For each computed velocity field, temperature profiles and heat transfer on the wall problems were solved for Prandtl numbers 0.7 and 4.0. In order to distinguish each case of computation, the geometry of the heat exchanger is coded by, for example, C3F1.5. Here, C3 denotes configuration 3, which has four staggered fins in the heat exchanger, and F1.5 denotes that the fin height is 1.5 times that of the inlet or exit height. The fins in the heat exchanger are installed in pairs and equally spaced, as shown in Figure 1. Three basic configurations are investigated, namely zero pair of fins (C1), one pair (C2), and two pairs (C3).

RESULTS AND DISCUSSION

Flow Field Analysis.

Figure 2 shows a series of dimensionless velocity vector plots for some of the configurations analyzed. The magnitude and direction of the velocity in each point is given by the vector size and direction at that point. Since the reference velocity is the maximum velocity U_0 at the inlet, its corresponding vector size at the center of the inlet is the reference size.

It is seen that when there is no fin in the heat exchanger, i.e., C1F0.0, the flow enters the heat exchanger without encountering much resistance until it reaches the exit end of the wall where the flow is reflected upward before it goes out of the outlet. The flow shown in Fig. 2, C1F0.0 reveals that there is not much motion at the upper left and lower right corners. Although there is recirculation at the corners, the strength of recirculation is relatively weak. Since the flow encounters small resistance and also does not need to go around any fin, the residence time for fluid in the heat exchanger is small. When fins are installed in the heat exchanger, such as two fins and four fins, the flow in these configurations is forced to change its path in the heat exchanger due to the presence of the fins. However, when the fin height is one half of the entry opening, i.e., C2F0.5, installations of fins in the heat exchanger do not affect or alter the flow pattern much when compared with the flow in the zero fin configuration. On the other hand, when the fin height is equal to the entry width F1.0, or greater F1.5, the flow pattern in the heat exchanger is significantly changed. The flow greatly accelerates near the narrow passage. The separation and recirculation are clearly seen. The recirculation strength is now much stronger than that in the heat exchanger without fins.

Figure 3 shows the profiles for the components u and v of velocity at different equally spaced locations. As in Fig. 2, the reference magnitude of u=1.0 is given at the maximum value of the profile of u at the inlet location. A careful examination of Fig. 3, C2F1.5 and C3F1.5 configurations, reveal that the flow does separate at the top of the second fin. There, the velocity profile of the u component shows a small reverse flow at the fin facing downward.

It should be pointed out that the boundary condition at the outlet is such that the gradients of u, v, θ are set equal to zero to approximately simulate that the flow exit to a large ambient.

Local Heat Transfer Analysis.

In order to study the heat transfer characteristics of various configurations of the finned heat exchanger, the wall and fin temperatures in the heat exchanger are set to a constant temperature Tw, which is different from the inlet fluid temperature of To. The heat transfer analysis is then concentrated on the effect of the flow parameter, Reynolds numbers varying from 100 to 500, the fluid property, Prandtl numbers from 0.7 (gas) and 4.0 (water with an average temperature of 40°C), the number of staggered fins 0, 2 and 4, and the fin height to the entry width of 0, 0.5, 1.0 and 1.5.

Figure 4 shows the isotherms corresponding to dimensionless temperatures of 0.1, 0.5, and 0.75 for several configurations at the flow conditions for Reynolds number of 150 and Prandtl number of 0.7. Since, in dimensionless plot, the inlet fluid temperature is zero while the wall and fin temperatures in the heat exchanger are set equal to one, the isotherm with dimensionless temperature 0.75 is closest to the walls.

Examination of the isotherm $\theta = 0.1$ reveals that fluid in heat exchangers with configurations of C1F0.0 and C3F0.5 flows through the exchanger without acquiring much energy from the heat exchanger and exits with a large portion of fluid at a temperature of $\theta=0.1$ or less. This immediately leads to the conclusion that the heat exchangers with fin heights smaller than the entry width are not effective heat exchangers, even though the pressure drop between the inlet and outlet may be small compared to the heat exchangers with higher fins. The remaining configurations with higher fins all show that sufficient heat transfer is made between the fluid and heated walls and fins so as to raise the fluid temperature at the exit higher than 0.1. Judging from the fluid exit temperature, it seems that the heat exchangers with configurations C3F1.0 and C3F1.5 are the most effective heat transfer devices in the seven configurations studied.

Figure 5 shows the local Nusselt number for all seven configurations when the flow is computed for Re = 150 and Pr= 0.7. The magnitude of local Nusselt number is plotted normal to the wall. The maximum Nusselt number or heat transfer occurs at the inlet wall due to entrance effect. It is interesting to see that the local Nusselt number at each fin tip facing the upstream flow has relatively high value, signifying much heat transfer is produced there. On the other hand, the heat transfer at the base of the fin and corners of the heat exchanger is quite small and indeed has minimal contribution to the heat exchange between the fluid and the wall. This is because the flow at the base of the fin or corner is not only very slow, but also is within the recirculating zone. Therefore, the heat cannot be easily carried away by the flow and out of the heat exchanger in the convection mode.

The heat transfer from the wall of the heat exchanger depends greatly on how the fins are arranged and the choice of the fin height. The configurations with the fin height of one-half of the entry width, C2F0.5 and C3F0.5, are not very effective in producing much heat transfer from the wall of heat exchanger, other than at the tip of the fins. It is also seen that configurations C2F1.5 and C3F1.5 are quite effective in channeling the fluid around the fin so as to create good convective heat transfer from the wall when the fluid is accelerated around the fin. The local Nusselt number on the wall may be as high as that at the tip of the fin.

Overall Heat Transfer

While the local heat transfer analysis may reveal details of physics as to why and where the heat transfer is low or high, it is important to the design and application of heat exchangers that the overall heat transfer is known for a given configuration of heat exchanger. The total heat transfer Q in the present case is given by equation (9), i.e., $Q = h_{av} A(T_w - T_o) = \int_A q dA = kA(T_w - T_o), Nu_{av}/L$ where A is the total surface area of the heat exchanger, including the walls and fins, k is the thermal conductivity, and Nu_{av} is the average Nusselt number.

The average Nusselt number for each configuration studied in the present investigation is given, respectively, in Figs. 6, 7 and 8, for Reynolds numbers from 100 to 1000 and Prandtl numbers 0.7 and 4.0. Figure 6 gives the average Nusselt number versus Reynolds number for heat exchanger of C1F0.0. This is the configuration without fins in the heat exchanger. The average Nusselt number increases gradually with Reynolds numbers, and is higher for higher value of Prandtl number. Figure 7 shows the average Nusselt number versus Reynolds numbers for two fin heat exchangers at different fin heights of F=0.5L, 1.0L, and 1.5L, where L is the entry width. It is seen that when the fin height is one-half of the entry width, F=0.5L, the Nusselt number at high Reynolds number may indeed become less than that of heat exchanger without fins, as shown in Fig. 6. However, this does not mean that the finned heat exchanger has a smaller total heat transfer. Indeed, the finned heat exchanger may still have a larger overall heat transfer because the total surface area is larger for the finned heat exchanger than that without fins. However, it reveals that with short fin height such as that of F=0.5L, the heat transfer efficiency is not good.

As an example, the heat exchanger with configuration C2F1.5 gives Nu_{av} =10.3 for Re=500 and Pr=4.0. This value of 10.3 is about twice the

average Nusselt number obtained from heat exchanger C1F0.0. Since the total surface area for C2F1.5 is 1.5 times of C1F0.0, the net heat transfer rate for C2F1.5 is close to three times of C1F0.0. For the same Prandtl and Reynolds numbers, but with four fins, the heat exchanger with C3F1.5 has an average Nusselt number 1.9 times that of C1F0.0 and a total heat transfer of 3.7 times of C1F0.0. Since the heat transfer capability for C3F1.5 and C2F1.5, with reference to the base configuration of C1F0.0, is respectively, 3.7 times for C3F1.5 and 3.0 times for C2F1.5, one may conclude that the configuration C2F1.5 is perhaps a more efficient heat exchanger in light of fewer number of fins being needed and less pressure drop created, which will be considered next.

Pressure Drop

The pressure drop across the heat exchanger is also an important factor in the design and selection of heat exchangers since it specifies the power required to move the fluid past the exchanger. The pressure drop for all seven configurations studied is plotted in Fig. 9. The dimensionless pressure drop is the difference between the inlet and outlet pressures divided by ρU_0^2, where U_0 is the inlet maximum velocity. Figure 9 shows that the pressure drop increases with the number of fins installed in the heat exchanger and also increases with the increases of the fin height. The pressure drop can increase almost tenfold from C1F0.0 to C3F1.5. Obviously, the presence of fins in the heat exchanger compels the fluid to go around the fins. As a result, it creates a substantial pressure drop. However, the benefit from consuming the pressure drop is the increase in heat transfer. For example, in configuration C3F1.5 with Re=100 and Pr= 4.0, the increase in pressure drop is 59 times that of C1F0.0, while for Re=500 and Pr=4.0, the pressure drop is 160 times.

CONCLUSIONS

Finite analytic numerical solutions are obtained for the flow and heat transfer in seven configurations of heat exchangers. The fins in the heat exchanger are arranged in staggered fashions of zero, two, and four fins, with fin heights of 0.5, 1.0, and 1.5 of the entry height. The fluid flows simulated are for air (Pr=0.7) and water (Pr=4.0) and Reynolds numbers from 100 to 1500. It is found that substantial increase in heat transfer can be obtained for a given size heat exchanger when fins are inserted in the exchanger. For the present investigation, the fin height of 1.5L is the most effective in increasing the heat transfer from the fins, as well as from walls of the heat exchanger. However, the penalty of increasing heat transfer by inserting fins is the increase of pressure drop for an order of magnitude. The proper selection of heat exchanger configuration will eventually depend on the cost of energy transfer and the cost of power required to create the necessary pressure drop.

APPENDIX

Finite Analytic Method

When the problem is well posed with the governing equations and appropriate boundary and initial conditions, the numerical solution can be obtained by a proper numerical method. In the present study, the Finite Analytic Method [8, 9] is used to obtain algebraic representation of the governing equations, Eqs. (2), (3) and (4). In the Finite Analytic Method the momentum and energy equations can be arranged as, for example,

$$\phi_{xx} + \phi_{yy} = R_e (\phi_t + u\phi_x + v\phi_y) + s \qquad (A1)$$

where ϕ is the dependent variable either u or v, Re is the Reynolds number, and s is the respective pressure gradient term. The central idea of the Finite Analytic Method is to obtain the algebraic representation of the governing Eq. (A1) by analytic solution of locally linearized governing equation in a small element. To derive the algebraic representation for Eq. (A1) with the Finite Analytic Method, the solution domain is first decomposed into many small elements. A typical element 2h x 2k, in which an interior node P is surrounded by eight boundary nodal points. Here, h=Δx and k=Δy.

The convective transport Eq. (A1) can be locally linearized by approximating the convective velocities u, v and the source term s with u_p, v_p and s_p which are constants over the small element and evaluated at node P. The unsteady term may be replaced by a backward finite difference formula between the time steps t^n and t^{n-1}, $\phi_t^n = (\phi_P^n - \phi_P^{n-1})/\Delta t$, where ϕ_P^{n-1} is the initial condition for the element.

Under these approximations, Eq. (A1) is reduced to steady like convective transport equation and can be written as

$$\phi_{xx} + \phi_{yy} = 2A\,\phi_x + 2B\,\phi_y + g \qquad (A2)$$

$$A = .5Re\,u_P, \quad B = .5Re\,v_P, \quad g = Re\,(\phi_P^n - \phi_P^{n-1}/\Delta t + s_P = \text{constant}$$

For Eq. (A2) to be well posed in the local element, 2h x 2k and a time interval Δt, the boundary conditions on north, south, east, and west sides of the local elements must be specified. The three nodal points at each side of the local element and suitable interpolation functions may be used to approximate the boundary conditions. In the present study, the functions used to approximate boundary conditions at t^n are, a constant, a linear and an exponential form [11]. For example, the south boundary function ϕ_s is specified as

$$\phi_s = a_S(\exp(2AX)-1) + b_S X + c_S \qquad (A3)$$

where ϕ_s is the boundary profile for the south boundary.

$$a_S = \frac{\phi_{SE} + \phi_{SW} - 2\phi_{SC}}{4\sinh^2 Ah}$$

$$b_S = \frac{1}{2h}\left[\phi_{SE} - \phi_{SW} - \coth Ah(\phi_{SE} + \phi_{SW} - 2\phi_{SC})\right]$$

$$c_S = \phi_{SC}$$

A similar function is used for the boundary functions, ϕ_E, ϕ_w, and ϕ_N, on the east, west and north sides.

Eq.(A2) is solved analytically by the method of separation of variable with boundary condition, Eq. (A3), expressed as the combinations of linear and exponential functions involving the nodal values. Evaluation of the analytic solution at the center node, ϕ_P, then provides a nine point algebraic discretization formula.

$$\phi_P = \frac{\left(\sum_{n=1}^{8} C_{nb}\phi_{nb} + C_P\,\dfrac{Re}{\Delta t}\phi_P^{n-1} - C_P s_P\right)}{(1 + C_P Re/\Delta t)} \qquad (A4)$$

where

$$C_{EC} = \frac{\exp(-Ah)}{2\cosh(Ah)}P_{B'} \qquad C_{WC} = \frac{\exp(Ah)}{2\cosh(Ah)}P_{B'}$$

$$C_{NC} = \frac{\exp(-Bk)}{2\cosh(Bk)}P_{A'} \qquad C_{SC} = \frac{\exp(Bk)}{2\cosh(Bk)}P_{A'}$$

$$C_{SW} = \frac{\exp(Ah+Bk)}{4\cosh(Ah)\cosh(Bk)}(1-P_A-P_B)$$

$$C_{SE} = \frac{\exp(-Ah+Bk)}{4\cosh(Ah)\cosh(Bk)}(1-P_A-P_B)$$

$$C_{NW} = \frac{\exp(Ah-Bk)}{4\cosh(Ah)\cosh(Bk)}(1-P_A-P_B)$$

$$C_{NE} = \frac{\exp(-Ah-Bk)}{4\cosh(Ah)\cosh(Bk)}(1-P_A-P_B)$$

$$C_P = \frac{h\tanh(Ah)}{2A}(1-P_A) = \frac{k\tanh(Bk)}{2B}(1-P_B)$$

$$P_A = 4Ah\cosh(Ah)\cosh(Bk)\coth(Ah)\,E_2$$

$$P_B = 1 + \frac{Bh\coth(Bk)}{Ak\coth(Ah)}\bullet(P_A-1)$$

and

$$E_2 = \sum_{m=1}^{\infty}\frac{-(-1)^m \lambda_m h}{[(Ah)^2 + (\lambda_m h)^2]^2 \cosh\sqrt{(A^2 + B^2 + \lambda_m^2)k^2}}$$

For large cell Reynolds numbers, i.e., Reh or Rek >100, the series summation in E_2 can be expressed by the following asymptotic expressions of P_A and P_B based on the theory of characteristics such that

For Ak coth Ah > Bh coth Bk,

$$P_A = 0, \quad P_B = 1 - (Bh\coth Bk)/(Ak\coth Ah)$$

For Ak coth Ah < Bh coth Bk,

$$P_B = 0, \quad P_A = 1 - (Ak\coth Ah)/(Bh\coth Bk)$$

Details for solution procedure for uniform and nonuniform grid spacing are given by Chen [9].

Algebraic Pressure Equation

To solve the Navier-Stokes equations, Eqs. (1), (2), and (3), with primitive variable approach, the continuity equation, Eq.(1), must first be converted into the equation for pressure variable. Following Patankar's [10] control volume formu-lation, the staggered arrangement for continuity equation, Eq (1), is used to derive the following equation for the pressure.

$$B_P P_P = B_E P_E + B_W P_W + B_N P_N + B_S P_S + D \qquad (A5)$$

where

$$B_E = \frac{ReC_{pE}}{(1+\dfrac{DC_{pE}}{\Delta t}h)} \qquad\qquad B_N = \frac{ReC_{pN}}{(1+\dfrac{DC_{pN}}{\Delta t}k)}$$

$$B_W = \frac{ReC_{pw}}{(1+\dfrac{DC_{pw}}{\Delta t}h)} \qquad\qquad B_s = \frac{ReC_{ps}}{(1+\dfrac{DC_{ps}}{\Delta t}k)}$$

$$B_P = B_E + B_W + B_N + B_S \qquad\qquad D = \frac{\tilde{u}_w - \tilde{u}_E}{\Delta x} + \frac{\tilde{v}_N - \tilde{v}_s}{\Delta y}$$

$\tilde{u}$, $\tilde{v}$ are the pseudo velocities and can be derived from Eq.(A4) by rewriting Eq.(A4) as

$$u_E = \tilde{u}_e - d_E(P_E - P_P), \qquad\qquad v_N = \tilde{v}_N - d_N(P_N - P_P)$$

with

$$\tilde{u}_E = \left(\sum_{n=1}^{8} C_{nb}u_{nb} + C_{pE}\,Re\,u_E^{n-1}/\Delta t\right)/(1+C_{pE}Re/\Delta t)$$

$$\tilde{v}_N = \left(\sum_{n=1}^{8} C_{nb}v_{nb} + C_{pN}\,Re\,v_N^{n-1}/\Delta t\right)/(1+C_{pN}Re/\Delta t)$$

$$d_N = \frac{C_{pN}}{(1+C_{pN}Re/\Delta t)k}, \qquad\qquad d_E = \frac{C_{pE}}{(1+C_{pE}Re/\Delta t)h}$$

Eqs.(A4) and (A5) can be solved for u, v, and p with proper boundary and initial conditions.

REFERENCES

1. Kays,W.M., London, H.L., Compact Heat Exchangers, McGraw-Hill, N.Y. 1984.

2. Taborek,J., Hewitt, G.F., Afgan, N., Heat Exchangers, McGraw-Hill, N.Y. 1983.

3. Kakac, S., Shah, R.K., Bergles, A.E., Low Reynolds Number Flow Heat Exchangers, Springer-Verlag, 1983.

4. Knudsen, J.G., and Katz, D.L., Fluid Dynamics and Heat Transfer, McGraw-Hill, New York, 1958.

5. Gunter, A.Y., Sennstrom, H.R., and Kopp, S., A Study of Flow Patterns in Baffled Heat Exchangers, <u>ASME Paper 47-A-103</u>, 1947.

6. Kelkar,K.M., Patankar,S.V., "Numerical Prediction of Flow and Heat Transfer in a Parallel Plate Channel With Staggered Fins". <u>ASME Journal of Heat Transfer</u>, Vol. 109, February 1987, pp. 25-30.

7. Berner, C., Durst, F., and McEligot, D.M., "Flow Around Baffles", <u>ASME Journal of Heat Transfer</u>, Vol. 106, 1984, pp. 743-749.

8. Chen,C.J., Bravo,R.H., Haik,Y.S., Sheikhoslelami, Z.M., "Numerical Flow Visualization of Two Dimensional Viscous Flow In Complex Internal Geometries", ASME Winter Annual Meeting, Nov 27- Dec 2, 1988, Chicago. FED-Vol.74, pp.1-9.

9. Chen,C.J. "Finite Analytic Method", Chapter 17, <u>Handbook of Numerical Heat Transfer.</u> Minkowycz,W.J., Sparrow,E.M., Pletcher,R.H., Schneider,G.E., Editors, John Wiley & Sons, Inc., 1988, pp. 723-746.

10. Patankar, S.V., <u>Numerical Heat Transfer and Fluid Flow</u>, McGraw-Hill, NY, 1980.

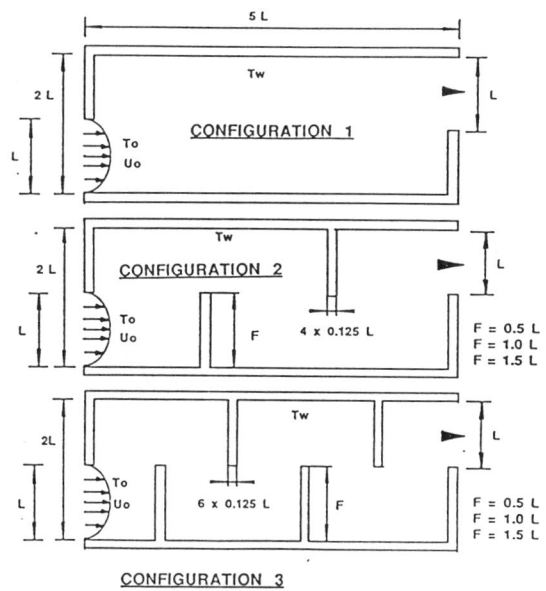

Figure 1   Heat Exchanger Geometries and Boundary Conditions

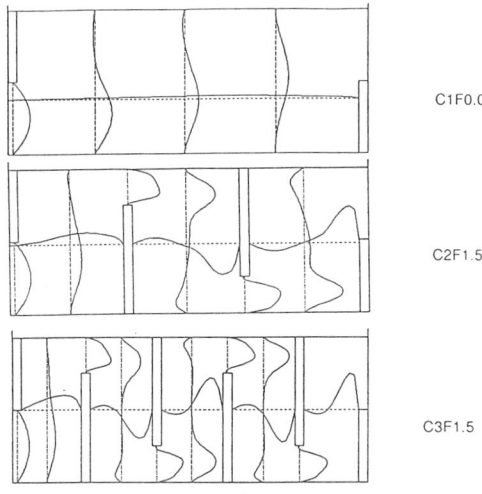

Figure 3   Velocity Profiles at Selected Locations for Re = 150

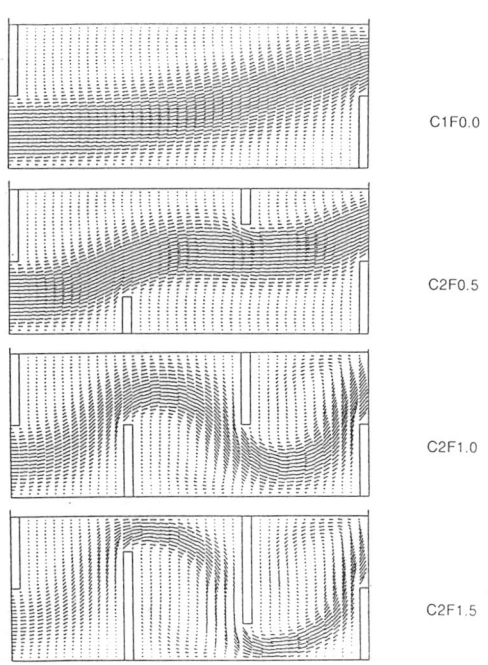

Figure 2   Velocity Vectors for Re = 150

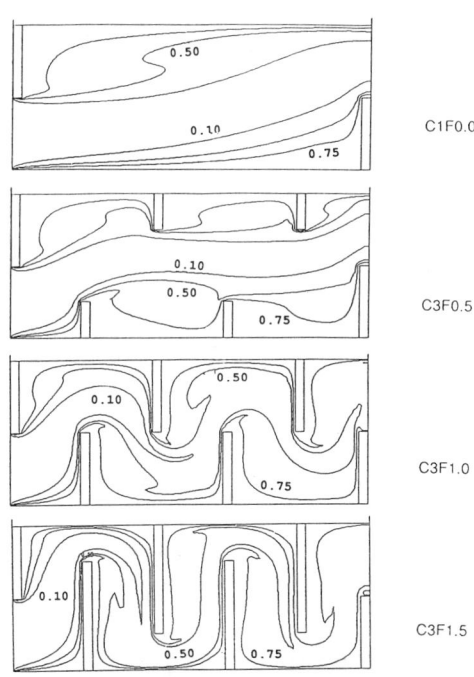

Figure 4   Isotherms for Dimensionless Temperature
$\theta = 0.1, 0.5, 0.75$ for Re=150 and Pr = 0.7

189

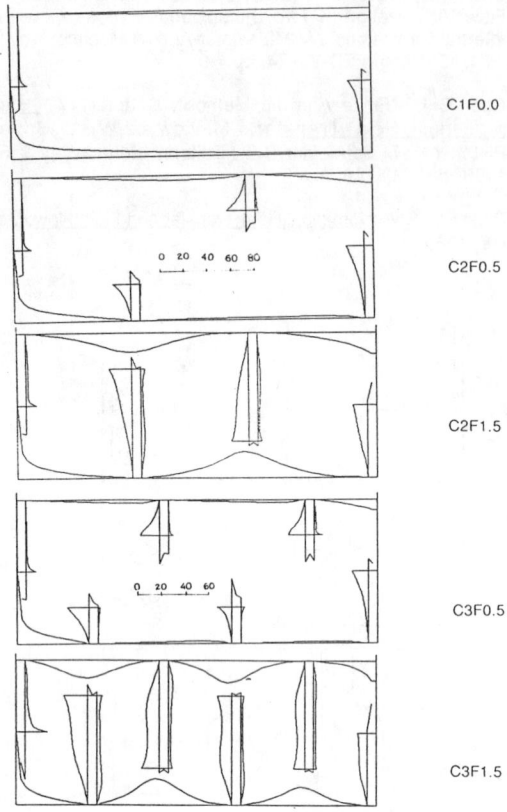

Figure 5   Local Nusselt Number for Re = 150 and Pr = 0.7

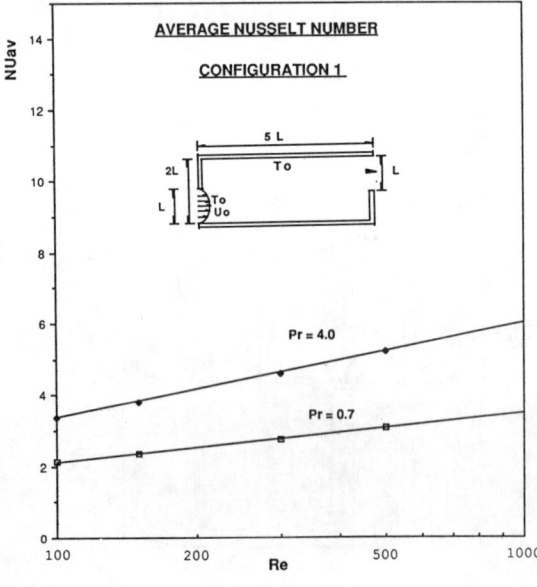

Figure 6   Average Nusselt Number for Configuration 1

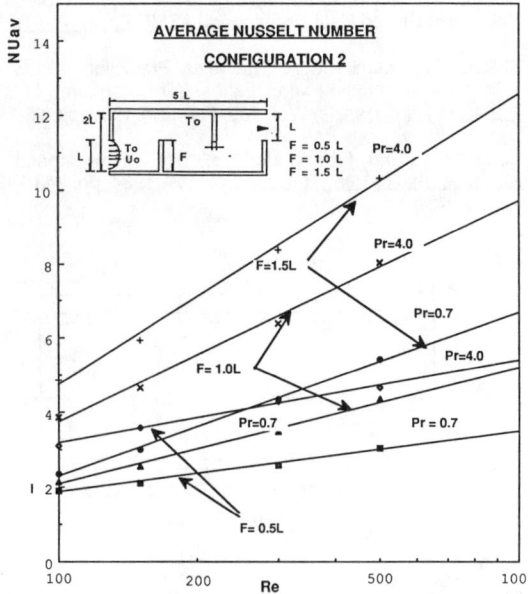

Figure 7   Average Nusselt Number for Configuration 2

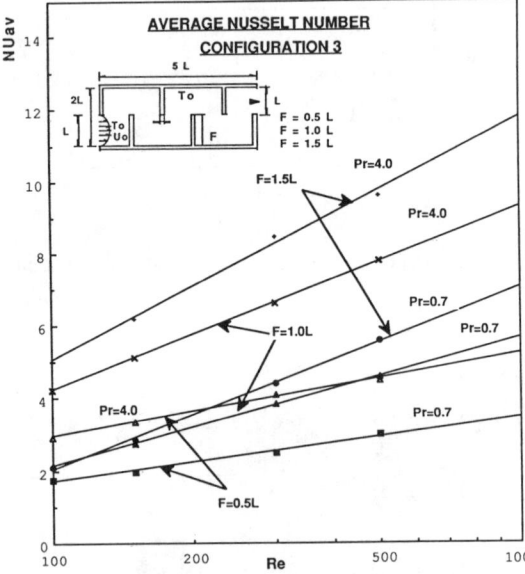

Figure 8   Average Nusselt Number for Configuration 3

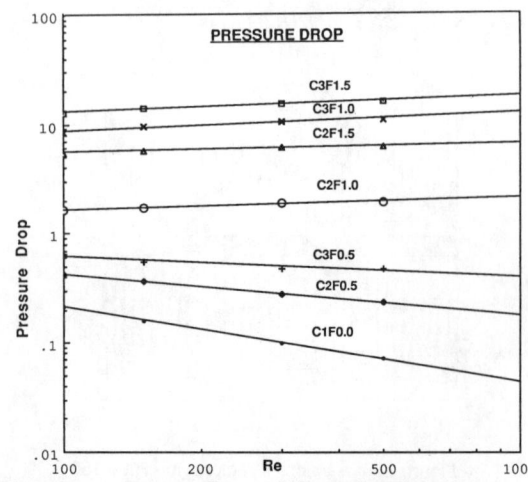

Figure 9   Pressure Drop at Different Reynolds Numbers

190

1989 National Heat Transfer Conference
HTD-Vol. 107, Heat Transfer in Convective Flows

TURBULENT FLOW CHARACTERISTICS IN A RECTANGULAR CHANNEL WITH REPEATED RIB ROUGHNESS

H. Sato, K. Hishida, and M. Maeda
Department of Mechanical Engineering
Keio University
Yokohama, Japan

Abstract

The turbulent flow structure over a repeated-rib geometry rough-walled surface was investigated in connection with a two dimensional square rib roughness which was set on the internal surface of a rectangular channel. The upper plate with ribs of the same pitch can be shifted to arbitrary positions. Reynolds number of duct flow was set to 2.0×10^4, which was similar to the condition of practical heat exchangers. Flow measurements were performed using a compact fiber laser Doppler velocimeter (LDV) probe. The results of the detailed flow measurements indicated that a recirculating flow existed behind the rib on symmetric and staggered rib geometries. A remarkable large peak value of Reynolds shear stress was observed behind the rib on the staggered rib geometry.

Nomenclature

$2d$ = distance between plates
d_h = hydraulic diameter
e = rib height
f = friction factor = $\{(p/L)d_h\}/(U_m^2/2)$
L = length of test section
p = rib pitch (mm)
ΔP = pressure drop
Re = Reynolds number = $U_m d/\nu$
U = mean velocity component in X direction
U_m = average air velocity; bulk velocity (m/s)
u' = velocity fluctuation in X direction
V = mean velocity component in X direction
v' = velocity fluctuation in Y direction

Greek symbols

ν = kinematic viscosity (m²/s)
ρ = fluid density (kg/m³)

1 Introduction

The development for the heat transfer have been carried out using artificial wall configurations such as sand grains or regular geometric roughness elements on the surface. However, the increase in heat transfer is accompanied by an increase in the resistance to fluid flow. Many investigators have studied this problem in an attempt to develop accurate predictions of the behavior of a given roughness geometry which gives a better heat transfer performance for a given flow friction.

The early study of this field was performed by Nikuradse (1933) who conducted a series of experiments with circular pipes roughened by sand grains. Heat transfer inside the tubes are enhanced by turbulence promoters on the internal surface of the tubes, and various types of the turbulence promoters are examined in previous studies. The heat transfer performance of tubes with fins on the internal surface were examined by Carnavos (1979). A number of friction and heat transfer measurements have been performed for helically rib roughened tubes. Gee and Webb (1980) indicated that helical rib roughness yielded greater heat transfer than transverse ribs, and the preferred helix angle was approximately 50 degree. Withers (1980a, 1980b) performed experimental studies on single and multiple helix tubes, and correlations were obtained by using a heat-momentum analogy, which enabled both types of rib to be handled in a common frame work. Li et al. (1982) investigated friction factors and heat transfer of helical ribbed tubes, and obtained a general correlations for the friction and Stanton number. In addition, separation and spiral flow were found in tubes by flow visualization, and the promotion of separated flow was of more benefit than that of spiral flow.

Webb et al. (1971) performed experiments on tubes having a repeated rib roughness. The data of Webb covered a wide range of rib height to hydraulic diameter (e/d_h), and a pitch to height ratio ranged from 10 to 40. The friction correlation was extended to a wider range of e/d_h. Han et al.

(1978) examined a heat transfer and friction of the rib roughened surface with a parallel plate geometry, i.e. a two dimensional geometry which simulated a rib roughened tube. The experiments were performed not only with a symmetric rib, but a staggered rib, and various cross-sections of rib and flow attack angles. The experimental results showed that the rib cross-section and flow attack angle had more effect on the friction factor than the rib geometries. Those series of works on enhancement of heat transfer were summarized by Nakayama (1982).

In the area of a practical heat exchanger, the two dimensional ribbed tubes were often used to upgrade the performance. Nakayama et al. (1983) investigated the heat transfer and friction factor of the tube having two dimensional spiral ribbing and indicated the best performance for the reduction of heat exchanger volume when the pumping power was held constant.

Although the turbulent heat transfer and friction in tubes with repeated rib roughness or another type of turbulent promoter has been investigated extensively, most of the investigators have concentrated on overall heat transfer or friction factor. The flow measurements, however, have hardly been performed for complex configurations of the ribbed surface involving a recirculating flow. A conventional measuring technique such as a hot wire anemometry could hardly be applied to those flow field. A laser Doppler velocimeter is powerful instrument for such flow configurations. However, for simultaneous measurements of two component velocity, the ordinary three or four beam LDV optical systems need huge traversing devices for their weight, and have a difficulty for the fine traverse of the optical unit. To avoid this kind of difficulties, a compact LDV optical probe was developed by the authors' grope (Maeda et al., 1985, Sato et al. 1985) and has been applied for the measurement of the flow field around the ribs.

Most of the previous investigations were performed for high Reynolds numbers, of 10^4-10^6, and with a small ratio of rib height to hydraulic diameter, between 10^{-2} and 10^{-3}. However, a few experimental data for a lower Reynolds number and a large ratio of rib height to hydraulic diameter were obtained. In addition, when the rib height to hydraulic diameter ratio is large, the rib location in the opposite side may influence the flow field between the ribs. The detailed information of such flow conditions are more useful to clarify the mechanism of heat transfer and turbulent flow characteristics, and the data of the detailed flow pattern around the rib have been needed in the fundamental data for the computational analysis.

In the present investigation, a two color, four beam fiber LDV has been applied to the two component velocity measurement around two dimensional ribs in a rectangular channel at a Reynolds number of the channel of 2.0×10^4. The relative position of each ribbed surface is set at 0, $p/4$ and $p/2$ and the influence of each ribbed surface on turbulent flow characteristics are discussed with experimental results.

2 Experimental Apparatus and Measuring System

2.1 Experimental Apparatus. Figure 1 shows a schematic drawing of the experimental apparatus. The air of room temperature and pressure was sucked into the rectangular stilling section of 850 mm long, 320 mm high and 480 mm

wide in streamwise, transverse and spanwise direction. The test section has the size of 780 mm long, 250 mm high and 50 mm wide and a rectangular form where the ribbed plates are set on both top and bottom sides. The stilling section and a contraction nozzle were used to insure that the air entering the test section had a uniform velocity profile. At the end of the test section, the air was exhausted by a blower driven by a servo motor.

The measuring point was 510 mm (about 7 pitches) from the inlet of test section. The glass with an optical flat surface was set on the side wall of measuring area in order to ensure optical access for the application of optical velocity measurement. The atomized fine oil particles of $d_p < 1\,\mu$m were seeded at the inlet as tracer particles.

Figure 2 illustrates a geometry of the test section which consists of two parallel ribbed plates of acrylate resins. The ribs with 10×10 mm square cross section were arranged on the plate in 70 mm rib pitch, and the ribbed plates were set on bottom- and top-walls spaced in 50 mm. The test channel width is 250 mm ensuring the two-dimensional condition so that the shear layer thickness of the lateral walls can be ignored in the middle of the channel. The top side plate could be shifted to an arbitrary location and the relative position of opposite ribbed plates was changed between 0 to 35 mm (a half pitch). The relative position of the ribbed surface was set at 0 (Geometry I; symmetry), $p/2$ (Geometry II; staggered) and $p/4$ (Geometry III; unsymmetry).

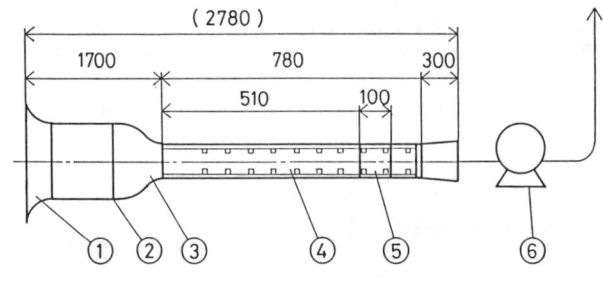

① Bell Mouth　　　④ Test Section
② Stilling Chamber　⑤ Measuring Section
③ Contraction Nozzle ⑥ Blower

Fig. 1 Schematic diagram of the experimental apparatus

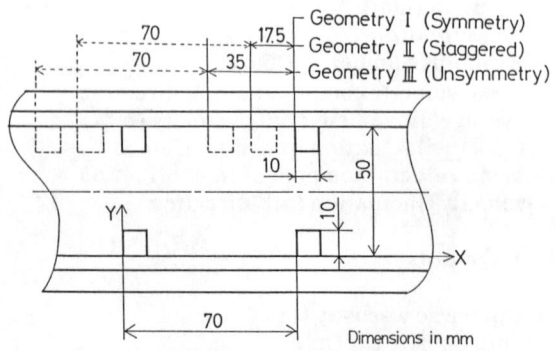

Fig. 2 Rib geometry

192

Arrangement of the pressure taps is shown in Figure 3. Pressure taps for static pressure at the wall were located on the bottom side plate. Reynolds number based on the bulk velocity and distance between the ribbed plates was set at 2.0×10^4 for all experimental conditions, where the mean air velocity in the duct was correspondingly set to 6.00 m/s.

2.2 Measuring System. The flow measurement around the rib has a difficulty in traversing the measuring points with the conventional three or four beam LDV optical systems, therefore the measurements were performed by use of an Ar-Ion four beam Fiber LDV optical probe developed by authors (Sato et al. 1985). Figure 4 illustrates an arrangement of optical modules of the two-color four-beam fiber LDV system. The optical probe head, of which dimensions are $40 \times 40 \times 200$ mm and weight is only 0.8 kg, is easier to position the measuring point precisely using a small traversing unit. Two component velocity signals from the photomultipliers were processed by counter type signal processors checking the coincidence by Doppler envelope pulses and the data were transferred to a micro computer. Mean velocities, turbulence intensities and Reynolds shear stresses were calculated using 4000-10000 data with the ϵ-σ data reduction method for two component velocity data proposed by Ueda et al. (1982).

The maximum uncertainty in the static pressure data was estimated to be 1.0% for the 95% confidence interval. Generally, for the velocity measurements in turbulent flow, there must be a bias error on mean velocity data. For the present study, the velocity data are checked by a long range of sampling time with seeding the tracer particle constantly in order to avoid the biasing error. As a result of the repeated measurements, the resultant error for the normalized mean velocity profiles was estimated to be 5.0% for the 95% confidence interval.

3 Experimental Results and Discussions

3.1 Pressure Drops. Figure 5 illustrates the static pressure on the bottom side surface measured around the rib of 8 pitches from the inlet. The static pressure at $X/e = 0$ was measured on the side wall of the duct. Pressure drops are presented in terms of the difference from the static pressure at $X/e = 0$ and normalized by dynamic pressure based on the bulk velocity. Behind the rib ($3 < X/e$), the pressure show a smaller value for Geometry III. Pressure gradient (dP/dX) was almost similar for all rib geometry on $3 < X/e$. On the rib ($0 < X/e < 1$), the sudden drop in pressure, which may be due to a separation bubble on the rib, was shown on the top of the rib at the leading edge. The pressure recovers as the flow reattaches to the top rib surface. Friction factor, estimated by total pressure losses ΔP at $Re = 2.0 \times 10^4$, were 0.763, 0.697 and 0.637, corresponding to symmetric rib, staggered rib and 0.25 pitch shifted unsymmetric rib. Those friction factors were exceedingly larger than the friction factor of a smooth tube ($f = 0.0266$).

3.2 Mean Velocity. Detailed velocity measurements were performed along eight vertical lines in the central plane. The variations of the normalized mean velocity profile of the U component around the rib at 510 mm (7 pitches) downstream from the inlet is shown in Figure 6. A broken line on each figure denotes a dividing streamline calculated by U component velocity profile of the measured plane.

For Geometry I (Fig. 6 (a)), the flow detaches at the rib edges. Behind the rib a recirculation (separation babble) arises, which is stretched downstream until the reattachment point. For Geometry II (Fig. 6 (b)), the flow pattern is almost similar to that of symmetric ribs. On the symmetric ribs, however, the free stream is accelerated over the rib and has a larger value than that over the staggered ribs. Because the

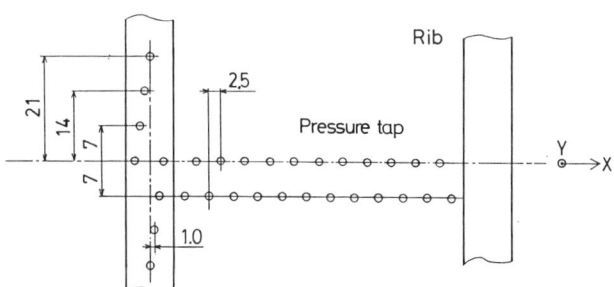

Fig. 3 Configuration of pressure taps

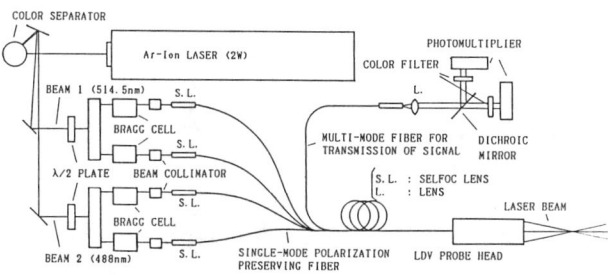

Fig. 4 Optical arrangement of four beam fiber LDV system

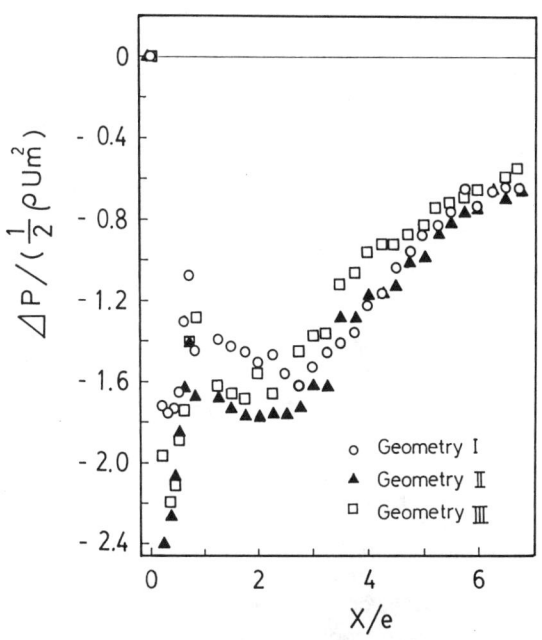

Fig. 5 Static pressure distribution

cross sectional area of the channel at $X/e=0$ for Geometry I was 25% smaller than that of Geometry II, and the acceleration is caused by this reduction of cross sectional area of the duct. For Geometry III (Fig. 6 (c)), the flow pattern around the top side ribbed plate has almost the same as the other two types of geometry, except the recirculating region is stretched downstream. On the contrary, the flow pattern around the bottom side rib shows a remarkable difference, i.e. a reattachment does not occur behind the rib. On the top of the rib, a separation bubble was not defined clearly by the velocity profile on $X/e=0.5$ for all rib geometries. The separation bubble in front of each rib can hardly be defined based on the results of the mean flow profile. These results indicate that the influence to the flow by the opposite rib exists more or less in each rib geometry. When considering the V component velocity, the effect of the opposite rib are made more clear. V component velocity profiles are shown in Fig. 7. In the recirculation region, a remarkable difference can not be seen on each rib geometry. Upward flow existed in the recirculation region just behind the rib (up to $X/e=2$). Near the center line ($Y/d=1.0$), V component velocities were different for each geometry. For Geometry I, the free stream flows parallel to the horizontal line (Fig 7 (a)). On the other hand, for Geometry II (Fig. 7 (b)), the free stream over the rib ($-0.5<X/e<1.5$) flows slightly upward, and the free stream between the rib ($2<X/e<4$) flows downward and is influenced by the rib of the opposite wall. On Geometry III (Fig 7 (c)), an upward

flow is observed on $0<X/e<2$, downward flow on $4<X/e<5.25$. For Geometry III, the position of the opposite rib affects significantly the recirculating zone in bottom and top plates.

3.3 Velocity Fluctuations. Turbulent fluctuations of the U and V components normalized by U_m are illustrated in Figure 8. For Geometry I, the peak of the turbulence level of U component was generated on the rib edge due to a separation at the trailing edge, and the peak spread and diffused to the center of the duct with increasing X/e. The turbulence levels of V component show smaller value than those of U component for whole range of X/e. However, the profile of V component has a peak value on each measuring cross section behind the rib, while the fluctuations in Y direction decreases in the near wall region. On Geometry II, the turbulence level in U component generated in the shear layer has a peak on every measured plain (Fig. 8 (b)). In the main flow region ($Y/d>0.5$) the turbulence level of U component have smaller values than those of symmetric rib. And the values of V component are almost the same as U component. For Geometry III (Fig. 8(c)), the profiles of the turbulence level are almost same as those of the staggered rib. These results indicate that the acceleration of the main flow velocity over the rib generated the turbulence fluctuations in U component, on the other hand the V component fluctuations are suppressed on the symmetrical rib geometry.

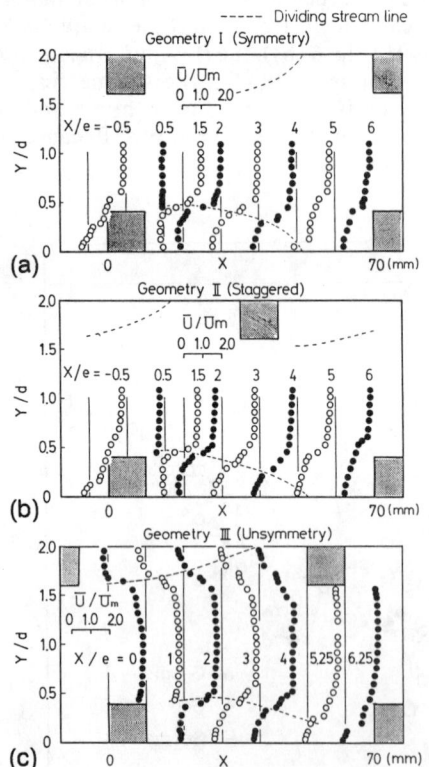

Fig. 6 Mean velocity profile of U component

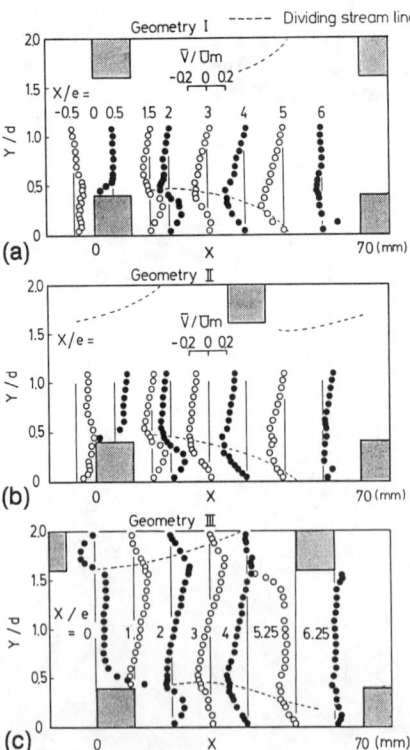

Fig. 7 Mean velocity profile of V component

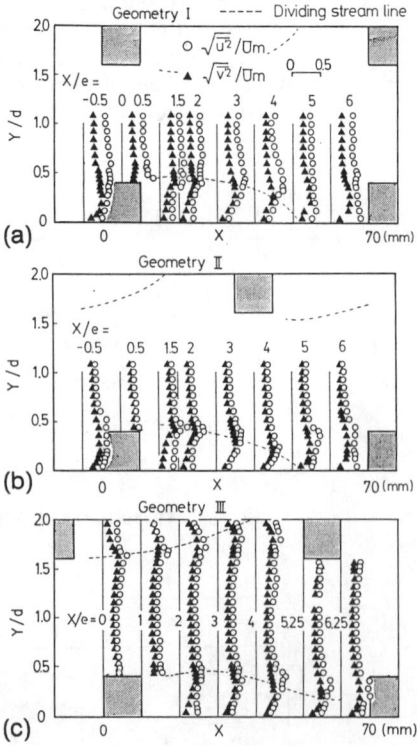

Fig. 8 Turbulence intensities of U and V component

3.4 Turbulent Shear Stress. Figure 9 denotes the Reynolds shear stress ($-\overline{u'v'}$) normalized by the square of bulk velocity (U_m^2). For all rib geometries, the profiles of the Reynolds shear stress have a peak in the shear layer between ribs. On Geometry I (Fig. 9 (a)), the large peak value of Reynolds shear stress remains between ribs and is observed around the rib height. On Geometry II (Fig. 9 (b)), the point of the peak value of Reynolds shear stress moves toward the wall in the region of $3 < X/e < 5$, and the Reynolds shear stress has larger values in the near wall region. For Geometry III (Fig. 9 (c)), a large peak value is shown on the near wall region of the upper side. On the bottom side, however, the profile is almost similar to that of Geometry II on $X/e > 3$ and the peak value is observed on about $Y/d = 0.3$.

Figure 10 shows a streamwise distribution of the Reynolds shear stress ($-\overline{u'v'}$) near the wall region ($Y/d = 0.12$). The profiles are classified to two types. The first type, which is denoted by symbols of triangles and filled squares, corresponded to staggered and top wall of unsymmetrical rib, has a peak on reattachment point. The second, denoted by symbols of circle and empty square, corresponds to symmetric rib and bottom wall of unsymmetrical rib, has a profile with no significant peak value. The value of turbulent energy and Reynolds shear stress of near wall region are strongly relevant to the heat transfer enhancement rate. It is satisfactory to consider that the staggered rib geometry may be more effective in order to upgrade the performance of heat transfer enhancement.

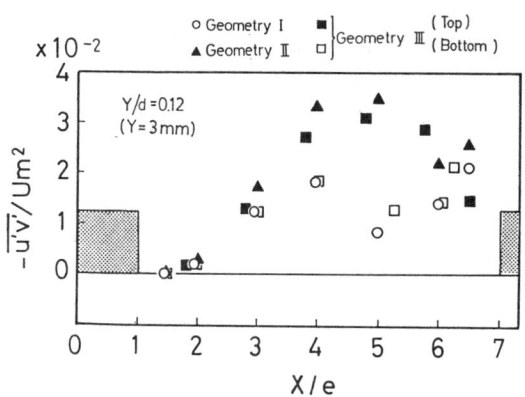

Fig. 10 Reynolds shear stress of near the wall

3.5 Correlation of Velocity Fluctuations. Figure 11 illustrates profile of the correlations coefficient of the velocity fluctuation defined by following equation.

$$R = -\overline{u'v'}/(\sqrt{\overline{u'^2}}\sqrt{\overline{v'^2}}) \qquad (1)$$

The larger value of correlation coefficients are observed in the recirculation zone for all rib geometries, but the remarkable differences are observed in main flow region ($0.5 < Y/e < 1.5$). For Geometry I (Fig. 11 (a)), the correlation coefficients have large value in the shear layer

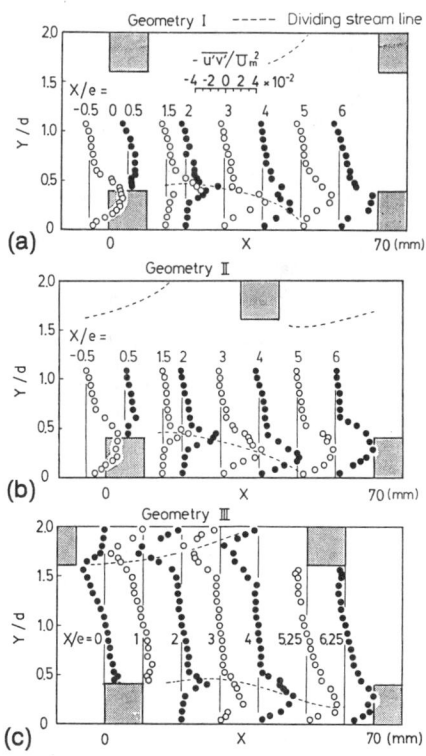

Fig. 9 Reynolds shear stress

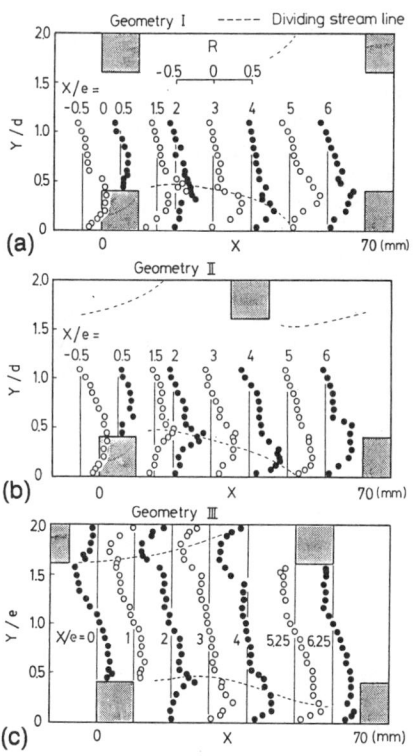

Fig. 11 Correlation coefficient of velocity fluctuations

and the recirculation zone. In the main flow region, however, the value of the correlation coefficients decreases in the central region where the zero velocity gradient due to symmetry. For Geometry II (Fig. 11 (b)), a large value of the correlation coefficient is observed in shear layer and in the near the wall region around reattachment point. In the main flow region the value is larger than those of Geometry I. For Geometry III (Fig. 11 (c)), a large value is shown in the recirculation zone, but the value is small in the middle of the rib pitch ($2 \leq X/e \leq 4$). Large values are observed between ribs on $X/e = 0$, 1, 5.25 and 6.25 from the wall to the center of the duct.

These results indicates that the rib may influence the generation of turbulence energy and generate a strong correlation of velocity fluctuations. In other words, when the rib is arranged staggered or unsymmetric, the velocity fluctuation of strongly correlated might be generated entirely in the duct.

4 Concluding Remarks

The flow field measurements were performed on the repeated rib roughness on the internal surface of rectangular channel with three types of rib geometries. Reynolds number based on the average velocity in the duct and the distance between ribbed plates was at 2.0×10^4. The results of the present investigation are summarized as follows.

For the profile of the static pressure distribution, a symmetrical arrangement of the plates gives approximately the same results as a staggered and a unsymmetrical arrangement.

A change of the rib arrangement has a remarkable effect on the turbulent characteristics and a modest effect on the mean velocity.

A strong turbulent shear stress has been generated in the shear layer for staggered and unsymmetrical rib arrangement. Correlation of the velocity fluctuations have larger value not only in the shear layer but also central area of the duct and near the wall region.

It is pointed out from the results of the friction factor for the each rib arrangement that the unsymmetrical rib arrangement is more suitable for upgrading the performance of the heat transfer enhancement involving the pumping power.

Acknowledgement

The authors would like to acknowledge Messrs. M. Motozuna and Y. Aikawa for their help in performing experiments.

References

Carnavos, T.C., "Heat Transfer Performance of Internally Finned Tubes in Turbulent Flow," *Advances in Enhanced Heat Transfer, ASME*, 1979, pp. 61.

Gee, D.L. and Webb, R.L., "Forced Convection Heat Transfer in Helically Rib-Roughened Tubes," *International Journal of Heat and Mass Transfer*, Vol. 23, 1980, pp. 1127-1136.

Han, J.C., Glicksman, L.R., and Rohsenow, W.M., "An Investigation of Heat Transfer and Friction for Rib-Roughened Surface," *International Journal of Heat and Mass Transfer*, Vol. 21, 1978, pp. 1143-1156.

Li, H.M., Ye, K.S., Tan, Y.K., and Deng, S.J., "Investigation on Tube-Side Flow Visualization, Friction and Heat Transfer Characteristics of Helical Ridging Tubes," *Proceedings of the 7th International Heat Transfer Conference*, Munich, Vol. 3, 1982, pp. 75-80.

Maeda, M., Hishida, K., and Ono, M,. "Development of Fiber LDV and Measurements of Turbulence in Boundary Layer on a Flat Plate," *Bulletin of JSME*, Vol. 28, No. 240, 1985, pp. 1050-1053.

Nakayama, W., "Enhancement of Heat Transfer," *Proceedings of the 7th International Heat Transfer Conference*, Munich, Vol. 1, 1982, pp. 223-240.

Nakayama, W, Takahashi, K., and Daikoku, T., "Spiral Ribbing to Enhance Single-Phase Heat Transfer Inside Tubes," *ASME/JSME Thermal Engineering Joint Conference Proceedings*, Vol. 1, 1983, pp. 365-372.

Nikuradse, J., "Stromungsgesetze in rauhen Rohren," *Forschg. Arb. Ing.-Wes. No. 361*, 1933. (in German)

Sato, H., Hishida, K., and Maeda, M., "Development of Fibre LDA and Two-Component Velocity Measurements in Boundary Layer on an Inclined Heated Plate," *Fluid Control and Measurement 1985*, Pergamon Press., Vol. 2, 1985, pp. 1065-1070.

Ueda, T., Mizomoto, M., and Ikai, S., "Velocity and Temperature Fluctuations in a Flat Plate Boundary Layer Diffusion Flame," *Combustion Science and Technology*, Vol.27, 1982, pp. 133-142.

Webb, R.L., Eckert, E.R.G., and Goldstein, R.J., "Heat Transfer and Friction in Tubes with Repeated-Rib Roughness," *International Journal of Heat and Mass Transfer*, Vol. 14, 1971, pp. 601-617.

Withers, J.G., "Tube-Side Heat Transfer and Pressure Drop for Tube Having helical Internal Ridging with Turbulent/ Transitional Flow of Single-Helix Ridging, " *Heat Transfer Engineering*, Vol. 2, No. 1, 1980, pp. 48-58.

Withers, J.G., "Tube-Side Heat Transfer and Pressure Drop for Tube Having Helical Internal Ridging with Turbulent/ Transitional Flow of Single-Helix Ridging," *Heat Transfer Engineering*, Vol. 2, No. 2, 1980, pp. 43-50.

1989 National Heat Transfer Conference
HTD-Vol. 107, Heat Transfer in Convective Flows

MODIFIED LOCAL SIMILARITY FOR NATURAL CONVECTION ALONG A NONISOTHERMAL VERTICAL FLAT PLATE INCLUDING STRATIFICATION

S. W. Webb
SPR Geotechnical Division 6257
Sandia National Laboratories
Albuquerque, New Mexico

ABSTRACT

An approximate method has been developed to analyze natural convection along a vertical flat plate with variable surface conditions and temperature stratification. This method uses the boundary layer velocity and temperature profiles from the local similarity method and imposes explicit conservation of energy along the plate resulting in required relationships for the similarity parameters for energy conservation. The results from this Modified Local Similarity (MLS) method are compared to those from other methods for a number of nonsimilar natural convection problems. Based on these comparisons, the MLS method is a significant improvement to the local similarity approach and is a useful approximate tool for analyzing natural convection on vertical surfaces for nonsimilar conditions.

NOMENCLATURE

A   area
c_p specific heat
f' velocity similarity variable
g   gravitational constant
Gr_x Grashof number based on x
J   stratification similarity parameter
LS   local similarity
$\dot{m}$ mass flow rate per unit width
n   temperature difference similarity parameter
N   temperature difference constant
PL   Power Law Distribution
Pr   Prandtl number
q'' heat flux
Q   integrated heat flux per unit width
ΔT temperature difference, $T_w - T_f$
T   temperature

u   x-direction velocity
v   y-direction velocity
W   width of plate
Δx difference in x, $x_2 - x_1$
x   distance along plate surface
y   distance normal to plate surface

Greek
α thermal diffusivity
β coefficient of thermal expansion
δ boundary layer thickness
η dimensionless coordinate
μ viscosity
ν kinematic viscosity
ρ density
θ dimensionless temperature

Subscripts
1   value at position x_1
2   value at position x_2
12 value between x_1 and x_2
f   fluid
r   reference
w   wall

Superscripts
−   average value
$'$ derivative with respect to η
*   entrainment or ejected value

INTRODUCTION

Natural convection along vertical surfaces occurs in the more than 50 oil-filled caverns in the Strategic Petroleum Reserve (SPR). These caverns are located in a number of large salt domes where the geothermal temperature difference over the cavern height of up to 600 m can be 15°C or more. The hotter salt is located at the bottom of the cavern; this configuration causes natural convection in the enclosed fluids as a result of buoyancy forces. Due to the large length scale, highly turbulent boundary layer conditions will be encountered with Rayleigh numbers up to approximately

* Prepared by Sandia National Laboratories, Albuquerque, New Mexico 87185 and Livermore, California 94550 operated for the United States Department of Energy under contract DE-AC04-76DP00789.

10^{16}. Since the heat transfer between the salt and the fluids in the cavern is coupled, heat transfer to the oil and the resulting natural convection can occur during the entire anticipated storage period of up to 30 years. SPR cavern wall conditions are nonuniform due to the geothermal temperature difference. In addition, the fluid temperature is nonuniform owing to the thermal stratification of the oil. Thus, the wall conditions and the ambient fluid temperature are both variable. In order to efficiently evaluate the natural convection boundary layer behavior in each cavern, a rapid analysis technique is needed.

The methods in general use for the analysis of natural convection are the integral (Sparrow, 1955), similarity, local similarity, local nonsimilarity (Sparrow, et al., 1970, 1971 and Minkowycz and Sparrow, 1974), and finite difference approaches (Cebeci and Bradshaw, 1984). In addition, approximate methods have been developed by Raithby, et al. (1975, 1977, 1978), Kao, et al. (1977), Yang, et al. (1982), and Lee and Yovanovich (1987, 1988).

The integral method could be used, although the assumed profiles are a problem for turbulent flow conditions. The wall and fluid temperature variations preclude direct use of the similarity solutions. The local similarity method, which applies the similarity solutions based only on the local boundary conditions, would be appropriate for SPR since the boundary layer results can be tabulated for use at each time step; therefore, the resulting calculations would be fast. However, the method does not consider the history of the boundary layer, and errors in the heat transfer rate can be significant even for simple cases. Local nonsimilarity and finite difference methods are impractical due to long estimated computing times for the 30 year transient involved.

Approximate methods have been proposed by a number of authors. However, the methods developed by Raithby, et al. (1975, 1977, 1978) and Lee and Yovanovich (1987, 1988) were not considered for use in SPR since neither method reduces to the similarity solutions for similar boundary conditions. Differences of up to 20% have been noted. The methods developed by Kao, et al. (1977) and by Yang, et al. (1982), approach the local nonsimilarity method in complexity and were therefore not considered.

If the heat transfer rate errors noted for the local similarity approach can be significantly reduced, the method would be ideal for SPR. The present study attempts to minimize this problem by modifying the local similarity approach to explicitly conserve energy as the boundary layer develops along the surface. This Modified Local Similarity (MLS) approach is developed and compared to results from other methods in this paper. This method is used in the SPR velocity model developed by Webb (1988a).

FORMULATION

Consider natural convection boundary layer flow along a flat plate as depicted in Figure 1. The boundary layer energy equation can be integrated along the plate using the local boundary layer velocity and temperature profiles. This equation must be satisfied for global energy conservation. In the present study, the boundary layer profiles used in this equation are calculated by the local similarity method. The local similarity method has two parameters which are mathe-

matical descriptions of the temperature variation along the plate and in the surrounding fluid. In addition to being mathematical parameters, these variables have physical significance with regard to conservation of energy. The global energy conservation equation, which must be satisfied, can be written in terms of the local similarity parameters. Relationships for the two local similarity parameters can then be developed to explicitly satisfy energy conservation.

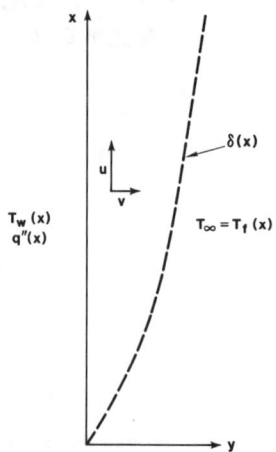

Fig. 1. Boundary layer coordinates.

For example, if the wall heat flux and fluid temperature variation are known, the use of the local similarity profiles in the global energy conservation equation results in a required variation of the two similarity parameters. If these relationships are satisfied, global energy conservation is achieved. This approach is called the MLS method and is detailed below.

Global Energy Equation

Conservation of energy along the plate per unit width can be written as

$$\dot{m}_1 \, c_p \, (\bar{T}_1 - T_f^*) + \bar{q}_{12}^{''} \, \Delta x = \dot{m}_2 \, c_p \, (\bar{T}_2 - T_f^*) \qquad (1)$$

where

$$\bar{q}_{12}^{''} = \frac{1}{\Delta x} \int_{x_1}^{x_2} q^{''} \, dx \qquad (2)$$

and T_f^* is the average temperature of the fluid entrained into or ejected from the boundary layer between x_1 and x_2. The value of T_f^* is then equal to T_f at location x^*. In the present analysis, the MLS approach results in a relationship for x^* based on energy conservation considerations. Note that the fluid specific heat, c_p, is assumed to be constant.

The average temperature of the entrained or ejected fluid will be assumed equal to the local environmental fluid temperature for this analysis. For Prandtl number fluids of order 1.0 and higher, such as air, water, and oil, the velocity boundary layer thickness is larger than the the thermal boundary layer, so

198

any fluid exchange will be at the environmental temperature. This assumption breaks down for low Prandtl number fluids such as liquid metals where the similarity solution gives a larger thermal boundary layer than velocity boundary layer (Gebhart, 1985).

The average boundary layer temperature, $\bar{T}$, to be used in equation (1) is simply the bulk fluid temperature at that location, or

$$\bar{T} = \frac{\int u\, T\, dy}{\int u\, dy} \qquad (3)$$

which for the present analysis is rewritten as

$$\bar{T} = T_f(x) + \frac{\int u\, (T - T_f(x))\, dy}{\int u\, dy}. \qquad (4)$$

Combining the equations (1) and (4) results in

$$\dot{m}_1\, c_p\, (\bar{T}_1 - T_f^*) + \bar{q}_{12}''\, \Delta x$$
$$= \dot{m}_2\, c_p\, (T_{f_2} - T_f^* + \frac{\int u\, (T - T_f(x))\, dy}{\int u\, dy}). \qquad (5)$$

where the integrals in equation (5) are evaluated at x_2. The above equation is general; any restrictions as to the orientation, etc. are from evaluation of the boundary layer velocity and temperature profiles. These profiles will be based on local similarity.

MLS Method

The boundary layer profiles in this study are for laminar natural convection over a nonisothermal vertical flat plate in a variable temperature fluid medium. Invoking the Boussinesq approximation with otherwise constant properties and neglecting viscous dissipation and the pressure-work term, the steady-state conservation equations are (Jaluria, 1980)

Continuity

$$\frac{\partial u}{\partial x} + \frac{\partial v}{\partial y} = 0 \qquad (6)$$

x-Momentum

$$u\, \frac{\partial u}{\partial x} + v\, \frac{\partial u}{\partial y} = g\, \beta\, (T - T_f(x)) + \nu\, \frac{\partial^2 u}{\partial y^2} \qquad (7)$$

Energy

$$u\, \frac{\partial T}{\partial x} + v\, \frac{\partial T}{\partial y} = \alpha\, \frac{\partial^2 T}{\partial y^2}. \qquad (8)$$

The above conservation equations can be integrated across the boundary layer resulting in

Momentum

$$\frac{d}{dx} \int u^2\, dy = g\, \beta \int (T - T_f(x))\, dy - \nu\, \frac{\partial u}{\partial y}\bigg|_w \qquad (9)$$

Energy

$$\frac{d}{dx} \int u\, (T - T_f(x))\, dy + \frac{dT_f}{dx} \int u\, dy$$
$$= -\alpha\, \frac{\partial T}{\partial y}\bigg|_w \qquad (10)$$

where the second term on the LHS of the energy equation accounts for temperature stratification.

Two energy equations are considered in the present analysis. The global energy equation (5) is concerned with the energy in the boundary layer as it develops along the plate. The local energy equation (10) is related to the energy in the boundary layer at location x only. Both equations must be satisfied. Similarity variables will be used to rewrite the energy equations. These equations will then be combined to lead to relationships for the similarity variables that must be satisfied for global and local energy conservation.

According to Sparrow and Gregg (1958) and Yang (1960), similarity exists for two temperature distributions: the power-law and the exponential distributions. The power-law distribution is the more useful case and is discussed in this paper. Results for the exponential distribution are given by Webb (1988b).

For the power-law distribution, the temperature difference between the wall and fluid is a function of the distance x to a power, or

$$\Delta T(x) = T_w(x) - T_f(x) = N\, x^n. \qquad (11)$$

For similarity, the fluid temperature variation must be of the same form, or (Jaluria, 1980)

$$T_f(x) - T_r = \frac{J\, N}{4\, n}\, x^n = \frac{J}{4\, n}\, \Delta T(x) \qquad (12)$$

where the reference temperature, T_r, is the fluid temperature at $x = 0$. If the fluid temperature is constant, J is equal to 0.

The similarity variables for this case are (Gebhart and Mollendorf, 1969)

$$f' = \frac{x}{2\, \nu}\, Gr_x^{-1/2}\, u. \qquad (13)$$

$$\eta = \frac{y}{x} \left[\frac{Gr_x}{4}\right]^{1/4} \qquad (14)$$

$$\theta(\eta) = \frac{T(x) - T_f(x)}{T_w(x) - T_f(x)} \qquad (15)$$

where

$$Gr_x = \frac{g\, \beta\, x^3\, (T_w(x) - T_f(x))}{\nu^2} \qquad (16)$$

Using the similarity variables and assuming local similarity, the boundary layer partial differential equations (6)-(8) reduce to a set of coupled ordinary differential equations which are (Jaluria, 1980)

$$f''' + (n + 3)\, f\, f'' - 2\, (n + 1)\, f'^2 + \theta = 0 \qquad (17)$$

199

$$\frac{\theta''}{Pr} + (n + 3) f \theta' - 4 n f' \theta - J f' = 0 \qquad (18)$$

so the boundary layer velocity and temperature profiles are a function of the similarity parameters n and J. The above equations have been solved by a finite difference method as summarized by Webb (1989).

Expressing the fluid temperature difference in terms of the similarity parameter J gives

$$T_{f_2} - T_f^* = \frac{J \, \Delta T_2}{4 n} \left(1 - \left(\frac{x^*}{x_2}\right)^n\right). \qquad (19)$$

Using the similarity parameters along with equation (19), the global energy equation (5) becomes

$$\dot{m}_1 \, c_p \, (\bar{T}_1 - T_f^*) + \bar{q}_{12}'' \, \Delta x$$

$$= \dot{m}_2 \, c_p \, \left(\frac{J}{4 n} \left(1 - \left(\frac{x^*}{x_2}\right)^n\right) + \frac{\int f' \theta \, d\eta}{\int f' \, d\eta}\right) \Delta T_2. \qquad (20)$$

The mass flow rate per unit width can be expressed in terms of the local similarity variables as

$$\dot{m} = \rho \, \bar{u} \, A / W = \rho \, \bar{u} \, \delta$$

$$= 4 \mu \int f' \, d\eta \, \left(\frac{g \, \beta}{4 \nu^2}\right)^{1/4} x^{3/4} \, \Delta T^{1/4}. \qquad (21)$$

The heat flux relationship

$$q'' = - k \left.\frac{\partial T}{\partial y}\right|_w = - k \, \theta_w' \, \left(\frac{g \, \beta}{4 \nu^2}\right)^{1/4} \Delta T^{5/4} \, x^{-1/4} \qquad (22)$$

can be used to get the temperature difference as a function of x, and the mass flow rate per unit width is

$$\dot{m} = 4 \mu \int f' \, d\eta \, \left(\frac{g \, \beta}{4 \nu^2}\right)^{0.2} x^{0.8} \left(\frac{q''}{- k \, \theta_w'}\right)^{0.2} \qquad (23)$$

The global energy equation (20) then becomes

$$\dot{m}_1 \, c_p \, (\bar{T}_1 - T_f^*) + \bar{q}_{12}'' \, \Delta x$$

$$= 4 \, Pr \int f' \, d\eta \, \left(\frac{q_2''}{- \theta_w'}\right)$$

$$\left(\frac{J}{4 n} \left(1 - \left(\frac{x^*}{x_2}\right)^n\right) + \frac{\int f' \theta \, d\eta}{\int f' \, d\eta}\right) x_2. \qquad (24)$$

Using the similarity variables in the integrated local boundary layer equations (9) and (10) results in the following equations for the boundary layer quantities under the assumption of local similarity

Momentum

$$(5 + 3n) \int f'^2 \, d\eta = \int \theta \, d\eta - f_w'' \qquad (25)$$

Energy

$$(5n + 3) \int f' \theta \, d\eta = - \frac{\theta_w'}{Pr} - J \int f' \, d\eta. \qquad (26)$$

Rearranging the local boundary layer energy equation (26) and substituting it into the global energy equation (24) results in

$$(\dot{m}_1 \, c_p \, (\bar{T}_1 - T_f^*) + \bar{q}_{12}'' \, \Delta x) / (q_2'' \, x_2)$$

$$= \frac{\frac{J}{n} \left(1 - \left(\frac{x^*}{x_2}\right)^n\right) + 4 \int f' \theta \, d\eta / \int f' \, d\eta}{J + (5n + 3) \int f' \theta \, d\eta / \int f' \, d\eta}. \qquad (27)$$

The similarity parameter n is independent of J. Therefore, from equation (27), the expression for n is

$$n = \frac{1}{5} \left[4 \, q_2'' \, x_2 / (\dot{m}_1 \, c_p \, (\bar{T}_1 - T_f^*) + \bar{q}_{12}'' \, \Delta x) - 3\right]. \qquad (28)$$

Taking x_1 at the leading edge of the plate ($x_1 = 0.$) with no initial mass flow rate, which is usually the case, then $\Delta x = x_2$, and the equation for n simplifies to

$$n = \frac{1}{5} \left[4 \, q_2'' / \bar{q}_{12}'' - 3\right]. \qquad (29)$$

The value of the similarity parameter n is just a function of the ratio of the local to the average heat flux up to that point.

Similarly, from equations (27) and (28),

$$x^* = x_2 \left[1 - \frac{4n}{(5n + 3)}\right]^{1/n}. \qquad (30)$$

The fluid temperature evaluated at x* is that required for global energy conservation.

Surprisingly, the stratification parameter, J, is independent of global conservation of energy. Instead, the value of J is determined by the local value of the heat flux, q_2''. Equating equations (1) and (28) gives

$$\dot{m}_2 \, c_p \, (\bar{T}_2 - T_f^*) = \frac{4}{(5n + 3)} \, q_2'' \, x_2 \qquad (31)$$

where the values on the RHS are known. This equation includes the effect of temperature stratification on the local energy balance. Using the relationships developed above for $\dot{m}$ and ΔT, the equation can be written as

$$A_1 \frac{\int f' \, d\eta}{(- \theta_w')^{0.2}} \left(A_2 \frac{\int f' \theta \, d\eta}{\int f' \, d\eta \, (- \theta_w')^{0.8}} + T_{f_2} - T_f^*\right) = A_3 \qquad (32)$$

where

$$A_1 = 4 \, c_p \, \mu \, \left(\frac{g \, \beta}{4 \nu^2}\right)^{0.2} x_2^{0.8} \left(\frac{q_2''}{k}\right)^{0.2} \qquad (33)$$

$$A_2 = \left(\frac{g \, \beta}{4 \nu^2}\right)^{-0.2} x_2^{0.2} \left(\frac{q_2''}{k}\right)^{0.8} \qquad (34)$$

200

$$A_3 = \frac{4}{(5n + 3)} \, q_2'' \, x_2. \tag{35}$$

For uniform fluid conditions, T_{f2} is equal to T_f^*, and the above expression reduces to the local integrated energy equation with J equal to zero. As discussed earlier, the boundary layer parameters in the above expression are dependent on the similarity parameters n and J. Since the value of n is determined by equation (28) or (29), the only undefined parameter is J.

Solution of the equation (32) for J initially looks difficult. In practice, however, solution is straightforward and, for the present investigation, has been accomplished by iterating on the form

$$\frac{\int f' \theta \, d\eta}{-\theta_w'} = \frac{A_3 - \dot{m}_2^{i-1} \, c_p \, (T_{f_2} - T_f^*)}{4 \, Pr \, q_2'' \, x_2} \tag{36}$$

where $\dot{m}_2^{i-1}$ is the value of $\dot{m}_2$ from the previous iteration and T_f^* is evaluated from conservation of energy. The ratio of the LHS of the equation is a strong function of J for a given value of n, and convergence has not been a problem.

In summary, for a specified heat flux problem, the similarity parameter n is determined directly from equation (28) or (29). For a uniform environmental fluid temperature, the similarity parameter J is equal to 0. Otherwise, the value of J is determined by iterating on equation (36). All the boundary layer parameters are uniquely determined by these values of n and J. For situations where similarity conditions are imposed, the similarity solutions are obtained. This is not the case for the approximate methods developed by Raithby, et al. (1975, 1977, 1978) and by Lee and Yovanovich (1987, 1988). For variable conditions where an exact similarity solution does not exist, the MLS method provides an estimate of "equivalent" similarity conditions including velocity and temperature profiles by requiring global conservation of energy and the same local heat flux at position x_2.

In the above development, the heat flux variation is assumed to be specified. This situation is not always the case, as the temperature distribution is sometimes given. In order to calculate the similarity parameters, energy consistency between the specified problem and the MLS method is required. The integrated heat flux per unit width for constant properties is proportional to the following integral

$$Q = \int q'' \, dx \propto \int \theta_w' \, \Delta T^{5/4} \, x^{-1/4} \, dx \tag{37}$$

and the expression for n becomes

$$n = \frac{1}{5} \left[4 \, q_2'' \, x_2 \, / \, (\dot{m}_1 \, c_p \, (\bar{T}_1 - T_f^*) + Q) - 3 \right] \tag{38}$$

which, for x_1 and $\dot{m}_1$ equal to 0. can be written as

$$n = \frac{1}{5} \left[\frac{4 \, \theta_w' \, \Delta T_2^{5/4} \, x_2^{3/4}}{\int \theta_w' \, \Delta T^{5/4} x^{-1/4} dx} - 3 \right]. \tag{39}$$

The temperature gradient for a given Prandtl number is only a function of n and J, so iteration is required on this equation and, when necessary, equation (39) for J.

For specified surface temperatures, the MLS method is not a local similarity approach since the answer at x depends on the results at the upstream locations. Iteration is required for the variation of the similarity parameters with x. However, this iteration is easily accomplished since the only term that depends on n and J is θ_w', and convergence is rapid for the cases analyzed in this report.

While specified temperatures are a convenient analytical case, the wall temperature and wall heat fluxes are usually coupled to each other through heat conduction, and either the wall temperature or the heat flux can be used in the solution scheme. For the MLS method, heat fluxes are considerably more convenient than temperatures since no iteration is involved.

EVALUATION

The Modified Local Similarity (MLS) method derived above has been applied to a number of nonsimilar wall temperature and heat flux cases with uniform fluid temperature and to an isothermal plate in a stratified fluid environment. The results in this section compare the predictions from the MLS method with those from other approaches and, for the case of an isothermal plate in a stratified fluid, to experimental data. The results from another possible implementation of the local similarity approach in addition to the MLS method are also given. While the MLS method is based on conservation of energy as the boundary layer develops and matching the local heat flux, another reasonable approach would be matching the local value of the specified parameter (temperature difference or heat flux) as well as the local slope of that parameter. The predictions from this method will be referred to as the LS* approach.

Uniform Fluid Temperature

For uniform fluid temperature conditions, the MLS and LS* methods have been applied to specified wall temperature and specified heat flux cases. Results of these cases for a number of other methods are summarized by Yang, et al. (1982) for a Prandtl number of 0.7 where the property term is assumed equal to 1.0, or

$$\left(\frac{g \, \beta}{4\nu^2} \right) = 1.0. \tag{40}$$

In all these cases, the stratification parameter, J, is equal to 0. since the fluid temperature is uniform.

The results from the MLS method and the LS* approach will be compared to the following predictions.

1. Numerical - as given by Kao, et al. (1977).
2. Kao LS - Kao, et al. (1977) local similarity.
3. Kao method - The method of Kao, et al. (1977) which is basically a perturbation approach.
4. Yang method - The method of Yang, et al. (1982) which is a series expansion approach.

Predictions from other methods, such as the integral approach, will also be included where available.

Specified Surface Temperature. The comparisons are based on the temperature gradient at the surface which is related to the local heat transfer coefficient. In addition, the approximation of the temperature difference behavior is presented.

1) $\Delta T = e^x$. Figure 2a shows the desired temperature difference as well as the variation predicted by the MLS and LS* approaches. The predictions depend on n which itself is a function of x. Therefore, in Figure 2a, two curves for the appropriate value of n corresponding to the two x values of 0.5 and 2.0 are shown for each approach. In general, the variation of the temperature difference is reasonably close to the desired behavior. The temperature difference variation is well represented by both methods. The surface temperature gradient as a function of x is depicted in Figure 2b. The gradient is underpredicted by the MLS method by approximately 5%. While the error is larger than the other methods, the magnitude is still relatively small. For the LS* approach, a slight overprediction of the gradient, especially near the front of the plate, is noted. This behavior is also seen for the Kao LS method. Predictions for the Kao and Yang methods are not shown in this figure since both approaches yield predictions indistinguishable from the numerical results.

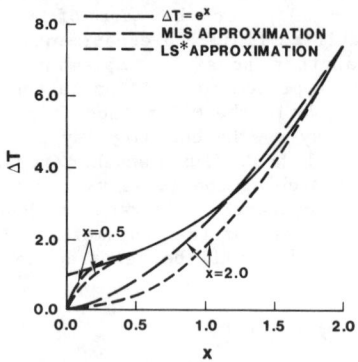

Fig. 2a. Approximation of ΔT for $\Delta T = e^x$.

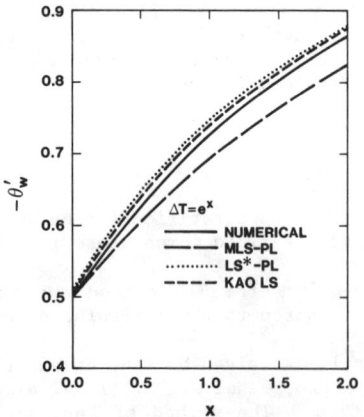

Fig. 2b. Surface temperature gradient for $\Delta T = e^x$.

2) $\Delta T = \sin x$. The predicted surface temperature gradient as a function of x for a number of other methods is depicted in Figure 3. The Yang method gives excellent results up to an x value of 2.2 after which the method has convergence problems. The Kao method also gives good results out to an x value of 2.3; after

this point, the Kao method also no longer converges. The Kao local nonsimilarity (LNS) results are surprisingly poor; only results to an x value of 2.0 are given by Kao (1976). The Kao LS results show reasonable agreement for the entire problem, although a systematic underprediction is evident.

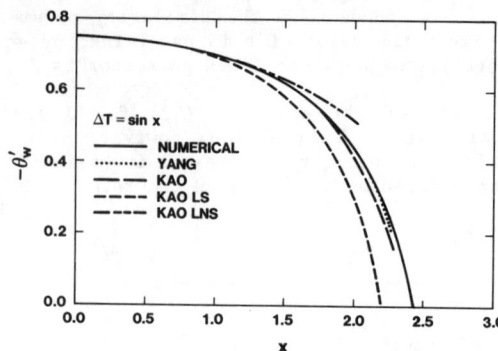

Fig. 3. Surface temperature gradient for $\Delta T = \sin x$. Results from other methods.

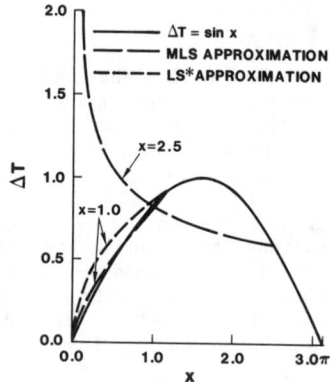

Fig. 4a. Approximation of ΔT for $\Delta T = \sin x$.

Results for the MLS and LS* power-law approaches are shown in Figure 4. The temperature difference is given in Figure 4a for n corresponding to x values of 1.0 and 2.5. For small x values ($< \sim \pi/2$), both methods give a good approximation of the sinusoidal temperature difference. For larger x values, the MLS distribution becomes increasingly poor since n becomes negative for x values greater than about 2.16. When n changes sign, the shape changes dramatically as shown in Figure 4a. No temperature difference variation is presented for the LS* method at an x value of 2.5 since the zero heat flux location is x~1.88 for this approach. From Figure 4b, the temperature gradient is well predicted for small x values. For large x values, the MLS method overpredicts the temperature gradient and the location of zero heat flux, while the LS* approach underpredicts the results. While not as good as some of the other methods, the MLS approach is better than the LS* method and about the same as the Kao LS approach. This discrepancy is not unexpected due to the poor approximation of the temperature difference behavior by the MLS method at large x values.

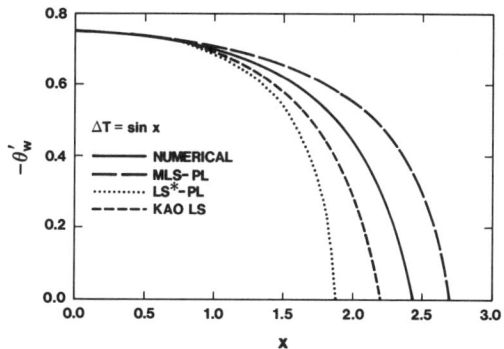

Fig. 4b. Surface temperature gradient for $\Delta T = \sin x$.

Specified Wall Heat Flux. The comparisons between the various methods are based on the wall to fluid temperature difference. In addition, the approximation of the heat flux behavior is presented.

1) $q''/k = e^x$. The heat flux distribution for the MLS and LS* approaches is given in Figure 5a for n corresponding to x values of 0.5 and 2.0. In general, the heat flux variation is well represented by both methods. These conclusions are similar to those for the exponential temperature difference case discussed earlier. The surface temperature as a function of x is depicted in Figure 5b. The MLS and LS* methods both successfully predict the surface temperature variation with x. The predictions of the Kao and the Yang methods are not shown since they are indistinguishable from the numerical results. All the methods perform well for this case.

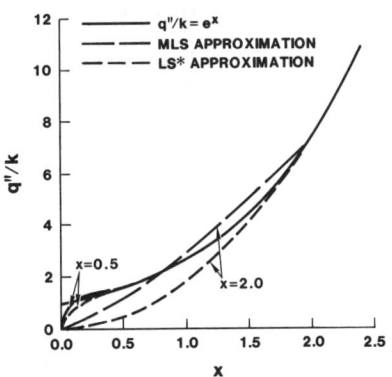

Fig. 5a. Approximation of q''/k for $q''/k = e^x$.

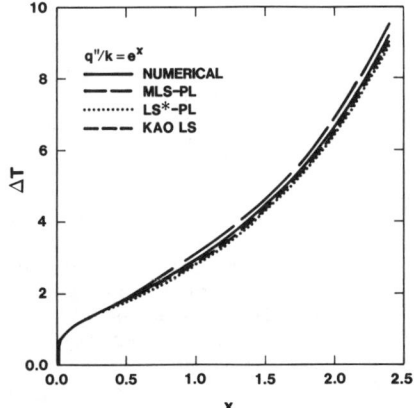

Fig. 5b. Temperature difference for $q''/k = e^x$.

2) $q''/k = 1 + x$. Figure 6a compares the heat flux variation for the power-law distribution to the desired variation for n corresponding to x values of 0.5 and 3.0. As with a number of the previous cases, the heat flux behavior is reasonably well represented by both approaches. Figure 6b shows the temperature difference variation along the plate. The answers from the Kao method and the Yang method are not given since they essentially coincide with the numerical results. The results from the integral analysis as given by Sparrow (1955) are also shown. All methods give good predictions for this case including the integral method.

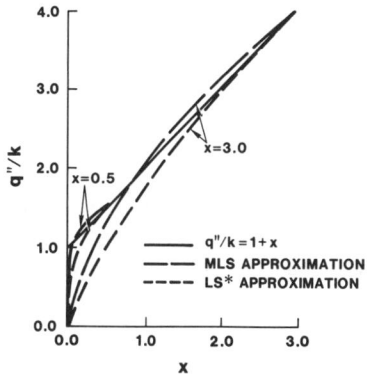

Fig. 6a. Approximation of q''/k for $q''/k = 1 + x$.

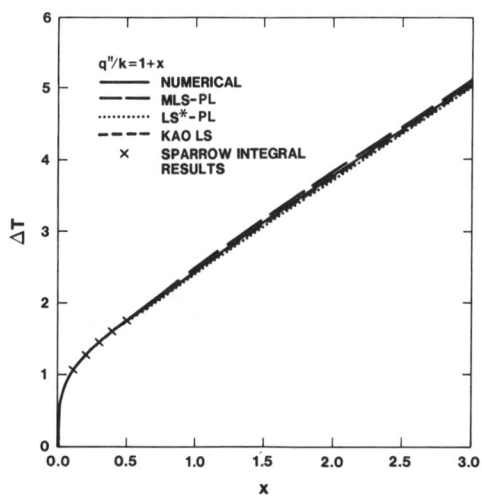

Fig. 6b. Temperature difference for $q''/k = 1 + x$.

3) $q''/k = 1 - x$. The predicted temperature difference variation along the plate for a number of different methods is given in Figure 7. The Kao and Yang methods diverge from the numerical solution for x values greater than about 0.5. The Kao LS method gives widely different results. The integral results from Sparrow (1955) seem to be well behaved, although the results are only provided out to an x value of 0.5 due to the limited information presented by Sparrow.

Figure 8a gives the heat flux predictions for the MLS and LS* methods for n corresponding to x values of 0.1 and 0.5. The behavior of both methods is not unreasonable, although significant differences can be seen between the approximation and the desired variation. Figure 8b shows the temperature difference results. The MLS method provides a reasonable prediction for the surface temperature behavior; the results are superior to all the other methods based on

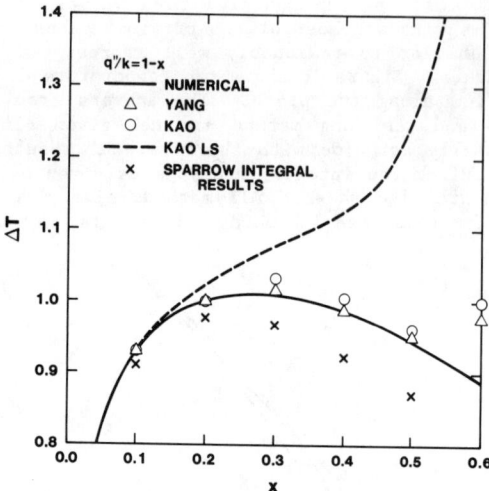

Fig. 7. Temperature difference for $q''/k = 1 - x$. Results from other methods.

comparison to the numerical predictions. The LS* predictions diverge like the Kao LS results.

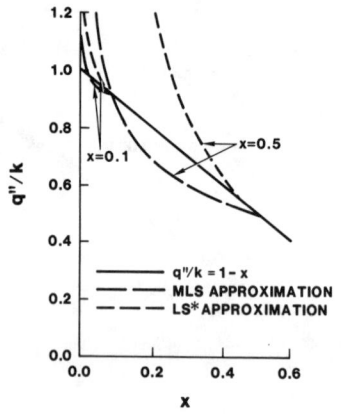

Fig. 8a. Approximation of q''/k for $q''/k = 1 - x$.

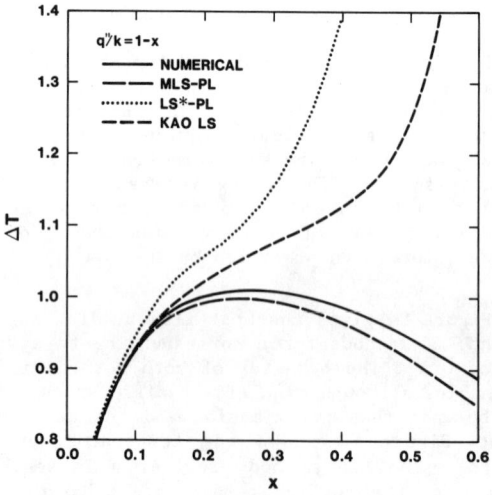

Fig. 8b. Temperature difference for $q''/k = 1 - x$.

Stratified Fluid Temperature

The stratified fluid temperature problem is an isothermal plate in a linearly stratified fluid as shown in Figure 9. The temperature difference between the plate and the fluid decreases linearly up the plate. A similar solution is not available for this problem. Chen and Eichhorn (1976) present an analysis of this problem using local similarity and local non-similarity methods for their coordinate transformation. In addition, Chen and Eichhorn (1976) acquired heat transfer data for water with a nominal Prandtl number of 6.0. Raithby and Hollands (1978) have applied their approximate technique (Raithby, et al. (1975, 1977)) to this problem with good results.

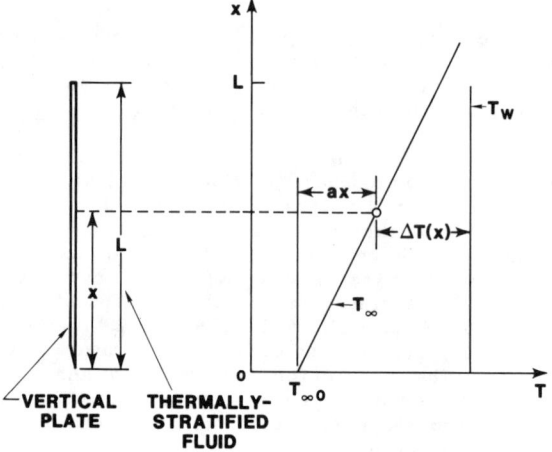

Fig. 9. Stratified fluid temperature problem.

Results for this problem are given in terms of the ratio of Nusselt numbers for the stratified fluid to that for an isothermal fluid as a function of the stratification parameter S, which is

$$S = \frac{L}{\Delta T} \frac{dT_f}{dx}. \tag{41}$$

When $S \leq 2$, the entire plate is hotter than the fluid. For $S > 2$, the bottom portion of the plate is hotter than the fluid while the top is colder.

The MLS and LS* approaches have been used to analyze this case for Prandtl numbers of 0.7 and 6.0. Only the results for a Prandtl number of 6.0 are included in this paper since this is the only case where data are available. The results for a Prandtl number of 0.7 are given in Webb (1988b). For the LS* approach, the value of n is determined by matching the local temperature difference value and the local slope; the value of J is calculated by the appropriate fluid temperature variation equation.

Figure 10 gives the variation of the temperatures for n corresponding to an x value of 0.5. For both methods, the reference value of the fluid temperature, T_r, is calculated by matching the temperatures at an x value of 0.5. The temperature variation for the MLS method is more reasonable than the LS* approach as the temperature difference is closer to the desired behavior. Both methods give an adequate, though certainly not perfect, prediction of the temperature variation.

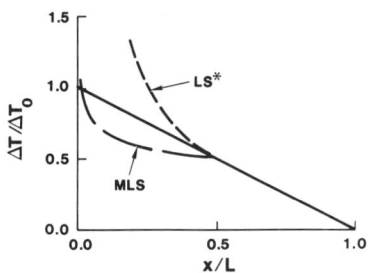

Fig. 10a. Approximation of ΔT for Stratified Fluid Case. (S=2).

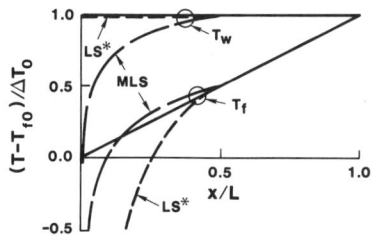

Fig. 10b. Approximation of T_w and T_f for Stratified Fluid Case. (S=2).

Figure 11 shows the predicted value of the average Nusselt number for a stratified fluid over that for an isothermal fluid with the same average temperature difference for a Prandtl number of 6.0. The MLS predictions are shown on this figure; results from the LS* method are not included as discussed below. The local similarity and local non-similarity (LNS) results are shown as well as the experimental data from Chen and Eichhorn (1976). The predictions by Raithby and Hollands (1978) are also included in the figure.

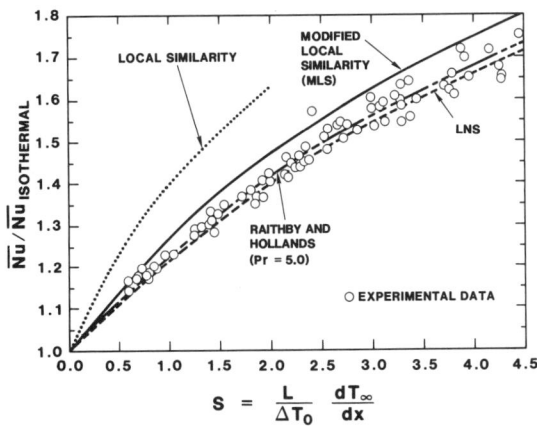

Fig. 11. Variation of Nusselt Number with Stratification for Pr=6.0.

The MLS results were calculated for a number of discrete S values out to 2.0. For S>2, the predictions are based on an $S^{1/4}$ dependence as used by Chen and Eichhorn (1976) and Raithby and Hollands (1978). The MLS predictions show reasonable agreement with the data with a consistent overprediction of about 4%. The local similarity results by Chen and Eichhorn (1976)

are much higher than the data with an error of about 16%. The Raithby and Hollands predictions go right through the data, although their results are for a Prandtl number of 5.0, not 6.0. The LNS results show good agreement with the data with a small consistent underprediction. Overall, the MLS, Raithby and Hollands, and LNS results are in good agreement with the data. The maximum difference between these methods is about 5%, while the uncertainty in the data is of this order, or ±3.2% for Nu and ±3.5% for S (Chen and Eichhorn, 1976).

The LS* method performs poorly for this case. For a Prandtl number of 6.0 and an S value of 2.0, the wall is always as hot or hotter than the fluid. The LS* method predicts that the wall temperature gradient will change sign about 1/4 up the plate. For the first 1/4 of the plate, heat is transferred from the hotter wall to the fluid. However, for the last 3/4 of the plate, heat is predicted to flow from the colder fluid to the hotter plate, which is unreasonable. Therefore, the LS* predictions are not shown on the figure.

SUMMARY AND CONCLUSIONS

The MLS method has been developed and evaluated for a number of nonsimilar temperature and heat flux cases for the power-law similarity distribution. For variable conditions where an exact similarity solution does not exist, the MLS method provides an estimate of "equivalent" similarity conditions including velocity and temperature profiles. This estimate is achieved by requiring global conservation of energy and the same local heat flux at position x_2. In addition, another possible application of the local similarity approach has been evaluated. This method, designated the LS* approach, matches the local value and local slope of the prescribed parameter whether it be temperature difference or heat flux. The MLS and LS* results have been compared to those from a number of other methods, including a numerical approach.

The predictions from the LS* approach vary from reasonable to absurd, so the LS* method is not a reliable technique. The MLS method is not the most accurate approach as expected but is superior to the traditional local similarity approach. In addition, many of the other more complex approaches, such as the methods of Kao, et al. (1977) and of Yang, et al. (1982), have problems with certain cases such as the linearly decreasing heat flux situation and have not been applied to a nonuniform fluid temperature case. In contrast, all the MLS predictions are reasonable even where the more complex methods fail or no longer apply. Through the introduction of global conservation of energy, the MLS method has significantly improved the predictive capability of the local similarity approach and may be superior to more complex methods.

The MLS method is not without its problems. For specified temperature cases, iteration is required which violates the local similarity assumption. However, in most practical cases, temperatures and heat fluxes are related through heat conduction in the wall, and the more convenient variable can be used. Use of heat flux information permits use of the MLS method on a local basis consistent with the local similarity approach.

Where computing times are a major constraint, as in the analysis of natural convection in SPR caverns (Webb (1988a)), standard techniques such as finite differences and local nonsimilarity are impractical. In this case, the approximate results provided by the MLS method should usually provide reasonable results with minimal computing time. Thus, the MLS method is a useful approximate tool for natural convection analysis for analyzing vertical plates for nonsimilar conditions.

REFERENCES

Cebeci, T., and Bradshaw, P., 1984, Physical and Computational Aspects of Convective Heat Transfer, Springer-Verlag, New York.

Chen, C. C., and Eichhorn, R., 1976, "Natural Convection From a Vertical Surface to a Thermally Stratified Fluid," Trans. ASME, J. Heat Transfer, pp. 446-451.

Gebhart, B., 1985, "Similarity Solutions for Laminar External Boundary Region Flows," Natural Convection - Fundamentals and Applications, S. Kakac, W. Aung, and R. Viskanta, eds., pp. 3-35, Hemisphere Publishing Corporation, Washington.

Gebhart, B., and Mollendorf, J., 1969, "Viscous dissipation in external natural convection flows," J. Fluid Mech., Vol. 38, Part 1, pp. 97-107.

Jaluria, Y., 1980, Natural Convection Heat and Mass Transfer, Pergamon Press, Oxford.

Kao, T-T., 1976, "Locally Nonsimilar Solution for Laminar Free Convection Adjacent to a Vertical Wall," Trans. ASME, J. Heat Transfer, pp. 321-322.

Kao, T-T., Domoto, G. A., and Elrod, Jr., H. G., 1977, "Free Convection Along A Nonisothermal Vertical Flat Plate," Trans. ASME, J. Heat Transfer, pp. 72-78.

Lee, S., and Yovanovich, M. M., 1987, "Laminar Natural Convection From a Vertical Plate With Variations in Wall Temperature," ASME HTD-Vol. 82, Convective Transport, ASME Winter Annual Meeting, pp. 111-119, Boston, MA.

Lee, S., and Yovanovich, M. M., 1988, "Laminar Natural Convection From a Vertical Plate With Variations in Surface Heat Flux," ASME HTD-Vol. 96, ASME Proceedings of the 1988 NHTC, Vol. 2, pp. 197-205, Houston, TX.

Minkowycz, W. J., and Sparrow, E. M., 1974, "Local Nonsimilar Solutions for Natural Convection on a Vertical Cylinder," Trans. ASME, J. Heat Transfer, pp. 178 -183.

Raithby, G. D., and Hollands, K. G. T., 1975, "A General Method of Obtaining Approximate Solutions to Laminar and Turbulent Free Convection Problems," Advances in Heat Transfer, Vol. 11, Academic Press.

Raithby, G. D., and Hollands, K. G. T., 1978, "Heat Transfer by Natural Convection Between a Vertical Surface and a Stably Stratified Fluid," Trans. ASME, J. Heat Transfer, pp. 378-381.

Raithby, G. D., Hollands, K. G. T., and Unny, T. E., 1977, "Analysis of Heat Transfer by Natural Convection Across Vertical Fluid Layers," Trans. ASME, J. Heat Transfer, pp. 287-293.

Sparrow, E. M., 1955, "Laminar Free Convection On a Vertical Plate With Prescribed Nonuniform Wall Heat Flux or Prescribed Nonuniform Wall Temperature," NACA TN 3508.

Sparrow, E. M., and Gregg, J. L., 1958, "Similar Solutions for Free Convection From a Nonisothermal Vertical Plate," Trans. ASME, pp. 379-386.

Sparrow, E. M., Quack, H., and Boerner, C. J., 1970, "Local Nonsimilarity Boundary-Layer Solutions," AIAA J. Vol. 8, No. 41, pp. 1936-1942.

Sparrow, E. M., and Yu, H. S., 1971, "Local Non-Similarity Thermal Boundary-Layer Solutions," Trans. ASME, J. Heat Transfer, pp. 328-334.

Webb, S. W., 1988a, Development and Validation of the SPR Cavern Fluid Velocity Model, Sandia National Laboratories Report SAND88-2711.

Webb, S. W., 1988b, Modified Local Similarity for Natural Convection Along a Nonisothermal Vertical Flat Plate Including Stratification, Sandia National Laboratories Report SAND88-2710.

Webb, S. W., 1989, Calculation of Natural Convection Boundary Layer Profiles Using the Local Similarity Approach Including Turbulence and Mixed Convection, Sandia National Laboratories Report SAND89-0821.

Yang, J., Jeng, D. R., and DeWitt, K. J., 1982, "Laminar Free Convection From a Vertical Plate With Nonuniform Surface Conditions," Numerical Heat Transfer, Vol. 5, pp. 165-184.

Yang, K. T., 1960, "Possible Similarity Solutions for Laminar Free Convection on Vertical Plates and Cylinders," Trans. ASME, J. Applied Mech., pp. 230-236.

THE CAVITY WIDTH EFFECT ON IMMERSION COOLING DUE TO DISCRETE FLUSH-HEATERS ON ONE VERTICAL WALL OF AN ENCLOSURE COOLED FROM THE TOP

R. Carmona and M. Keyhani
Department of Mechanical and Aerospace Engineering
The University of Tennessee
Knoxville, Tennessee

ABSTRACT

The width effect on natural convection heat transfer due to discrete flush-heated sections of equal height in an enclosure cooled from the top is experimentally investigated. Five heated sections are uniformly distributed along a vertical side wall, where the height of the unheated sections is equal to that of the heated sections. All other vertical surfaces and the bottom plate are insulated. The experiments are conducted for six values of cavity width resulting in a variation in the cavity height-to-width ratio (aspect ratio) of 3.67 to 12.22. Ethylene glycol and FC-75 (a dielectric fluid) are used as the convective media. The flow visualization results with ethylene glycol reveal a fairly inactive core flow at low power inputs and small cavity width. Higher values of the power input and cavity width transforms a rather well structured core flow into a time dependent one with higher horizontal velocities toward the "hot" wall. For a given cavity width, it is found that the heat transfer results for all the heated sections can be unified and presented by a single correlation through use of a local height as the length scale. The local height of a given heated section is measured from the bottom of the cavity to the mid-height of that section. Based on the local height length scale the data for all the cavity widths is correlated and an explicit relation for the aspect ratio effect on local Nusselt number is reported. The data indicate that an increase in the cavity width results in an increase in the heat transfer coefficients of the heated sections. A comparison between ethylene glycol and FC-75 revealed no appreciable Prandtl number effect which is in agreement with a previously reported prediction. A heat transfer coefficient for the top surface is defined based on the total convective heat flux at this surface and an average temperature difference between the heated sections and the top plate. The result shows that for a given heat flux at the top, the highest heat transfer coefficient on this surface can be obtained with the lowest cavity width.

NOMENCLATURE

A	aspect ratio, H/W
A_h	surface area of heater, m^2
A_T	surface area of top sink surface, m^2
C	isobaric specific heat, $J/kg\text{-}K$
D	Depth of the enclosure in the z direction, m

g	acceleration due to gravity, m/s^2
h	local heat transfer coefficient $q/(T_h\text{-}T_c)$, $W/m^2\text{-}K$
h_T	heat transfer coefficient at the top sink surface $q_T/(T_h\text{-}T_c)$, $W/m^2\text{-}K$
H	height of the enclosure, m
k	thermal conductivity, $W/m\text{-}K$
L_1	height of the heated section, m
L_2	height of the unheated section, m
N	heater row number (starting from the bottom)
Nu	Nusselt number, $h[\text{length scale}]/k$
$Nu_{W,T}$	Nusselt number based on the top sink surface
Pr	Prandtl number, ν/α
q	convective heat flux per heater section, $(Q\text{-}Q_l\text{-}Q_{sc})/A_h$, W/m^2
q_T	total convective heat flux to the sink surface, $(Q_{To}\text{-}Q_{l,To})/A_T$, W/m^2
Q	power input per heater section, W
Q_l	heat loss per heater section, W
$Q_{l,To}$	total heat loss from the enclosure to the enviroment, W
Q_{sc}	substrate conduction heat transfer per heated section, W
Q_{To}	total power input, W
Ra	Rayleigh number, $g\beta(T_h\text{-}T_c)[\text{length scale}]^3/\alpha\nu$
Ra^*	modified Rayleigh number, $NuRa$
$Ra^*_{W,T}$	modified Rayleigh number based on the top sink surface, $g\beta q_T W^4/k\nu\alpha$
T	temperature, K
t	exposure time of photographs, seconds
x, y, z	Cartesian coordinates, m
W	width of the cavity, m

Greek Symbols

α	thermal diffusivity, $k/\rho C$, m^2/s
ρ	density, kg/m^3
ν	kinematic viscosity, m^2/s
σ	standard deviation

Subscripts

c   cold wall
h   heated section
L_1 vertical length scale of a heated section.
y   based on the local height, measured from the bottom of the cavity to the mid-height of a heated or unheated section.

Superscripts

−   average value

INTRODUCTION

The evolution in miniaturization of electronic components, which unquestionably started with the invention of the transistor in 1947, has not come to an end yet. The desire to decrease time of electronic signals as well as the interest in having higher performance from printed circuits and solid state devices has resulted in the design of more densely packed electronic components. As a consequence, heat removal requirements have risen not only because power dissipation has increased but also because of the lower temperature requirements of integrated circuits. Techniques such as using higher velocities to enhance forced convection of the electronic arrays have become less satisfactory due to excessive acoustical noise generated. Natural convection, on the other hand, has taken a new dimension with the acceptance of liquid immersion cooling as a mean of maintaining thermal control of electronic devices.

Chu, Churchill, and Patterson (1976) were among the first investigators to conduct experimental and numerical studies of laminar natural convection from a flush-mounted heat source on an insulated vertical wall of an air-filled rectangular cavity. Their calculations and experiments were for aspect ratios (H/W) from 0.4 to 5.0 and a range of Grashof numbers up to 10^5. However, as pointed out by Turner and Flack (1980), their range of Grashof numbers was considerably low for actual electronic cooling applications. Turner and Flack experimentally expanded the numerical study made by Chu et al. to Grashof numbers as high as 9×10^6.

The numerical study conducted by Kuhn and Oosthuizen (1986) on three-dimensional transient natural convection flow in a rectangular enclosure with localized heating had demostrated that the proximity of the heater elements to the horizontal walls (especially the top wall) has a strong effect on the heat transfer coefficient while the side walls have neglegible effects. Ozoe (1987) presented a two-dimensional numerical study of natural convection in a square cavity with nine point sources of uniform heat flux. He reported results for the two cases of cooling from the top and cooling from the side. Keeping in mind that the side and the top areas were kept the same size, the conclusion was that cooling from the top was more effective and resulted in lower temperatures. Thus, the cooling of the bottom surface may not be required in the packaging considerations. Other studies deserving credit for their contribution to natural convection heat transfer in rectangular enclosures with protruding heaters are due to Kelleher, Knock and Yang (1987); Lee, Liu, Yang and Kelleher (1987); and Liu, Yang and Kelleher (1987). Liu et al. (1987) reported numerical analysis for the case of three-dimensional, natural convection cooling of an array of heated protrusions in an enclosure filled with fluorinert FC-75. The wall opposing the protruding elements was insulated while the top and bottom were isothermal. Similarly as Ozoe (1987) did, they concluded that the top sink surface is responsible for about all the heat dissipated from the enclosure.

Sparrow, Vemury and Kadle (1983) have indicated that in order to facilitate the design of digital computers, sophisticated computer programs are already in existence to trace the heat flow through complex paths. The imminent problem in the computer simulation is the unavailability of heat transfer coefficients which may be appropriate to the complex geometries typically encountered in a computer cooling passage. Liu et al. (1987) pointed out that no studies dealing with the cooling of arrays of chips in a closed enclosure were available for comparison with their numerical results. The only experimental data for multiple in-line heaters in an enclosure have been reported by Keyhani, Prasad and Cox (1988a); Chen, Keyhani and Pitts (1988); and Keyhani, Prasad, Shen, and Wong (1988b).

Review of the relevant literature has lead to the conclusion that experimental information is extremely scarce for the design and analysis of the thermal performance of electronic packages when the geometric considerations require that the problem be modeled as an enclosure. The aim of this work was to expand the limited understanding of natural convection in fluid-filled enclosures with multiple heaters by experimentally studying the cavity width effects (while keeping the cavity height constant) on the heat transfer coefficients and the flow regimes.

EXPERIMENTAL APPARATUS AND PROCEDURE

The test cell consisted of a liquid filled, rectangular enclosure isothermally cooled from the top. Five identical rows of isoflux heated strips were flush-mounted on one vertical side wall. These heated strips were distributed such that identical rows of unheated and heated sections were obtained. All other surfaces of the cavity were insulated. The schematic of the enclosure and the details of the experimental apparatus are given in Figs.1 and 2, respectively. Data were collected at steady state with ethylene glycol ($64 < Pr < 138$), and Fluorinert FC-75 ($22 < Pr < 25$) as the working fluids. The actual test section height and depth were 165 mm and 141 mm, respectively.

In order to have sufficient space for a wide variation in the width of the actual enclosure under study, an outer rectangular box with inner dimensions of H=171.35 mm, D=141 mm and W=139.7 mm was made of 25.4 mm thick plexiglass sheet. The vertical end-walls of the box (z-direction) were made out of 6.35 mm thick, ground plate glass. These plates were bolted to the outer plexiglass box by using aluminum frames. To make the cavity leak proof, a continuous groove was cut at the edges of the outer enclosure walls to accomodate a 3.18 mm silicone O-ring which was sandwiched between the outer plexiglass enclosure and the glass plates.

The top surface covering the cavity was a copper heat exchanger plate. Two 6.35 mm thick, rectangular pieces of copper formed the heat exchanger. A 3.18 mm deep semicircular groove was milled on the surface of each copper plate in the form of a sinusoidal wave line. The groove in one plate was the mirror image of the other and they met to form a channel when the plates were silver-soldered together. Water was circulated through this channel to maintain the plate at constant temperature. A constant temperature water bath circulator maintained the cooling water within ±0.15 °C. Seven 30 AWG (copper-constantan) thermocouples were placed in wells which were within 1 mm of the bottom surface of the plate to verify that the copper heat exchanger was maintained at a constant temperature. Eight holes of 3.97 mm diameter were drilled through the top plate so that sealed probes could be inserted into the enclosure to measure the core flow temperature.

The actual experimental enclosure was formed by placing a "movable" vertical plate and the "hot" plate inside the rectangular box enclosure formed by the outer plexiglass plates and the two glass end-walls (please see Fig. 2 for details). In order to minimize heat losses, the movable vertical wall was instrumented with four silicone-rubber thermofoil heaters which were sandwiched between 6.35 mm thick and 19.1 mm thick plexiglass plates. The thicker plate was instrumented with differential thermopiles to determine the amount of heat losses. The voltage supplied to the "guard" heaters was adjusted such that the output of the differential thermopiles were reduced to a level that the heat loss was below 2 percent of the total power input.

The heaters were placed such that a seven mm gap was available on the "movable" wall. This gap was used to introduce a thin plane of light during flow visualization experiments. A 3.18-mm deep groove was milled on the surface of the 19.1 mm thick plate to facilitate channeling of the thermocouple leads. All the thermocouple leads were secured in their channel grooves with a 8-hour cure time epoxy. A 3.18-mm O-ring was placed on the edge of the movable plate to maintain it firmly inside the experimental enclosure. With this movable plate the actual test cell width could be varied from 0 to 88 mm.

The heated phenolic wall was 177.7 mm high and 141 mm wide. The complete "hot" wall was formed by bolting together two 12.7 mm thick phenolic sheets. One sheet was instrumented with "guard" heaters similar to those of the "movable" plexiglass plate. Five slots with dimensions of 15 by 141 by 3.425 mm deep were cut in the face of the frontal phenolic sheet. Silicon rubber thermofoil heaters with a thickness of 0.25 mm were placed in each slot and covered with heat sink compound to enhance heat transfer. Aluminum plates with the same dimensions as the slots, but with a thickness of 3.125 mm, were placed over the strip heaters. Each aluminum plate was instrumented with at least one 30 AWG thermocouple (type T) which was inserted in a well drilled on the side of the aluminum plate. These thermocouples were located at the mid-height of the aluminum plates and at the midspan of the cavity. Each unheated section of the phenolic plate was also instrumented with 30 AWG thermocouples for temperature measurements. The thermocouple and heater leads were routed through holes made in the phenolite plate and then directed out of the cavity through channels milled on the back of the plate. The face of the hot wall was sanded to produce a smooth-flat surface between the aluminum plates and the phenolite wall. The bottom horizontal wall was also instrumented with "guard" heaters and differential thermopiles to measure and control the heat losses.

Visualization experiments were conducted with ethylene glycol to investigate the flow structure. A thin plane of light was projected through the existing transparent gap on the "movable" wall to illuminate the midspan section of the test enclosure. Aluminum powder (5 to 20 μm in size) was used as tracer particles. Still photographs were taken for each cavity width with at least two nominal power inputs.

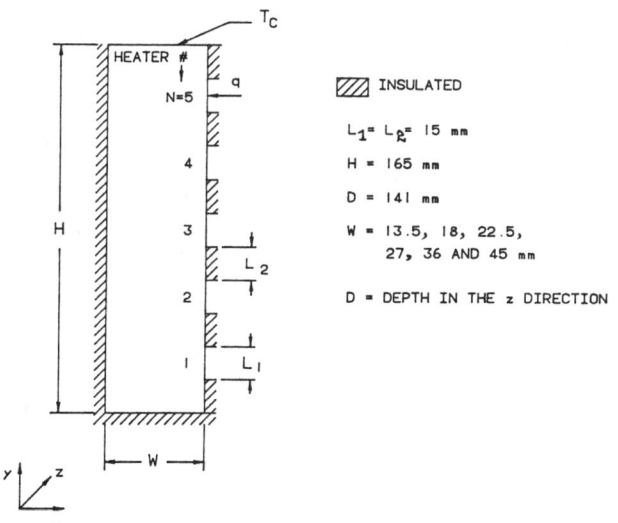

Fig. 1. Schematic of the enclosure.

A Sorencen[TM] power panel containing a voltage regulator, a step-down transformer, and five variacs was used to measure and control the voltage and current in each circuit. These units are compatible with an existing data acquisition package consisting of a Hewlett-Packard control unit (HP 3497A), a digital voltmeter (HP 3456A), an extender (HP 3498A) and a system voltmeter (HP 3437A). All digital information were collected via a Hewlett-Packard 9817A desktop computer and stored for subsequent data reduction and analysis.

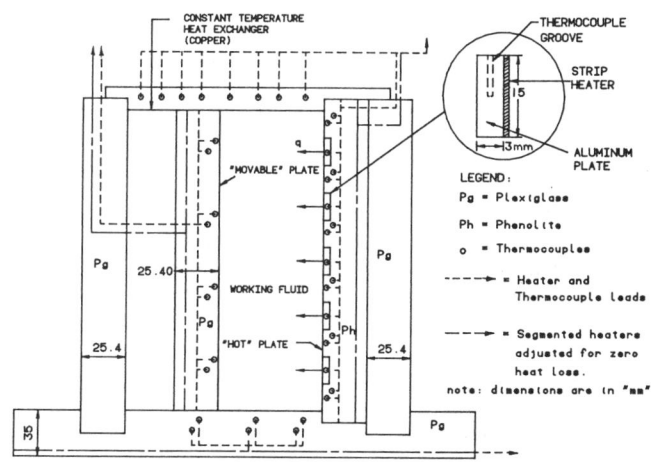

Fig. 2. Details of the experimental apparatus.

DATA REDUCTION

The Nusselt number and modified Rayleigh number used in the presentation of the data were calculated by

$$Nu = h[\text{length scale}]/k \qquad (1)$$

$$Ra^* = g\beta q[\text{length scale}]^4/k\upsilon\alpha \qquad (2)$$

Three length scales of local height (Nu_y, Ra^*_y), heater height (Nu_{L1}, Ra^*_{L1}) and cavity width ($Nu_{W,T}$, $Ra^*_{W,T}$) are used in the presentation of the results.

When using the length scales of L_1 and y, the local values of the thermophysical properties for ethylene glycol and fluorinert (FC-75) were evaluated at the average temperature,

$$T_f = (T_h + T_c)/2 \qquad (3)$$

where T_h is the local mid-height temperature of a given heater and T_c is the constant sink temperature. Kelleher et al. (1987) and Lee et al. (1987) suggest the use of this simple arithmetic mean temperature for evaluating the thermophysical properties. The local convective heat transfer coefficient h (of a given heated section) is defined as

$$h = (Q - Q_l - Q_{sc})/A_h(T_h - T_c) \qquad (4)$$

where $(T_h - T_c)$ is the excess local temperature between each heater and the cold plate, Q is the power input to each heater, Q_l is the heat loss per heater to the laboratory environment by conduction through the enclosure walls, Q_{sc} is the substrate conduction heat dissipated to the fluid from the unheated phenolite surfaces, and A_h is the heater surface area. A two-dimensional unit cell conduction heat transfer model was numerically solved to calculated the heat convected to the fluid through the unheated phenolic sections. The unit cell modeling heaters 2, 3, and 4 represented the details of the phenolite plate from the midheight of an unheated section to the midheight of the next unheated section. Hence, this unit cell comprised half the vertical length of two unheated regions which were adjacent to the full vertical length of a heated section (with a line source 3 mm from the surface). In order to model the asymmetrical unheated sections of heaters 1and 5, a simple alteration was done on the unit cell, just mentioned, to include the full vertical length of one of the unheated sections. The measured temperatures were used to specify the boundary conditions of the unit cell. It was found that the convection heat transfer to the fluid through substrate conduction ranged from about 2 to 20 percent of the power input per heater for glycol and 16 to 38 percent for experiments conducted with FC-75. The FC-75 requires a higher substrate correction due to its lower thermal conductivity (~0.063 W/m K) than that of ethylene glycol (~0.25 W/m K).

For the results presented using the length scale W, the thermophysical properties were evaluated at an overall film average temperature of

$$\overline{T}_f = (\overline{T}_h + T_c)/2 \qquad (5)$$

where $\overline{T}_h$ is the average temperature of the heaters. The heat transfer coefficient for this scaling is defined as

$$h_T = (Q_{To} - Q_{l,To})/A_T(\overline{T}_h - T_c) \qquad (6)$$

Where Q_{To} is the total power input, $Q_{l,To}$ is the total heat loss to the laboratory environment and A_T is the top sink surface. It should be noted that the convective energy at the top surface includes the energy convected from the unheated sections.

The standard deviation of the correlations presented in this report was calculated by using :

$$\sigma^2 = (\quad \Sigma(Nu_{y,i} - \overline{Nu}_y)^2)/(n-1) \qquad (7)$$

where n is the total number of data points (based on the "y" length scale), $Nu_{y,i}$ is the experimental local Nusselt number and $\overline{Nu}_y$ is the mean of the experimental local Nusselt number.

UNCERTAINTY ANALYSIS

The thermocouple outputs were measured to $\pm 0.1 \mu V$ which rendered a sensitivity of 0.0025 $^{\circ}$C. A reasonable limit of the accuracy of the temperature measurement is within ± 0.15 $^{\circ}$C. This conservative limit contains errors associated with reference junction compensation (± 0.01 $^{\circ}$C), temperature difference along terminals, thermal offset, voltage to temperature conversion error, and DVM accuracy. Measurements of the voltage input to each heater circuit were done with a sensitivity of 1mV and an accuracy of 0.1 percent. The input voltage and current to each circuit were measured for each run to verify the experimental stability of each circuit resistance. A maximum uncertainty of ± 1.5 percent was obtained when comparing the original resistance accurately measured before the experiments with the average experimental resistance. The uncertainty associated with the length scale for the enclosure width was ± 0.25 mm. The thermophysical properties of ethylene glycol and fluorinert (FC-75) were assumed to have an uncertainty of ± 2 percent.

This judgment was based on the observed variations on the reported values in the literature.

The uncertainty in the Nusselt and modified Rayleigh numbers has been found to be strongly dependent on the power input. Low power input results in higher uncertainties due to small temperature differences. As the power input is increased, the uncertainty in the experimental data is decreased. The estimated uncertainties in the Nusselt and modified Rayleigh numbers are 3.1 to 10.0 and 4.2 to 10.9 percent respectively, for glycol. Likewise, for fluorinert the corresponding percentages are 5.3 to 10.2 and 7.8 to 11.6.

RESULTS AND DISCUSSION

The experiments with ethylene glycol were conducted for cavity widths of 13.5, 18, 22.5, 27, 36 and 45 mm, and for nominal power inputs ranging from 2 to 18 watts per heater. The data presented for FC-75 were collected for a cavity width of 18 mm and nominal power inputs of 2 and 4 watts per heater. It should be noted that the absence of a freeboard height above the dielectric fluid level (to keep the vapor in place) results in a rather high rate of evaporation. Due to other constraints on the design of the apparatus, no provisions were made to provide a freeboard height above the liquid level. Thus, a rather high rate of evaporation occurred during the experiments. The objective of our experiments with FC-75 was to verify that the ethylene glycol results are applicable to immersion cooling as reported by Keyhani et al. (1988b). Considering the fact that FC-75 is rather expensive, the experiments were limited to two power levels which were sufficient to achieve the stated objective.

The experiments conducted with ethylene glycol as convective medium covered a range of local modified Rayleigh numbers Ra_y^* from 4.1×10^6 to 1.37×10^{11}. This corresponds to a local Prandtl number variation between 137.8 and 63.7. The range of modified Rayleigh numbers covered with FC-75 was from 1.02×10^9 to 5.24×10^{12}, with a corresponding local Prandtl number variation between 24.88 and 22.03.

Flow Visualization

Photographs of the flow patterns for the entire enclosure are shown in Figs. 3, 4, 5 and 6. A primary flow traveling along the vertical side walls of the cavity in a counterclockwise sense constitutes the main mechanism for the transport of energy to the top surface. The following general observations can be made from a detailed study of the flow visualization results for all the cavity widths. At low power inputs a very well structured core flow is evident (Figs. 3 and 6). As the power input is increased the following two events take place : (i) the bottom of the core flow moves upward (Figs. 3 and 6); and (ii) a stable and structured core flow at a lower power input is transformed into a time dependent one (Figs.3 to 6).

The cavity width effect on the flow structure can be summarized as following : (i) For W>13.5 mm, a large secondary flow cell (counterclockwise motion) at the top portion of the cavity is observed (Figs. 3 to 6). As the power input is increased the center of rotation of this cell moves toward the upper right-hand corner of the cavity ("hot" wall), thus, causing the creation of a tertiary cell (clockwise motion) in the upper left-hand corner of the cavity (Fig. 6). (ii) It is observed that, for a given power input, an increase in the cavity width results in a much stronger horizontal movement of the fluid in a larger lower portion of the cavity (please see the photographs for Q=2 watts, and cavity width of 13.5, 27 and 45 mm). The combined effects of increasing power input and cavity width are : (i) a transition from a fairly inactive core to one where secondary and tertiary cells are created at the top of the cavity; and (ii)

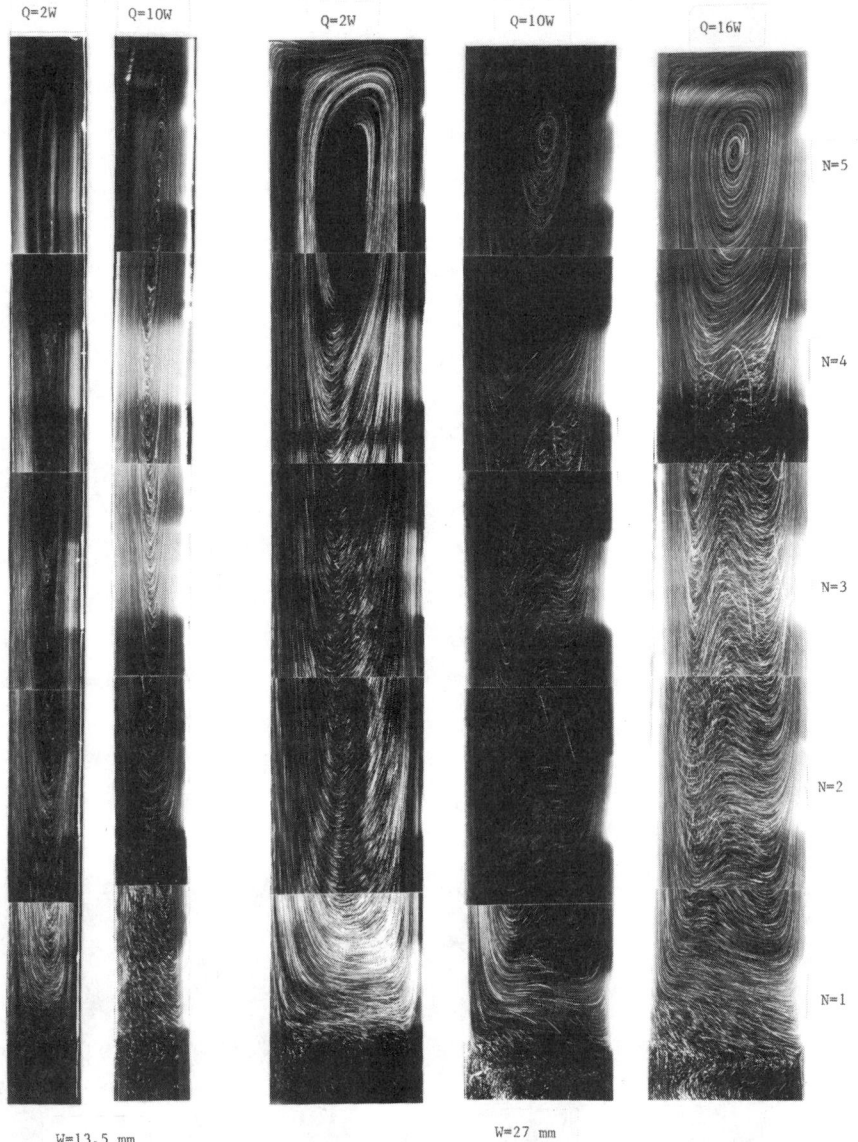

Fig. 3.   Streak photographs for the entire cavity : (a) W = 13.5 mm; (b) W = 27 mm.

development of a time dependent core flow (below the secondary and tertiary cells), with increasing horizontal velocities toward the "hot" wall.

Sample photographs of the time dependent core flow are shown in Figs. 4 and 5. Figure 4 shows the core flow adjacent to heaters 3 and 4 for the case of 12 watts per heater and cavity width of 18 mm. The time sequence of the close-up photographs taken is from left to right in the figure. Figure 5 shows the core flow adjacent to heater 4 for the case of 16 watts per heater and cavity width of 36 mm. The time sequence of the close-up photographs shown in this figure is from top to bottom. The exposure time of each photograph is noted in the figures. Sporadic extension and subsequent shrinkage of the secondary flow cell adjacent to heater 5 into the region of heater 4 can be seen in Fig. 4. The photographs of the core flow adjacent to heater 3 (N = 3) indicate a sporadic formation and subsequent vanishing of a secondary flow cell in this region. At a higher power input and a larger cavity width, one can observe sporadic formation and subsequent vanishing of a tertiary flow cell below the top secondary cell next to heater 4 (Fig.5).

Temperature Distribution

The temperature distributions $(T-T_c)$ along the hot wall (unheated and heated sections) for various cavity widths and power inputs with ethylene glycol as the working fluid are presented in Figs. 7 and 8. The locations of the heated sections are indicated via the heater number (N) in the graphs. Figure 7 shows the effect of the variation in the cavity width on the temperature profile, while Fig. 8 demonstrates the dependence of the profile on the power input.

It should be emphasized that the magnitude of the $(T-T_c)$ of the unheated sections does not imply substantial rate of transfer of energy to the fluid from these regions. Indeed, the conduction

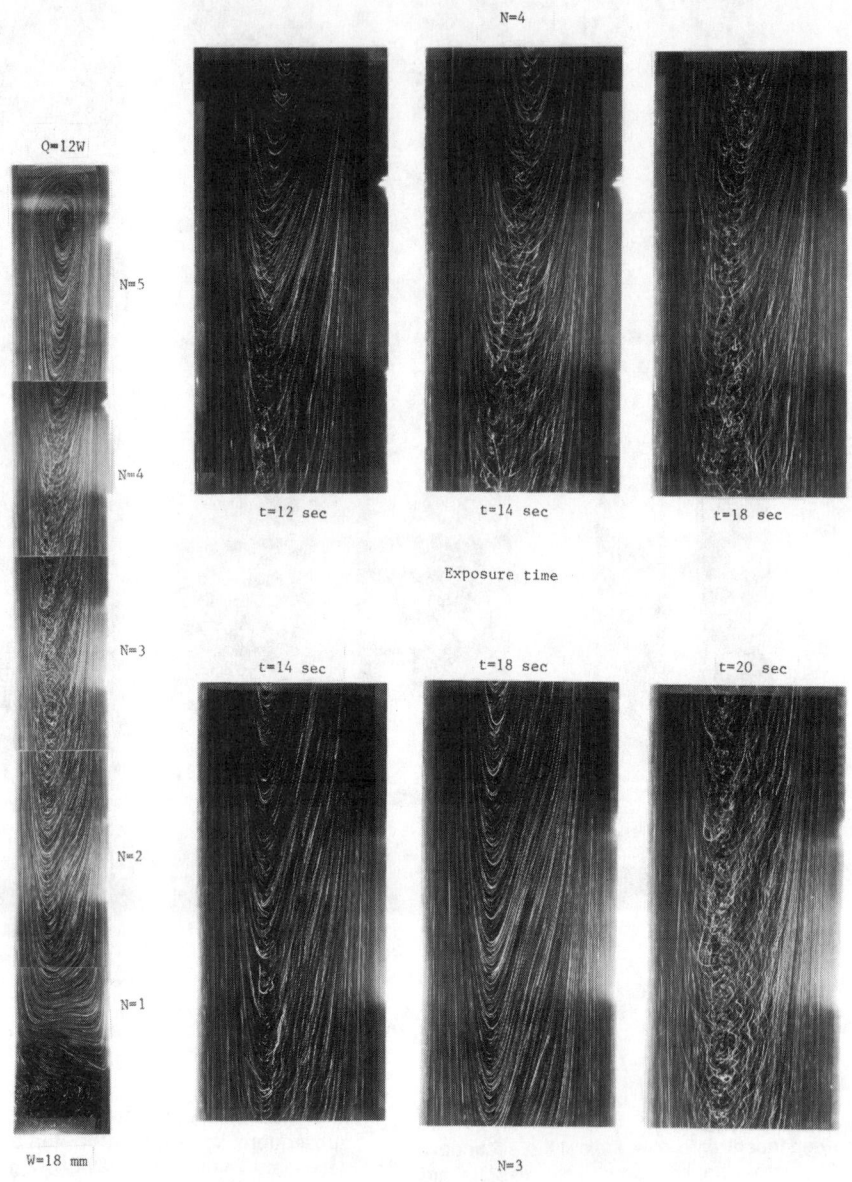

Fig. 4.   Streak photographs of the time dependent core flow at heaters 3 and 4 for W = 18 mm and Q = 12 Watts.

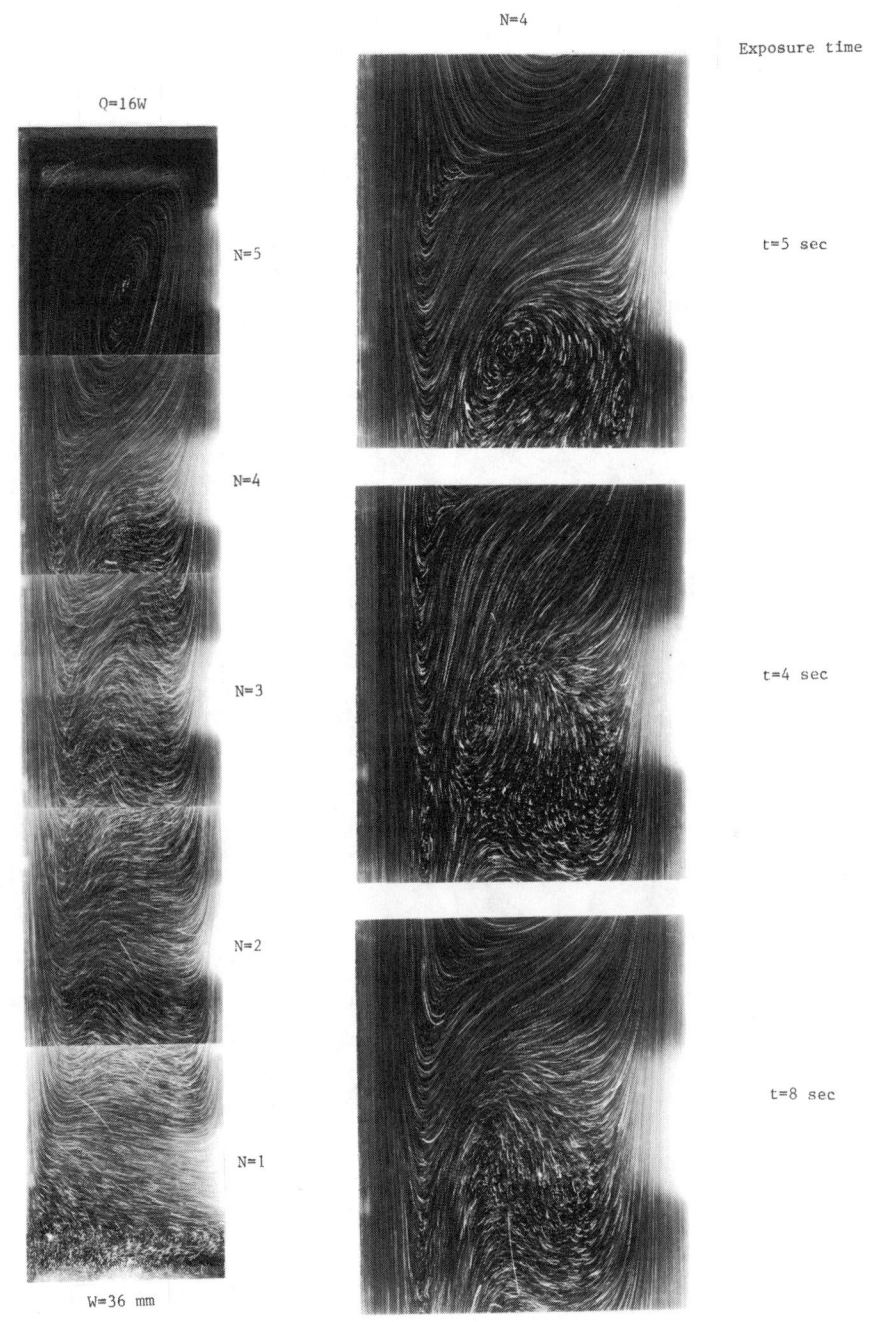

Fig. 5.   Streak photographs of the time dependent core flow below heater 5 (N = 4) for W = 36 mm and Q = 16 W.

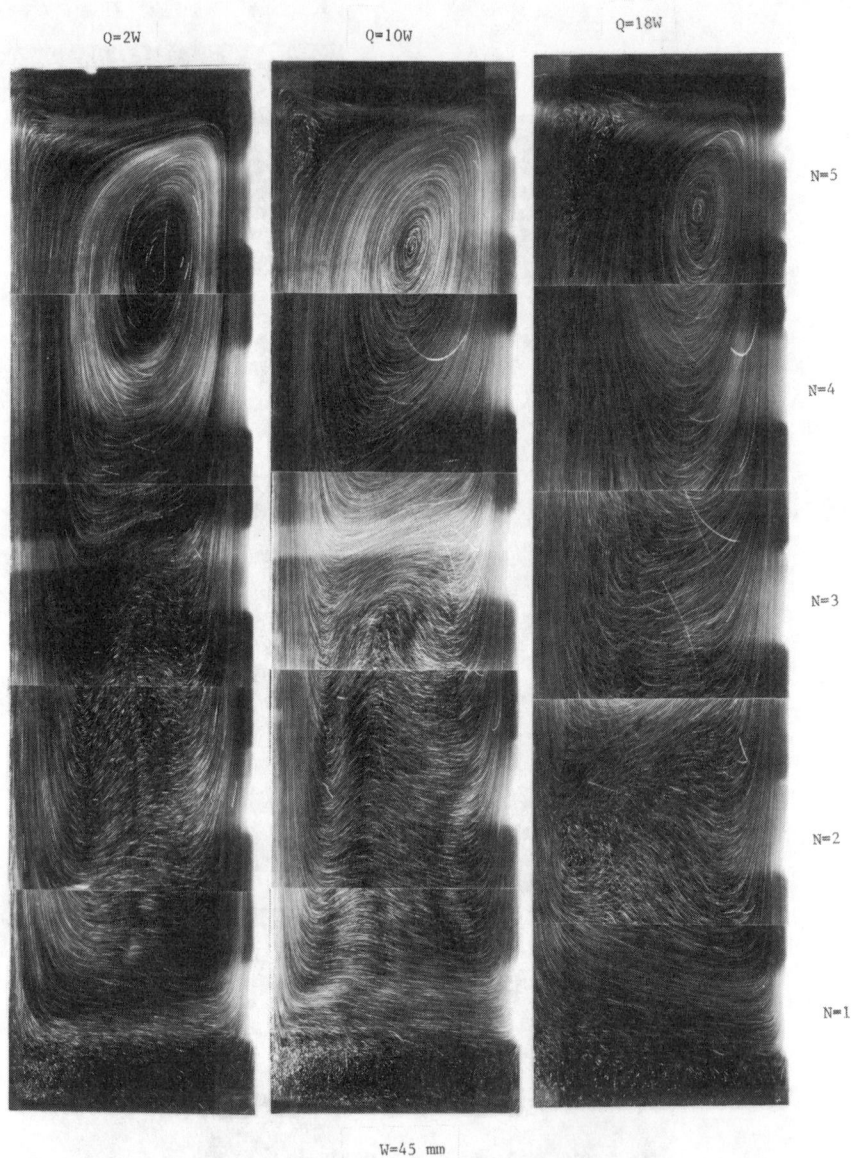

Fig. 6.   Streak photographs for the entire cavity for W = 45 mm.

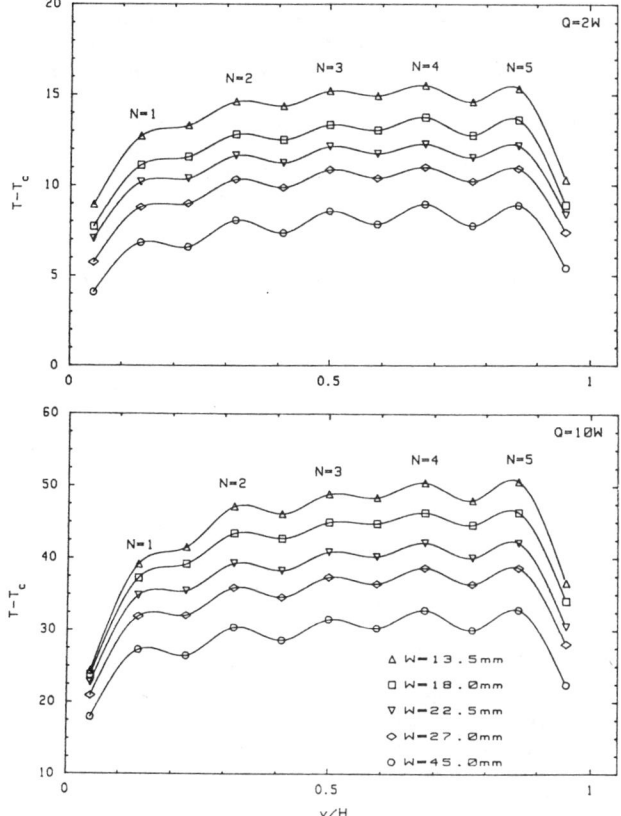

Fig. 7. Cavity width effect on the temperature profile along the "hot" wall.

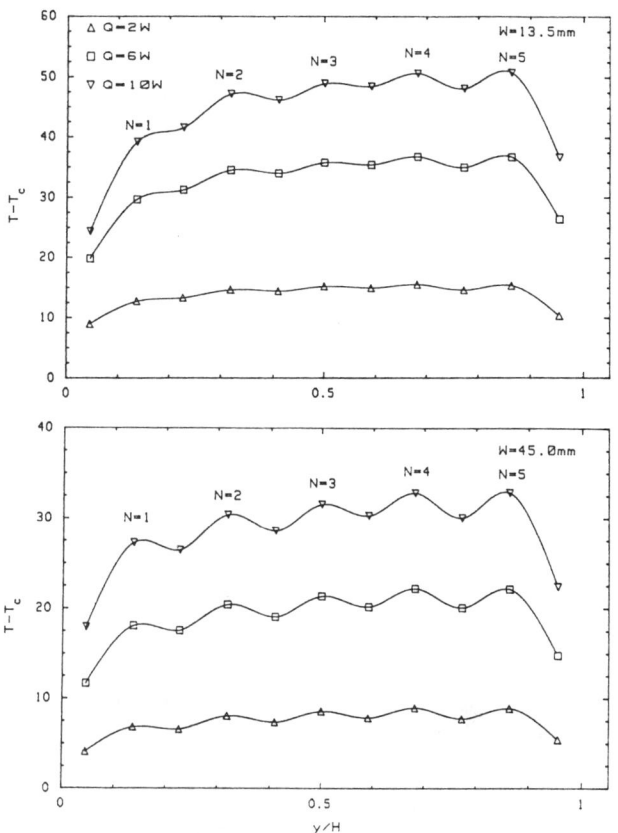

Fig. 8. Power input effect on the temperature profile along the "hot" wall.

solutions of the substrate show that the highest rate of convection heat transfer from the unheated regions occurs in the section with the lowest temperature. Thus, the highest substrate conduction corrections are applied to heaters 1 and 5, respectively.

The lowest temperature for the heated sections occurs at heater 1, while the highest value fluctuates between heaters 4 and 5. The difference between the temperature of the heaters 4 and 5, however, is less than 0.2 °C. Although there is a steady increase in the temperature of the heated sections from heater 1 to 4, the magnitude of the temperature rise is not substantial. The data suggest that the temperature of the heated sections, for each power input and cavity width, can be represented with an average value $\overline{T}_h \pm 5$ percent. Based on the temperature difference, the data can be approximated with an average value $(\overline{T}_h - T_c) \pm 18$ percent. The variation of the maximum temperature difference from the average value, however, is less than 8 percent. A rather small difference between the maximum observed temperature and the average value can have important practical applications in design and analysis of cooling problems with multiple heated sections.

The sensitivity of the "hot" wall temperature distribution to variation in the power input for a fixed cavity width is illustrated in Fig. 8. For W = 13.5 mm, the maximum rises in $T_h - T_c$ above those of heater 1 for Q = 2 and 10 watts are 22% and 29.5%, respectively. Therefore, the rate of rise in the temperature of the heated sections increases with power input. A reverse trend is observed for W = 45 mm. At this cavity width,

the maximum rises in $T_h - T_c$ above those of heater 1 for Q = 2 and 10 watts are 31.5% and 20.6%, respectively. It should be noted that the maximum rise in the $T_h - T_c$ above that of heater 1 is nearly constant for W = 18 mm (23.8% at Q=2 watts and 24.5% at Q=10 watts). Thus, the reversal in the trend begins with W = 22.5 mm. As discussed earlier a simultaneous increase in the power input and cavity width results in a more active core flow as well as formation of a tertiary cell at the top of the cavity. Thus, for W = 45 mm and Q = 10 watts we observe a relatively smaller rise in the temperature of the heated sections above that of heater 1.

The temperature profiles at x/W = 0.167, 0.186, 0.583 and 1 (the "hot" wall) are presented in Fig. 9 for a power input of 12 watts per heater and cavity widths of 18 and 36 mm. For both cavity widths, the fluid temperature, with the exception of the end regions, is nearly constant. Excluding the end regions, the fluid temperature $T - T_c$ for W = 18 mm and 36 mm can be approximated as 32.58 ± 0.5 °C and 23.36 ± 0.5 °C, respectively. A measure of comparison of the core flow at these two cavity widths is the ratio of the average core-fluid temperature to that of the heated sections. The ratios of the average core-fluid temperature (excluding the end regions) to that of the heated sections for W = 18 mm and 36 mm are 0.69 and 0.594, respectively. The higher cavity width has resulted in a relatively "cooler" core-fluid temperature. It may be noted that the fluid temperature distributions at x/W = 0.186 and 0.583 are nearly identical. This behavior is attributed to the insulated boundary condition at x=0.

215

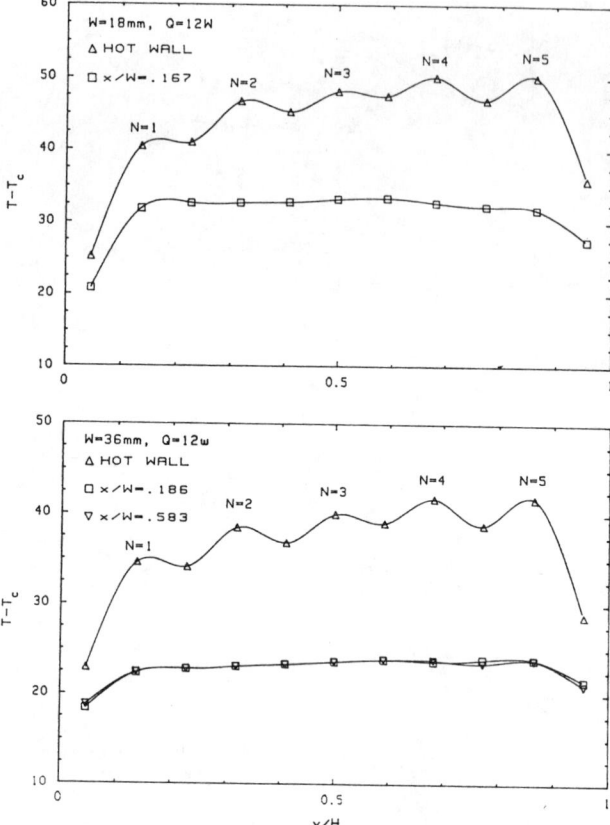

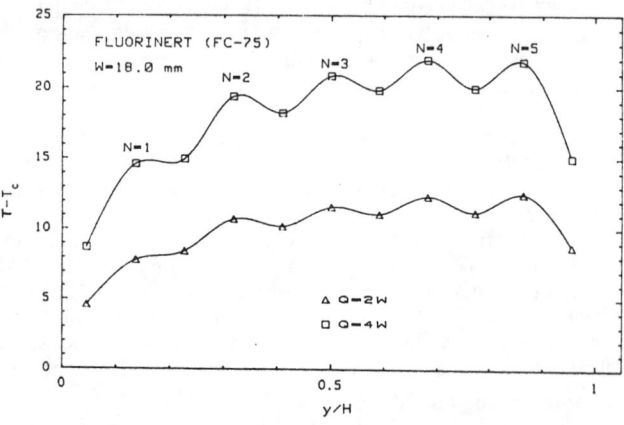

Fig. 9. Core flow temperature distribution; (a) W=18 mm, Q=12Watts; and (b) W=36 mm, Q=12 Watts.

Fig. 10. Temperature profile along the "hot" wall with FC-75 as the working fluid.

The temperature distributions along the "hot" wall with FC-75 as the working fluid for W = 18 mm and power inputs of 2 and 4 watts per heater are shown in Fig. 10. It should be mentioned that the thermophysical properties group of $(g\beta)/(\alpha\upsilon)$ for FC-75 at 25 °C is 105 times larger than that of ethylene glycol. Thus, the FC-75 temperature profiles shown in Fig. 10 are for substantially higher Rayleigh numbers than those given in Fig. 7 for glycol. The general trend of the temperature variation for FC-75 is similar to that discussed for glycol data. The heated section temperatures can be represented with an average value $\overline{T}_h \pm 12$ percent (7% higher deviation than the glycol data). However, the deviation of the maximum temperature from the average value is only 5.2%. Based on the temperature difference, the data can be approximated with an average value $(\overline{T}_h - T_c) \pm 29\%$ (11% higher deviation than the glycol data). The maximum rises in $T_h - T_c$ above those of heater 1 for Q = 2 and 4 watts are 60% and 51%, respectively.

Heat Transfer

The Nusselt numbers Nu_{L1} for heaters 1 to 5 for cavity widths of 13.5 and 45 mm are presented in Fig. 11. As can be seen from the figure, heater 1 has the highest heat transfer coefficient, while heater 5 has the lowest. For W = 13.5 mm, the difference between the Nusselt numbers of heaters 1 and 2 is rather small at low Ra^*_{L1} values and increases with Ra^*_{L1}. A reverse trend is observed for W = 45 mm. This behavior is due to the changes in the flow patterns caused by variations in the power input and cavity width (discussed earlier).

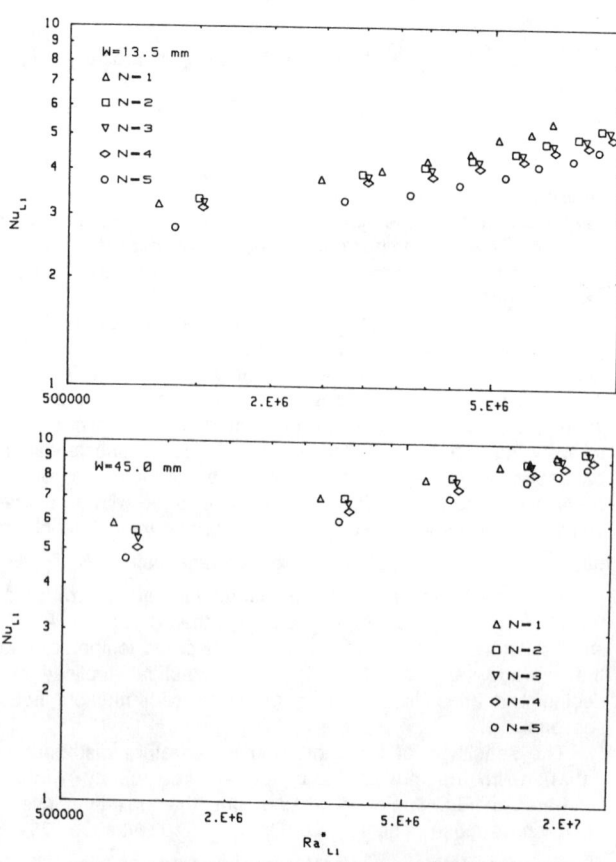

Fig. 11. Heated section Nusselt numbers (Nu_{L1}) as a function of Ra^*_{L1} for W=13.5 mm and W=45 mm.

The presentation of the Nusselt numbers in terms of Nu_{L1} for all of the heated sections and cavity widths would require several graphs and thirty correlations. Chen et al. (1988) have shown that the choice of length scale of local height "y" substantially simplifies the task of the presentation and discussion of the results. The present data based on the length scale of "y" are presented in Fig. 12. It is interesting to note that for each cavity width, the log-log plot of the Nu_y versus Ra^*_y for all the heated sections forms a straight line. Moreover, the data for each cavity width can be correlated with a standard deviation of less than 0.83. It is obvious that the task of analysis of cooling problems is substantially simplified when one can unify the data via local height scaling. The local Nusselt number of the heated sections Nu_y for each cavity width is correlated in terms of its respective local modified Rayleigh number Ra^*_y in the form of

$$Nu_y = C(Ra^*_y)^m \qquad (8)$$

and the constants C, exponents m, and the standard errors are tabulated in Table 1. These correlations represent the data with an average deviation of less than 4.6 percent.

Table 1. Heat transfer correlation coefficients for each cavity width (equation (8)).

W (mm)	C	m	σ	Applicable range (min < Ra^*_y < max)
13.5	0.156	0.226	0.657	4.1E6 - 7.23E10
18.0	0.154	0.234	0.830	4.1E6 - 5.24E12
22.5	0.180	0.230	0.682	3.8E6 - 1.07E11
27.0	0.210	0.228	0.676	3.7E6 - 1.22E11
36.0	0.229	0.226	0.684	3.7E6 - 1.22E11
45.0	0.276	0.223	0.677	3.6E6 - 1.40E11

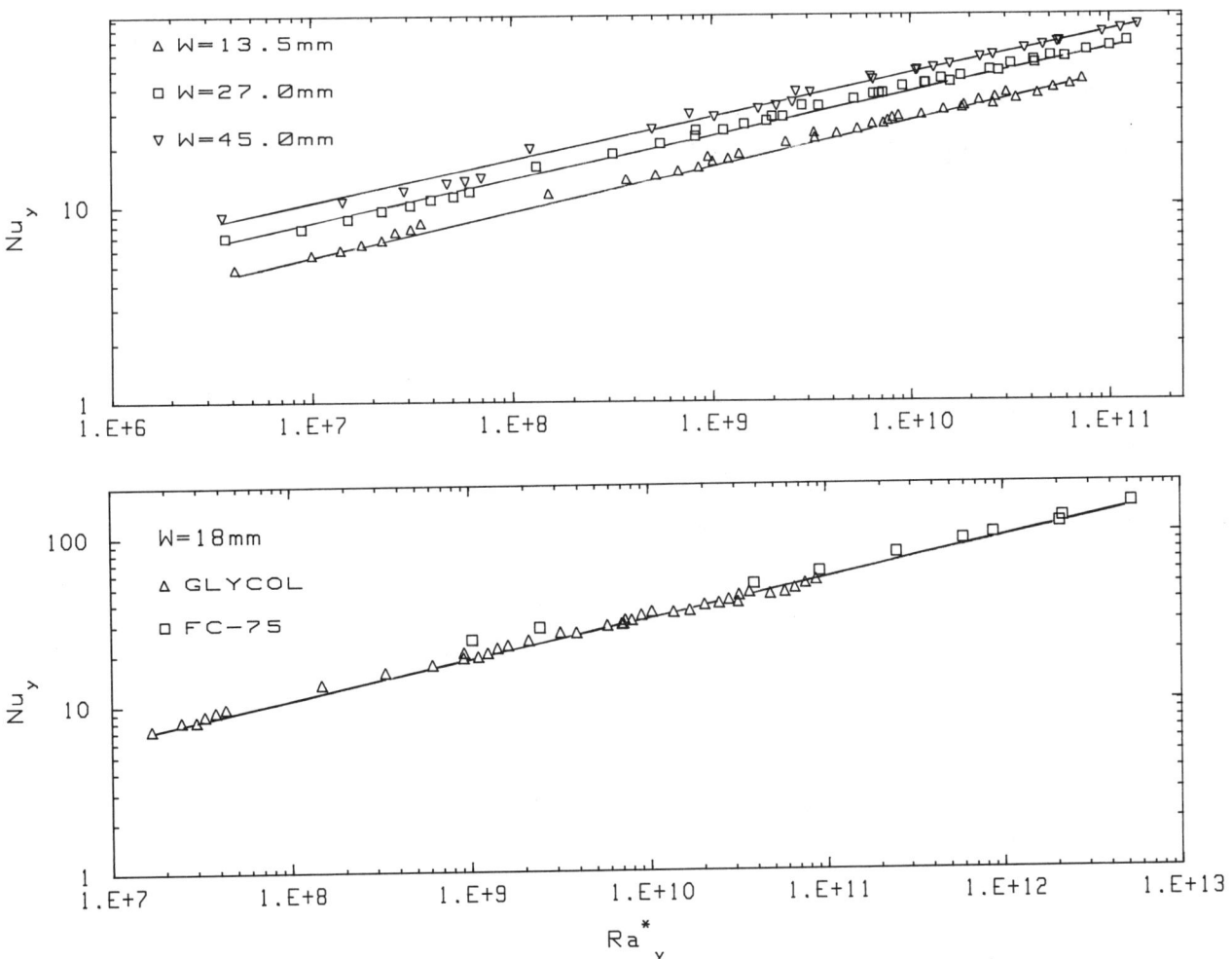

Fig. 12. Local Nusselt number (Nu_y) as a function of modified Rayleigh number (Ra^*_y).

All the experimental data collected with ethylene glycol and FC-75 for the six cavity widths can be correlated into the following general equation :

$$Nu_y = 0.437(A)^{-0.416}Ra^*_y{}^{0.227} \qquad (9)$$

Equation (9) predicts the experimental data with an average deviation of 4.6 percent and it is applicable for the range of aspect ratios (A) between 3.67 and 12.22.

The effect of increasing the cavity width on the heat transfer coefficient can clearly be seen in Fig. 12. As expected, the increase in W, i. e., heat sink surface, results in higher heat transfer coefficients for all the heaters. The data indicates that at $Ra^*_y = 10^9$ an increase in W from 13.5 to 45 mm results in a 66% higher heat transfer coefficient. The ethylene glycol and FC-75 data for W = 18 mm are presented in a separate plot in Fig. 12. A single correlation as given in Table 1 represents the two fluid data with an average deviation of 4.6 percent. This agreement obviously indicates a negligible Prandtl number effect for the range covered (23 < Pr < 124). Keyhani et al. (1988b) conducted experimental and numerical studies of natural convection in a vertical cavity with an isothermal vertical wall and three flush-mounted heated sections on the opposing vertical wall. Based on the numerical results, they reported that although the flow structure varies significantly when the Prandtl number is reduced from 25 to 1, the variations are negligible when Pr is increased from 25 to 166.

The total energy convected from the heated and unheated sections is transferred to the top sink surface. The heat transfer coefficient h_T at the top surface as defined by equation (6) is plotted against the total convective heat flux at the top surface q_T

in Fig. 13. This figure shows that h_T increases with decreasing cavity width. It should be noted that for a given power input the highest fluid velocity was observed at the smallest cavity width during the flow visualization experiments. Liu et al. (1987) reported a similar finding in their numerical study dealing with convection cooling of an array of heated protrusions in a cavity. The FC-75 data for W = 18 mm are even higher than the glycol data for W=13.5 mm. This is due to the fact that the corresponding $Ra^*_{W,T}$ for the FC-75 data are more than 300 times larger than those for glycol data. The non-dimensional ($Nu_{W,T} = h_T W/k$ versus $Ra^*_{W,T} = (g\beta/\alpha\upsilon)W^3(q_T W/k)$) version of the data in Fig. 13 are presented in Fig. 14. It appears that the data for all the cavity widths converge into a single line at low $Ra^*_{W,T}$ values. It is evident that for all cavity widths as the $Ra^*_{W,T}$ is decreased a change in flow regime is taking place. This may be an indication of transition to a conduction dominated flow regime. For a given fluid, the data at low $Ra^*_{W,T}$ suggests that for a fixed modified Rayleigh number one obtains the same value for $h_T W$ product for the present range of aspect ratios. Thus, a decrease in W is off-set by a corresponding increase in h_T. However, as the $Ra^*_{W,T}$ is increased the data for each W form distinct separate lines. The slope of $Nu_{W,T}$ versus $Ra^*_{W,T}$ on the log-log plot increases with decreasing W. At high $Ra^*_{W,T}$ the data indicate that, for a given fluid and $Ra^*_{W,T}$, a decrease in W results in higher value for $h_T W$ product. Thus a decrease in W is compensated for via a proportionately higher increase in the h_T. Unfortunately, we do not have sufficient data at either end of the $Ra^*_{W,T}$ range in order to report any correlations.

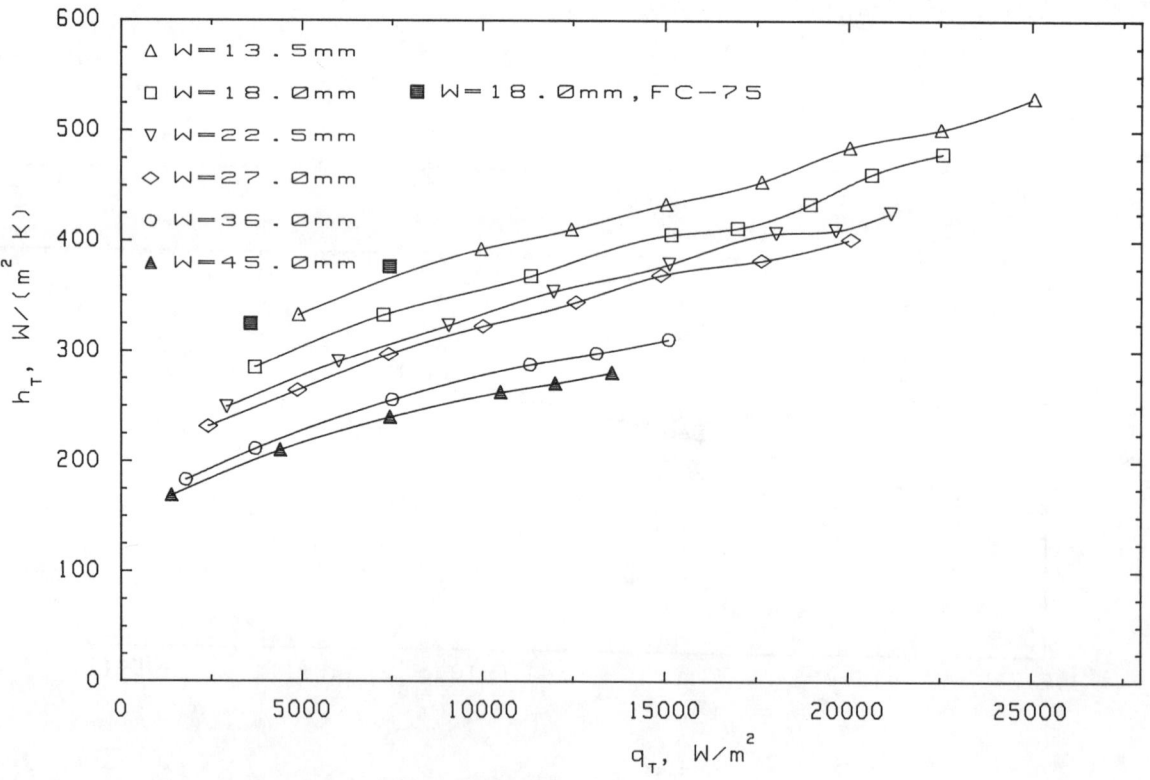

Fig. 13.   Heat transfer coefficient (based on the top heat sink surface) as a function of the heat flux at the top surface.

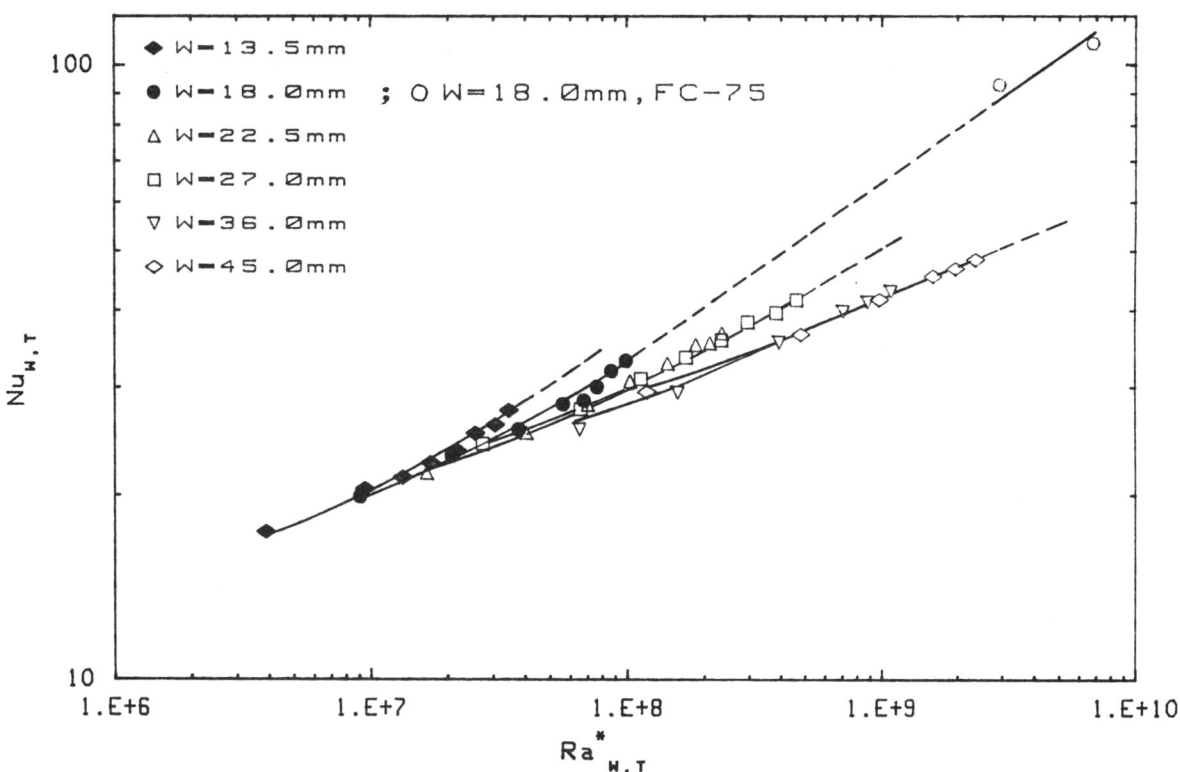

Fig. 14.   Nusselt number as a function of modified Rayleigh number (based on the top heat
sink surface).

REFERENCES

Chen, L., Keyhani, M., and Pitts, D. R., 1988, "An experimental Study of Natural Convection Heat Transfer in a Rectangular Enclosure with Protruding Heaters," Proceedings of The 1988 National Heat Transfer Conference, ASME HTD-Vol. 2, pp. 125-133.

Chu, H. H., Churchill, S. W., and Patterson, C. V. S., 1976, "The Effects of Heater Size, Location, Aspect Ratio, and Boundary Conditions on Two-Dimensional, Laminar, Natural Convection in Rectangular Channels," Journal of Heat Transfer, Vol.98, pp. 194-201.

Kellerher, M. D., Knock, R. H., and Yang, K. T., 1987, "Laminar Natural Convection in a Rectangular Enclosure due to a Heated Protrusion on one Vertical Wall-Part I: Experimental Investigation," Proc. 2nd. ASME/JSME Thermal Engineering Joint Conference, Honolulu, Hawaii, Vol.2, pp.169-177.

Keyhani, M., Prasad, V., and Cox, R., 1988a, "An Experimental Study of Natural Convection in a Vertical Cavity with Discrete Heat Sources," ASME Journal of Heat Transfer, Vol. 110, pp. 616-624.

Keyhani, M., Prasad, V., Shen, R., and Wong, T. T., 1988b "Free Convection Heat Transfer from Discrete Heat Sources in a Vertical Cavity," in Natural and Mixed convection in Electronic Equipment Cooling, Ed. R. A. Wirtz, ASME HTD-Vol. 100, pp. 13-24.

Kuhn, D., and Oosthuizen, P. H., 1986, "Three-Dimensional Transient Natural Convective Flow in a Rectangular Enclosure with Localized Heating," Natural Convection in Enclosures-1986, ASME HTD-Vol. 63, pp. 55-62.

Lee, J. J., Liu, K. V., Yang, K. T., and Kelleher, M. D., 1987, "Laminar Natural Convection in a Rectangular Enclosure Due to a Heated Protrusion on One Vertical Wall - Part II: Numerical Simulations," Proc. 2nd. ASME/JSME Thermal Engineering Joint Conference, Honolulu, Hawaii, Vol. 2, pp. 179-185.

Liu, K. V., Yang, K. T., and Kelleher, M. D., 1987, "Three Dimensional Natural Convection Cooling of an Array of Heated Protrusions in a Enclosure Filled with a Dialectric Fluid,"Proc. Int. Symposium on Cooling Technology for Electronic Equipment, Honolulu, Hawaii, pp. 486-497.

Ozoe, H., 1987, "The Development of a Computational Scheme for Natural Convection in an Enclosure with Multiple Isolated Heat Sources," Proceedings, International Symposium of Cooling Technology for Electronic Equipment, Honolulu, Hawaii, pp. 522-534.

Sparrow, E. M., Vemury, S. B., and Kadle, D. S., 1983, "Enhanced and Local Heat Transfer, Pressure Drop, and Flow Visualizations for Arrays of Block-Like Electronic Components," International Journal of Heat and Mass Transfer, Vol.26, pp. 689-699.

Turner, B. L., and Flack, R. D., 1980, "The Experimental Measurement of Natural Convective Heat Transfer in Rectangular Enclosures with concentraded Energy Sources," Journal of Heat Transfer, Vol.102, pp. 236-241.

EXPERIMENTAL STUDY OF COMBINED CONVECTIVE HEAT TRANSFER FROM TANDEM CYLINDERS IN A HORIZONTAL AIR FLOW

C. Henderson and P. H. Oosthuizen
Department of Mechanical Engineering
Queen's University
Kingston, Ontario
Canada

ABSTRACT

Mean combined (or mixed) convective heat transfer rates from a pair of tandem horizontal cylinders mounted normal to a horizontal forced air flow have been experimentally measured. The experimental models used in the present study all had a diameter of 25.4 mm. The tests were carried out in an open jet blower type wind tunnel that had a horizontal working stream. The working stream velocity was varied in the present tests between 0 and approximately 1 m/s. The models were made from solid aluminum and the heat transfer rates were determined by the transient method, i.e. by heating the models and then measuring their temperature-time variations while they cooled when inserted into the working stream of the wind-tunnel. Because the variation in model temperature during the test was kept small and because only one model diameter was considered, the Grashof number was essentially the same in all tests and equal to approximately 90,000. Furthermore, over the range of temperatures used in the present study, the Prandtl number for air remains effectively constant. In the present study, therefore, the variation of Nu with Re for various values of S/D for fixed values of Gr and Pr was measured. Here Nu, Re and Gr are the Nusselt, Reynolds and Grashof numbers respectively Pr is the Prandtl number and S is the distance between the center-lines of the cylinders. Re values of between 0 and about 2000 were covered in the present tests and results were obtained for S/D values of 1.3, 2.6, 3.9 and 6.5 in the mixed convection region. Additional tests covering S/D values of between 1 and 12 were carried out under forced convective conditions.

NOMENCLATURE

A = surface area of model
c = specific heat of material from which model is made
D = diameter of cylinder
Gr = Grashof number based on D
h = convective heat transfer coefficient
h_t = total heat transfer coefficient
L = heated length of cylinder
Nu = mean Nusselt number based on D
Nu_D = mean Nusselt number for downstream cylinder
Nu_F = mean Nusselt number that would exist in forced convection
Nu_N = mean Nusselt number that would exist in free convection
Nu_U = mean Nusselt number for upstream cylinder

Pr = Prandtl number
Re = Reynolds number based on D
S = distance between the center-lines of the cylinders
t = time
T = model temperature
T_i = initial model temperature
T_∞ = ambient air temperature

INTRODUCTION

Mean convective heat transfer rates from a pair of tandem horizontal cylinders mounted normal to a horizontal forced air flow have been experimentally measured. The flow situation considered is thus as shown in Fig. 1. The conditions covered in the tests were such that, despite the presence of the forced velocity, the buoyancy forces resulting from the temperature differences had an influence on the heat transfer rate, i.e. the heat transfer rate was by mixed- or combined forced and free convection.

There have been a number of previous studies of mixed convective heat transfer from single circular cylinders in a horizontal cross flow, e.g. see Hatton et al (1970), Oosthuizen and Madan (1971) and Jackson and Yen (1971). These results and those for forced and free convection from a single cylinder in a cross flow have been reviewed by Morgan (1975). There appear to be no previous experimental studies of combined convection from tandem cylinders. There have, of course, been many studies of flow without heat transfer over tandem cylinders, e.g. see Zdravkovich (1977) and Igrashi (1981, 1984), and the results obtained in these studies are useful in interpreting heat transfer behaviour in tandem cylinder flow. There have also been a number of experimental studies of forced convective heat transfer from tandem cylinders, e.g. see Zukauskas (1987), Kostic and Oka (1972), Oosthuizen and Paul (1979), Hiwada et al (1979, 1982) and the results of Baughn et al (1986) are also relevant. Several experimental studies of free convection from multiple tubes are also available and these are relevant to the interpretation of combined convective results for tandem cylinders, e.g. see Marsters (1972), Tsubouchi and Saito (1978), Dutton and Welty (1975), Sparrow and Boessneck (1983), and Sparrow and Niethammer (1981). All the studies mentioned above have been experimental. There have been some numerical studies of flow in tube banks, e.g. see Launder and Massey (1978), Fujii et al (1984), and Dhaubhadel (1986) although most of these have assumed a fully developed type flow. There have also been some numerical studies of natural convective heat transfer from multiple cylinders, e.g. see Wong and Chen (1987), Parsons and

Beach (1987), Farouk and Güçeri (1983) and Paykoç et al (1988). The last of these references also includes some experimental results. Some numerical studies of combined convection from multiple cylinders has also been undertaken by Wong and Chen (1986, 1987) for circular cylinders. A direct comparison between the results of the latter two numerical studies and the present experimental results is difficult because of the difference in the Reynolds number ranges considered, e.g. Wong and Chen (1987) considered only Reynolds numbers of 20, 40 and 80.

EXPERIMENTAL PROCEDURE

The models used in the present study all had a diameter of 25.4 mm and a heated length of 305 mm. The models were made from solid aluminum and had nylon insulating end pieces of the same diameter as the model attached to each end. The ends of the models were internally chamfered in order to reduce the contact area between the model and the end pieces in order to reduce the conduction heat transfer from the model to the end pieces. A series of small diameter holes was drilled longitudinally to various depths into the models and thermocouples inserted into these holes were used to measure the temperature of the models. The thermocouples were held in place mechanically and measurements were undertaken to ensure that the the time response of the thermocouples was adequate for the present study.

The tests were carried out in an open jet blower type wind tunnel that had a horizontal working stream that was 600 mm high by 400 mm wide. The open jet working stream was surrounded by a very large ventilated chamber which minimized the effects of disturbances in the laboratory in which the wind tunnel was housed. The working stream velocity was essentially uniform over the area in which the tests were undertaken and the turbulence level was low (less than 0.5 percent). The working stream velocity was varied in the present tests between 0 and approximately 1 m/s by varying the speed of the tunnel fan. The models were mounted horizontally in a frame in this tunnel at right angles to the flow direction. A series of holes in the frame allowed the models to be mounted at a variety of distances from each other.

The heat transfer rates from the models were determined by the transient method, i.e. by heating the models and then measuring their temperature - time variations while they cooled when inserted into the working stream of the wind-tunnel. In an actual test, the tunnel speed was set at the value selected for that test. The models were heated to a mean temperature of approximately $105°C$ and then inserted into the tunnel. They were then allowed to stand until their mean temperatures had dropped to approximately $100°C$ to ensure that any non-uniformities in temperature that had arisen during heating had been evened out by internal heat conduction. The temperature-time variations were then measured using an automatic data acquisition system while the models cooled from $100°C$ to $95°C$.

Tests and approximate calculations indicated that, because the Biot number existing during the tests was very small, the temperature of the aluminum models remained effectively uniform during the cooling. The overall heat transfer coefficient, h_t, could then be determined for each of the models from its measured temperature - time variation using the usual procedure, i.e. by assuming that because of the relatively small temperature range covered during a test, the overall heat transfer coefficient, h_t, remained constant during the test and that the temperature variation of the model was then related to the time by:

$$\left(\frac{h_t A}{m c} \right) t = \ln \left(\frac{T_i - T_\infty}{T - T_\infty} \right)$$ (1)

Hence, by fitting a straight line to the measured variation of $\ln (T_i - T_\infty)/(T - T_\infty)$ with t, h_t could be determined using the known value of $(A/m c)$. The measurement and data reduction was all done by an automatic data-acquisition system. The value of h_t so determined is, of course, made up of contributions to the heat transfer due to convective heat transfer to the surrounding air, the radiant heat transfer to the surroundings and conduction from the model to the end pieces. Calculations and subsidiary tests using different forms of

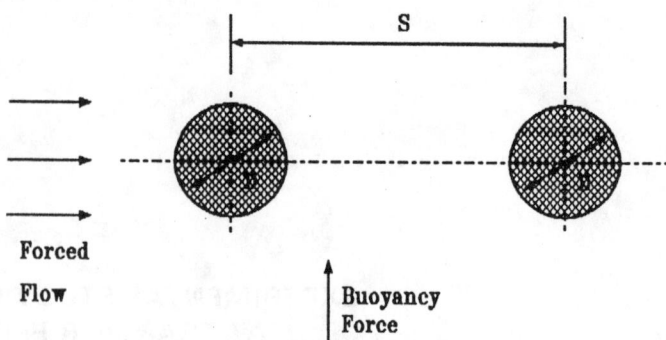

Fig. 1   Flow situation considered.

contact between the model and the end pieces indicated that the conduction heat transfer was negligible. The radiant heat transfer rate could be allowed for by calculation using the known emissivity of the polished surface of the aluminum models, the models surfaces being cleaned prior to every test. Thus by allowing for radiant heat transfer in this way, it being less than 10% of the total heat transfer in all cases, and neglecting conduction heat transfer, the convective heat transfer coefficient, h, could be determined. The time taken for the models to cool was, in all cases, sufficiently long to ensure that unsteady effects on the convective heat transfer were negligible, this conclusion being reached by considering the order of magnitude of the unsteady terms in the governing equations in comparison with the orders of magnitude of the other terms in these equations. The experimental procedure, therefore, allowed the convective heat transfer coefficients that would exist in steady flow under the same circumstances to be found. Estimates of the uncertainties involved in the experimental procedure indicate that the maximum uncertainty in the measured heat transfer rates resulting from uncertainties in the measurements is about 1%. However, when the uncertainties in the correction for radiation and those that arise from the neglect of conduction losses are considered, the overall uncertainty in the measured heat transfer coefficients is about 3%.

In expressing the experimental results in dimensionless form, all air properties have been evaluated at the mean film temperature existing during the test, this being approximately $60°C$ in all cases.

RESULTS

Under the conditions covered in the present tests:

$$Nu = f (Re, Gr, S/D, Pr)$$ (2)

where Nu, Re and Gr are the Nusselt, Reynolds and Grashof numbers based on the cylinder diameter D respectively. Pr is the Prandtl number and S is the distance between the center-lines of the cylinders. Because the variation in model temperature during the test was kept small (about $5°C$) compared to the overall temperature difference (about $75°C$) and because only one model diameter was considered, the Grashof number was essentially the same in all tests and equal to approximately 90,000 (the variation in Grashof number arising from the temperature change during a test was approximately 6%) Furthermore, over the range of temperatures used in the present study, the Prandtl number for air remains effectively constant. In the present study, therefore, the variation of Nu with Re for various values of S/D for a fixed value of Gr and of Pr was measured. Re values of between 0 and about 2000 were covered in the present tests and results were obtained for S/D values of 1.3, 2.6, 3.9 and 6.5 in the mixed convection region. Additional tests covering S/D values of between 1 and 12 were carried out under purely forced convective conditions.

Beside the main tandem cylinder tests, additional tests were carried out with a single cylinder for comparison. These single cylinder results will first be considered. The variation of Nusselt number with Reynolds number for this situation is shown in Fig. 2. At larger

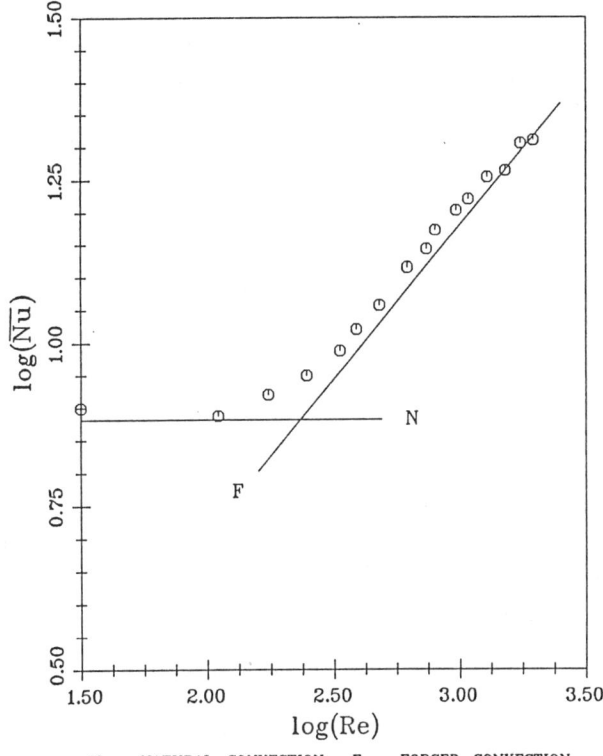

Fig. 2 Variation of Nusselt number with Reynolds number for a single cylinder. The lines marked F and N indicate standard variations for purely forced and purely free convection respectively.

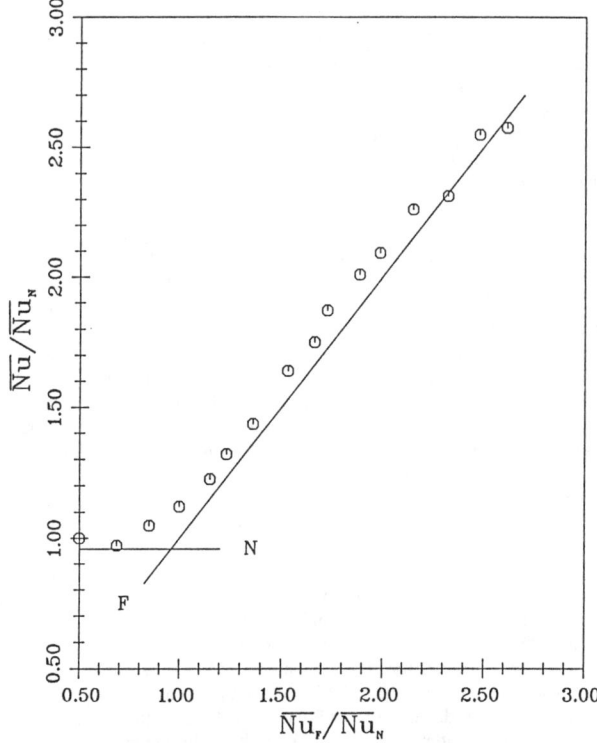

Fig. 3 Attempted correlation of results for a single cylinder. The lines marked F and N indicate standard variations for purely forced and purely free convection respectively.

Reynolds numbers the heat transfer rate becomes, of course, effectively equal to the rate that would exist in purely forced convection while at small Reynolds the heat transfer rate becomes effectively equal to the rate that would exist in purely free convection. The measured heat transfer rates in these two limiting states are effectively the same as those obtained in previous studies, e.g. see Oosthuizen and Mansingh (1986). In purely forced convection, for example, the Nusselt number is relatively well described by:

$$Nu = 0.45 \, Re^{0.5}$$

The transition from very nearly forced convective flow to very nearly free convective flow will be seen to be quite sharp, the combined convection region covering a fairly narrow range of Reynolds numbers. In a number of previous studies, e.g. see Jackson and Yen (1971), Oosthuizen and Chow (1986) and Oosthuizen and Paul (1988), an attempt has been made to correlate combined convection heat transfer rates by assuming that:

$$\frac{Nu}{Nu_N} = \text{function} \left(\frac{Nu_F}{Nu_N} \right) \qquad (3)$$

where Nu_N is the Nusselt number that would exist under the same conditions in purely free convection and Nu_F is the Nusselt number that would exist under the same conditions in purely forced convection. The single cylinder results are presented in this form in Fig. 3. If it is assumed that the effectively purely free convection region ends when the Nusselt number differs by more than 5% from the purely free convection value and that the effectively purely forced convection region begins when the Nusselt number is within 5% of the purely forced convection value, then the values of Nu_F / Nu_N at which the combined convection region begins and ends can be determined from the results given in Fig. 3. It will be seen from this figure that in the combined convection region:

$$0.8 < Nu_F / Nu_N < 2.4$$

The results for the tandem cylinder situation will now be considered. Typical variations of Nusselt number with Reynolds number for this situation are shown in Figs. 4, 5, 6 and 7 for the four cylinder spacings considered. The heat transfer rates in purely free convection were also measured by undertaking tests with no velocity in the tunnel. The curves labeled F and N in Figs. 4 to 7 represent the purely forced and purely free convective results for a single isolated cylinder. It will be seen that the Nusselt number variation for the downstream cylinder is in particular strongly effected by the cylinder spacing and that a sharp almost discontinuous change in this variation occurs at the smaller spacings. This is associated with a change in the flow pattern that occurs as the flow goes from one that is dominated by the buoyancy forces to one that is dominated by forced flows. In the buoyancy dominated flow, the flow near the cylinders is upwards at right angles to the direction of the forced flow and little interaction between the cylinders occurs. When the forced flow effects begin to dominate, however, the wake from the upstream cylinder impinges directly on the downstream cylinder causing a reduction in the heat transfer rate from this downstream cylinder.

Although the Nusselt number from the downstream cylinder is still changing significantly with increasing Reynolds number at the smaller spacings considered (e.g. see Figs. 4 and 5), the Nusselt numbers at the highest Reynolds considered in the present study, i.e. approximately 2000, have been taken as effectively those for purely forced convection for both the upstream and downstream cylinders and, as previously mentioned, tests at a larger number of cylinder spacings under these conditions have been undertaken. The results of these forced convective tests are shown in Fig.8. Because of the Reynolds number range being considered and the results obtained with a single cylinder, it is to be expected that the Nusselt number will depend on the square root of the Reynolds number. The ratio $Nu / Re^{0.5}$ for each cylinder has, therefore, been given in Fig. 8. The single cylinder result given in this figure was that obtained as part of the present study and discussed earlier. It will be seen from Fig. 8, that at the smaller spacings the heat transfer rate from both cylinders is reduced by the interaction between the cylinders and the heat transfer rate from the

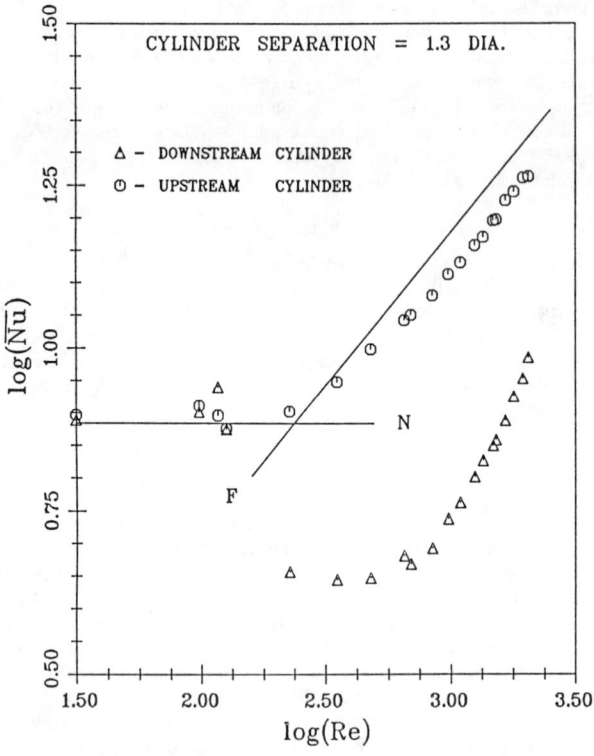

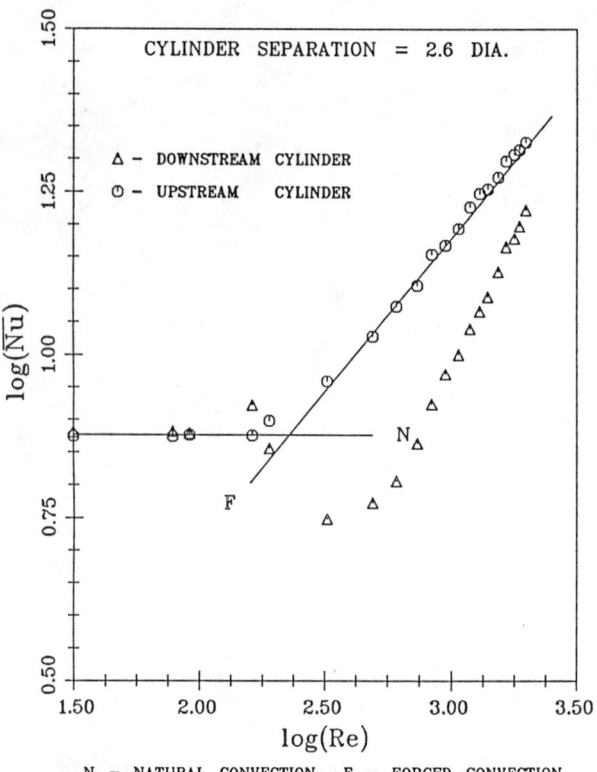

Fig. 4 Variations of Nusselt numbers for upstream and downstream cylinders with Reynolds number for *S/D* = 1.3. The lines marked F and N indicate standard variations for purely forced and purely free convection for a single cylinder respectively.

Fig. 5 Variations of Nusselt numbers for upstream and downstream cylinders with Reynolds number for *S/D* = 2.6. The lines marked F and N indicate standard variations for purely forced and purely free convection for a single cylinder respectively.

downstream cylinder is smaller than that from the upstream cylinders. At the larger cylinder spacings considered, however, the heat transfer rate from the upstream cylinder becomes almost independent of cylinder spacing and the heat transfer rate for the downstream cylinder rises above that for the upstream cylinder. At the smaller spacings, the heat transfer rate from the downstream cylinder is dominated by the temperature rise and velocity defect in the wake of the upstream cylinder, there being, as observed in a number of previous studies, a closed recirculation zone between the cylinders when the cylinder spacing is less than about 3.8. At the larger spacings, however, the heat transfer rate from the downstream cylinder is dominated by the unsteadiness in the wake from the upstream cylinder. The variation of the ratio of the Nusselt number for the downstream cylinder to that for the upstream cylinder with cylinder spacing for these "forced" convection conditions is shown in Fig.9. It will be seen from this figure that, at small spacings, the heat transfer rate from the downstream cylinder is only about half of that from the upstream cylinder. The ratio then rises rapidly with increasing cylinder spacing, becoming greater than one at a spacing greater than about 3.8. This is the value beyond which previous studies, e.g. see Kostic and Oka (1972), have indicated a closed vortex ceases to exist between the two cylinders.

The variations of the ratio of the Nusselt number for the downstream cylinder to that for the upstream cylinder with Reynolds number for the various cylinder spacings considered are shown in Figs. 10 to 13. At small values of the Reynolds number, when the flow is effectively purely free convective, this ratio is one. For the three smaller spacings considered, increasing the Reynolds number beyond this region produces a sharp drop in the Nusselt number ratio, this ratio passing through a minimum with further increase in Reynolds number and then rising to the forced convective value which is less than one for the two smaller spacings and approximately equal to one at a spacing of 3.9 diameters. The minimum value of the Nusselt number ratio is lowest at the smallest cylinder spacing considered, having a value of about 0.4 in this case, and rises with increasing

cylinder spacing. At the largest cylinder spacing considered, there is only a small drop in the Nusselt number ratio as the combined convection region is entered, the ratio then rising above one as the forced convection region is approached.

As with a single cylinder, the tandem cylinder results for combined convection have been expressed in terms of Nu_N, the Nusselt number that would exist under the same conditions in purely free convection, and Nu_F, the Nusselt number that would exist under the same conditions in purely forced convection using the form indicated in eq.(3). These variations are shown in Figs. 14 to 17 for the four cylinder spacings considered. In deriving the results shown in these figures it has been assumed that the heat transfer rates in the purely forced convection region for both the upstream and the downstream cylinder can be described by assuming that $Nu/Re^{0.5}$ is a constant, the value of the constants for each cylinder for each cylinder spacing being deducible from Fig. 8. It will be seen from these figures that the variation for the upstream cylinder is basically very similar to that for a single cylinder except that the transition from the purely free convection region to the purely forced convection region is sharper at the smaller cylinder spacings. The variations for the downstream cylinder are, however, strongly dependent on the cylinder spacing except at the largest cylinder spacing considered where the variation is again similar to that for a single cylinder. In summary, then, the heat transfer rate in the combined convection region for the upstream cylinder can be correlated by a similar curve to that which applies to a single cylinder unless the cylinder spacing is very small. For the downstream cylinder, however, if the spacing is less than that which gives a closed vortex in purely forced convective flow the results are not well correlated, the form of the correlation curve varying quite strongly with cylinder spacing. When the cylinder spacing is greater than that required to given a closed vortex in purely forced convection, the heat transfer rate in the combined convection region for the downstream cylinder is again correlated quite well by a curve that applies to a single cylinder.

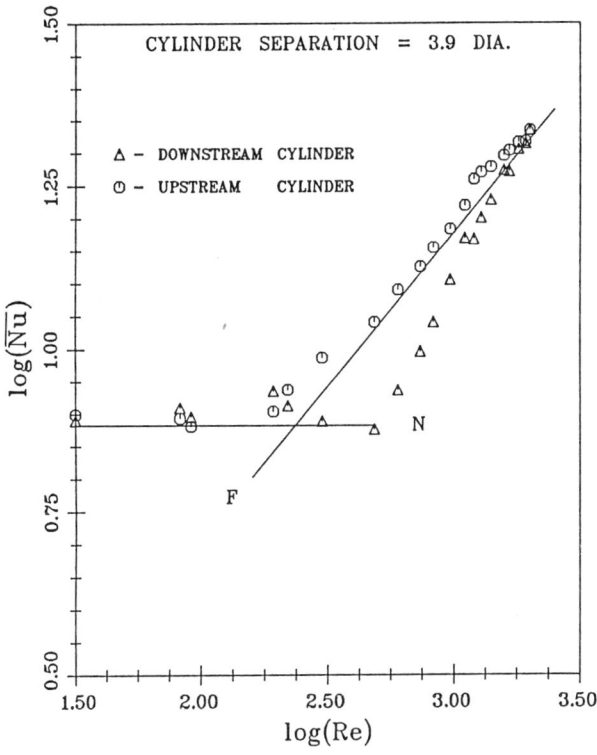

N — NATURAL CONVECTION   F — FORCED CONVECTION

Fig. 6   Variations of Nusselt numbers for upstream and downstream cylinders with Reynolds number for *S/D* = 3.9. The lines marked F and N indicate standard variations for purely forced and purely free convection for a single cylinder respectively.

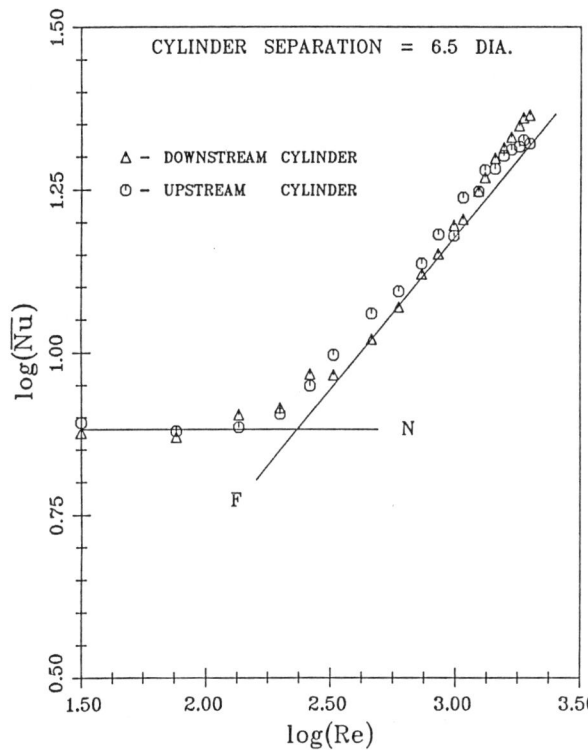

N — NATURAL CONVECTION   F — FORCED CONVECTION

Fig. 7   Variations of Nusselt numbers for upstream and downstream cylinders with Reynolds number for *S/D* = 6.5. The lines marked F and N indicate standard variations for purely forced and purely free convection for a single cylinder respectively.

Using the curves given in Figs. 14 to 17 for the upstream cylinder and again assuming that the free convection region ends when the heat transfer rate differs by more than 5% from the purely free convective value and that the forced convection region begins when the heat transfer rate is within 5% of the assumed variation in purely forced convection the limits on the combined convection region for the upstream cylinder can be deduced as a function of cylinder spacing. The results so obtained are shown in Fig. 18. From this figure it will be seen that, as previously mentioned, the combined convection region is narrow at small cylinder spacings but becomes essentially the same as that for a single cylinder at cylinder spacings beyond about four diameters.

CONCLUSIONS

The results of the present experimental study lead to the following conclusions, it being realized that some of these conclusions may only apply to the Reynolds number range covered by the present tests:

- Except when the cylinder spacing is very small (roughly less than 2 diameters), the variation of Nu / Nu_N with Nu_F / Nu_N where Nu_N is the Nusselt number that would exist under the same conditions in purely free convection and Nu_F is the Nusselt number that would exist under the same conditions in purely forced convection, for the upstream cylinder is approximately the same as for a single cylinder under the same conditions.
- When the cylinder spacing is very small, the value of Nu_F / Nu_N at which the flow becomes effectively forced convective is significantly lower than the value for a single cylinder.

- The heat transfer variation for the downstream cylinder in the combined convection region for cylinder spacings of less than about 5 diameters is very different from that for a single cylinder. The heat transfer rate for the downstream cylinder compared to that for the upstream cylinder for these smaller cylinder spacings is lower than in purely forced convection and much lower than that for a single cylinder.
- It is not possible to correlate the downstream cylinder results for small cylinder spacings (less than about 4 diameters) using the approach that has been successful for the upstream cylinder and for many other combined convection situations.
- For spacings greater than that necessary for a closed vortex between the cylinders in forced convection, i.e. greater than a dimensionless spacing of about 4, the results for the downstream cylinder can be correlated in a similar way to those for the upstream cylinder.

The most important conclusion, then, is that with small cylinder spacings the heat transfer rate from the downstream cylinder in the combined convection region can be significantly lower than it would be in either purely forced or purely free convection.

ACKNOWLEDGEMENTS

This work was supported by the Natural Sciences and Engineering Research Council of Canada.

225

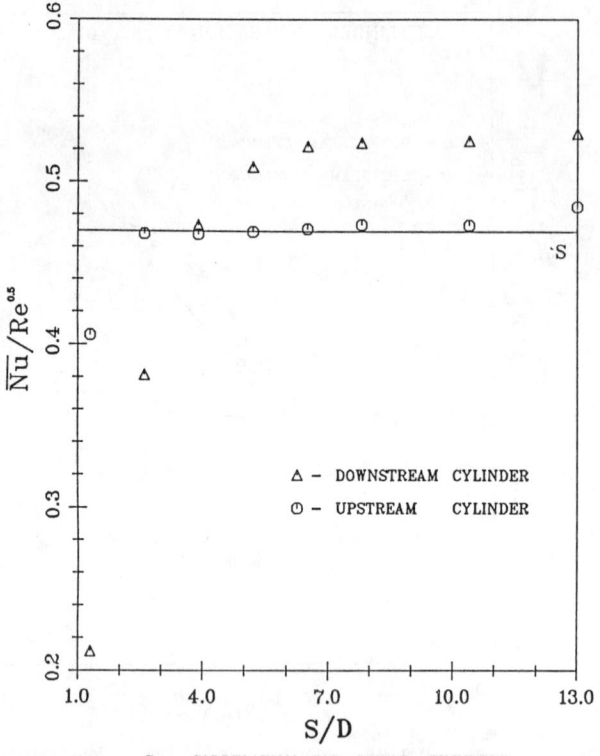

S — CORRELATION FOR SINGLE CYLINDERS

Fig. 8 Variation of the ratio of Nusselt number to the square root of the Reynolds number with cylinder separation distance, *S/D*, for upstream and downstream cylinders in purely forced convection.

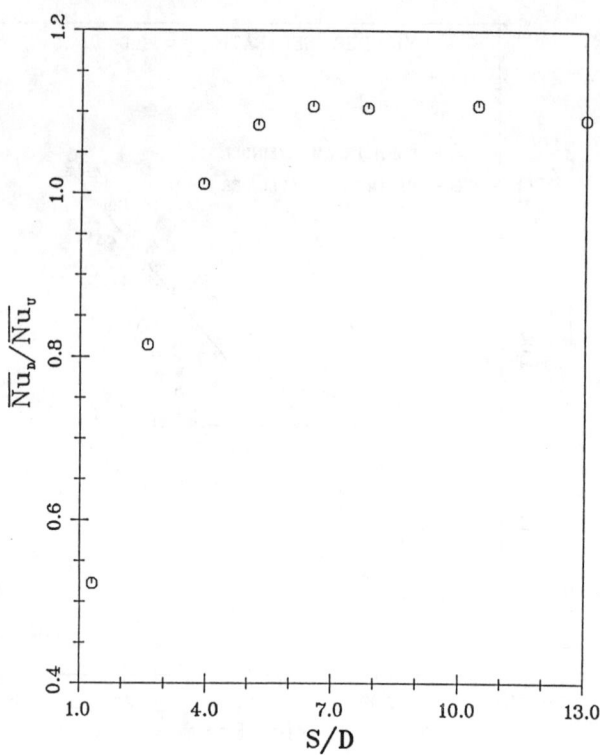

Fig. 9 Variation of the ratio of the Nusselt number for the downstream cylinder to that for the upstream cylinder with cylinder separation distance, *S/D*, cylinders in purely forced convection.

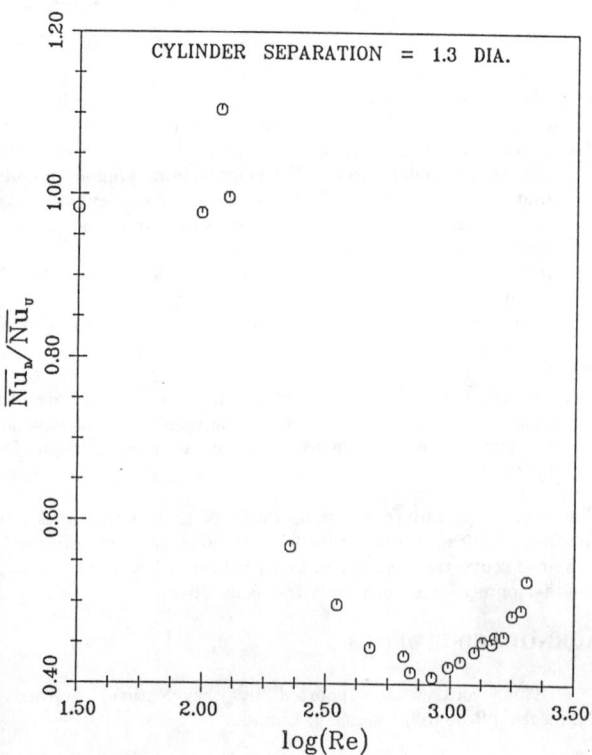

Fig. 10 Variation of the ratio of the Nusselt number for the downstream cylinder to that for the upstream cylinder with Reynolds number for *S/D* = 1.3.

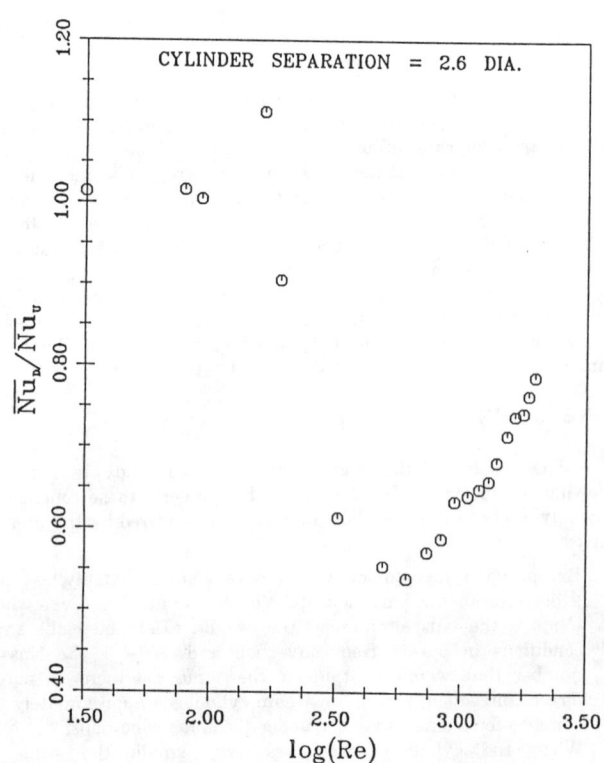

Fig. 11 Variation of the ratio of the Nusselt number for the downstream cylinder to that for the upstream cylinder with Reynolds number for *S/D* = 2.6.

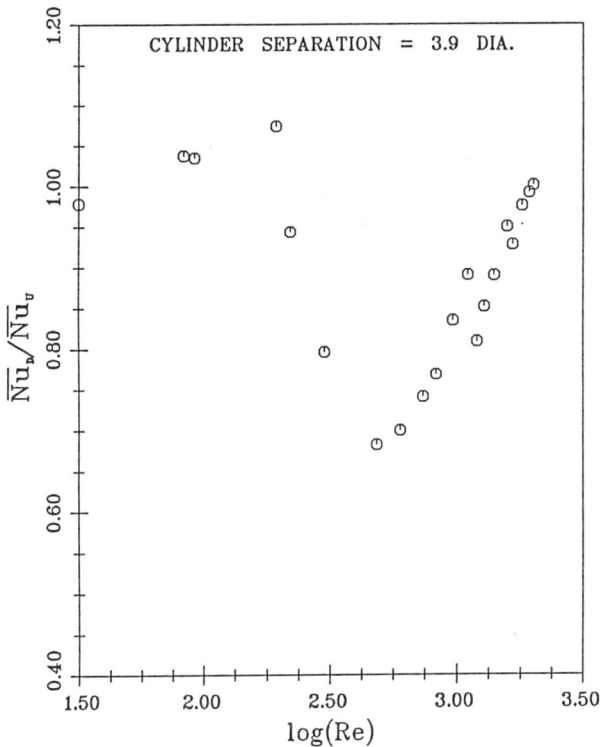

Fig. 12 Variation of the ratio of the Nusselt number for the downstream cylinder to that for the upstream cylinder with Reynolds number for *S/D* = 3.9.

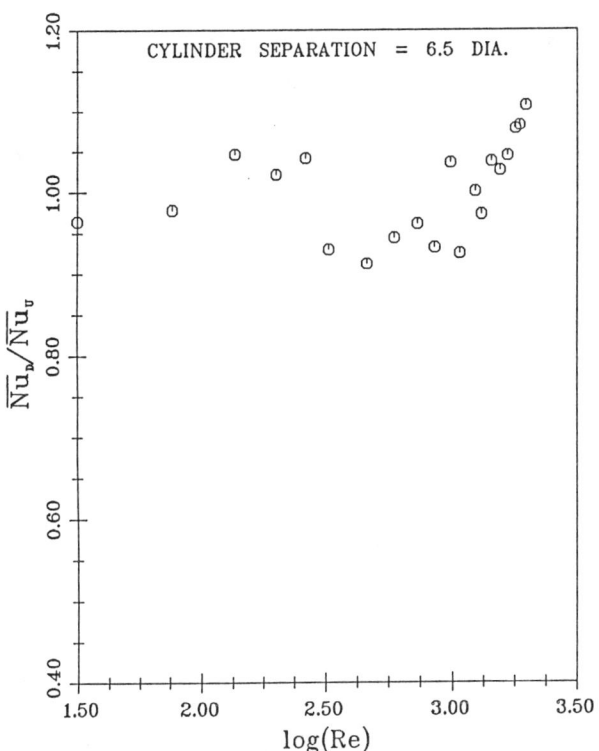

Fig. 13 Variation of the ratio of the Nusselt number for the downstream cylinder to that for the upstream cylinder with Reynolds number for *S/D* = 6.5.

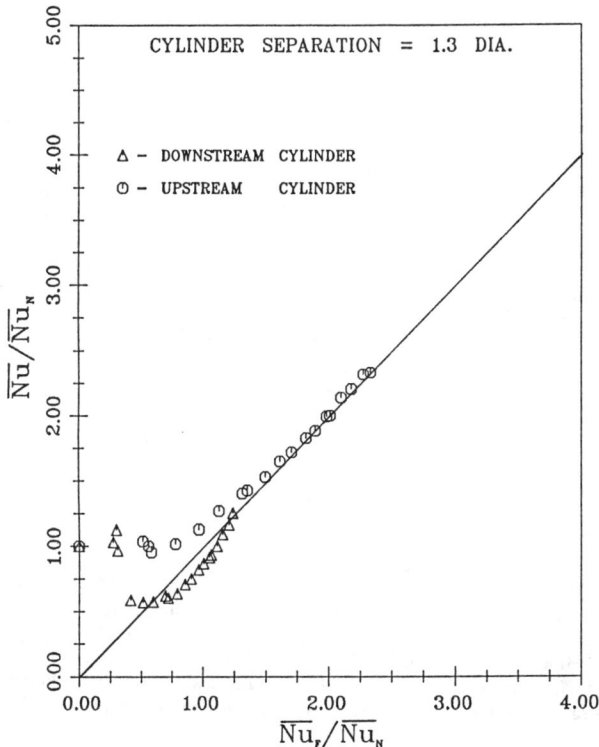

Fig. 14 Attempted correlation of results for *S/D* = 1.3.

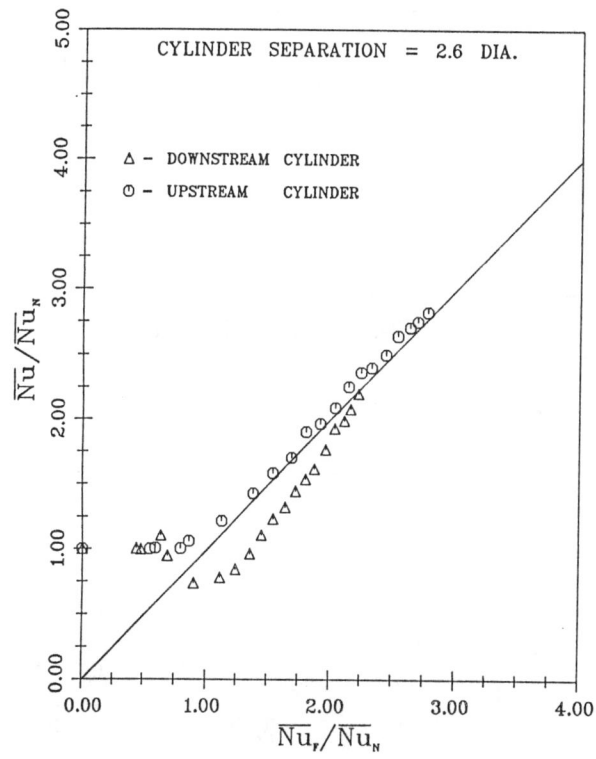

Fig. 15 Attempted correlation of results for *S/D* = 2.6.

227

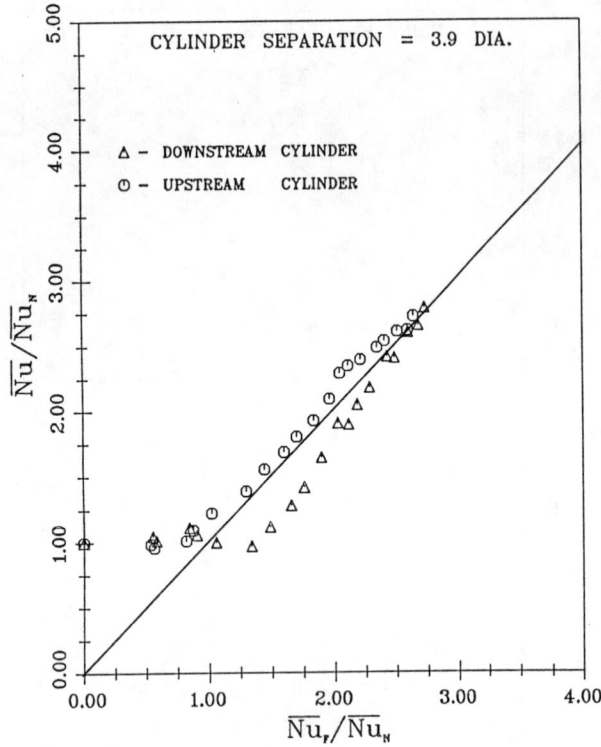

Fig. 16 Attempted correlation of results for *S/D* = 3.9.

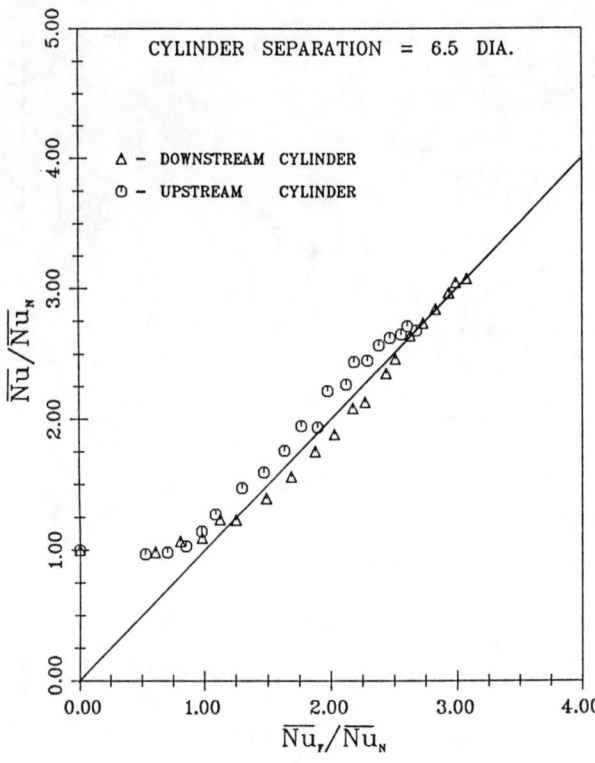

Fig. 17 Attempted correlation of results for *S/D* = 6.5.

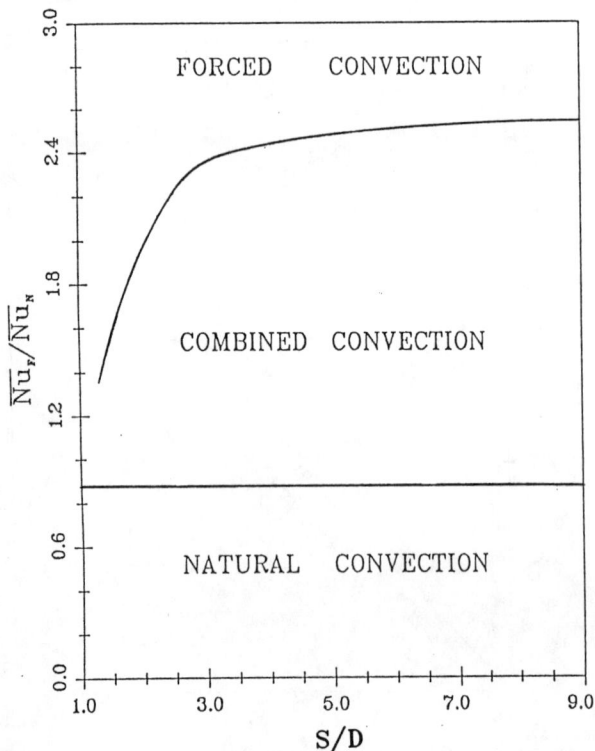

Fig. 18 Limits on mixed convection region for upstream cylinder.

REFERENCES

Baughn, J. W., Elderkin, M. J. and McKillop, A. A., 1986, "Heat Transfer From a Single Cylinder, Cylinders in Tandem, and Cylinders in the Entrance Region of a Tube Bank With a Uniform Heat Flux", *Journal of Heat Transfer, Trans. ASME*, Vol. 108, pp. 386 - 391.

Dhaubhadel, M. N., Reddy, J. N. and Telionis, D. P., 1986, "Penalty Finite - Element Analysis of Coupled Fluid Flow and Heat Transfer for In-Line Bundle of Cylinders in Cross Flow", *International Journal of Non-Linear Mechanics*, Vol. 21, pp. 361 - 373.

Dutton, J. C. and Welty, J. R., 1975, "An Experimental Study of Low Prandtl Number Natural Convection in an Array of Uniformly Heated Vertical Cylinders", *Journal of Heat Transfer, Trans. ASME*, Vol. 97, pp. 372 - 377.

Farouk, B. and Güçeri, S. I., 1983, "Natural Convection from Horizontal Cylinders in Interacting Flow Fields", *International Journal of Heat and Mass Transfer*, Vol. 26, pp. 231 - 243.

Fujii, M., Fujii, T. and Nagata, T., 1984, "A Numerical Analysis of Laminar Flow and Heat Transfer of Air in an In-Line Tube Bank", *Numerical Heat Transfer*, Vol. 7, pp. 89 - 102.

Hatton, A.P., James, D.D. and Swire, H.W., 1970, "Combined Forced and Natural Convection with Low-Speed Air Flow Over Horizontal Cylinders", *J. Fluid Mech.*, Vol. 42, pp. 17-31.

Hiwada, M., Taguchi, T., Mabuchi, I. and Kumada, M., 1979, "Fluid Flow and Heat Transfer Around Two Circular Cylinders of Different Diameters in Cross Flow", *Bulletin of the JSME*, Vol. 22, pp. 715 - 723.

Hiwada, M., Mabuchi, I. and Yanagihara, H., 1982, "Fluid Flow and Heat Transfer Around Two Circular Cylinders", *Bulletin of the*

JSME, Vol. 25, pp. 1737 - 1745.

Igarashi, T., 1981, "Characteristics of the Flow Around Two Circular Cylinders Arranged in Tandem (1st Report)", *Bulletin of the JSME*, Vol. 24, pp. 323 - 331.

Igarashi, T., 1984, "Characteristics of the Flow Around Two Circular Cylinders Arranged in Tandem (2nd Report, Unique Phenomenon at Small Spacing)", *Bulletin of the JSME*, Vol. 27, pp. 2380 - 2387.

Jackson, T. J. and Yen, H. H., 1971, "Combining Forced and Free Convective Equations to Represent Combined Heat-Transfer Coefficients for a Horizontal Cylinder", *Journal of Heat Transfer, Trans. ASME*, Vol. 93, pp. 247 - 249.

Kostic, Z. G. and Oka, S. N., 1972, "Fluid Flow and Heat Transfer with Two Cylinders in Cross Flow", *International Journal of Heat and Mass Transfer*, Vol. 15, pp. 279 - 299.

Launder, B. E. and Massey, T. H., 1978, "The Numerical Prediction of Viscous Flow and Heat Transfer in Tube Banks", *Journal of Heat Transfer, Trans. ASME*, Vol. 100, pp. 565 - 571.

Marsters, G. F., 1972, "Arrays of Horizontal Cylinders in Natural Convection", *International Journal of Heat and Mass Transfer*, Vol. 15, pp. 921 - 933.

Morgan, V. T., 1975, "The Overall Convective Heat Transfer from Smooth Cylinders", *Advances in Heat Transfer*, Vol. 11, Academic Press, pp. 199 - 264.

Oosthuizen, P.H. and Madan, S., 1971, "Combined Convective Heat Transfer From Horizontal Cylinders in Air", *J. Heat Transfer*, Vol. 93, pp. 240-242.

Oosthuizen, P. H. and Mansingh, V., 1986, "Free and Forced Convection Heat Transfer from Short Inclined Circular Cylinders", *Chem. Eng. Comm.*, Vol. 42, pp. 333 - 348.

Oosthuizen, P. H. and Chow, K., 1986, "An Experimental Study of Free Convective Heat Transfer from Short Cylinders with "Wavy" Surfaces", *Proc. 8th Int. Heat Transfer Conf.*, Vol. 3, pp. 1311 - 1316.

Oosthuizen, P. H. and Paul, J. T., 1979, "Heat Transfer with Cross Flow Over Two In-Line Cylinders", *Trans. CSME*, Vol. 5, pp. 121 - 124.

Oosthuizen, P. H. and Birk, A. M., 1988, "Mixed Convective Heat Transfer from Inclined Circular Cylinders", *Experimental Heat Transfer, Fluid Mechanics, and Thermodynamics 1988-* Proceedings of the First World Conference on Experimental Heat Transfer, Fluid Mechanics, and Thermodynamics, pp. 200 - 207, Elsevier.

Parsons, J. R. and Beach, T. A., 1987, "A Numerical Investigation of Natural Convection from Two Vertically Aligned Horizontal Cylinders", Proceedings of the 1987 ASME/JSME Thermal Engineering Joint Conference, Honolulu, pp. 17 - 24.

Paykoç, E., Yüncü, H. and Bezzazoglu, M., 1988, "Laminar Natural Convective Heat Transfer Over Two Vertically Spaced Isothermal Horizontal Cylinders", *Experimental Heat Transfer, Fluid Mechanics, and Thermodynamics 1988-* Proceedings of the First World Conference on Experimental Heat Transfer, Fluid Mechanics, and Thermodynamics, pp. 208 - 216, Elsevier.

Sparrow, E. M. and Boessneck, D. S., 1983, "Effect of Transverse Misalignment on Natural Convection from a Pair of Parallel, Vertically Stacked, Horizontal Cylinders", *Journal of Heat Transfer, Trans. ASME*, Vol. 105, pp. 241 - 247.

Sparrow, E.M. and Niethammer, J. E., 1981, "Effect of Vertical Separation Distance and Cylinder to Cylinder Temperature Imbalance on Natural Convection for a Pair of Horizontal Cylinders", *Journal of Heat Transfer, Trans. ASME*, Vol. 103, pp. 638 - 644.

Tsubouchi, T. and Saito, E., 1978, "An Experimental Study of Natural Convection Heat Transfer in Banks of Uniformly Heated Horizontal Circular Cylinders", *Heat Transfer - Japanese Research*, Vol. 7, pp. 44 - 58.

Wong, K. and Chen, C., 1986, "The Finite - Element Solution of Laminar Combined Convection from Two Horizontal Cylinders in Tandem Arrangement", *AIChE Journal*, Vol. 32, pp. 557 - 565.

Wong, K. and Chen, C., 1987, "The Finite Element Solutions of Laminar Flow and Combined Convection of Air from Three Horizontal Cylinders in Staggered Arrangement", Proceedings of the 1987 ASME/JSME Thermal Engineering Joint Conference, Honolulu, Vol. 5, pp. 9 - 15.

Zdravkovich, M. M., 1977, "Review of Flow Interference Between Two Circular Cylinders in Various Arrangements", *Journal of Fluids Engineering, Trans. ASME* , Vol. 99, pp. 618 - 633.

Zukauskas, A., 1987, "Heat Transfer from Tubes in Cross Flow", in *Advances in Heat Transfer*, Vol. 18, J. P. Hartnett and T. F. Irvine, eds., Academic, pp. 87 - 159.

FREE CONVECTIVE FLOW IN AN INCLINED SQUARE CAVITY WITH A PARTIALLY HEATED WALL

P. H. Oosthuizen and J. T. Paul
Department of Mechanical Engineering
Queen's University
Kingston, Ontario
Canada

ABSTRACT

Free convective flow in a cavity which has one wall partially heated to a uniform temperature and another parallel wall cooled over its entire surface to a uniform lower temperature has been considered. All remaining wall surfaces are adiabatic. The cavity is, in general, inclined at an angle to the vertical. The flow has been assumed to be laminar and two-dimensional. Fluid properties have been assumed constant except for the density change with temperature that gives rise to the buoyancy forces, this being treated by means of the Boussinesq approximation. The governing equations, expressed in terms of stream function and vorticity, have been written in dimensionless form. The resultant equations, subject to the assumed boundary conditions, have been solved using the finite-element method. Because of the possible applications that motivated the study, results have only been obtained for a Prandtl number of 0.7. Attention has been restricted to a cavity with an aspect ratio of 1, i.e. a square cavity. Results have then been obtained for Rayleigh numbers between 1,000 and 100,000 for a variety of heated wall portion sizes and positions for angles of inclination between $-90°$ and $+90°$.

NOMENCLATURE

A = aspect ratio, h'/w'
C = c'/w'
c' = position of center-line of heated wall section
h' = height of "vertical" walls of cavity
k = thermal conductivity
Nu_L = mean Nusselt number based on s' for hot wall
Nu_{LM} = maximum value of Nu_L
Nu_{L0} = value of Nu_L at an angle of inclination of 0°
Nu_W = mean Nusselt number based on w' for hot wall
Nu_{WM} = maximum value of Nu_W
Nu_{W0} = value of Nu_W at an angle of inclination of 0°
n = coordinate measured normal to surface
Pr = Prandtl number
q = local dimensionless heat transfer rate
$\bar{q}$ = mean heat transfer rate
Ra = Rayleigh number based on w'
S = s'/w'
s' = size of heated wall section
T = dimensionless temperature, see equation (9)
T' = temperature
T_H' = temperature of hot wall
T_C' = temperature of cold wall

u' = velocity component in x' direction
v' = velocity component in y' direction
w' = width of cavity
x = dimensionless x' coordinate, see equation (9)
x' = horizontal coordinate position
y = dimensionless y' coordinate, see equation (9)
y' = vertical coordinate position
ψ = dimensionless stream function, see equation (9)
ψ_m = maximum absolute value of ψ in cavity
ψ' = stream function
ω = dimensionless vorticity, see equation (9)
ω' = vorticity
ϕ = angle of inclination of cavity as defined in Fig. 1

INTRODUCTION

Free convective flow in a cavity which has one wall partially heated to a uniform temperature, T'_H, and another parallel wall cooled over its entire surface to a uniform lower temperature, T'_C, has been considered. All remaining wall surfaces are adiabatic. The cavity is, in general, inclined at an angle to the vertical. The flow situation considered is, therefore, as shown in Fig. 1. Interest in this situation arises because it is a highly idealized model of some situations that arise in the electrical industry and because previous studies of heat transfer across a vertical cavity with a partially heated wall have shown that when the heated portion of the wall is near the top or near the bottom of the wall, a significant reduction in the heat transfer rate, compared to that which exists when the heated portion of the wall is in the middle of the wall, can arise. It has been suggested that this heat transfer rate reduction can be avoided if the cavity is inclined to the horizontal. The purpose of the present study was, basically, to investigate whether this was, in fact, the case.

Two-dimensional free convective flow in a rectangular cavity with one vertical wall heated to a uniform temperature and the other vertical wall cooled to a uniform temperature and with the horizontal walls adiabatic has been the subject of many numerical and some experimental studies, e.g., Catton (1978), Ostrach (1972), Ostrach (1982), Wong and Raithby (1979) and de Vahl Davis (1986). The interest in this type of flow arises because it is widely accepted as a flow that can be used for testing and evaluating numerical solution procedures for fluid flow, e.g. see de Vahl Davis and Jones (1983), and because it is an approximate model of a number of practically important flow situations. Two quite extensive studies of steady state two-dimensional free convective flow in a rectangular cavity with a partially heated vertical wall and with the opposite wall fully cooled to a lower temperature are available. Chu et al (1976) undertook a numerical study of this problem while Turner and Flack (1980) studied the problem experimentally. Both of these studies were concerned with a vertical cavity. Chu et al (1976) noted that as the heated element location moves from the top to the bottom the heat transfer from the heated element increases to a maximum and then decreases. Kuhn and Oosthuizen (1987) studied the same geometrical problem with unsteady heat transfer. There appear to be no previous studies of the effect of inclining the cavity on the heat transfer rate with a partially heated wall. However, Oosthuizen and Paul (1987) numerically studied the effect of inclination on heat transfer in a cavity with a heated square element on one wall.

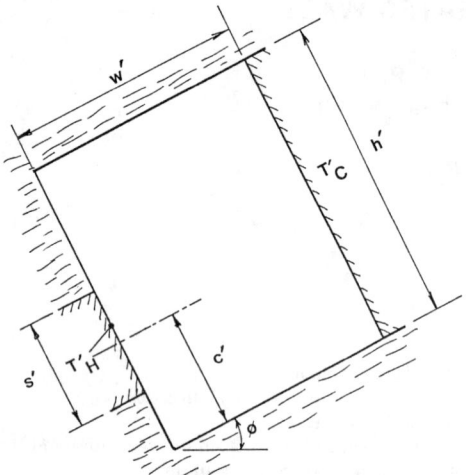

Fig. 1   Form of cavity considered

GOVERNING EQUATIONS AND SOLUTION PROCEDURE

The flow has been assumed to be steady, laminar and two-dimensional. The solution is based, of course, on the use of the steady, two-dimensional Navier-Stokes, continuity and energy equations. Fluid properties have been assumed constant except for the density change with temperature that gives rise to the buoyancy force, this being treated by means of the Boussinesq approximation. The governing equations are then:

$$\frac{\partial u'}{\partial x'} + \frac{\partial v'}{\partial y'} = 0 \tag{1}$$

$$u' \frac{\partial v'}{\partial x} + v' \frac{\partial v'}{\partial y} = \beta g (T' - T_C') \cos \phi$$

$$- \frac{1}{\rho} \frac{\partial p'}{\partial y'} + \nu \left(\frac{\partial^2 v'}{\partial x'^2} + \frac{\partial^2 v'}{\partial y'^2} \right) \tag{2}$$

$$u' \frac{\partial u'}{\partial x} + v' \frac{\partial u'}{\partial y} = \beta g (T' - T_C') \sin \phi$$

$$- \frac{1}{\rho} \frac{\partial p'}{\partial x'} + \nu \left(\frac{\partial^2 v'}{\partial x'^2} + \frac{\partial^2 v'}{\partial y'^2} \right) \tag{3}$$

$$u' \frac{\partial T'}{\partial x'} + v' \frac{\partial T'}{\partial y'} = \left(\frac{k}{\rho c} \right) \left(\frac{\partial^2 T'}{\partial x'^2} + \frac{\partial^2 T'}{\partial y'^2} \right) \tag{4}$$

The prime (′) denotes a dimensional quantity and the coordinates are as defined in Fig. 2.

The solution has been obtained in terms of the stream function and vorticity defined, as usual, by:

$$u' = \frac{\partial \psi'}{\partial y'} , \qquad v' = - \frac{\partial \psi'}{\partial x'}$$

$$\omega' = \frac{\partial v'}{\partial x'} - \frac{\partial u'}{\partial y'} \tag{5}$$

In terms of these variables, the governing equations become:

$$\frac{\partial^2 \psi'}{\partial x'^2} + \frac{\partial^2 \psi'}{\partial y'^2} = - \omega' \tag{6}$$

$$\frac{\partial \psi'}{\partial y'} \frac{\partial \omega'}{\partial x'} - \frac{\partial \psi'}{\partial x'} \frac{\partial \omega'}{\partial y'}$$

$$= \beta g \left(\frac{\partial T'}{\partial x'} \cos \phi + \frac{\partial T'}{\partial y'} \sin \phi \right) + \nu \left(\frac{\partial^2 \omega'}{\partial x'^2} + \frac{\partial^2 \omega'}{\partial y'^2} \right) \tag{7}$$

$$\frac{\partial \psi'}{\partial y'} \frac{\partial T'}{\partial x'} - \frac{\partial \psi'}{\partial x'} \frac{\partial T'}{\partial y'} = \left(\frac{k}{\rho c} \right) \left(\frac{\partial^2 T'}{\partial x'^2} + \frac{\partial^2 T'}{\partial y'^2} \right) \tag{8}$$

The following dimensionless variables have been introduced:

$$x = x'/L' , \quad y = y'/L' , \quad \omega = \omega' L'^2 Pr/\nu$$

$$\psi = \psi' Pr/\nu , \quad T = (T' - T_C')/(T_H' - T_C') \tag{9}$$

the cavity width, w', thus being used as the characteristic length scale. In terms of these variables, the governing equations become:

$$\frac{\partial^2 \psi}{\partial x^2} + \frac{\partial^2 \psi}{\partial y^2} = - \omega \tag{10}$$

$$\frac{\partial^2 \omega}{\partial x^2} + \frac{\partial^2 \omega}{\partial y^2} = \frac{1}{Pr} \left(\frac{\partial \psi}{\partial y} \frac{\partial \omega}{\partial x} - \frac{\partial \psi}{\partial x} \frac{\partial \omega}{\partial y} \right) \tag{11}$$

$$- Ra \left(\frac{\partial T}{\partial x} \cos \phi + \frac{\partial T}{\partial y} \sin \phi \right)$$

$$\frac{\partial^2 T}{\partial x^2} + \frac{\partial^2 T}{\partial y^2} = \frac{\partial \psi}{\partial y} \frac{\partial T}{\partial x} - \frac{\partial \psi}{\partial x} \frac{\partial T}{\partial y} \tag{12}$$

where Ra is the Rayleigh number based on the cavity width, w' :

$$Ra = \beta g (T_H' - T_C') w'^3 / \nu \alpha \tag{13}$$

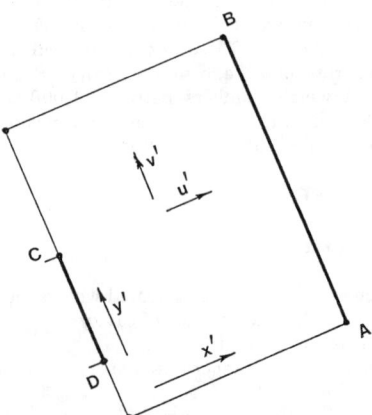

Fig. 2   Coordinate system and wall segments used in defining boundary conditions

The boundary conditions on the solution are as follows, the lettered wall segments being as defined in Fig. 2:

on all walls:

$$\psi = 0$$

on A B:

$$T = 1 \tag{14}$$

on C D:

$$T = 0$$

on AD and CB (which are assumed adiabatic):

$$\partial T / \partial n = 0$$

where n is the coordinate measured normal to the surface.

The above dimensionless governing equations, subject to these boundary conditions, have been solved using the finite element method. Simple linear triangular elements were used. The procedure adopted has been successfully used before in a number of studies of natural convective cavity flows, e.g. Oosthuizen and Paul (1987).

The solution directly gives the local dimensionless heat transfer rate distributions on the walls of the cavity, i.e. the local Nusselt number distributions on the hot and cold walls. These distributions can then be integrated to give the mean dimensionless heat transfer rates on the hot and the cold walls. This mean heat transfer rate has been expressed in the form of a mean Nusselt number based on the cavity width, w' :

$$Nu_W = \frac{\bar{q} w'}{k (T_H' - T_C')}$$

and in terms of a mean Nusselt number based on the length of the heated wall element, s':

$$Nu_L = \frac{\bar{q} s'}{k (T_H' - T_C')}$$

The values of Nu_W for the hot and cold walls will, in general, be different, the values of Nu_W being dimensionless measures of the mean heat transfer rates from the two walls. The value of Nu_L for the hot wall element will be equal to the value of Nu_W for the opposite cold wall, the value of Nu_L being a dimensionless measure of the total heat transfer rate from the element, i.e., of the product of the mean heat transfer rate and the length of the element. It should, perhaps, be noted that neither of the Nusselt numbers so defined is, in general, equal to the ratio of the mean heat transfer rate to the heat transfer rate that would exist with pure conduction. This ratio can, however, be deduced from the results presented. The results given by the present computer program were compared where possible with previous results and good agreement was obtained.

RESULTS

The solution to the equations given in the previous section has, in general, the following parameters:

- the Rayleigh number, Ra
- the Prandtl number, Pr
- the aspect ratio of the cavity, $A = h' / w'$
- the size of the heated portion of the wall relative to the width of the cavity, $S = s' / w'$
- the distance of the center-line of the heated portion of the wall along the wall relative to the width of the cavity $C = c' / w'$
- the angle of inclination of the cavity, ϕ

Only a square cavity ($A = 1$) has been considered here. Because of the possible applications that motivated the present study, results have only been obtained for $Pr = 0.7$, leaving as parameters Ra, S, C and ϕ.

Results have been obtained for Rayleigh numbers, Ra, between 1000 and 100,000 for values of S of 0.25, 0.5, 0.75, and 1.0, for angles of inclination, ϕ, between $-90°$ and $+90°$ for various values of C. The case of $S = 1$ corresponds to the classical case of a square cavity with fully heated hot and cold walls while the case of $\phi = 0°$ corresponds to the case of a vertical cavity. The results for $S = 1$ and $\phi = 0°$ have been compared with previous results for this case, e.g., Catton (1978), Wong and Raithby (1979) and de Vahl Davis and Jones (1983). Excellent agreement with these previous results was found to exist in all cases considered.

Typical variations of the mean Nusselt number based on the heated element length with angle of inclination for various Rayleigh numbers are shown in Fig. 3 for the case of $S = 0.5$ and $C = 0.5$, i.e. for a centrally mounted element. It will be seen that at

$\phi = -90°$ the results all coincide, the heat transfer rate in this case being by pure conduction. As ϕ increases, the Nusselt numbers rise and pass through a maximum at a value of ϕ between approximately 30° and 60°. The Nusselt numbers then decrease with increasing angles of inclination. The reason that the Nusselt number is a maximum at an angle of inclination greater than 0° is basically, of course, because the flow must rise up the wall containing the heated element, flow across the "upper" adiabatic wall, down the cold wall and across the "lower" adiabatic wall. When the cavity is vertical ($\phi = 0°$) the buoyancy forces act in the same direction as the flow near the hot and cold walls but act at right angles to the direction of flow near the "upper" and "lower" walls. When the wall is at a positive angle of inclination, components of the buoyancy force act in the flow direction along all four walls thus leading to a higher heat transfer rate than exists with a vertical cavity. However, as the angle of inclination nears 90°, the components of buoyancy force parallel to the heated and cooled walls tend to zero leading to a reduction in the heat transfer rate. When the wall is inclined at a negative angle of inclination, the buoyancy force components act in the opposite direction to that of the flow along the "upper" and "lower" walls, thus leading to a decrease in the heat transfer rate compared to that for a vertical cavity. In all cases, then, the maximum Nusselt number, Nu_{LM} is significantly higher than the Nusselt number for a vertical cavity, Nu_{L0} . The variation of these two Nusselt numbers and of their ratio for $S = 0.5$ and $C = 0.5$ is shown in Fig. 4, from which it will be seen that the maximum Nusselt number is about 30% higher than the Nusselt number for a vertical cavity at a Rayleigh number of about 10,000, the maximum percentage difference occurring at around this Rayleigh number.

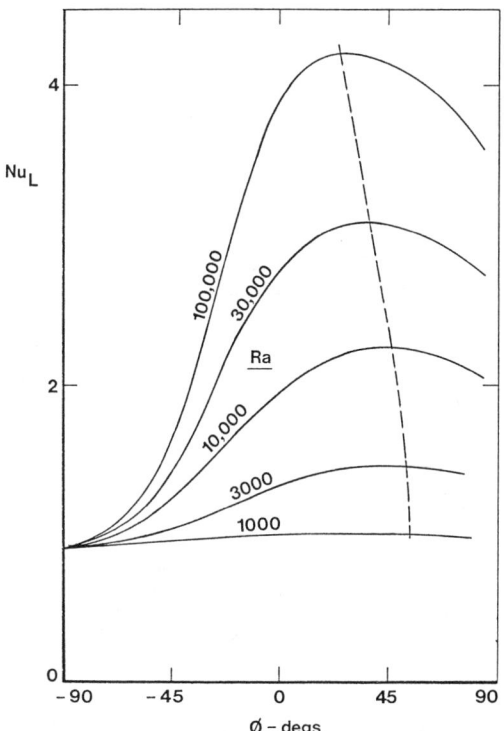

Fig. 3   Variation of mean Nusselt number based on heated element length with angle of inclination for various values of Rayleigh number for $S = 0.5$ and $C = 0.5$

The effect of the position of the heated element on the results is illustrated by the results given in Fig. 5 which shows the variation of mean Nusselt number based on heated element length with angle of inclination for $S = 0.5$ and $Ra = 30,000$ for C values of 0.25, 0.5 and 0.75. The C values correspond to the cases where the element is touching the "lower" wall of the cavity, is in the middle of the heated wall and is touching the "upper" wall of the cavity respectively. The general form of the variation is similar in all cases. For the vertical cavity case, the Nusselt number for $C = 0.75$, is considerably less than the values for $C = 0.5$ and $C = 0.25$. The maximum Nusselt number for $C = 0.75$ is, however, quite close to the Nusselt numbers for the other two values of C for $\phi = 0$. This means that while mounting a heated element high on the wall in the vertical cavity case leads to a lower heat transfer rate, a fact that has been noted by Chu et al (1976) and Kuhn and Oosthuizen (1987), if the cavity is inclined a significant improvement in the heat transfer rate can be achieved. This point is further illustrated by the results given in Fig. 6 which shows the maximum value of the Nusselt number, the value of the Nusselt for a vertical cavity and the ratio of these two Nusselt numbers with element position, C, for $S = 0.5$ and $Ra = 30,000$. It will be seen that the decrease in the maximum value of the Nusselt number at higher values of C is much less than the decrease in the value of the Nusselt number for a vertical cavity. Similar conclusions can be drawn from the results given in Fig. 7 which shows the variation of mean Nusselt number with angle of inclination for $S = 0.75$ and $Ra = 30,000$ for C values of 0.5 and 0.625, these values correspond to the case where the element is in the centre of the wall and where it is touching the "upper" wall of the cavity. The magnitude of the effect in this case is less than that for the case of $S = 0.5$ due to the larger size of the heated element.

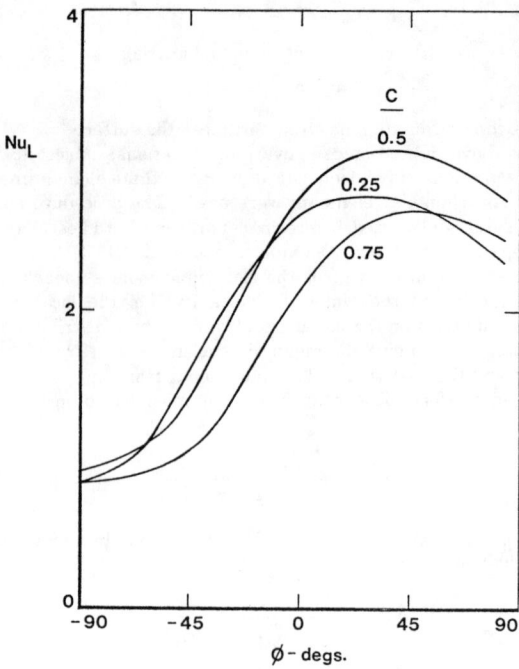

Fig. 5　Variation of mean Nusselt number, based on heated element length, with angle of inclination for various values of C for $S = 0.5$ and $Ra = 30,000$

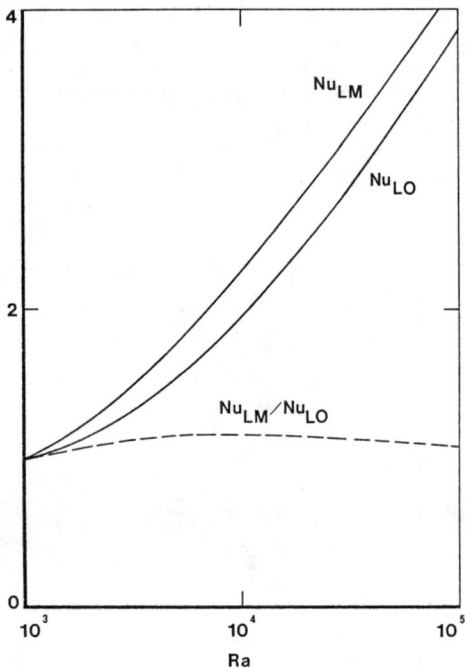

Fig. 4　Variation of mean Nusselt number, based on heated element length, at an angle of inclination of 0° and the maximum value of this Nusselt number with Rayleigh number for $S = 0.5$ and $C = 0.5$

The effect of element size on the results is illustrated in Fig. 8. This shows the variation of the mean Nusselt number based on element length with angle of inclination for various values of S for $C = 0.5$ and $Ra = 30,000$. The same results are presented in Fig. 9 which shows the variation of the mean Nusselt number based on cavity width with angle of inclination. As mentioned before, Nu_L is a dimensionless measure of the total heat transfer rate from the element while Nu_W is a dimensionless measure of the mean heat transfer rate from the element. The Nusselt number based on cavity width decreases at all angles of inclination with increasing element size, S. The Nusselt number based on element length, however, tends to decrease with decreasing element size except when S is near 1. The maximum Nusselt numbers and the Nusselt numbers for a vertical cavity for the case considered in Figs. 8 and 9 are shown in Fig. 10.

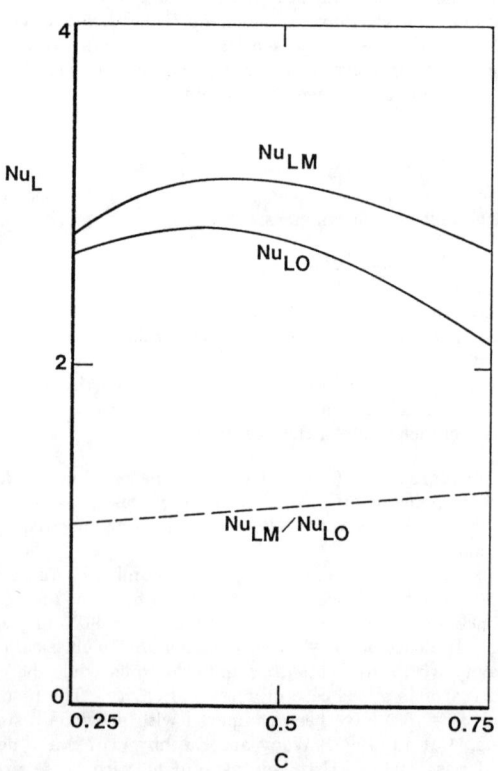

Fig. 6　Variation of mean Nusselt number, based on heated element length, at an angle of inclination of 0° and the maximum value of this Nusselt number with C for $S = 0.5$ and $Ra = 30,000$

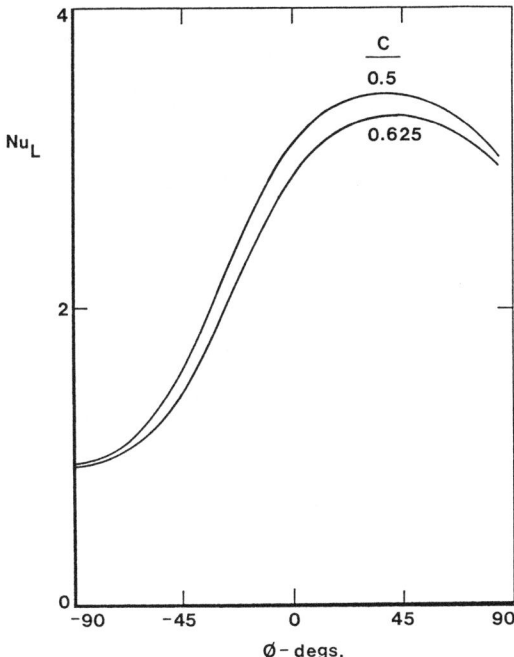

Fig. 7 Variation of mean Nusselt number, based on heated element length, with angle of inclination for various values of C for $S = 0.75$ and $Ra = 30,000$

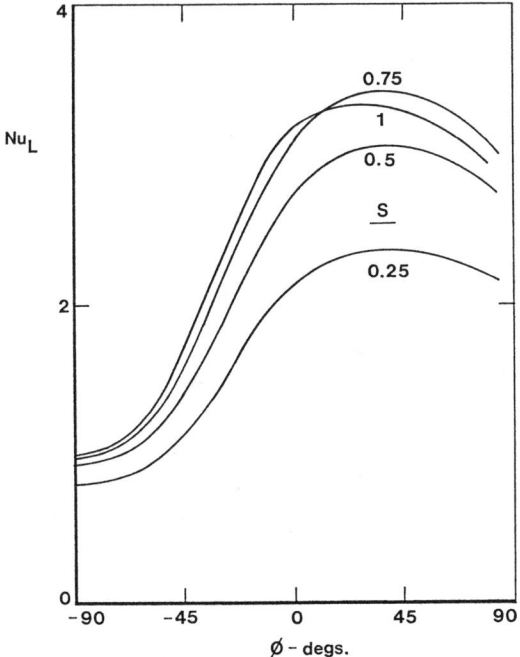

Fig. 8 Variation of mean Nusselt number, based on heated element length, with angle of inclination for various values of S for $C = 0.5$ and $Ra = 30,000$

The effect of angle of inclination on the streamline and isotherm patterns is illustrated by the results given in Fig. 11 which shows the patterns for $S = 0.5$, $C = 0.5$ and $Ra = 30,000$ for various angles of inclination. The effect of Rayleigh number on the streamline and isotherm patterns is illustrated by the results given in Fig. 12 and 13 which shows the patterns for $S = 0.5$ and $C = 0.5$ for various values of Ra for angles of inclination of $-30°$ and $+30°$ respectively.

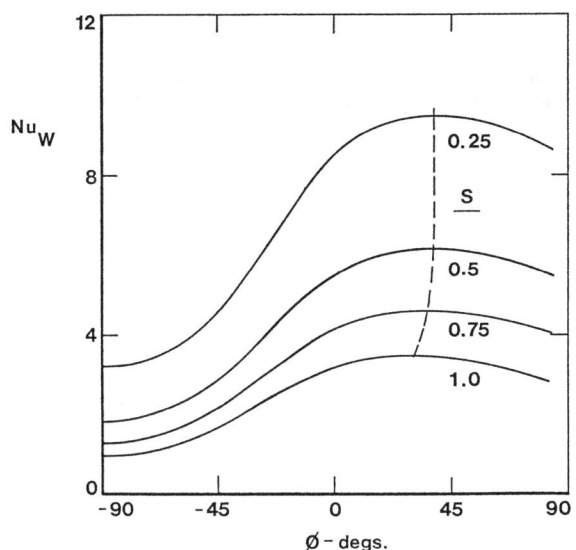

Fig. 9 Variation of mean Nusselt number, based on cavity width, with angle of inclination for various values of S for $C = 0.5$ and $Ra = 30,000$

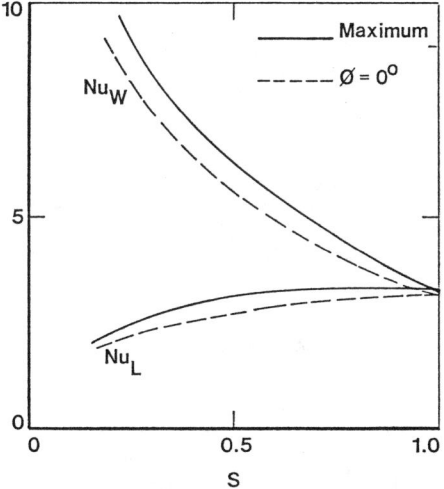

Fig. 10 Variations of mean Nusselt numbers, based on heated element length and cavity width, at an angle of inclination of $0°$ and the maximum values of these Nusselt numbers with S for $C = 0.5$ and $Ra = 30,000$

Drastic changes in the streamline pattern occur at angles of inclination near $+90°$, there being a single rotational cell when the angle is less than $90°$ but two identical but oppositely rotating cells at $90°$. This is illustrated by the results given in Fig. 14. The effect of element position on the streamline and isotherm patterns is illustrated by the results given in Fig. 15. This shows the patterns for $S = 0.75$ and $C = 0.625$ for a Rayleigh number of $30,000$ and for various angles of inclination. These patterns may be compared with those given earlier for a centrally located element. The effect of element size on the streamline and isotherm pattern is illustrated by comparing the results given in Fig. 16 for $S = 0.25$, $C = 0.5$ and $Ra = 30,000$ with those given in Fig. 11 for $S = 0.5$.

CONCLUSIONS

The results obtained in the present study indicate that the decreases in the heat transfer rate from heated elements that arise when these elements are located near the top of the wall of a cavity may be partly avoided by inclining the cavity to an angle of between approximately 30 and 60 degrees to the vertical, the optimum angle depending on the actual element size, element position and Rayleigh number.

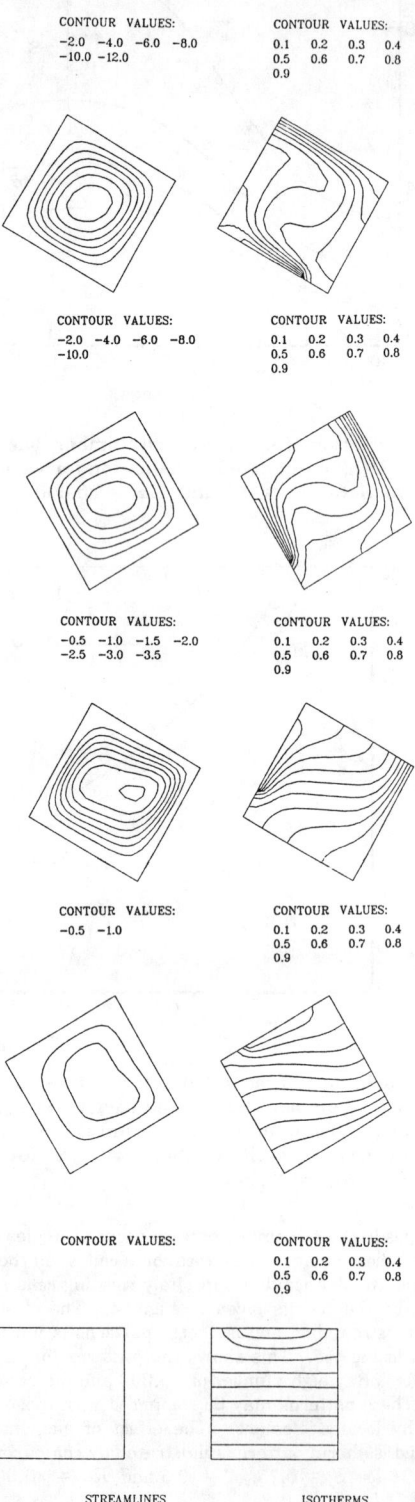

Fig. 11   Streamline and isotherm patterns for S = 0.5, C = 0.5 and Ra = 30,000 for angles of inclination of, from the top, 60, 30, -30, -60 and -90 degrees. The hot face is on the "left".

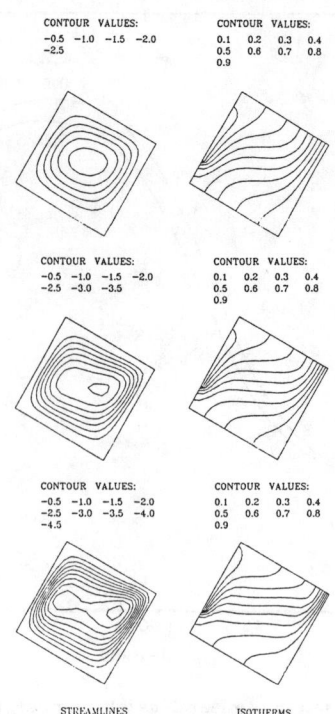

Fig. 12   Streamline and isotherm patterns for S = 0.5 and C = 0.5 for an angle of inclination of -30 degrees for Rayleigh numbers of, from the top, 10,000, 30,000 and 100,000. The hot face is on the "left".

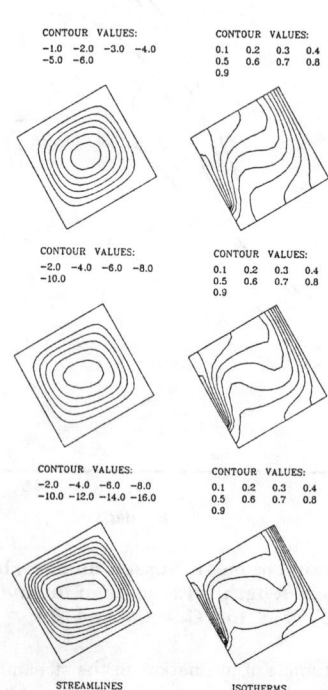

Fig. 13   Streamline and isotherm patterns for S = 0.5 and C = 0.5 for an angle of inclination of 30 degrees for Rayleigh numbers of, from the top, 10,000, 30,000 and 100,000. The hot face is on the "left".

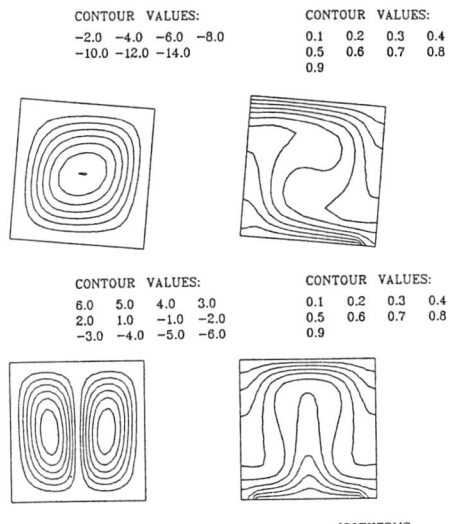

CONTOUR VALUES:
−2.0 −4.0 −6.0 −8.0
−10.0 −12.0 −14.0

CONTOUR VALUES:
0.1 0.2 0.3 0.4
0.5 0.6 0.7 0.8
0.9

CONTOUR VALUES:
6.0 5.0 4.0 3.0
2.0 1.0 −1.0 −2.0
−3.0 −4.0 −5.0 −6.0

CONTOUR VALUES:
0.1 0.2 0.3 0.4
0.5 0.6 0.7 0.8
0.9

STREAMLINES   ISOTHERMS

Fig. 14 Streamline and isotherm patterns for S = 0.75, C = 0.5 and Ra = 30,000 for angles of inclination of, from the top, 85 and 90 degrees. The hot face is on the "left".

ACKNOWLEDGEMENTS

This work was supported by the Natural Sciences and Engineering Research Council of Canada.

REFERENCES

Catton, I., 1978, "Natural Convection in Enclosures," *Proceedings of the 6th International Heat Transfer Conference*, Vol. 6, pp. 13-43.

Chu, H. H. S., Churchill, S. W. and Patterson, C. V. S., 1976, "The Effect of Heater Size, Location, Aspect Ratio and Boundary Conditions on Two-Dimensional, Laminar, Natural Convection in Rectangular Channels," *ASME J. Heat Transfer*, Vol. 98, pp. 194-201.

de Vahl Davis, G., 1986, "Finite Difference Methods for Natural and Mixed Convection in Enclosures," *Proceedings of the 8th International Heat Transfer Conference*, Vol. 1, pp. 101-109.

de Vahl Davis, G. and Jones, I. P., 1983, "Natural Convection in a Square Cavity: A Comparison Exercise," *Int. J. Num. Meth. Fluids*, Vol. 3, pp. 227-248.

Kuhn, D. and Oosthuizen, P. H., 1987, "Unsteady Natural Convection in a Partially Heated Rectangular Cavity," *ASME J. Heat Transfer*, Vol. 109, pp. 798-801.

Oosthuizen, P. H. and Paul, J. T., 1987, "Natural Convective Heat Transfer From a Square Element Mounted on the Wall of an Inclined Square Enclosure," Paper AIAA-87-1588, AIAA 22nd Thermophysics Conference.

Ostrach, S., 1972, "Natural Convection in Enclosures," *Advances in Heat Transfer*, Vol. 8, Academic Press, New York, pp. 161-227.

Ostrach, S., 1982, "Natural Convection Heat Transfer in Cavities and Cells," *Proceedings of the 7th International Heat Transfer Conference*, Vol. 1, pp. 365-379.

Turner, B. L. and Flack, R. D., 1980, "The Experimental Measurement of Natural Convective Heat Transfer in a Rectangular Enclosure with Concentrated Energy Sources," *J. Heat Transfer, Trans. ASME*, Vol. 102, pp. 236-241.

Wong, H. H. and Raithby, G. D., 1979, "Improved Finite-Difference Methods Based on a Critical Evaluation of the Approximation Errors," *Numerical Heat Transfer*, Vol. 2, pp. 139-163.

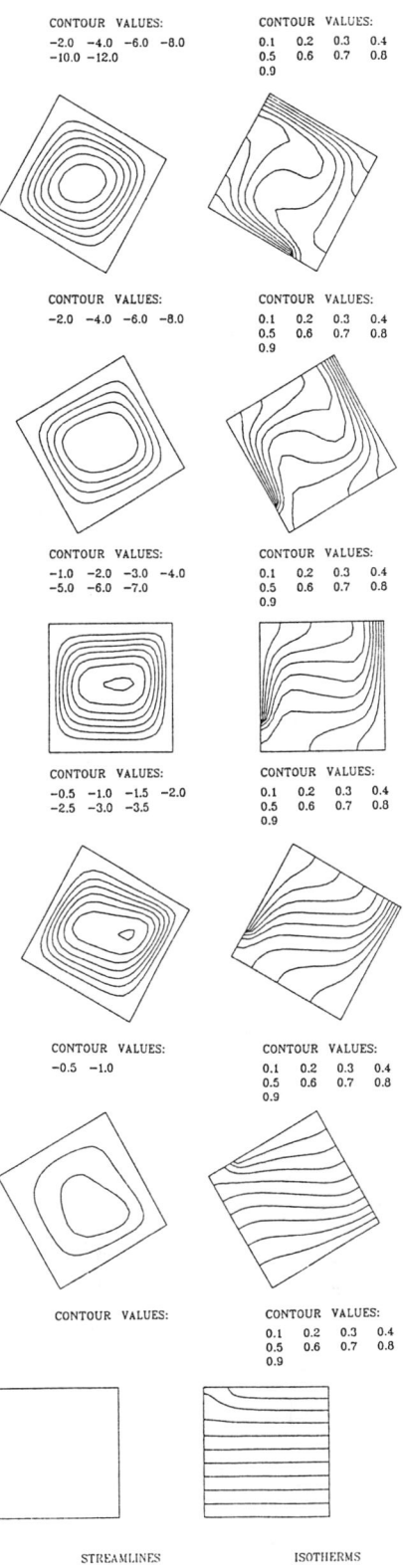

CONTOUR VALUES:
−2.0 −4.0 −6.0 −8.0
−10.0 −12.0

CONTOUR VALUES:
0.1 0.2 0.3 0.4
0.5 0.6 0.7 0.8
0.9

CONTOUR VALUES:
−2.0 −4.0 −6.0 −8.0

CONTOUR VALUES:
0.1 0.2 0.3 0.4
0.5 0.6 0.7 0.8
0.9

CONTOUR VALUES:
−1.0 −2.0 −3.0 −4.0
−5.0 −6.0 −7.0

CONTOUR VALUES:
0.1 0.2 0.3 0.4
0.5 0.6 0.7 0.8
0.9

CONTOUR VALUES:
−0.5 −1.0 −1.5 −2.0
−2.5 −3.0 −3.5

CONTOUR VALUES:
0.1 0.2 0.3 0.4
0.5 0.6 0.7 0.8
0.9

CONTOUR VALUES:
−0.5 −1.0

CONTOUR VALUES:
0.1 0.2 0.3 0.4
0.5 0.6 0.7 0.8
0.9

CONTOUR VALUES:

CONTOUR VALUES:
0.1 0.2 0.3 0.4
0.5 0.6 0.7 0.8
0.9

STREAMLINES   ISOTHERMS

Fig. 15 Streamline and isotherm patterns for S = 0.75, C = 0.625 and Ra = 30,000 for angles of inclination of, from the top, 60, 30, 0, -30, -60 and -90 degrees. The hot face is on the "left".

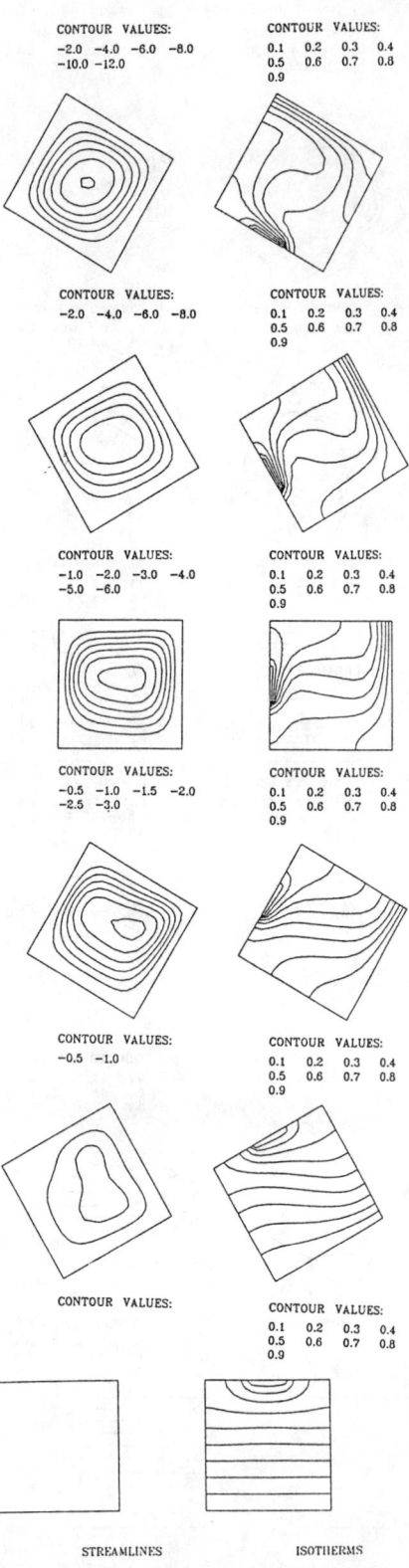

CONTOUR VALUES:
-2.0 -4.0 -6.0 -8.0
-10.0 -12.0

CONTOUR VALUES:
0.1 0.2 0.3 0.4
0.5 0.6 0.7 0.8
0.9

CONTOUR VALUES:
-2.0 -4.0 -6.0 -8.0

CONTOUR VALUES:
0.1 0.2 0.3 0.4
0.5 0.6 0.7 0.8
0.9

CONTOUR VALUES:
-1.0 -2.0 -3.0 -4.0
-5.0 -6.0

CONTOUR VALUES:
0.1 0.2 0.3 0.4
0.5 0.6 0.7 0.8
0.9

CONTOUR VALUES:
-0.5 -1.0 -1.5 -2.0
-2.5 -3.0

CONTOUR VALUES:
0.1 0.2 0.3 0.4
0.5 0.6 0.7 0.8
0.9

CONTOUR VALUES:
-0.5 -1.0

CONTOUR VALUES:
0.1 0.2 0.3 0.4
0.5 0.6 0.7 0.8
0.9

CONTOUR VALUES:

CONTOUR VALUES:
0.1 0.2 0.3 0.4
0.5 0.6 0.7 0.8
0.9

STREAMLINES   ISOTHERMS

Fig. 16   Streamline and isotherm patterns for S = 0.25, C
= 0.5 and Ra = 30,000 for angles of inclination of,
from the top, 60, 30, 0, -30, -60 and -90 degrees. The
hot face is on the "left".

238

NATURAL CONVECTION FROM A VERTICAL PLATE
WITH STEP CHANGES IN SURFACE HEAT FLUX

S. Lee and M. M. Yovanovich
Microelectronics Heat Transfer Laboratory
Department of Mechanical Engineering
University of Waterloo
Waterloo, Ontario, Canada

ABSTRACT

An approximate analytical model has been developed to predict heat transfer and flow characteristics of a steady state, two dimensional laminar boundary layer in the vicinity of a vertical flat plate under natural convection. The plate dissipates heat, with multi-step changes in surface heat flux of an arbitrary pattern, into an extensive, stagnant fluid which is maintained at uniform temperature. Wall temperature and maximum velocity variations are predicted and compared with existing numerical data, which were obtained by solving the boundary layer equations for cases with a number of uniform heat flux sources mounted flush on a vertical adiabatic plate in air using finite difference methods. The agreement is good. In order to examine the validity and accuracy of the present model, additional test computations were carried out for situations in which either the exact solutions or the proper behavior of the solutions are known. The model is shown to be valid and discussions are included herein as a result.

NOMENCLATURE

$\tilde{C}$ parameter, $\tilde{C} = (\sqrt{\mathrm{Pr}}\,\tilde{C}_u/4)^{2/3}$ or $\tilde{C} = \mathrm{Pr}/(4\tilde{C}_\eta^2)$

$\tilde{C}_u, \tilde{C}_\eta$ parameters defined in Eqs. (18), and (19) or (20)

$\mathrm{erfc}\,\eta$ Complementary Error function, $\dfrac{2}{\sqrt{\pi}}\displaystyle\int_\eta^\infty e^{-\tau^2}\,d\tau$

f_T, f_u functions defined by Eqs. (11) and (12)

g gravitational acceleration

G_e, G_m functions given in Table 1

$\tilde{h}_e, \tilde{h}_m$ functions given in Table 2

$\mathcal{H}_e, \mathcal{H}_m$ functions expressed by Eqs. (45) and (44)

$\mathrm{i}^n\mathrm{erfc}\,\eta$ Complementary Error function integrated n-times, $\displaystyle\int_\eta^\infty \mathrm{i}^{n-1}\mathrm{erfc}\,\tau\,d\tau$, $\mathrm{i}^1\mathrm{erfc}\,\eta = \mathrm{ierfc}\,\eta$, $\mathrm{i}^0\mathrm{erfc}\,\eta = \mathrm{erfc}\,\eta$

k thermal conductivity of fluid

p square root of Prandtl number, $\sqrt{\mathrm{Pr}}$

q local heat-flux

q^* dimensionless local heat-flux, q/q_{w_0}

q_e^*, q_m^* expressions given by Eqs. (41) and (40)

t time variable defined in $t-y$ plane

T temperature excess over ambient fluid temperature

u local velocity parallel to plate, in x-direction

u_c characteristic velocity across the boundary layer

U dimensionless u-velocity defined by Eq. (47)

v local velocity normal to plate, in y-direction

x vertical coordinate measured from leading edge

y horizontal coordinate measured from plate surface

Greek Symbols

α thermal diffusivity of fluid, $k/(\rho c_p)$

β thermal expansion coefficient, $-(\partial\rho/\partial T)_P/\rho$

γ function defined by Eq. (38)

η_0 variable defined by Eq. (13) or Eq. (48)

η_i variable defined by Eq. (14) or Eq. (48)

ν kinematic viscosity, μ/ρ

ξ dimensionless x-coordinate, x/x_0

$\tilde{\phi}, \tilde{\varphi}, \tilde{\psi}$ modifying functions

$\tilde{\Phi}$ function defined by Eq. (39)

θ dimensionless temperature defined by Eq. (46)

Subscripts

0 parameters at leading section

i parameters at i-th step

r references

w wall conditions

Dimensionless Groups

G^* modified Grashof number, $g\beta q_r x_r^4/k\nu^2$

Gr_x Grashof number, $g\beta T_{w_0} x^3/\nu^2$

Gr_x^* modified Grashof number, $g\beta q_{w_0} x^4/k\nu^2$

Nu_x Nusselt number, $q_{w_0} x/T_{w_0} k$

Pr Prandtl number, ν/α

Introduction

The natural convection heat transfer from a vertical flat plate has been a subject of numerous investigations in the past few decades. The plate with thermal conditions that allow similarity transformations have been examined by Ostrach (1953), Sparrow and Gregg (1956, 1958), and Jaluria and Gebhart (1977). They have considered steady state, two dimensional laminar boundary layer equations for uniform wall temperature, uniform surface heat flux, excess wall temperature variations of the power and exponential forms, and a line source on an adiabatic plate. Yang (1960) found that there are no other types of boundary conditions which would make similarity solutions possible for steady natural convection from a vertical plate. Numerous studies have been carried out to expand available solutions for non-similar boundary conditions by using various methods. Unfortunately, analytical solution techniques for problems with arbitrary boundary conditions, that are frequently expected from most practical applications, are not available.

The modeling of an isolated vertical flat plate with arbitrary surface thermal conditions would be useful in many technological applications such as the thermal design of Printed Circuit Boards (PCBs) on which a number of finite sized heat sources are mounted. Complete thermal phenomena involved in the final product of a PCB are too complex and are impractical to analyze as a whole. A PCB is often modeled as a flat plate in thermal analyses with heat sources mounted flush with its surface. Bar-Cohen (1985) has discussed and found that it is possible to apply heat transfer relations developed for a smooth wall to non-smooth component carrying PCBs. Nonetheless, the sources are discrete and randomly distributed in general. The heat transfer associated with PCB applications is usually a conjugate heat transfer, in which all three modes of heat transfer, namely heat conduction through the board, heat convection in the ambient fluid and surface radiation to the surroundings, may occur simultaneously. It becomes apparent that neither the resulting wall temperature nor convective heat flux variations would be known *a priori*, and similarity transformations would rarely be allowed in the analyses.

During the early stages of an ongoing development towards conjugate thermal modeling of a PCB cooled by natural convection, the present authors adopted the flat plate approximation but recognized a need for a model that can predict wall temperature variations in a vertical plate when arbitrary surface heat flux variations are prescribed. Only then, by means of an iterative procedure, would the model become capable of simulating conjugate heat transfer which is ubiquitous in situations involving PCBs (Lee and Yovanovich, 1989). As previously mentioned, an analytical model capable of dealing with problems under the present consideration does not exist in the literature. Other techniques to obtain solutions for cases with the arbitrary thermal conditions may include experimental investigations and fully numerical methods. The data from experiments are, however, those of case-by-case studies and cannot be manipulated to predict results of other cases, since their range of reliable application is mostly limited within the range of parameters that are examined. Utilization of fully numerical methods such as finite difference or finite element methods usually requires a main-frame computer with associated high computing expenses, and may become impractical in some situations. Many applications call for simpler and inexpensive solution techniques, though they may be less accurate than more rigorous and time consuming methods (Schetz, 1963).

In this paper, an approximate analytical model is presented which can be used to predict two dimensional laminar natural convection flow and heat transfer about a vertical flat plate dissipating energy into a quiescent medium with multi-step changes in surface heat flux of an arbitrary pattern. With sufficient discretization, the model can be applied to most surface heat flux variations that can be expected from a flat plate model of PCBs.

Other approximate methods are available in the literature for cases with a continuous variation in the thermal boundary conditions (Tribus, 1958, Raithby and Hollands, 1975). However, applicability of these results is limited to problems that closely maintain the form of the specified polynomials in characterizing the temperature and velocity profiles across the boundary layer. This deficiency in Tribus's solutions was pointed out and further discussed by Sparrow and Gregg in their "Authors Closure" following Tribus's discussion. In addition, Zinnes (1970) has shown that, when the surface of a plate experiences abrupt thermal variations, the results obtained by using the model of Tribus considerably deviate from those obtained by using finite difference methods.

The downstream wall temperature and maximum velocity variations are obtained by using the present model and compared with existing numerical data of Jaluria (1982), who solved the boundary layer equations by using finite difference methods for cases with a number of strip thermal sources of uniform surface heat flux mounted flush on an adiabatic plate in air. The comparison resulted in satisfactory to good agreement. Since the present model is developed based on an approximate method, computations are carried out to examine the validity and accuracy of the model. A complete verification over all ranges of parameters is obviously neither practical nor necessary for the present purpose. The validity of the model is, therefore, demonstrated within the range of parameters for comparisons with the results of the aforementioned numerical data. The model is further validated by examining its behavior using the cases for which either the exact solutions or the proper behavior of the solutions are known.

Problem Statement

The geometric configuration and coordinate system of the problem are depicted in Fig. 1, where a vertical flat plate is shown with a possible step variation in surface heat flux. The problem is two dimensional in the $x - y$ plane, having a large transverse dimension in the z-direction normal to the page. The plate is dissipating heat into an extensive, quiescent fluid which is assumed to be maintained at uniform temperature. As shown in the figure, the value of the sectional heat flux, except q_{w_0} in the leading section, may be zero; the plate is insulated, or negative; the fluid is heating the plate, so long as the resulting temperature excess over the ambient fluid temperature at any point within the boundary layer is maintained positive.

A set of boundary layer equations that governs two dimensional, steady state momentum and energy transport in natural convection is written below.

$$\frac{\partial u}{\partial x} + \frac{\partial v}{\partial y} = 0 \tag{1}$$

$$u\frac{\partial u}{\partial x} + v\frac{\partial u}{\partial y} = \nu\frac{\partial^2 u}{\partial y^2} + g\beta T \tag{2}$$

$$u\frac{\partial T}{\partial x} + v\frac{\partial T}{\partial y} = \alpha\frac{\partial^2 T}{\partial y^2} \tag{3}$$

The boundary conditions associated with the above equations are

at $y = 0$, $\quad u = v = 0$,

$$-k\frac{\partial T}{\partial y} = q_{w_0} \text{ for } 0 < x \leq x_0$$

$$-k\frac{\partial T}{\partial y} = q_{w_i} \text{ for } x_{i-1} < x \leq x_i \,;\, i = 1, 2, 3, \dots$$

as $y \to \infty$, $\quad u \to 0$, $\quad T \to 0$

at $x = 0$, $\quad u = T = 0 \tag{4}$

where x and y are the coordinates parallel and normal to the plate, u and v are the corresponding components of the velocity, and T is the local temperature excess over the ambient fluid temperature. All q_{w_i} including q_{w_0} are uniform. The usual assumptions and approximations, such as those of constant fluid properties except the density in the derivation of the buoyant term and negligible viscous heating, are made in deriving these equations. The present problem includes the cases with a step change in surface heat flux when $i = 1$, which were inclusively examined by the present authors in an earlier study (Lee and Yovanovich, 1988).

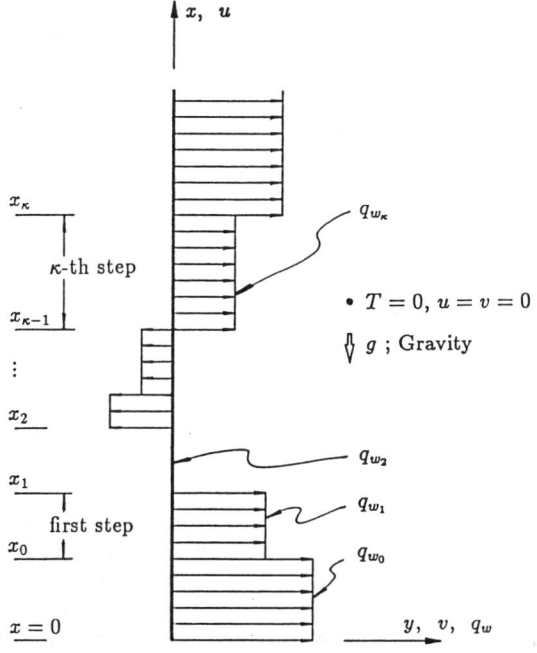

Figure 1: Geometric configuration shown with a schematic surface heat flux variation.

ANALYSIS

An exact analytical solution to the above problem does not exist, and obtaining such solutions in the near future seems improbable. As was done in the earlier study, the non-linear convection terms in the left hand side of the momentum and energy equations, Eqs. (2) and (3), are linearized through two stages as follows.

Firstly, a transient coordinate t is introduced through a $t - x$ transformation defined as

$$t = \frac{x}{u_c} \tag{5}$$

where u_c is called the characteristic velocity. By the above definition, it can be said that the distance, measured from the leading edge of the plate at $x = 0$, to a location along the x-coordinate corresponds to a specific lapse in time. The original x-coordinate is hence transformed into a transient coordinate specified by the time variable t. The characteristic velocity is presently an unknown function of x. It may be viewed as an effective mean flow velocity of the fluid in the boundary layer initiated at the leading edge of the plate.

Secondly, upon this transformation, an assumption is made such that diffusion is dominant in transporting both heat and momentum across the boundary layer in the y-direction at fixed time t. The effect of convection transport in the $t - y$ plane is thereby neglected, and profiles of the temperature and velocity distributions in the original $x - y$ plane will be determined approximately by transient diffusion equations in the $t - y$ plane. Subsequently, the above governing differential equations are linearized in the $t - y$ plane and the transformed momentum and energy equations take forms of

$$\frac{\partial u}{\partial t} = \nu\frac{\partial^2 u}{\partial y^2} + g\beta T \tag{6}$$

$$\frac{\partial T}{\partial t} = \alpha\frac{\partial^2 T}{\partial y^2} \tag{7}$$

respectively. This set of equations is identical to the governing differential equations which describe *real time* transient natural convection heat transfer from an infinite plate. The continuity equation, Eq. (1), is obsolete since the u-velocity becomes one dimensional in y at fixed time t. The above assumption of dominant diffusion in the y-direction in the $t - y$ plane can only be validated indirectly through comparisons of the resulting temperature and velocity distributions with existing and known data, as will be carried out in the following section. The corresponding transient boundary conditions that are compatible with those given by Eq. (4) can now be given as

at $y = 0$, $\quad u = 0$,

$$-k\frac{\partial T}{\partial y} = q_{w_0} \text{ for } 0 < t \leq t_0$$

$$-k\frac{\partial T}{\partial y} = q_{w_i} \text{ for } t_{i-1} < t \leq t_i \,;\, i = 1, 2, 3, \dots$$

as $y \to \infty$, $\quad u \to 0$, $\quad T \to 0$

at $t = 0$, $\quad u = T = 0 \tag{8}$

where t_0 and t_i are the times corresponding to the locations at x_0 and x_i, respectively.

The transient solutions to the above equations are found for the temperature and velocity distributions by means of similarity transformations and the method of superposition (Lee, 1988). They are, for the κ-th step defined for $t_{\kappa-1} < t \leq t_\kappa$,

$$T = \frac{2\sqrt{\alpha}}{k}\left[q_{w_0}t^{1/2}f_T(\eta_0) + \sum_{i=1}^{\kappa}(q_{w_i} - q_{w_{i-1}})(t - t_{i-1})^{1/2}f_T(\eta_i)\right]$$

(9)

$$u = \frac{4\sqrt{\alpha}g\beta}{k}\left[q_{w_0}t^{3/2}f_u(\eta_0) + \sum_{i=1}^{\kappa}(q_{w_i} - q_{w_{i-1}})(t - t_{i-1})^{3/2}f_u(\eta_i)\right]$$

(10)

where

$$f_T(\eta) = \text{ierfc}\,\eta$$

(11)

$$f_u(\eta) = \begin{cases} \eta\,\text{i}^2\text{erfc}\,\eta & \text{for } \text{Pr} = 1 \\ \dfrac{2}{1 - \text{Pr}}(\text{i}^3\text{erfc}\,\eta - \text{i}^3\text{erfc}\,\dfrac{\eta}{\sqrt{\text{Pr}}}) & \text{for } \text{Pr} \neq 1 \end{cases}$$

(12)

$$\eta_0 = \frac{y}{2\sqrt{\alpha t}}$$

(13)

$$\eta_i = \frac{y}{2\sqrt{\alpha(t - t_{i-1})}}$$

(14)

These solutions are exact to the set of transient equations, Eqs. (6) to (8), which approximately represents the set of original steady state equations, Eqs. (1) to (4), through the $t - x$ transformation. Determination of t, as well as $t - t_{i-1}$, in terms of x would, therefore, render these equations to become approximate solutions to the original problem in the $x - y$ plane. Although the present analysis is based on the linearization of the equations in the $t - y$ plane, it is to be emphasized that the non-linearity of the steady state problem in the $x - y$ plane will be restored by treating each and every $t - x$ transformation, that is associated with the time lapse corresponding to the displacement within each step, to be specific as detailed below.

Consider the case of uniform surface heat flux, or, equivalently, the leading section of the present problem between $x = 0$ and $x = x_0$. Owing to the boundary layer approximations, solutions at the upstream locations are unaffected by the thermal variations imposed at the downstream locations. The solutions in this section, where $\kappa = 0$, are thus given solely by the first terms in Eqs. (9) and (10), in which t is the only unknown parameter in terms of the steady state variables.

It was found in the earlier study (Lee and Yovanovich, 1988) that the required $t - x$ transformation for $0 < x \leq x_0$ is

$$t = \frac{x}{u_c} = \tilde{C}\,\frac{x}{\frac{\nu}{x}\text{Gr}_x^{*2/5}}$$

(15)

where $\tilde{C}$ is a function only of the Prandtl number, Pr, and is given by two different expressions. They are

$$\tilde{C} = \frac{\text{Pr}}{4\tilde{C}_\eta^2}$$

(16)

and

$$\tilde{C} = \left(\frac{\sqrt{\text{Pr}}\tilde{C}_u}{4}\right)^{2/3}$$

(17)

where $\tilde{C}_u$ and $\tilde{C}_\eta$ were determined by using the integral method.

$$\tilde{C}_u^{-5} = \frac{\text{Pr}}{\pi^{3/2}}G_e\left(\frac{7G_m}{5} + \frac{2G_e}{1 + 1/\sqrt{\text{Pr}}}\right)^2$$

(18)

Table 1 : G_e and G_m

	$\text{Pr} = 1$	$\text{Pr} \neq 1$ $(p = \sqrt{\text{Pr}})$
G_e	$\dfrac{7 - 4\sqrt{2}}{120}$	$\dfrac{2p^5 + (8\sqrt{2} - 9)p^3 + 5p^2 + 2 - 2(p^2 + 1)^{5/2}}{60\,p^3(1 - p^2)}$
G_m	$\dfrac{19 - 13\sqrt{2}}{105}$	$\dfrac{2p^7 + 7p^5 + (8\sqrt{2} - 9)(p^4 + p^3) + 7p^2 + 2 - 2(p^2 + 1)^{7/2}}{52.5\,p^3(1 - p^2)^2}$

$$\tilde{C}_\eta^2 = \frac{\text{Pr}}{\pi^{1/2}}G_e\tilde{C}_u$$

(19)

where G_e and G_m are tabulated in Table 1.

The two different expressions, as given by Eqs. (16) and (17), for $\tilde{C}$ are necessary in order to conserve concurrently the energy and momentum flows in the boundary layer. The $t - x$ transformation, Eq. (15), with $\tilde{C}$ given by Eq. (16) transforms t in the similarity variable η_0 and $t^{1/2}$ in the temperature solution, Eq. (9). Equation (15) with $\tilde{C}$ given by Eq. (17) transforms $t^{3/2}$ occurring in the velocity solution, Eq. (10).

A value of $\tilde{C}_\eta$ may alternatively be found based on any natural convection heat transfer results obtained for a vertical plate with uniform surface heat flux. For example, using the correlation equation presented by Fujii and Fujii (1976) for Nu_x as a function of Pr and Gr_x^*, $\tilde{C}_\eta$ can be expressed as

$$\tilde{C}_\eta = \frac{1}{\sqrt{\pi}}\frac{\text{Nu}_x}{\text{Gr}_x^{*1/5}} = \frac{1}{\sqrt{\pi}}\left(\frac{\text{Pr}^2}{4 + 9\sqrt{\text{Pr}} + 10\,\text{Pr}}\right)^{1/5}$$

(20)

For the transformation of the time variables in the succeeding step sections, the $t - x$ transformation defined by Eq. (5) can further be generalized for the time lapse $t - t_{i-1}$ associated within the i-th step as

$$t - t_{i-1} = \frac{x - x_{i-1}}{u_c} \qquad \text{for } i = 1, 2, 3, \ldots$$

(21)

Upon this transformation, the problem is now reduced to determining the characteristic velocities introduced at every section of the discretized domain. For the κ-th step change, introduced beyond $x = x_{\kappa-1}$, all the existing characteristic velocities, that are determined and used to define the $t - x$ transformations up to the $(\kappa - 1)$-th step, have to be modified. Furthermore, an additional characteristic velocity has to be determined which would represent the flow within the new boundary layer evolved from the surface at the beginning of the κ-th step. The number of unknown modifying functions required to determine the characteristic velocities, and in turn, complete the solutions in the κ-th step, becomes $\kappa + 1$. This number can be seen also from the above transient solutions, Eqs. (9) and (10), as the total number of functions required to transform t and $t - t_{i-1}$, for i from 1 to κ, is $\kappa + 1$.

In order to clarify this, consider a problem with two step changes in surface heat flux. The $t - x$ transformations within the first step, $x_0 < x \le x_1$, are defined as

$$t = \tilde{C}\,\tilde{\phi}_1\,\frac{x}{\frac{\nu}{x}\mathrm{Gr}_x^{*2/5}} \tag{22}$$

$$t - t_0 = \tilde{C}\,\tilde{\psi}_1\,\frac{x - x_0}{\frac{\nu}{x}\mathrm{Gr}_x^{*2/5}} \tag{23}$$

A subscript 1 is used to denote $\tilde{\phi}_1$ and $\tilde{\psi}_1$ as the modifying functions incurred by the first step and $\tilde{C}$ is given previously. $\tilde{\phi}_1$ is the function that modifies the primary characteristic velocity established from the leading edge of the plate, whereas $\tilde{\psi}_1$ is the function required in determining the characteristic velocity within the secondary boundary layer evolved at the beginning of the first step, $x = x_0$. They were determined by using the integral method in the earlier study (Lee and Yovanovich, 1988).

As soon as the second step is introduced, another boundary layer evolves from the surface at the beginning of the second step. This additional boundary layer grows within the existing secondary boundary layer in much the same way as the secondary boundary layer itself evolved when the first step was introduced. Not only do the two existing characteristic velocities as found in the above equations, namely $\frac{\nu}{x}\mathrm{Gr}_x^{*2/5}/\tilde{C}\tilde{\phi}_1$ and $\frac{\nu}{x}\mathrm{Gr}_x^{*2/5}/\tilde{C}\tilde{\psi}_1$, have to be modified, but a new characteristic velocity for the flow in the additional boundary layer has to be determined also. Therefore, two additional functions, say $\tilde{\phi}_2$ and $\tilde{\varphi}_2$, are required to modify the two existing characteristic velocities, and a third function, $\tilde{\psi}_2$, is introduced to determine the characteristic velocity within the new boundary layer. The resulting transformation functions that are associated with the second step, $x_1 < x \le x_2$, may then be written as

$$t = \tilde{C}\,\tilde{\phi}_1\tilde{\phi}_2\,\frac{x}{\frac{\nu}{x}\mathrm{Gr}_x^{*2/5}} \tag{24}$$

$$t - t_0 = \tilde{C}\,\tilde{\psi}_1\tilde{\varphi}_2\,\frac{x - x_0}{\frac{\nu}{x}\mathrm{Gr}_x^{*2/5}} \tag{25}$$

$$t - t_1 = \tilde{C}\,\tilde{\psi}_2\,\frac{x - x_1}{\frac{\nu}{x}\mathrm{Gr}_x^{*2/5}} \tag{26}$$

The above set of transformation functions show that there are three unknown modifying functions in all, due to the introduction of the second step.

It was found (Lee, 1988) that the heat transfer characteristic of the flow within the first step is insensitive to induced changes in the $\tilde{\phi}_1$ variations. This observation may be postulated to yield a general interpretation, and as such, the resulting heat transfer phenomena are not strongly dependent on those modifying functions that modify *existing* characteristic velocities. Moreover, the $\tilde{\phi}_1$ variations are shown to respond gradually to the step changes at the immediate downstream locations beyond $x = x_0$, in the region where the greatest portion of the changes in heat transfer rate take place. The abrupt changes in the heat transfer rate, observed right after the step, are mostly attributed to the changes in $\tilde{\psi}_1$ variations. Further modifying functions, $\tilde{\phi}_2$ and $\tilde{\varphi}_2$, that are introduced to modify *existing* characteristic velocities due to the addition of the second step, are expected to behave not only qualitatively similar to $\tilde{\phi}_1$, but also quantitatively similar to each other. In all, they are assumed to be unique, $\tilde{\varphi}_2 = \tilde{\phi}_2$, and hence, the above set

of transformations become

$$t = \tilde{C}\,\tilde{\phi}_1\tilde{\phi}_2\,\frac{x}{\frac{\nu}{x}\mathrm{Gr}_x^{*2/5}} \tag{27}$$

$$t - t_0 = \tilde{C}\,\tilde{\psi}_1\tilde{\phi}_2\,\frac{x - x_0}{\frac{\nu}{x}\mathrm{Gr}_x^{*2/5}} \tag{28}$$

$$t - t_1 = \tilde{C}\,\tilde{\psi}_2\,\frac{x - x_1}{\frac{\nu}{x}\mathrm{Gr}_x^{*2/5}} \tag{29}$$

Two functions, $\tilde{\phi}_2$ and $\tilde{\psi}_2$, are now required to be determined.

Although the approximation leading to the reduction of the number of unknown modifying functions to two was based on the observation that primarily concerns temperature distributions, the same may be applied for the velocity distributions. Similar approximations are introduced as additional step changes are considered. As a result, $(\kappa + 1)$ $t - x$ transformations, required due to the induction of the κ-th step between $x = x_{\kappa-1}$ and $x = x_\kappa$, are defined as

$$t = \tilde{C}\,\tilde{\phi}_1\tilde{\phi}_2\tilde{\phi}_3\cdots\tilde{\phi}_\kappa\,\frac{x}{\frac{\nu}{x}\mathrm{Gr}_x^{*2/5}} \tag{30}$$

$$t - t_0 = \tilde{C}\,\tilde{\psi}_1\tilde{\phi}_2\tilde{\phi}_3\cdots\tilde{\phi}_\kappa\,\frac{x - x_0}{\frac{\nu}{x}\mathrm{Gr}_x^{*2/5}} \tag{31}$$

$$t - t_1 = \tilde{C}\,\tilde{\psi}_2\tilde{\phi}_3\cdots\tilde{\phi}_\kappa\,\frac{x - x_1}{\frac{\nu}{x}\mathrm{Gr}_x^{*2/5}} \tag{32}$$

$$\vdots$$

$$t - t_{i-1} = \tilde{C}\,\tilde{\psi}_i\tilde{\phi}_{i+1}\cdots\tilde{\phi}_\kappa\,\frac{x - x_{i-1}}{\frac{\nu}{x}\mathrm{Gr}_x^{*2/5}} \tag{33}$$

$$\vdots$$

$$t - t_{\kappa-1} = \tilde{C}\,\tilde{\psi}_\kappa\,\frac{x - x_{\kappa-1}}{\frac{\nu}{x}\mathrm{Gr}_x^{*2/5}} \tag{34}$$

Each additional κ-th step requires the determination of two additional modifying functions, $\tilde{\phi}_\kappa$ and $\tilde{\psi}_\kappa$, that are dependent on the history of the surface thermal condition up to and including the step. These functions are determined, again, by applying the integral method. After manipulating, simplifying and rearranging expressions, the integral method results in a set of equations as

$$\frac{d\gamma_\kappa^3}{d\xi} = \frac{\frac{3}{5\xi q_e^*}\left(q_m^*\frac{\mathcal{H}_{e_\kappa}}{\mathcal{H}_{m_\kappa}} - q_{w_\kappa}^*\right) - \sum_{i=1}^{\kappa-1}\left[\left(\frac{\partial\mathcal{H}_{m_\kappa}/\partial\gamma_i}{7\gamma_i^2\mathcal{H}_{m_\kappa}} - \frac{\partial\mathcal{H}_{e_\kappa}/\partial\gamma_i}{5\gamma_i^2\mathcal{H}_{e_\kappa}}\right)\frac{d\gamma_i^3}{d\xi}\right]}{\frac{\partial\mathcal{H}_{m_\kappa}/\partial\gamma_\kappa}{7\gamma_\kappa^2\mathcal{H}_{m_\kappa}} - \frac{\partial\mathcal{H}_{e_\kappa}/\partial\gamma_\kappa}{5\gamma_\kappa^2\mathcal{H}_{e_\kappa}}} \tag{35}$$

$$\tilde{\Phi}_\kappa = \left(\frac{q_e^*}{\mathcal{H}_{e_\kappa}}\right)^{2/5} \tag{36}$$

where the initial condition for γ_κ^3 is

$$\lim_{\xi \to \xi_{\kappa-1}^+}\gamma_\kappa^3 = 0 \tag{37}$$

and

$$\gamma_i = \sqrt{\frac{\tilde{\psi}_i}{\tilde{\Phi}_i}\left(1 - \frac{\xi_{i-1}}{\xi}\right)} \tag{38}$$

$$\tilde{\Phi}_i = \prod_{j=1}^{i}\tilde{\phi}_j \tag{39}$$

$$q_m^* = 1 + \sum_{i=1}^{\kappa}(q_{w_i}^* - q_{w_{i-1}}^*)\gamma_i^2 \tag{40}$$

Table 2 : $\tilde{h}_m(\gamma)$ and $\tilde{h}_e(\gamma)$

$Pr = 1$	$\tilde{h}_m(\gamma)$	$\dfrac{12\gamma^7 + 7\gamma^5 + 7\gamma^2 + 12 + 35\gamma^2(\gamma^2+1)^{3/2} - 12(\gamma^2+1)^{7/2}}{19 - 13\sqrt{2}}$
	$\tilde{h}_e(\gamma)$	$\dfrac{2\gamma^5 + 5\gamma^3 + 5\gamma^2 + 2 - 2(\gamma^2+1)^{5/2}}{7 - 4\sqrt{2}}$
$Pr \neq 1$	$\tilde{h}_m(\gamma)$	$\begin{aligned}&\big[(p^2-1)(p-1)\{7p^2\gamma^2(\gamma^3+1)+2(p^2+1)(p^2+p+1)(\gamma^7+1)\}\\&+2p^3(p+1)(\gamma^2+1)^{7/2}-2(\gamma^2+p^2)^{7/2}-2(p^2\gamma^2+1)^{7/2}\big]\\&\div\big[2p^7+7p^5+(8\sqrt{2}-9)(p^4+p^3)+7p^2+2-2(p^2+1)^{7/2}\big]\end{aligned}$
	$\tilde{h}_e(\gamma)$	$\begin{aligned}&\big[4p^3(\gamma^2+1)^{5/2}-2(p^2\gamma^2+1)^{5/2}-2(\gamma^2+p^2)^{5/2}\\&+2(p^5-2p^3+1)(\gamma^5+1)-5p^2(p-1)(\gamma^3+\gamma^2)\big]\\&\div\big[2p^5+(8\sqrt{2}-9)p^3+5p^2+2-2(p^2+1)^{5/2}\big]\end{aligned}$

$$q_e^* = 1 + \sum_{i=1}^{\kappa}(q_{w_i}^* - q_{w_{i-1}}^*)\Big(1 - \frac{\xi_{i-1}}{\xi}\Big) \tag{41}$$

$$q_{w_i}^* = \frac{q_{w_i}}{q_{w_0}} \tag{42}$$

$$\xi_i = \frac{x_i}{x_0} \tag{43}$$

$\mathcal{H}_{m_\kappa}$ and $\mathcal{H}_{e_\kappa}$ are given by recurrence relations as

$$\mathcal{H}_{m_\kappa} = \mathcal{H}_{m_{\kappa-1}} + (q_{w_\kappa}^* - q_{w_{\kappa-1}}^*) \tag{44}$$
$$\times\left[(q_{w_\kappa}^* - q_{w_{\kappa-1}}^*)\gamma_\kappa^7 + \tilde{h}_m(\gamma_\kappa) + \sum_{i=1}^{\kappa-1}(q_{w_i}^* - q_{w_{i-1}}^*)\gamma_i^7\tilde{h}_m(\gamma_\kappa/\gamma_i)\right]$$

$$\mathcal{H}_{e_\kappa} = \mathcal{H}_{e_{\kappa-1}} + (q_{w_\kappa}^* - q_{w_{\kappa-1}}^*) \tag{45}$$
$$\times\left[(q_{w_\kappa}^* - q_{w_{\kappa-1}}^*)\gamma_\kappa^5 + \tilde{h}_e(\gamma_\kappa) + \sum_{i=1}^{\kappa-1}(q_{w_i}^* - q_{w_{i-1}}^*)\gamma_i^5\tilde{h}_e(\gamma_\kappa/\gamma_i)\right]$$

with $\mathcal{H}_{m_0} = \mathcal{H}_{e_0} = 1$, and algebraic expressions for $\tilde{h}_m(\cdot)$ and $\tilde{h}_e(\cdot)$ can be found in Table 2.

This first order ordinary differential equation, Eq. (35), is independent of $\tilde{\Phi}_\kappa$ and is, therefore, decoupled from Eq. (36). With the above initial condition, it represents a complete problem that can be solved numerically for all γ_i^3 in succession for i from 1 to κ with given uniform $q_{w_i}^*$ and Pr. Upon finding γ_κ^3, $\tilde{\Phi}_\kappa$ can be determined from Eq. (36). The modifying functions $\tilde{\phi}_i$ and $\tilde{\psi}_i$ can be subsequently obtained by considering Eqs. (39) and (38), respectively. Readers who are interested in the detailed derivations and complete expressions of the differential terms appearing in the above differential equation may find the full context in Lee, 1988, where additional discussions on the modifying functions are also included.

RESULTS AND DISCUSSION

Temperature and Velocity Distributions. By substituting the $t - x$ transformations, defined by Eqs. (30) through (34), into Eqs. (9) and (10), a set of dimensionless approximate solutions to the original steady state problem may be obtained as

$$\begin{aligned}\theta &= \frac{T}{q_{w_0}x_0/k\mathrm{Gr}_{x_0}^{*\,1/5}}\\&= \frac{\xi^{1/5}\sqrt{\tilde{\Phi}_\kappa}}{\tilde{C}_\eta}\left[f_T(\eta_0) + \sum_{i=1}^{\kappa}(q_{w_i}^* - q_{w_{i-1}}^*)\gamma_i f_T(\eta_i)\right]\end{aligned} \tag{46}$$

$$\begin{aligned}U &= \frac{u}{\frac{\nu}{x_0}\mathrm{Gr}_{x_0}^{*\,2/5}}\\&= \xi^{3/5}\tilde{C}_u\tilde{\Phi}_\kappa^{3/2}\left[f_u(\eta_0) + \sum_{i=1}^{\kappa}(q_{w_i}^* - q_{w_{i-1}}^*)\gamma_i^3 f_u(\eta_i)\right]\end{aligned} \tag{47}$$

for $x_{\kappa-1} < x \le x_\kappa$, where $f_T(\cdot)$ and $f_u(\cdot)$ are as defined by Eqs. (11) and (12), respectively, and

$$\eta_0 = \gamma_i\eta_i = \frac{\tilde{C}_\eta}{\xi^{1/5}\sqrt{\tilde{\Phi}_\kappa}}\mathrm{Gr}_{x_0}^{*\,1/5}\frac{y}{x_0} \tag{48}$$

For an additional κ-th step, γ_κ and $\tilde{\Phi}_\kappa$ found in the previous section are required to complete the above solutions.

The above expressions for the temperature and velocity distributions are given in dimensionless forms with the non-dimensionalizing parameters which are invariant of x. They are evaluated for the case in air with a number of thermal sources mounted flush with the surface of an adiabatic plate, and the results are plotted in Fig. 2. All eight sources shown in this figure have uniform surface heat flux of equal strength and size, with non-source spaces equal to the source size. The value of $\tilde{C}_\eta$ from Eq. (20) was used in evaluating local temperatures throughout the present study, as it was recommended by Lee and Yovanovich (1988) for Pr > 0.1.

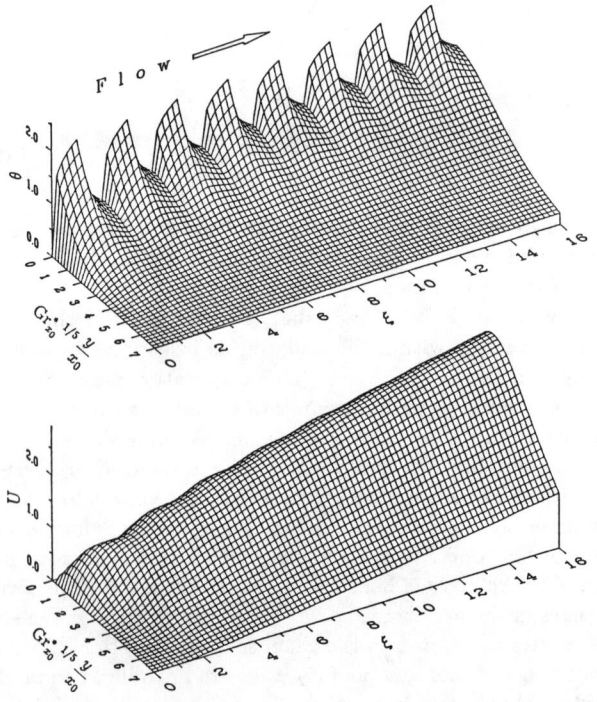

Figure 2: Dimensionless temperature and velocity distributions due to alternating positive uniform and zero surface heat fluxes of equal size : Pr = 0.7.

244

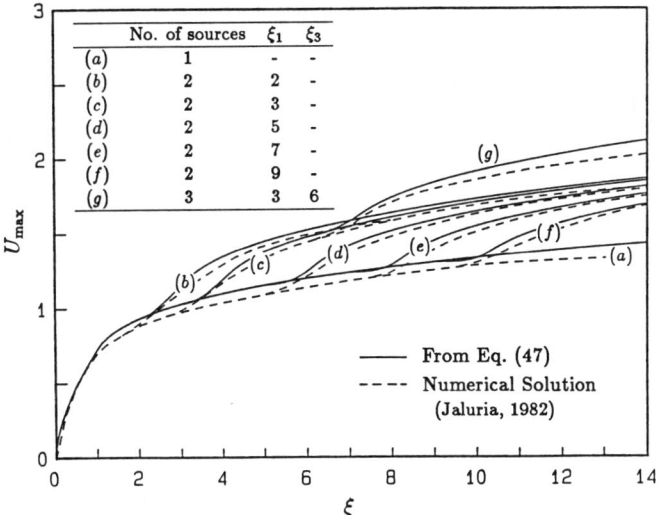

Figure 3: Comparison of maximum velocity variation with identical thermal sources of uniform surface heat flux on an adiabatic plate : Pr = 0.7.

The dimensionless local maximum flow velocity, U_{max}, is plotted and compared with the numerical data of Jaluria (1982) in Fig. 3 for cases with identical thermal sources and different source spacings. The sources are again mounted flush on an adiabatic plate in air, and they all have uniform surface heat flux. The figure shows excellent agreement of the velocity variations with the numerical results.

Wall Temperature Variation. The local wall temperature for $x_{\kappa-1} < x \leq x_\kappa$ can be obtained by substituting $\eta_0 = \eta_i = 0$ into Eq. (46). The resulting expression is

$$\theta_w = \frac{T_w}{q_{w_0} x_0 / k \mathrm{Gr}_{x_0}^{*\,1/5}} = \frac{\xi^{1/5}}{\tilde{C}_\eta} \sqrt{\frac{\tilde{\Phi}_\kappa}{\pi}} \left[1 + \sum_{i=1}^{\kappa} (q_{w_i}^* - q_{w_{i-1}}^*) \gamma_i \right]$$

(49)

where $\mathrm{ierfc}\, 0 = 1/\sqrt{\pi}$ was used.

Comparisons of the wall temperature variations evaluated by using the present model are made with the results of Jaluria (1982) who used finite difference methods. The plots shown in Fig. 4 are obtained for two identical strip thermal sources of uniform surface heat flux, mounted flush on an adiabatic plate in air with a source spacing between the leading edges of the strips equal to ξ_1. Satisfactory agreement of the present results, particularly in the trend of the local peak temperatures with increasing ξ_1, with the numerical solutions is obtained. Figure 5 depicts a similar situation with three sources for Pr = 0.1, 0.7 and 6, showing the effect of the Prandtl number on the surface temperature variations.

An examination of Fig. 4 reveals that the local peak temperature over the downstream source decays and becomes lower than that over the upstream source as the source spacing increases. The wall temperature due to the single source also decreases asymptotically to the ambient temperature as ξ increases, and the fluid temperature excess within the boundary layer always remains positive.

The positive fluid temperature excess represents an upward buoyant force on the fluid. A portion of this force will be used to overcome the friction, and the remainder will be used to accelerate the flow, resulting in a perpetual increase in the overall downstream flow velocity. Despite the fact that there is a higher fluid temperature at the beginning of the second source than at the leading edge of the first one, this fluid flow results in a lower temperature over the second source when the source spacing is large enough. As far as the downstream source is concerned, the flow becomes an induced free stream flow and thus, enhances the heat transfer rate.

Jaluria's numerical prediction overestimates the peak temperature over the first source by approximately 5% when it is compared to the result of Eq. (49), whose overall scale of the magnitude is based on the correlation equation of Fujii and Fujii (1976) through $\tilde{C}_\eta$ from Eq. (20).

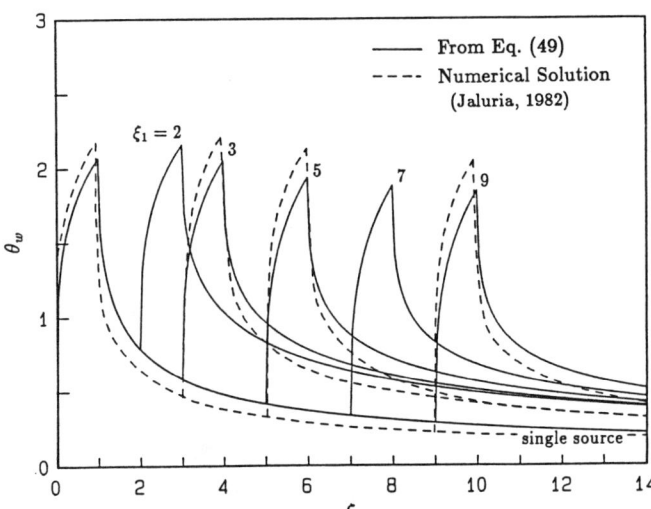

Figure 4: Comparison of dimensionless wall temperature variation due to two identical thermal sources of uniform surface heat flux on an adiabatic plate with various source spacings : Pr = 0.7.

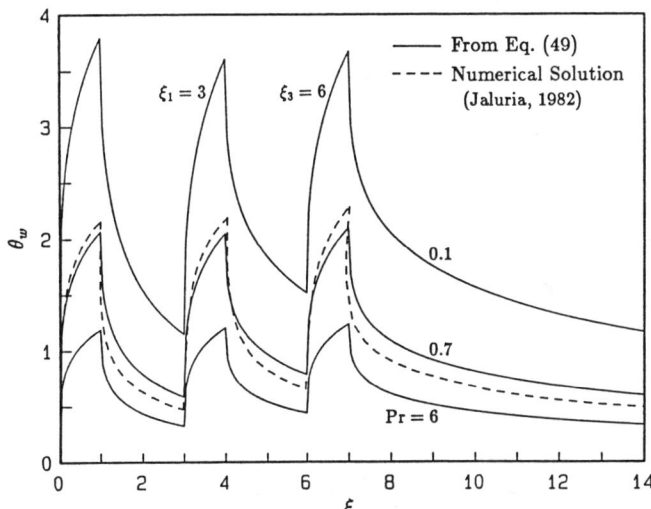

Figure 5: Comparison of dimensionless wall temperature variation due to three identical thermal sources of uniform surface heat flux on an adiabatic plate and the effect of Prandtl number.

The rest of this section exhibits some results obtained for cases that are intended to examine the validity and accuracy of the present model. A couple of figures for selected cases are presented herein for discussion purposes.

An expression for the dimensionless wall temperature variation based on an arbitrary reference heat flux q_r and a reference length scale x_r can be obtained from Eq. (49), as

$$\frac{T_w}{q_r x_r / kG^{*1/5}} = \left(\frac{q_{w_0}}{q_r}\right)^{4/5} \left(\frac{x}{x_r}\right)^{1/5} \frac{1}{\tilde{C}_\eta} \sqrt{\frac{\tilde{\Phi}_\kappa}{\pi}} \left[1 + \sum_{i=1}^{\kappa} (q_{w_i}^* - q_{w_{i-1}}^*)\gamma_i\right]$$ (50)

where G* is the modified Grashof number based on q_r and x_r. This equation supersedes the previous equation for θ_w, as they become identical when q_{w_0} and x_0 are chosen to be the references. Although the above equation carries no additional information from Eq. (49), the usefulness of this expression is apparent in Fig. 6, as it allows to compare different wall temperature variations for various combinations of q_{w_0} and x_0 on the same plot.

In Fig. 6, the line denoted by (a) is the result of alternating thermal sources of equal size dissipating at two different levels of uniform heat fluxes. The line denoted by (b) is for the case with a continuous and uniform surface heat flux whose magnitude is equal to the average value of the alternating heat fluxes. The lines denoted by (c) through (f) are the results of other combinations of heat flux distributions as shown at the top of the figure. All cases dissipate the same total amount of energy into the fluid over the length of the plate. They are evaluated for Pr = 0.7.

As can be seen from Fig. 6, the effect of the step changes in the heat flux input made at the upstream decays rather

quickly, and it may be considered to penetrate no more than approximately five to six step distances into the downstream. The flow characteristics far enough downstream become indifferent to the specifics of the upstream heat input variation, as long as the total energy dissipated into the flow over the upstream section is kept the same. More examinations with different combinations of surface heat flux exhibited the same behavior.

Further testing, which involves another model, named a temperature model and developed by Lee (1988) for cases with a step change in surface temperature, is carried out as follows.

A step change in uniform wall temperature is denoted by (a) in Fig. 7 for $T_{w_1}^* = T_{w_1}/T_{w_0} = 0.75$ and Pr = 0.7. Using this as an input to the temperature model, the corresponding surface heat flux variation denoted by (b) is obtained. This surface heat flux variation is then discretized to yield multi-step changes denoted by (c). Although the step sizes can be arbitrary, they are determined such that every step before $x/x_r = 1$ dissipates roughly the same amount of energy into the fluid. The step sizes after $x/x_r = 1$ are determined simply by overlapping those from $x/x_r \leq 1$. The magnitude of heat flux in each step is equal to the average heat flux of the original variation over the step. This multi-step change in surface heat flux is used in the present model, Eq. (49), to produce the wall temperature variation denoted by (d). The reference heat flux, q_r, was sized to yield $T_{w_0}/(q_r x_r / kGr_{x_r}^{1/4}) = 1$ at the beginning of the process.

As can be seen from Fig. 7, excellent agreement between the input and the resulting temperature variations, lines (a) and (d), is obtained. Peaks observed in the resulting temperature variation at the leading edge, $x/x_r = 0$, and at the point of a step, $x/x_r = 1$, are due to the finite representation, line (c), of the unbounded original heat flux variation, line (b), at the locations. Performing the same process with other values of $T_{w_1}^*$ as an input wall temperature variation resulted in the same level of agreement.

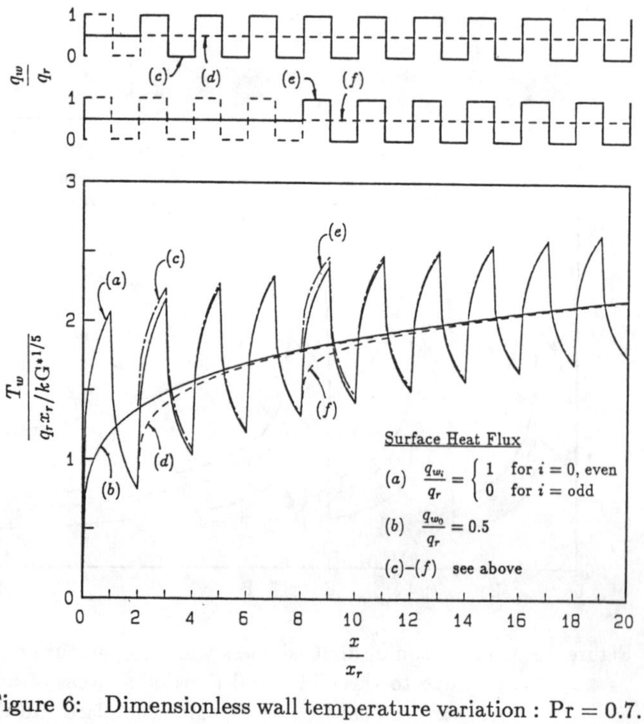

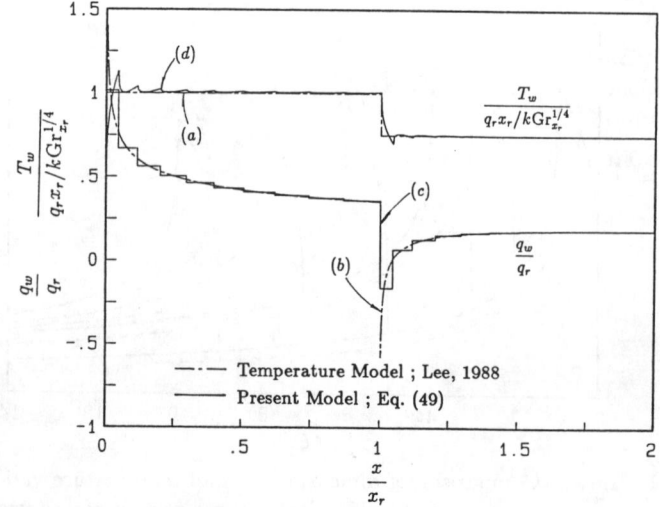

Figure 6: Dimensionless wall temperature variation : Pr = 0.7.

Figure 7: Dimensionless wall temperature and surface heat flux variations : Pr = 0.7.

246

Results of the present model are less accurate in the vicinity of the leading edge and locations of a discontinuity in surface thermal variations, due to the boundary layer approximations. Jaluria (1985), using finite difference methods, investigated this subject and presented the level of inaccuracies introduced by the approximations in both temperature and maximum velocity variations for different modified Grashof numbers.

A pertinent application of the model includes the thermal design of printed circuit boards (PCBs). In view of PCB applications, conjugate analyses would be required which include conduction heat transfer in the board substrate. When air is used as the coolant fluid as in most cases, the thermal conductivity of a typical substrate is usually orders of magnitude greater than that of the air, and the natural convection heat transfer coefficients are usually small. As a result, the in-plane conduction heat transfer would be characterized predominantly by the thermal properties of the substrate, and the effect of the fluid conduction heat transfer in the directions parallel to the plate surface may be neglected. This further supports the boundary layer approximation in the energy equation which already ignores the streamwise conduction heat transfer in the fluid.

Since the model is capable of dealing with arbitrary surface heat flux variations, it can be used in conjunction with a heat conduction analysis in the solid plate for a two dimensional flat plate modeling of conjugate heat transfer problems (Lee and Yovanovich, 1989). Further, as the fluid conduction in the direction across the plate width may be neglected as aforementioned, the present two dimensional model may also be used in solving full three dimensional conjugate problems involving an isolated vertical flat plate.

ACKNOWLEDGEMENTS

The authors wish to acknowledge the financial support of the Natural Sciences and Engineering Research Council of Canada under a Postgraduate Scholarship for Dr. Lee, and the operating grant to Dr. Yovanovich.

REFERENCES

Bar-Cohen, A., 1985, "Bounding Relations for Natural Convection Heat Transfer from Vertical Printed Circuit Boards," *Proceedings of IEEE*, Vol. 73, No. 9, pp. 1388–1395.

Fujii, T., and Fujii, M., 1976, "The Dependence of Local Nusselt Number on Prandtl Number In the Case of Free Convection Along a Vertical Surface With Uniform Heat Flux," *Int. J. Heat Mass Transfer*, Vol. 19, pp. 121–122.

Jaluria, Y., 1982, "Buoyancy-Induced Flow Due to Isolated Thermal Sources on a Vertical Surface," *J. Heat Transfer*, Vol. 104, pp. 223–227.

Jaluria, Y., 1985, "Interaction of Natural Convection Wakes Arising From Thermal Sources on a Vertical Surface," *J. Heat Transfer*, Vol. 107, pp. 883–892.

Jaluria, Y., and Gebhart, B., 1977, "Buoyancy-Induced Flow Arising From a Line Thermal Source on an Adiabatic Vertical Surface," *Int. J. Heat Mass Transfer*, Vol. 20, pp. 153–157.

Lee, S., 1988, "Laminar Natural Convection From a Vertical Plate With Variations in Thermal Boundary Conditions," Ph.D. Thesis, Department of Mechanical Engineering, University of Waterloo, Ontario.

Lee, S., and Yovanovich, M.M., 1988, "Laminar Natural Convection From a Vertical Plate With Variations in Surface Heat Flux," ASME HTD-Vol. 96, ed. H. R. Jacobs, Vol. 2, pp. 197–205, The 25th National Heat Transfer Conf., July 24–27, Houston, Texas.

Lee, S., and Yovanovich, M.M., 1989, "Conjugate Heat Transfer From a Vertical Plate With Discrete Heat Sources Under Natural Convection," - submitted for presentation at the ASME Winter Annual Meeting, December 10–15, San Francisco, California.

Ostrach, S., 1953, "An Analysis of Laminar Free-Convection Flow and Heat Transfer About a Flat Plate Parallel to the Direction of the Generating Body Force," NACA Report No. 1111.

Raithby, G.D., and Hollands, K.G.T., 1975, "A General Method of Obtaining Approximate Solutions to Laminar and Turbulent Free Convection Problems," *Advances in Heat Transfer*, Vol. 11, pp. 265–315, Academic Press, NY.

Schetz, J.A., 1963, "On the Approximate Solution of Viscous-Flow Problems," *J. Appl. Mech.*, Vol. 30, *Trans. ASME*, Vol. 85, pp. 263–268.

Sparrow, E.M., and Gregg, J.L., 1956, "Laminar Free Convection From a Vertical Plate With Uniform Surface Heat Flux," *Trans. ASME*, Vol. 78, pp. 435–440.

Sparrow, E.M., and Gregg, J.L., 1958, "Similar Solutions for Free Convection From a Nonisothermal Vertical Plate," *Trans. ASME*, Vol. 80, pp. 379–386.

Tribus, M., 1958, "Discussion on Similar Solutions for Free Convection From a Nonisothermal Vertical Plate," *Trans. ASME*, Vol. 80, pp. 1180-1181.

Yang, K.T., 1960, "Possible Similarity Solutions for Laminar Free Convection on Vertical Plates and Cylinders," *Trans. ASME*, Vol. 82, pp. 230–236.

Zinnes, A.E., 1970, "The Coupling of Conduction With Laminar Natural Convection From a Vertical Flat Plate With Arbitrary Surface Heating," *J. Heat Transfer*, Vol. 92, pp. 528–535.

1989 National Heat Transfer Conference
HTD-Vol. 107, Heat Transfer in Convective Flows

THERMOSOLUTAL NATURAL CONVECTION IN A RECTANGULAR ENCLOSURE: NUMERICAL RESULTS

C. Benard, D. Gobin and J. Thevenin
Université Pierre et Marie Curie
Orsay, Cedex, France

ABSTRACT

This study deals with natural convection in rectangular enclosures due thermosolutal buoyancy forces. A numerical method previously tested for thermal natural convection (hybrid discretization scheme, SIMPLE algorithm) has been extended to situations where the density variation of the fluid is due to horizontal temperature and concentration gradients. The flow field calculation requires the solution of the Navier-Stokes and continuity equations and of two equations of convection-diffusion. Two new parameters are introduced in the dimensionless equations, the Schmidt number and a solutal Grashof number.

Numerical results for the steady state solution are presented for Dirichlet boundary conditions. The parametric study deals with the influence of two ratios on heat and mass transfer in the enclosure : the buoyancy ratio and the Lewis number. The results show that the influence of the dominating term in the buoyancy force is largely dependent on the relative importance of diffusion of mass and heat. The case of large Lewis numbers, which is relevant in many practical situations is examined.

INTRODUCTION

In many industrial processes involving solid-liquid phase change, both heat transfer and diffusion of constituents (mass transfer) in the fluid phase have to be considered. This is the case in a number of techniques of crystal growth [1] or solidification of alloys, for example. The competition between buoyancy effects due to thermal and solutal gradients leads to very complex situations for the fluid flow and the stability of the liquid-solid interface.

In this paper we will only consider the fluid domain with non-moving boundaries and study natural convection due to both thermal and solutal buoyancy forces in a 2D closed cavity with imposed horizontal temperature and concentration gradients. We will analyse the structure of the temperature, concentration and velocity fields for a wide range of the parameters characterizing the relative roles of concentration and temperature.

The problem of natural convection due to these two cooperating (or opposing) buoyancy forces has received some attention in the last few years [2]. For theoretical simplicity, the assumption is always made that the temperature and concentration boundary conditions are perfectly defined at all the surfaces of the enclosure. To our knowledge, very few experimental results are available in the litterature for horizontal temperature and concentration gradients, due to the difficulty of imposing well-known boundary conditions in such a situation [3,4]. Some numerical results have been proposed in this case for given boundary conditions [5], and the scope of this study is to provide and analyse numerical results for other boundary conditions over a wide range of parameters, in the following precise situation : 2D rectangular geometry with laminar convection in the fluid due to cooperating thermal and solutal buoyancy effects.

The first part of this paper is devoted to setting the equations and defining the underlying hypotheses. The numerical method is then briefly described : a numerical method previously tested for thermal natural convection has been extended to situations where the density variation of the fluid is due to temperature and concentration gradients. Compared to the purely thermal case, two new parameters are introduced in the dimensionless equations, the Schmidt number and a solutal Grashof number.

In the second section, the results of the parametric study are analysed, the varying parameters being the buoyancy ratio and the Lewis number. They are respectively the ratio of the solutal and thermal Grashof numbers and the ratio of the Schmidt and Prandtl numbers.

PROBLEM DEFINITION AND SOLUTION PROCEDURE

Definition of the problem.

A 2D rectangular cavity (height H, width L : aspect ratio A = H/L) is bounded by two vertical walls at different uniform temperatures and concentrations, respectively T_1 and T_2 (c_1, c_2), (Fig. 1). The top and bottom walls of the enclosure are defined by zero heat and mass transfers. No-slip dynamic boundary conditions are imposed at the four walls.

The fluid in the cavity is assumed to be newtonian and incompressible and to satisfy the Boussinesq approximation. The thermo-physical properties are constant, except for the density in the buoyancy term where a linear dependence on temperature and concentration is assumed :

$$\rho(T,c) = \rho_0 [1 - \beta_T(T - T_0) + \beta_c(c - c_0)] ,$$

$$(\beta_T > 0, \quad \beta_c > 0) .$$

Soret and Dufour effects are assumed to have negligible influence on heat and mass diffusion. Under these hypotheses, the dimensionless expression of the equations of mass, energy and momentum conservation is :

$$\nabla . V = 0 \tag{1}$$

$$\partial V/\partial t + (V.\nabla)V = \nabla^2 V - \nabla P + (Gr_T \theta + Gr_c C) \, g/|g| \tag{2}$$

$$\partial\theta/\partial t + V.\nabla\theta = Pr^{-1} \nabla^2\theta \tag{3}$$

$$\partial C/\partial t + V.\nabla C = Sc^{-1} \nabla^2 C \tag{4}$$

where V, θ and C are the dimensionless velocity, temperature and concentration :

$$V = v^* \, H/\nu , \qquad (\nu : \text{kinematic viscosity}) ,$$

$$\theta = -(T-T_m)/ \Delta T,$$

$$T_m = (T_2 + T_1)/2 , \quad \Delta T = |T_2 - T_1| , \tag{5-a}$$

$$C = (c-c_m)/ \Delta c,$$

$$c_m = (c_2 + c_1)/2 , \quad \Delta c = |c_2 - c_1| , \tag{5-b}$$

P being the pressure and g the gravity.
The characteristic dimensionless parameters of the problem are :

$$Gr_T = \beta_T . g . \Delta T . H^3 / \nu^2 ,$$

$$N = Gr_c/Gr_T = \beta_c . \Delta C / \beta_T . \Delta T ,$$

$$\text{with } Gr_c = \beta_c . g . \Delta c . H^3 / \nu^2 ,$$

$$Pr = \nu / \alpha \quad (\alpha : \text{thermal diffusivity}) , \tag{6}$$

$$Le = Sc / Pr = \alpha / D,$$

$$\text{with } Sc = \nu / D \quad (D : \text{solutal diffusivity}) .$$

The two ratios N and Le characterize the relative importance of heat and mass transfers.

The boundary conditions that will be considered in this work correspond to cooperating thermal and solutal buoyancy forces :

at the vertical wall x = 0 :

$$\theta_1 = +0.5 , \quad C_1 = -0.5 , \quad u = v = 0 ,$$

at the vertical wall x = 1/A :

$$\theta_2 = -0.5 , \quad C_2 = +0.5 , \quad u = v = 0 , \tag{7}$$

at the horizontal walls y = 0 and y = 1 :

$$\partial\theta/\partial y = 0 , \quad \partial C/\partial y = 0 .$$

Numerical method.

The set of coupled equations (1-4) with the boundary conditions detailed above are solved using a finite difference technique similar to the method described by Patankar [7]. A non-uniform grid is defined on the rectangular domain (Fig. 1), to take account of the thin boundary layers near the walls of the cavity, and the equations are integrated over each control volume defined by the grid. The temperatures, concentrations and pressures are defined at the main nodes of the grid and the velocity components are calculated at the nodes of two staggered sub-grids. The integrated equations are discretized using the hybrid scheme proposed by Patankar, which we have proved, by comparing the two schemes, to be less time-consuming than the power-law scheme for the same precision in the range of parameters considered in this study.

The pressure-velocity coupling is solved using the SIMPLE pressure correction algorithm [7]. The set of uncoupled linearized equations is solved iteratively and an ADI procedure is used to solve each pentadiagonal system. The convergence criterion retained here uses the ratio of the pressure correction term to the calculated pressure : when the maximum of this ratio over the whole domain is less than 10^{-3}, the solution is considered to be converged.

The choice of the optimal number of space nodes has been obtained by comparing a 23 x 23 and a 53 x 53 grid for A = 1 : a 0.1% difference only is obtained on the resulting fields, while the CPU time on a MicroVax II increases from about 9' to more than 2 hours. As a consequence a 23 x 23 grid is used for A = 1 and similarly a 37 x 23 grid is used for A = 0.5 .

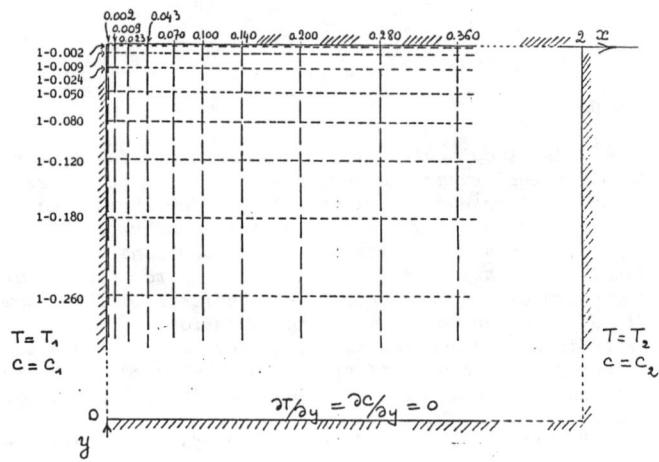

Figure 1. 2D irregular main grid.

The other aspects of the numerical code that have been checked, apart from the obvious case Le = 1 and from symmetry properties between thermal and solutal effects, deal with the influence of the initial conditions : we could show that starting from zero fields or from a pure steady thermal regime leads to the same final coupled solution .

The numerical results presented in the next section have been obtained for an enclosure of aspect ratio A = 0.5 with a 37 x 23 grid. The nodes are irregularly spaced. In the boundary layers they are distributed according to a geometrical law, 4 to 5 nodes being present in each boundary layer (Fig. 1). Calculations were performed on a Micro-VAX II and a standard run requires between 200 and 500 iterations.

ANALYSIS AND RESULTS.

Guidelines for the analysis of the results.

Our purpose in this paragraph is to compare the orders of magnitude characterizing the two independent problems, solutal convection and thermal convection. This comparison will be used as a basis for proposing some hypotheses on the coupled problem. The validity

N \ Le	1	2	5	10
0.1	4.42	4.38	4.34	4.32
1	5.34	5.02	4.69	4.53
2	6.04	5.54	4.92	4.64
5	7.43	6.61	5.05	4.42
10	8.88	7.78	5.28	3.35

Table 1. Nusselt Numbers
$(Pr = 10 , Gr_T = 10^4 , A = 0.5)$

N \ Le	1	2	5	10
0.1	4.42	5.90	8.19	10.39
1	5.34	6.87	9.28	11.58
2	6.04	7.65	10.17	12.54
5	7.43	9.24	12.09	14.57
10	8.88	10.91	14.17	17.12

Table 2. Sherwood Numbers.
$(Pr = 10 , Gr_T = 10^4 , A = 0.5)$

of these hypotheses will be checked by analyzing the numerical results in the next section.

The analysis that will be developped here is based on the orders of magnitude characterizing the underlined thermal convection and solutal convection problems when Grashof numbers are high enough for the left and right wall boundary layers to be separated. More precisely, it will use, as a reference frame, the study by Patterson [6] on the growth of the thermal and viscous boundary layers in thermal convection problems.

In each independent problem ($Pr \geq 1$, $Sc \geq 1$), the diffusion length δ_T (δ_C) grows as :

$$\delta_T(t) = (\alpha t)^{1/2} \qquad (\delta_C(t) = (Dt)^{1/2}) , \qquad (8)$$
$$\delta_T(t)/\delta_C(t) = Le^{1/2} . \qquad (9)$$

In the growing boundary layer, we may admit that buoyancy and viscous forces balance one another :

$$\nu v_T(t)/\delta_T(t)^2 \approx g.\beta_T.\Delta T , \qquad (\nu v_C(t)/\delta_C(t)^2 \approx g.\beta_C.\Delta C)$$

that is:

$$v_T(t) \approx g.\beta_T \Delta T.t / Pr , \qquad (v_C(t) \approx g.\beta_C \Delta C.t / Sc) \quad (10)$$
$$v_T(t)/v_C(t) \approx Le/N . \qquad (11)$$

Stabilization of the thermal (solutal) boundary layer occurs when advection and diffusion of heat (mass) compensate one another :

$$v_T \Delta T / H \approx \alpha \Delta T/\delta_T^2 \qquad (v_C \Delta C / H \approx D \Delta C/\delta_C^2) ,$$

that is, according to Eq.(10), when :

$$t_T \approx H^2(Pr.Gr_T)^{-1/2}/\alpha \qquad (t_C \approx H^2(Sc.Gr_C)^{-1/2}/D) \quad (12)$$

which (Eq. 8) corresponds to :

$$\delta_T \approx H (Pr.Gr_T)^{-1/4} \qquad (\delta_C \approx H (Sc.Gr_C)^{-1/4}) . \quad (13)$$

Thus, the steady state regime is not reached at the same time for the two boundary layers:

$$t_T / t_C \approx (N/Le)^{1/2} , \qquad (14)$$

and the steady boundary layers are characterized by :

$$\delta_T(t_T) / \delta_C(t_C) \approx (Le.N)^{1/4} , \qquad (15)$$
$$v_T / v_C \approx (Le/N)^{1/2} . \qquad (16)$$

From the foregoing analysis , we know that, if Le $\gg$ 1, the transient regime leading to the permanent one is characterized by δ_T growing much more quickly than δ_C. Then, in the coupled problem, the effect of the thermal buoyancy term will be felt on a much wider zone. Moreover, if N is smaller than one , the thermal buoyancy force in every point of this zone is dominant in the momentum equation, and our hypothesis will be that, in this case, the temperature and velocity fields are not strongly disturbed by the concentration field. On the contrary, if N is larger than one and reaches the same order of magnitude as Le, the previous analysis shows that a more complicated situation arises : the velocity fields created in the wide thermal zone and in the narrower solutal zone are of the same order of magnitude (Eq. 11 and 16). As a consequence our hypothesis will be that a strong coupling occurs that changes completely the resulting fields.

These two hypotheses for N < 1 and N > 1 will be checked and their domain of validity more precisely defined by numerical analysis ; moreover, since the resulting fields for N > 1, Le > 1 cannot be inferred from the previous parametric analysis, the role of the numerical analysis below will be also to quantify these fields and the corresponding boundary heat and mass transfers.

Analysis of the numerical results.

The results that are analysed here correspond to the following situations :
- a given geometry : A = 0.5
- given thermal characteristics : Pr = 10 , $Gr_T = 10^4$.
- solutal parameters ranging from Le = 1 to Le = 10 , and N = 0.1 to N = 10 .

with cooperating thermal and solutal buoyancy forces.

The results for Le varying from 0.1 to 1 are not considered since they can be obtained from the previous cases by symmetry between the temperature and concentration fields.

Temperature, concentration and vertical velocity profiles in the horizontal mid-plane (y = 0.5) are represented in Figures 2 to 4 and 6 to 9 as functions of the horizontal coordinate x. Isotherms, isoconcentration lines and stream lines are given in Figures 5 and 10.

The corresponding values of the average Nusselt and Sherwood numbers (Nu, Sh) at the vertical walls are given in Tables 1 and 2. Nu is the ratio of the average heat flux at the vertical wall to a reference conductive heat flux $k.\Delta T/H$ (k is the thermal conductivity of the fluid), and Sh is defined accordingly for the solute mass flux.

Let us first analyse the influence of the Lewis number for a given N (Fig. 2 to 6). When N = 0.1, the solutal buoyancy force is negligible compared to the thermal buoyancy force. As a consequence, the increase of the Lewis number will only reduce the width of the zone where the small solutal buoyancy force acts and will influence very little the thermal boundary layer thickness (Fig. 7 and 8). Indeed, the latter depends mainly on the constant Rayleigh number ($Ra_T = Pr.Gr_T$) as in the purely thermal case. On the contrary, Le will act strongly on the solute concentration profile, which depends on the solutal diffusivity D. As a consequence Nu varies very little with Le, while Sh increases with Le (see Tables 1 and 2).

For $N \geq 1$, the solutal buoyancy force is of the same order of magnitude as or larger than the thermal buoyancy force. As a consequence, increasing the Lewis number - that is, acting on the concentration field and reducing the solutal boundary layer thickness (Fig. 2 and Table 2) - acts on the dominant buoyancy force and modifies the driving term in Eq. (2), especially in the boundary layer region. The influence of Le on the vertical velocity in the horizontal mid-plane can be seen for N = 2 (Fig. 3) and N = 5 (Fig. 4). The effect on the streamlines appears for N = 10 on Fig. 5. It can also be noticed that though the thermal boundary layer thickness is not much affected by the Lewis number for N = 2 (Fig. 6), the Nusselt number at N = 2 is significantly smaller for higher values of Le (Table 1), because the velocity profile in the boundary layer is lower for higher Le (Fig. 3). For N = 10, the Nusselt number is increased by the effect of the cooperating solutal driving force when the thermal and solutal boundary layers are similar (Le = 1), and decreases strongly with increasing Le, due to a very strong reduction of the boundary layer velocities (Fig. 4). This will even partly destroy the thermal boundary layer structure. This phenomenon corresponds to the case that we pointed out in our preliminary approach : the independent problems lead for these values of N and Le to the same order of magnitude for the maximum velocities (Eq. 11 and 16) in two boundary layers of very different widths (Eq. 15).

Let us now consider the influence of N for a given value of the Lewis number (Fig. 7 to 10). For Le = 1, obvious results are obtained (Tables 1 and 2) : the Nusselt and Sherwood numbers are the same since the concentration and temperature fields are identical. Increasing N leads to increase in the same way Nu and Sh, since the source term in Eq. (2) is globally increased.

For Le = 10, increasing N from 0.1 to 1 will have a weak influence (Fig. 7 to 9, Tables 1 and 2). Indeed in both cases the dominant buoyancy and viscous forces balancing one another are the ones present in the widest diffusing layer, that is in the whole thermal diffusion layer. Final balance of the temperature field will be reached when the thermal advection and diffusion terms are of the same order of magnitude in this layer, as in the pure thermal case.

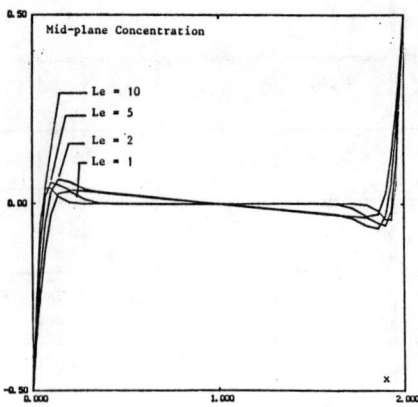

Figure 2. Concentration in the horizontal mid-plane $Gr_T = 10^4$, Pr = 10, A = 0.5, N = 2 (Le ranging from 1 to 10)

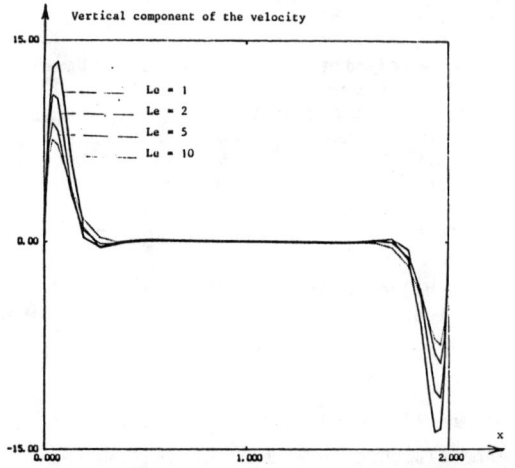

Figure 3. Vertical component of V in the y=0.5 plane $Gr_T = 10^4$, Pr = 10, A = 0.5, N = 2 (Le ranging from 1 to 10)

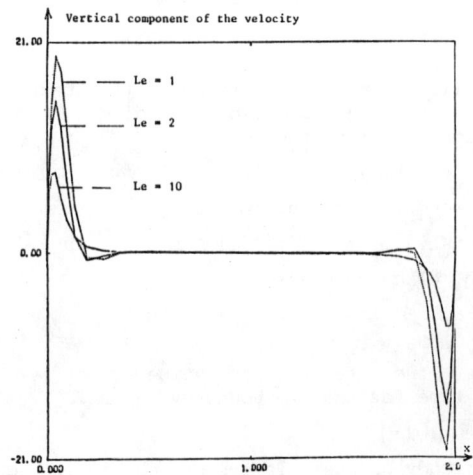

Figure 4. Vertical component of V in the y=0.5 plane $Gr_T = 10^4$, Pr = 10, A = 0.5, N = 5 (Le ranging from 1 to 10)

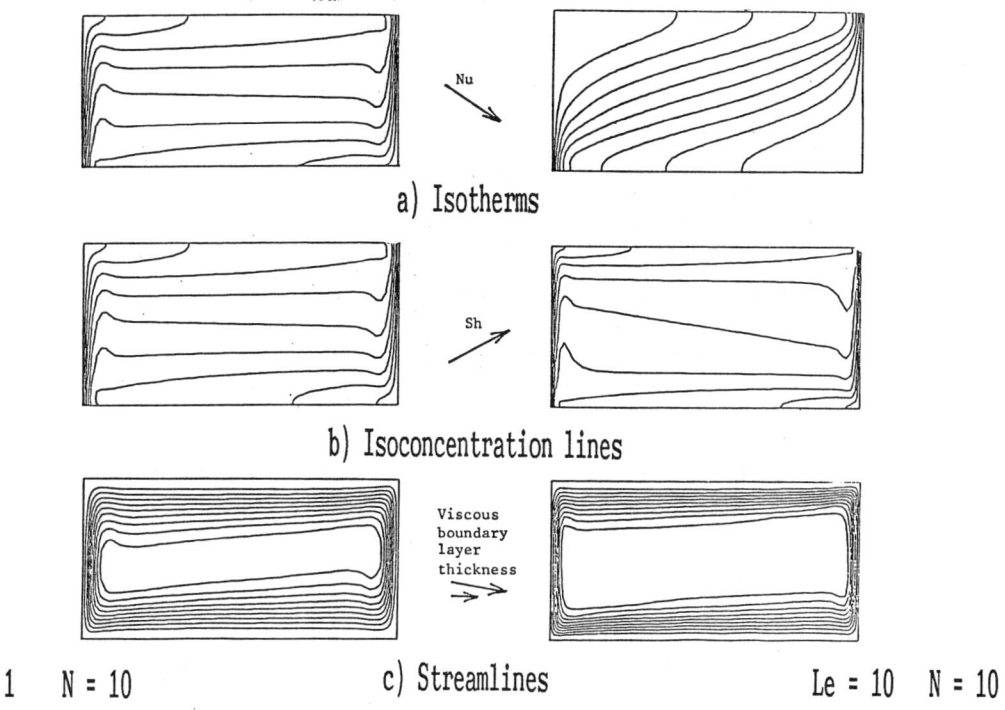

a) Isotherms

b) Isoconcentration lines

c) Streamlines

Le = 1   N = 10

Le = 10   N = 10

Figure 5. Isotherms, isoconcentration lines and streamlines.
$Gr_T = 10^4$, Pr = 10, A = 0.5, N = 10
Left Column : Le = 1
Right Column : Le = 10.

On the contrary, increasing N from 1 to 10 will significantly change the field structure (Fig. 7 to 10) and the value of Sh (Table 2). In fact, for high values of N and Le, the balance forces in the wider diffusion layer (that is, the thermal diffusion layer) are still the viscous forces and the thermal buoyancy force. But in the solutal diffusion layer, which is much thinner and grows more slowly with time, the solutal buoyancy force grows as N grows and must be balanced by a growing viscous force creating increasing local velocities, which may finally reach the same order as the the velocities in the thermal diffusion zone. This leads to the strongly coupled case already mentionned above (Fig. 10).

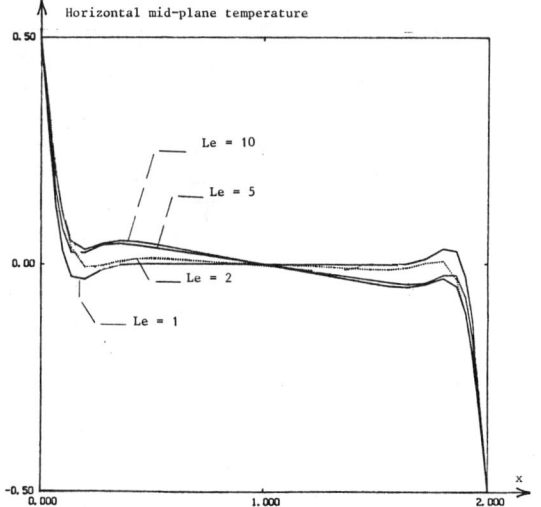

Figure 6. Temperature in the horizontal mid-plane.
$Gr_T = 10^4$, Pr = 10, A = 0.5, N = 2
(Le ranging from 1 to 10)

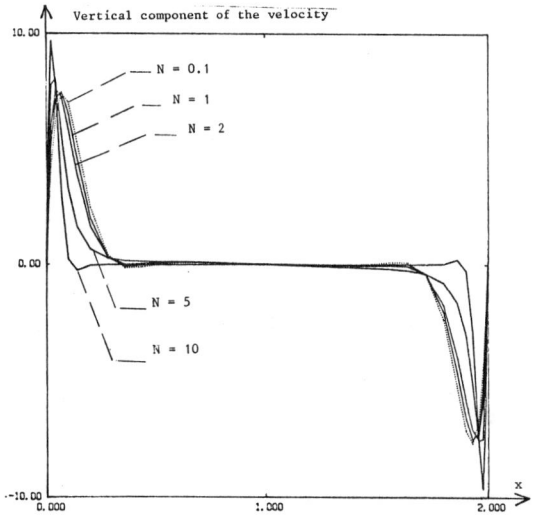

Figure 7. Vertical component of V in the y=0.5 plane
$Gr_T = 10^4$, Pr = 10, A = 0.5, Le = 10
(N ranging from 0.1 to 10)

253

CONCLUSION

In real situations, the solutal diffusivity is much weaker than the thermal one . Thus Le >> 1. Our results have been obtained for moderatly high Lewis numbers (Le ≈ 10). Nevertheless, the trends that we have pointed out give us an insight on what is going on in real cases :

The boundary heat transfer is not very sensitive to the cooperating solutal buoyancy force, even if it is ten times stronger than the thermal buoyancy force, (last column of Table 1).

On the contrary, the boundary mass transfer is extremely sensitive to the solutal buoyancy force even if it is ten times weaker than the thermal one, (last column of Table 2).

REFERENCES.

[1] OSTRACH S. (1983) : Fluid Mechanics in Crystal Growth - The 1982 Freeman Scholar Lecture. *J. Fluids Engin.*, 105, 5-20.

[2] OSTRACH S. (1980) : Natural Convection with Combined Driving Forces. *Physico-Chemical Hydrodynamics*, 1, 233-247.

[3] KAMOTANI Y., WANG L.W., OSTRACH S., JIANG H.D. (1985) : Experimental Study of Natural Convection in Shallow Enclosures with Horizontal Temperature and Concentration Gradients. *Int. Journal Heat Mass Transfer*, 28 n° 1, 165-173.

[4] JIANG H.D., OSTRACH S., KAMOTANI Y. (1988) : Thermosolutal convection with opposed buoyancy forces in shallow enclosures. ASME Winter Meeting, Nov.27-Dec.2 1988, Chicago, USA. (HTD, Vol.99).

[5] TREVISAN O.V., BEJAN A. (1987) : Combined Heat and Mass Transfer by Natural Convection in a Vertical Enclosure. *J. Heat Transfer*, 109, 104-112.

[6] PATTERSON J., IMBERGER J. (1980) : Unsteady Natural Convection in a Rectangular Cavity. *J. Fluid Mech.*, 100, 65-86.

[7] PATANKAR S.V. (1980) : Numerical Heat Transfer and Fluid Flow. Hemisphere Corp. New-York.

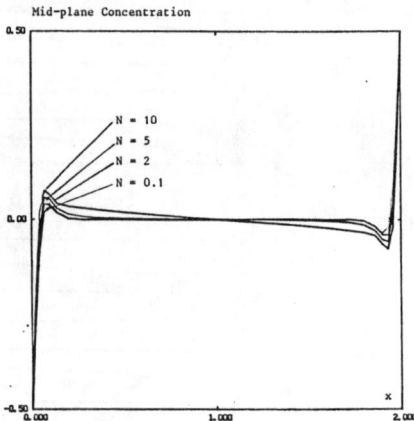

Figure 8. Concentration in the horizontal mid-plane. $Gr_T = 10^4$, Pr = 10, A = 0.5, Le = 10 (N ranging from 0.1 to 10)

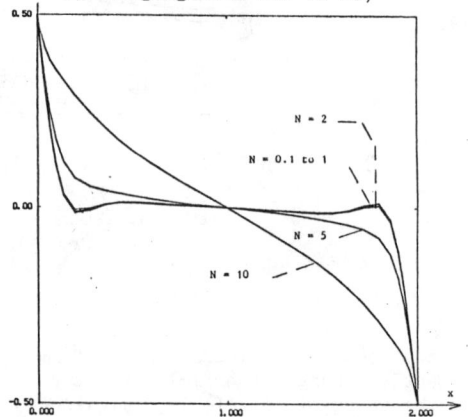

Figure 9. Temperature in the horizontal mid-plane. $Gr_T = 10^4$, Pr = 10, A = 0.5, Le = 10 (N ranging from 0.1 to 10)

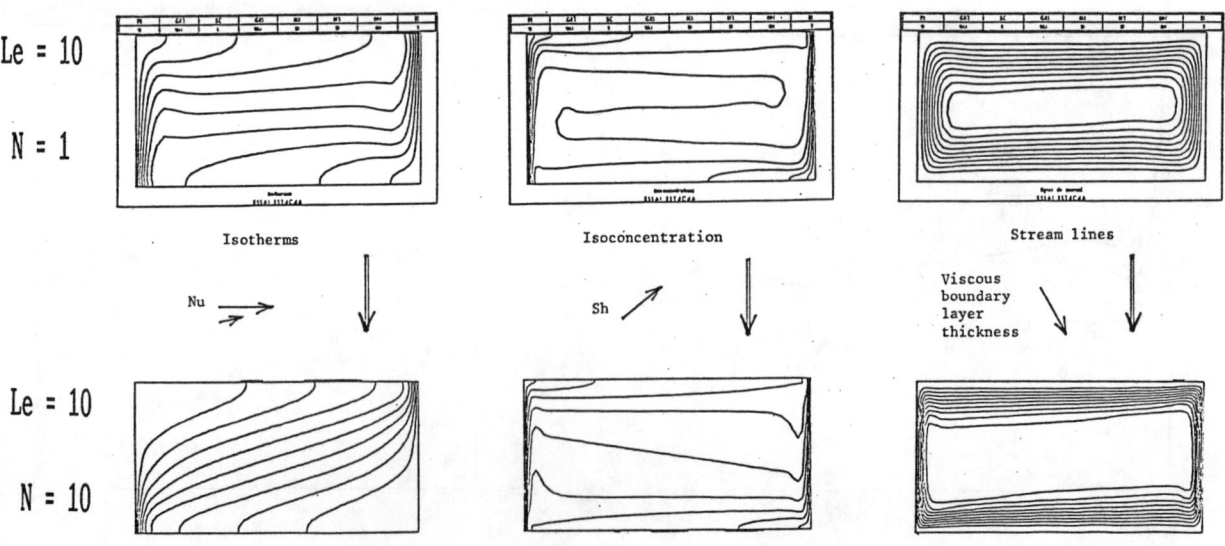

Figure 10. Isotherms, isoconcentration lines and streamlines. $Gr_T = 10^4$, Pr = 10, A = 0.5, Le = 10

First Line : N = 1   Second Line : N = 10.

1989 National Heat Transfer Conference
HTD-Vol. 107, Heat Transfer in Convective Flows

VISCOUS DISSIPATION EFFECT ON SOLIDIFICATION
OF A LIQUID ONTO A TUBE WALL

M. T. Ahmadian and L. C. Burmeister
Mechanical Engineering Department
University of Kansas
Lawrence, Kansas

ABSTRACT

The effect of viscous dissipation on solidification onto the inner surface of a straight circular tube of a liquid flowing laminarly with a specified volumetric rate is analytically determined. Steady conditions and a constant tube wall temperature below the freezing temperature of the liquid are assumed. Solutions for the thickness of the frozen shell and for the pressure drop are obtained. The pressure gradient versus volumetric flow rate curve exhibits a minimum, suggesting that it might be possible to design runners for injection molding machines to have a minimum pressure drop if the volumetric flow rate is specified.

NOMENCLATURE

C_p liquid specific heat, kJ/kg K

E Eckert number, $E = \left(\dfrac{Q}{\pi R^2}\right)^2 \dfrac{1}{C_p (T_0 - T_m)}$

Pr Prandtl number, $Pr = \dfrac{\nu}{\alpha}$

Q volumetric flow rate, $\dfrac{m^3}{s}$

$Q*$ dimensionless flow rate, $Q* = \dfrac{Q}{R^2}\left[\dfrac{\mu}{k_s(T_m - T_w)}\right]^{1/2}$

R tube radius or function of dimensionless radius, m

T temperature of the liquid, C

T_m melting temperature, C

T_0 temperature of the liquid at the entrance, C

T_s temperature of the solid, C

T_w temperature of the wall, C

$T*$ dimensionless tube wall temperature, $T* = \dfrac{T - T_m}{T_0 - T_m}$

V_z axial fluid velocity, m/s

V_r radial fluid velocity, m/s

$z*$ dimensionless axial distance, $z* = \dfrac{\alpha\pi}{Q} z$

Greek Symbols

α liquid thermal diffusivity, m^2/s

δ thickness of the frozen layer, m

Δ radius of unfrozen opening, m

$\Delta*$ dimensionless unfrozen opening radius, $\Delta* = \dfrac{\Delta}{R}$

η dimensionless radius, $\dfrac{r}{\Delta}$

λ_n separation constant

INTRODUCTION

Solidification of flowing liquids onto the chilled walls of the ducts through which they flow is a phenomenon that is observed in such applications as chemical processing, heat exchangers, and injection molding machines. Many of these liquids are nonNewtonian and an analytical study of their flows is further complicated by the coupling of the motion and energy equations caused by the temperature dependence of viscosity. If the geometry is also complex, numerical simulation is of especially great help.

However, analytical approximation often leads to results that are close to the real solution. In this manner, the trend of the real solution is obtained and avenues of detailed investigation are suggested. Zerkle and Sunderland (1968) applied such an approach in 1968 to the problem of determining the entrance region thickness of the frozen shell that forms on the inner surface of a straight circular duct maintained at a constant subfreezing wall temperature from a Newtonian fluid in forced laminar flow of specified volumetric flow rate. They showed that if natural convection effects are unimportant, use of the fully developed parabolic velocity profile that is expected for a fluid of high Prandtl number for the local axial velocity puts the problem into the form of the classical Graetz problem studied by Sellars et al. (1956). Tago and Fukasaka (1988) recently reported measurements for water freezing onto the chilled interior wall of a curved duct but did not include the effect of viscous dissipation in their discussion. Their work demonstrates that this topic is of contemporary interest.

In the present study the analysis of Zerkle and Sunderland is extended by considering the effects of viscous dissipation. The results obtained are applicable to circular ducts of any length, not just to the thermal entrance region.

DETERMINATION OF THE TEMPERATURE PROFILE

Consider a circular duct with the boundary temperature lower than the solidification temperature of the polymer that steadily flows through it. The effect of viscous dissipation on freezing of the fluid onto the wall is approximated. The pattern of the flow is represented in Figure 1 with δ as the thickness of the frozen layer and Δ as the radius of the unfrozen opening of the tube. Assuming $\text{Pr} \gg 1$,

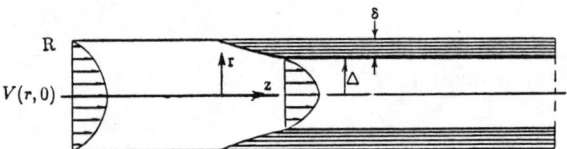

Figure 1. Flow pattern with solidification

the velocity profile develops much faster than does the temperature profile. Accordingly, assume the profile of the local velocity V_z in the stream direction to be the fully developed profile

$$V_z = \frac{2Q}{\pi \Delta^2} \left[1 - \left(\frac{r}{\Delta} \right)^2 \right] \tag{1}$$

The continuity equation is

$$\frac{1}{r} \frac{\partial r V_r}{\partial r} = - \frac{\partial V_z}{\partial z} \tag{2}$$

Substituting for V_z from equation (1) into equation (2) gives

$$\frac{1}{r} \frac{\partial r V_r}{\partial r} = \frac{4Q}{\pi \Delta^3} \left(1 - \frac{2r^2}{\Delta^2} \right) \frac{d\Delta}{dz} \tag{3}$$

Integration of equation (3) gives the radial velocity V_r as

$$V_r = \frac{2Q}{\pi \Delta^2} \, \eta \, (1 - \eta^2) \frac{d\Delta}{dz} \tag{4}$$

where $\eta = \frac{r}{\Delta}$.

The energy equation is

$$V_r \frac{\partial T}{\partial r} + V_z \frac{\partial T}{\partial z} = \alpha \frac{1}{r} \frac{\partial}{\partial r} \left(r \frac{\partial T}{\partial r} \right) + \frac{r \Phi}{\rho C_p} \tag{5}$$

where

$$\Phi = 2 \left[\left(\frac{\partial V_r}{\partial r} \right)^2 + \left(\frac{V_r}{r} \right)^2 + \left(\frac{\partial V_z}{\partial z} \right)^2 \right] + \left[\frac{\partial V_r}{\partial z} + \frac{\partial V_z}{\partial r} \right]^2 \tag{6}$$

Execution of the coordinate transformation $r, z \rightarrow \eta = r/\Delta$, $z^* = f(z)$ gives

$$\frac{\partial}{\partial r} = \frac{\partial}{\partial \eta} \frac{\partial \eta}{\partial r} + \frac{\partial}{\partial z^*} \frac{\partial z^*}{\partial r} = \frac{1}{\Delta} \frac{\partial}{\partial \eta}$$

since

$$\frac{\partial \eta}{\partial r} = \frac{1}{\Delta} \qquad \frac{\partial z^*}{\partial r} = 0$$

Similarly,

$$\frac{\partial^2}{\partial r^2} = \frac{\partial}{\partial r} \left(\frac{\partial}{\partial r} \right) = \frac{1}{\Delta^2} \frac{\partial^2}{\partial \eta^2}$$

and

$$\frac{\partial}{\partial z} = \frac{\partial}{\partial \eta} \frac{\partial \eta}{\partial z} + \frac{\partial}{\partial z^*} \frac{\partial z^*}{\partial z} = \frac{-\eta}{\Delta} \frac{d\Delta}{dz} \frac{\partial}{\partial \eta} + f'(z) \frac{\partial}{\partial z^*}$$

where

$$\frac{\partial \eta}{\partial z} = \frac{r}{\Delta^2} \frac{d\Delta}{dz} = - \frac{\eta}{\Delta} \frac{d\Delta}{dz} \qquad \frac{\partial z^*}{\partial z} = f'(z)$$

The energy equation then is

$$\frac{1}{\Delta} \left(V_r - V_z \eta \frac{d\Delta}{dz} \right) \frac{\partial T}{\partial \eta} + V_z f'(z) \frac{\partial T}{\partial z^*} = \frac{\alpha}{\Delta^2} \frac{1}{\eta} \frac{\partial}{\partial \eta} \left(\eta \frac{\partial T}{\partial \eta} \right) + \frac{\mu \Phi}{\rho C_p} \tag{7}$$

Then substituting for V_r from equation (4) and V_z from equation (1) gives the energy equation (7) as

$$\frac{2}{\alpha \pi} Q (1 - \eta^2) f'(z) \frac{\partial T}{\partial z^*} = \frac{1}{\eta} \frac{\partial}{\partial \eta} \left(\eta \frac{\partial T}{\partial \eta} \right) + \frac{\mu \Delta^2 \Phi}{k} \tag{8}$$

Let

$$f'(z) = \frac{\alpha \pi}{Q} z = 4 \frac{z}{D} \frac{1}{\text{Re}_D \text{Pr}} = z^* \tag{9}$$

Substitution of equation (9) into equation (8) results in

$$2(1 - \eta^2) \frac{\partial T}{\partial z^*} = \frac{1}{\eta} \frac{\partial}{\partial \eta} \left(\eta \frac{\partial T}{\partial \eta} \right) + \frac{\mu \Delta^2}{k} \Phi \tag{10}$$

Substituting V_z and V_r from equations (1) and (4) into equation (6) for Φ gives

$$\Phi = 2 \left[\frac{4}{\pi^2} \frac{Q^2}{\Delta^6} \left(\frac{d\Delta}{dz} \right)^2 (1 - 3\eta^2)^2 + \frac{4}{\pi^2} \frac{Q^2}{\Delta^6} \left(\frac{d\Delta}{dz} \right)^2 (1 - \eta^2)^2 + \right.$$

$$\frac{16}{\pi^2} \frac{Q^2}{\Delta^6} \left(\frac{d\Delta}{dz} \right)^2 (1 - 2\eta^2)^2 \left. \right] + \frac{4}{\pi^2} Q^2 \eta^2 \left[\left(\frac{d^2 \Delta}{dz^2} \right) \frac{1}{\Delta^2} (1 - \eta^2) - \right.$$

$$\left(\frac{d\Delta}{dz} \right)^2 \frac{1}{\Delta^3} (3 - 5\eta^2) - \frac{2}{\Delta^3} \right]^2$$

Neglecting curvature effects, all terms involving a derivative of Δ, gives

$$\Phi = \frac{16}{\pi^2 \Delta^6} Q^2 \eta^2$$

The energy equation (10) then is

$$2 (1 - \eta^2) \frac{\partial T}{\partial z^*} = \frac{1}{\eta} \frac{\partial}{\partial \eta} \left(\eta \frac{\partial T}{\partial \eta} \right) + \frac{16}{\pi^2} \frac{\mu}{k \Delta^4} Q^2 \eta^2 \tag{11}$$

The initial condition

$$T(\eta, 0) = T_o$$

for a uniform inlet temperature and the boundary condition

$$T(1, z^*) = T_m$$

for a well defined freezing temperature are imposed. The interfacial boundary condition for the conservation of energy is more detailed, however. As illustrated in Figure 2, it can be expressed as

$$- k \frac{\partial T}{\partial r} 2\pi \Delta dz - k \frac{\partial T}{\partial z} 2\pi \Delta dr = - k_s \frac{\partial T_s}{\partial r} 2\pi \Delta dz -$$

$$k_s \frac{\partial T_s}{\partial z} 2\pi \Delta dr$$

Note that at the interface $r = \Delta$. Hence,

$$-k \left[\frac{\partial T}{\partial r} + \frac{\partial T}{\partial z} \frac{d\Delta}{dz} \right] = -k_s \left[\frac{\partial T_s}{\partial r} + \frac{\partial T_s}{\partial z} \frac{\partial \Delta}{\partial z} \right]$$

256

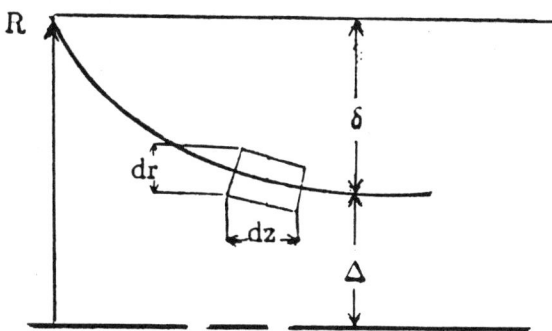

Figure 2. Interface control volume

Or, making use of the previous coordinate transformation,

$$-k\left[\frac{1+\left(\frac{d\Delta}{dz}\right)^2}{\Delta}\frac{\partial T}{\partial \eta} + \frac{\alpha\pi}{Q}\frac{d\Delta}{dz}\frac{\partial T}{\partial z*}\right] =$$

$$-k_s\left[\frac{1+\left(\frac{d\Delta}{dz}\right)^2}{\Delta}\frac{\partial T_s}{\partial \eta} + \frac{\alpha\pi}{Q}\frac{d\Delta}{dz}\frac{\partial T_s}{\partial z*}\right]$$

Neglecting curvature effects results in

$$k\frac{\partial T}{\partial \eta} = k_s\frac{\partial T_s}{\partial \eta}$$

Finally, the dimensionless energy equation and boundary conditions are

$$2(1 - \eta^2)\frac{\partial T*}{\partial z*} = \frac{1}{\eta}\frac{\partial}{\partial \eta}\left(\eta\frac{\partial T*}{\partial \eta}\right) + 16\ EPr\left(\frac{R}{\Delta}\right)^4\eta^2 \quad (12)$$

$$T*(\eta,0) = 1 \quad (13)$$

$$T*(1,z*) = 0 \quad (14)$$

$$\frac{\partial T*(1,z*)}{\partial \eta} = \frac{k_s}{k}\frac{\partial T*_s(1,z*)}{\partial \eta} \quad (15)$$

where

$$T* = \frac{T - T_m}{T_0 - T_m}$$

and

$$E = \left(\frac{Q}{\pi R^2}\right)^2\frac{1}{C_p(T_0 - T_m)}$$

A particular solution $T*_p$ to equation (12) is

$$T*_p = -EPr\left(\frac{R}{\Delta}\right)^4(\eta^4 - 1) \quad (16)$$

The complementary solution $T*_c$ is the solution to

$$2(1 - \eta^2)\frac{\partial T*}{\partial z*} = \frac{1}{\eta}\frac{\partial}{\partial \eta}\left(\eta\frac{\partial T*}{\partial \eta}\right) \quad (17)$$

which is assumed to be $T* = Z(z*)R(\eta)$. Substitution of this assumed solution into equation (17) gives

$$2(1 - \eta^2)\ R\ \frac{dZ}{dz*} = \frac{1}{\eta}\frac{d}{d\eta}\left(\eta\frac{dR}{d\eta}\right)Z$$

Separation of variables gives

$$\frac{2}{Z}\frac{dZ}{dz*} = \frac{1}{R\eta(1 - \eta^2)}\frac{d}{d\eta}\left(\eta\frac{dR}{d\eta}\right) = -\lambda_n^2$$

in which λ_n is a separation constant. The solution for Z is

$$Z = \exp\left(\frac{-\lambda_n^2 z*}{2}\right)$$

The η-dependent equation is

$$\frac{1}{\eta}\frac{d}{d\eta}\left(\eta\frac{dR}{d\eta}\right) + \lambda_n^2(1 - \eta^2)R = 0 \quad (18)$$

with the solution

$$R = C_n R_n(\eta)$$

Then the complementary solution to $T*_c$ is

$$T*_c = \sum_{n=0}^{\infty}C_n R_n(\eta)\exp\left(-\frac{\lambda_n^2 z*}{2}\right)$$

where λ_n is such as to make $R(\eta = 1) = 0$. The complete solution $T*$ is

$$T* = T*_p + T*_c$$

Thus

$$T* = EPr\left(\frac{R}{\Delta}\right)^4(1 - \eta^4) + \sum_{n=0}^{\infty}C_n R_n(\eta)\exp\left(-\frac{\lambda_n^2 z*}{2}\right) \quad (19)$$

The constants C_n are evaluated by satisfying the initial condition of equation (13). Then,

$$1 - EPr\left(\frac{R}{\Delta}\right)^4(1 - \eta^4) = \sum_{n=0}^{\infty}C_n R_n(\eta)$$

A Sturm-Liouville system such as is described by

$$\frac{d}{dx}\left[r(x)\frac{dX}{dx}\right] + \left[g(x) + \lambda p(x)\right]X = 0$$

and

$$\alpha_1 X'(a) + \alpha_2 X(a) = 0$$

$$\beta_1 X'(b) + \beta_2 X(b) = 0$$

enables a function $f(x)$ to be represented by the series

$$f(x) = \sum_{n=0}^{\infty}C_n X_n(x)$$

where

$$C_n = \frac{\int_a^b p(x)f(x)X_n(x)dx}{\int_a^b p(x)X_n^2(x)dx}$$

For equation (18) one has

$$\frac{d}{d\eta}\left(\eta\frac{dR}{d\eta}\right) + \lambda_n^2(\eta - \eta^3)R = 0$$

where $R(0) = 0$ and $R(1) = 0$. Hence

$$C_n = \frac{\int_0^1\left[1 - EPr\left(\frac{R}{\Delta}\right)^4(1 - \eta^4)\right]\eta(1 - \eta^2)R_n d\eta}{\int_0^1\eta(1 - \eta^2)R_n^2 d\eta}$$

which can be cast into the form

$$C_n = C_{1n} - C_{2n}EPr\left(\frac{R}{\Delta}\right)^4$$

where

$$c_{1n} = \frac{\int_0^1 \eta(1 - \eta^2) R_n d\eta}{\int_0^1 \eta(1 - \eta^2) R_n^2 d\eta} \qquad (20a)$$

$$c_{2n} = \frac{\int_0^1 \eta(1 - \eta^2)(1 - \eta^4) R_n d\eta}{\int_0^1 \eta(1 - \eta^2) R_n^2 d\eta} \qquad (20b)$$

Then T^* can be written as

$$T^* = \sum_{n=0}^{\infty} c_{1n} R_n(\eta) \exp\left(-\frac{\lambda_n^2 z^*}{2}\right) +$$

$$EPr\left[1 - \eta^4 - \sum_{n=0}^{\infty} c_{2n} R_n(\eta) \exp\left(-\frac{\lambda_n^2 z^*}{2}\right)\right]\left(\frac{R}{\Delta}\right)^4 \quad (20c)$$

PRESSURE GRADIENT AND RADIUS OF THE OPENING

The variation of Δ with z^* will be determined next. The temperature distribution in the frozen shell is approximated by

$$T_s = T_m + (T_m - T_w)\frac{\ln\left(\frac{r}{\Delta}\right)}{\ln\left(\frac{\Delta}{R}\right)} \qquad (21)$$

In dimensionless form this is

$$T^*_s = -T^*_w \frac{\ln(\eta)}{\ln\left(\frac{\Delta}{R}\right)}$$

Use of equation (21) in equation (15) gives

$$\frac{\partial T^*(1,z^*)}{\partial \eta} = -\frac{k_s}{k} T^*_w \frac{1}{\ln\left(\frac{\Delta}{R}\right)} \qquad (22)$$

Evaluation of $\frac{\partial T^*(1,z^*)}{\partial \eta}$ from equation (20c)

enables equation (22) to be put into the form

$$\ln\left(\frac{\Delta}{R}\right)\left\{\sum_{n=0}^{\infty} c_{1n} R'_n(1)\exp\left(-\frac{\lambda_n^2 z^*}{2}\right)\right.$$

$$\left. -EPr\left[4 + \sum_{n=0}^{\infty} c_{2n} R'_n(1)\exp\left(-\frac{\lambda_n^2 z^*}{2}\right)\right]\left(\frac{R}{\Delta}\right)^4\right\} = -\frac{k_s}{k} T^*_w$$

$$(23)$$

The axial pressure drop required to force the fluid to flow at the specified volumetric rate is obtained from the axial equation of motion

$$V_r \frac{\partial V_z}{\partial r} + V_z \frac{\partial V_z}{\partial z} = -\frac{1}{\rho}\frac{\partial p}{\partial z} + \nu\left[\frac{1}{r}\frac{\partial}{\partial r}\left(r\frac{\partial V_z}{\partial r}\right) + \frac{\partial^2 V_z}{\partial z^2}\right]$$

$$(24)$$

Multiplication of equation (24) by r and integration with respect to r from $r = 0$ to $r = \Delta$, assuming $p = p(z)$, gives

$$\int_0^{\Delta} r\left[V_r \frac{\partial V_z}{\partial r} + V_z \frac{\partial V_z}{\partial z}\right] dr =$$

$$-\frac{1}{\rho}\frac{\Delta^2}{2}\frac{dp}{dz} + \nu\int_0^{\Delta}\frac{\partial}{\partial r}\left(r\frac{\partial V_z}{\partial r}\right)dr + \nu\int_0^{\Delta} r\frac{\partial^2 V_z}{\partial z^2} dr \quad (25)$$

Recall that

$$V_r \frac{\partial V_z}{\partial r} + V_z \frac{\partial V_z}{\partial z} = \frac{\partial(V_z V_r)}{\partial r} + \frac{\partial(V_z V_z)}{\partial z} - V_z\left(\frac{\partial V_r}{\partial r} + \frac{\partial V_z}{\partial z}\right)$$

$$(26)$$

But the continuity equation, equation (2), gives

$$\frac{\partial V_z}{\partial z} = \frac{1}{r}\frac{\partial}{\partial r}(rV_r)$$

which enables equation (26) to be rewritten as

$$V_r \frac{\partial V_z}{\partial r} + V_z \frac{\partial V_z}{\partial z} = \frac{\partial(V_z V_r)}{\partial r} + \frac{\partial(V_z V_z)}{\partial z} + \left(\frac{V_z V_r}{r}\right)$$

Thus the left hand side of equation (25) is

$$\int_0^{\Delta(z)} r\left[V_r \frac{\partial V_z}{\partial r} + V_z \frac{\partial V_z}{\partial z}\right] dr =$$

$$\int_0^{\Delta(z)} \left[r\frac{\partial(V_z V_r)}{\partial r} + V_z V_r\right] dr + \int_0^{\Delta(z)} r\frac{\partial V_z^2}{\partial z} dr$$

But

$$\frac{\partial}{\partial z}\int_0^{\Delta(z)} rV_z^2 dr = \int_0^{\Delta(z)}\frac{\partial}{\partial z}(rV_z^2)dr + rV_z^2(\Delta)\frac{d\Delta}{dz} - rV_z^2(0)\frac{\partial 0}{\partial z}$$

The last two terms vanish, the first one because $V_z(\Delta) = 0$. Also

$$\int_0^{\Delta(z)} r\left(\frac{\partial V_z V_r}{\partial r}\right)dr = rV_z V_r\Big|_0^{\Delta} - \int_0^{\Delta(z)} V_z V_r dr$$

The first term on the right hand side is zero since $V_r(\Delta) = 0 = V_r(0)$. Consideration of these facts reveals that the axial momentum equation (25) becomes

$$\frac{\Delta^2}{2\rho}\frac{dp}{dz} = \nu\Delta\frac{\partial V_z(\Delta,z)}{\partial r} + \nu\int_0^{\Delta(z)} r\frac{\partial^2 V_z}{\partial z^2}dr - \frac{\partial}{\partial z}\int_0^{\Delta(z)} rV_z^2 dr$$

$$(27)$$

Substituting the assumed profile of equation (1) into equation (27) gives

$$\frac{dp^*}{dz^*} = -\frac{16Pr}{\Delta^{*4}}\left[1 + \frac{4}{Re^2 Pr^2}\left(\frac{d\Delta^*}{dz^*}\right)^2\right] + \frac{16}{3}\frac{1}{\Delta^{*5}}\frac{d\Delta^*}{dz^*} \quad (28)$$

where

$$p^* = \frac{2p\pi^2 R^4}{\rho Q^2} = \frac{p}{\left(\frac{\rho V_{av}^2}{2}\right)}$$

and

$$z^* = \frac{\alpha\pi}{Q} z$$

The unity term in the brackets on the right hand side of this equation represents the contribution of viscous shear.

The pressure gradient for the fully developed case ($z^* \to \infty$) is obtained from equation (28) by setting $\frac{d\Delta^*}{dz^*} = 0$. Also, the unfrozen radius Δ is related to the volumetric flow rate Q for the fully developed case ($z^* \to \infty$) by

$$4EPr\left(\frac{R}{\Delta}\right)^4 \ln\left(\frac{\Delta}{R}\right) = \frac{k_s}{k} T^*_w$$

from equation (23). From this it is found that

$$Q^2 = -\frac{\pi^2}{4} R^4 \frac{k_s}{\mu} (T_m - T_w) \frac{\left(\frac{\Delta}{R}\right)^4}{\ln\left(\frac{\Delta}{R}\right)} \quad (29)$$

and the pressure gradient $p' = \frac{dp}{dz}$ for the fully developed condition is related to the unfrozen radius Δ by equation (28) as

$$-\frac{dp}{dz} = \frac{8}{\pi} \frac{\mu Q}{R^4 \left(\frac{\Delta}{R}\right)^4}$$

Substitution of this relationship for Q into equation (29) gives

$$-p' = \frac{4}{R^2} \left[\mu k_s (T_m - T_w)\right]^{1/2} \left(\frac{\Delta}{R}\right)^{-2} \left(-\ln\left(\frac{\Delta}{R}\right)\right)^{\frac{-1}{2}} \quad (30)$$

To find the condition that gives the minimum pressure gradient, the minimum of equation (30) is found by equating its derivative with respect to $\frac{\Delta}{R}$ to zero. After manipulation it is found that

$$-p'_{minimum} = \frac{8e^{1/2}}{R^2} \left[\mu k_s (T_m - T_w)\right]^{1/2}$$

or, in dimensionless form,

$$\left(-\frac{dp*}{dz*}\right)_{min} = 16ePr$$

and

$$\Delta* = \frac{\Delta}{R} = \exp\left(-\frac{1}{4}\right) = 0.7788$$

It is found from equation (30) that the volumetric flow rate Q_{min} associated with this minimum pressure gradient is related to the parameters of the problem by

$$Q_{min} = \frac{1}{e^{1/2}} R^2 \left[\frac{k_s}{\mu}(T_m - T_w)\right]^{1/2}$$

or, in dimensionless terms,

$$Q*_{min} = \frac{1}{e^{1/2}}$$

The existence of the minimum shown in Fig. 3 suggests that the cross-sectional dimensions of a tube or a runner in an injection molding machine can be selected

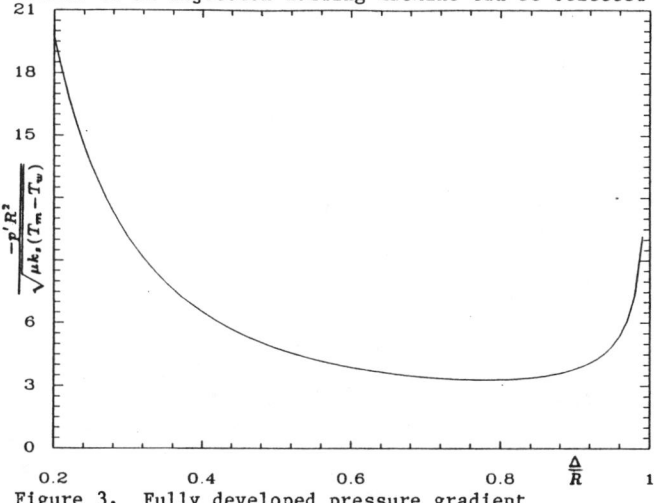

Figure 3. Fully developed pressure gradient from equation (30)

for minimum pressure drop if the volumetric flow rate Q is specified. If a pressure gradient greater than the minimum is imposed, it is seen that two different flow rates are possible, a perplexing situation for a designer to face. It is possible, as a consequence, that a flow that divides into the two branches of a tee might not divide equally.

Consider the problem of computing $\frac{\Delta}{R}$ versus $z*$ for the case where $z* \to \infty$. The dimensionless derivative $\frac{d\Delta*}{dz*}$ can be obtained from equation (23) for use in equation (28) to obtain the dimensionless pressure gradient as

$$-\frac{dp*}{dz*} = \frac{16Pr}{\Delta*^4}\left[1 + \frac{4}{Re^2 Pr^2}\left(\frac{d\Delta*}{dz*}\right)^2\right] + \frac{4}{3}\frac{d\Delta*^{-4}}{dz*}$$

Thus

$$p*_0 - p*(z*) = 16Pr \int_0^{z*} \left(\frac{1}{\Delta*^4}\right)$$
$$\left[1 + \frac{4}{Re^2 Pr^2}\left(\frac{d\Delta*}{dz*}\right)^2\right]dz + \frac{4}{3}\left(1 - \frac{1}{\Delta*^4}\right) \quad (30a)$$

Taking the derivative of equation (23) gives

$$\frac{d\Delta*}{dz*} = \left\{ \frac{\Delta*}{2}\ln\Delta*\left(\sum_{n=0}^{\infty} C_{1n}R'_n(1)\lambda_n^2\exp\left(-\lambda_n^2 z*/2\right)\right.\right.$$

$$\left.\left. - \frac{EPr}{\Delta*^4}\sum_{n=0}^{\infty} C_{2n}R'_n(1)\lambda_n^2\exp\left(-\lambda_n^2 z*/2\right)\right)\right\}$$

$$\left/ \left\{\sum_{n=0}^{\infty} C_{1n}R'_n(1)\exp(-\lambda_n^2 z*/2) - \frac{EPr}{\Delta*^4}(1 - 4\ln\Delta*)\right.\right.$$

$$\left.\left[4 + \sum_{n=0}^{\infty} C_{2n}R'_n(1)\exp(-\lambda_n^2 z*/2)\right]\right\}$$

In order to calculate the derivative $\frac{d\Delta*}{dz}$ at any point, coefficients C_{1n} and C_{2n} must be known. However, values of C_{1n}, λ_n, and $R'_{1n}(1)$ are given by Jakob (1949) and Sellars (1956) as represented in Table 1. The values of C_{2n} are calculated from equation (20b)

Table 1. Eigenfunctions and eigenvalues occurring in the Graetz problem from Sellars et al.(1956)

n	λ_n	C_{1n}	$C_{1n}R'_n(1)$
0	2.704364	1.466220	0.748790
1	6.679032	-0.802476	0.544240
2	10.673380	0.587094	0.462880
3	14.671080	-0.474897	0.415180
4	18.669870	0.404402	0.382370

as follows.

From Jakob (1949)

$$R_n(\eta) = \sum_{m=0}^{\infty} B_{2m}\beta^{2m}\eta^{2m} \quad (31)$$

where

$$B_0 = 1$$

$$B_2 = -\frac{1}{2^2}$$

$$B_4 = \frac{1}{4^2}\left(\frac{1}{\beta^2} + \frac{1}{2^2}\right)$$

$$B_6 = -\frac{1}{6^2}\left(\frac{1}{2^2\beta^2} + \frac{1}{4^2}\left(\frac{1}{\beta^2} + \frac{1}{2^2}\right)\right)$$

$$*$$
$$*$$
$$*$$

$$B_{2m} = -\frac{1}{(2m)^2}\left[\frac{B_{2m-4}}{\beta^2} - B_{2m-2}\right] \qquad (32)$$

and $\beta = \lambda$ is the eigenvalue for the differential equation (18). Let

$$b_{2m} = B_{2m}\beta^{2m}$$

Hence $b_0 = 1$

$$b_2 = B_2\beta^2 = -\frac{\beta^2}{2}$$

$$b_4 = B_4\beta^4 = \frac{1}{4^2}\left(\beta^2 + \frac{\beta^4}{2^2}\right)$$

$$b_6 = B_6\beta^6 = -\frac{1}{6^2}\left(\beta^4\left(\frac{1}{2^2} + \frac{1}{4^2}\right) + \frac{\beta^6}{2^2 \times 4^2}\right)$$

$$*$$
$$*$$
$$*$$

$$b_{2m} = \frac{\beta^2}{(2m)^2}(b_{2m-4} - b_{2m-2})$$

$$*$$
$$*$$

Now define I as

$$I = \int_0^1 \eta(1 - \eta^2)(1 - \eta^4)R_n\,d\eta$$

where

$$R_n(\eta) = \sum_{m=0}^{\infty} b_{2m}\eta^{2m}$$

thus

$$I = \sum_{m=0}^{\infty} b_{2m}\int_0^1 \eta^{(2m+1)}(1 - \eta^2)(1 - \eta^4)d\eta$$

Jakob (1949) gives

$$\int_0^1 \eta(1 - \eta^2)R_n^2\,d\eta = \frac{1}{2\beta}\left[\left(\frac{dR(1)}{d\beta}\right)_n R'_n(1)\right]$$

and

$$C_{1n} = -\frac{2}{\beta\left(\frac{dR(1)}{d\beta}\right)_n}$$

With these relationships and the result for I it can be shown that

$$C_{2n} = \left(\frac{\beta^2 C_{1n}^2}{2}\right)\frac{\sum_{m=0}^{\infty} b_{2m}\dfrac{(2m+5)}{(m+1)(m+2)(m+3)(m+4)}}{-\frac{1}{2}C_{1n}R'_n(1)}$$

With C_{1n} and $-\frac{1}{2}C_{1n}R'_n(1)$ available as given in Table 1, the calculated C_{2n} values in Table 2 were used in equation (23) to obtain the Δ^* versus z^* results plotted in Figure 4. It is seen that viscous dissipation has an important influence for long ducts. The asymptotic value of the frozen annulus thickness is seen in Figure 4 to be achieved for $z^* \sim 1$, a distance that can be taken to be the entrance region. Equation (23) can be used to develop a more general entrance region estimate.

Table 2. Coefficients C_{1n}, C_{2n}, and $\dfrac{dR'(1)}{d\beta}$ at different β values

n	β	$\left(\dfrac{dR'}{d\beta}\right)_n$	C_{1n}	C_{2n}
0	2.704364	-0.5043891	1.466220	1.3347197
1	6.679032	0.3731508	-0.802476	-0.4990974
2	10.673380	-0.3191688	0.587094	0.2308420
3	14.671080	0.2870572	-0.474897	-0.1323893
4	18.669870	-0.2648960	0.404402	0.0814687

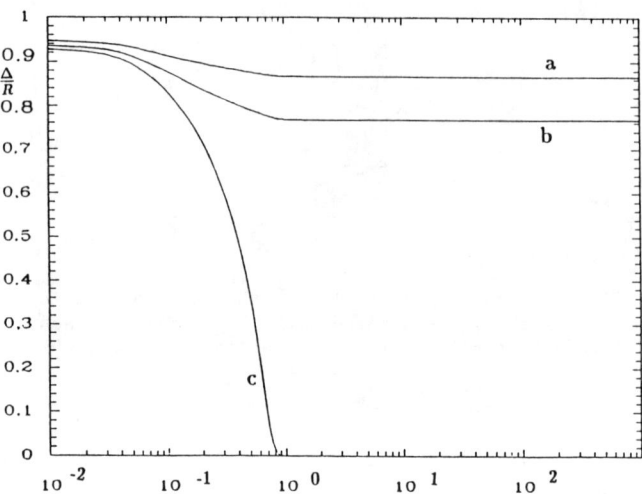

Figure 4. Variation of $\dfrac{\Delta}{R}$ along the duct from equation (23) for $\dfrac{k_s}{k}T^*_w = -0.5625$: (a) with viscous dissipation (EPr = 0.6); (b) with viscous dissipation (EPr = 0.2) and (c) no viscous dissipation (EPr = 0.0)

REFERENCES

Zerkle, R.D. and Sunderland, J.E., 1968, "The Effect of Liquid Solidification in a Tube Upon the Laminar-Flow Heat Transfer and Pressure Drop," Trans. ASME, J. Heat Transfer, Vol. 90, pp. 183-190.

Sellars, J.R., Tribus, M., and Klein, J.S., 1956, "Heat Transfer to Laminar Flow in a Round Tube or Flat Conduit--The Graetz Problem Extended," Trans. ASME, Vol. 78, pp. 441-448.

Tago, M., and Fukasako, S., 1988, "Freezing Heat Transfer Characteristics In Return Bend With A Rectangular Cross Section," Proceedings of the 1988 National Heat Transfer Conference, editor: H.R. Jacobs, ASME HTD-96, Vol. 3, pp. 215-224.

Jakob, M., 1949, Heat Transfer, Vol. 1, John Wiley, pp. 451-459.

1989 National Heat Transfer Conference
HTD-Vol. 107, Heat Transfer in Convective Flows

THE USE OF RHEOLOGICAL FLUIDS IN CONTINUOUS FLOW ELECTROPHORESIS

I. Kim and T. F. Irvine Jr.
Department of Mechanical Engineering
State University of New York
Stony Brook, New York

Abstract

A continuous flow electrophoresis device which stabilizes the flow field by using advantageous characteristics of non-Newtonian fluids has been developed. Various factors including the geometry and orientation of the flow channel with respect to the gravity vector have been considered to mitigate free convection currents due to Joule heating. The viscosity increase in the carrier fluid allows an increase of the carrier velocity and the migrant throughput without increasing the carrier Reynolds number, thus decreasing the Richardson number. The use of non-Newtonian fluids flattens the velocity profile, which is advantageous in keeping the migrant fluid from spreading due to different residence times in the electric field.

A horizontal continuous flow electrophoresis device using optimum design conditions has been constructed and the electrophoretic separation of a two-dye mixture was performed. Hydrodynamic, thermal and electrophoretic measurements are presented showing that the use of rheological fluids offers many advantages including flow field stability and migrant throughput increase.

Nomenclature

a channel height (m)
C concentration (%)
g gravitational acceleration (m/s^2)
Gr_g generalized Grashof number
K fluid consistency (Ns^n/m^2)
k thermal conductivity (W/mK)
k_e electrical conductivity (S/m)
n flow index
p pressure (Pa)
Q flow rate (m^3/s)
Q''' heat generation rate per unit volume (W/m^2)
$\overline{Q'''}$ cross section average heat generation rate per unit volume (W/m^3)
Re Reynolds number
Re_g generalized Reynolds number
Ri Richardson number

T temperature (K)
T_r reference temperature (K)
u velocity component in flow direction (m/s)
$\overline{u}$ average velocity (m/s)
v velocity component in gravity direction (m/s)
x coordinate in flow direction (m)
y coordinate in gravity direction (m)

Greek Letters

β thermal expansion coefficient (1/K)
$\dot{\gamma}$ shear rate (1/s)
η_a apparent viscosity (Ns/m^2)
ρ density (Kg/m^3)

Introduction

Continuous Flow Electrophoresis

Electrophoresis is an electrokinetic phenomenon wherein the charged particles migrate in an electric field either toward the positive or negative electrode according to the sign of the charge which they carry. A variety of methods for implementing electrophoretic separation have been developed for various applications in the fields of biology, biochemistry,etc [1,2,3,4]. Most of the methods developed until the late 1950s were batchwise but after that the need for continuous operation for many applications has been favored and the development of continuous flow electrophoresis (CFE) devices has accelerated.

The technique of CFE is illustrated in Fig.1. The carrier fluid flows in a direction normal to the lines of an electric field, and the mixture to be separated is added continuously in the flowing medium. Components of the mixtures are deflected in many different directions according to their electrophoretic mobilities and charges which they carry and, after passage through the electrode section, can be collected continuously at a variety of exit positions. The electrodes are usually separated from the main flow by permeable membranes to allow cooling and the removal of electrolysis products.

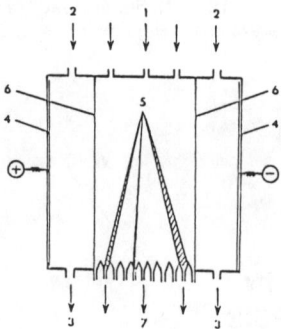

Fig.1. Schematic representation of CFE
1. Carrier in   2. Electrolyte in   3. Electrolyte out
4. Electrodes   5. Migrant in   6. Membranes
7. Migrant and carrier out

Maintaining the stability of CFE against convective disturbances caused by Joule heating is a key problem in scaling up electrophoresis systems. A larger migrant throughput - amount of migrant which can be separated with maximal purity in a unit time - seems to be possible by increasing the gap size in the Hannig type CFE [2,5] which is widely used, but it also increases the electric voltage to achieve sufficient separation and consequently increases the amount of Joule heating, which causes an excessive temperature rise and local density differences in the separation channel and thus distorts the separated migrant bands due to free convection and hydrodynamic instabilities.

Even if the amount of the migrant throughput is increased up to 20 ml/min by a rotating CFE apparatus [6], most devices still have too large a temperature increase during separation and some method to lower this temperature increase is required. This problem is commonly minimized either by pumping cold electrolyte into the electrode channels or by cooling the incoming carrier fluid to 2-3 °C thus reducing the outlet temperature [6]. This method however requires several tons of extra refrigeration power and consequently high capitol and operating costs. Thus another way to solve this temperature increase problem is desirable.

General Description of non-Newtonian Fluids

Fluids can be divided into Newtonian and non-Newtonian fluids. For Newtonian fluids, the dynamic viscosity (η) is a fluid property and is independent of shear rate ($\dot{\gamma}$). If shear stress (τ) is plotted against shear rate, a linear relation exists between two quantities and the slope of the curve is the dynamic viscosity. Fluids which do not obey this relation are non-Newtonian fluids and classification of such fluids is well described in Irvine and Karni [7]. In non-Newtonian fluids, the viscosity is not a fluid property but is a function of shear rate and so it is called the apparent viscosity (η_a). Non-Newtonian fluids are often prepared in the laboratories by adding soluble powders to a liquid such as water. Because of their large apparent viscosity, non-Newtonian fluids have a tendency to operate at low Reynolds and Grashof numbers. Accordingly, in engineering practice, laminar flow situations are encountered more often than with Newtonian fluids.

Non-Newtonian Fluids in CFE

To the authors' knowledge, Dobry [8] was the first to utilize the viscous properties of a fluid in CFE. He mixed a water soluble viscosity enhancing material - Methocel or Dexatrin - into water and suppressed the free convection eddies by means of high viscosity (10 cp). Recently, the laminar mixed convection problem with internal heat generation was considered by Brewster and Irvine [9] who proposed a theoretical CFE model with flow between two vertical parallel plates with non-uniform heat generation in the transverse direction. A power-law fluid was taken as the carrier fluid and it was shown that the use of pseudo-plastic non-Newtonian fluids with high viscosity can effectively reduce or prevent the convection.

The use of non-Newtonian fluids in CFE has several advantages over Newtonian fluids including:
1. An increased carrier fluid viscosity allows an increase of the average carrier and migrant velocities without unduly increasing the carrier Reynolds number thus avoiding the transition to turbulence.
2. The use of a non-Newtonian fluid can flatten the velocity profile which is advantages in keeping the migrant fluid from spreading due to different residence times in the electric field.

Aims of the Present Work

Development of a CFE device stabilizing the flow field and providing large migrant throughput by using the advantageous characteristics of non-Newtonian fluids and basic heat transfer principles is the main purpose of the present work. It also includes in particular a consideration of the design philosophy including the following most important aspects:
1. To choose the geometry and orientation with respect to the gravity vector to minimize free convection effects.
2. To reduce the necessary heat generation as much as possible, and to reduce the temperature increase during electrophoresis. The latter is important from the standpoint of inhibiting free convection.
3. To avoid, if possible, the necessity of refrigeration in order to keep the exit carrier temperature within allowable limit.
4. To design an apparatus which is simple to construct and operate with reduced capitol and operating costs.
5. To utilize the properties of non-Newtonian carrier fluid in order to
   (i) Make use of the increased apparent viscosity to be able to increase the average velocity together with the migrant throughput without unduly increasing the carrier Reynolds number thus avoiding the possibility of transition to turbulent flow.
   (ii) Flatten the carrier velocity profile to reduce the velocity profile dispersion due to unequal residence times along various streamlines of the carrier fluid.

Analysis

Consideration of Free Convection Effects in CFE

The situation depicted in Fig. 2 is considered as a model for the CFE apparatus. A fluid flows through a horizontal parallel plate channel. The fluid entering the duct is assumed to be fully developed, steady and laminar. It flows with an average velocity u and has uniform inlet temperature and density. Both plates are thermally insulated and, as an electric current passes through the duct in vertical direction, heat is generated arbitrary inside the fluid (Q''') in y direction. The fluid under consideration is a power-law fluid, $\eta_a = K \dot{\gamma}^{n-1}$, where K and n are fluid properties called fluid consistency and flow index respectively.

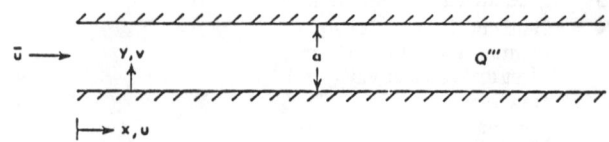

Fig. 2. Schematic of horizontal parallel-plate channel

Using the Boussinesq approximation and dimensionless variables defined below, the gravitational direction momentum equation can be written as:

$$u^+\frac{\partial v^+}{\partial x^+} + v^+\frac{\partial v^+}{\partial y^+} = -\frac{\partial p^+}{\partial y^+} + \frac{1}{Re_g}\left[\frac{\partial}{\partial x^+}\left(\frac{\partial v^+}{\partial x^+}\right)^n + \frac{\partial}{\partial y^+}\left(\frac{\partial v^+}{\partial y^+}\right)^n\right] + Ri\,T^+$$

where $u^+ = \dfrac{u}{\bar{u}}$, $v^+ = \dfrac{v}{\bar{u}}$, $x^+ = \dfrac{x}{a}$, $y^+ = \dfrac{y}{a}$, $p^+ = \dfrac{p}{\rho\bar{u}^2}$, $T^+ = \dfrac{T - T_r}{\bar{Q}''\,a^2/k}$

$$Re_g = \frac{\rho\,\bar{u}^{2-n}\,a^n}{K} \quad , \quad Gr_g = \frac{\rho^2\,g\,\overline{Q''}\,a^{2n+3}}{k\,K^2\bar{u}^{2-n}} \quad , \quad Ri = \frac{Gr_g}{Re_g^2} = \frac{g\,\beta\overline{Q''}\,a^3}{k\,\bar{u}^2}$$

The existence of free convection depends upon the Richardson number (Ri), i.e., the lower its value the more the free convection is reduced. Thus one way to decrease free convection is to decrease the Ri. It is of interest to note that the separate quantities Gr_g and Re_g contain viscous properties but the Ri does not. Thus, the only practical way to reduce the Ri is to decrease the channel height (a) or increase the average fluid velocity ($\bar{u}$). Decreasing the channel dimension in the gravity direction can be achieved by having the device in a horizontal position instead of a vertical position, so that the length scale in the Ri can be the channel width instead of channel height, which could be usually of the order of 100 times smaller. An increase in velocity is bounded by the fact that the Re cannot exceed a certain value or the flow becomes turbulent. It is here that the fluid viscosity becomes a factor indirectly. If the viscosity is increased, the average velocity can also be increased without increasing the Re, thus no turbulent flow. However, by increasing the average velocity, the Ri has been decreased which further suppresses the free convection occurs. Thus, viscosity plays an indirect role in allowing the average carrier velocity to be increased therefore increasing the throughput and decreasing the free convection. The use of non-Newtonian fluids usually brings a lower n value with a flatter velocity profile and a higher viscosity. Therefore, by using non-Newtonian fluids, the main problems in CFE - free convection disturbances, velocity profile dispersion and migrant throughput - can be reduced.

Stabilization of the flow field against free convection effects by using non-Newtonian fluids has been studied by Brewster and Irvine [9]. They specified a power-low fluid as the carrier fluid in a theoretical CFE model consisting of two vertical parallel plates. Figure 3 shows their numerical results where the dimensionless exit velocity profiles for n = 1.0(pure water), 0.807(CMC 100 wppm) and 0.735(CMC 300 wppm), which correspond to the Richardson numbers of 2428.32, 74.51 and 17.13 respectively, are plotted. An increase in viscosity (decrease in n) at constant Reynolds number allows an increase in the fluid velocity and subsequent decrease of Richardson numbers. In this figure, it is seen that as Richardson number decreases, the resulting asymmetry of the velocity decreases indicating that the free convection effects can be suppressed by increasing the fluid viscosity and velocity, which reduce Richardson number.

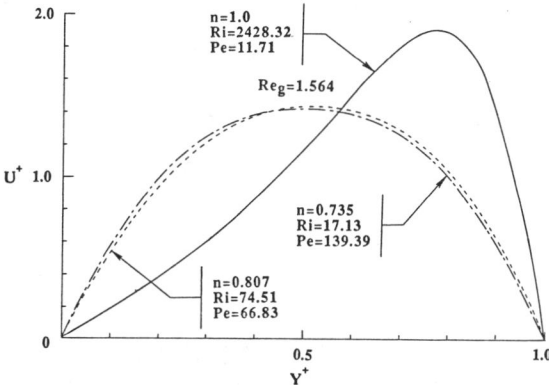

Fig.3. Exit velocity profiles for different n's at constant Re [9]

Electrophoretic Factors

In addition to the hydrodynamic and thermal constraints discussed above, there are also limitations imposed by the electrophoretic process which must be taken into account in the design of a CFE using the stabilized laminar flow including:
1. Migrant displacement, voltage gradient and migrant residence time
2. Velocity profile dispersion
3. Joule heating
These factors are not independent but most of them are interrelated. For example, if the voltage gradient is increased to have larger migrant separation, the heat generation is also increased thus the carrier temperature is increased and, if low voltage is applied to avoid the temperature rise, the system length should be very long to achieve necessary migration thus, the migrant stays in the system for a long time, etc. To obtain the optimum design and operating conditions to satisfy all the factors, a computer program was written and acceptable conditions were selected, and the CFE device was constructed and operated based on these results [10].

Experiment and Results

A schematic and flow path views of the apparatus are shown in Fig.4. The overall dimensions are approximately 200 cm x 12 cm x 6 cm with the short sides parallel to the gravity vector. It consists of three rectangular channels - a carrier channel, and top and bottom electrode channels. A more complete description including the details of the migrant collecting system, materials, etc. can be found from [10].

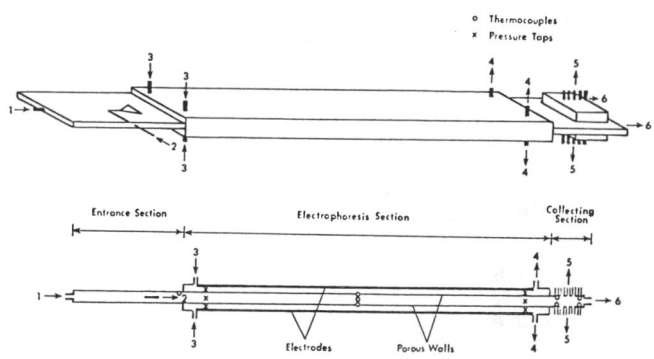

Fig.4. Schematic and flow path views of the CFE apparatus
1.Carrier in   2.Migrant in   3.Electrolyte in   4.Electrolyte out
5.Migrant and carrier out   6.Carrier out

Methocel A4M (manufactured by the Dow Chemical Co.) 12,000 wppm was selected as the carrier fluid through a series of preliminary tests with many different polymers and concentrations. Small amounts of buffers (Citric acid and Tris) were added to the carrier fluid to stabilize the pH and electrical conductivity and the rheological properties were measured before and after the experiment with a falling needle viscometer (manufactured by J & L instrument Co.) and the flow curve is shown in Fig,5. . The properties of the final buffered carrier fluid for electrophoresis were:

Flow index (n) = 0.88   Consistency (K) = 0.54 Ns^n/m^2
pH= 7.3   Electrical conductivity = 1.21 mS/cm

Equal amount of two color dyes which were known to have different electrical charges and electrophoretic mobilities - Naphthol Green B and Orange II by Aldrich Chemical Co. - were well mixed with the same fluid as the carrier and used as a migrant. To prevent any deleterious electrolysis penetrating into the separation channel from the electrode channels and to provide high electrical conductivity in the electrolyte, the viscosity of the electrolyte was adjusted to be the same as that of the carrier by mixing the same concentration of the Methocel polymer to Isoton II electrolyte (by Coulter Diagnostics).

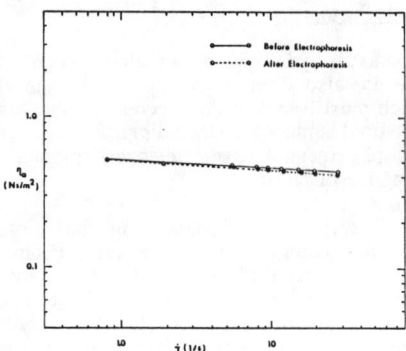

Fig.5. Flow curve of the carrier fluid (Methocel A4M 12,000wppm)

Figure 6 is another sketch of the apparatus showing various flow loops and collecting stations. The carrier fluid was pumped to the carrier inlet with a flow rate of 500 - 600 ml/min. The electrolyte was also pumped into the electrode channels and the flow rate was controlled until no pressure difference between the separation and electrode channel was found. The migrant was then introduced in the center of the carrier by a triangular shaped injector with a form of thin ribbon with a flow rate of 7 - 10 ml/min.

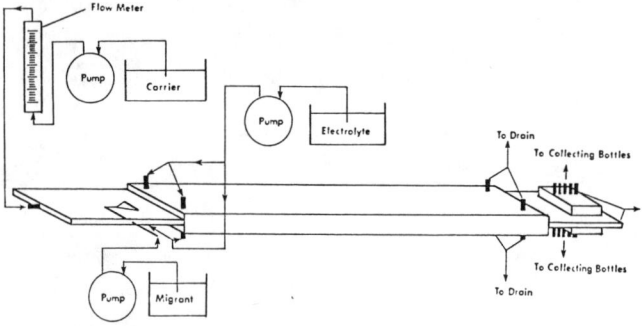

Fig.6. Sketch of the CFE apparatus showing various flow loops and collecting stations

As the power supply was turned on and voltage was applied to the apparatus, the migrant band which was initially a dark brown color started being separated into its constituents up and down. After passing the separation section, the migrant consisted of two individual color bands of about 2 mm thick and 1.5 cm wide while maintaining stable laminar conditions and no free convection disturbance could be observed. The separated dyes were then removed through ten exit ports and uncontaminated portion of the carrier was also collected separately. The temperatures of the incoming and outgoing carrier, and the temperatures of the top, middle and bottom of the channel were measured during electrophoresis by using thermocouples to determine the overall flow direction and gravity direction temperature differences and they were found to be about 6 °C for both cases.

The CFE device developed in this study utilized no refrigeration system. A relatively low voltage (20-30 V) was possible by reducing the length scale between the electrodes and by keeping a relatively long migrant residence time (150-180 s) to achieve sufficient separation without excessive carrier temperature rise. As a result, the gravity directional temperature increase (6 °C) was not enough to create any perceptible free convection currents and the maximum temperature of the carrier (30 °C) was far below the limit for many protein separations [11]. However, since some biological migrants are to be kept and separated at very low temperature, performing the

electrophoresis separation at low temperature for example 4 °C which gives the maximum water density, can further decrease the free convection effect due to the decrease in the thermal expansion coefficient and thus the Grashof number.

pH and electrical conductivity were also measured independently after every electrophoresis run from the fluids in the collecting bottles and Fig.7 shows the results. Reasonably flat pH profile (7.8 ± 0.5) and electrical conductivity profile (1.3 ± 0.2 mS/cm in exits 1 to 8) were maintained during electrophoresis.

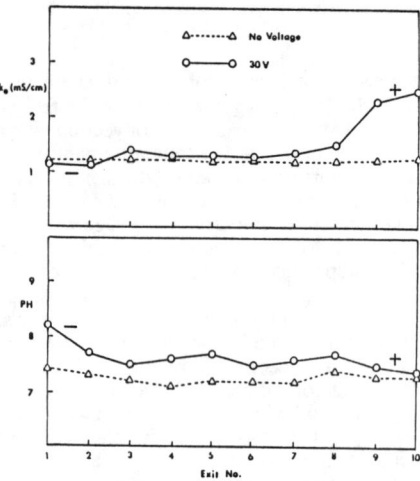

Fig.7. pH and electrical conductivity profiles for dye separations

It is of interest and also important to see how much of the injected migrant is collected and how much of the migrant is diluted by the carrier. After the separated migrant mixed with carrier fluid was collected in ten collecting bottles, the optical densities were measured with a UV/VIS spectrophotometer. Figure 8 shows the concentration measurement results for single and two-dye mixtures. As seen in Fig. 8-a , the green dye was collected through exits 8 and 9 when there was no electricity applied. With an original migrant

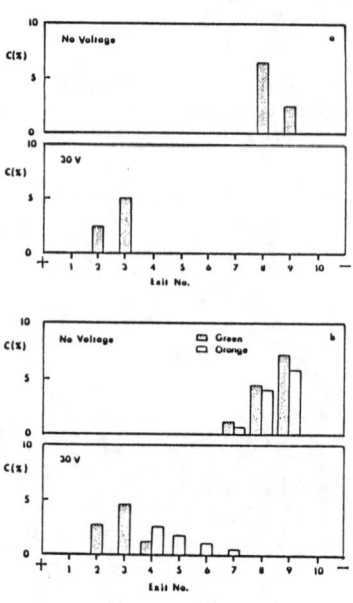

Fig.8. Results of concentration measurements

concentration of a 100%, this represents about 15 fold dilution in exit 8. When 30 V were applied, the dye migrated towards the positive electrode and was collected through exits 2 and 3 with about a 20 fold dilution. Similar phenomena can be seen in Fig.8-b which shows a two-dye mixture separation. Without electricity, it was collected through exits 7 to 9 and with 30 V, the migrant mixture was separated and collected through exits 2 to 7, which shows about the same dilution as Fig.8-a . From Fig.8, the electrophoresis efficiency was determined considering flow rate and concentration of each exit. Generally about 70 - 80% of the injected migrant was collected with no electricity while about 60% of the separated species were collected with an applied voltage.

Table 1 summarizes some of the operating characteristics and the results of the electrophoresis runs for dye separations, and several deserve special comments :

1. The apparent viscosity of the carrier fluid is approximately 540 times more viscous than water.
2. The migrant throughput is about half of the rotating apparatus [6] which is known to give the highest migrant throughput in the CFE market but, considering much lower construction and operating costs, it is quite satisfactory.
3. A relatively low electric power was required by reducing the length scale between electrodes and, as a result, the temperature rises without any refrigeration were quite acceptable, and no free convection disturbance could be observed.
4. The Richardson number is decreased to the order of one by means of the increased carrier fluid viscosity and reduction of the length scale in the gravity direction, which indicates the suppression of the free convection.

Table 1. Operating characteristics and results of the dye-separation electrophoresis

Carrier fluid	Methocel 12000 wppm
n	0.88
$K(Ns^n/m^2)$	0.54
$\bar{u}$ (cm/s)	0.8 - 1.0
Carrier Q (ml/min)	500 - 600
Migrant Q (ml/min)	7 - 10
Dilution (fold)	15 - 20
Voltage(V) / Current(A)	20-30 / 5-8
Electric power (W)	100 - 240
Flow direction ΔT (^{o}C)	6
Gravity direction ΔT (^{o}C)	6
Re_g	0.15 - 0.19
Gr_g	0.073 - 0.149
Ri	2.189 - 4.345

Conclusions

The application of rheological fluids to CFE has been studied including geometry, orientation and careful heat transfer considerations in designing a CFE device to suppress free convection currents during electrophoresis. A horizontal CFE device was constructed to reduce the length scale in the gravity direction and the electrophoresis separation of a two-dye mixture was performed with a non-Newtonian carrier fluid. Without any refrigeration, the temperature increases in the separation channel due to Joule heating were found to be insufficient to create free convection disturbances and, due to the high viscosity of the carrier fluid, the Richardson number was decreased to the order of one. The CFE device developed in the present study using a non-Newtonian carrier fluid and in a horizontal position illustrates the possibilities of improving CFE devices including higher throughput, flow field stability and reduction of construction and operating costs.

Acknowledgement

The authors are grateful for the financial support for this investigation from the United Kingdom Atomic Energy Authority, Harwell, England and for the technical advice and encouragement of Drs. G.F.Hewitt and C.Lambe from Harwell.

References

1. D.Freifelder, "Electrophoresis", Physical Biology - Application to Biochemistry and Molecular Biology, Ch.9, W.H.Freeman and Company, 1976

2. K.Hannig, "Preparative Electrophoresis", Electrophoresis : Theory, Methods and Applications, M.Bier (ed), Ch.9, Academic Press, 1967

3. Z.Deyl, F.M.Everaerts, Z.Prusic and P.J.Svendsen (eds), "Electrophoresis : A survey of Techniques and Applications", Elsevier Scientific Publishing Co., 1979

4. D.J.Shaw, "Electrophoresis", Academic Press, 1969

5. K.Hannig, "Die Trageerfreie Kontinuierliche Elektrophorese und iher Awnwendung", Z.Anal. Chem., 1181, pp 244-254, 1961

6. P.Mattock, G.F.Aitchison and A.R.Thomson, "Velocity Gradient Stabilized, Continuous, Free Flow Electrophoresis. A Review", Separation and Purification Methods, 9(1), pp 1-68, 1980

7. T.F.Irvine Jr. and J.Karni, "Non-Newtonian Fluid Flow and Heat Transfer", Ch.20, Handbook of Single-Phase Convective Heat Transfer, S.Kakac, R.K.Shaw and W.Aung (eds), John Wiley & Sons, 1987

8. R.Dobry and R.K.Finn, "New Approach to Continuous Flow Electrophoresis", Science, 127, pp 697-698, 1958

9. R.A.Brewster and T.F.Irvine Jr., "Laminar Mixed Convection in Power Law Fluids", Accepted for Publication, Int. J. Heat and Mass Transfer, 1989

10. I.Kim, "The Use of Rheological Fluids in Continuous Flow Electrophoresis", Ph.D. Thesis, State University of New York at Stony Brook, 1988

11. R.A.Yoshisato, L.M.Korndorf, G.R.Carmichael and R.Datta, "Performance Analysis of a Continuous Rotating Electrophoresis Column", Separation Science and Technology, 21(8), pp 727-753, 1986

1989 National Heat Transfer Conference
HTD-Vol. 107, Heat Transfer in Convective Flows

ANALYSIS OF MULTIPLE SOLUTIONS IN PLANE POISEUILLE FLOW WITH VISCOUS HEATING AND TEMPERATURE DEPENDENT VISCOSITY

M. Sen
University of Notre Dame
Notre Dame, Indiana

P. Vasseur
École Polytechnique
Montréal, Canada

ABSTRACT

The two-dimensional problem of plane Poiseuille flow with viscous heating is analyzed. The flow is assumed to be laminar and unidirectional. The plane walls are kept isothermal. In the first case the thermal conductivity of the fluid is considered to vary linearly with temperature while in the second the viscosity is assumed to decrease exponentially with temperature. Some characteristics relating to the existence of solutions in certain parameter ranges are pointed out. In the latter case two solutions are found to exist, for each of which numerical and approximate results are presented.

NOMENCLATURE

A,B constants of integration
a coefficient for variation of thermal conductivity with temperature
h half the separation between plates
p pressure
P^* pressure gradient
Q ratio of volumetric flow rate
T temperature
T_i temperature perturbation of order i
T_a average temperature
T_b bulk temperature
T_m centerline temperature
T_w wall temperature
u,v Cartesian components of velocity
u_i velocity perturbation of order i
x,y Cartesian coordinates

Greek letters

α dimensionless parameter related to viscous heating
β dimensionless parameter related to temperature effect on viscosity
γ $= \alpha\beta$
θ $= T - 1$
θ_m centerline value of θ
λ thermal conductivity of the fluid
μ dynamic viscosity of the fluid
ϕ $= \beta\theta$

Subscripts and superscripts

* dimensional quantities
w evaluated at the temperature of the wall
1 solution with lower temperature and flow rate
2 solution with higher temperature and flow rate

INTRODUCTION

It has been known for some time that some problems of flow of a Newtonian fluid with viscous heating and temperature dependent viscosity admit two steady solutions, leading to issues that are of importance in processes such as polymer processing and extrusion. Typically there are two possible flow rates for the same pressure gradient with a liquid for which the viscosity depends exponentially with temperature. A textual description of such solutions is given in Platten and Legros (1984) and a brief history of previous work in Sukanek and Laurence (1974). Geometries which have been studied include Poiseuille flow in a cylindrical pipe (Kearsley, 1962; Nihoul, 1971; Sukanek and Laurence, 1972), and plane and circular Couette flows (Gavis and Laurence, 1968). The solution for a power-law fluid has also been given by Sukanek (1971). In all these cases, the corresponding nonlinear equations could be analytically solved and the multiplicity of steady states thus demonstrated.

Of the two flow rates found, early investigators tended to doubt whether the higher one was physically possible. This flow rate tends to infinity as the parameter representing fluid viscosity variation with temperature is changed so the viscosity tends towards a constant value. The lower flow rate solution approaches, of course, a standard Poiseuille flow with constant viscosity. A related concern with respect to the higher flow rate solution is its stability, a problem addressed in the inviscid analysis of Joseph (1965). Linear stability of the cylindrical pipe problem to axisymmetric perturbations have been studied by Vanderborck and Platten (1977) and Vanderborck *et al.* (1979) using a Galerkin method. In the earlier paper they neglected temperature fluctuations which they included in the latter. Both flows were found to be stable for the parameter values studied. Ho *et al.* (1977) carried out stability analyses of both the plane Couette and axisymmetric Poiseuille flows using direct numerical integration.

The major experimental investigation of this problem has been by Sukanek and Laurence (1974). They used a high viscosity Newtonian fluid with extreme sensitivity of its viscosity to temperature. Plane and circular Couette flows were studied along with cylindrical Poiseuille flow. Some qualitative similarity of theory with experiment was discovered, though quantitative agreement was not found.

The objective of the present work is to extend the theoretical results to the case of plane Poiseuille flow in which the fluid is driven by a pressure gradient through the gap between two infinite stationary flat plates. In this case analytical solutions cannot be found so that the number of solutions has to be numerically determined. Two solutions, both symmetrical with respect to the centerline, are obtained. The characteristics of both solutions are examined for different parameter values. A perturbation technique is used to approximate the lower flow rate solution and a weighted residual moment method the higher one.

BASIC EQUATIONS

Consider the steady laminar flow between flat plates of a viscous incompressible fluid with temperature dependent properties. The plates are kept at a constant temperature T_w and are separated by a distance 2h. The coordinate system is shown in Fig. 1. We make the usual assumptions of $p^* = p^*(x^*)$, $u^* = u^*(y^*)$, $v^* = 0$ and $T^* = T^*(y)$, where p^* is the pressure, u^* and v^* are the Cartesian velocity components in the x^* and y^* directions, and T^* is the temperature. The two-dimensional momentum and energy equations reduce to

$$\frac{d}{dy^*}\left(\mu^* \frac{du^*}{dy^*}\right) = -P^* \tag{1}$$

$$\frac{d}{dy^*}\left(\lambda^* \frac{dT^*}{dy^*}\right) + \mu^* \left(\frac{du^*}{dy^*}\right)^2 = 0 \tag{2}$$

where $P^* = -dp^*/dx^*$ is the pressure gradient, $\lambda^*(T^*)$ is the coefficient of thermal conductivity and $\mu^*(T^*)$ is the dynamic viscosity. The nondimensional variables

$$y = y^*/h \tag{3a}$$
$$u = \frac{\mu_w}{Ph^2}u^* \tag{3b}$$
$$T = T^*/T_w \tag{3c}$$
$$\mu = \mu^*/\mu_w \tag{3d}$$
$$\lambda = \lambda^*/\lambda_w \tag{3e}$$

are without asterisks. Here λ_w and μ_w are the values of λ and μ at the temperature of the wall T_w. The nondimensional equations which describe the phenomena are

$$\frac{d}{dy}\left(\mu \frac{du}{dy}\right) = -1 \tag{4}$$

$$\frac{d}{dy}\left(\lambda \frac{dT}{dy}\right) + \alpha \mu \left(\frac{du}{dy}\right)^2 = 0 \tag{5}$$

where the parameter α is given by

$$\alpha = \frac{P^2 h^4}{\mu_w \lambda_w T_w} \tag{6}$$

which is the Brinkman number (Platten and Legros, 1984) if the characteristic velocity is taken to be Ph^2/μ_w as in equation (3b). This is twice the maximum velocity for the constant viscosity case.

We do make any assumptions *a priori* about the symmetry of the velocity or temperature fields. Thus the boundary conditions are taken to be

$$u = 0 \text{ and } T = 1 \text{ at } y = \pm 1 \tag{7}$$

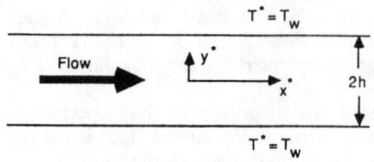

Fig. 1: Parallel plate geometry

CONSTANT THERMAL CONDUCTIVITY AND VISCOSITY

The simplest solution is, of course, for constant thermal conductivity and viscosity with which we will compare the other cases. Taking $\lambda = \mu = 1$, the solutions of equations (4) and (5) are

$$u = \frac{1}{2}(1 - y^2) \tag{8}$$

and

$$T = 1 + \frac{\alpha}{12}(1 - y^4) \tag{9}$$

The main effect of the viscous heating is near the centerline $y = 0$.

CONSTANT VISCOSITY AND LINEARLY VARYING THERMAL CONDUCTIVITY

Usually the variation of the thermal conductivity is neglected. However, there are some interesting results to be obtained if we consider the viscosity to be constant and the thermal conductivity linearly varying with temperature. So in this section we take $\mu = 1$ and $\lambda = 1 + aT$. Gases normally have a positive a, while there are many liquids with a negative a over a certain temperature range. Equations (4) and (5) then become

$$\frac{d}{dy}\left(\frac{du}{dy}\right) = -1 \tag{10}$$

and

$$\frac{d}{dy}\left\{(1 + aT)\frac{dT}{dy}\right\} + \alpha \left(\frac{du}{dy}\right)^2 = 0 \tag{11}$$

The momentum equation (10), now decoupled from the energy equation (11), has a solution given by equation (8). We introduce this velocity field in the energy equation, solve and determine the constants of integration from the temperature boundary conditions. Then we get

$$T + \frac{a}{2}T^2 = 1 + \frac{a}{2} + \frac{\alpha}{12}(1 - y^4) \tag{12}$$

For a = 0, the temperature field reduces to equation (9). Otherwise, the quadratic nature of the relation seems to indicate the possibility of two solutions. Specifically, let us consider the centerline temperature T_m at $y = 0$ given by

$$T_m^2 + \frac{2}{a}T_m - (1 + \frac{2}{a} + \frac{\alpha}{6a}) = 0 \tag{13}$$

Real solutions are not obtained in the (α, a) region shown shaded in Fig. 2. As an example, the real solutions for $\alpha = 1$ are indicated in Fig. 3. It can be observed that for $-3/2 < a < -2/3$, no steady state temperature distribution is possible. Outside of this range, for each value of a there are two real roots for T_m. The solutions with negative temperature or λ are not physical. It is more important to note that the temperature distributions corresponding to values indicated by broken lines do *not* satisfy the thermal boundary conditions at the walls. In other words, for $y = \pm 1$ in equation (12) we have $T = 1$ and $-(1 + 2/a)$ as the two solutions of the quadratic equation, of which the latter must be discarded. Thus it turns out that if a solution exists, it is unique.

The one possible solution in the left branch of Fig. 3 has a centerline temperature $T_m < 1$. Due to the strong decrease in thermal conductivity with temperature, viscous heating reduces the temperature of the fluid. The other branch has $T_m > 1$ with the fluid temperature higher than that of the wall. The qualitative difference between these two temperature profiles is illustrated in Fig. 4 for the two turning points $a = -3/2$ and $-2/3$ of Fig. 3.

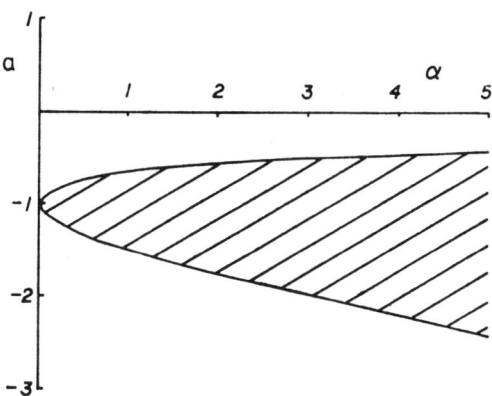

Fig. 2: Steady solutions for variable thermal conductivity not possible in shaded region

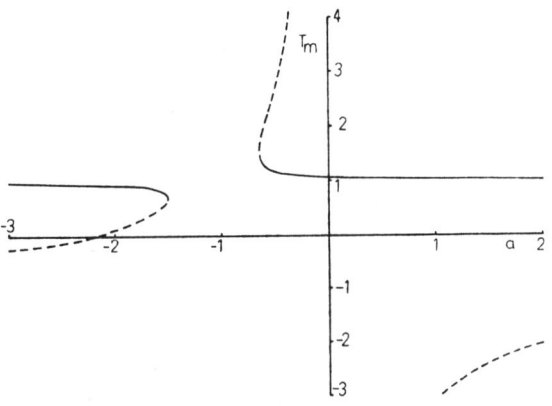

Fig. 3: Centerline temperatures for variable thermal conductivity

CONSTANT THERMAL CONDUCTIVITY AND EXPONENTIALLY VARYING VISCOSITY

For this case we know that two steady states have been found for Couette flows and cylindrical Poiseuille flow. We take

$$\lambda = 1 \tag{15}$$

and

$$\mu = \exp\{-\beta(T-1)\} \tag{16}$$

Such an exponential decrease of viscosity with temperature is characteristic of many liquids. Substituting in the basic equations, we get

$$\frac{d}{dy}\left(e^{-\beta(T-1)}\frac{du}{dy}\right) = -1 \tag{17}$$

and

$$\frac{d^2T}{dy^2} + \alpha\, e^{-\beta(T-1)}\left(\frac{du}{dy}\right)^2 = 0 \tag{18}$$

Equation (17) can be integrated once to give

$$\frac{du}{dy} = (A - y)\, e^{\beta(T-1)} \tag{19}$$

where A is a constant. For the moment we will not invoke symmetry of the velocity field to set A = 0. The controlling parameters of the system are a and b.

NUMBER OF SOLUTIONS

Equations (18) and (19) with boundary conditions (7) constitute a two-point nonlinear boundary value problem. In what follows a fourth order Runge-Kutta method with an integration step of 0.01 is used to obtain numerical solutions. Only two of the boundary conditions, those at y = −1, can be used to start the integration, and the other two at y = 1 must be satisfied by the shooting technique. Letting dT/dy = B at y = −1, the value of the constants A and B must be varied until the other two boundary conditions at y = 1 are satisfied. For parameter values α = β = 1, Fig. 5 shows the two curves representing the locus of (A,B) such that T(1) = 1 and u(1) = 0 respectively. At the intersection of the two curves, both conditions at y = 1 are satisfied. This happens at two points, both with A = 0. Thus there are two solutions to the boundary value problem, each one being symmetrical about the centerline.

Defining θ = T − 1, the two temperature profiles represented by θ(y) are shown in Fig. 6, the smaller temperature being indicated by a subscript 1 and the larger by subscript 2. $θ_1$ is multiplied by a factor of ten to enable it to be seen clearly on the graph. It is observed that there is a large temperature rise in solution 2 due to heat generation, especially near the center of the channel. The corresponding velocity profiles are shown in Fig. 7. Again, for the second solution the velocity at the center is large.

The variation of the overall temperature and flow characteristics with change in the two parameters a and b can be analyzed. We define the volumetric flow rate ratio Q, the average temperature T_a and the bulk temperature T_b by

$$Q = \frac{3}{2}\int_{-1}^{1} u(y)\, dy \tag{20a}$$

$$T_a = \frac{1}{2}\int_{-1}^{1} T(y)\, dy \tag{20b}$$

$$T_b = \frac{\int_{-1}^{1} u(y)T(y)\, dy}{\int_{-1}^{1} u(y)\, dy} \tag{20c}$$

Without viscous heating (i.e. with α = 0), we would have Q = 1, $T_b = T_a = 1$. Figure 8 shows the variation of these quantities as a function of β with α = 1. The lower branch corresponds to solution 1 while the upper branch is solution 2. Solutions are found to exist only for β < 5.7. As β → 0, the flow rate and temperatures for solution 2 become unbounded while solution 1 tends to the constant viscosity case. Figure 9 shows a similar variation with respect to α, keeping β = 1.

There is some similarity in the α and β dependence of the results of Figs. 8 and 9 which can be demonstrated by a suitable change of variables. Writing

$$\beta\,\theta(y) = \phi(y) \tag{21}$$

for nonzero β, equations (18) and (19) can be simplified to

$$\frac{d^2\phi}{dy^2} + \gamma\, y^2\, e^\phi = 0 \tag{22}$$

where γ = αβ, with the boundary conditions dφ/dy = 0 at y = 0, and φ = 0 at y = 1 (on using the symmetry condition A = 0). We can see that α and β have similar roles in the determination of φ even though θ from equation (21), and hence the other flow quantities, may be different on varying α and β with γ constant.

269

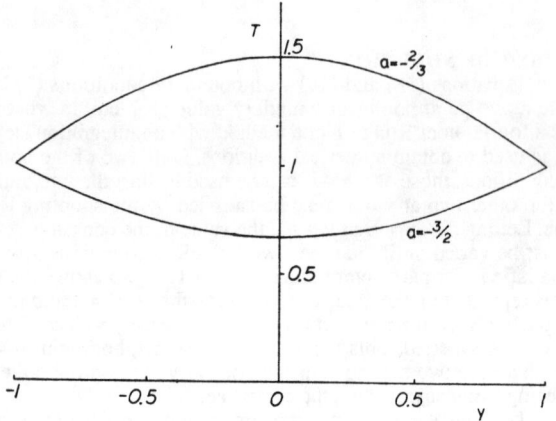

Fig. 4: Two temperature profiles for variable thermal conductivity

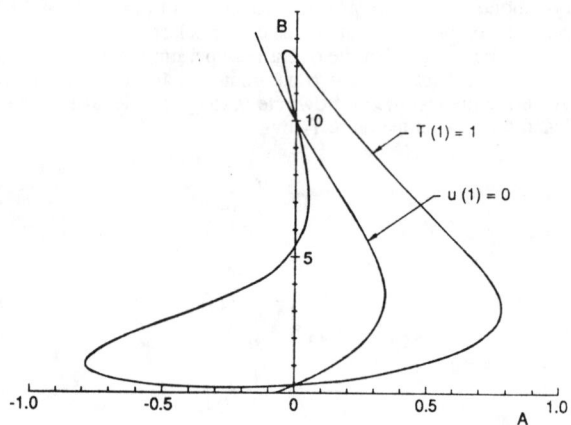

Fig. 5: A,B values such that T(1) = 1 and u(1) = 0.
Parameters $\alpha = \beta = 1$

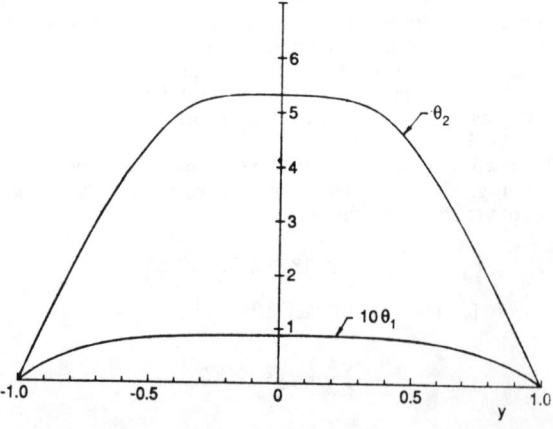

Fig. 6: Two θ profiles with parameters $\alpha = \beta = 1$

PERTURBATION ANALYSIS OF LOWER SOLUTION

The lower solution can be conveniently analyzed through a regular perturbation series. Assuming $\alpha \ll 1$, we can expand

$$T = T_0 + \alpha T_1 + \alpha^2 T_2 + \ldots \tag{23a}$$
$$u = u_0 + \alpha u_1 + \alpha^2 u_2 + \ldots \tag{23b}$$

These are substituted into equations (17) and (18) to obtain the following sequence of equations:

Order α^0:
$$\frac{d}{dy}\left\{\exp[-\beta(T_0 - 1)]\frac{du_0}{dy}\right\} = -1 \tag{24a}$$
$$\frac{d^2 T_0}{dy^2} = 0 \tag{24b}$$

Order α^1:
$$\frac{d}{dy}\left\{-\beta T_1 \frac{du_0}{dy} + \frac{du_1}{dy}\right\} = 0 \tag{24c}$$
$$\frac{d^2 T_1}{dy^2} + \left(\frac{du_0}{dy}\right)^2 = 0 \tag{24d}$$

Order α^2:
$$\frac{d}{dy}\left\{\frac{du_2}{dy} - \beta T_1 \frac{du_1}{dy} - \beta\left(T_2 - \frac{\beta}{2}T_1^2\right)\frac{du_0}{dy}\right\} = 0 \tag{24e}$$
$$\frac{d^2 T_2}{dy^2} + 2\frac{du_0}{dy}\frac{du_1}{dy} - \beta T_1\left(\frac{du_0}{dy}\right)^2 = 0 \tag{24f}$$

and so on. The leading terms of the expansions satisfy the boundary conditions while the higher order terms are zero at $y = \pm 1$. After some algebra carried out with the help of the symbolic manipulation program REDUCE, the following velocity and temperature fields are obtained

$$u = \frac{1}{2}\left(1 - y^2\right) + \alpha\beta\frac{1}{36}\left\{1 - \frac{3}{2}y^2 + \frac{1}{2}y^6\right\}$$
$$+ \alpha^2\beta^2\frac{1}{378}\left\{1 - \frac{27}{16}y^2 + \frac{7}{8}y^6 - \frac{3}{16}y^{10}\right\} + \ldots \tag{25}$$

$$T = 1 + \frac{\alpha}{12}\left(1 - y^4\right) + \alpha^2\beta\frac{11}{2016}\left\{1 - \frac{14}{11}y^4 + \frac{3}{11}y^8\right\} + \ldots \tag{26}$$

From these relations, using definitions (20), we can deduce the corresponding volumetric flow rate ratio and the average and bulk temperatures

$$Q = 1 + \frac{1}{21}\alpha\beta + \frac{1}{231}\alpha^2\beta^2 + \ldots \tag{27}$$

$$T_a = 1 + \frac{1}{15}\alpha + \frac{4}{945}\alpha^2\beta + \ldots \tag{28}$$

$$T_b = 1 + \frac{8}{105}\alpha + \frac{122}{24255}\alpha^2\beta + \ldots \tag{29}$$

up to order α^2. Due to viscous heating, the temperature is seen to increase, especially near the centerline. As a consequence the viscosity decreases and the flow rate increases. However, the net effect is small, the first correction to the bulk and average temperature being of the order of about 7% of α and is independent of β. For the volumetric flow rate the temperature rise of the fluid is only important in so far as it affects the viscosity. The first correction of the flow rate is thus a function of β, being about 5% of the product $\alpha\beta$.

If, on the other hand, instead of small α we assume that $\beta \ll 1$, the perturbation analysis gives

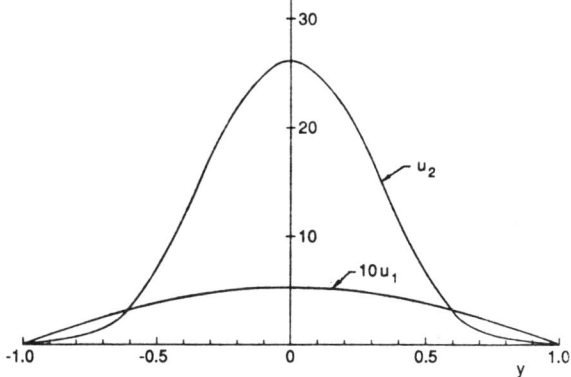

Fig. 7: Two velocity profiles with parameters $\alpha = \beta = 1$

$$u = \frac{1}{2}\left(1 - y^2\right) + \alpha\beta\frac{1}{36}\left\{1 - \frac{3}{2}y^2 + \frac{1}{2}y^6\right\}$$
$$+ \alpha^2\beta^2\frac{1}{378}\left\{1 - \frac{27}{16}y^2 + \frac{7}{8}y^6 - \frac{3}{16}y^{10}\right\} + \ldots \quad (30)$$

$$T = 1 + \frac{\alpha}{12}\left(1 - y^4\right) + \alpha^2\beta\frac{11}{2016}\left\{1 - \frac{14}{11}y^4 + \frac{3}{11}y^8)\right\}$$
$$+ \alpha^3\beta^2\frac{71}{133056}(1 - \frac{99}{71}y^4 + \frac{33}{71}y^8 - \frac{5}{71}y^{12}) + \ldots \quad (31)$$

$$Q = 1 + \frac{1}{21}\alpha\beta + \frac{1}{231}\alpha^2\beta^2 + \ldots \quad (32)$$

$$T_a = 1 + \frac{1}{15}\alpha + \frac{4}{945}\alpha^2\beta + \frac{166}{405405}\alpha^3\beta^2 + \ldots \quad (33)$$

$$T_b = 1 + \frac{8}{105}\alpha + \frac{122}{24255}\alpha^2\beta + \frac{9809}{19864845}\alpha^3\beta^2 + \ldots \quad (34)$$

to order β^2. There is agreement where the two expansions overlap, though the latter has an additional term of order $\alpha^3\beta^2$ in the temperatures.

Quantitative comparison between the analytical and numerical results is made in Fig. 10 for $\alpha = 1$. The exact numerical result of Q and T_b is shown with continuous lines and the corresponding approximate relations (32) and (34) with broken lines. The average temperature T_a gives similar results and is not shown. The small β expansion is found to give reasonable values even for values of β as large as 2. The first order correction in β is an order of magnitude smaller for T_b than it is for Q, so that the error in the bulk temperature is seen to be generally smaller than that in the volumetric flow rate.

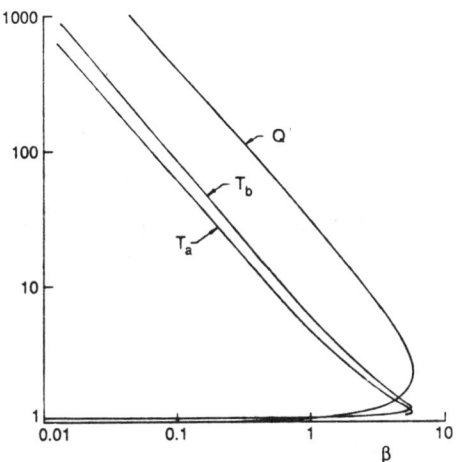

Fig. 8: Variation of Q, T_a and T_b with β. Parameter $\alpha = 1$

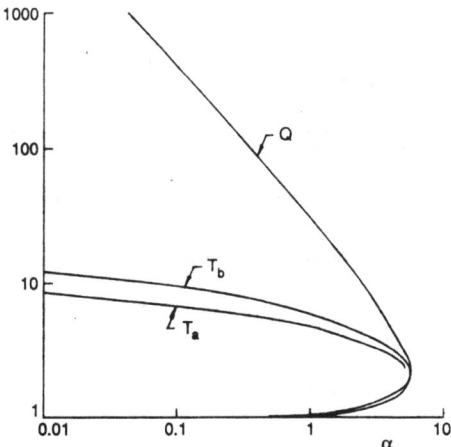

Fig. 9: Variation of Q, T_a and T_b with α. Parameter $\beta = 1$

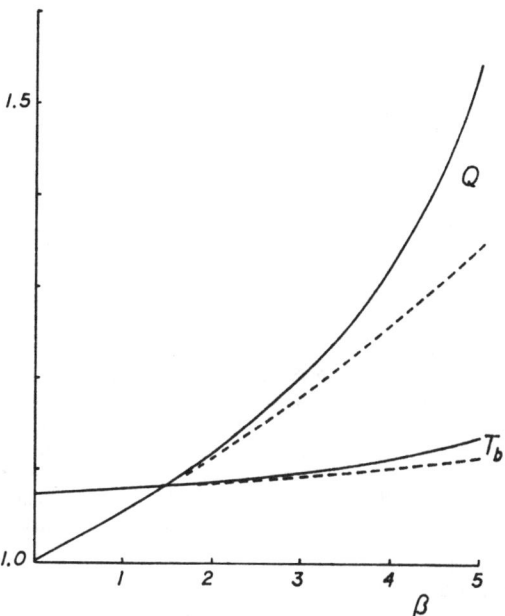

Fig. 10: Numerical (continuous lines) and perturbation (broken lines) results for Q and T_b. Parameter $\alpha = 1$

Weighted Residual Analysis of Upper Solution

The upper solution is more difficult to approximate through a perturbation analysis since it tends to infinity as either one of the parameters α or β tends to zero from above. Here we have chosen a weighted residual method based on moments for the purposes of approximation, the purpose being to find the large temperature solution for small values of the parameter α or β. Thus we attempt to solve equation (22) approximately for small γ. The simplest trial function similar in shape to the exact profiles and which fits the boundary conditions is a parabola. With this in mind we assume that

$$\phi(y) = \phi_m (1 - y^2) \qquad (35)$$

Equating the zeroth moment of the residual of equation (22) to zero, we have

$$2\,\phi_m - \gamma \exp(\phi_m) \int_0^1 y^2 \exp(-\phi_m\, y^2)\, dy = 0 \qquad (36)$$

For large ϕ_m, the integral

$$\int_0^1 y^2 \exp(-\phi_m\, y^2)\, dy$$

can be approximated by

$$\int_0^\infty y^2 \exp(-\phi_m\, y^2)\, dy$$

the difference being about $1.6 \times 10^{-2}\%$ for $\phi_m = 10$. This integral can be equated to $\frac{1}{4\phi_m}\sqrt{\frac{\pi}{\phi_m}}$ (Abramowitz and Stegun, 1964), so that

$$\frac{8}{\sqrt{\pi}}\phi_m^{5/2} - \gamma \exp(\phi_m) = 0. \qquad (37)$$

For $\gamma < 3.7$, there are two solutions of this equation (compare this with the critical value of 5.7 for the exact equation (22)). Only the larger one is applicable because of the large ϕ_m assumption made. It gives an estimate of the peak temperature ϕ_m, from which the other overall quantities can be derived. Figure 11 shows a comparison between the exact numerical solution of equation (22) and that of the approximate equation (37). The qualitative features are the same, and even the quantitative comparison is good.

Conclusions

The steady state problem of plane Poiseuille flow with viscous heating and variation of properties with temperature has been considered. For constant viscosity and linearly varying thermal conductivity it is seen that steady state solutions do not exist under certain circumstances, otherwise it is unique. On the other hand, for constant thermal conductivity and exponentially decreasing viscosity, there are two steady states within a certain parameter range. Outside this range there are no steady solutions once again. The present study has been directed towards the existence of the solutions rather than their stability. The linear stability analysis will be taken up in the future. For the present it seems appropriate to remark that previous authors working on different geometries have found parts of both solutions to be stable.

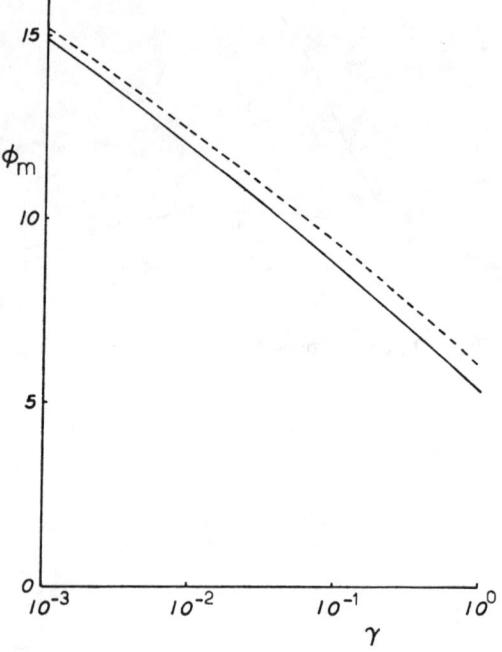

Fig. 11: Numerical (continuous line) and approximate (broken line) results for ϕ_m.

References

Abramowitz, M. and Stegun, I.A., 1964, *Handbook of Mathematical Functions,* Dover Publ., New York

Gavis, J. and Laurence, R.L., 1968, "Viscous Heating in Plane and Circular Flow between Moving Surfaces", *I&EC Fund.* **7**, 232-239

Ho, T.C., Denn, M.M. and Anshus, B.E., 1977, "Stability of Low Reynolds Number Flow with Viscous Heating", *Rheol. Acta* **16**, 61-68

Joseph, D.D., 1965, "Stability of Frictionally-Heated Flow", *Phys. Fluids* **8**, 2195-2200

Kearsley, E.A., 1962, "The Viscous Heating Correction for Viscometric Flows", *Trans. Soc. Rheology* **6**, 253-261

Nihoul, J.C.J., 1971, "Non-Linear Channel Flows with Temperature Dependent Viscosity", *Ann. Soc. Sci. Bruxelles* **85**, I, 18-28.

Platten, J.K. and Legros, J.C., 1984, *Convection in Liquids*, Springer-Verlag, Berlin.

Sukanek, P.C., 1971, "Poiseuille Flow of a Power-Law Fluid with Viscous Heating", *Chem. Engng. Sci.* **26**, 1775-1776

Sukanek, P.C. and Laurence, R.L., 1972, "The Uniqueness of Solutions for Poiseuille Flow with Viscous Heating", *Ann. Soc. Sci. Bruxelles* **86**, II, 201-209

Sukanek, P.C. and Laurence, R.L., 1974, "An Experimental Investigation of Viscous Heating in Some Simple Shear Flows", *AIChE J.* **20**, 474-484

Vanderborck, G. and Platten, J.K., 1977, "Stabilité Hydrodynamique de l'Ecoulement de Poiseuille Cylindrique incluant la Dissipation Visqueuse", *Lett. Heat Mass Transfer* **4**, 453-463

Vanderborck, G., Platten, J.K. and Cornet, P., 1979, "Stabilité Hydrodynamique de l'Eloulement de Poiseuille Cylindrique incluant la Dissipation Visqueuse, II. Influence des Fluctuations de Temperature, *Lett. Heat Mass Transfer* **6**, 83-91

1989 National Heat Transfer Conference
HTD-Vol. 107, Heat Transfer in Convective Flows

INFLUENCE OF APERTURE HEIGHT AND WIDTH
ON INTERZONAL NATURAL CONVECTION
IN A FULL-SIZE, AIR-FILLED ENCLOSURE

C. R. Boardman III and A. Kirkpatrick
Mechanical Engineering Department
Colorado State University
Fort Collins, Colorado

R. Anderson
Solar Energy Research Institute
Golden, Colorado

ABSTRACT

The topic of this paper is the influence of aperture height and width on interzonal high Rayleigh number natural convection heat transfer. Experiments were conducted in an eight-foot, air-filled cube divided into two zones by a vertical partition which was centered between a constant flux hot wall and an isothermal cold wall. The partition was configured to form doorway-like apertures. The aperture height relative to test cell height ranged from 1/8 to 1 and the aperture width relative to test cell width ranged from 0.009 to 1. The zone to zone temperature difference and the overall Nusselt number were determined experimentally, and correlated with the overall Rayleigh number, aperture and enclosure geometry, using a series resistance model for the enclosure. A turbulent boundary layer resistance was used to represent the hot and cold wall boundary layer flow, while an orifice resistance was used to represent the aperture flow. For flux Rayleigh numbers between $5*10^{11}$ and $5*10^{12}$, the enclosure Nusselt numbers ranged between 15 and 165 with a strong dependence on aperture height.

NOMENCLATURE

a	Zone temperature at mid-height of doorway
A	Wall Nusselt-Rayleigh number correlation coefficient
Ap	Dimensionless aperture area, hw/HW
b_h, b_c	Temperature stratification in hot and cold zones
B, C, D, F	Coefficients used to fit model to data
c_p	Specific heat at constant pressure
dT	Zone to zone temperature difference, $T_h - T_c$
DT	Wall to wall temperature difference, $T_H - T_C$
Eu	Euler number, $Eu \cong 0.6$
g	Gravitational constant
h	Height of aperture
h_T	Convection heat transfer coefficient
H	Height of enclosure, 8 feet (2.44 m)
k	Thermal conductivity
L	Length of enclosure, 8 feet (2.44 m)

Nu	Nusselt number, $h_T H/k$, $Q/kW\Delta T$
Pr	Prandtl number, ν/α
q"	Heat transfer rate per unit area
Q	Heat transfer rate
R	Thermal resistance
Ra	Rayleigh number, $g\beta\Delta T H^3/\alpha\nu$
Ra^*	Flux Rayleigh number, $g\beta q'' H^4/\alpha\nu k$, $Nu * Ra$
T	Temperature
T_c	Temperature of cold zone at mid-height of aperture
T_C	Temperature of isothermal cold wall
T_h	Temperature of hot zone at mid-height of aperture
T_H	Characteristic temperature of constant flux hot wall
u	Velocity in horizontal or x direction
w	Width of aperture, variable
W	Width of enclosure, 8 feet (2.44 m)
z	Distance from mid-height of doorway
α	Thermal diffusivity, k/c_p
β	Coefficient of volume expansion
ΔT	Characteristic temperature difference
ν	Kinematic viscosity
ρ	Density

INTRODUCTION

The goal of this research is to experimentally determine the effect of simple geometric changes in height and width to a doorway-like aperture in an enclosure on the interzonal natural convection within the enclosure. Natural convection flow in a building is usually characterized as a high Rayleigh number flow in a complex geometry, so that the convective transport is governed by both wall boundary layer and aperture effects. This work extends previous research on natural convection in buildings by taking into account the far wall boundary layers of a full-size, air-filled enclosure, and developing heat transfer correlations for a wide range of aperture heights and widths.

A review of research related to natural convection heat transfer in buildings is given in Anderson (1986) and Barakat (1987). Significant early work was done on natural convection in partitioned enclosures by Brown and Solvason (1962). They examined the heat transfer through an aperture resulting from a zone to zone bulk density difference across the aperture. They also carried out experiments in an air-filled eight foot cube divided in half by a vertical partition. Recently, air velocities and temperatures have been measured in the doorways of passive solar houses, to determine the convective transport from the sunspace to the rest of the building, for example, see Balcomb and Jones (1985), Mahajan (1986), and Hill, Kirkpatrick, and Burns (1986). The correlation of the thermal transport with the zone temperatures is accomplished with the aid of a bulk density transport model which uses an empirically determined discharge coefficient or Euler number, Eu. The interzonal heat transfer measured by Hill et al. (1986) was successfully correlated with the zone temperature differences using a aperture Euler number of 0.7. Recent experiments by Mahajan (1986) indicate that the aperture Euler number can have a range from 0.45 to 0.75 depending on aperture flow rate, which is a complex function of aperture size, enclosure geometry, and zone to zone temperature difference. However, in all of these full scale studies, the relationship between the wall temperatures and the zone temperatures were not studied, so it was not possible to develop a an overall wall to wall heat transfer correlation equation.

Other investigators have used scale models with water as the working fluid and correlated the cross cavity transport with the enclosure wall temperatures, for example, see Bauman et al. (1980), Nansteel and Greif (1984), and Lin and Bejan (1983). Using wall boundary layer scaling, Lin and Bejan (1983) correlated their results with $Nu = 0.336 Ra^{1/4}/(Ap^{-3/4} + 0.5)$. The aperture size, Ap, was included in this correlation due to the wall boundary layer scaling of both h and H when separation was taken into account. However, due to relatively large aperture areas, the aperture resistances were negligible in the above experiments. Scott, Anderson, and Figliola (1987) performed experiments using a scale model with small apertures, so that both boundary layer flow and bulk density flow would occur. The experimental results indicated that boundary layer flow would dominate for high Rayleigh numbers and relatively large apertures, with separation of the flow from the hot wall at half-height. For small aperture sizes the bulk density flow regime dominated. For this case the boundary layer flow could not travel through the doorway without a significant rise in zone to zone temperature difference needed to overcome the increased aperture resistance. Neymark, Boardman, Kirkpatrick and Anderson (1988) conducted experiments in both scale model water-filled and full size air-filled enclosures with a fixed half height partition. A decrease in the Nusselt number and increase in the zone to zone temperature difference indicated the onset of boundary layer flow blockage. Again, separation of the boundary layer flow along the hot wall at a height approximately equal to the doorway height was observed in both water and air, leaving the top section of fluid in the hot zone trapped.

This paper extends the full-size air cell work of Neymark et al. (1988) to include the effects of aperture height as well as width, as shown in Figure 1, and develops a simple model to correlate the influence of Rayleigh number, aperture height, and aperture width on the Nusselt number and zone to zone temperature difference of the enclosure. In the present study, the hot wall has a constant flux boundary condition, while the cold wall is isothermal, as shown in Figure 2. These boundary conditions were chosen to closely match a solar irradiated hot wall and the cold wall of a massive building interior. The exact correlations which will be presented are accurate only for the cubic geometry studied (aspect ratio of one), and

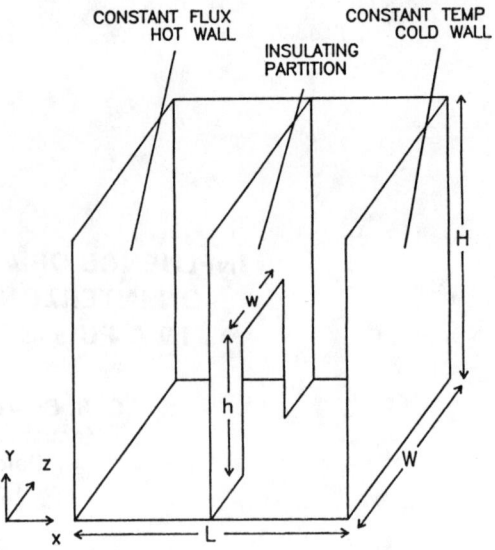

Figure 1. Representative Geometry.

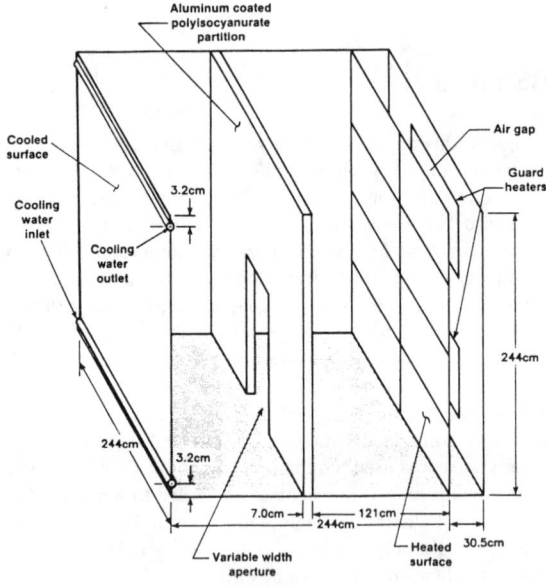

Figure 2. Schematic of test cell.

for the range of flux Rayleigh numbers investigated, but the modeling approach should be useful outside this range. An application for the research in this paper can be seen by reference to a standard computer program, SUNCODE, to model the heat transfer characteristics of passive solar buildings, DeLahunt (1985). This program uses a simple user-specified resistance to model the interzonal convection between two zones, with little guidance provided for choosing the resistance. There is a need to accurately model the aperture resistance for improved calculation of the interzonal heat transfer due to convection, and to be able to base the interzonal flow on enclosure internal surface temperatures. Another application is in the development of passive solar design guidelines for boundary layer convection heat transport through doorways. In seeking to provide guidelines for bulk density driven transport, Balcomb and Jones (1988) recommend a doorway area of 15% of the glazing area for an 80 inch (2 meter) high door between a sunspace and cold interior room. The results from our research indicate that a smaller

274

doorway can be chosen if boundary layer flow is used for the heat transport.

CORRELATION EQUATIONS

There are three significant resistances to high Rayleigh number natural convection heat transfer in a differentially heated partitioned enclosure: the hot wall boundary layer resistance, the cold wall boundary layer resistance, and the aperture resistance. If the resistances are assumed to be connected in series as shown in Figure 3, scaling relationships for enclosure heat transfer and temperatures can be developed. The wall to wall temperature difference, $DT = T_H - T_C$, or the enclosure Rayleigh number Ra, is assumed known, and the resulting enclosure Nusselt number, Nu, and zone to zone temperature difference, dT, are to be determined. The hot and cold wall heat transfer are modeled using the empirical correlation of Churchill and Chu (1975): $Nu_H = A Ra_H^{1/3}$, where A is a constant, for turbulent boundary layer vertical plate heat transfer. Olson, et al. (1986), and Cheesewright and Zial (1986), have found that the boundary layer flow in air-filled enclosures is turbulent at high Rayleigh numbers ($Ra > 10^{10}$). Churchill and Chu (1975) have shown that the average Nusselt number dependence on Rayleigh number is the same for both constant flux and isothermal walls, so the same equation can be used to model both the hot and cold walls. For isothermal walls, A=0.107, which is the value used in this paper. An additional constant coefficient is used in the correlation equations in this paper. The wall resistances can be written as

$$R_{hw} = 1/(kWARa_h^{1/3}) \qquad (1)$$
$$R_{cw} = 1/(kWARa_H^{1/3}) \qquad (2)$$

The above Rayleigh number characteristic temperature difference is that between a wall and its zone, representing a local heat transfer resistance. The characteristic length for the cold wall is H, and for the hot wall is h, the height of the aperture, since the boundary layer separates from the hot wall at about this height. Equating the hot and cold wall heat transfer leads to a scaling relationship for cross cavity heat transfer:

$$Q_{hw} = Q_{cw}$$
$$(T_H - T_h)/R_{hw} \approx (T_c - T_C)/R_{cw} \qquad (3)$$

which upon substitution of equations (1) and (2) yields:

$$T_c - T_C \approx \frac{DT - dT}{1 + (H/h)^{3/4}} \qquad (4)$$

The aperture heat transfer is modeled using a stratified bulk density flow equation developed by Jones and Otis (1986). The velocity profile through the aperture is assumed to be parabolic:

$$u(z) = \text{Eu}(2gz\beta(T_h - T_c)^{1/2}, \qquad (5)$$

where z is the vertical distance in upward or downward direction from the mid- height of the aperture. The Euler number, Eu, for apertures is about $\pi/(\pi + 2)$, Rouse (1945), which is the value used in this paper. This number is modified by a constant coefficient in the final correlation equations. The temperature stratification in each zone is assumed linear and is of the form $T = a + bz$, where a is the temperature of the appropriate zone, T_h or T_c, and b is the stratification. Integration of the velocity and temperature profiles from z=0 to h/2 and to -h/2 respectively results in the aperture heat transfer:

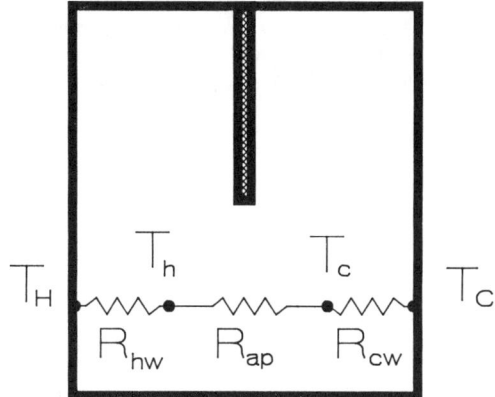

Figure 3. Schematic of resistance model.

$$Q_{ap} = (\text{Eu}/3)c_p\rho wh^{3/2}(g\beta dT)^{1/2}[dT + .3h(b_h + b_c)] \qquad (6)$$

If we now assume that the stratification scales linearly with the overall length scales and temperature differences, then $b_h + b_c = F * (DT/h + DT/H)$, where F is an undetermined constant. An equation for the dimensionless zone temperature difference, dT/DT, can be obtained by equating the aperture and cold wall heat transfer:

$$Q_{ap} = Q_{cw} \qquad (7)$$

Upon substitution of equations (2), (4), and (6), and subsequent rearrangement, a nonlinear equation of the form $f(dT/DT) = 0$ results:

$$(dT/DT)^{3/2} + 0.3F(1 + h/H)(dT/DT)^{1/2} -$$
$$\frac{ARa^{1/3}[(1 - dT/DT)/(1 + (H/h)^{3/4}]^{4/3}}{(\text{Eu}/3)(w/W)(h/H)^{3/2}(PrRa)^{1/2}} = 0 \qquad (8)$$

The dimensionless heat transfer across the enclosure is

$$Nu = \frac{Q}{kW \, DT} \qquad (9)$$

which in resistance form is:

$$Nu = \frac{1}{kW(R_{hw} + R_{ap} + R_{cw})} \qquad (10)$$

The two wall resistance terms can be combined to give:

$$R_w = R_{hw} + R_{cw}$$
$$= \frac{(1 + (h/H)^{3/4})^{4/3}}{kWA(h/H)(1 - dT/DT)^{1/3}Ra^{1/3}} \qquad (11)$$

The aperture resistance term is:

$$R_{ap} = 1/[kW(Eu/3)(w/W)(\frac{h}{H})^{3/2}(PrRa)^{1/2}$$
$$(dT/DT)^{1/2}(1 + 0.3F(DT/dT))(1 + (h/H))] \qquad (12)$$

The characteristic length and temperature for the Rayleigh number are the overall height and temperature, H and DT respectively, for equations (8), (11) and (12). Since the Nusselt number is a function of the dimensionless zone temperature difference, the solution of equation (8) for dT/DT is required for determination of the overall Nusselt number. With the different aperture and wall velocity, length, and temperature scales indicated by the scaling relationships, the enclosure Nusselt number cannot be expressed by a simple power law function of Rayleigh number and geometry.

The series resistance scaling equations, equations (8) and (10), will be used as a template for the correlation of experimental results. The dimensionless temperature will be fit with two coefficients: B, a leading coefficient, and F, the stratification coefficient:

$$\left(\frac{dT}{DT}\right)_{measured} = B \left(\frac{dT}{DT}\right)_{model} \qquad (13)$$

The enclosure Nusselt number will be fit with two coefficients: C, a wall coefficient, and D, an aperture coefficient:

$$Nu_{measured} = \frac{1}{kW(C*R_w + D*R_{ap})} \qquad (14)$$

As the aperture height and width decrease, the zone temperature difference dT/DT will increase and the Nusselt number will decrease, both in a nonlinear fashion, due to the aperture height dependence of both the wall and aperture terms.

DESCRIPTION OF EXPERIMENT

An existing test cell was used for the measurement of velocity and temperature profiles, and enclosure heat transfer. The test cell was a 2.44 m (8 foot) cube with a vertical partition extending from the ceiling centered between the hot and cold walls. The partition was configured to form doorway- like apertures of variable height and width. A table of the geometries tested is provided in Table 1, giving the relative height and width, h/H and w/W, and the dimensionless aperture area, Ap. All interior walls, except the hot and cold walls, were constructed of 2.54 cm (1 inch) Celotex, a poly-isocyanurate foam board, and covered with aluminum to minimize radiation exchange. The exterior walls were further insulated with fiberglass to minimize external heat loss. The partition was formed with two layers of Celotex separated by a 2 cm air gap. The cold wall temperature was maintained within 0.5 C from inlet to outlet by running tap water through modified copper solar collector absorber plates. The constant flux hot wall used eight graphite based electrical resistance heaters for input power distribution. A separately heated air gap on the outside of the hot wall was used to minimize direct losses from the hot wall to the environment. The different temperature stratifications in the air gap and on the hot wall resulted in less than 7 percent variation in local heat flux from the hot wall. The power supplied to the hot wall, which varied between 60 and 400 watts, was measured with a volt- ammeter to better than 2 percent accuracy.

A heat balance applied to the hot zone was used to indirectly determine the convection heat transfer across the enclosure. The input power was measured and the conduction losses and radiation heat transfer were calculated. Conduction losses and effective thermal conductivity were empirically determined. Temperatures in the enclosure walls and in the air were measured using 79 copper-constantan (Type-T) thermocouples, calibrated to within 0.2 C. The thermocouple voltages were read by a HP3497A data acquisition unit controlled by a computer which subsequently calculated the envelope, air, and partition temperatures. Steady state conditions, needed to apply the heat balance, were usually obtained after a twenty-four hour period, and were determined by checking for negligible change (less than 0.05 C in one hour) in the hot wall and air gap temperatures. Radiation exchange with the cold zone was approximated for each test point using a three-dimensional exchange factor Monte Carlo method, MONTE3D, developed by Burns and Maltby (1986). The absorptivity of the Celotex aluminum surfaces was 0.04 ± 0.02, and the black chrome surface of the cold wall was 0.11 ± 0.02, as measured by a Gier-Dunkle D9-100 infrared interferometer.

TABLE 1

Height	Width	Ap
0.125	1.0	0.125
0.125	0.04	0.005
0.25	1.0	0.25
0.25	0.40	0.10
0.25	0.10	0.025
0.25	0.04	0.01
0.333	0.40	0.133
0.50	1.0	0.50
0.50	0.20	0.10
0.50	0.10	0.05
0.50	0.04	0.02
0.50	0.02	0.01
0.50	0.009	0.0045
0.666	0.04	0.0266
0.75	0.40	0.30
0.75	0.04	0.03
1.0	0.40	0.40
1.0	0.10	0.10
1.0	0.04	0.04

The zone to zone temperature difference, dT, was measured with a shielded thermocouple in the center of each zone at a height equal to the midheight of the doorway. The average temperature of the hot wall was chosen for T_H, the characteristic temperature of the constant flux wall. It is not clear what temperature to choose to characterize constant flux walls in partitioned enclosures with variable height apertures, so the average was chosen. The fluid properties needed to calculate the flux Rayleigh number, Ra^*, and and the Nusselt number, Nu, were evaluated at a bulk temperature which was the average between the cold wall temperature and the hot wall temperature at the height of the aperture. Both the constant temperature Rayleigh number, Ra, and the constant flux Rayleigh number $Ra^* \equiv Nu*Ra$, are used in this paper.

The resulting enclosure Nusselt number had a typical uncertainty of less than 15%, but extremes of up to 27% for low Nusselt numbers. These resulted from the uncertainty in envelope and partition conduction losses of less than 8% and 12% respectively, and uncertainty in radiation exchange of 50% because of the lack of precision in measuring the absorptivity. While the above uncertainties are important because they are related to the final accuracy of the results, the precision of the data was much better. Repeated testing for the same aperture and power input yielded Nusselt numbers within 3% of each other, as indicated in Table 2.

TABLE 2

Ra^*	Nu	Accuracy	Geometry
$.135*10^{13}$	41.5	20.7%	h/H=1/8
$.136*10^{13}$	41.8	20.5%	w/W=1
$.295*10^{13}$	97.0	11.1%	h/H=1/2
$.292*10^{13}$	97.0	11.1%	w/W=1
$.352*10^{13}$	127.0	3.1%	h/H=3/4
$.354*10^{13}$	126.7	3.1%	w/W=0.04

VELOCITY MEASUREMENTS

Flow visualization and hot wire anemometer measurements were used to confirm turbulent boundary layer flow along the hot wall. Consequently, turbulent flow was assumed for the correlation model. Further details on the flow visualization and anemometer measurements can be found in Neymark et al. (1989) and Boardman (1988). Detailed doorway velocity profile measurements were also taken. The velocities at nine heights in a half height, w/W = 0.1 doorway were sampled. using a thermistor velocity probe, Guire (1987). The uncertainty in velocity measurement was ±0.025 m/s.

Doorway velocity profiles are shown in Figure 4 with hot wall $Ra^* = 1.6 * 10^{12}$ and in Figure 5 with hot wall $Ra^* = 3.6 * 10^{12}$. The neutral height, where the pressure difference between the two zones was zero, was about 7 cm above the mid-height of the doorway. The predictions of the bulk density velocity profile equation, equation (5), for an Euler number, Eu, of 0.6 are shown as a solid line, and the experimental data as a dashed line. The upper dashed line in both plots represents the flow in the upper portion of the doorway flowing out of the hot zone. The lower dashed line represents the flow in the lower portion of the doorway flowing into the hot zone. The large jump in velocity at the top of the door in Figure 4 is probably a result of the hot wall boundary layer flow which separated from the wall and moved through the top of the doorway into the cold zone. This effect was neglected by the bulk density model which assumes the doorway velocity profile results only from zone temperature differences. Reference to Figure 5 shows that the deviation from the bulk density velocity profile prediction becomes greater as the boundary layer velocity increases because of increased wall heat flux. The drop in velocity at the bottom of the door is due to the no slip condition at the floor.

These velocity measurements were used in conjunction with seven doorway temperature measurements to directly calculate the heat flux convected through the doorway. A comparison of the indirect flux measurements with direct temperature and velocity derived flux measurements is provided in Figure 6. The uncertainty in the direct measurements was primarily the result of the fluctuations in velocity at each height. The direct measurement results shown were the average of many individual measurements taken when the test cell had reached steady state. Within the uncertainty of the measurements, Figure 6 shows the agreement of this direct heat flux calculation with the indirect calculation used to fit the correlation equations.

HEAT TRANSFER RESULTS AND CORRELATION

A total of 80 test points were obtained. The heat transfer results include two model graphs, three data graphs, and two correlation graphs. The model graphs in Figures 7 and 8 indicate how the dimensionless zone temperature, dT/DT, and the enclosure Nusselt number, Nu, depend on the dimensionless height, h/H, and width, w/W, of the aperture. The bulk temperature is assumed to be 30 C , and the overall enclosure temperature difference, DT, is assumed to be 30 C, so the enclosure Rayleigh number is $2.67 * 10^{10}$ for these graphs. Equation (8) was solved numerically for dT/DT using a Raphson-Newton method given w/W, h/H, DT, Ra, and Pr, using B=0.91 and F=0.27. The Nusselt number was calculated from (14) using C=0.83 and D=0.69. The B, C, D, and F coefficients were determined empirically from the experimental data as explained below. As shown in Figure 7, the boundary lay er regime, c haracterized

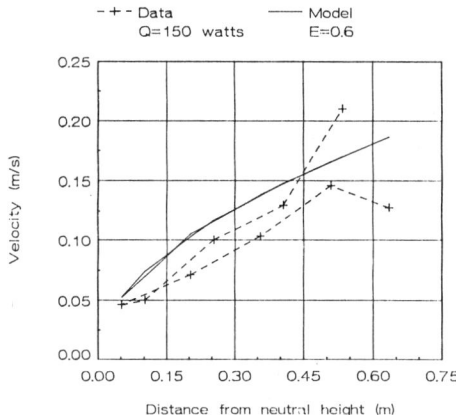

Figure 4. Doorway velocity profile for Q = 150 watts, $Ra^* = 1.6 * 10^{12}$.

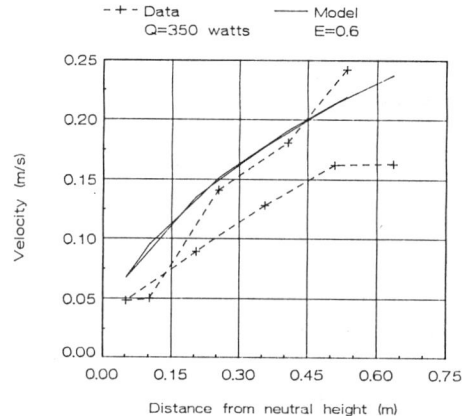

Figure 5. Doorway velocity profile for Q = 350 watts, $Ra^* = 3.6 * 10^{12}$.

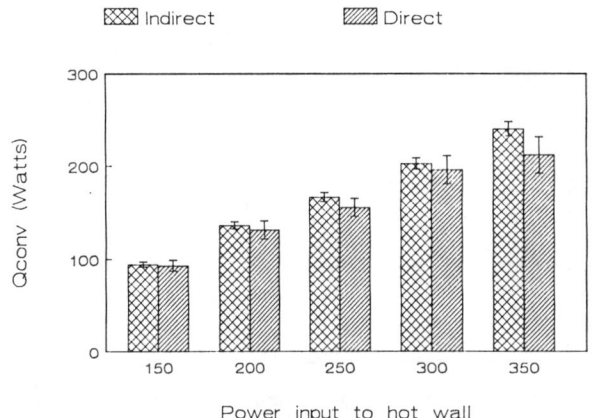

Figure 6. Comparison of methods to calculate heat convection through the aperture.

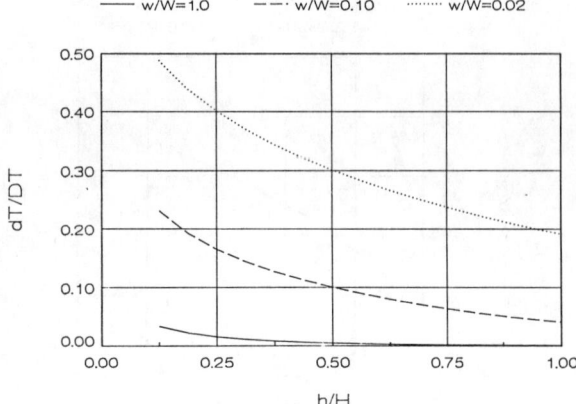

Figure 7. Model predictions for dT/DT vs. h/H and w/W at DT = 30 C.

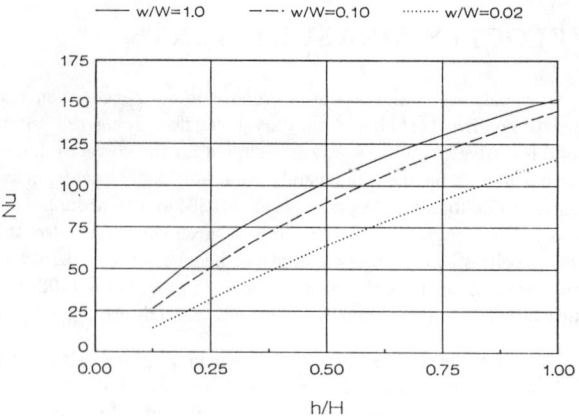

Figure 8. Model predictions for Nu vs. h/H and w/W at DT = 30 C.

here as when dT/DT is less than 0.1, dominates when $w/W > 0.1$. The same effect can be seen in Figure 8 as the Nusselt number drops significantly when $w/W < 0.1$.

The experimental results are shown in Figures 9, 10 and 11. The experimental data covers both the boundary layer dominated regime ($dT/DT < 0.1$) and the bulk density dominated regime ($dT/DT > 0.1$). Figure 9 shows a plot of Nu versus Ra^* for various aperture heights. The Nusselt number increases as the flux Rayleigh number increases and as the height increases. The effect of aperture height on the dimensionless temperature difference is shown in Figure 10 for a fixed flux Rayleigh number. The effect of aperture height on the enclosure Nusselt number is shown in Figure 11 for a fixed flux Rayleigh number. The experimental data is not plotted on the same graph as the model since that would involve interpolation of the various flux Rayleigh data points to obtain a constant temperature Rayleigh number used in the model. Instead, the data points are used to find the leading correlation coefficients B, F, C, and D in equations (13) and (14) in order to convert the scaling relationships to actual equations. The goodness of fit between the data points and the correlation predictions provides a measure for the accuracy of the model.

The correlation equations used to fit the data with the model employ leading correlation coefficients which were obtained using the ASYSTANT software package curve-fitting module. This package allows the user to define simple arbitrary functions with numerous undetermined coefficients to be fit using a least squares approach. The result of fitting the leading coefficient B in equation (13) was B=0.91 which had a goodness of fit $R^2 = .9947$, with a standard deviation of 6%. The stratification coefficient F=0.27 was found by trial and error. The excellent agreement of this correlation with the dimensionless zone temperature data is shown in Figure 12. Equation (10) is the Nusselt number scaling equation. The curve fit of equation (14) yields

$$Nu = 1/[kW(0.83 * R_w + 0.69 * R_{ap})] \qquad (15)$$

with $R^2 = 0.9865$ for a standard deviation of 13%. The data and the correlation are compared in Figure 13. The correlation tends to slightly underpredict the large Nusselt numbers and overpredict the low Nusselt numbers. This is presumably the result of physical effects neglected in this simple model. These effects could include a higher velocity through the doorway than the bulk density model

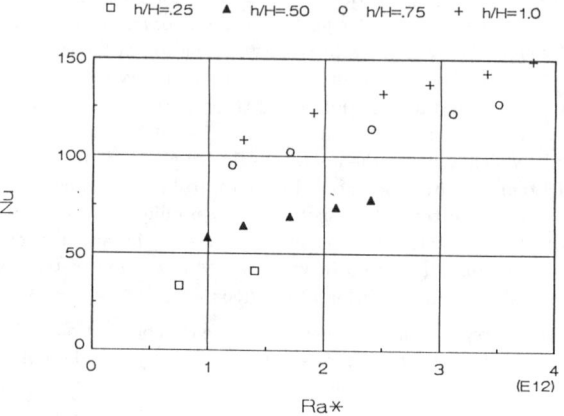

Figure 9. Nu vs. Ra^* for selected w/W = 0.04 data.

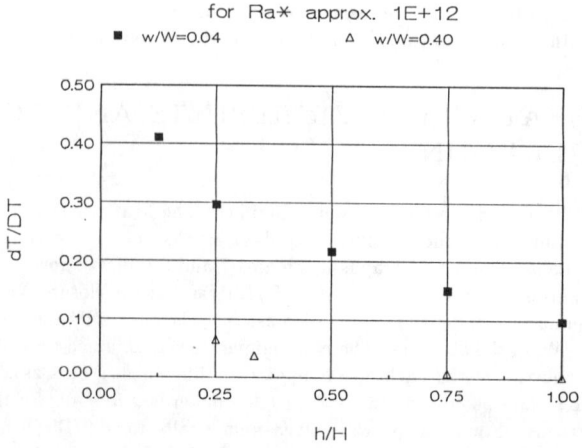

Figure 10. dT/DT vs. h/H for selected experimental data.

predicts for large wall to wall temperature differences, and the neglect of wall friction on the boundary layer flow. The accuracy of the correlation predictions will decrease as deviations from a cubic geometry with Ra^* between $5*10^{11}$ and $5*10^{12}$ increase.

SUMMARY AND CONCLUSIONS

Correlation equations, based on a simple resistance model of a differentially-heated partially-divided enclosure, have been presented which are able to predict the zone to zone temperature difference, and the enclosure Nusselt number, within the accuracy of the experimental results. The correlation structure takes into account both wall and aperture resistance effects. These results are useful for computer modeling of interzonal natural convection driven by boundary layer flow, and for designing passive solar houses.

The results of this research indicate that a 10% doorway area relative to glazing area would be sufficient for good heattransfer, given the 80 inch (2 m) high doorway for which Balcomb (1988) gives his recommendation. This research shows the relative advantage of using a vertical collecting surface for solar radiation if the designer wishes to minimize doorway area or improve convective heat transport. The bulk density flow, which results when energy is captured on the floor or far walls , is a less efficient heat transfer mechanism than boundary layer dominated flow.

ACKNOWLEDGEMENTS

This work was supported by the Solar Energy Research Institute and by funding provided to the Energy Teaching and Training Laboratory by the Office of Energy Conservation of the State of Colorado.

REFERENCES

Anderson, R., 1986, "Natural Convection Research and Solar Buildings Applications," *Passive Solar Journal*,Vol. 3, pp. 33-76.

Balcomb, J.D., and Jones, G.F., 1985, "Natural Convection Airflow and Heat Transport in Buildings: Experimental Results," *Proc. 10th Passive Solar Conf.*, Rayleigh, North Carolina, ASES.

Balcomb, J.D., and Jones R.W., 1988, "Workbook on Advanced Passive Solar Design," Balcomb Solar Associates, Santa Fe. New Mexico.

Barakat, S.A., 1987, "Inter-Zone Convective Heat Transfer in Buildings: A Review," *J. Solar Energy Eng.*, Vol. 109, pp. 71-78.

Bauman, F., Gadgil, A., Kammerud, R., and Greif, R., 1980, "Buoyancy-Driven Convection in Rectangular Enclosures: Experimental Results and Numerical Calculations," ASME paper 80-HT-66.

Boardman, C.R., 1988, "Influence of Aperture Height and Width on Interzonal Natural Convection," M.S. Thesis, Mechanical Engineering Department, Colorado State University.

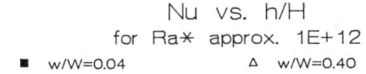

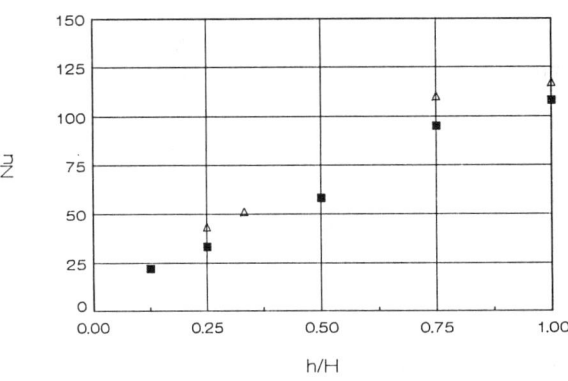

Figure 11. Nu vs. h/H for selected experimental data.

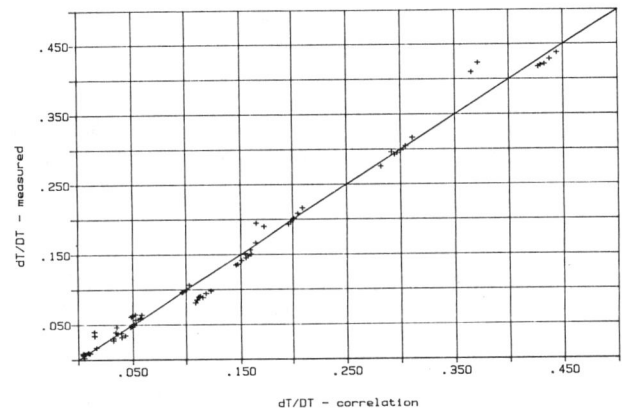

Figure 12. dT/DT correlation fit.

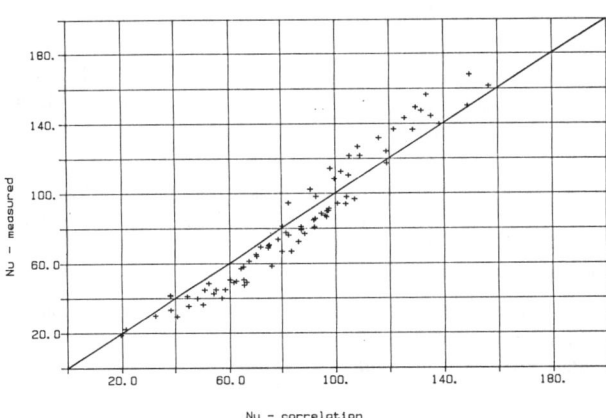

Figure 13. Nu correlation fit.

Brown, W.G., and Solvason, K.R., 1962, "Natural Convection through Rectangular Openings in Partitions," *Int. J. Heat Mass Transfer*, Vol. 5, pp. 859-868.

Burns, R. and Maltby, J., 1986, "MONTE User's Manual for MONT2D and MONT3D," Department of Mechanical Engineering, Colorado State University.

Cheesewright, R. and Zial, S., 1986, "Distributions of Temperature and Local Heat Transfer Rate in Turbulent Natural Convection in a Large Rectangular Cavity," *Proc. 8th Int. Heat Transfer Conf.*, Vol. 4, pp. 1465-1470.

Churchill, S.W., and Chu, H.H.S., 1975, "Correlating Equations for Laminar and Turbulent Free Convection From a Vertical Plate," *Int. J. Heat Mass Transfer*, Vol. 18, pp. 1323-1329.

DeLaHunt, M.J., 1985, "Suncode-PC: A Program User's Manual," Ecotope Inc., Seattle, WA.

Gadgil, A., Bauman, F., and Kammerud, R., 1982, "Natural Convection in Passive Solar Buildings: Experiments, Analysis, and Results," *Passive Solar Journal*, Vol. 1, pp. 28-40

Guire, J.L., 1987, "A multi-channel, low velocity, hot film anemometry system for measuring air flows in buildings," M.S. Thesis, Mechanical Engineering Department, Colorado State University.

Hill, D., Kirkpatrick, A., and Burns, P., 1986, "Analysis and Measurements of Interzonal Natural Convection Heat Transfer in Buildings," *J. Solar Energy Eng.*, Vol. 108, pp. 178-184.

Jones, G.F., Balcomb, J.D., and Otis, D.R., 1985, "A Model for Thermally Driven Heat and Air Transport in Passive Solar Buildings," *Proc. ASME Winter Annual Meeting*, Miami Beach, FL.

Jones, G.F., and Otis, D.R., 1986, "On the Correlation of Natural Convection Heat Transfer in Divided Enclosures," *Int. Comm. Heat and Mass Transfer*, Vol. 13, pp.109-113.

Lin, N.N., and Bejan, A., 1983, "Natural Convection in a Partially Divided Enclosure," *Int. J. Heat Mass Transfer*, Vol. 26, pp. 1867-1878.

Mahajan, B.M., 1986, "Inter-room Air Flow by Natural Convection via a Doorway Opening," *Proc. 1986 ASME Solar Energy Division Conf.*, Anaheim, CA.

Nansteel, M.W., and Greif, R., 1984, "An Investigation of Natural Convection in Enclosures with Two- and Three-Dimensional Partitions," *Int. J. Heat Mass Transfer*, Vol. 27, pp. 561-571 .

Neymark, J., Boardman, C.R., Kirkpatrick, A., and Anderson, R., 1989, "High Rayleigh Number Natural Convection in Partially Divided Air and Water Filled Enclosures," *Int. J. Heat Mass Transfer*, in press.

Olson, D.A., Glicksman, L.R., and Ferm, H.M., 1986, "Scale Model Studies of Natural Convection in Enclosures with Turbulent Vertical Boundary Layers," *Proc. ASME Winter Annual Meeting*, Anaheim, CA.

Rouse, H., 1946, *Elementary Mechanics of Fluids*, Wiley, New York.

Scott, D., Anderson R., and Figliola, R.S., 1988, "Blockage of Natural Convection Boundary Layer Flow in a Multizone Enclosure, *Int. J. Heat and Fluid Flow*, Vol. 9, pp. 208-214.

1989 National Heat Transfer Conference
HTD-Vol. 107, Heat Transfer in Convective Flows

NATURAL CONVECTION ON A HORIZONTAL SURFACE WITH HIGH GRAVITY

M. E. Ulucakli
Department of Mechanical Engineering
Lafayette College
Easton, Pennsylvania

H. Merte, Jr.
Department of Mechanical Engineering and Applied Mechanics
University of Michigan
Ann Arbor, Michigan

ABSTRACT

The purpose of this paper is to report non-boiling natural convection results obtained during an experimental investigation of nucleate pool boiling under high gravity.

Natural convection plays an important role in nucleate pool boiling in several respects. First, it precedes the boiling process, and therefore is an integral part of the incipient boiling process. Second, after the onset of boiling the buoyancy driven flows continue to make significant contributions to the over-all heat transfer process, combining with the bubble induced flows until the boiling becomes dominant.

Natural convection heat transfer measurements are presented for water in an enclosure heated from below, in an accelerating system with various subcoolings and heat flux. The range of experimental variables include: heat flux between 0.19 MW/m^2 and 1.5 MW/m^2, accelerations normal to the heating surface from 1 to 100 times earth gravity, and liquid subcoolings between 0 K and 89 K. The Rayleigh numbers varied between 1.0×10^8 and 2.11×10^{10}, with corresponding Nusselt numbers between 78 and 436.

The test results were correlated by an equation in the form of $Nu_D = 0.16 (Ra_D)^{1/3}$. The considerable influence of temperature dependent properties on the heat transfer correlation is discussed.

1. INTRODUCTION:

The purpose of this paper is to report the non-boiling natural convection heat transfer results obtained during study of nucleate pool boiling under high gravity, reported by Ulucakli [1987] and Ulucakli and Merte [1988]. Gebhart et al. [1988] provide a recent survey of an existing state of knowledge of the natural convection processes. Increased acceleration perpendicular to the heating surface was obtained by rotating the test vessel. The centrifugal acceleration is treated as the radial component of the body force.

The single phase natural convection process plays an important role in pool boiling in several respects. First, it precedes the boiling process, and therefore is an integral part of incipient boiling. Second, after the onset of boiling the buoyancy driven flows continue to make a significant contribution to the over-all heat transfer, combining with the bubble-induced flows until the boiling becomes dominant. At low heat fluxes, where bubbles nucleate at active sites relatively isolated from each other, natural convection is postulated to occur in the areas outside of the "area of influence" of the bubble. As the heat flux is increased the boiling process becomes quite vigorous, as exemplified by the lateral coalescence of bubbles and the formation of bubble columns, and the natural convection contribution to the total heat transfer diminishes.

Another aspect of natural convection is observed in fully developed pool boiling systems when subcooling and acceleration, or both, are significantly increased with constant imposed heat flux. It was observed [Ulucakli and Merte, 1988] that increasing the level of acceleration and/or decreasing the bulk liquid temperature could result in the complete suppression of the boiling process. This is attributed to the associated enhancement of natural convection.

Mikic and Rohsenow [1968] presented a model of nucleate pool boiling process, expressing the total heat flux as:

$$q'' = q''_b + f\, q''_{nc} \qquad (1)$$

where q''_b and q''_{nc} are the boiling and non-boiling natural convection components of the total heat flux, respectively. f represents the mean area fraction of the heating surface not influenced by the bubbles. The model proposed by Mikic and Rohsenow [1968] provides an expression for q''_b. Clearly, an appropriate expression for the natural convection process is also needed.

The influence and significance of natural convection for incipient boiling has been summarized [Rohsenow, 1985]. According to the so-called Hsu-Bergles-Rohsenow model of nucleation presented therein, incipient boiling occurs when the temperature profile in the liquid adjacent to the heating surface becomes tangent to the curve representing the local liquid superheat necessary for activation of the wide range of vapor-filled cavities in the surface. The resulting predictions for the heat flux and heating surface superheat are given by Equations (2) and (3), respectively, as functions of subcooling, the non-boiling heat transfer coefficient, and other parameters:

$$q''_i = \frac{h}{2\Gamma} (1 + \sqrt{1 + 4\Gamma(T_{sat} - T_b)} + h (T_{sat} - T_b) \quad (2)$$

$$(T_w - T_{sat})_i = (1 + \sqrt{1 + 4\Gamma (T_{sat} - T_b)})/2\Gamma \quad (3)$$

The natural convection heat transfer coefficient is defined as:

$$h = q''_{nc} / (T_w - T_b) \quad (4)$$

The natural convection heat transfer coefficient (or Nusselt number) is a function of a number of variables, including the Rayleigh and Prandtl numbers, the aspect and heater ratios of the enclosure, and the thermal stratification ratio [Torrance, 1979]. The thermal stratification ratio is defined as the ratio of horizontal and vertical temperature differences in the liquid space. The aspect and heater ratios of the enclosure are shown on Figure 1.

2. PREVIOUS WORK:

Extensive reviews of natural convection in enclosures are presented by Catton [1978] and Ostrach [1982], but a brief review will be included here for the objectives of this investigation.

Natural convection in an enclosure is influenced by the geometry and orientation of the heating surface as well as that of the surrounding surfaces. For example, heated, cooled or insulated confining lateral boundaries introduce horizontal temperature and velocity gradients which would be quite different were the lateral boundaries not significant.

Catton and Edwards [1967] presented experimental data for natural convection in enclosures of hexagonal shape which were heated from below and cooled above. Various aspect ratios were employed, defined as the ratio of the distance between the plates to the diameter of the cell. The vertical walls of the cells were constructed from either an insulating or a conducting material. As the Rayleigh number based on the distance between the plates was increased to 10^6 and above, the Nusselt numbers for various aspect ratios asymptotically reached the limit corresponding to the smallest aspect ratio used of less than one. In other words, for large Rayleigh numbers extending into the turbulent regime, the size and thermal conductivity of the horizontally confining surfaces did not matter. Water and silicone oils were used.

Other investigators, such as Globe and Dropkin [1959] and Silveston [1958] did not observe any significant effects of the lateral surfaces when a large Rayleigh number was combined with a low aspect ratio.

Kamotani et al. [1983] reported experimental results for natural convection in an enclosure heated from below by a disk-shaped heating surface smaller in size than the horizontal area of the enclosure. The Nusselt number was expressed as:

$$Nu = C \, Gr^{0.5} \, Ar^{-1.25} \quad (5)$$

where the aspect ratio Ar was the ratio of the depth of fluid to half the horizontal dimension of the enclosure. Water was used as the test fluid and the heater ratio was not varied. The Prandtl number for water is incorporated into the coefficient in Equation (5). It is noted that the exponent of the Grashof number in Equation (5) is 1/2 rather than the usual 1/3 for the turbulent natural convection or 1/4 for the laminar case. The measurements correspond to Grashof numbers on the order of 4×10^6 with an aspect ratio of 0.37, and the flow regime appears to be in the laminar range. Torrance [1979] reported computed results of natural convection in thermally stratified enclosures with localized heating from below. For the case of strong buoyancy with weak stratification, the theoretical results showed a Gr dependency of the Nusselt number for large Grashof numbers. Ishiguro et al. [1978] also reported a 1/2 power on the Rayleigh number for rectangular heating surfaces submerged in water in large test vessels.

3. EXPERIMENTAL APPARATUS AND PROCEDURE:

A flat circular horizontal heating surface was used, and the geometry and orientation relative to the enclosure are shown in Figure 1. The aspect and heater ratios were 6 and 1, respectively, and were not varied in the experimental results reported here. The heating surface and related equipment are contained in a test vessel as shown in Figure 2, which, in turn, is attached to the horizontal arm of a centrifuge, as shown in Figure 3. As the vertical shaft is rotated the test vessel and the identical counterweight swing outward such that the vector sum of the centrifugal and gravitational accelerations is normal to the heating surface at all times.

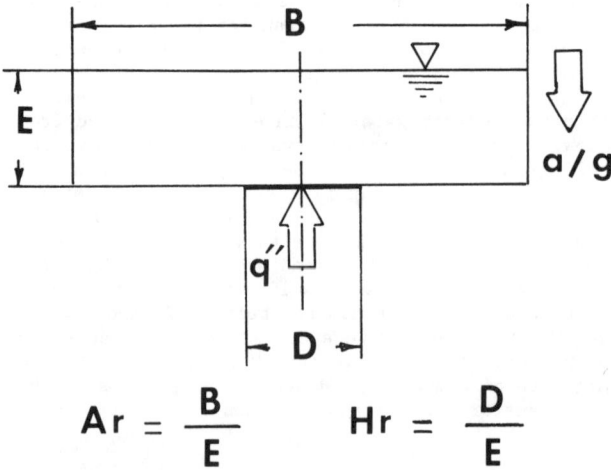

Fig. 1 - Heating Surface and its Surrounding Surfaces.

Five cartridge heaters are embedded at the bottom of a tellurium copper cylinder 5 cm in diameter and 10 cm in length. The top portion is reduced to 2.5 cm in diameter and covered with a stainless steel type 347 foil by silver brazing in vacuum to provide the heating surface. Three thermocouples embedded in the linear portion of the cylinder are used to determine the heat flux and heating surface temperature by extrapolation, taking into account the thicknesses and thermal conductivities of the stainless steel, silver braze and copper layers. Liquid temperatures were measured at three locations: the main thermo-couple is 2.5 mm above the center of the heating surface and provides the bulk temperature reported here, and the other two thermocouples are located as shown in Figure 2. The thermocouple probe at the centerline was located close to the surface to measure the liquid temperatures in boiling experiments as well. The other two thermocouples were used to monitor the liquid pool temperatures. Typical measurements for given condition of subcooling are shown in Figure 6. The data indicate a horizontal temperature gradient at a/g = 1. Further details are given by Ulucakli [1987].

Water was used as the test fluid and the liquid depth was maintained constant at 25.4 mm. The bulk temperature was varied using a submerged heat exchanger as shown in Figure 2, and adjusting the mass flow rate of the cooling water. A detailed description of the experimental apparatus is given by Ulucakli [1987].

The experimental procedure involved varying the bulk liquid temperature at a given heat flux and acceleration. Nondimensional accelerations of 1, 10, and 100 were employed. Natural convection results were obtained in the course of the study of boiling when the heater surface superheat was below of that required for nucleation.

4. RESULTS:

The heat transfer results for non-boiling natural convection are shown in Figure 4 in the conventional Nusselt number Rayleigh number format while the numerical results are listed in Table 1. All data were obtained under steady-state conditions.

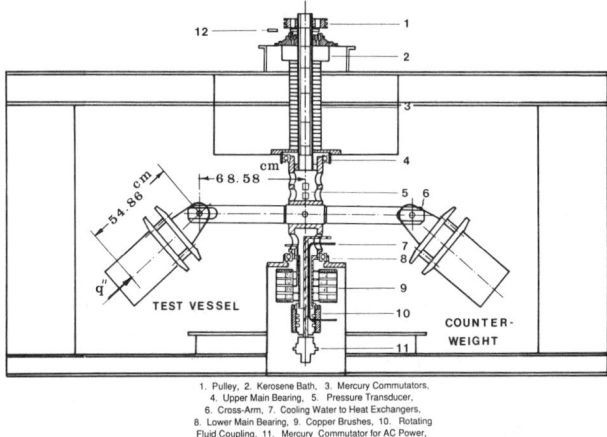

1. Pulley, 2. Kerosene Bath, 3. Mercury Commutators, 4. Upper Main Bearing, 5. Pressure Transducer, 6. Cross-Arm, 7. Cooling Water to Heat Exchangers, 8. Lower Main Bearing, 9. Copper Brushes, 10. Rotating Fluid Coupling, 11. Mercury Commutator for AC Power, 12. Magnetic Pick-up.

Fig. 3 - Drawing of Centrifuge.

The characteristic dimension used is the diameter of the heating surface, 2.54 cm (1 inch). Goldstein et al. [1973] and Lloyd and Moran [1974] pointed out that the correlation of experimental results may be improved for a variety of heater surface planform geometries by using a characteristic length defined by $L = A_p/P$. The reference temperature used to calculate the thermodynamic and transport properties of water was taken as the average of the heating surface and bulk liquid. Since the heating surface temperatures change considerably, a wide variation of properties occurs in the dimensionless parameters, and these will be discussed later here. The thermodynamic and transport properties of water are taken from Reynolds [1979] and Irvine and Lilley [1984].

Since the relatively large variation in Rayleigh numbers occur as a result of body force changes, the heat transfer results are presented and distinguished for the nondimensional accelerations of 1, 10, and 100. The gravitational acceleration term in the Rayleigh number is multiplied with the nondimensional acceleration a/g when the gravity is larger than 1. The data includes results obtained at various levels of heat flux. A best fit line that includes all the data was obtained in the form of:

$$Nu_D = 0.16 \ (Ra_D)^{1/3} \tag{5}$$

The data for the 1- and 10-g accelerations correlated better with the coefficient of 0.18 rather than 0.16. While the data for the 100-g acceleration appear to be slightly lower than the trend indicated by

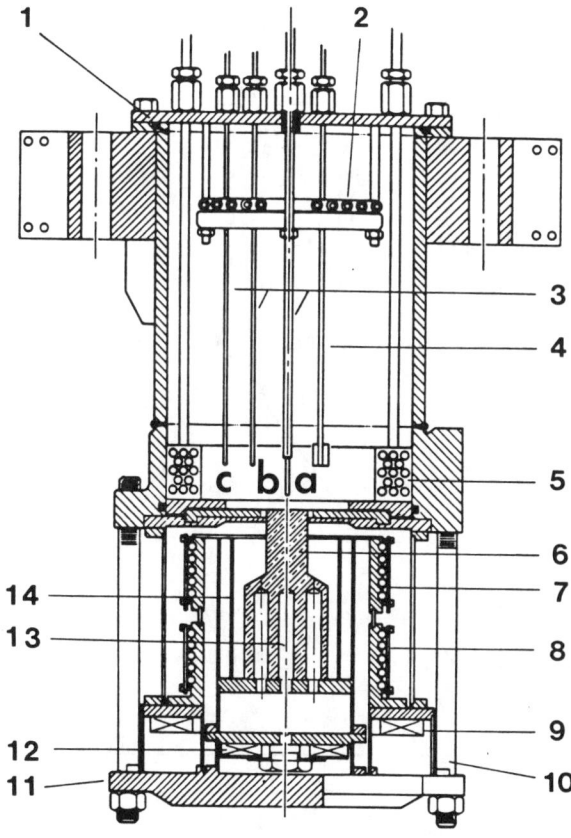

1. Cover Plate, 2. Condensing Heat Exchanger, 3. Liquid Temperature Probes, 4. Liquid Depth Control Tube, 5. Subcooling Heat Exchanger, 6. Copper Cylinder, 7. Upper Cylindrical Guard Heater, 8. Lower Cylindrical Guard Heater, 9. Bottom Ring Guard Heater, 10. Support Bolts, 11. Support Plate, 12. Bottom Disc Guard Heater, 13. Cartridge Heaters, 14. Radiation Shields

Fig. 2 - Test Vessel Schematic.

Equation (6). The 1/3 exponent of the Rayleigh number indicate the existence of a turbulent flow regime. A linear least-squares regression analysis of all the data yielded an equation in the form of:

$$Nu_D = 0.20 \ (Ra_D)^{0.325} \qquad (6)$$

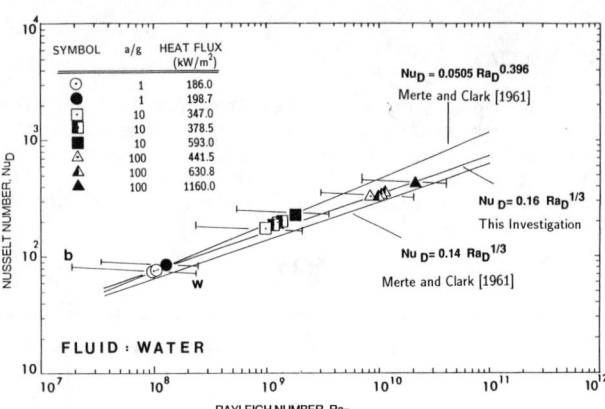

Fig. 4 - Natural Convection Heat Transfer Results.

with a regression coefficient of r = 0.992, which is slightly different than that given by Equation (6). Arpaci [1986] and Chu and Goldstein [1973] present theoretical and experimental evidence, respectively, as to why the exponent of the Rayleigh number should be less than 1/3 for fluids having a Prandtl number on the order of 1. The coefficient in the correlations above varies widely among various investigators. Most recent investigations show that this coefficient is strongly influenced by various aspect and heater ratios existing in the heat transfer enclosure. Another factor incorporated into this coefficient is the Prandtl number. A proper correlation of heat transfer for turbulent natural convection should include the Prandtl number [Arpaci, 1986]. However, for investigations performed with only one fluid, as in the work presented here, the influence of the Prandtl number can not be explicitly correlated.

Natural convection data obtained by Merte and Clark [1961] are also shown in Figure 4 for comparison. The heating surface-enclosure geometry used is similar to that of this investigation. The correlation plotted as the lower line corresponds to natural convection in a pool 14.5 cm in diameter, where the heating surface diameter was 7.6 cm, and the liquid depth 6.4 cm. These correspond to an aspect ratio of 2.31 and heater ratio of 1.2, respectively. The correlation plotted as the upper line corresponds to the case where a flow guide 7.6 cm in diameter was used to restrict the convection flow above the heating surface, reducing the aspect ratio to 1.2, and the heater ratio to 1. The natural convection heat transfer results obtained in the present study at 1-, 10-, and 100-g acceleration levels fall between these two lines.

The results obtained are also plotted in a Nu.Ra vs Ra format, after Globe and Dropkin [1959], in Figure 5. The Nu.Ra product is proportional to the heat flux, and does not amplify errors made in the measurement of temperatures. The Rayleigh number is proportional to the temperature difference between the heating surface and bulk liquid temperatures, and is influenced by the uncertainties in temperature measurements. The log-log plot appears to be linear.

Bonilla and Sigel [1961] observed a gradual reduction of the slope of the Nusselt number vs Rayleigh number correlation line as the Rayleigh number was increased to very large values, as in Figure 4 where a gradual decrease in the slope of the line is observed in the 100-g cluster.

5. INFLUENCE OF THERMODYNAMIC AND TRANSPORT PROPERTIES:

The thermodynamic and transport properties of water change considerably as result of the large temperature differences between the heating surface and the bulk liquid. The temperature-dependent properties of water that contriute to the calculation of Rayleigh number are volumetric thermal expansion coefficient, kinematic viscosity, and thermal diffusivity. The thermal expansion coefficient and kinematic viscosity of water are known to be strongly dependent on the temperature. The thermal conductivity and thermal diffusivity are also dependent on the temperature, but to a lesser extent.

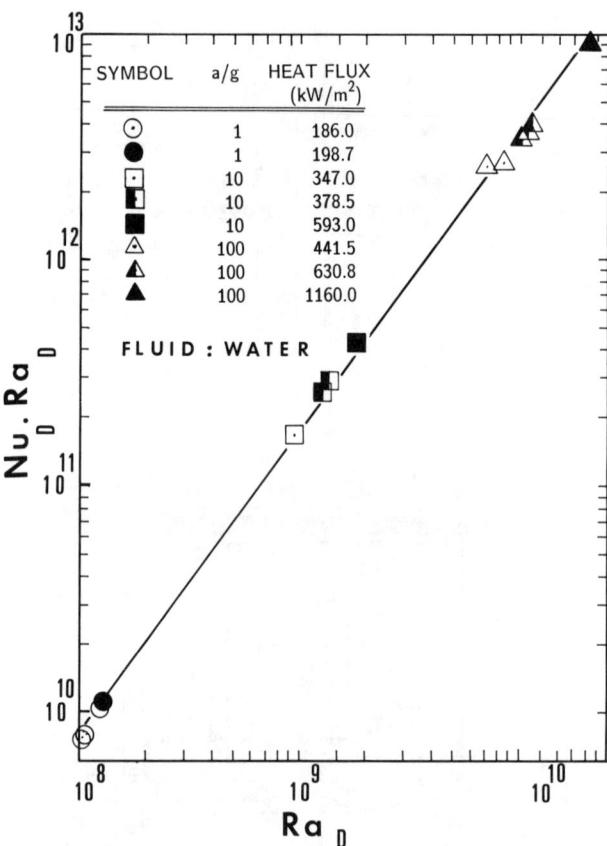

Fig. 5 - Nu.Ra Product vs Rayleigh Number Plot

The temperature dependent property problem has been given some attention by investigators. A recent survey is presented by Kakac et al. [1985]. In order to account for the variation of properties two major methods were used: the reference temperature method and the property ratio method. The method used depends on the phase of the fluid, and the degree of temperature dependency of a particular property. It appears that Sparrow and Gregg [1958] introduced a reference temperature in the form of $T_r = T_w + m \ (T_w$

- T_b) to calculate the properties, where m is a factor less than 1. The fluid under investigation was air. Later, this method was applied to liquids in various forms. It also appears that there is no general agreement upon the reference temperature to be used for enclosure problems. Fujii and Imura [1972] used a reference temperature $T_r = T_w - 0.25$ ($T_w - T_b$) to evaluate all the properties except the volumetric expansion coefficient which was evaluated at a reference temperature of 0.5 ($T_b + 0.5$ ($T_b + T_w$)). Ishiguro et al. [1978] used a reference temperature of $T_r = 0.1$ ($T_w - T_b$) $+ T_b$ to evaluate all the properties.

In order to determine the sensitivity of the property variations with temperature level, the Rayleigh numbers were evaluated using the same driving temperature differences as before, but with the properties evaluated at different temperatures. The temperatures used for evaluation of all the properties were that at the heating surface, bulk liquid, and four intermediate levels. The results are shown in Figure 4. The almost horizontal bars represent the variation of Rayleigh numbers over these temperature levels, and demonstrate the importance of choosing the appropriate reference temperature for property evaluation. The letters b and w indicate the reference temperatures equal to bulk and heating surface temperatures. The slight deviation of the bars from the horizontal is due to the variation of the liquid thermal conductivity with temperature. The choice of the reference temperature would determine the slope and the position of the correlation as shown in Figure 4. The deviation from the 1/3 slope becomes smaller as reference temperatures close to the surface temperature are chosen.

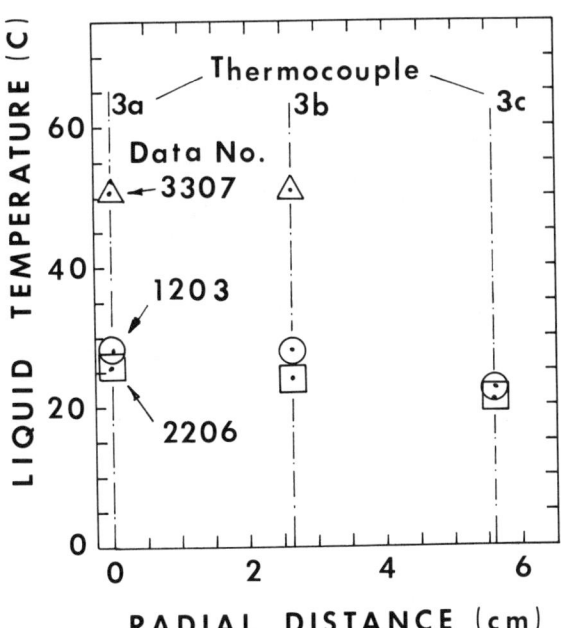

Fig. 6 - Typical Liquid Temperatures in the Enclosure.

6. **CONCLUSIONS:**

a. Natural convection heat transfer measurements were presented for water in an enclosure heated from below in an accelerating system with very large subcoolings and heat flux.

b. The measurements made in the experiments reported here show that, for very large differences between the heating surface temperature and the bulk liquid temperature, overall heat transfer rates could be expressed in terms of conventional correlations.

c. Since the properties of the liquid changed considerably for the large temperature differences employed in these experiments, the choice of the reference temperature influenced the coefficient and the exponent in Equation (6).

TABLE 1

NATURAL CONVECTION HEAT TRANSFER RESULTS AT VARIOUS ACCELERATIONS

Fluid : Water

DATA NO.	a/g	ΔT_{sat} K	ΔT_{sub} K	q" kW/m²	T_w C	Ra_D x 10^{-8}	Nu_D
1203	1	6.6	88.9	182	123.5	1.045	78
1204	1	7.2	84.5	178	122.9	1.029	76
1205	1	6.0	81.9	172	124.1	1.064	76
2301	1	0.2	82.9	188	130.4	1.228	87
2206	1	2.1	86.4	199	130.4	1.283	86
3105	10	- 9.9	88.9	349	120.2	9.68	171
3209	10	19.0	74.7	593	149.1	18.11	239
3307	10	6.7	61.2	378	136.5	13.56	210
3308	10	2.0	69.2	373	131.8	12.78	199
3309	10	- 4.5	81.8	347	125.4	11.54	186
3310	10	1.9	69.5	376	131.7	12.57	200
5005	102	- 7.7	55.3	407	120.6	84.18	324
5006	102	-16.7	65.2	467	111.8	70.94	368
5107	101	0.9	71.9	650	127.3	112.5	350
5108	102	- 4.0	77.0	646	124.3	107.2	340
5109	102	- 6.1	81.4	652	122.2	103.0	334
5206	101	28.8	66.4	1107	157.5	211.8	436

T_{sat} = 112.23° C at 154.4 kPa pressure at the heating surface.

NOMENCLATURE:

a	Acceleration
A_p	Area
Ar	Aspect ratio, B/E
a/g	Nondimensional acceleration
b	Symbol in Fig. 4 indicating typical data point where the properties are calculated at the bulk liquid temperature
B	Diameter of the enclosure
C	Coefficient in Equation (5)
D	Diameter of the heating surface
E	Liquid depth
f	Area fraction
g	Standard gravitational acceleration
G_r	Grashof number, Ra/Pr
h	Heat transfer coefficient

h_{fg}	Enthalpy of evaporation
Hr	Heater ratio, D/E
k	Thermal conductivity of liquid
L	Characteristic length (Equal to the heater diameter D)
Nu_D	Nusselt number, h D/k
P	Perimeter of the heating surface
Pr	Prandtl number, α/υ
Ra_D	Rayleigh number, $(a/g)g\ \beta D^3\ (T_w - T_b)/\alpha\ \upsilon$
q"	Heat Flux
T	Temperature
Δ Tsat	Heating surface superheat, $T_w - T_{sat}$
Δ Tsub	Liquid subcooling, $T_{sat} - T_b$
w	Symbol in Fig. 4 indicating a typical data point where the properties are evaluated at the wall temperature

Greek Letters:

α	Thermal diffusivity of liquid
β	Volumetric thermal expansion coefficient of liquid
Δ	Difference
Γ	Nucleation parameter, $k\ h_{fg}\ \rho/8\ \sigma\ T_{sat}\ h$
ν	Kinematic viscosity of liquid
ρ	Density of liquid
σ	Surface tension

Subscripts:

b	Bulk, boiling
D	Diameter
i	Incipient
nc	Natural convection
r	Reference
sat	Saturation
sub	Subcooling
w	Heating surface

REFERENCES:

Arpaci, V. S., "Microscales of Turbulence and Heat Transfer Correlations", 1986, Int. J. Heat Mass Transfer, Vol. 29, pp. 1071-1078.

Bonilla, C. F. and Sigel, L. A., 1961, "High Intensity Natural Convection Heat Transfer Near Critical Point", Chemical Engineering Progress Symposium Series, Vol. 32, No. 32, pp.87-95.

Catton, I. and Edwards, D. K., "Effect of Side Walls on Natural Convection Between Horizontal Plates and Heated from Below", 1967, J. Heat Transfer, Vol. 89, pp. 295-299.

Catton, I., 1978, "Natural Convection in Enclosures", Proc. 6th Int. Heat Transfer Conference, Vol. 6, pp 13-31.

Chu, T. Y. and Goldstein, R. J., 1973, "Turbulent Convection in a Horizontal Layer of Water", Journal of Fluid Mechanics, Vol. 60, Part 1, pp. 141-159.

Fujii, T. and Imura, H., 1972, "Natural Convection Heat Transfer from a Plate with Arbitrary Inclination", Int. J. Heat Mass Transfer, Vol. 15, pp. 755-767.

Gebhart, B. et al., 1988, Buoyancy-Induced Flows and Transport, Hemisphere Publishing Corp., New York.

Globe, S. and Dropkin, D., 1959, "Natural Convection Heat Transfer in Liquids Confined by Two Horizontal Plates and Heated from Below", J. Heat Transfer, Vol. 81, pp. 24-27.

Goldstein, R. J., Sparrow, E. M. and Jones, D. C., 1973, " Natural Convection Mass Transfer Adjacent to Horizontal Plates", Int. J. Heat Mass Transfer, Vol. 16, pp. 1025-1034.

Irvine, T. F., Jr. and Lilley, P. E., 1984, "Steam and Gas Tables with Computer Equations", Academic Press, Inc.

Ishiguro, R. et al., "Heat Transfer and Flow Instability of Natural Convection Over Upward-Facing Horizontal Surfaces", Proceedings 6th International Heat Transfer Conf., Vol. 2, pp. 229-234.

Kakac, S. et al., 1985, "The Effects of Temperature Dependent Fluid Properties on Natural Convection - Summary and Review", in "Natural Convection: Fundamentals and Applications", Ed. by Kakac, S. et al., pp. 721-773.

Kamotani, Y. et al., 1983, "Natural Convection Heat Transfer in a Water Layer with Localized Heating from Below", "Natural Convection in Enclosures-1983", ASME HTD. Vol. 26.

Lloyd, J. R. and Moran, W. R., 1974, "Natural Convection Adjacent to Horizontal Surfaces of Various Planforms", ASME Paper No.: 74-WA/HT-66.

Merte, H., Jr. and Clark, J. A., 1961, "Pool Boiling in an Accelerating System", J. Heat Transfer, Vol. 83, pp. 234-242.

Mikic, B. and Rohsenow, W. M., 1968, " A New Correlation of Pool Boiling Data Including the Effect of Heating Surface", ASME Paper No.: 68-WA/HT-22.

Ostrach, S., 1982, "Natural Convection Heat Transfer in Cavities and Cells", Proceedings 7th International Heat Transfer Conference, Vol. 1, pp. 365-379.

Reynolds, W. H., 1979, "Thermodynamic Properties in SI", Department of Mechanical Engineering, Stanford University, Stanford CA.

Rohsenow, W. M., 1985, "Handbook of Heat Transfer Fundamentals", 2nd Edition, Hemisphere Publishing Co.

Silveston, P. L., 1958, "Heat Transfer in Horizontal Fluid Layers", in German, Forschung Gebiete Ingenieurwes., Vol. 24, Part 1.: pp. 29-32, Part 2: 59-69.

Sparrow, E. M. and Gregg, J. L., 1958, "The Variable Fluid Property Problem in Free Convection", Trans. American Society of Mechanical Engineers, Vol. 80, pp. 869-886.

Torrance, K. E. et al., 1969, "Experiments on Natural Convection in Enclosures with Localized Heating from Below", Journal of Fluid Mechanics, Vol. 36, Part I, pp. 21-31.

Torrance, K. E., 1979, "Natural Convection in Thermally Stratified Enclosures with Localized Heating from Below", Journal of Fluid Mechanics, Vol. 95, Part 3, pp. 477-495.

Ulucakli, M. E., 1987, "Nucleate Pool Boiling with Increased Acceleration and Subcooling", Ph. D. Thesis, Dept. of Mechanical Engineering, University of Michigan, Ann Arbor, Michigan.

Ulucakli, M. E. and Merte, H., Jr., 1988, "Nucleate Pool Boiling with High Gravity and Subcooling", Proceedings 1988 National Heat Transfer Conference, H. R. Jacobs, ed., Vol. 2, pp. 415-422, ASME HTD-96, Houston, Texas.

1989 National Heat Transfer Conference
HTD-Vol. 107, Heat Transfer in Convective Flows

NATURAL CONVECTION HEAT TRANSFER
IN A SQUARE ENCLOSURE
CONTAINING AN OBSTRUCTION

J. M. House, C. Beckermann and T. F. Smith
Department of Mechanical Engineering
The University of Iowa
Iowa City, Iowa

ABSTRACT

A numerical study was performed of natural convection in a vertical square enclosure containing a centered square obstruction. The analysis reveals that the fluid flow and heat transfer processes are governed by the Rayleigh and Prandtl numbers, dimensionless obstruction size, and the ratio of the thermal conductivity of the obstruction to that of the fluid. For relatively large ranges of these parameters, results are reported in terms of streamlines, isotherms, and the overall heat transfer across the enclosure as described by the Nusselt number. It is found that the heat transfer across the enclosure, in comparison to that for no obstruction, may be enhanced (reduced) by an obstruction with a thermal conductivity ratio less (greater) than unity. Furthermore, the heat transfer may attain a minimum as the obstruction size is increased. These and other findings are justified through a careful examination of the local heat and fluid flow phenomena.

NOMENCLATURE

c_p	isobaric specific heat
g	gravitational acceleration
h	convection coefficient
k	thermal conductivity of fluid
k_s	thermal conductivity of solid
k^*	thermal conductivity ratio, k_s/k
L	spacing between hot and cold walls
Nu	Nusselt number, $h\,L/k$
P	pressure
$\bar{P}$	dimensionless pressure, $P\,L^2/\rho\,\alpha^2$
Pr	Prandtl number, ν/α
Ra	Rayleigh number, $g\,\beta\,\Delta T\,L^3/\nu\,\alpha$
T	temperature
u	velocity in x direction
$\bar{u}$	dimensionless velocity in x direction, $u\,L/\alpha$
v	velocity in y direction
$\bar{v}$	dimensionless velocity in y direction, $v\,L/\alpha$

Greek symbols

α	thermal diffusivity, $k/\rho\,c_p$
β	coefficient of thermal expansion
ζ	dimensionless obstruction size, W/L
η	dimensionless y coordinate, y/L
θ	dimensionless temperature, $(T-T_c)/(T_h-T_c)$
μ	dynamic viscosity
ν	kinematic viscosity, μ/ρ
ξ	dimensionless x coordinate, x/L
ρ	density

Subscripts

c	cold surface
h	hot surface
s	solid

INTRODUCTION

In several engineering applications such as building energy components, electronic cooling, and manufacturing processes, natural convection heat transfer is of importance. As indicated in the reviews by Yang (1987) and Ostrach (1988), analytical, numerical, and experimental studies have been conducted in an attempt to quantify natural convection heat transfer processes. One particular system that has received considerable attention is the vertical rectangular enclosure, both as a practical engineering system as well as a convenient system for studying natural convection phenomena. In some instances, an obstruction may be located in the enclosure, thereby impeding the natural convection flow. Considerable research has been performed with obstructions in the form of partitions or partial baffles (Yang, 1987). However, other than the study reported by Emery and Chu (1969) for a baffle centered vertically in a vertical enclosure, there is lacking information about the natural convection processes when an obstruction is placed at the

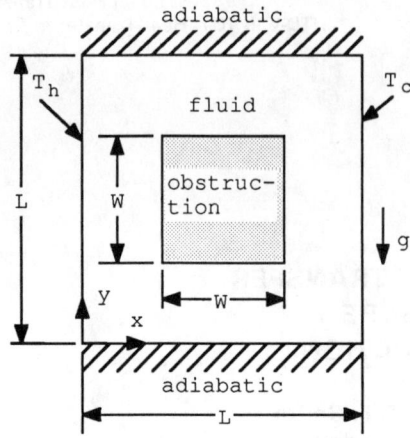

Figure 1 Schematic of enclosure with obstruction.

center of the enclosure (see Fig. 1). In the case of building energy components such as walls or windows or an electronic enclosure, the obstruction may reduce the flow thereby lowering the heat transfer across the enclosure. In a metal casting process, an obstruction may be placed in the mold to form a hole in the casting. On a smaller scale, understanding natural convection in an enclosure with an obstruction may provide needed effective heat transfer data for porous media studies. It is, therefore, of interest to examine natural convection in enclosures containing an obstruction.

The purpose of this investigation is to examine natural convection heat transfer in a square enclosure containing a square obstruction located at the center of the enclosure. The system shown schematically in Fig. 1 consists of a square enclosure with sides of length L. The left and right side walls are isothermal at respective temperatures of T_h and T_c, while the bottom and top walls are adiabatic. The obstruction is centered at L/2, has sides of length W, and is solid with thermal conductivity k_s. The flow within the enclosure is laminar, gravitational acceleration acts parallel to the isothermal walls, and radiation effects are taken to be negligible. Except for the density in the buoyancy term, the fluid properties are assumed to be constant, and the Boussinesq approximation applies. Steady-state conditions prevail. Results of interest include the effect of the size and thermal conductivity of the solid obstruction on the flow and temperature distributions within the the enclosure and the heat transfer across the enclosure.

ANALYSIS

The steady-state conservation equations for the fluid in dimensionless form are as follows:
Continuity:

$$\frac{\partial}{\partial \xi}(\bar{u}) + \frac{\partial}{\partial \eta}(\bar{v}) = 0 \tag{1}$$

ξ-momentum:

$$\frac{\partial}{\partial \xi}(\bar{u}\,\bar{u}) + \frac{\partial}{\partial \eta}(\bar{u}\,\bar{v})$$

$$= Pr\left[\frac{\partial}{\partial \xi}\left(\frac{\partial \bar{u}}{\partial \xi}\right) + \frac{\partial}{\partial \eta}\left(\frac{\partial \bar{u}}{\partial \eta}\right)\right] - \frac{\partial \bar{P}}{\partial \xi} \tag{2}$$

η-momentum:

$$\frac{\partial}{\partial \xi}(\bar{v}\,\bar{u}) + \frac{\partial}{\partial \eta}(\bar{v}\,\bar{v})$$

$$= Pr\left[\frac{\partial}{\partial \xi}\left(\frac{\partial \bar{v}}{\partial \xi}\right) + \frac{\partial}{\partial \eta}\left(\frac{\partial \bar{v}}{\partial \eta}\right)\right] - \frac{\partial \bar{P}}{\partial \eta} + Ra\,Pr\,\theta \tag{3}$$

Energy:

$$\frac{\partial}{\partial \xi}(\bar{u}\,\theta) + \frac{\partial}{\partial \eta}(\bar{v}\,\theta) = \frac{\partial}{\partial \xi}\left(\frac{\partial \theta}{\partial \xi}\right) + \frac{\partial}{\partial \eta}\left(\frac{\partial \theta}{\partial \eta}\right) \tag{4}$$

For the solid obstruction, the energy equation is

$$0 = \frac{\partial}{\partial \xi}\left(k^\star \frac{\partial \theta_s}{\partial \xi}\right) + \frac{\partial}{\partial \eta}\left(k^\star \frac{\partial \theta_s}{\partial \eta}\right) \tag{5}$$

where $k^\star$ is the ratio of the thermal conductivity of the obstruction to that of the fluid.

The flow boundary conditions are zero velocity at all solid surfaces. The thermal boundary conditions are

at $\xi = 0$, $\theta = 1$,   at $\xi = 1$, $\theta = 0$

at $\eta = 0$, $\dfrac{\partial \theta}{\partial \eta} = 0$,   at $\eta = 1$, $\dfrac{\partial \theta}{\partial \eta} = 0$

at $\xi = \dfrac{1-\zeta}{2}$, $\theta_s = \theta$, $\dfrac{\partial \theta}{\partial \xi} = k^\star \dfrac{\partial \theta_s}{\partial \xi}$

at $\xi = \dfrac{1+\zeta}{2}$, $\theta_s = \theta$, $\dfrac{\partial \theta}{\partial \xi} = k^\star \dfrac{\partial \theta_s}{\partial \xi}$ $\quad(6)$

at $\eta = \dfrac{1-\zeta}{2}$, $\theta_s = \theta$, $\dfrac{\partial \theta}{\partial \eta} = k^\star \dfrac{\partial \theta_s}{\partial \eta}$

at $\eta = \dfrac{1+\zeta}{2}$, $\theta_s = \theta$, $\dfrac{\partial \theta}{\partial \eta} = k^\star \dfrac{\partial \theta_s}{\partial \eta}$

where $\zeta = W/L$.

The average Nusselt number at the hot wall is expressed by

$$Nu_h = -\int_0^1 \left.\frac{\partial \theta}{\partial \xi}\right|_{\xi=0} d\eta \tag{7}$$

This expression without the minus sign and evaluated at $\xi = 1$ applies to the cold wall.

By conservation of energy across the enclosure, the average Nusselt numbers at the hot and cold walls must be equal, that is, $Nu_h = Nu_c$.

Equations (1) - (5), and the boundary conditions, Eq. (6), reveal that there are four dimensionless parameters (Ra, Pr, k^*, ζ) that govern natural convection heat transfer in a square enclosure containing a centered square obstruction.

NUMERICAL PROCEDURE

Numerical solutions of the governing equations were obtained using the SIMPLER algorithm developed by Patankar (1980). In a similar manner as Beckermann, Viskanta, and Ramadhyani (1988), one set of conservation equations is solved over the entire domain. By setting the Prandtl number to infinity in the region occupied by the solid obstruction, the velocities automatically approach zero in this region. At the same locations, the dimensionless diffusion coefficient in the energy equation is changed from unity to k^*. By combining the energy equations in this manner, the matching conditions at the fluid/obstruction interface stated by Eq. (6) are satisfied automatically. The algorithm is based on a control-volume formulation that ensures continuity of the heat fluxes across all control surfaces and, thus, the fluid/obstruction interface. The harmonic mean formulation adopted for the interface diffusion coefficients between two control volumes yields physically realistic results for abrupt changes in these coefficients (for example, if $k^* \neq 1$) without requiring an excessively fine grid in the neighborhood of the fluid/obstruction interface.

Numerical experiments were performed to establish the number of control volumes required to produce accurate results within a reasonable computational effort. Numerical solutions presented in this paper were acquired using 38 control volumes in both the ξ- and η-directions. Ten control volumes where placed in the obstruction and 14 control volumes were placed in each passage way on either side of the obstruction in both the ξ- and η- directions. The distribution of the control volumes was slightly skewed along all solid surfaces in order to resolve accurately large velocity and temperature gradients. Convergence of the numerical solution was checked by performing overall mass and energy balances. The calculations were performed on a PRIME 9955 computer and typically required 6,000 central processor seconds. As shown in Table 1, the Nusselt number results for the limiting case of no obstruction agree favorably with the accepted values (de Vahl Davis, 1983). It is realized that the comparisons in Table 1 do not validate

necessarily the entire code when an obstruction is present, but the good agreement and the noted numerical experiments established confidence in the solutions.

RESULTS AND DISCUSSION

It is first instructive to examine the ranges of the governing dimensionless parameters Ra, Pr, ζ, and k^* considered in the present studies. To retain laminar flow, the Rayleigh number is kept less than or equal to 10^6. The Prandtl number was varied from 0.01, corresponding to a liquid metal, to 0.71, which is representative of a gas. Results for the latter value of the Prandtl number could be applied to non-metallic liquids since in natural convection in enclosures the Nusselt numbers are known to be relatively insensitive to Prandtl numbers above unity (Yang, 1987). The dimensionless size, ζ, of the obstruction was varied from 0 to 1, where the limit of zero corresponds to pure natural convection, while unity represents pure conduction in the obstruction with a resultant Nusselt number equal to k^*. Emphasis is placed on results for a thermal conductivity ratio, k^*, equal to 0.2 and 5.0. Physically, the former value of k^* represents, for example, a solid obstruction of wood in a liquid with properties similar to those of water, while the latter is for a solid obstruction of wood in a gas with properties similar to those of air.

Streamlines and Isotherms

Streamlines and temperature fields for the six cases identified in Table 2 are displayed in Figs. 2 to 7. These contour plots are presented in an attempt to demonstrate typical flow and heat transfer characteristics and to assist in explaining the findings for the overall Nusselt numbers (see next section). In order to illustrate important details of the flow structure, the increments between some of the streamlines in Figs. 4a and 5a were chosen not to be equal. In all cases, the streamlines show a large cell rotating clockwise about the obstruction. The isotherms in the fluid are somewhat similar to those found in natural convection in cavities without an obstruction, in that they are clustered near the lower-left and upper-right corners of the enclosure.

Figures 2 and 3 show the streamlines and isotherms for Cases 1 and 2, where the obstruction size, ζ, is equal to 0.5. The only difference between these two cases is the value of k^*. The streamlines (Figs. 2a and 3a) are quite similar to each other. In fact, a comparison with streamlines for the case without an obstruction (not shown here) reveals that

Table 1 Nusselt number comparisons
for Pr = 0.71 and AR = 1.0.

Ra	Nu	
	Present	Exact
10^3	1.118	1.118
10^4	2.254	2.243
10^5	4.561	4.519
10^6	8.923	8.800

Table 2 Dimensionless parameters for the
streamlines and isotherms plots.

Case	Ra	Pr	ζ	k^*	Nu
1	10^5	0.71	0.5	0.2	4.624
2	10^5	0.71	0.5	5.0	4.324
3	10^5	0.01	0.5	0.2	2.866
4	10^5	0.01	0.5	5.0	2.539
5	10^6	0.71	0.9	0.2	2.402
6	10^6	0.71	0.9	5.0	3.868

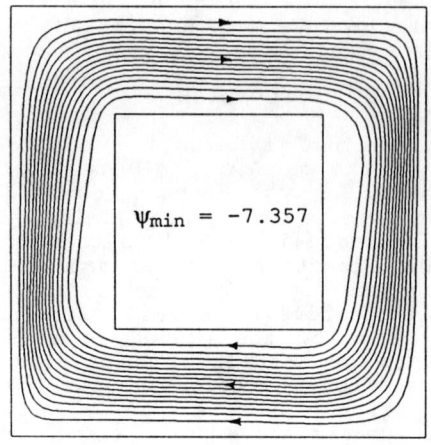

$\Psi_{min} = -7.357$

(a) Streamlines (equally spaced).

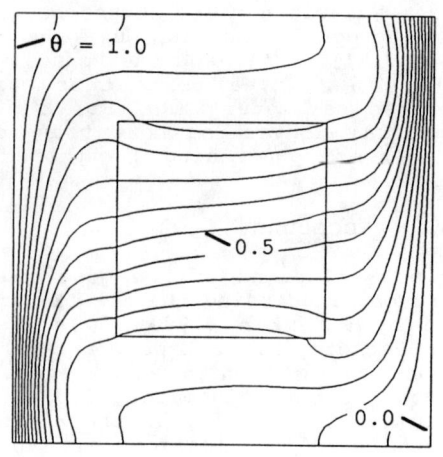

$\theta = 1.0$

0.5

0.0

(b) Isotherms (equally spaced).

Figure 2   Streamlines and temperature contours for Case 1.

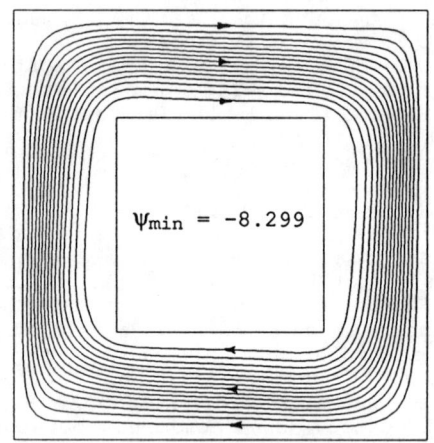

$\Psi_{min} = -8.299$

(a) Streamlines (equally spaced).

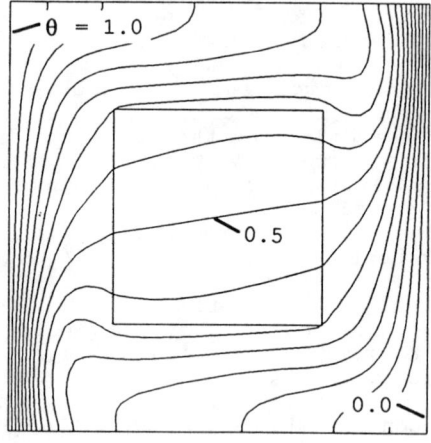

$\theta = 1.0$

0.5

0.0

(b) Isotherms (equally spaced).

Figure 3   Streamlines and temperature contours for Case 2.

the obstruction has relatively little influence on the flow. This can be attributed to the fact that in pure natural convection at this Rayleigh number, the central core of the fluid region is relatively stagnant. This stagnant region approximately coincides with the obstruction of Cases 1 and 2. As a result, the Nusselt numbers for Cases 1 and 2 (see Table 2) do not differ significantly from the corresponding pure natural convection value (see also Fig. 8).

In both cases, the isotherms (Figs. 2b and 3b) in the obstruction are almost horizontal, indicating that heat is conducted vertically through the obstruction, from the higher temperature fluid in the upper passage to the lower temperature fluid in the lower passage. The spacing of the isotherms in the obstruction shows, as expected, that this effect is particularly important for a high thermal conductivity obstruction. Thus, for $k^* = 5$ (Fig. 3b), the hot fluid flowing in the upper passage transfers a significant portion of its sensible

heat through the obstruction to the cold fluid flowing in the lower passage, instead of carrying it all the way to the cold wall of the enclosure. In turn, the cold fluid flowing in the lower passage is significantly heated by the obstruction, instead of by the hot wall of the enclosure. This "short-circuiting" by the high thermal conductivity obstruction effectively reduces the overall natural convection heat transfer between the hot and cold walls of the enclosure. The opposite is true for $k^* = 0.2$ (Fig. 2b). Here, the relatively low thermal conductivity obstruction acts as an insulator between the hot and cold fluid streams in the upper and lower passages. The fluid flowing between the adiabatic wall and the obstruction experiences virtually no temperature change (see Fig. 2b). Thus, for $k^* = 0.2$, the horizontally flowing fluid advects heat more effectively between the vertical walls of the enclosure, instead of losing it to or gaining it from the obstruction as for $k^* = 5$.

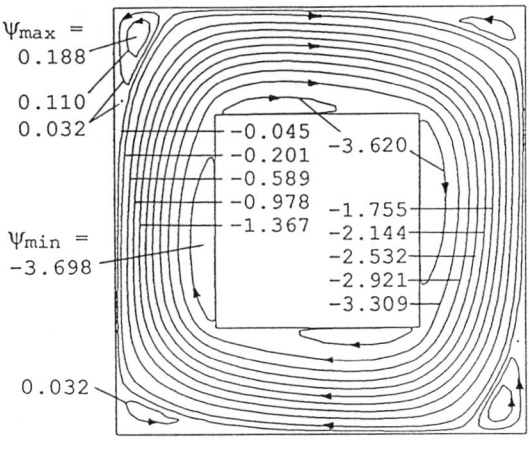

(a) Streamlines.

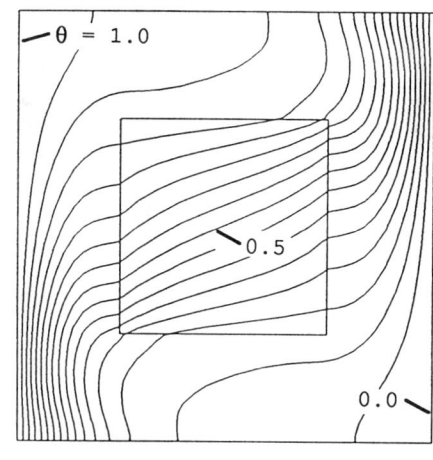

(b) Isotherms (equally spaced).

Figure 4   Streamlines and temperature contours for Case 3.

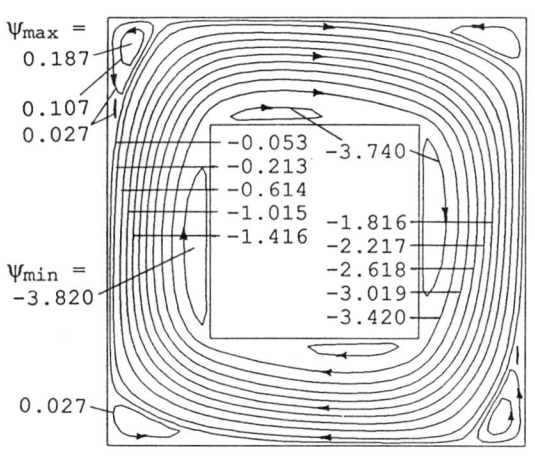

(a) Streamlines.

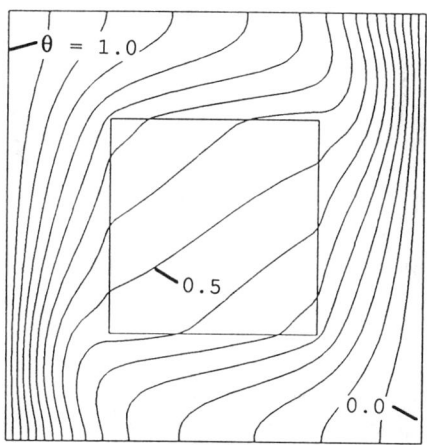

(b) Isotherms (equally spaced).

Figure 5   Streamlines and temperature contours for Case 4.

The different heat transfer properties of the obstructions also have an effect, albeit small, on the flow. For $k^\star$ = 0.2 (Fig. 2a), the streamlines near the lower-left and upper-right corners of the obstruction are more separated from the vertical obstruction walls than for $k^\star$ = 5 (Fig. 3a), indicating that the fluid has a higher momentum when making the 90 deg turn. This is due to the above mentioned fact that the horizontally flowing fluid experiences a smaller temperature change for $k^\star$ = 0.2 than for $k^\star$ = 5, resulting in larger temperature differences between the fluid leaving the horizontal passages and the vertical walls and, hence, larger buoyancy forces. As a consequence of this separation from the obstruction for $k^\star$ = 0.2, the flow is forced into more narrow regions between the obstruction and the vertical enclosure walls, resulting in locally higher velocities and heat transfer rates at the heated and cooled walls.

The above discussion clearly shows that the combined effects of the obstruction on the heat transfer and fluid flow are that the overall Nusselt number for $k^\star$ = 0.2 is higher than for $k^\star$ = 5 (see Table 2). In other words, a relatively low thermal conductivity obstruction enhances the heat transfer across the enclosure, while a high thermal conductivity obstruction reduces it. This result is somewhat unexpected and is only true up to a certain obstruction size. For relatively large obstructions (see, for example, Figs. 6 and 7, as well as Fig. 8) the natural convection is effectively suppressed, and heat transfer is mostly by conduction through the obstruction. The Nusselt number results for other obstruction sizes, as well as comparisons with pure natural convection and conduction values, are presented in the next section.

The effect of the Prandtl number of the fluid on the flow and heat transfer characteristics can be seen by comparing Figs. 4 and 5 (Pr = 0.01) to Figs. 2 and 3 (Pr = 0.71), respectively. As opposed to Cases 1 and 2, the streamlines in Figs. 4a and 5a show weak

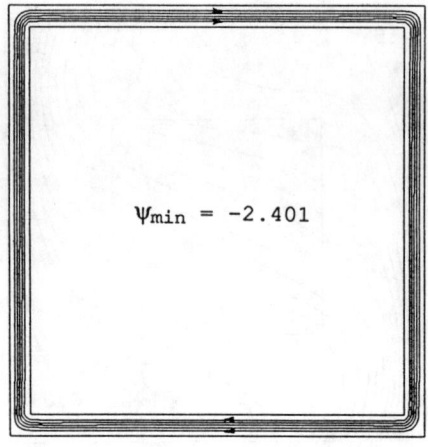

$\Psi_{min} = -2.401$

(a)   Streamlines (equally spaced).

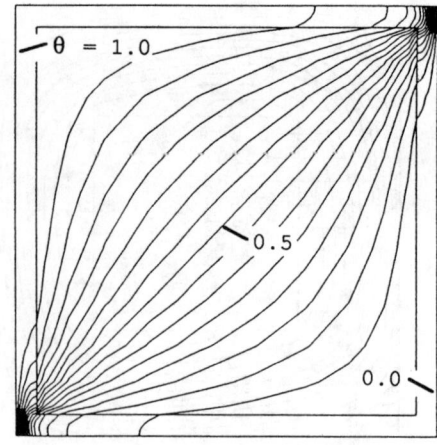

$\theta = 1.0$

0.5

0.0

(b)   Isotherms (equally spaced).

Figure 6 Streamlines and temperature contours for Case 5.

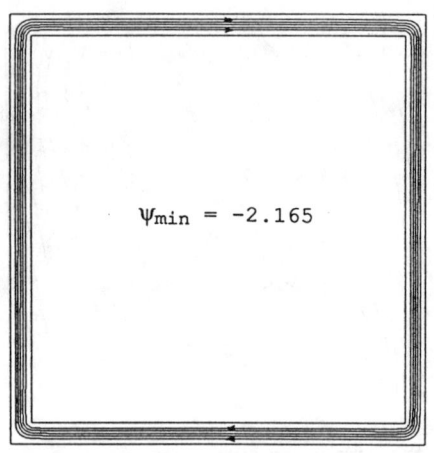

$\Psi_{min} = -2.165$

(a)   Streamlines (equally spaced).

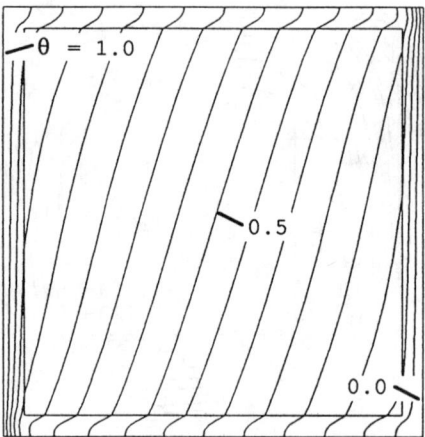

$\theta = 1.0$

0.5

0.0

(b)   Isotherms (equally spaced).

Figure 7 Streamlines and temperature contours for Case 6.

recirculations in the corners of the enclosure and along the walls of the obstruction. Such recirculations are typical for natural convection of liquid metals and can be attributed to the small value of the Prandtl number (Wolff et al., 1988). In addition, the low Prandtl number in Cases 3 and 4 causes the natural convection flow to be weaker and, hence, the Nusselt numbers to be lower (see Table 2) than in Cases 1 and 2. As can be seen from Figs. 4b and 5b, the effect of the obstruction on the heat transfer is similar as for Cases 1 and 2. Again, a relatively low thermal conductivity obstruction ($k^* = 0.2$, Fig. 4b) enhances the heat transfer across the enclosure by providing an insulation between the hot and cold fluids, while a relatively high thermal conductivity obstruction ($k^* = 5$, Fig. 5b) reduces the heat transfer across the enclosure by "short-circuiting" the two fluid streams. However, for Pr = 0.01 (Cases 3 and 4), the isotherms in the obstruction (Figs. 4b and 5b) are more vertical than for Pr = 0.71 (Figs. 2b and 3b), which can

be attributed to the weaker natural convection in the fluid (i.e., for no natural convection, the isotherms would be completely vertical). A comparison of Figs. 4b and 5b reveals that the isotherms in the obstruction for $k^* = 0.2$ are slightly more horizontal than for $k^* = 5$, indicating that, as for Pr = 0.71, the flow intensity and, thus, Nu is higher for $k^* = 0.2$ than for $k^* = 5$. Furthermore, the effect of the obstruction on the flow is similar to Cases 1 and 2, in that for $k^* = 0.2$ the streamlines near the lower left and upper right corners of the obstruction are more separated from the vertical obstruction walls than for $k^* = 5$. Consequently, the recirculations regions along the obstruction walls are slightly larger for $k^* = 0.2$ than for $k^* = 5$ (see Figs. 4a and 5a).

Streamlines and isotherms for Cases 5 and 6 with $\zeta = 0.9$ are displayed in Figs. 6 and 7, respectively. Since the obstruction occupies a significant portion of the enclosure, the effect of the fluid properties is negligible and the results for both cases are similar to those for $Pr = 0.01$. Due to the large size of the obstruction the natural convection flow is much weaker than in Cases 1 to 4 (see Figs. 6a and 7a). In fact, the streamlines for Cases 5 ($k^* = 0.2$) and 6 ($k^* = 5$) appear to be almost identical. The different thermal conductivities of the obstructions in Cases 5 and 6 have, however, a significant influence on the isotherms in both the fluid and the obstruction (see Figs. 6b and 7b).

For $k^* = 0.2$ (Fig. 6b), the isotherms in the fluid are highly concentrated in the lower-left and upper-right corners of the enclosure. The cold fluid arriving at the hot wall is heated within a short distance from the lower-left corner. Then, it flows almost isothermally through the left and upper passages until it is cooled again within a short distance from the upper-right corner. The cold fluid then flows almost isothermally through the right and lower passages until it reaches the lower-left corner. These peculiar heat transfer patterns can be explained by considering the isotherm patterns in the obstruction (Fig. 6b). As in Cases 1 and 3 ($k^* = 0.2$), the relatively low thermal conductivity obstruction acts as an insulator between the hot and cold fluid in the upper and lower (as well as left and right) passages. However, there is significant heat transfer through the obstruction from the fluid in the left (upper) passage to the fluid in lower (right) passage, particularly near the lower-left (upper-right) corners. This is due to the fact that the thermal resistance (i.e., conduction path) across a corner is much less than across the entire obstruction. Because of this conduction across the lower-left and upper-right corners of the obstruction, the cold (hot) fluid in the lower (upper) passage is already significantly heated (cooled) while it is approaching the hot (cold) enclosure wall. The resulting temperature increase (decrease) can clearly be seen in Fig. 6b. After flowing around the lower-left (upper-right) corner, the fluid then attains the hot (cold) wall temperature within a short distance from the bottom of the hot (cold) vertical wall. Although the remainder of the heated (cooled) wall has no further effect on the fluid in the left (right) passage, there is significant heat transfer <u>across</u> the isothermal fluid into the obstruction. In other words, in the regions where the fluid flows isothermally, the heat loss (gain) of the fluid to (from) the obstruction is balanced by the heat gain (loss) from (to) the vertical wall. Although the overall heat transfer across the enclosure may be expected to be highly dominated by heat conduction through the (large) obstruction, the fluid significantly augments the heat transfer by advecting heat from the lower-left to the upper-right corners of the enclosure. Consequently, the Nusselt number is far greater than the pure conduction value of 0.2 (see Table 2).

For $k^* = 5$ (Case 6), the isotherms (Fig. 7b) indicate that heat transfer is almost one-dimensional across the enclosure. Due to the relatively high thermal conductivity of the obstruction, the fluid at each position in the horizontal passages has almost the same temperature as the obstruction bounding it. Because

of the advecting nature of the horizontally flowing fluid, the isotherms in the obstruction are slightly tilted from the vertical. On the other hand, the (relatively low thermal conductivity) fluid in the vertical passages acts as a thermal barrier for the heat transfer across the enclosure. This can clearly be seen from the different spacings of the isotherms in the obstruction and the vertically flowing fluid. Overall, it can be seen that the fluid advects virtually no heat across the enclosure, and heat transfer is dominated by conduction through the vertical fluid passages and the obstruction. Due to the large thermal resistance of the fluid in the vertical passages, the Nusselt number is significantly below the pure conduction value of 5 (see Table 2).

Nusselt Numbers

Nusselt numbers as a function of the dimensionless size of the obstruction, ζ, are shown in Fig. 8 for $Pr = 0.71$. Results are presented for $Ra = 10^3$, 10^4, 10^5, and 10^6 and $k^* = 0.2$ and 5.0. As expected, the Nusselt numbers for $k^* = 0.2$ and 5 collapse into a single value as the obstruction size approaches zero, corresponding to pure natural convection in a vertical square enclosure. Similarly, as the obstruction size approaches unity, the Nusselt numbers for all Rayleigh numbers collapse into a single value of k^*, corresponding to pure conduction through the obstruction. As discussed earlier, the obstruction has relatively little influence on the natural convection flow, as long as it is not much larger than the almost stagnant central core present during natural convection in a vertical square enclosure without an obstruction. Since the size of this stagnant core increases with the Rayleigh number, the value of ζ at which the obstruction begins to significantly suppress the natural convection flow increases with Ra. As a result, the Nusselt number for $Ra = 10^3$ stays fairly close to the pure natural convection value up to only about $\zeta = 0.2$, while for $Ra = 10^6$ the obstruction has only a small effect on Nu up to almost $\zeta = 0.7$ (see Fig. 8). The above is true for both thermal conductivity ratios, although small differences can be observed. For $k^* = 0.2$, the Nusselt number first increases slightly above the pure natural convection value, reaches a maximum, and then decreases to the pure conduction value of k^*. This maximum in Nu corresponds to an obstruction size smaller than the size at which the obstruction significantly suppresses the natural convection flow. The maximum can be directly attributed to the insulating nature of the obstruction between the hot and cold fluid streams, as discussed in detail in connection with Cases 1 and 3. This finding implies that natural convection heat transfer for a vertical enclosure can be enhanced by inserting a relatively low thermal conductivity (and not too large) obstruction in the center. It should be noted that for $Ra = 10^3$, this effect is not observable due to the low intensity of the natural convection flow.

For $k^* = 5$, no such maximum is present. With increasing ζ, the Nusselt number decreases continually from the pure natural convection value until it reaches a minimum. Below the value of ζ at which the obstruction begins to

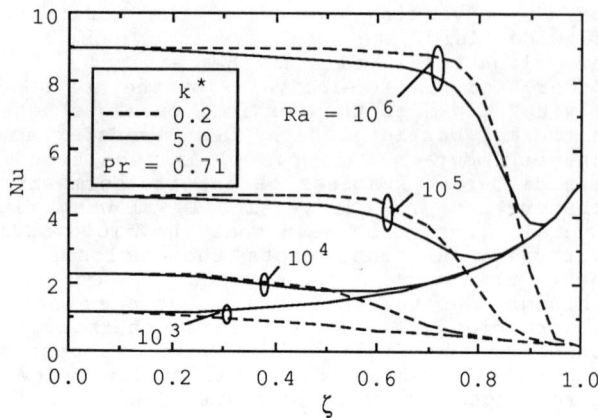

Figure 8   Effect of Rayleigh number and thermal
conductivity ratio on Nusselt number.

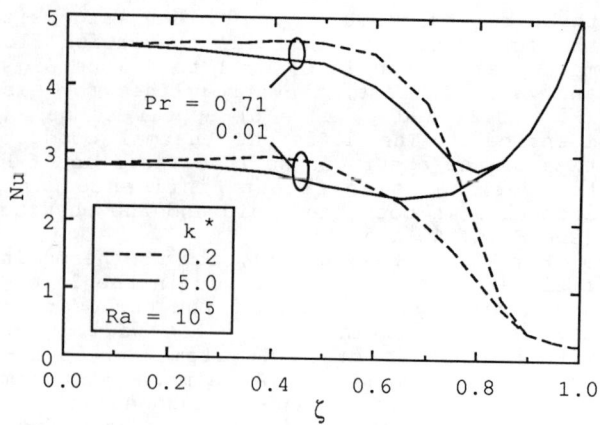

Figure 9   Effect of Prandtl number and thermal
conductivity ratio on Nusselt number.

significantly suppress the natural convection
flow, the decrease can be directly attributed
to the "short-circuiting" of the heat transfer
by the relatively high thermal conductivity
obstruction, as discussed in detail in connection with Cases 2 and 4.   Above that value of
ζ, the decrease is more steep and is predominantly due to the suppression of the natural
convection flow (that is present regardless of
the value of k^*).   The minimum in Nu corresponds to a value of ζ at which the natural
convection flow is relatively unimportant, but
the vertical fluid passages still represent a
major thermal resistance.   Case 6 (for Ra =
10^6) discussed in the previous section corresponds closely to that minimum.   Beyond the
minimum, the Nu number increases with ζ until
it reaches the pure conduction value of $k^* = 5$,
because the thickness and, thus, the thermal
resistance of the vertical fluid passages
decreases.   As expected, for smaller Rayleigh
numbers, the minimum in Nu shifts to lower
values of ζ and, at Ra = 10^3, a minimum is not
observable.   It is interesting to note that the
Nusselt numbers for $k^* = 5$ are consistently
below the ones for $k^* = 0.2$, up to a value of ζ
at   which   the   obstruction   has   already
significantly suppress the natural convection
flow (see Fig. 8).   This finding is generalized
for other values of k^* in the discussion of
Fig. 10 (see below).

Figure 9 shows the effect of the Prandtl
number on the variation of Nu with ζ, for Ra =
10^5 and $k^* = 0.2$ and 5.   In general, the trends
for Pr = 0.01 are very similar to the ones for
Pr = 0.71.   Essentially, a lower Prandtl number
reduces the flow intensity, so that the results
for Pr = 0.01 and Ra = 10^5 apparently
correspond to Pr = 0.71 and some lower value of
Ra.   This finding is consistent with previous
studies of natural convection of low Prandtl
number liquid metals (Wolff et al., 1988).

The effect of k^* on the Nusselt number is
shown in Fig. 10.   The values of Ra (=10^5),
Pr (=0.71), and ζ (=0.5) in Fig. 10 were
chosen to illustrate the effect of k^* for an obstruction that does not significantly suppress the
natural convection flow.   In order to demonstrate the corresponding variation of Nu with

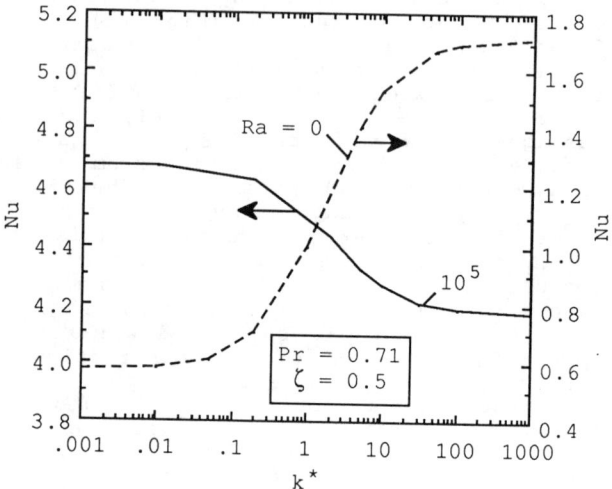

Figure 10   Effect of thermal conductivity ratio
on Nusselt number.

k^* that can be expected in the absence of natural convection flow, a curve for Ra = 0 is also
included in Fig. 10.   As expected, for Ra = 0,
the Nusselt number increases monotonically with
an increasing thermal conductivity ratio.   The
opposite is, however, true in the presence of
natural convection (i.e., for Ra = 10^5).   Here,
the Nusselt number continually decreases with
increasing k^*.   In the limits of $k^* \to 0$ and ∞,
the Nusselt number asymptotically approaches
values that are about 2.4 percent higher and
8.6 percent lower, respectively, than the
corresponding   Nusselt   number   for   no
obstruction, while most of the change takes
place between thermal conductivity ratios of
0.1 and 10.   Again, the increase of Nu above
the pure convection value can be explained by
the fact that for small k^*, the obstruction
provides an insulation between the horizontally
flowing hot and cold fluid streams.   On the
other hand, the decrease below the pure natural
convection value for large k^* is mainly due to
the "short-circuiting" of the two fluid streams
by conduction vertically through the obstruction.   Another reason for a lower Nusselt num-

ber is, of course, the suppression of the flow by the obstruction that is present regardless of the value of k^*. The latter effect is primarily responsible for the asymmetry of the Nu vs. k^* curve about the pure natural convection value. It can be seen from Fig. 10 that the Nusselt number (for $\zeta = 0.5$) is equal to the pure natural convection value (i.e., for $\zeta = 0$) for a thermal conductivity ratio of approximately 0.45.

CONCLUSIONS

A numerical study has been performed of natural convection in a vertical square enclosure containing a centered square obstruction. From an examination of the heat and fluid flow phenomena revealed by the numerical experiments, the following major conclusions can be drawn:

(i) The Nusselt number is not significantly different from the one for pure convection without an obstruction at the same Rayleigh and Prandtl numbers, up to an obstruction size that approximately coincides with the relatively stagnant fluid core present in a vertical enclosure without an obstruction.

(ii) For larger obstruction sizes, the variation of Nusselt number with obstruction size is significantly influenced by the ratio of the thermal conductivity of the obstruction to that of the fluid. For relatively large ratios, the Nusselt number displays a minimum at an obstruction size at which the natural convection flow is "successfully" suppressed, but the vertical fluid passages still represent a major thermal resistance. On the other hand, for relatively low ratios, natural convection in the relatively small fluid passages does play an important role and the Nusselt numbers are always above the value of the ratio.

(iii) For smaller obstruction sizes, the present study conclusively shows that the presence of a relatively low thermal conductivity obstruction can enhance the heat transfer rate across the enclosure to values that are above the ones for pure natural convection without the obstruction at the same Rayleigh and Prandtl numbers.

(iv) Prandtl numbers corresponding to liquid metals cause the formation of weak recirculations in the corners of the enclosure and along the obstruction walls. For such low Prandtl numbers, the Nusselt number results approximately coincide with the ones for a higher Prandtl number but a somewhat lower Rayleigh number.

In summary, the results of the present study are, in some respects, unexpected and can have important implications in many applications. Although this fundamental study revealed some of the basic heat transfer mechanisms, natural convection in enclosures containing an obstruction needs considerable more research attention. In particular, three-dimensional effects may be important in many practical systems, for example, if the obstruction does not extend over the entire depth of the enclosure.

ACKNOWLEDGEMENTS

One of the authors (C.B.) wishes to acknowledge partial support of the work reported in this paper by the National Science Foundation under grant No. CBT-8808888. Computer facilities were made available by The University of Iowa WEEG Computing Center.

REFERENCES

Beckermann, C., Viskanta, R., and Ramadhyani, S., 1988, "Natural Convection in Vertical Enclosures Containing Simultaneously Fluid and Porous Layers," J. Fluid Mechanics, Vol. 186, pp. 257-284.

de Vahl Davis, G., "Natural Convection of Air in a Square Cavity: A Bench Mark Numerical Solution," Int. J. Num. Methods Fluids, Vol. 3, pp. 249-264, 1983.

Emery, A. F., 1969, "Exploratory Studies of Free-Convection Heat Transfer Through an Enclosed Vertical Liquid Layer With a Vertical Baffle," J. Heat Transfer, Vol. 91, pp. 163-165.

Ostrach, S., 1988, "Natural Convection in Enclosures," J. Heat Transfer, Vol. 110, pp. 1175-1190.

Patankar, S. V., 1980, Numerical Heat Transfer and Fluid Flow, McGraw-Hill Book Co., New York, 1980.

Wolff, F., Beckermann, C., and Viskanta, R., 1988, "Natural Convection of Liquid Metals in Vertical Cavities," Experimental Thermal Fluid Sciences, Vol. 1, pp. 83-91.

Yang, K. T., 1987, "Natural Convection in Enclosures," Chap. 13, Handbook of Single-Phase Convective Heat Transfer, ed. S. Kakac, R. K. Shah, and W. Aung, Wiley, New York.

1989 National Heat Transfer Conference
HTD-Vol. 107, Heat Transfer in Convective Flows

NATURAL CONVECTION IN A VERTICAL HEATED TUBE ATTACHED TO A THERMALLY INSULATED CHIMNEY OF A DIFFERENT DIAMETER

Y. Asako and H. Nakamura
Department of Mechanical Engineering
Tokyo Metropolitan University
Tokyo, Japan

M. Faghri
Department of Mechanical Engineering
and Applied Mechanics
University of Rhode Island
Kingston, Rhode Island

ABSTRACT

Natural convection in a vertical heated tube attached to a thermally insulated chimney of a different diameter has been investigated numerically. The chimney located over the vertical heated tube enhances natural convection in the tube and leads to a higher heat transfer rate, which is the well known chimney effect. If the chimney diameter is larger than the heated tube diameter, the friction loss in the chimney region decreases with increasing chimney diameter. This induces an increase in the mass flow and leads to a higher heat transfer rate. However, from a geometrical consideration it is evident that the chimney effect diminishes when the chimney diameter exceeds its height. Therefore, there exists an optimum diameter which corresponds to the mavimum heat transfer rate. To investigate the chimney effect, computations are carried out for Rayleigh number in the range of 12.5 to 1250, based on the heated tube radius, and for a Prandtl number of 0.7. The numerical results are based on the control volume finite difference method. The average Nusselt number results are compared with the numerical results obtained for a chimney attached toa tube of the same diameter.

NOMENCLATURE

a	thermal diffusivity
C_p	specific heat of the fluid
H_1	dimensionless height of the heated tube, $=h_1/r_1$
H_2	dimensionless height of the chimney, $=h_2/r_1$
h	average heat transfer coefficient, equation (9)
h_1	height of the heated tube
h_2	height of the chimney
k	thermal conductivity
Nu	average Nusselt number, equation (11)
Pr	Prandtl number
P	dimensionless pressure difference
p'	pressure difference
Q	total heat transfer rate
Ra	Rayleigh number $[= g\beta r_1^3(t_W-t_\infty)/a\nu]$
r_1	radius of the heated tube
r_2	radius of the chimney
T	dimensionless temperature
t_W	tube temperature
t_∞	ambient temperature
U, V	dimensionless velocity components, equation (1)
u, v	velocity components
X, R	dimensionless coordinates [$= x/r_1$, $= r/r_1$]
x, r	coordinates
μ	viscosity
ν	kinematic viscosity
ρ	density of the fluid

INTRODUCTION

Natural convection is often a convenient and inexpensive mode of heat transfer. It is commonly employed in the cooling of electronic equipment and many other applications. Since the initial work by Bodoia and Osterle (1962) first succeeded on finite difference solutions of natural convection between vertical isothermal plates, many other researchers have studied natural convection in vertical channels. Davis and Perona (1971) studied natural convection in vertical heated tubes whereas Aung et. al. (1972a, 1972b) studied natural convection in vertical channels heated asymmetrically. Aihara (1973) and Nakamura et. al. (1982) included the entrance effect in their study of natural convection between vertical plates.

A thermally insulated chimney attached to a vertical heated channel induces an increase in the natural convection in the channel and leads to a higher heat transfer rate. This is the well known chimney effect discussed in the paper by Haaland and Sparrow (1983). If the chimney diameter is larger than the heated tube diameter, the friction loss in the chimney region decreases with increasing the chimney diameter. This induces an increase in the mass flow rate and leads to a higher heat transfer rate than the case for a chimney tube of the same diameter. However, from a geometrical consideration it is evident that the chimney effect diminishes in the limiting case of an extremely large chimney diameter compared with its height. Therefore, there exist an optimum diameter where the heat transfer is maximum. To investigate the chimney effect computations are carried out for Rayleigh number in the range of 12.5 to 1250, based on the heated tube radius, and for a Prandtl number of 0.7. The numerical results are based on a control volume finite difference method. The average Nusselt number results are compared with the numerical results obtained for a chimney attached to a tube of the same diameter.

FORMULATION

Description of the Problem

The problem to be considered in this study is schematically depicted in Fig. 1. It involves the determination of two-dimensional heat transfer for laminar natural convection in a vertical heated tube at a uniform temperature t_W attached to a thermally insulated chimney of a different diameter. The geometry of the heated tube is specified by the height h_1 and the radius r_1 and that of the chimney by the height h_2 and the radius r_2. This is a two-dimensional problem and because of the symmetry only half of the domain need to be considered for the numerical analysis.

The Conservation Equations

The governing equations to be considered are the continuity, momentum and energy equations. Laminar flow is assumed to prevail and the only fluid property variation considered is the density difference needed to establish the buoyancy term for which the Bussinesq density-temperature relation is employed. The following dimensionless variables are used:

$$X = x/r_1, \quad R = r/r_1, \quad U = ur_1/a, \quad V = vr_1/a,$$
$$P = p'/\rho(a/r_1)^2, \quad T = (t-t_\infty)/(t_W-t_\infty),$$
$$Ra = g\beta r_1^3(t_W-t_\infty)/a\nu \tag{1}$$

where r_1 is the radius of the heated tube, p' is the pressure difference between the local values within the channel and the ambient values at the same elevation. Then, upon introduction of the dimensionless variables, and parameters, the governing equations have the following forms:

$$\partial(RU)/\partial X + \partial(RV)/\partial R = 0 \tag{2}$$
$$U(\partial U/\partial X) + V(\partial U/\partial R) = -\partial P/\partial X + Pr \nabla^2U + Ra\, Pr\, T \tag{3}$$
$$U(\partial V/\partial X) + V(\partial V/\partial R) = -\partial P/\partial R + Pr \nabla^2V - Pr(V/R^2) \tag{4}$$
$$U(\partial T/\partial X) + V(\partial T/\partial R) = \nabla^2T \tag{5}$$

where $\nabla^2 = \partial^2/\partial X^2 + \partial^2/\partial R^2 + (1/R)\partial/\partial R \tag{6}$

In these equations, the streamwise second derivatives and pressure variations transverse to the streamwise direction are also present. To complete the formulation of the problem, it remains to discuss the boundary conditions. These are

on all walls	:	$U = V = 0$
on symmetry line (R = 0)	:	$\partial U/\partial R = V = 0$	(7)
on heated tube wall	:	$T = 1$
on chimney wall	:	$\partial T/\partial N = 0$	(8)

where N is the dimensionless normal vector to the chimney wall. The inflow and outflow boundary conditions will be discussed later.

Numerical Solutions

To avoid the specification of the inflow and outflow boundaries, which are not known, the solution domain is extended from $25r_1$ to $150r_1$ in the X-direction and $70r_1$ in the R-direction as depicted in Fig. 2. These limits were obtained by numerical experimentation so that further extension of the solution boundaries had no

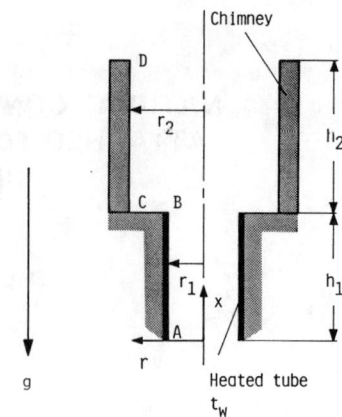

Fig. 1   A schematic diagram of a heated vertical tube with a chimney

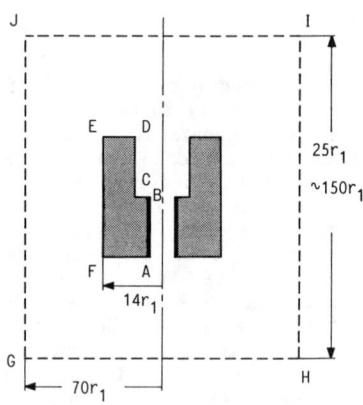

Fig. 2   Solution domain

effect on the results to within plotting accuracy. The boundary of this domain is denoted by GHIJ and is maintained at an ambient temperature t_∞.

The shaded domain ABCDEF in Fig. 2 consists of the heated tube and the chimney. The boundaries AB and CD are the inner surfaces of the heated tube and the chimney, respectively. For the application of the general-purpose computer program in cylindrical coordinates, the viscosity is taken to have a large value and the thermal conductivity is taken to be zero. These choices ensure that the shaded domain ABCDEF acts as a solid wall inside of which is thermally insulated. Only the boundary AB adjacent control volumes are maintained at constant temperature t_W. Note the boundary EF is fixed to $14r_1$ for all computations. The large step change in the fluid properties is handled by using harmonic-mean practice by Patankar (1980).

The discretization procedure of the equations is based on the power-law scheme of Patankar (1981), and the discretized equations are solved by using a line-by-line method. The pressure and velocities are linked by the SIMPLER algorithm of Patankar (1980).

The final computations were carried out for a grid containing 28X25 nodal points. The points in the X and

R coordinate directions are distributed in a nonuniform manner with a higher concentration of grids closer to the center of the solution domain except in the channel. The grid points in the channel are distributed with a higher concentration of grids closer to the wall. Each interior control volume contains one grid point, while the boundary adjacent control volume contains two grid points. A typical distribution for the main control volumes corresponding to $h_1/r_1=10$, $h_2/r_1=10$, $r_2/r_1=8.2$ is illustrated in Fig. 3.

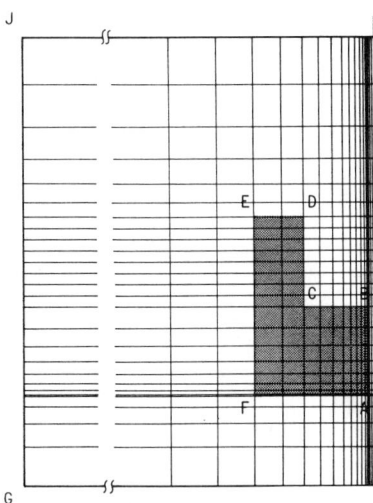

Fig. 3   A typical grid distribution for the main control volume

Supplementary runs for a channel without a chimney for $h_1/r_1=10$, Ra=12.5 and for a number of extended solution domains (height X radius : $25r_1$ X $70r_1$), ($50r_1$ X $85r_1$) and ($82r_1$ X $105r_1$) were performed to investigate size of the solution domain. A comparison of the Nusselt number under various size of the solution domain is given in Table 1. Thus, the solution domain ($25r_1$ X $70r_1$) is considered to be large enough for all computaions to maintain relatively moderate computing cost.

Supplementary runs for a channel without chimney, for $h_1/r_1=10$, Ra=12.5, with the grid points of (36X34), and (44X40) were also performed to investigate the grid size effect. A comparison of the Nusselt number with various grid size is given in Table 2. The (28X25) mesh are chosen for all computations to maintain relatively moderate computer time.

The convergence criterion that is used in this computations is that the value of the mass flux residuals (mass flow) divided by the total mass flow in each control volume takes a value under 10^{-5}. Since the present solution domain for the computation is very large, this convergence criterion is larger than that used in the general two-dimensional computations. The under-relaxation factors for the velocity and pressure are set to .2 and 1, respectively. About 400 to 1200 iterations are required to obtain a converged solution. This number of iterations depends on the geometric parameter and Rayleigh number. The computation for a higher Rayleigh number required a large number of iterations.

From an examination of the governing equations (2) to (5), it can be seen that there are two parameters whose values have to be specified prior to the initiation of the numerical solutions. These are the Prandtl number, Pr, and the Rayleigh number, Ra. The natural convection in this configuration can be applied for cooling of the micro-electronic components. Most of the micro-electronic components are small in size, and therefore their Ra number can be considered to be low. In this paper a value of 0.7 is chosen for Pr, and the Rayleigh number ranges from 12.5 to 1250. Aside from Pr and Ra, there are four geometric parameters which have to be specified. These are the radius of the heated tube and the chimney, r_1 and r_2, and heights of the heated tube and the chimney, h_1 and h_2. If the r_1 is used as a reference length, then $H_1=h_1/r_1$, $H_2=h_2/r_1$, and r_2/r_1 needs to be specified as geometric parameter. The selected values of H_1 are 5, 10, 20, and 40 and those of H_2 are 5, 10, and 20. The radius ratio, r_2/r_1 ranges from 1 to 10.7.

Nusselt Numbers
Attention will now be focused on the calculation of Nusselt numbers. The average heat transfer coefficients on the heated wall is defined as

$$h = Q/[A(t_W-t_\infty)] \qquad (9)$$

where A is the area of the heated wall, $2\pi r_1 h_1$, and Q is the total heat transfer rate from the heated wall. It can be expressed as

$$Q = 2\pi\rho C_p \left[\int_{exit} rut\ dr - \int_{ent.} rut\ dr\right]$$

$$+ 2\pi \left[\int_{ent.} rk(\partial t/\partial x)\ dr - \int_{exit} rk(\partial t/\partial x)dr\right] \qquad (10)$$

Therefore, the Nusselt number is expressed as follows:

$$Nu = hr_1/k$$

$$= (1/H_1)\left[\int_{exit} RUT\ dR - \int_{ent.} RUT\ dR\right.$$

$$+ \left.\int_{ent.} R(\partial T/\partial X)\ dR - \int_{exit} R(\partial T/\partial X)\ dR\right] \qquad (11)$$

RESULTS AND DISCUSSIONS

Nusselt Number for Channel without Chimney
The Nusselt number for a channel without a chimney is plotted as a function of Ra/H_1 for ($25r_1$ X $70r_1$) solution domain and for a grid containing 28X25 nodal

Table 1   The effect of the solution domain size on the average Nusselt number

height X radius	Nu
$25r_1$ X $70r_1$	0.0870
$48r_1$ X $84r_1$	0.0862
$82r_1$ X $105r_1$	0.0858

Table 2 Grid size effect on the average Nusselt number

Meshes (height X radius)	Nu
28 X 25	0.0870
36 X 34	0.0891
44 X 40	0.0911

points in Fig. 4. This results is for H_1=5, 10, 20, 40 and for Ra=12.5, 125 and 1250. The experimental data of Elenbaas (1942) and the analytical data of Davis and Perona (1971) are also depicted on this figure for comparison. The latter authors solved the boundary layer form of the conservation equations and also plotted their Nusselt number results as a function of Ra/H_1. As seen from this figure, slight discrepancies exist between the experimental and numerical results.

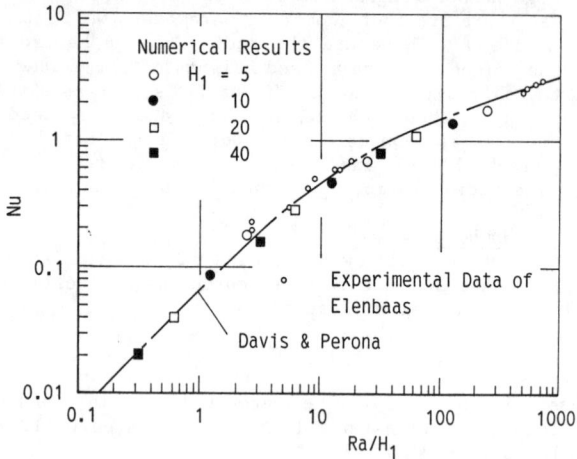

Fig. 4   Comparison of Nusselt number for a heated tube without a chimney

Nusselt Number for channel with Chimney

The representative Nusselt number results for H_1=10 and for H_2=10 are plotted as a function of Ra/H_1 in Fig. 5 with the radius ratio r_2/r_1 as curve parameters. In this figure, the limiting case of H_2=0 represents the result for the channel without the chimney. The dashed line indicates the result of Davis and Perona (1971). As seen from this figure, the Nusselt number results increase with increasing the radius ratio, r_2/r_1, as expected.

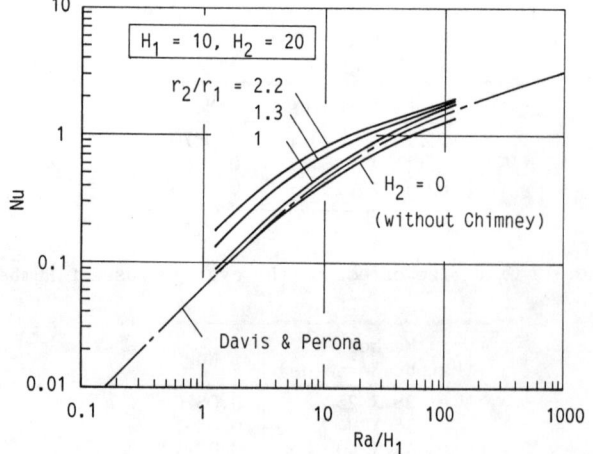

Fig. 5   Average Nusselt number for H_1=10 and for H_2=20

Effect of Radius Ratio on Nusselt Number

The representative Nusselt number ratios, Nu/Nu_0, for the nondimensional heated tube height, H_1=10, and for Ra=12.5 are plotted as a function of the radius ratio, r_2/r_1, in Fig. 6 with the nondimensional chimney height, H_2, as a curve parameter. Nu_0 represents the value of the Nusselt number for a channel without a chimney.   As seen from this figure, the Nusselt number ratio increases with increasing the radius ratio until it reaches a peak value.   Then, it decreases with increasing radius ratio until it approaches the value for the channel without a chimney.   In the computations, in the range of r_2/r_1=3 to 8, it was hard to obtain a converged solution. therefore, the converged values outside this range are connected by dashed lines.

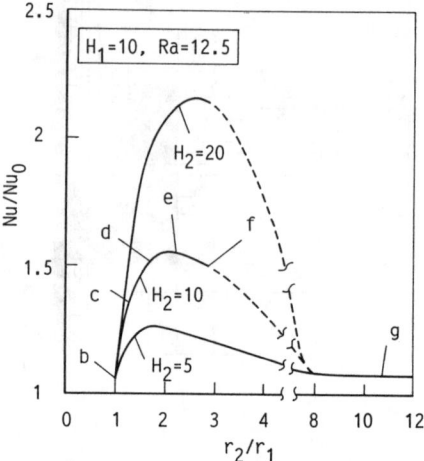

Fig. 6   Nusselt number ratio, Nu/Nu_0, for H_1=10, Ra=12.5 as a function of radius ratio, r_2/r_1

Streamline Diagrams

The representative streamline maps obtained from the numerical solutions for H_1=10, H_2=10 and for Ra=12.5 are presented in Fig. 7.   The Figs. 7(b) to (g) are for r_2/r_1=1, 1.3, 1.7, 2.2, 2.9, and 10.7, respectively. The streamline map for the channel without a chimney is presented in Fig. 7(a).   The streamlines Ψ/Ψ_W are spaced in increment of 0.2 and their values on the symmetric axis and on the wall are 0 and 1, respectively.   The line for H_2=10 in Fig. 6 are marked with letters b, c, d,•••g.   These letters correspond to radius ratios in Figs. 7(b) to (g).   Namely, Fig. 7(b) represents the streamline for a channel with a chimney of same diameter. Fig. 7(e) represents the streamlines for the case where maximum heat is transferred.   In fig. 7(f), a separation bubble can be seen at the bottom of the chimney.   Fig. 7(g) represents a streamline for the channel with a large radius chimney, where the inflow from the upper end of the chimney can be seen.   The chimney effect diminishes for such an extremely large radius ratio.

Flow Rate

The flow rate obtained by the evaluation of the stream function at the heated tube, Ψ_W, is plotted as a function of the radius ratio, $r_2/r1$, in Fig. 8 with the nondimensional chimney height, H_2, as a curve parameter for the case of H_1=10 and Ra=12.5.   The dashed line in the figure represents the value for the channel without the chimney. The streamfunction and the dimensionless streamfunction are defined as

$$\psi_W = \int_0^{r_1} ru\ dr \qquad (12)$$

$$\Psi_W = \psi_W/Ra \qquad (13)$$

In case of an infinitely long heated tube, the terms $\partial/\partial X$, $\partial^2/\partial X^2$ in equation (3) become zero, and the dimensionless temperature T becomes 1. Then, the momentum equation reduces to

$$\partial^2 U/\partial R^2 + (1/R)(\partial U/\partial R) + Ra = 0 \qquad (14)$$

Integrating this equation twice, one obtained U as

$$U = (Ra/4)(1-R^2) \qquad (15)$$

Therefore, the streamfunction Ψ_W can be obtained as

$$\Psi_W = \int_0^1 RU\ dR = (1/16)\ Ra \qquad (16)$$

The streamfunction for Ra=12.5 takes a value of 0.781. The streamfunction for the channel with a chimney of the same diameter cannot exceed this value. However, as seen from Fig. 8, the maximum value of the streamfunction for the channel with the chimney of $H_2=20$ is about 2.5 times higher than this value.

Optimum Diameter

As seen from Fig. 6, there exists an optimum diameter where the heat transfer is maximum. The optimum radius ratio, $(r_2/r_1)_{opt}$, is listed in Table 3, which was obtained by a curve fitting technique. Three numerically obtained values near the maximum are chosen for the curve fittings. Namely, a parabolic curve which passes through these three points, is obtained. The optimum radius ratio decreases with increasing nondimensional heated tube height, H_1, and it increases with an increase in the nondimensional chimney height, H_2.

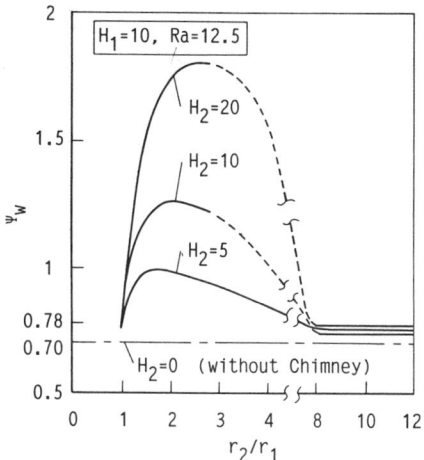

Fig. 8   Flow rate, Ψ_W, for $H_1=10$, Ra=12.5 as a function of radius ratio, r_2/r_1

Maximum Heat Transfer Rate

The maximum value of the Nusselt number ratio, $(Nu/Nu_0)_{max}$, is plotted as a function of the nondimensional chimney height, H_2, in Fig. 9 with the nondimensional heated tube height, H_1, as a curve parameter. Nu_0 represents the value for the channel without the chimney. As seen from the figure, the chimney effect is accentuated for a lower Rayleigh number, for lower H_1, and for a higher H_2. In the range of the present computations, the maximum Nusselt number ratio reaches about 2.5. This value is for the case of Ra=12.5, $H_1=5$, and $H_2=20$.

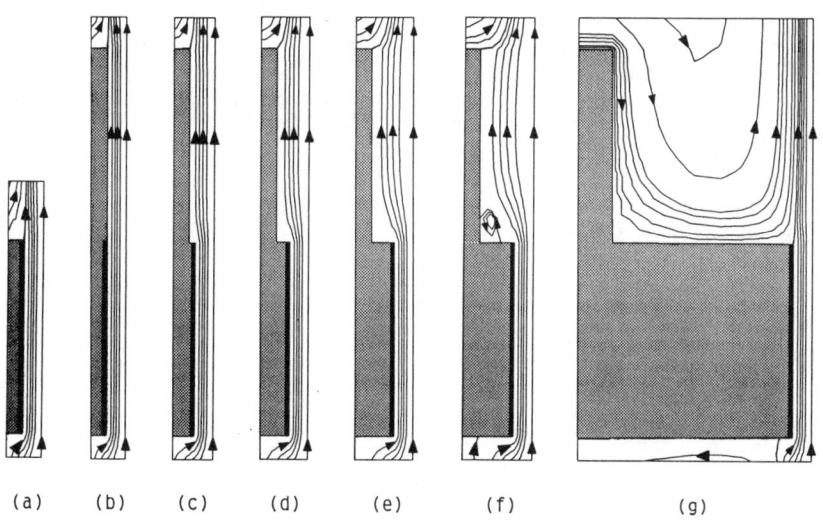

Fig. 7   Streamline diagrams for $H_1=10$, $H_2=10$, and for Ra=12.5 and radius ratio, r_2/r_1, as parameter: (a) ($H_2=0$), (b) $r_2/r_1=1$ (c) $r_2/r_1=1.3$ (d) $r_2/r_1=1.7$ (e) $r_2/r_1=2.2$ (f) $r_2/r_1=2.9$ (g) $r_2/r_1=10.7$

Table 3 Optimum diameter ratio $(r_2/r_1)_{opt}$

(Ra=12.5)

H_1	\(r_2/r_1\)$_{opt}$		
	$H_2=5$	$H_2=10$	$H_2=20$
5	1.95	2.41	2.73
10	1.90	2.24	2.63
20	1.84	2.05	2.51
40	1.77	1.95	2.08

(Ra=125)

H_1	$H_2=5$	$H_2=10$	$H_2=20$
5	1.80	2.01	2.51
10	1.73	1.99	2.46
20	1.70	1.98	2.07
40	1.70	1.97	2.06

(Ra=1250)

H_1	$H_2=5$	$H_2=10$	$H_2=20$
5	1.99	2.22	2.64
10	1.98	2.06	2.66
20	1.70	2.06	2.62
40	1.44	1.59	1.98

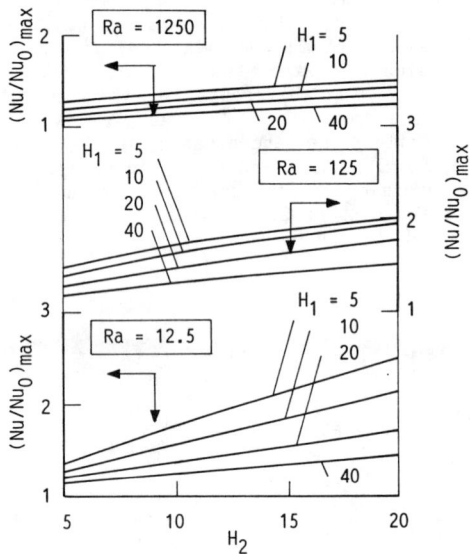

Fig. 9 Maximum Nusselt number ratio, $(Nu/Nu_0)_{max}$ as a function of H_2

CONCLUDING REMARKS

The heat transfer characteristic of natural convection in a vertical heated tube attached to a thermally insulated chimney of different diameter has been investigated by a numerical study to find the optimum chimney diameter where the maximum heat is transferred. The optimum diameter ratio, r_2/r_1 ranges from 1.4 to 2.7 depending on the geometric parameters, H_1 and H_2, and the Rayleigh number. The heat transfer rate for the channel with a chimney of optimum diameter is about 2.5 higher than the rate for the channel without a chimney. This effect is accentuated for lower Rayleigh numbers.

REFERENCES

Aihara, T., 1973, "Effects of Inlet Boundary Condition on Numerical Solutions of Free Convection between Vertical Parallel Plates," Rep. Inst. High Speed Mech., Tohoku Univ., Vol. 28, pp. 1-27.

Aung, W., 1972a, "Fully Developed Laminar Free Convection Between Vertical Plates Heated Asymmetrically," Int. Journal of Heat and Mass Transfer, Vol. 15, pp.1577-1580.

Aung, W., Fletcher, L.S. and Sernas, S., 1972b, "Developing Laminar Free Convection Between Vertical Plates with Asymmetric Heating," Int. Journal of Heat and Mass Transfer, Vol. 15, pp. 2293-2308.

Bodoia, J.R. and Osterle, J.F., 1962, "The Development of Free Convection Between Heated Vertical Plates," Journal of Heat Transfer, Vol. 84, pp. 40-44.

Davis L.P. and Perona, J.J., 1971, "Development of Free Convection Flow of a Gas in a Heated Vertical Open Tube," Int. Journal of Heat and Mass Transfer, Vol. 14, pp.889-903.

Elenbaas, W., 1942, "The Dissipation of Heat by Free Convection. The inner surface of Vertical tubes of Different Shapes of Cross-Section," Physica, Vol.9, pp. 865-874.

Haaland S.E. and Sparrow, E.M., 1983, "Solutions for The Channel Plume and The Parallel-Walled Chimney," Numerical Heat Transfer, Vol. 6, pp.155-172.

Nakamura, H., Asako, Y. and Naitou, T., 1982, "Heat Transfer by Free Convection between Two Parallel Flat Plates," Numerical Heat Transfer, Vol. 5, pp. 95-106.

Patankar, S.V., 1980, Numerical Heat Transfer and Fluid Flow, McGraw-Hill, New York.

Patankar, S.V., 1981, "A Calculation Procedure for Two-Dimensional Elliptic Situations," Numerical Heat Transfer Vol. 4, pp.409-425.

TRANSIENT NATURAL CONVECTION IN A VERTICAL CYLINDER
WITH A SPECIFIED WALL FLUX

J. Sun and P. H. Oosthuizen
Department of Mechanical Engineering
Queen's University, Kingston,
Ontario, Canada

ABSTRACT

Transient laminar natural convection inside a vertical cylinder containing a fluid following a sudden change in the wall heat flux has been numerically studied. The two-dimensional governing equations have been solved numerically using a finite difference iterative implicit technique with under relaxation. Solutions have been obtained for Prandtl numbers of from 0.7 to 10.0, Rayleigh numbers of 5000 to 30000 and for cylinder aspect ratios, i.e, the ratio of the cylinder length to its radius, of from 1 to to 6. The effects of the changes in these parameters on the variations of the temperature and flow patterns within the cylinder have been studied.

NOMENCLATURE

A = aspect ratio H/R
R = Radius of cylinder
H = height of cylinder
r z= dimensionless r and z coordinate
T = dimensionless temperature
t = dimensionless time
U V= dimensionless component of velocity in r and z directions
ψ = dimensionless stream function
ϕ = dimensionless vorticity

Superscripts

n = iteration step
'= indicates the dimensional variables

INTRODUCTION

Transient laminar natural convection inside a vertical cylinder following the sudden application of a uniform heat flux at the surface has been numerically studied. It has been assumed that the fluid in the cylinder is initially at a uniform temperature and at rest. Then, at time zero, a constant heat flux is applied to the vertical cylindrical wall, the top and bottom walls being assumed to be adiabatic. The flow patterns and temperature distributions within the cylinder and the transient wall temperature following this change have been computed and the effects of the cylinder aspect ratio, Prandtl number and Rayleigh number on the flow and temperature fields have been studied.

Steady state two-dimensional natural convection within a vertical cylindrical enclosure with various boundary conditions and aspect ratios has been studied quite extensively. Huang and Hsieh (1987) numerically studied steady laminar natural convection in a vertical enclosure with an aspect ratio of 1.0 The fluid was heated from the side wall and from the top wall and was cooled from the bottom wall. Rayleigh numbers of 10^2 to 2×10^5 and Prandtl numbers of 1 to 200 were considered. Lin and Akins (1986) used the SIMPLER numerical method to study the pseudo-steady-state laminar convection inside a cylinder with the aspect ratios of from 0.2 to 2 and Rayleigh numbers of from 10^3 to 10^7. Both the side wall and the end wall were heated. They indicated that the numerical results became unstable as the Rayleigh number increased over 10^7 due to the transition from laminar to weak turbulence. Lin and Akins also stated that the steady state natural convection can be described by a geometric parameter and Rayleigh number alone when the Prandtl number is greater than 5. This disagreed with the conclusion advanced by Huang and Hsieh (1987). The characteristic length used by Lin and Akins was defined as 6×volume/surface area. Concerning the transient case, Barakat and Clark (1966) undertook a numerical and experimental study for 2-D laminar natural convection in a vertical cylinder, which was partly filled with liquid. The thermal boundary condition used were such that side wall temperature was a function of axial location and time and the top and the bottom of the cylinder were adiabatic. They concluded that the heat transfer coefficient decreases rapidly following the start of transient, after which its value oscillates. The time average quasi-stead state value of the mean Nusselt number for laminar flow was found to be represented $Nu=0.54Ra^{0.25}$ for Rayleigh numbers from 10^4 to 10^9. A numerical study was conducted by Park and Lee (1985) for a similar geometry. Both the adiabatic top wall and the isothermal top wall cases were examined. Their results indicate that the average fluid temperatures in the cylinder are basically the same for the isothermal top case and adiabatic top case. A secondary motion was observed at later times due to the positive temperature gradient at the bottom wall. They also noted that the mean Nusselt number decreases at the beginning and then raises and then oscillates after the initial transient period. The mean Nusselt number was given $Nu=0.55\ Ra^{025}$ for Ra number from 9×10^7 to 6×10^8.

MATHEMATICAL FORMULATION

The flow is assumed to be laminar and the fluid properties are assumed constant except for the density changes with temperature. The resultant buoyancy forces have been dealt with using the Boussinesq approximation. The flow is described by the unsteady two dimensional continuity, Navier-Stokes and energy equations in cylindrical coordinate, i.e. by,

Continuity:

$$\frac{U'}{r} + \frac{\partial U'}{\partial r} + \frac{\partial W'}{\partial z} = 0 \qquad (1)$$

Momentum:

$$\rho\left(\frac{\partial U'}{\partial t'} + U'\frac{\partial U'}{\partial r'} + W'\frac{\partial U'}{\partial z'}\right)$$

$$=Br-\frac{\partial P'}{\partial r'}+\mu\left(\frac{1}{r'}\frac{\partial U'}{\partial r'} - \frac{U'}{r^2} + \frac{\partial^2 U'}{\partial r'^2} + \frac{\partial^2 U'}{\partial z'^2}\right) \qquad (2)$$

$$\rho\left(\frac{\partial W'}{\partial t'} + U'\frac{\partial W'}{\partial r'} + W'\frac{\partial W'}{\partial z'}\right)$$

$$=-\frac{\partial P'}{\partial z'}+ \mu\left(\frac{1}{r'}\frac{\partial W'}{\partial r'} + \frac{\partial^2 W'}{\partial r'^2} + \frac{\partial^2 W'}{\partial z'^2}\right) \qquad (3)$$

Energy:

$$\rho C_P\left(\frac{\partial T'}{\partial t'} + U'\frac{\partial T'}{\partial r'} + W'\frac{\partial T'}{\partial z'}\right)$$

$$= k\left(\frac{1}{r'}\frac{\partial T'}{\partial r'} + \frac{\partial^2 T'}{\partial r'^2} + \frac{\partial^2 T'}{\partial z'^2}\right) \qquad (4)$$

Where the buoyancy term is given, because of use the Boussinesq approach by:

$$Br=\beta g\rho(T'-T'_{in})$$

A stream function-vorticity formulation of the above set of equations has been adopted. These quantities have, as usual, been defined by:

$$U'= -\frac{\partial\psi'}{\partial z'} \qquad W'= \frac{\partial\psi'r'}{r'\partial r'}$$

$$\phi'= \frac{\partial U'}{\partial z'} - \frac{\partial W'}{\partial r'} \qquad (5)$$

The use of these variables allows the pressure terms to be eliminated from the governing equations.

The following dimensionless variables have then been introduced:

$$\psi = \frac{\psi'}{\alpha R} \qquad \phi = \frac{\phi'R^3}{\alpha} \qquad r = \frac{r'}{R}$$

$$z = \frac{z'}{R} \qquad t = \frac{t'}{R^2}\nu \qquad T=\frac{(T'-T_{in})k}{q_w R} \qquad (6)$$

The governing equations become in terms of these dimensionless variables

$$- \frac{\psi}{r^2} + \frac{1}{r^2}\frac{\partial\psi}{\partial r}+ \frac{\partial^2\psi}{r\ \partial r^2} + \frac{\partial^2\psi}{r\partial z^2} = - \phi \qquad (7)$$

$$\frac{\partial\phi}{\partial t} +\frac{1}{Pr}\frac{1}{r\partial r}\left(\frac{\partial\psi r}{\partial z}\frac{\partial\phi}{\partial z} - \frac{\partial\psi}{\partial z}\frac{\partial\phi}{\partial r} +\phi\frac{\partial\psi}{r\partial z}\right)$$

$$= \frac{\partial\phi}{r\partial r} + \frac{\partial^2\phi}{\partial r^2} + \frac{\partial^2\phi}{\partial z^2} - \frac{\phi}{r^2} -Ra\frac{\partial T}{\partial r} \qquad (8)$$

$$Pr\frac{\partial T}{\partial t} + \frac{\partial(\psi r)}{r\partial r}\frac{\partial T}{\partial z} - \frac{\partial\psi}{\partial z}\frac{\partial T}{\partial r}$$

$$= \frac{1}{r}\frac{\partial T}{\partial r} + \frac{\partial^2 T}{\partial r^2} + \frac{\partial^2 T}{\partial z^2} \qquad (9)$$

Where the Rayleigh number has as usual been defined by:

$$Ra = \frac{g\beta q_w R^4}{\nu\alpha} \qquad (10)$$

The flow is assumed to be symmetrical about the vertical center line, so the boundary conditions on the solution are:

on side wall:

$$r=1: \quad \psi=0, \quad \phi= - \frac{\partial^2\psi}{r^2\partial r^2} \quad \frac{\partial T}{\partial r} = 1$$

on center line:

$$r=0: \quad \psi=0, \quad \phi = - \frac{\partial^2\psi}{r\partial z^2} \quad \frac{\partial T}{\partial r} = 0 \qquad (11)$$

The initial condition at time zero are: everywhere in the cylinder

306

$\psi = 0, \qquad \phi = 0,$

$r \neq 1, \; z \neq 0, \; z \neq A; \qquad T = 0$

$r = 1; \qquad \dfrac{\partial T}{\partial r} = 1 \qquad (12)$

$z = 0, \; z = A; \qquad \dfrac{\partial T}{\partial z} = 0$

NUMERICAL TECHNIQUE

The computations were only carried out for one half of the cylindrical space since symmetry about the axis was assumed. The numerical solution was obtained using a method that is similar to that used by Oosthuizen and Kuhn (1984), which is an extension of that used by Han in (1979). The method is an iterative semi-implicit finite difference technique with under-relaxation to insure the stability of numerical solution. The spatial derivatives in the governing equations, boundary and initial conditions were approximated using second order finite difference approximations while the time derivatives were approximated using first order finite-difference approximations. A central difference scheme was used for all interior grid points and a backward or forward difference scheme was used for grid points on the wall. A uniform grid spacing was used in both the r and z directions.

Equations (6) to (13) can be written in following finite difference form:

$$\frac{\psi_{i+1,j} - 2\psi_{i,j} + \psi_{i-1,j}}{\Delta r^2} + \frac{\psi_{i,j+1} - 2\psi_{i,j} + \psi_{i,j-1}}{\Delta z^2}$$

$$+ \frac{\psi_{i+1,j} - \psi_{i-1,j}}{2r\Delta r} - \frac{\psi_{i,j}}{r^2} = \phi_{i,j} \qquad (13)$$

$$\frac{\phi^2 - \phi^1}{\Delta t} + \frac{1}{Pr} \left[\left(\frac{\psi_{i+1,j} r_{i+1} - \psi_{i-1,j} r_{i-1}}{2r\Delta r} \right) \right.$$

$$\left(\frac{\phi_{i,j+1} - \phi_{i,j-1}}{2\Delta z} \right)$$

$$+ \left(\frac{\psi_{i,j+1} - \psi_{i,j-1}}{2\Delta z} \right) \left(\frac{\phi_{i+1,j} - \phi_{i-1,j}}{2\Delta r} \right)$$

$$+ \phi_{i,j} \left(\frac{\psi_{i,j+1} - \psi_{i,j-1}}{2r\Delta z} \right) \right]$$

$$= \frac{\phi_{i+1,j} - 2\phi_{i,j} + \phi_{i-1,j}}{\Delta r^2} +$$

$$\frac{\phi_{i,j+1} - 2\phi_{i,j} + \phi_{i,j-1}}{\Delta z^2} - \frac{\phi_{i,j}}{r^2}$$

$$+ \frac{(\phi_{i+1,j} - \phi_{i-1,j})}{r2\Delta r} - Ra \left[\frac{T_{i+1,j} - T_{i-1,j}}{2\Delta r} \right] \qquad (14)$$

$$Pr \frac{T^2 - T^1}{\Delta t} + \left(\frac{\psi_{i+1,j} r_{i+1} - \psi_{i-1,j} r_{i-1}}{2r\Delta r} \right)$$

$$\left(\frac{T_{i,j+1} - T_{i,j-1}}{2\Delta z} \right)$$

$$- \left(\frac{\psi_{i,j+1} - \psi_{i,j-1}}{2\Delta z} \right) \left(\frac{T_{i+1,j} - T_{i-1,j}}{2\Delta r} \right)$$

$$= \frac{T_{i+1,j} - 2T_{i,j} + T_{i-1,j}}{\Delta r^2}$$

$$+ \frac{T_{i,j+1} - 2T_{i,j} + T_{i,j-1}}{\Delta z^2} + \frac{T_{i+1,j} - T_{i-1,j}}{r \, 2\Delta r} \qquad (15)$$

where superscripts 1 and 2 refer to conditions at the beginning and the end of the time step. Similarly the boundary conditions become a finite diffidence form:

$$\psi_{1,j} = 0, \qquad \phi = - \frac{-2\psi_{2,j} + \psi_{3,j}}{\Delta r^2}$$

$$\psi_{i,1} = 0, \qquad \phi = - \frac{-2\psi_{i,2} + \psi_{i,3}}{\Delta z^2}$$

$$\psi_{i,n} = 0, \qquad \phi = - \frac{-2\psi_{i,n-1} + \psi_{i,n-1}}{2}$$

$$T_{1,j} = \frac{4T_{2,j}}{3} - \frac{T_{3,j}}{3} + \frac{2\Delta r}{3}$$

$$T_{n,j} = \frac{4T_{n-1,j}}{3} - \frac{T_{n-2,j}}{3}$$

$$T_{i,1} = \frac{4T_{i,2}}{3} - \frac{T_{i,3}}{3}$$

$$T_{i,n} = \frac{4T_{i,n-1}}{3} - \frac{T_{i,n-2}}{3} \qquad (16)$$

initial conditions
for all points

$$\psi_{i,j} = 0, \quad \phi_{i,j} = 0, \quad T_{i,j} = 0$$

at i=1, any j:

$$T_{1,j} = \frac{4T_{2,j}}{3} - \frac{T_{3,j}}{3} + \frac{2\Delta r}{3} \qquad (17)$$

As was mentioned above, equations (13) through (17) form as a set of the semi implicit difference equations, i.e the values of the unknowns in the equations (14) and (15) are taken at the time level t+Δt, but the nonlinear terms $\partial\psi/\partial z \partial\phi/\partial r$ and $\partial\psi/\partial z \partial T/\partial r$ are linearized by considering the derivatives of stream function to be known and are taken equal to their values at time t when solving the vorticity and temperature field. With this numerical technique, numerical instabilities may occur when the dimensionless time increment becomes large. For the present study, the time step was varied according to the Ra, values 10^{-2} and 10^{-3} were used.

In the actual numerical solution, iteration was employed at each time step. The energy equation was first solved, then the vorticity equation was solved and then the stream function at each grid point was computed using the updated vorticity. Each step has been repeated twice before continuing to the next one. The iterative procedure was terminated when the following relative-error convergence criterion on temperature was satisfied:

$$\text{Max} \left| \frac{T_{i,j}^{(n+1)} - T_{i,j}^{(n)}}{T_{i,j}^{(n)}} \right| \leq 0.001$$

where (n) indicates the temperature T at the nth step of the iteration. Selection of the optimum value for the under relaxation factor to obtain a fast and stable solution was based on experience. It was found that values between 0.6 to 1.0 were suitable for use with the present numerical method.

Grids of 21×11, 21×21 and 21×31 were generally used depending on the values Ra and aspect ratio used. Calculations were carried out using a number of different grid sizes. The results showed that the grid independence was achieved when the number of grid points in

radial direction was ≥21.

RESULTS AND DISCUSSION

The solution has as parameters the Rayleigh number, the Prandtl number and the aspect ratio of cylinder, H/R. Calculations were carried out for Rayleigh numbers of between 5000 and 30000, for Prandtl numbers of between 0.7 to 10.0 and for aspect ratios of between 1 to 6.

Flow Pattern and Temperature Distribution

Typical isotherms and streamline patterns are shown in Figs. 1 to 4. These results are for aspect ratios of 1, 2, 4 and 6 respectively for Ra of 10^4 and Pr of 1.0. Each figure shows contours at dimensionless times of 0.01, 0.02 and 0.03. Since the flow is assumed to be symmetrical, isotherms are shown on the left hand side and streamlines are shown on the right hand side. t will be noted from the results given in these figures that, unlike what happens in the constant wall temperature case (Sun and Oosthuizen 1988), only single vortex forms in the present case for all the time and at all the aspect ratios considered. From these figures it will also be noted that at early times, the horizontal temperature gradient across the fluid is large as shown by the results given in the diagrams labeled (a) in each figure due to sudden application in the wall heat flux. The fluid motion is generated by this horizontal temperature

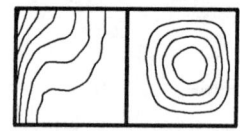 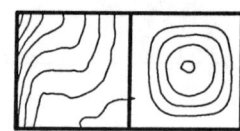

isotherms streamlines
$T_{max} = 0.4$ $\psi_{max} = 4.7$
$\Delta T = 0.05$ $\Delta \psi = 1.0$

(a) time=0.1

isotherms streamlines
$T_{max} = 0.5$ $\psi_{max} = 4.1$
$\Delta T = 0.05$ $\Delta \psi = 1.0$

(b) time=0.2

isotherms streamlines
$T_{max} = 0.6$ $\psi_{max} = 4.0$
$\Delta T = 0.05$ $\Delta \psi = 1.0$

(c) time=0.3

Fig.1 Isotherms and Streamlines for Ra=10000, Pr=1.0 A=1.0

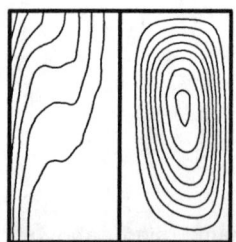 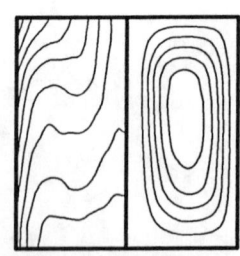 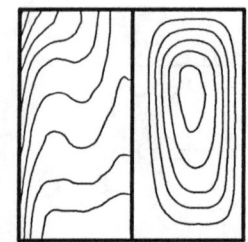

isotherms streamlines
$T_{max} = 0.4$ $\psi_{max} = 8.2$
$\Delta T = 0.05$ $\Delta \psi = 1.0$

(a) time=0.1

isotherms streamlines
$T_{max} = 0.6$ $\psi_{max} = 5.9$
$\Delta T = 0.05$ $\Delta \psi = 1.0$

(b) time=0.2

isotherms streamlines
$T_{max} = 0.7$ $\psi_{max} = 5.4$
$\Delta T = 0.05$ $\Delta \psi = 1.0$

(c) time=0.3

Fig.2 Isotherms and Streamlines for Ra=10000, Pr=1.0 A=2.0

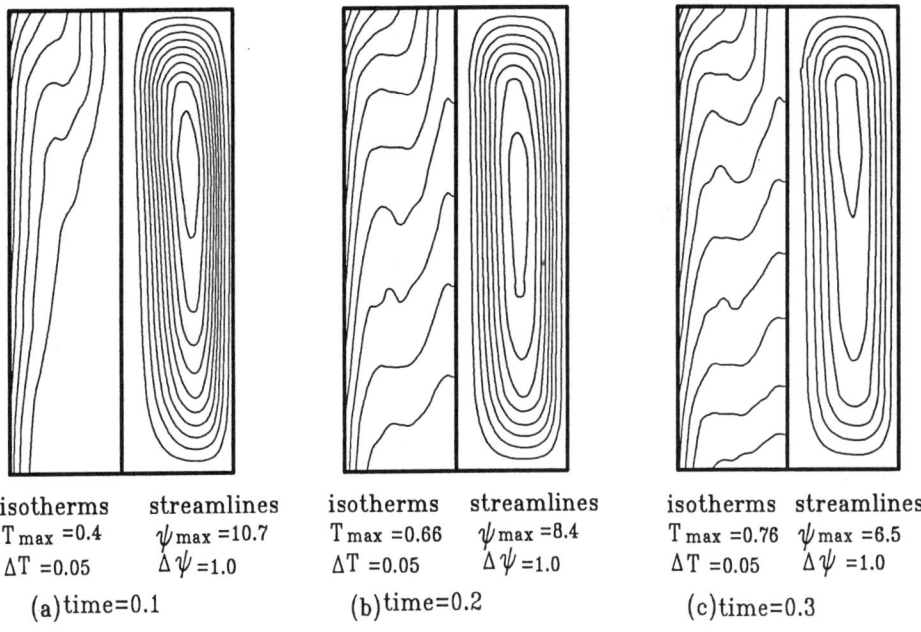

isotherms streamlines
$T_{max} = 0.4$ $\psi_{max} = 10.7$
$\Delta T = 0.05$ $\Delta\psi = 1.0$
(a)time=0.1

isotherms streamlines
$T_{max} = 0.66$ $\psi_{max} = 8.4$
$\Delta T = 0.05$ $\Delta\psi = 1.0$
(b)time=0.2

isotherms streamlines
$T_{max} = 0.76$ $\psi_{max} = 6.5$
$\Delta T = 0.05$ $\Delta\psi = 1.0$
(c)time=0.3

Fig.3 Isotherms and Streamlines for Ra=10000, Pr=1.0 A=4.0

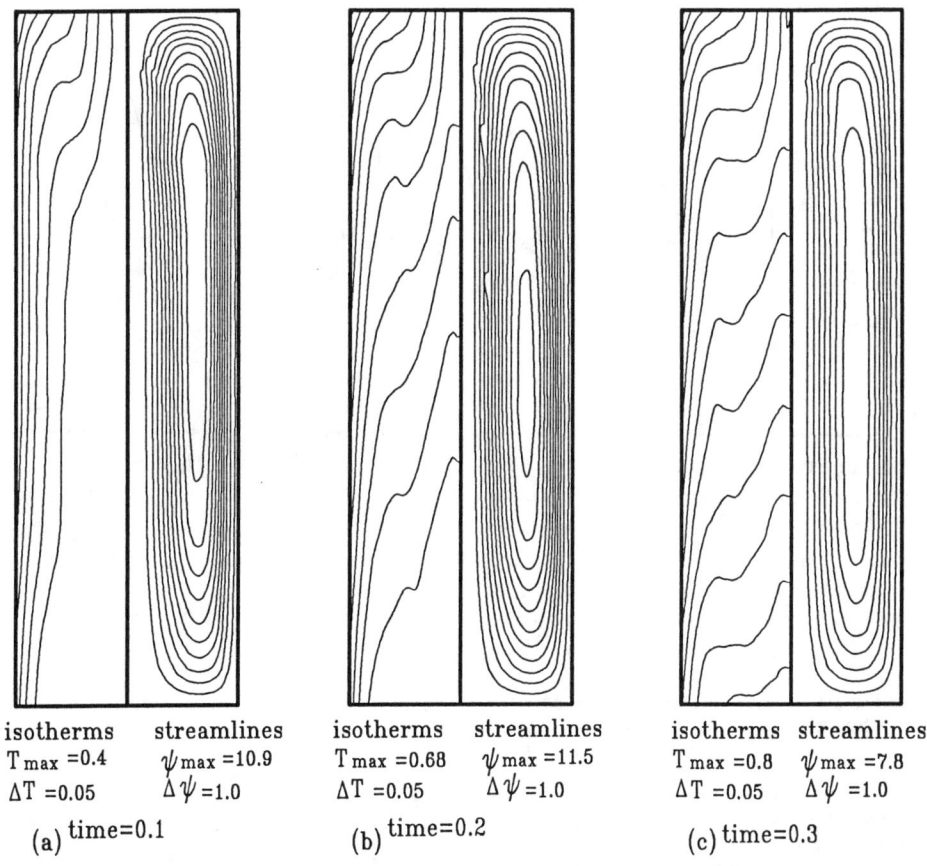

isotherms streamlines
$T_{max} = 0.4$ $\psi_{max} = 10.9$
$\Delta T = 0.05$ $\Delta\psi = 1.0$
(a) time=0.1

isotherms streamlines
$T_{max} = 0.68$ $\psi_{max} = 11.5$
$\Delta T = 0.05$ $\Delta\psi = 1.0$
(b) time=0.2

isotherms streamlines
$T_{max} = 0.8$ $\psi_{max} = 7.8$
$\Delta T = 0.05$ $\Delta\psi = 1.0$
(c) time=0.3

Fig.4 Isotherms and Streamlines for Ra=10000, Pr=1.0 A=6.0

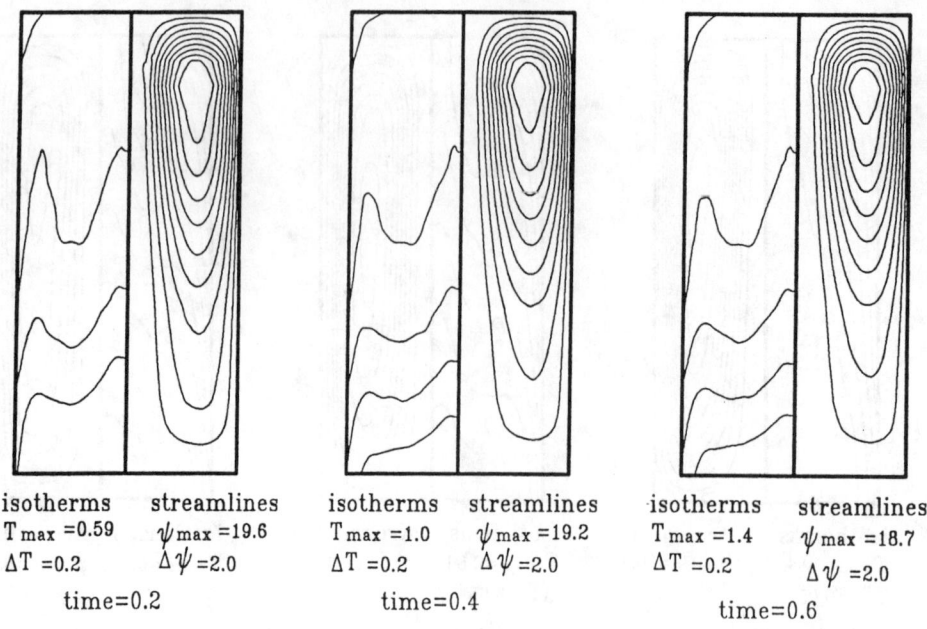

isotherms   streamlines
$T_{max} =0.59$ $\psi_{max} =19.6$
$\Delta T =0.2$ $\Delta\psi =2.0$
time=0.2

isotherms   streamlines
$T_{max} =1.0$ $\psi_{max} =19.2$
$\Delta T =0.2$ $\Delta\psi =2.0$
time=0.4

isotherms   streamlines
$T_{max} =1.4$ $\psi_{max} =18.7$
$\Delta T =0.2$ $\Delta\psi =2.0$
time=0.6

Fig.5 Isotherms and Streamlines for Ra=10000, Pr=0.7 A=4.0

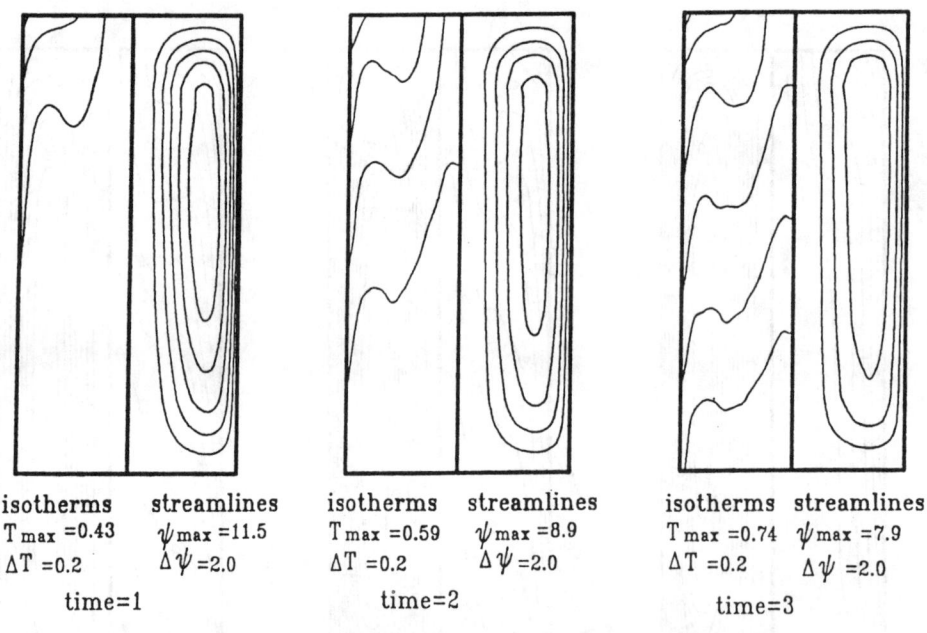

isotherms   streamlines
$T_{max} =0.43$ $\psi_{max} =11.5$
$\Delta T =0.2$ $\Delta\psi =2.0$
time=1

isotherms   streamlines
$T_{max} =0.59$ $\psi_{max} =8.9$
$\Delta T =0.2$ $\Delta\psi =2.0$
time=2

isotherms   streamlines
$T_{max} =0.74$ $\psi_{max} =7.9$
$\Delta T =0.2$ $\Delta\psi =2.0$
time=3

Fig.6 Isotherms and Streamlines for Ra=10000, Pr=1.0 A=4.0

gradient. The maximum temperature occurs at top corner of heated wall and minimum temperature is at the bottom near the centerline. As time progresses, the horizontal temperature gradients are reduced and a quasi-steady state is approached. In this quasi-steady state, the temperature gradient in the cylinder remains a constant with the time. Comparing the results shown in Figs. 1 to 4, for aspect ratios are of 1 to 6, it will be seen that the changes in aspect ratio do not have a major influence on the flow structure but that the value of the maximum stream function increases as with increasing aspect ratio, i.e, as the aspect ratio increases, the intensity of the fluid motion increases. A stronger thermal stratification is also observed for the larger aspect ratio.

The effect of Pr on the isotherm and streamline patterns is illustrated in the results given in Figs 5 and 6 which given results for Prandtl number of 0.7 and 10.0

respectively. At the lower Prandtl number, the isotherms tend to spread into the core region at higher rate and entire temperature field moves towards its final quasi-steady state at an earlier time. This is because the Prandtl number is a measure of the relative rate of diffusion of momentum and thermal energy and the spread of temperature field tends, therefore to be greater at the lower Pr. It will be seen from the figures that the maximum stream function reaches at a higher value at lower Pr.

The effect of Ra on the isotherms and streamlines is illustrated in the results given in Figs.7, 8 and 9, which show the results for Ra=5×10^3, 10^4 and 3×10^4 respectively. As Rayleigh number increases, the intensity of the fluid motion increases due to a stronger buoyancy force. Although the maximum wall temperatures in the three cases are only slightly different, the isotherms penetrate more into the unheated portion of the fluid and temperature field inside the

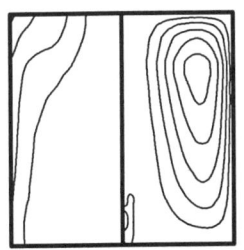 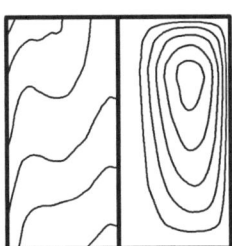 

isotherms   streamlines
T_{max} =0.43 ψ_{max} =5.4
ΔT =0.1 $\Delta\psi$ =1.0
 (a) time=0.1

isotherms   streamlines
T_{max} =0.67 ψ_{max} =5.3
ΔT =0.1 $\Delta\psi$ =1.0
 (b) time=0.3

isotherms   streamlines
T_{max} =0.92 ψ_{max} =5.0
ΔT =0.1 $\Delta\psi$ =1.0
 (c) time=0.5

Fig.7 Isotherms and Streamlines for Ra=5000, Pr=1.0 A=2.0

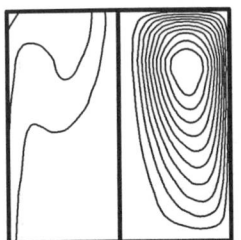 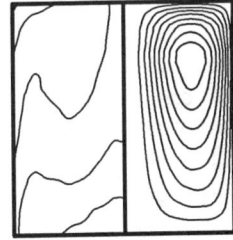

isotherms   streamlines
T_{max} =0.38 ψ_{max} =10.8
ΔT =0.1 $\Delta\psi$ =1.0
 (a) time=0.1

isotherms   streamlines
T_{max} =0.65 ψ_{max} =9.2
ΔT =0.1 $\Delta\psi$ =1.0
 (b) time=0.3

isotherms   streamlines
T_{max} =0.93 ψ_{max} =8.6
ΔT =0.1 $\Delta\psi$ =1.0
 (c) time=0.5

Fig.8 Isotherms and Streamlines for Ra=10000,Pr=1.0 A=2.0

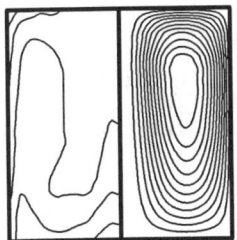 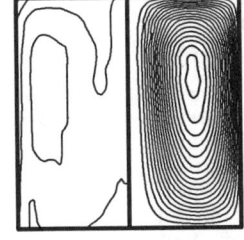 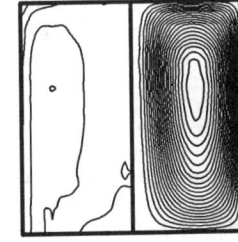

isotherms   streamlines
T_{max} =0.31 ψ_{max} =23.9
ΔT =0.05 $\Delta\psi$ =2.0
 (a) time=0.1

isotherms   streamlines
T_{max} =0.62 ψ_{max} =36.6
ΔT =0.05 $\Delta\psi$ =2.0
 (b) time=0.3

isotherms   streamlines
T_{max} =0.95 ψ_{max} =43.5
ΔT =0.05 $\Delta\psi$ =2.0
 (c) time=0.5

Fig.9 Isotherms and Streamlines for Ra=30000, Pr=1.0 A=2.0

cylinder becomes more uniform, i.e., there is less thermal stratification at the higher Ra due to the stronger convection. The final quasi-steady state will also be seen to be approached at earlier times at the higher Ra values.It will also be seen from the figures that at the highest Ra value considered, i.e., Ra≈3×10⁴, the strong circulation causes the colder region to occur in the upper portion of the cylinder near the wall rather than at the bottom.

Wall temperature and maximum stream function

Typical variation of the local temperature on the surface of the cylinder side wall with time and of the average side wall temperature with time are presented in Figs.10 to 14. Typical variations of the maximum stream function with time are given in Figs. 15 to 16.

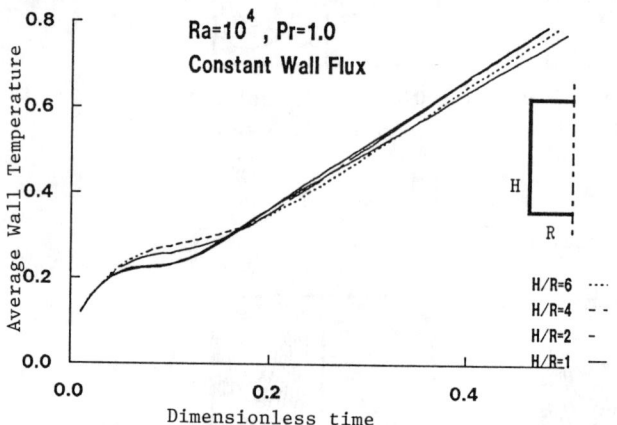

Fig. 10 Effect of aspect ratio on average wall temperatures

The effect of aspect ratio on the average wall temperature is illustrated by the results given in Fig.10. As will be seen, the average wall temperature for all the aspect ratios increases at the same rate with time at small times when conduction is the dominant mechanism. As time increases, there is a period where the temperature increment slows down slightly. The average wall temperatures for A=6 and A=1 are slightly higher than those for A=2 and A=4 during this period. With further increase in time, the average wall temperatures for different aspect ratios became identical. As will be seen from Fig.11 and Fig.12, the thermal stratification caused by convection is stronger at the higher aspect ratio. It will be noted from Fig.11, which gives results for A=1.0, that the changes in wall temperature along the side wall are not as strong as those for A=6 which are shown in Fig.12. The temperature at the bottom of the side wall is higher at A=1.0 than it is at A=6.0. However, the average wall temperature remains essentially the same for the various aspect ratio. Fig.13 shows effect of Pr on the average wall temperature, results being shown for Pr=0.7, 2.0 and 10.0. As is to be expected, the wall temperature increases at much lower rate for the higher Pr values.

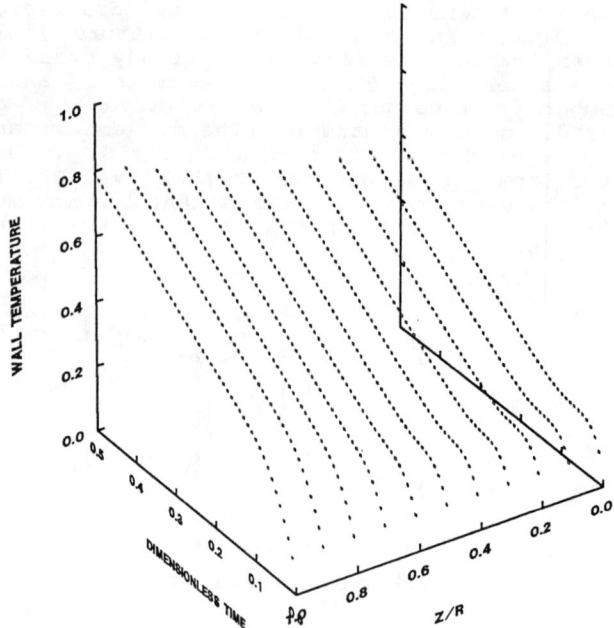

Fig.11 Local wall temperatures Ra=10⁴
Pr=1.0 A=1.0

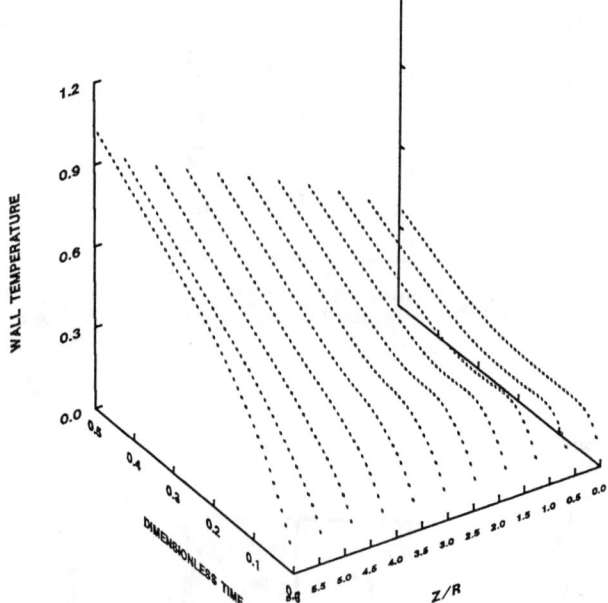

Fig.12 Local wall temperatures for Ra=10⁴
Pr=1.0 A=6.0

The variations of the average wall temperature with dimensionless time for various values of Ra are presented in Fig.14. It will be noted that there is a short period time, where following the initial rise the increase in average wall temperature slows

312

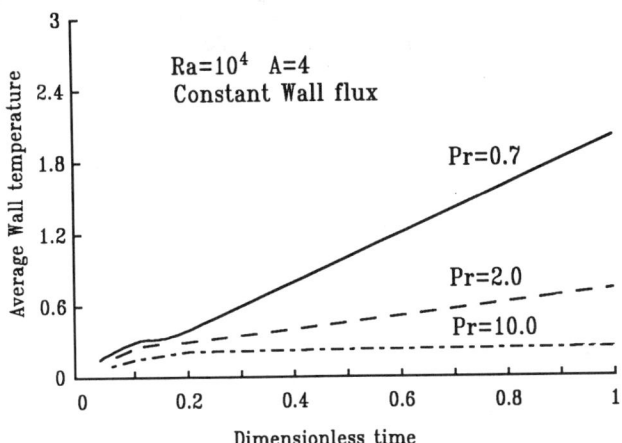

Fig.13 Effect of Prandtl number on
Average wall temperatures

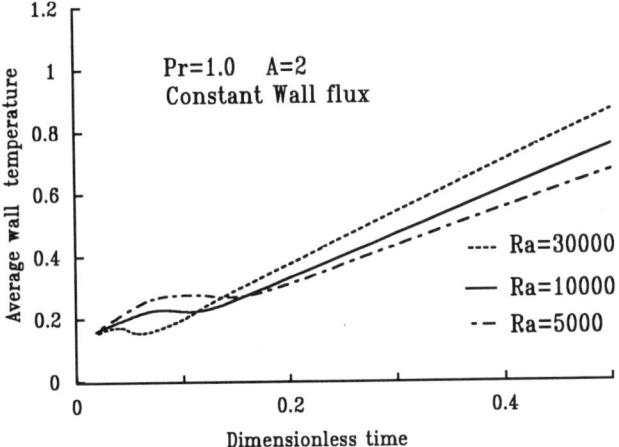

Fig.14 Effect of Rayleigh number on
Average wall temperatures

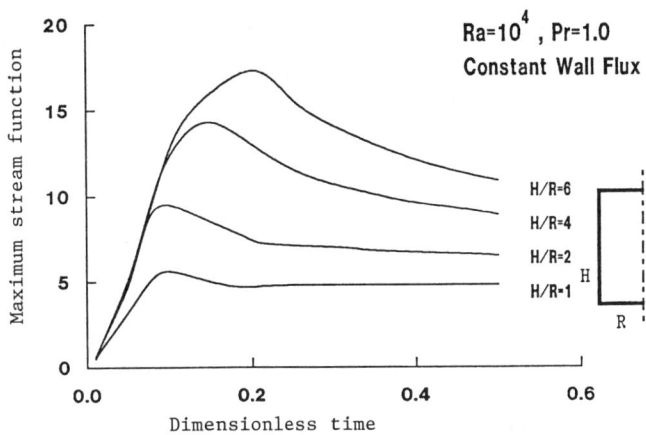

Fig.15 Effect of aspect ratio on the
maximum stream function

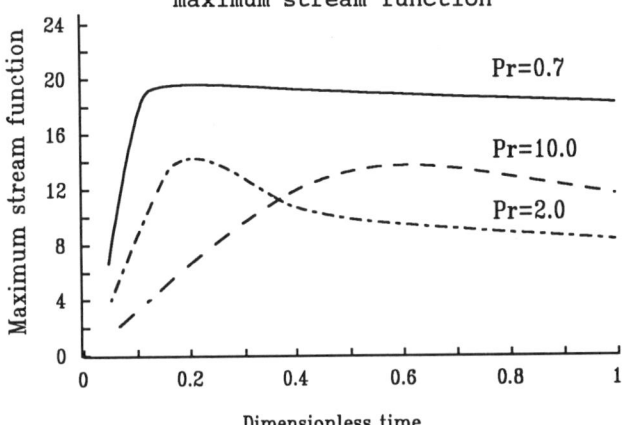

Fig.16 Effect of Prandtl number on the
maximum stream function

down. This is because as a result of the sudden change in wall heat flux the convective motion brings the cold fluid from the core region into the wall boundary layer region. As the Rayleigh number increases, this phenomenon occurs at smaller times and becomes more pronounced.

Typical variations of maximum stream function for the aspect ratios from 1 to 6 and for Prandtl numbers from 0.7 to 10 are shown in Fig. 15 and Fig. 16 respectively. It will be seen from these figures, the maximum stream function increases at small times, passes through a maximum value and then decreases towards a constant value as the flow approaches the quasi-steady state. At higher aspect ratios and higher values of Pr, the transient period is longer and the peak value of the maximum stream function is lower and occurs at later times.

CONCLUSIONS

The two dimensional flow in a closed vertical cylindrical enclosure following a sudden application of a uniform wall heat flux has been numerically studied. Following the change in the wall heat flux, a buoyancy induced fluid motion arises. However, as the fluid moves towards a new thermal balance this motion approaches a quasi-steady state. The main motion consists of a single vortex with upward flow along the hot wall and downward flow along the center line due to the symmetry. An examination of effect of aspect ratio, Pr number and Ra number on the flow and isotherm patterns as well as wall temperature has been conducted. The results indicate that the changes in the aspect ratio do not have a major influence on the flow structure and isotherm patterns. At higher Ra value, the average wall temperature decreases for a short period at small times. The value of the maximum stream function increases with increasing aspect ratio and Rayleigh number and decreases Prandtl number.. The peak value of stream function occurs at later dimensional time as aspect ratio increases and Pr decreases.

ACKNOWLEDGMENTS

This work was supported by the Natural Sciences and Engineering Research Council of Canada. The authors wish to thank Mr. S.J. Harrison and Ms. J. Paul for their assistance.

REFERENCES

Barakat, H. Z.and Clark, J. A., 1966, "Analytical and experimental Study of the Transient Convection Flows in Partially Filled Liquid Containers" Proceedings of 3rd International Heat Transfer Conference, Vol. 2 pp. 152-162.

Han, J. T. 1979, "A Computational Method to Solve Nonlinear Elliptic Equations of Natural Convection in Enclosures" *Numerical Heat Transfer*,Vol.2, pp. 165-175.

Huang, D. Y. and Hsieh, S. S 1987, "Analysis of Natural Convection in a Cylindrical Enclosure" *J. of Numerical Heat Transfer*, Vol.12, pp. 121-135.

Lee, J. H. and Park, W. H. 1985, "Theoretical Study of the Natural Convection Flows in a Partially Filled Vertical Cylinder Subjected to a Constant Wall Temperature", Korea Institute of Energy and Resources, Daejeon, Korea

Lin, Y. S. and Akins, R. G. 1986, "Pseudo-Steady-State Natural Convection Heat Transfer Inside a Vertical Cylinder" *J. of Numerical Heat Transfer*, Vol. 108, pp. 310-316.

Oosthuizen, P. H. and Kuhn, D 1984, "Unsteady Free Convective Flow in a Circular Container Half Filled with a Liquid and Half Filled with a Gas" ASME HTD Vol. 39, pp. 1-11, ASME Winter Annual Meeting.

Sun, J and Oosthuizen, P. H. 1988 "Transient Natural Convection in a Vertical Cylinder with a Specified Wall Temperature". ASME HTD-Vol. 96 Proceedings of the 1988 Natural Heat Transfer Conference, Vol. 2, pp. 107-113.

THE EFFECT OF PARTITIONS ON NATURAL CONVECTION IN ENCLOSURES

P. K.-B. Chao
Cotek Company
Paoli, Pennsylvania

N. Lior and S. W. Churchill
University of Pennsylvania
Philadelphia, Pennsylvania

H. Ozoe
Kyushu University
Fukuoka, Japan

ABSTRACT

This paper presents the results of a numerical and experimental study of the effect of partial baffles on natural convection in rectangular enclosures (aspect ratio 2 x 1 x 1) at angles of inclination between 0 and π rad. The analysis was performed for Ra =6000, and the experiments for Ra up to 2×10^4. There is good qualitative agreement between the numerical and experimental results, including the flow patterns. The baffles were shown to reduce the Nusselt number up to about 20% for all angles of inclination as compared to the enclosure without baffles. Increasing the conductivity of the baffles radically changes the flow pattern and heat transfer rate.

NOMENCLATURE

c   specific heat capacity of baffle, J/kg•K
g   acceleration due to gravity, m/s^2
H   height of enclosure, m
k   thermal conductivity, W/m•K
L   thickness of baffle, m
Nu $qH/L_f \Delta T$ = mean Nusselt number
Pr ν/α = Prandtl number
q   mean heat flux density through fluid, W/m^2
Ra $g\Delta\rho H^3/\nu\alpha\ \rho$
t   time, s
T   temperature, K
x   distance in the longer horizontal dimension from the non-elevated end, m
X   x/H
y   distance in the shorter horizontal dimension from the side to which the baffle is attached, m
Y   y/H
z   distance from the cooled plate, m
Z   z/H

Greek Symbols

α thermal diffusivity, m^2/s
δ_i finite-difference operator
θ angle of inclination of heated surface, rad

ν kinematic viscosity, m^2/s
ρ density, kg/m^3

Subscripts

f   fluid
i,j,k   indices for baffle grid
I,J,K   indices for fluid grid
NXF   location of baffle in X-direction
NYF   location of tip of baffle in Y-direction
w   baffle

INTRODUCTION

Motivated in large part by the desire to improve thermal insulation in solar collector windows and in building walls, and to improve heat distribution and transfer in passive solar houses, the "energy crisis" seventies have seen a major acceleration of existing studies of natural convection in enclosures, and an inception of new studies in this field (cf. references [1-28]). These studies continue to bear fruit not only as related to the original applications, but also in many other areas of current interest, such as crystal growth and building air quality control.

It is now understood more clearly that real natural convection flows in enclosures are not two-dimensional, and that a velocity component parallel to the familiar roll-cell axis is also present. The flow thus resembles a double helix, with fluid particles moving along both the circumference of the roll-cell and the direction of its principal axis, from the walls into the enclosure, up to a certain distance, and then back towards the walls. This third flow velocity component is due both to the drag at the end walls, and to thermal gradients generated at these walls because of the diminished rate of circulation.

The significance of the three-dimensionality of the flow

becomes even more pronounced when the enclosure is tilted, or when partial partitions are inserted. It was determined that steady laminar natural convection in horizontal enclosures with a bottom temperature higher than the top, is characterized by a train of roll-cells with axis parallel to the short side of the enclosure, an observation which also served to justify the many two-dimensional analyses of the phenomenon. This is no longer correct when the box is tilted: **inclination** about its longer side causes the axis of the roll cells to become oblique to the side, and at some critical angle all the roll cells form one large cell which has an axis perpendicular to the short side of the box. Tilting the box along its shorter side gradually merges the parallel roll cells into one large circulating cell, with its axis still parallel to the short side of the box. As the box is tilted from the horizontal position, the Nusselt number is first seen to decrease gradually to a minimum which coincides with the transition from one convective pattern to another, reaches a maximum at a higher angle of inclination (~60° to 90°) and then diminishes monotonically as the angle is increased to 180°. Both the minimum and maximum of the Nusselt number occur at slightly higher values for boxes inclined about the long side than for those inclined about the shorter side, but the values of Nu are about the same in both cases.

3-d calculations produce lower Nusselt number values than 2-d ones, for the same case, principally because the 3-d calculations account for the slow-down and redirection of the circulation by the solid ends.

The fact that the minimal Nusselt number occurred at angles of inclination at which the roll-cells were lined up with the longest axis, or the one along which the motion is most tortuous, indicated that the manipulation of roll-cell orientation by such means as **internal partial baffles** may result in the reduction of Nu. Studies by Chao et al. [29,30] for Ra up to 2 x 10⁴, Lin and Bejan [31] for $10^9 < Ra < 10^{10}$, and by Nansteel and Greif [32] for $10^{10} \leq Ra \leq 10^{11}$, have shown that such baffles indeed reduce the Nusselt number when compared to an enclosure which has no internal baffles.

This paper summarizes the results of a numerical and experimental study of the effects of partial baffles on natural convection in rectangular enclosures with arbitrary angle of inclination. It may be noted that a significant portion of this study was published in more detail by Chao et al. [29,30] in Chemical Engineering Fundamentals, but did not become known to the broader scientific community because this journal ceased publication after only two volumes. More detail can be found in [33].

NUMERICAL ANALYSIS

The Considered System and Range of Parameters

The enclosure, shown in Fig. 1, is a 2 x 1 x 1 rectangular box, with uniform dimensionless temperatures of 0.5 and -0.5 on the lower and upper 2 x 1 surfaces, respectively, and perfectly insulated side walls. The baffle extended perpendicularly from the longer side wall, and was always attached to the heated surface on one end and to the cooled one on the other. Solutions were obtained for two baffle breadths: 0.5 and 0.7 of the enclosure width; three baffle locations: at 0.25, 0.5, and 0.75 of the longer side wall; two baffle thicknesses: 0.01 and 0.02 of the enclosure height; a full range of baffle/fluid conductivity ratios from perfectly

insulating to perfectly conducting baffles; and for a series of inclinations of the enclosure about the short axis, from 0 to Π rad.

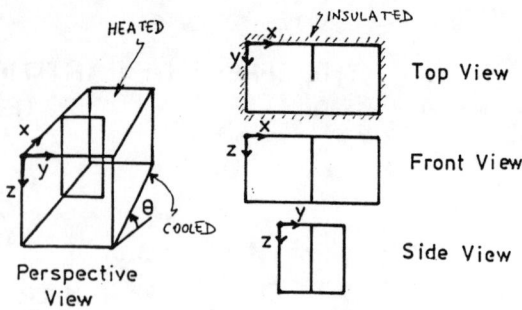

Fig. 1 The enclosure geometry, boundary conditions, and viewing angles for streakline displays

All the calculations were for Pr → ∞ and Ra = 6000. This implies that the inertia terms in the momentum equation are ignored, and the remaining terms are reatined. From the practical standpoint, it has been shown by many (cf. [34]) that the Prandtl number has a negligible effect on natural convection in enclosures for the range of approximately 1 < Pr → ∞. The significant solution simplification obtained by assuming Pr → ∞ thus still produces excellent results in that range. Computations for Ra > 6000 required excessive time on the computers that were available for this study.

Method

The mathematical model consists of the three-dimensional continuity, momentum, and energy equations employing the Boussinesq approximation, nondimensionalized and then transformed by the introduction of the vorticity and the vector potential. Transformed into finite-difference form, the equations were then solved by the ADI method. With exception of the existence of the baffles here, these are the same equations and technique used and published by Ozoe et al. (1976), and would not be repeated here. As there, a false transient term was introduced in the equation relating the vector potential to the vorticity, to transform this expression from elliptic form to the parabolic form of the vorticity and energy equations.

A uniform grid size of 0.1 was used for all of the calculations, resulting in a 21 x 11 x 11 = 2541 grid points. As compared to the boundary conditions used in the past (Ozoe et al. 1976) for the enclosure without baffles, more rapid convergence of the calculations to the steady state were obtained here by setting the X-component of the vector potential to equal zero on the baffle, and on the X = 0 and X = 2 surfaces as well.

Since further grid size reductions created in this 3-d an untenable problem for the computing facilities that were available, it was not possible to assess the error in this computation. This grid size, was, however carefully selected by performing computations of the 2-d case and examining the convergence of the results as the grid was grid approached zero, and selecting the size that produced errors of a few percent in Nu (more detail can be found in [33]).

The temperature of a perfectly conducting baffle was taken to vary linearly between the heated and cooled surfaces.

The temperatures on a baffle of finite conductivity were updated one step behind the energy, vorticity and velocity calculations using the following scheme.

The temperature variation across the thickness of the baffle was postulated to be negligible. The energy balance over a finite element of the baffle can then be expressed as

$$\frac{H^2}{x_w}\frac{\delta T}{\delta t} = \delta_Z^2 T + \delta_Y^2 T + \frac{\lambda_f H}{\lambda_w L}\Delta X \, \delta_X^2 T \qquad (1)$$

where here

$$\delta_X^2 T = (T_{i+j,k+1} - 2T_{i,j,k+1} + T_{i-i,j,k})/(\Delta X)^2 \qquad (2)$$

$$\delta_Z^2 T = (T_{i,j,k+1} - 2T_{i,j,k+1} + T_{i,j,k-1})/(\Delta Z)^2 \qquad (3)$$

The corresponding expression for the y-direction depends on j. For $j = 1$

$$\delta_Y^2 T = 2(T_{i,2,k} - T_{i,j,k})/(\Delta Y)^2 \qquad (4)$$

while for $1 < j < NYF$

$$\delta_Y^2 T = (T_{i,j+1,k} - 2T_{i,j,k} + T_{i,j-1,k})/(\Delta Y)^2 \qquad (5)$$

and for $j = NYF$

$$\delta_Y^2 T = 2[T_{i,j-1,k} - (1 + \frac{\lambda_f}{\lambda_w}) T_{i,j,k} + \frac{\lambda_f}{\lambda_w} T_{i,j+1,k}]/(\Delta Y)^2 \qquad (6)$$

In the above

$i = NXF$
$k = 2.3, \dots 10$
$j = NYF$

For the horizontal orientation of the enclosure, the stable mode of circulation is known from prior work to consist of two roll-cells with their axes parallel to the shorter horizontal dimension of the enclosure. For the mathematical model, circulation of these roll-cells in either direction is equally probable. In the experimental work, the circulation was always upward near the baffle and downward near the ends owing to a slight nonisothermality in the heated surface. Hence, a thermal shock was imposed as an initial condition for the calculations to induce this direction of circulation.

Results

The Flow Pattern. Streaklines were computed from the velocity field solutions, and displayed dynamically on the screen of a Vector General 3404 Display Unit using the procedure described in [35]. Photographs of these displays were taken for four different viewing angles: perspective, top, front, and side, as shown in Fig. 1. A few out of the many cases photographed are shown below, all for the enclosure with the centrally-located, half-extended baffle, to present the key phenomena observed.

Fig. 2 is for the horizontal enclosure (heated surface at bottom and cooled at top) with a non-conducting baffle. The flow consists of four roll-cells with their axes

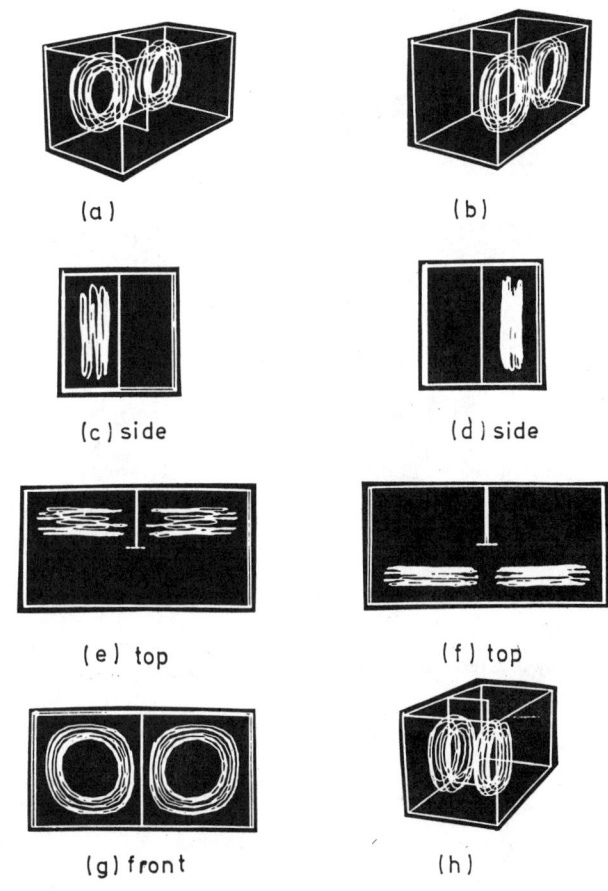

(a)   (b)

(c) side   (d) side

(e) top   (f) top

(g) front   (h)

Fig. 2 Streaklines for horizontal enclosure with centrally-located, half-extended, non-conducting baffle

parallel to the shorter horizontal side of the enclosure. The circulation is up near the central plane and down near the end walls, and the pattern is very similar to that observed in an enclosure without a baffle. For cases in which the baffle was not located centrally (not shown here) and flow symmetry about the long side was lost, the cells became oblique.

Inclination of the enclosure about the short side as axis (shown for 5II/180 rad in Fig. 3) shows a strengthening and enlargement of the lower (left-side) frontal roll-cell because the buoyancy forces associated with the elevation of the other side promote circulation in the same (counter-clockwise) direction. Due to the opposition of the buoyancy force to the original circulation for the higher (right-side) rear roll cell, it becomes oblique and its axis rotates by about 45II/180 rad, and the higher frontal roll cell has therefore apparently decayed completely. As the inclination angle increases, the frontal cell grows to occupy the entire frontal space, circulating counter-clockwise. Fig. 4 shows the streaklines for an inclination of 90II/180 rad

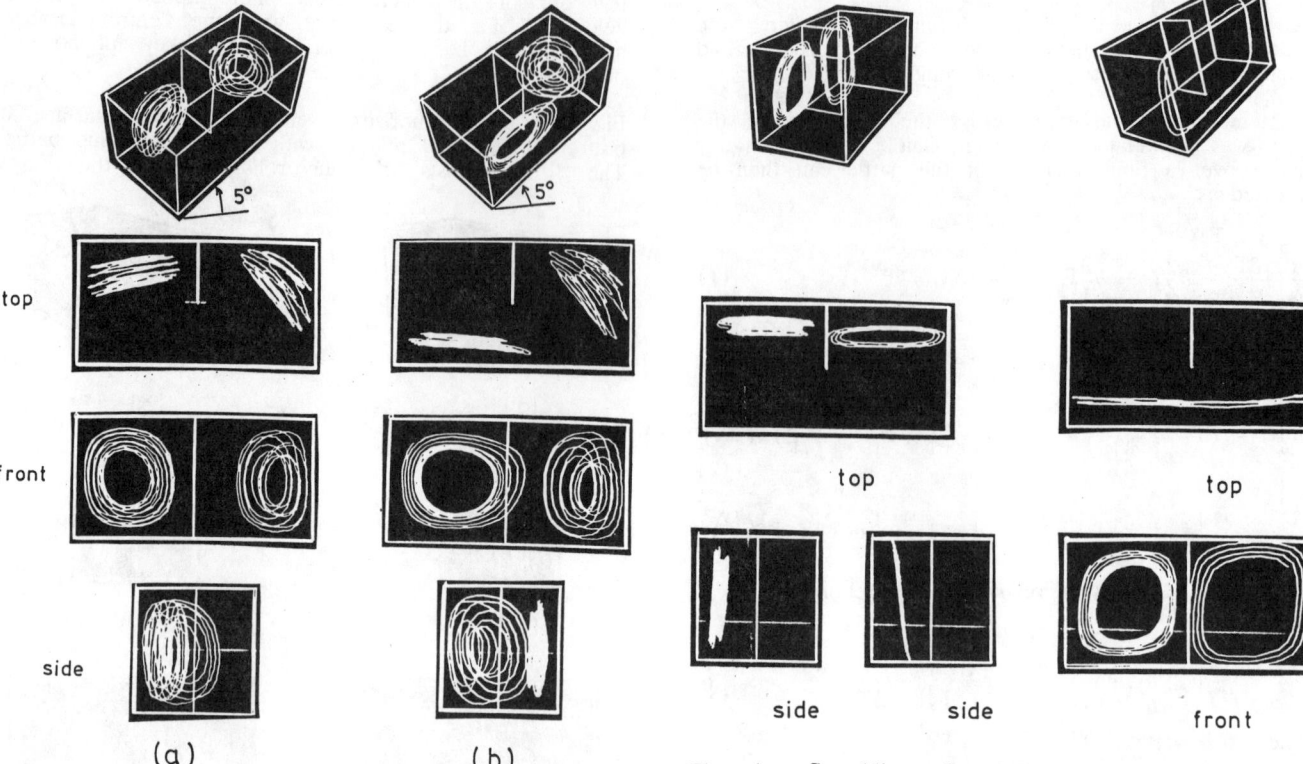

top

front

side

(a)   (b)

top   top

side   side   front

Fig. 4   Streaklines for 90Π/180 rad inclination with centrally-located, half-extended, non-conducting baffle

Fig. 3   Streaklines for 5Π/180 rad inclination with centrally-located, half-extended, non-conducting baffle
a)   rear roll-cells
b)   left-front and right-rear roll-cells

(enclosure heated from one side and cooled from the other).   The baffle is seen to cause a slight inclination of the roll-cells with respect to the insulated vertical walls.

The streaklines for the horizontal enclosure with a **perfectly-conducting baffle** are shown in Fig. 5.   In comparison with the circulation in a horizontal enclosure with a non-conducting baffle (Fig. 2), one can see that the effect of conduction in the baffle is to rotate the axes of the rear roll-cells significantly and of the frontal roll-cells slightly.   The differences between these two cases diminish for inclined enclosures.

Nusselt Numbers.   The variation of the mean Nusselt number with enclosure inclination is compared in Fig. 6 for no baffle, a non-conducting baffle, and a perfectly-conducting baffle.   The Nusselt number is always based on the heat flux through the fluid only. The baffle (half-extended, centrally-located in this case) is seen to reduce the Nusselt number by up to about 16% for all inclinations (except, of course for the stable case when the inclination is Π/180 rad).   Conduction in the baffle is seen to reduce the inclination angle at which the minimal Nusselt number occurs, from 10Π/180 to 8Π/180 rad, to reduce the Nusselt number for lower inclinations, and to increase it for higher ones.   Here and in the other cases computed, reduction in the Nusselt number is associated with obliqueness of the

front

side

top

(a)   (b)

Fig. 5   Streaklines for horizontal enclosure with centrally-located, half extended, perfectly-conducting baffle
a) rear roll-cells
b) front roll-cells

318

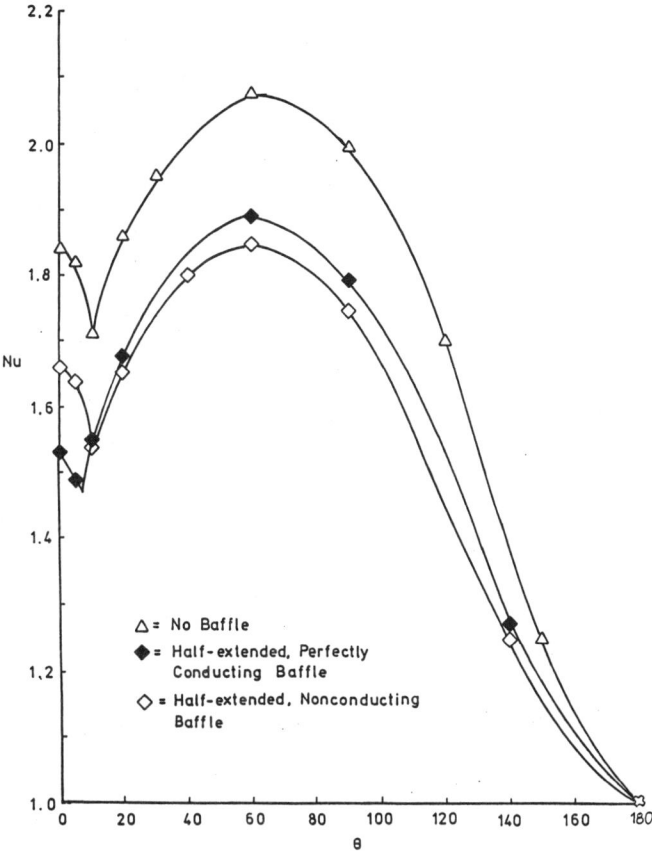

Fig. 6 Effect of baffle conduction on the mean Nusselt number in inclined enclosures

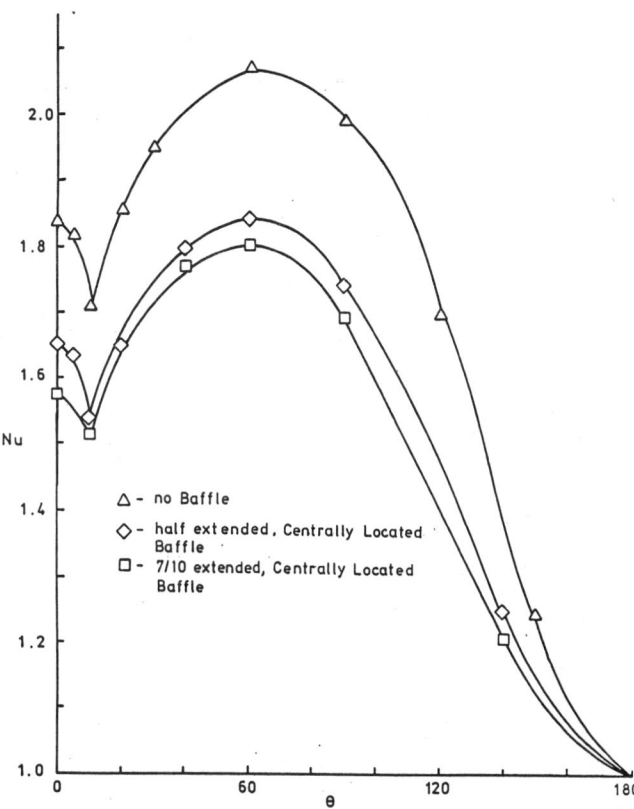

Fig. 7 Effect of extension of a non-conducting baffle on the mean Nusselt number in inclined enclosures

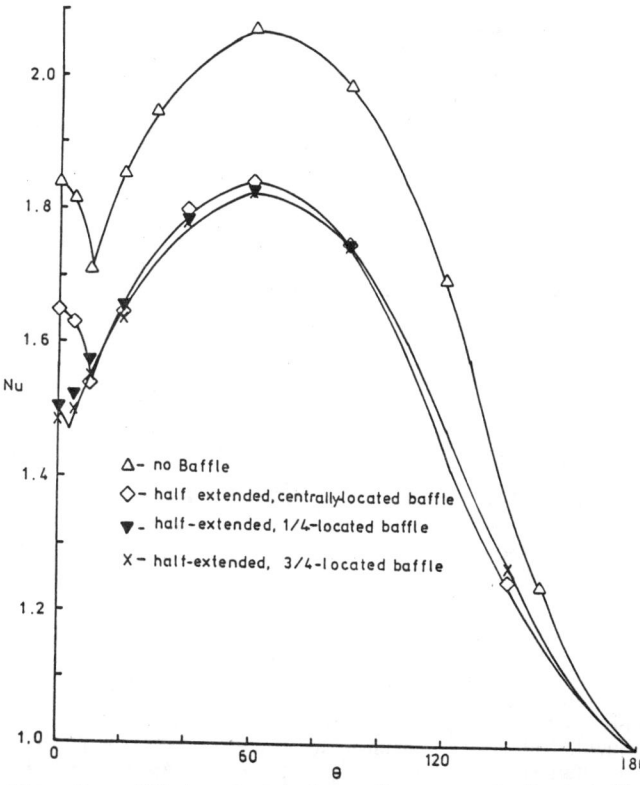

Fig. 8 Effect of location of non-conducting, half-extended baffle on the mean Nusselt number in inclined enclosures

roll-cells. The increase for higher inclinations is apparently the result of heat transfer from one roll-cell to the other by conduction normal to the thin dimension of the baffle. The computations have also shown no parametric dependence of Nu on L/H or k_w/k_f alone, but it does depend on the composite variable $k_w L/k_f H$ indicated by equation (1). For baffles of finite conductivity and thickness an additional heat flux passes through the baffle between the heated and cooled surfaces. For a thickness ratio $H/L = 100$, the computed increment in the overall Nusselt number is 0.05, 0.33, and 2.5 for k_w/k_f of 10, 100, and 1000, respectively.

Fig. 7 shows that extension of the breadth of a non-conducting baffle has the expected effect of reducing the rate of heat transfer, but the marginal effect of the increase from a 1/2 to a 7/10 baffle is small.

The effect of location of a non-conducting baffle on the Nusselt number is illustrated in Fig. 8. The two non-central locations of the baffle reduce Nu at small inclinations.

EXPERIMENTAL RESULTS

Method

The experiments were conducted in a rectangular enclosure with a cross section of 63.50 x 31.75 mm and height of 31.75 mm, i. e. with a width/height aspect ratio of 2/1 and depth/height aspect ratio of 1/1. The vertical side walls were Plexiglas, and 19.05 mm thick copper plates served as the top and bottom of the enclosure. The entire enclosure was surrounded laterally by a vacuum chamber (maintained at about 1 Pa), and the enclosure top and bottom periphery contained high reflection front-surface mirrors, to minimize heat losses and exchange between the top and bottom surfaces. Constant temperature water circulators were used to maintain the top and bottom copper plates at the desired temperatures. A heat flux meter was installed over each the entire top and bottom surfaces of the enclosure, and a thermocouple was imbedded into the center of each of the two heat flux meters. The baffles were made from transparent acetate, 1.5 mm thick.

The fluid in the enclosure was 99% glycerol. Relative to air, its high density and conductivity reduce the relative effects of conduction through and along the side-walls. Its greater viscosity and density provide a larger time constant and reduce the sensitivity to external perturbations, and impede the diffusion of the suspended aluminum flakes used for flow visualization. At the experimental conditions it has a Prandtl number of about 4261, providing, as discussed above, convection which is practically indistinguishable from the Pr → ∞ value used in the numerical computations.

All of the instruments were calibrated prior to the experiments, to within an accuracy of 2%. The experiments were typically conducted for one enclosure surface at 27.75 ±0.05°C and the opposite one at 31.55 ±0.05 °C, producing a Rayleigh number of about 2×10^4. Smaller temperature differences, to match the Ra = 6000 value of the numerical analysis, were impossible to obtain with that apparatus without introducing unacceptable experimental errors in the determination of the mean Nusselt number.

Twelve to twenty-four hours were required to establish a steady state at each inclination. Aluminum flakes were then introduced (by hypodermic needle), and about 40 minutes were then required for the flakes to establish good streakline patterns. A narrow slit of high intensity light was used to illuminate selected regions of the fluid, and photographs were taken.
The Nusselt numbers were defined as the ratio of the heat flux through the fluid, as measured by the heat flux meter, and the pure conduction heat flux through the fluid.

The experiments were conducted for seven enclosure/baffle configurations as shown in Fig. 9. Case 1: no baffle. Case 2: centrally-located, "quarter-baffle" (half height, half extension). Case 3: centrally-located, half-extended, full-height baffle. Case 4: 3/4-located (from axis), half-extended, full-height baffle. Case 5: 1/4-located (from axis), half-extended, full-height baffle. Case 6: centrally-located, 3/4-extended, full-height baffle. Case 7: centrally-located, fully-extended, 1/2-height baffle.

Results

The experimental results, showing the Nusselt number as a function of inclination for the seven cases, are presented in Fig. 9. The shape of this function is qualitatively the same for all cases, and also qualitatively the same as those observed in the numerical analysis described above, and in the past work by the authors and their co-workers. A minimum in the rate of heat transfer occurs between 5Π/180 and 10Π/180

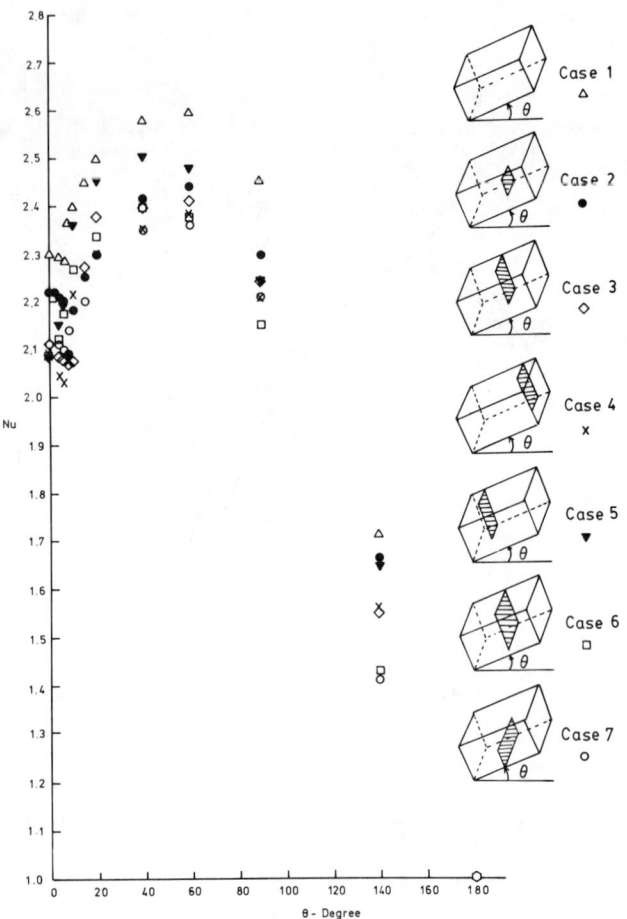

Fig. 9 Experimental Nusselt numbers at various angles of inclination and baffle configurations

rad, and a maximum between 40Π/180 and 60Π/180 rad.

Case 1, enclosure with no baffle, has the highest rates of heat transfer for all inclinations. Generally speaking, case 7 has the lowest heat transfer rates, but the minimum amongst the cases investigated is inclination-dependent. No one baffle-configuration was the best for all angles of inclination. For inclinations from 2Π/9 to Π rad, increasing the height and/or width of the baffle produced a monotonic decrease in the rate of heat transfer. Location of a baffle of half-width and full-height near the elevated end of the enclosure reduced the rate of heat transfer relative to the central or lower locations. A baffle of half-height and full-width produced the greatest reduction in the rate of heat transfer for all inclinations greater than Π/9 rad.

The maximum reduction in the rate of heat transfer which was achieved with these baffles varied from about 10% at no inclination, up to about 20% at 7Π/9 rad, and then down to zero at Π rad. It is likely that comparable reductions can be achieved in enclosures of greater aspect ratios.

A typical photograph of the streaklines, for case 3, is shown in Fig. 10. Because photographs were possible only from the sides, a sketch was also constructed of the presumed top view of the streaklines.

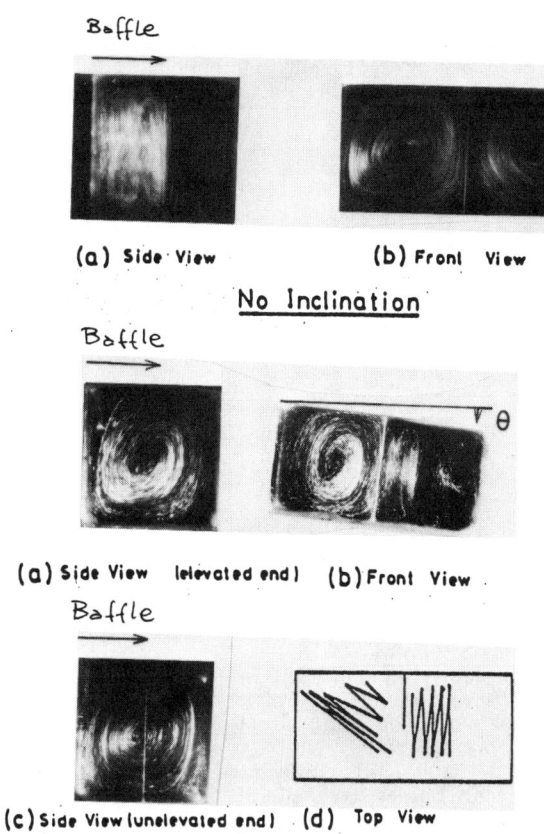

(a) Side View   (b) Front View

No Inclination

(a) Side View (elevated end)   (b) Front View

(c) Side View (unelevated end)   (d) Top View

5π/180 rad of Inclination

(a) Side View (elevated end)   (b) Front View   (c) Side View (unelevated end)

10π/180 rad of Inclination

Fig. 10 Streaklines for Case 3 with 0, 5Π/180, and 10Π/180 rad inclination

The circulation and its dependence on the inclination are essentially the same as those observed in Figs. 2 and 3 (obtained from the numerical analysis), confirming in principle the validity of the analysis. Quantitative validation requires close duplication of the boundary conditions, which is a very difficult task in the experimental work.

As shown in the numerical analysis, all of the baffle sizes and locations produced oblique roll-cells, particularly for low angles of inclination. As the angle of inclination of the heated surface was increased to Π/2 rad, the axes of the circulation again became parallel to the smaller horizontal dimension.

CONCLUSIONS

The good agreement between the computed and measured results indicates that such calculations, even with a relatively crude grid, can be relied upon to reveal the flow pattern.

Increasing the conductivity of the baffles radically changes the flow pattern, reduces the inclination for the minimum in the Nusselt number, reduces the heat flux through the fluid for inclinations lower than the critical value and increases it for greater inclinations. The heat flux between the baffle and the heated and cooled surfaces is negligible for small conductivity ratios, but becomes significant as k_w/k_f exceeds 100.

Partial baffles can be used to modify the flow pattern and thereby reduce the heat flux due to natural convection significantly. The size and location of the baffle which produces the lowest heat flux depend on the angle of inclination.

ACKNOWLEDGMENT

This work was supported in major part by the U.S. Department of Energy Solar Heating and Cooling R&D Branch, and in part by the Japanese Ministry of Education, Grants-in-Aid for Special Research Projects on Energy Problems. The participation of Professor Ozoe at the University of Pennsylvania was co-sponsored by the Japan Society for the Promotion of Science and the National Science Foundation.

REFERENCES

1. Hollands, K. G. T., "Honeycomb Devices in Flat-Plate Solar Collectors," Solar Energy, Vol. 9, No. 3, 1965, pp. 159-164.
2. Hollands, K. G. T., "Natural convection in horizontal thin-walled honeycomb panels", J. Heat Transfer, Vol. 95, 1973, pp. 439-444.
3. Buchberg, H., I. Catton and D. K. Edwards, "Natural Convection in Enclosed Spaces - A Review of Applications to Solar Energy Collection," J. Heat Transfer, Vol. 98, No. 2, 1976, pp. 182-188.
4. Hollands, K. G. T., T. E. Unny, G. D. Raithby, and L. Konicek, "Free Convection Heat Transfer Across Inclined Air Layers," J. Heat Transfer, Vol. 98, No. 2, 1976, pp. 189-193.
5. Ozoe, H., K. Yamamoto, S. W. Churchill, and H. Sayama, "Three-dimensional, numerical analysis of laminar natural convection in a confined fluid heated from below", J. Heat Transfer, Vol. 98C, 1976, pp. 202-207.
6. Ozoe, H., H. Sayama, and S. W. Churchill, "Natural Convection Patterns in along inclined rectangular box heated from below - Part I: Three dimensional photography", Int. J. Heat Mass Transfer, Vol. 20, 1977, pp. 123-129.
7. Ozoe, H., H. Sayama, and S. W. Churchill, "Natural Convection Patterns in along inclined rectangular box heated from below - Part II: Three dimensional numerical results", Int. J. Heat Mass Transfer, Vol. 20, 1977, pp. 131-139.
8. Cane, R. L. D., K. G. T. Hollands, G. D. Raithby, and T. E. Unny, "Free convection heat transfer across inclined honeycomb panels", J. Heat Transfer, Vol. 99, 1977, pp. 86-91.

9. Arnold, J. N., D. K. Edwards, and I. Catton, "Effect of tilt and horizontal aspect ratio on natural convection in rectangular honeycomb solar collectors", J. Heat Transfer, Vol. 99, 1977, pp. 120-122.

10. Chan, A. M. C. and S. Banerjee, "Three-dimensional numerical analysis of transient natural convection in rectangular enclosures", J. Heat Transfer, Vol. 101, 1979, pp.114-119.

11. Chan, A. M. C. and S. Banerjee, "A numerical study of three-dimensional roll cells within rigid boundaries", J. Heat Transfer, Vol. 101, 1979, pp. 233-237.

12. Meyer, B. A., M. M. El-Wakil, and J. W. Mitchell, "Natural Convection Heat Transfer in Small and Moderate Aspect Ratio Enclosures-An Application to Flat Plate Collectors," J. Heat Transfer, Vol. 101, 1979, pp. 655-659.

13. Chao, P., H. Ozoe, and S. W. Churchill, "The effect of a non-uniform surface temperature on laminar natural convection in a rectangular enclosure", Chem. Eng. Commun., Vol. 9, 1981, pp. 245-254.

14. H. Ozoe, K. Fujii, N. Lior and S. W. Churchill, "A Theoretically Based Correlation for Natural Convection in Horizontal Rectangular Enclosures Heated from Below with Arbitrary Aspect Ratios," Paper No. NC23, 7th International Heat Transfer Conference, 2, Munich, Germany, pp. 257-262, Hemisphere Publishing Corp., Washington, 1982.

15. Elsherbiny, S. M., G. D. Raithby, and K. G. T. Hollands, "Heat transfer by natural convection across vertical and inclined air layers", J. Heat Transfer, Vol. 104, 1982, pp. 96-102.

16. N. Lior, H. Ozoe, P. Chao, G.F. Jones, and S.W. Churchill, "Heat Transfer Considerations in the Use of New Energy Resources", chapter in Heat Transfer in Energy Problems, T. Mizushina and W.J. Yang, Editors, pp. 175-189, Hemisphere Publishing Co. and Springer-Verlag, 1983.

17. H. Ozoe, P. K-B Chao, S. W. Churchill and N. Lior, "Laminar Natural Convection in an Inclined Rectangular Box with the Lower Surface Half Heating and Half Insulated," ASME J. Heat Transfer, 105, 1983, pp. 425-442.

18. H. Ozoe, K. Fujii, N. Lior and S. W. Churchill, "Long Rolls Generated by Natural Convection in an Inclined Rectangular Enclosure", International Journal of Heat and Mass Transfer, 26, 10, 1983, pp. 1427-1438.

19. Kim, D. M., and R. Viskanta, "Effect of wall conductance and radiation on natural convection in a rectangular cavity", Num. Heat Transfer, Vol. 7, 1984, pp.449-470.

20. Kim, D. M., and R. Viskanta, "Effect of wall heat conduction on natural convection in a square enclosure", J. Heat Transfer, Vol. 107, 1985, pp. 139-146.

21. H. Ozoe, A. Mouri, M. Ohmuro, S. W. Churchill, and N. Lior, "Numerical Calculations of Laminar and Turbulent Natural Convection of Water in Rectangular Channels Heated and Cooled Isothermally on the Opposing Vertical Walls", Int. J. Heat and Mass Transfer, 28, 1, 1985, pp. 125-138.

22. Hoogendorn, C. J., "Natural convection supression in solar collectors", in Natural Convection Fundamentals and Applications, S. Kakac, W. Aung, and R. Viskanta, Eds., Hemisphere, Washington, 1985, pp. 940-960.

23. N. Lior, "Natural Convection in High Technology Applications", Chapter in Heat Transfer in High Technology and Power Engineering, W. J. Yang and Y. Mori, Editors, pp. 573-596, Hemisphere, Washington, 1986.

24. Ozoe, H., A. Mouri, M. Hiramitsu, S. W. Churchill, and N. Lior, "Numerical Calculation of Three-dimensional Turbulent Natural Convection in a Cubical Enclosure Using a Two-Equation Model for Turbulence", J. Heat Transfer, Vol. 108, 1986, pp. 806-813.

25. de Vahl Davis, G., "Finite difference methods for natural and mixed convection in enclosures", Heat Transfer 1986, Proc. Eighth International Heat Transfer Conf., Hemisphere, Washington DC, pp. 101-109.

26. Hoogendorn, C. J., "Natural Convection in Enclosures", Heat Transfer 1986, Proc. Eighth International Heat Transfer Conf., Hemisphere, Washington DC, Vol. 1, pp. 111-120.

27. Yang, K. T., "Natural Convection in Enclosures", Ch. 13, Handbook of Single-Phase Convective Heat Transfer, S. Kakac, R. K. Shah, and W. Aung, Eds., Wiley, New York, 1987.

28. Ostrach, S., "Natural Convection in Enclosures", J. Heat Transfer, Vol. 110, 1988, pp. 1175-1190.

29. P. K. Chao, H. Ozoe, N. Lior and S. W. Churchill, "The Effect of Partial Baffles on Natural Convection in an Inclined Rectangular Enclosure Part I: Experimental Observations", Chemical Engineering Fundamentals, 2, 2, 1983, pp. 23-37.

30. P. K. Chao, H. Ozoe, N. Lior and S. W. Churchill, "The Effect of Partial Baffles on Natural Convection in an Inclined Rectangular Enclosure Part II: Numerical Solution", Chemical Engineering Fundamentals, 2, 2, 1983, pp. 38-49.

31. Lin, N. N., and A. Bejan, "Natural convection in a partially divided enclosure", Int. J. Heat Mass Transfer, Vol. 26, 1983, pp. 1867-1878.

32. Nansteel, M, and R. Greif, "An investigation of natural convection in enclosures with two- and three-dimensional partitions, Int. J. Heat Mass Transfer, Vol. 27, 1984, pp. 561-571.

33. Chao, Paul, K.-B., "Effects of Non-Uniform Heating and Internal Baffles on Natural Convection in Inclined Rectangular Enclosures", Ph.D. Dissertation in Chemical Engineering, University of Pennsylvania, 1981.

34. Churchill, S. W., "Free convection in layers and enclosures", Ch. 2.5.8 in Heat Exchanger Design Handbook, Hemisphere, Washington DC, 1983.

35. Yamamoto, K., Ozoe, H., Chao, P. K.-B, and Churchill, S. W., "The computation of dynamic display of three dimensional streaklines for natural convection in enclosures" Computers & Chem. Engrg., Vol. 6, 1982, pp. 161-167.

A LINEAR STABILITY ANALYSIS OF A MIXED CONVECTION
PLUME:
FIRST ORDER MIXED CONVECTION EFFECTS

R. Krishnamurthy
Department of Mechanical Engineering
The Catholic University of America
Washington, D.C.

ABSTRACT

Aiding mixed convection flow resulting from the vertical flow of a uniform stream past a horizontal line source of heat is of importance in many practical situations such as hot-wire anemometry, etc. In this paper, the stability of such a flow to small disturbances is analyzed in terms of the linear stability theory. The analysis treats the presence of the free stream as a perturbation of a natural convection plume generated by the line source of heat. The base flow as well as the disturbance field are determined by means of a systematic perturbation expansion.

The results reveal that the free stream has a stabilizing effect. The reported results are valid at a large distance from the source where the flow field is dominated by buoyancy effects.

NOMENCLATURE

A   defined in equation (29)

c $\equiv \Omega / \alpha$

F_i terms in the expansion for $\bar{\psi}$ in equation (6), i=1,3

g   acceleration due to gravity

G $= (Gr_x)^{1/5}$

Gr_x $\equiv g\beta Q_0 x^3 /k\nu^2$

H_i terms in the expansion for $(\bar{T} - T_\infty)$ in equation (7), i=1,3

k   thermal conductivity

Pr   Prandtl number

Q(x)   local value of thermal convected energy

Q_0 thermal input per unit length of the line source

$\bar{R}$ parameter defined in equation (27)

R_i defined in equations (21) and (23), i=1,4

Re_x Reynolds number, $= U_\infty x / \nu$

S   defined in equation (8)

S_i defined in equation (14)

$\bar{T}$ base flow temperature

$\tilde{T}$ temperature of the disturbance

$\bar{u}$ base flow velocity component in x-direction

$\tilde{u}$ disturbance velocity component in x-direction

U   characteristic plume velocity in x-direction, $= \nu G^2 / x$

$\bar{v}$ base flow velocity component in y-direction

$\tilde{v}$ disturbance velocity component in y-direction

W   defined in equation (26)

x   vertical co-ordinate along the plume centerline

y   co-ordinate normal to x

Greek Symbols

α $= \delta\ d\Lambda/dx$

β coefficient of thermal expansion

γ $= \Omega \dfrac{d\Omega}{d\alpha_3}$

δ $= x / G$, Characteristic thickness of the plume

ΔT characteristic temperature difference between the centerline and edge of the plume, $\equiv Q_0 \delta /kx$

ϵ_H $= 1/G$

ϵ_M $= Re_x / G^2$

$\bar{\zeta}$ $= \dfrac{\partial \bar{v}}{\partial x} - \dfrac{\partial \bar{u}}{\partial y}$

$\tilde{\zeta}$ $= \dfrac{\partial \tilde{v}}{\partial x} - \dfrac{\partial \tilde{u}}{\partial y}$

η $= y / \delta$

Λ defined in equations (8) and (9)

ν kinematic viscosity

ϕ defined in equation (9)

ϕ_i terms defined in equation (15)

τ time

$\bar{\psi}$ base flow stream function

$\tilde{\psi}$ stream function for disturbance field

ω frequency of disturbance

Ω $= \omega\ \delta / U$

INTRODUCTION

Analyses of laminar mixed convection from a horizontal line source of heat have been reported in a number of recent studies. These include the earliest by Wood [1], followed by those of Wesseling [2], Afzal [3] and Krishnamurthy and Gebhart [4]. All these studies were primarily concerned with the predictions of velocity and temperature fields.

In this paper, the stability of such flows to small disturbances is investigated in terms of the Linear stability theory. The buoyancy force and the free stream flow are taken to be in the same direction. The region sufficiently downstream of the source is considered, where buoyancy effects dominate. This flow configuration, shown in Figure 1, is usually termed aiding mixed convection.

The effect of the free stream is considered as a perturbation in the far-field boundary condition on the tangential velocity component u (see Fig. 1), of the natural convection plume. This perturbation is termed the mixed convection effect and is characterized by the parameter ϵ_M. Also taken into account is the first order correction to the " Classical " boundary layer solution to the Natural convection plume. This correction results from the interaction of the plume with the irrotational flow outside the boundary layer. This perturbation is termed the higher-order effect and is characterized by ϵ_H . The base flow is taken to be the classical natural convection plume perturbed by ϵ_M and ϵ_H .The stability analysis is then performed by expanding the disturbance field too, in terms of these two perturbation parameters. These two perturbation parameters have been so chosen that at zero order, the governing equations reduce to that of the laminar natural convection plume. Computed results are presented and discussed for Pr = 0.7.

ANALYSIS

The mixed convection flow arising from an infinitely long horizontal line source of heat is considered as a two-dimensional steady flow. See Figure 1 for a sketch of the flow configuration. With the usual Boussinesq approximations, neglecting viscous dissipation and pressure terms in the energy equation, the full two-dimensional governing equations take the form,

$$\psi_y \frac{\partial}{\partial x}(\nabla^2 \psi) - \psi_x \frac{\partial}{\partial x}(\nabla^2 \psi) - \nu\nabla^4\psi -g\beta \frac{\partial T}{\partial y} = 0 \qquad (1)$$

$$\psi_y \frac{\partial T}{\partial x}- \psi_x \frac{\partial T}{\partial y} = \frac{\nu}{Pr}(T_{xx}+T_{yy}) \qquad (2)$$

where the stream function ψ has been so defined that,
$$u = \psi_y \qquad and \qquad v = - \psi_x$$
Boundary conditions are:

$$y = 0, \quad \psi = \psi_{yy} = T_y = 0 \; ; \; for \; all \; x \qquad (3)$$

$$y \to \infty \; , \; \psi_y \to U_\infty, \; T \to T_\infty \; ; \; for \; all \; x \qquad (4)$$

Also for $\underset{\sim}{x} > 0$, the convected energy is,

$$Q(x) = \int_{-\infty} \rho c_p \psi_y (T - T_\infty)dy = Q_0 = Constant \qquad (5)$$

where Q_0 is the thermal input per unit length of the line source.

In the region $y < O(\delta)$, the base flow can be represented as,

$$\bar\psi = U\delta (F_1(\eta) + \epsilon_M F_2(\eta) + \epsilon_H F_3(\eta)) \qquad (6)$$

and

$$\bar T - T_\infty = \Delta T(H_1(\eta) + \epsilon_M H_2(\eta) + \epsilon_H H_3(\eta)) \qquad (7)$$

where the governing equations and corresponding boundary conditions for F_i and H_i , $i = 1,3$, are given in the Appendix.

In the usual manner for Linear stability analyses, we superimpose on the base flow an arbitrarily small disturbance of the form,

$$\tilde\psi = U\delta \; S(\eta) \; exp \; (i \; (\Lambda(x) - \omega\tau \;)) \; + \qquad c.c. \qquad (8)$$

$$\tilde T = \Delta T \; \phi(\eta) \; exp \; (i(\Lambda(x)-\omega\tau)) \qquad + \qquad c.c. \qquad (9)$$

where "c.c." denotes complex conjugate and S, ϕ and Λ are complex and ω is taken to be real. Also, $\tilde u = \tilde\psi_y$ and $\tilde v = -\tilde\psi_x$.

Each flow variable is represented by the sum of its base flow component and the disturbance component. Then by subtracting the base flow equations from the complete two-dimensional, time-dependent governing equations and combining the x- and y-momentum equations to eliminate pressure terms, the vorticity and energy equations for the disturbance components are obtained. The linearized forms of these equations are given below.

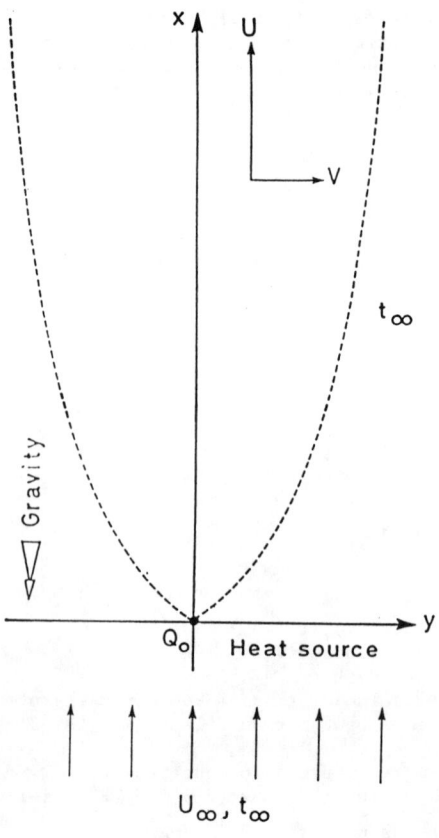

Fig. 1   Flow Configuration

324

$$\frac{\partial \tilde{\zeta}}{\partial \tau} + \bar{u}\frac{\partial \tilde{\zeta}}{\partial x} + \tilde{u}\frac{\partial \bar{\zeta}}{\partial x} + \bar{v}\frac{\partial \tilde{\zeta}}{\partial y} + \tilde{v}\frac{\partial \bar{\zeta}}{\partial y} = \nu\nabla^2\tilde{\zeta} - g\beta\frac{\partial \bar{T}}{\partial y} \quad (10)$$

$$\frac{\partial \tilde{T}}{\partial \tau} + \bar{u}\frac{\partial \tilde{T}}{\partial x} + \tilde{u}\frac{\partial \bar{T}}{\partial x} + \bar{v}\frac{\partial \tilde{T}}{\partial y} + \tilde{v}\frac{\partial \bar{T}}{\partial y} = \frac{\nu}{Pr}(\nabla^2\tilde{T}) \quad (11)$$

In stability analyses of natural convection boundary layers, an approach that has been sucessfully used in the past is to exploit the linearity of the disturbance equation by representing the disturbance field as

$$S = \bar{S}_1 + B_2\bar{S}_2 + B_3\bar{S}_3 \quad (12)$$

$$\phi = \bar{\Phi}_1 + B_2\bar{\Phi}_2 + B_3\bar{\Phi}_3 \quad (13)$$

where each $(\bar{S}_j, \bar{\Phi}_j)$ is an integral of the coupled Orr-Sommerfeld equations, with $j = 1$ corresponding to the inviscid limit and $j = 2,3$, being characterized by viscous effects. This very approach has also been successfully used by Carey and Gebhart [5] in analyzing the stability of an aiding mixed convection boundary layer flow adjacent to a vertical uniform-flux surface. However, in boundary-free flows such as plumes, B_2 and B_3 have to be identically zero as pointed out by Lin [6] and discussed by Hieber and Nash [7]. A more appropriate method is to expand the disturbance field in terms of the perturbation parameters as done in [7]. Thus,

$$S = S_1(\eta) + \epsilon_M S_2(\eta) + \epsilon_H S_3(\eta) \quad (14)$$

$$\phi = \phi_1(\eta) + \epsilon_M \phi_2(\eta) + \epsilon_H \phi_3(\eta) \quad (15)$$

$$\Lambda = \Lambda_1(x) + \epsilon_M \Lambda_2(x) + \epsilon_H \Lambda_3(x) \quad (16)$$

Additional quantities that arise are, non-dimensional frequency Ω, complex wave number α and the complex wave speed c, given by :

$$\Omega = \delta\omega / U$$

$$\alpha = \delta\frac{d\Lambda}{dx} = \alpha_1 + \epsilon_M\alpha_2 + \epsilon_H\alpha_3 \quad (17)$$

$$c = \Omega / \alpha = c_1 + \epsilon_M c_2 + \epsilon_H c_3 \quad (18)$$

Here, the value of ω will be taken as real.

Substituting the expressions for $\bar{u}, \tilde{u}, \bar{v}, \tilde{v}, \bar{T}, \tilde{T}, \bar{\zeta}$ and $\tilde{\zeta}$ into eqns.(10) and (11) and ordering the terms in terms of ϵ_M and ϵ_H, the following equations result.

At zero order:

$$\mathcal{L}(S_1) = (F_1'\alpha_1 - \Omega)(S_1'' - \alpha_1^2 S_1) - \alpha_1 F_1''' S_1 = 0 \quad (19)$$

$$\phi_1 = H_1 S_1 \alpha_1 / (F_1'\alpha_1 - \Omega) \quad (20)$$

At $O(\epsilon_M)$:

$$\mathcal{L}(S_2) = R_1 + \alpha_2 R_2 \quad (21)$$

where, $R_1 = -F_2\alpha_1(S_1'' - \alpha_1^2 S_1) + \alpha_1 S_1 F_2'''$

and $R_2 = 2\alpha_1 S_1(F_1'\alpha_1 - \Omega)$

$$\phi_2 = (\alpha_2(H_1 S_1 - F_1\phi_1) + \alpha_1(S_2 H_1 + S_1 H_2 - F_2'\phi_1))/(F_1'\alpha_1 - \Omega) \quad (22)$$

At $O(\epsilon_H)$:

$$\mathcal{L}(S_3) = R_3 + R_4 \quad (23)$$

where $R_3 = -i(S_1^{iv} - 2\alpha_1^2 S_1'' + \alpha_1^4 S_1 + \phi_1' - i\alpha_1 F_3(S_1'' - \alpha_1^2 S_1) + i\alpha_1 F_3'' S_1 - 4/5\,\alpha_1^2\eta F_1' S_1' + 3/5\,\alpha_1^2 F_1' S_1 + 3/5\,\alpha_1\gamma\,F_1 S_1 + 1/5\,F_1' S_1'' + 1/5$

$$S_1' F_1'' - 3/5\,\alpha_1^2 F_1 S_1' + 3/5\,F_1 S_1' + 3/5 F_1''' S_1$$
$$+ \Omega(4/5\,\alpha_1(\eta S_1' - S_1) - 1/5\,\gamma\,S_1))$$

$$R_4 = 2\,\alpha_1\Omega\,S_1 - 3\alpha_1^2 F_1' S_1 + F_1 S_1'' - F_1''' S_1$$

The boundary conditions are that, $S_j(0) = S_j(\infty) = 0$; $j=1,3$. The choice of the first boundary condition has been made on the basis of measurements in a natural convection plume reported by Pera and Gebahrt [8]. They found this mode of the disturbance to be more unstable than the symmetric mode. Such measurements in mixed convection plumes are not yet available. The equations at zero order and at $O(\epsilon_H)$ are essentially the same as those in [7]. Since the homogeneous problems for S_2 and S_3 are the same as those for S_1, it is required that,

$$\int_0^\infty (R_1 + \alpha_2 R_2)\,W\,d\eta = 0 \quad (24)$$

and

$$\int_0^\infty (R_3 + \alpha_3 R_4)\,W\,d\eta = 0 \quad (25)$$

where $W(\eta)$ is a non-trivial solution of the adjoint homogeneous problem,

$$(F_1'\alpha_1 - \Omega)(W'' - \alpha_1^2 W) + 2F_1''\alpha_1 W = 0 \quad (26)$$

with $W(0) = W(\infty) = 0$

From equations (24) and (25), it is easy to see that,

$$\alpha_2 = -\int_0^\infty R_1 W\,d\eta / \int_0^\infty R_2 W\,d\eta \text{ and } \alpha_3 = -\int_0^\infty R_3 W\,d\eta / \int_0^\infty R_4 W\,d\eta$$

chosen value of Ω, equation (19) is solved to determine S_1 and α_1. Then $W(\eta)$ is determined from equation (26). With α_1 and W known, α_2 can then be determined from equation (24). The procedure for obtaining α_3 is similar and is given in [7].

The two perturbation parameters ϵ_M and ϵ_H arise from distinct physical considerations. Yet the two can be related by,

$$\epsilon_M = \bar{R}\,\epsilon_H^{1/3} \quad (27)$$

where, $\bar{R} = U_\infty(\nu^2 k /g\beta Q_0)^{1/3} / \nu$

Clearly $\bar{R}$ is independent of x. If $\bar{R}$ is $O(1)$ or smaller, then the effect of mixed convection on the stability of the flow is inviscid in nature.

Computed values of α_1, α_2 and α_3 are listed in Table 1 at various values of Ω, for Pr = 0.7. Using these values neutral stability and amplification contours can be constructed. These are shown in Figure 2.

TABLE I

Computed eigenvalues of the disturbance flow field

Ω	α_1	α_2	α_3
0.02	0.0610 - 0.0582i	-0.0049 + 0.2291i	-0.033 + 0.779i
0.04	0.1150 - 0.0884i	-0.1058 + 0.3097i	-0.078 + 0.813i
0.06	0.1693 - 0.1069i	-0.2250 + 0.3198i	-0.123 + 0.842i
0.08	0.2213 - 0.1158i	-0.3142 + 0.2743i	-0.165 + 0.856i
0.10	0.2693 - 0.1181i	-0.3628 + 0.2113i	-0.198 + 0.860i
0.12	0.3130 - 0.1162i	-0.3819 + 0.1528i	-0.223 + 0.861i
0.14	0.3529 - 0.1120i	-0.3843 + 0.1050i	-0.242 + 0.863i
0.16	0.3897 - 0.1064i	-0.3785 + 0.0674i	-0.258 + 0.866i
0.18	0.4238 - 0.1000i	-0.3688 + 0.0382i	-0.271 + 0.870i
0.20	0.4558 - 0.0933i	-0.3576 + 0.0155i	-0.282 + 0.875i
0.24	0.5147 - 0.0794i	-0.3347 - 0.0164i	-0.300 + 0.888i
0.28	0.5685 - 0.0657i	-0.3139 - 0.0366i	-0.316 + 0.904i
0.32	0.6187 - 0.0525i	-0.2958 - 0.0496i	-0.329 + 0.925i

RESULTS AND DISCUSSION

Equation (19) was integrated inwards, that is, towards the centerline of the plume from its outer edge, by making use of the asymptotic form for $S_1(\eta)$ as $\eta \to \infty$. A similar procedure was used in obtaining α_2 and α_3 from equations (24) and (25) respectively. The neutral stability curve i.e. $\Omega(G)$ on which $\alpha_i = 0$, is obtained by solving,

$$\alpha_i = \alpha_{1,i} + \epsilon_M \alpha_{2,i} + \epsilon_H \alpha_{3,i} = 0 \qquad (28)$$

The value of G at a given Ω that one obtains from eqn.(28) depends on the value of $\bar{R}$. In all the computations here, $\bar{R}$ has been taken to be unity. The neutral curve so obtained is shown in Fig. 2 along with contours of constant amplification. These latter curves represent the exponential growth of a disturbance of fixed frequency i.e. ω , as it crosses the neutral curve and propagates downstream. If A_n is the amplitude of a disturbance at a downstream location corresponding to neutral stability and A_x is its amplitude further downstream, then,

$$A_x/A_n = e^A, \qquad A = -\int_{x_n}^{x} \alpha_i \, dx/\delta = -5/3 \left(\int_{G_n}^{G} \alpha_i \, dG\right) \qquad (29)$$

with α_i being the imaginary part of α. The neutral curve is A ≡ 0. Curves of constant amplification have been obtained by determining α_i at various values of G, keeping ω fixed. The integral in eqn.(29) is then evaluated by simple trapezoidal rule, with a step size in G of 2.5. Also shown for comparison, in Fig. 2 are neutral curve and contours of constant amplification for a natural convection plume.

It is clear from Fig. 2 by comparing the neutral curves and those for A = 2, that mixed convection effect stabilizes the flow considerably. The reason for this enhanced stability, lies in the changed nature of the velocity profile for the base flow, near the inflection point. This can be seen from Fig. 3. The slope of the velocity profile for mixed convection is reduced on either side of the inflection point, compared to that for pure natural convection plume. This observation taken together with Lin's [6] interpretationof the point of inflection criterion, gives a physical basis for the stabilizing influence of the free stream on the natural convection plume.Further details of this line of reasoning can be found in Krishnamurthy and Gebhart [9].

ACKNOWLEDGEMENT

The author would like to thank the Computer Center of The Catholic University of America for providing the facilities for computations.

REFERENCES

1. W. W. Wood, 1972, Free and Forced Convection from fine hot-wires, J. Fluid Mech., Vol. 55, 419 - 438

2. P. Wesseling, 1975,An asymptotic solution for slightly buoyant laminar plumes, J. Fluid Mech., Vol. 70, 81-87

3. N. Afzal, 1981, Mixed Convection in a Buoyant Plume, J. Fluid Mech.,Vol. 105, 347 - 368

4. R. Krishnamurthy and Gebhart,B., 1986, Mixed Convectionfrom a horizontal Line source of heat, Int. J. Heat Mass Transfer, Vol. 29, No. 2, 344-347

5. V. P. Carey and Gebhart, B., 1983, The stability and Disturbance amplification Characteristics of Vertical Mixed Convection Flow, J. Fluid Mech.,Vol. 127, 185 - 201

6. C. C. Lin, 1966, The Theory of Hydrodynamic Stability, Cambridge University Press, Cambridge

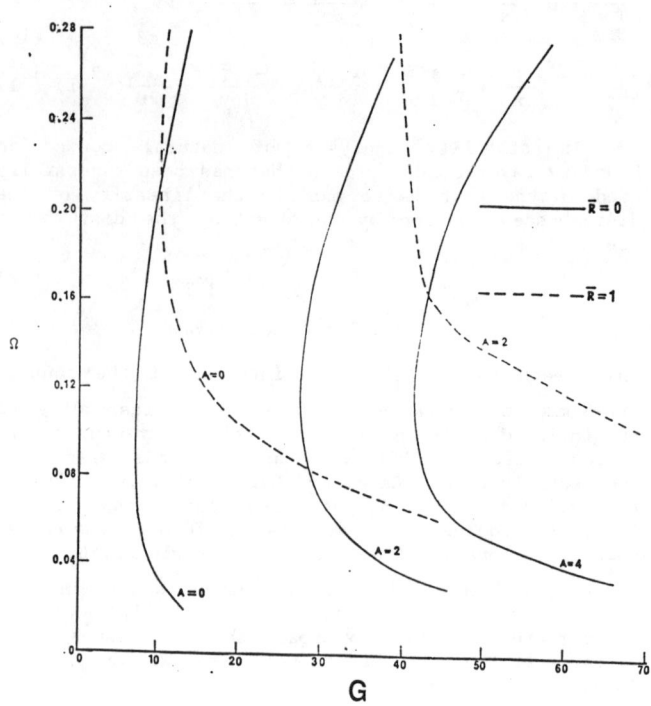

Fig. 2   Contours of Constant Amplification in Natural and Mixed Convection Flows

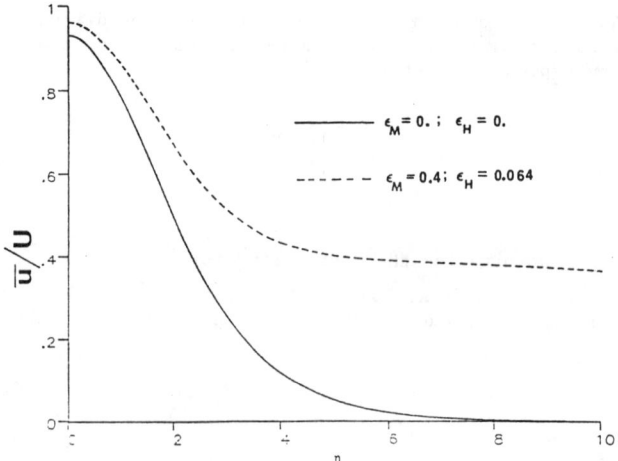

Fig. 3   A comparison of the base flow velocity profiles in Mixed(dashed curve) and Natural convection(solid curve)

7. C. A. Hieber and Nash, E.J., 1975,Natural Convection above a line source:Higher -Order Effects and Stability, Int. J. Heat Mass Transfer, Vol. 18, 1473 - 1479

8. L. Pera and Gebhart, B, 1971,On the stability of Laminar Plumes: some numerical solutions and experiments, Int. J. Heat Mass Transfer, Vol. 14, 975-984

9. R. Krishnamurthy and Gebhart, B, 1989,An experimental study of Transition to Turbulence in Vertical Mixed Convection Flows,ASME J. Heat Transfer, Vol. 111, 121-130

10. N. Riley, 1974, Free Convection from a Horizontal Line Source Of Heat, Z. angew. Math. Phys., Vol. 25, 817 - 828

APPENDIX

Governing Equations of the Base Flow:

At Zero Order:

$$F_1''' + 1/5 (3F_1F_1'' - F_1'^2) + H_1 = 0$$

$$H_1' + 3/5 \; Pr \; F_1 H_1 = 0$$

$$F_1(0) = F_1''(0) = F_1'(\infty) = 0 \; ; \; \int_0^\infty F_1' H_1 d\eta = \frac{1}{2 \; Pr}$$

At $0 \; (\epsilon_M)$:

$$F_2''' + 1/5 (3F_1F_2'' + 2F_2F_1'' - F_1'F_2') + H_2 = 0$$

$$H_2'' + 1/5 \; Pr(3F_1H_2' + 2F_2H_1' + 4H_2F_1' + 3H_1F_2') = 0$$

$$F_2(0) = F_2''(0) = F_2'(\infty) - 1 = H_2'(0) = H_2(\infty) = 0$$

At $0 \; (\epsilon_H)$:

$$F''' + 1/5 (3F F'' + F' F') + H = 0$$

$$H_3'' + 1/5 \; Pr (3F_1H_3' + 6F_1'H_3 + 3F_3'H_1) = 0$$

$$F_3(0) = F_3''(0) = H_3'(0) = H_3(\infty) = 0$$

$$F_3'(\infty) = (\frac{3}{5}\cot\frac{2\pi}{5})F_1(\infty)$$

The solution to these equations at zero order and at $0 \; (\epsilon_M)$ can be found in [3] and in [4] whereas [10] and [7] give the solution at $0 \; (\epsilon_H)$.However, the non-dimensionalization employed here differs from that used in [3], [4] and [10]. For the sake of completeness, all the unknown boundary conditions required to numerically integrate these equations, in their present form are listed below.

i	$F_i(0)$	$H_i(0)$
1	0.93273	0.49654
2	0.05982	-0.21831
3	0.09969	-0.25111

1989 National Heat Transfer Conference
HTD-Vol. 107, Heat Transfer in Convective Flows

CONVECTIVE INSTABILITY IN THE THERMAL ENTRANCE REGION
OF HORIZONTAL RECTANGULAR CHANNELS

F. C. Chou
Department of Mechanical Engineering
National Central University
Chung-Li, Taiwan

J. N. Lin
Department of System Engineering
Chung-Cheng Institute of Technology
Tashi, Taiwan

ABSTRACT

The present paper shows a systematic theoretical study on convective instability in the thermal entrance region of horizontal rectangular channels. By using the vorticity-velocity method, the variation of local Nusselt number is obtained and shown versus Rayleigh number and the dimensionless axial position. Two kinds of thermal boundary conditions, which are an isothermal channel and a channel heated from below, are considered. Based on 2% deviation of local Nusselt number from that of classical Graetz theory, the present analytical predictions of the onset of convective instability are compared with the previous experimental results with good agreement. The effect of specific parameters such as Prandtl number, aspect ratio, and thermal boundary conditions are also studied. Another criterion, based on the location of minimum local Nusselt number, is also considered to study the effect of the selection of criterion on the prediction of onset of instability.

NOMENCLATURE

A	= cross-sectional area of a channel
a,b	= height and width of a rectangular channel respectively
C	= constant, $-(D_e^2/\mu \bar{W}_f)\partial P_f/\partial Z$
D_e	= equivalent hydraulic diameter , $4A/S$
f	= friction factor , $2\bar{\tau}_w/(\rho \bar{W}'^2)$
g	= gravitational acceleration
Gr	= Grashof number, $g\beta\theta_c D_e^3/\nu^2$
$\bar{h}$	= average heat transfer coefficient
k	= thermal conductivity
M, N	= number of divisions in X and Y directions respectively
n	= outward normal direction to the wall
Nu	= local Nusselt number , $\bar{h}D_e/k$
P, P_f	= pressure deviation and pressur for fully developed laminar flow before thermal entrance respectively

p	= dimensionless quantity for P
$\bar{P}$	= mean pressure averaged over the passage cross-section
Pe	= Peclet number, PrRe
Pr	= Prandtl number, ν/α
Ra	= Rayleigh number, PrGr
Re	= Reynolds number, $\bar{W}_f D_e/\nu$
S	= circumference of cross-section
T	= temperature
T_0	= uniform fluid temperature at entrance
U,V,W	= velocity components in X, Y, Z directions
u,v,w	= dimensionless quantities for U, V, W
W_f	= fully developed axial velocity before thermal entrance
w_f	= dimensionless quantity for W_f
W	= axial velocity in the thermal entrance region, W_f+W
w'	= dimensionless axial velocity, $w_f+Ra\cdot w$
X,Y,Z	= rectangular coordinate
x,y,z	= dimensionless rectangular coordinates

GREEK LETTERS

α	= thermal diffusivity
β	= coefficient of thermal expansion
γ	= aspect ratio of a rectangular channel, a/b
θ	= dimensionless temperature, $(T-T_w)/\theta_c$
μ	= viscosity
τ	= shear stress
ν	= kinematic viscosity
ξ	= dimensionless vorticity in axial direction
ρ	= density

SUBSCRIPT

c	= characteristical quantity
f	= fully developed quantity before thermal entrance
w	= value at wall
O	= condition for purely forced convection

SUPERSCRIPT

⁻	= average value

1. INTRODUCTION

The flow and heat transfer characteristics of laminar forced convective flows are significantly affected by buoyancy-induced secondary flow. Thus the effect of buoyancy force on laminar forced convective flows with various geometrical shapes have been studied by many investigators in recent years because of the practical significance and theoretical interest. The buoyancy effect is particularly pronounced for horizontal ducts, where they may act to reduce the entry length, enhance heat transfer, and induce early transition to turbulence. The precursor to the effect is a convective instability which results in the development of longitudinal vortex rolls. Hence, to establish the existence of the buoyancy effect, conditions marking the onset of convective instability must be known.

The problem of convective instability for an infinite horizontal fluid layer is usually approached by employing a linear stability theory. Mori and Uchida(1966) applied the linear stability analysis to determine onset of an infinitesimally small disturbance for fully developed laminar flow between parallel plates. The convective instability problem concerning the onset of longitudinal columnar vortices due to buoyancy force for fully developed laminar forced convection between two horizontal plates was studied theoretically by Nakayama et al.(1970) and experimentally by Akiyama et al.(1971). In extending the analysis to the thermal entrance region, Hwang and Cheng(1973) determined theoretically the conditions marking the onset of convective instability in a hydrodynamically fully developed but thermally developing region. They calculated the critical Rayleigh number at different locations in the thermal boundary layer. A linear stability analysis was also performed to determine the onset of longitudinal rolls for the horizontal flat plate in parallel flow by Wu and Cheng(1976).

To verify the predictions of Hwang and Cheng(1973), Hwang and Liu(1976) visualized air flow in the thermal entrance region of a rectangular channel heated from below. A single stream of dye was injected along the midline of the bottom plate, and a side view was used to determine the longitudinal station at which buoyance force causes the dye to ascend or descend from the plate. It is noteworthy that the measured values of the critical Rayleigh number exceeded predictions by more than an order of magnitude. Kamotani and Ostrach(1976) and Kamotani et al.(1979) also performed experiments for air flow in the thermal entrance region of a rectangular channel and determined onset of the convective instability by detecting small spanwise temperature variations (Kamotani and Ostrach, 1976) and from visualization of the buoyancy induced flow(Kamotani et al., 1979). They also found that the values of critical Rayleigh number are larger than the theoretical values given by Hwang and Cheng(1973). Additional experiments had also been performed by Incropera and his coworkers for water flow at high Rayleigh number range in the thermal entry region of a rectangular channel. Correlations for the critical Rayleigh number, which is determined by the criterion of 10% departure of local Nusselt number from the forced convection results(Osborne and Incropera, 1985) or by flow visualization and the occurrence of minimum local Nusselt number (Incropera et al., 1986), are both found to be a function of Graetz number. But the results of Osborne and Incropera(1985) are found to suggest earlier onset of thermal instability when compared with that of Incropera et al.(1986). More recently experiments on mixed convection for airflow in a horizontal and inclined channel was performed by Maughan and Incropera(1987). The departure of spanwise-average longitudinal Nusselt number from the forced convection results was used to mark the onset point, and the instability data were found to be in reasonable agreement with the previous experimental results.

The possibility of finite difference solution for convective instability in channels using complete Navier-Stokes and energy equations instead of the conventional linear stability theory has been well known for sometime. However, the published works are rather limited and unable to predict accurately the relation of critical Rayleigh number versus the critical Graetz number for onset of convective instability. A solution was obtained by Ou et al.(1974) for laminar flow in a rectangular duct with uniform wall temperature for large Prandtl number fluids. A 2% deviation of local Nusselt number from the value for pure forced convection was used as the criterion for onset of the secondary flow. The critical Rayleigh number was correlated in terms of the Graetz number. Compared to the experimental data, it suggested earlier onset of convective instability about an order of magnitude. The onset of thermally driven secondary flow in horizontal rectangular ducts was also studied numerically by Incropera et al.(1986) for water flow in the combined and thermal entrance regions for different surface thermal conditions and aspect ratio. The onset of secondary flow is ascribed to an occurrence of reduction in bottom plate temperature near the side wall, which is due to buoyancy driven upward motion of warm fluid along the side wall and a descending motion of adjoining cooler fluid. The predictions of critical Rayleigh number for the thermal entry region also exceeded the measured values by about an order of magnitude. There are also other previous works concerning laminar mixed convection in the entrance region of horizontal ducts, including that for large Prandtl number fluids (Cheng et al., 1972) and that extending to moderate and small Prandtl number fluids with various thermal boundary condition (Ramakrishna et al., 1982; Abou-Ellail and Morcos, 1983; Coutier and Greif, 1985; Incropera and Schutt, 1985; Mahaney et al., 1987; Chou and Hwang, 1987).

While the foregoing studies contributed to the understanding of mixed convection, unanswered questions remain, particularly the influence of key parameters and conditions on the onset of thermally driven secondary flow in a horizontal rectangular duct. The present paper shows a systematic theoretical study on laminar mixed convection in the thermal entrance region of horizontal rectangular channels. A 2% deviation of local Nusselt number from the value for pure forced convection is used as the criterion for the prediction of the onset of convective instability, and the present results show a good agreement with the existing experimental data. The specific effects of parameters such as Prandtl number, aspect ratio, and thermal boundary conditions are also considered. Finally, a comparison between the results based on the criterion of 2% deviation of local Nusselt number and that based on the occurrence of local minimum local Nusselt number are shown for the study of the influence of selection of criterion.

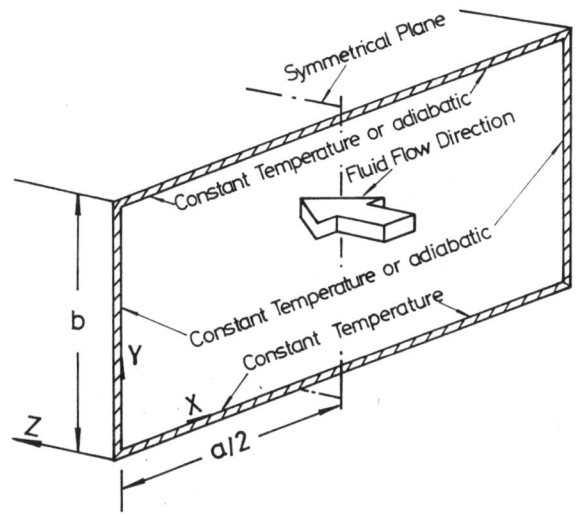

Fig.-1. Flow in a horizontal rectangular channel

2.THEORETICAL METHODS

The channel flow, which is shown in Fig. 1, is three-dimensional and is characterized by velocity components u, v and w in the spanwise x, vertical y and longitudinal z directions. The viscous dissipation and compressibility effects in the energy equation are neglected. The Boussinesq approximation is used to characterize the buoyancy effect. By introducing the dimensionless variables

$$x=X/D_e, \quad y=Y/D_e, \quad z=Z/[PrReD_e],$$
$$u=U/U_c, \quad v=V/U_c, \quad w_f=W_f/\bar{W}_f, \quad w=W/[Ra\bar{W}_f],$$
$$p=P/[\rho U_c \nu/D_e], \quad \theta=[T-T_w]/\theta_c, \quad Gr=g\beta\theta_c D_e^3/\nu^2, \qquad (1)$$
$$Pr=\nu/\alpha, \quad Ra=PrGr, \quad Re=\bar{W}_f D_e/\nu, \quad Pe=PrRe$$

where $D_e=4A/S$, $U_c=Gr\nu/D_e$ and $\theta_c=T_0-T_w$, the dimensionless governing equations can be obtained. By introducing a vorticity function in the axial direction, the vorticity-velocity formulation of the Navier-Stokes equations is employed as shown in Chou and Hwang(1987).

$$\nabla^2 u = \frac{\partial\xi}{\partial y} - \frac{\partial^2 w}{\partial x \partial z} \qquad (2)$$

$$\nabla^2 v = -\frac{\partial\xi}{\partial x} - \frac{\partial^2 w}{\partial y \partial z} \qquad (3)$$

$$Gr(u\frac{\partial\xi}{\partial x} + v\frac{\partial\xi}{\partial y} + w\frac{\partial\xi}{\partial z} + \frac{\partial u}{\partial x}\xi + \frac{\partial v}{\partial y}\xi + \frac{\partial w}{\partial y}\frac{\partial u}{\partial z}$$
$$-\frac{\partial w}{\partial x}\frac{\partial v}{\partial z}) + \frac{1}{Pr}(\frac{\partial w_f}{\partial y}\frac{\partial u}{\partial z} - \frac{\partial w_f}{\partial x}\frac{\partial v}{\partial z} + w_f\frac{\partial\xi}{\partial z})$$
$$= \nabla^2\xi - \frac{\partial\theta}{\partial x} \qquad (4)$$

$$Gr(u\frac{\partial w}{\partial x} + v\frac{\partial w}{\partial y} + w\frac{\partial w}{\partial z}) + \frac{1}{Pr}(u\frac{\partial w_f}{\partial x} + v\frac{\partial w_f}{\partial y}$$
$$+ w_f\frac{\partial w}{\partial z}) = -\frac{1}{Pe^2}\frac{\partial\bar{p}}{\partial z} + \nabla^2 w \qquad (5)$$

$$\nabla^2 w_f = C \qquad (6)$$

$$Ra(u\frac{\partial\theta}{\partial x} + v\frac{\partial\theta}{\partial y} + w\frac{\partial\theta}{\partial z}) + w_f\frac{\partial\theta}{\partial z} = \nabla^2\theta \qquad (7)$$

where $\xi = \partial u/\partial y - \partial v/\partial x$ is the axial vorticity and $C=-(D_e^2/\mu\bar{W}_f)\partial P_f/\partial Z=$constant. In the above equations the axial diffusion effect is already neglected under the high Peclet number assumption as employed in the previous works on convective instability by Incropera et al.(1986), Ou et al.(1974) and other previous numerical studies on laminar mixed convection in the entrance region of ducts by Ramakrishna et al.(1982), Abou-Ellail and Morcos(1983), Incropera and Schutt(1985), Chou and Hwang(1987), Patankar and Spalding(1972).

Because of symmetry, only half of the channel region is considered. The boundary conditions are

$$u=v=w_f=w=0 \quad \text{at } x=0, \ y=0, \text{ and } y=(\gamma+1)/2\gamma$$
$$u=\partial v/\partial x=\partial w/\partial x=\partial w_f/\partial x=0 \quad \text{at } x=(1+\gamma)/4 \qquad (8)$$
$$u=v=w=\xi=\theta=0 \text{ at thermal entrance } z=0$$

There are two kinds of thermal boundary conditions considered in the present work to study the effect of surface thermal conditions. In case 1, an isothermal channel is considered as that in Ou et al.(1974). In case 2 the channel is heated from below as in previous analytical or experimental works (for example: Hwang and Cheng, 1973; Kamotani and Ostrach, 1976; Incropera et al., 1986), and the other walls are subjected to an adiabatic condition. A uniform inlet fluid temperature is prescribed for both cases.

After the developing velocity and temperature fields are obtained, the computations of local Nusselt number are of practical and theoretical interest. The local Nusselt number Nu can be written based on overall energy balance for the axial length dZ as presented in Ou et al.(1974).

$$Nu = \overline{w'(\partial\theta/\partial z)}/\overline{4[w'(\theta_w-\theta_b)]} \qquad (9)$$

In case 2 the Nusselt number along the bottom plate can be written as

$$Nu = [(1+\gamma)\overline{w'(\partial\theta/\partial z)}]/[2\gamma(1-\theta_b)] \qquad (10)$$

Simpson's rule is used to compute the average quantities indicated above.

The computation procedures for the simultaneous solutions of equations (2)-(7) under thermal boundary conditions of uniform wall heat flux had been given in Chou and Hwang (1987). Since a step change in wall temperature is imposed at the entrance z=0 in the present work, an oscillatory behavior of local Nusselt number may arise in the region near the entrance when $z \leq 10^{-3}$ is used. This undesired oscillation can be avoided by using a small axial step size Δz and a simple forward difference for the axial derivative terms at the left hand side of equations (4), (5) and (7). After considerable numerical experiments, axial step Δz ranging from 10^{-6} near the entrance to 3.2×10^{-5} near the fully developed region was found to be satisfactory. The cross-sectional mesh size MxN= 60x12 was used. The average required computing time for each step was about 0.12 s on a VAX-8650 computer.

3. RESULTS AND DISCUSSION

To ensure the accuracy of the present numerical results, a numerical experiment was made on the mesh size MxN and axial step size Δz. Table 1 presents the values of local Nusselt number Nu calculated by using MxN= 60x12 and 80x16 and $\Delta z= 10^{-6}\sim 3.2\times10^{-5}$ at some selected axial positions for the case 1 with Pr= 0.7, Ra= 10^5 and $\gamma= 10$. It is seen that the deviations of Nu for MxN(Δz) = 60x12($10^{-6}\sim 3.2\times10^{-5}$) and 60x12($2\times10^{-6}\sim 6.4\times10^{-5}$) are all less than 0.04% and those for 80x16($10^{-6}\sim 3.2\times10^{-5}$) and 60x12($10^{-6}\sim 3.2\times10^{-5}$) are all less than 0.7%. Therefore MxN(Δz)= 60x12($10^{-6}\sim 3.2\times10^{-5}$) is used for the case 1. For case 2, a numerical experiment was also done and shown in Table 2. The maximum deviations of Nu for MxN(Δz)=80x16($10^{-6}\sim 3.2\times10^{-5}$) and 60x12($10^{-6}\sim 3.2\times10^{-5}$) are 1.12%, and those for 60x12 ($10^{-6}\sim 3.2\times10^{-5}$) and 60x12($2\times10^{-6}\sim 6.4\times10^{-5}$) are all less than 0.5%. Therefore, MxN(Δz)=60x12($10^{-6}\sim 3.2\times10^{-5}$) is also used for the case 2.

The development of secondary flow is of interest in understanding the flow characteristics and heat transfer mechanisms. For flow visualization the developing contours for velocity vector and isotherms are shown in

TABLE 2

Numerical experiment on mesh size (MxN) and axial step size (Δz) for case 2 with γ=10, Pr=0.7 and Ra=6x10^4

MxN (Δz) \ z → Nu	0.002	0.010	0.050	0.100	0.300
60x12 ($1\times10^{-6}\sim 3.2\times10^{-5}$)	9.4091	5.9866	5.1598	6.7766	4.1199
60x12 ($2\times10^{-6}\sim 6.4\times10^{-5}$)	9.4046	6.0084	5.1559	6.7713	4.0847
80x16 ($1\times10^{-6}\sim 3.2\times10^{-5}$)	9.5143	6.0051	5.1346	6.7583	4.1643

Figs. 2(a)-(e) for the case 2 with γ=10, Pr=0.7 and Ra=6x10^4. The secondary flow patterns shown in Fig. 2(a) at z=0.0131 is just the beginning of the secondary flow, and upward motion of warm fluid along the sidewall is accompanied by the motion of cooler fluid toward the bottom plate. At z= 0.0602 the vortex near the sidewall grows in size and strength and a weak plume ascending from the bottom plate at X/D$_e$=0.41 can be seen from both the contours of velocity vector and isotherms, and at z= 0.0800 the strength of the ascending plume at X/D$_e$=0.41 has increased and that of the vortex near the sidewall slightly decreased. At z=0.0903 a second weak plume is ascending from the bottom plate at X/D$_e$=0.77, and the plume at X/D$_e$=0.41 has become the strongest. At

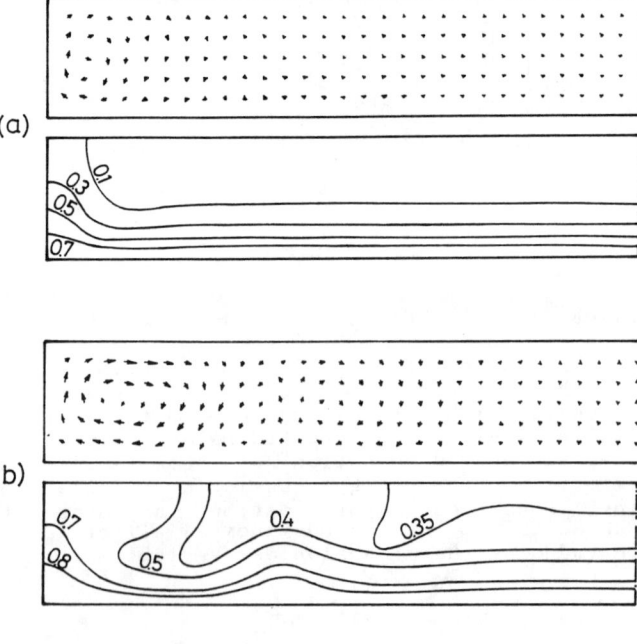

(a)

(b)

(c)

TABLE 1

Numerical experiment on mesh size (MxN) and axial step size (Δz) for case 1 with γ=10, Pr=0.7 and Ra=10^5

MxN (Δz) \ z → Nu	0.002	0.010	0.050	0.100	0.300
60x12 ($1\times10^{-6}\sim 3.2\times10^{-5}$)	9.5337	6.8647	6.2976	6.0532	5.8984
60x12 ($2\times10^{-6}\sim 6.4\times10^{-5}$)	9.5256	6.8669	6.3016	6.0653	5.9014
80x16 ($1\times10^{-6}\sim 3.2\times10^{-5}$)	9.5902	6.8506	6.5259	6.3119	5.9149

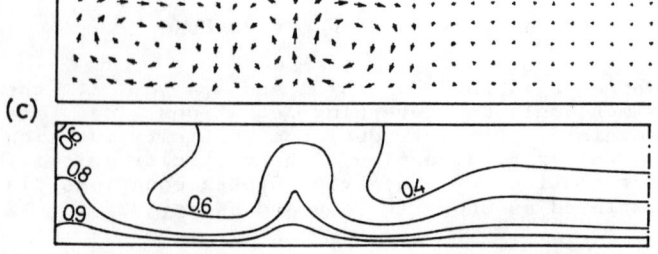

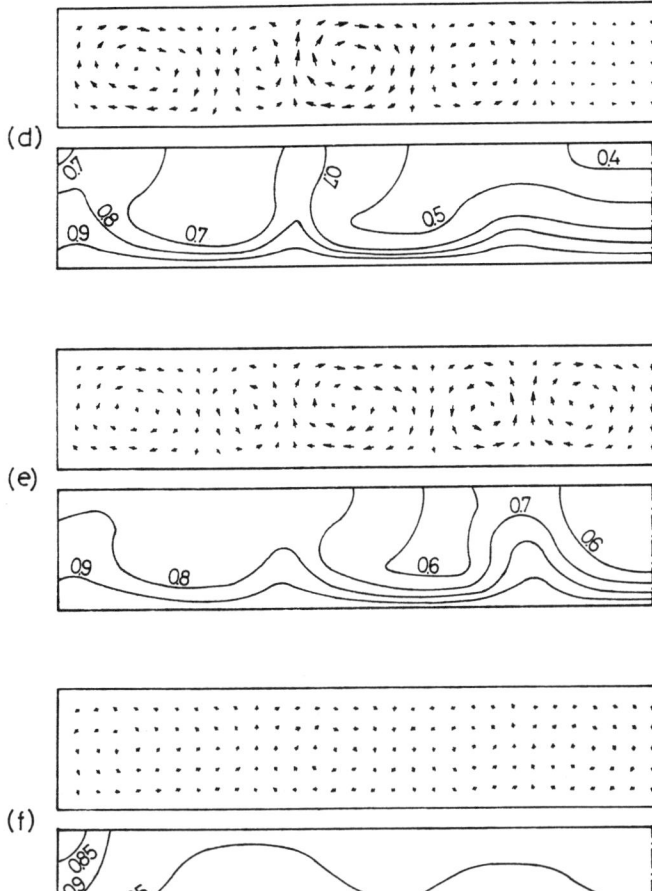

(d)

(e)

(f)

Fig.-2. Cross-stream velocity vectors and isotherms for case 2 with Υ=10, Pr=0.7 and Ra=6x10^4 ; (a)z=0.0131 (b)z=0.0602 (c) z= 0.0800 (d) z=0.0903 (e)z=0.1108 (f)z=0.3002

z=0.1108, the plume at X/D_e=0.77 has grown in strength and one can see clearly five independent vortices in the left half of the channel. From z=0.0131 to z=0.1108, the stronger circulation region shifts from the sidewall toward the central region. At z=0.3002, there are only two weak plumes since the bulk mean fluid temperature approaches the bottom plate temperature.

The heat transfer characteristics in the thermal entrance region of rectangular channels are usually presented by the spanwise average local Nusselt number Nu at each axial position. The variations of Nu versus axial distance z are shown in Figs. 3 and 4 for case 1 and 2 with Υ=5 and 10, Pr=0.7 and 5, and Ra=0 ~ 2x10^5, respectively. One can observe from the variations of local Nusselt number that the buoyancy effect is negligible up to a certain axial distance and this axial distance depends mainly on the magnitude of the Rayleigh number: the greater is Ra, the shorter is the distance. Each curve of Nu in Figs. 3-4 is seen to deviate from that for purely forced convection

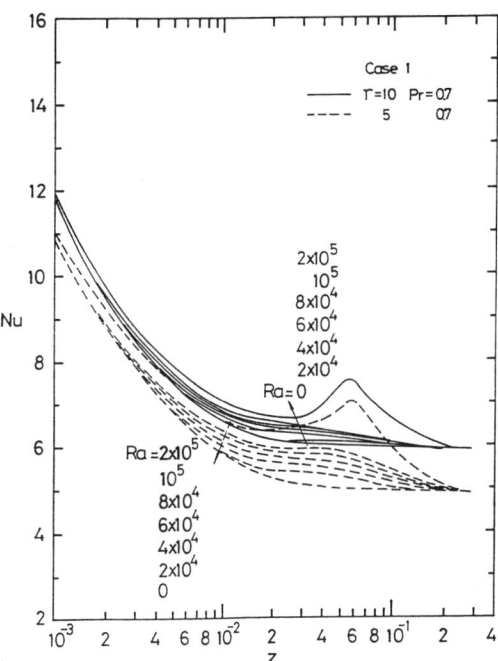

Fig.-3. Effect of Rayleigh number on local Nusselt number variation for case 1 with Υ =10, 5 and Pr= 0.7

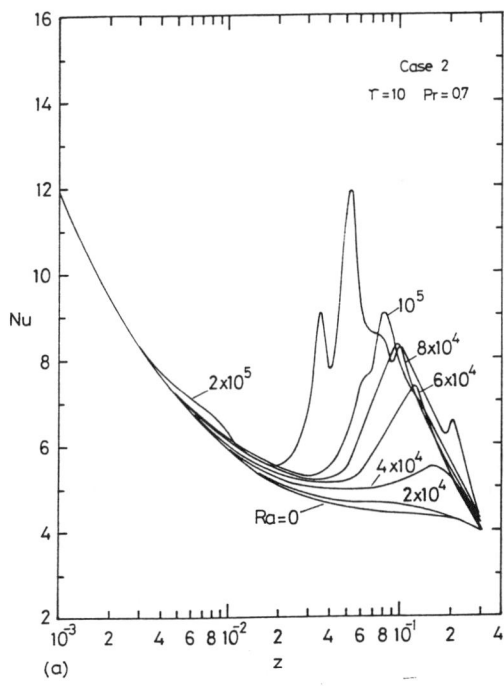

(a)

and after reaching a minimum and then maximum values for some curves, the curves gradually approach the asymptotic values when the fully-developed conditions are reached. It is well known that the occurrence of minimum local Nusselt number is caused by the combined entrance and secondary flow effect. It was found that the occurrence of maximum local Nusselt number for the cases of high Ra is

333

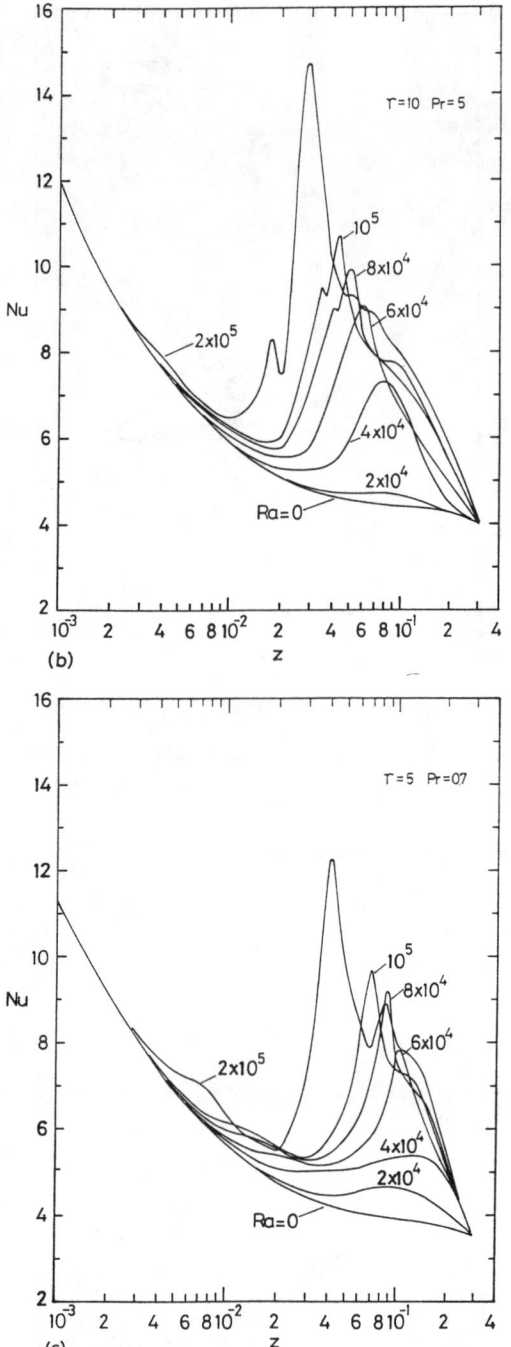

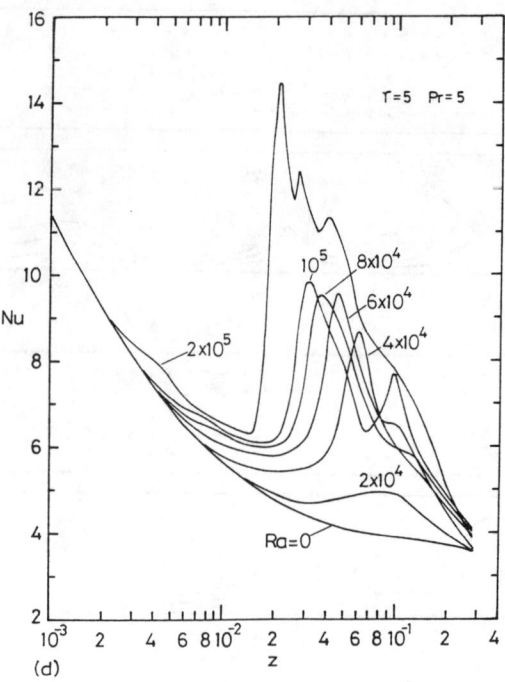

Fig.-4. Effect of Rayleigh number on
local Nusselt number variation
for case 2 with (a) ɤ=10 and Pr=
0.7 (b) ɤ=10 and Pr=5(c) ɤ=5 and
Pr=0.7 (d) ɤ=5 and Pr=5

closely related to the appearance of local
maximum secondary flow intensity(Chou and
Hwang, 1987).

It has been noted in the paper review that
several kinds of criteria have been suggested
for the onset of convective instability. In the
experimental studies, a criterion based on flow
visualization with an injection of dye was
usually used(Hwang and Liu, 1976; Kamotani et
al., 1979; Incropera et al., 1986). A criterion
based on the first longitudinal station at
which heat transfer enhancement begins to
deviate sharply from a gradual increase with
the longitudinal coordinate was also

used(Maughan and Incropera, 1987). Reasonable
agreement was obtained between these results
based on the above two criteria. Since where
and how the dye ascends from the plates or heat
transfer enhancement increases sharply is
directly caused by a finite amplitude of
buoyancy effect, the criterion based on 2%
deviation of local Nusselt number from that of
classical Graetz theory is used in the present
work. This criterion was also used by Ou et
al.(1974). Although numerical, or false,
diffusion effects existed for the cases of
larger Rayleigh number, which are about Ra
≧4x10^5 in the present study, the effects are
confined to downstream regions of the buoyancy
driven secondary flow and do not influence
conditions corresponding to onset of this flow,
and hence appearance of the convective
instability. The comparison of the present
theoretical results with the existing
theoretical and experimental results for onset
of convective instability in the horizontal
rectangular channels is shown in Fig.5. The
numerical results of Ou et al.(1976) correspond
to flow of a large Prandtl number fluid in a
channel of ɤ=2, while the numerical results of
Incropera et al.(1986) correspond to flow of
Pr=6.5 in both the combined entry and thermal
entry region of channels of ɤ=2 and 5. The
experimental results of Hwang and Liu(1976),
and Kamotani and Ostrach (1979) correspond to
air flow in channels of large aspect ratio, say
about 15 and 17 respectively. Finally, the
solid line is the correlation equation
suggested by Incropera et al.(1986) for the
data of water in the channel of ɤ=10. In the
numerical study of Incropera et al.(1986),
convective instability corresponds to onset of

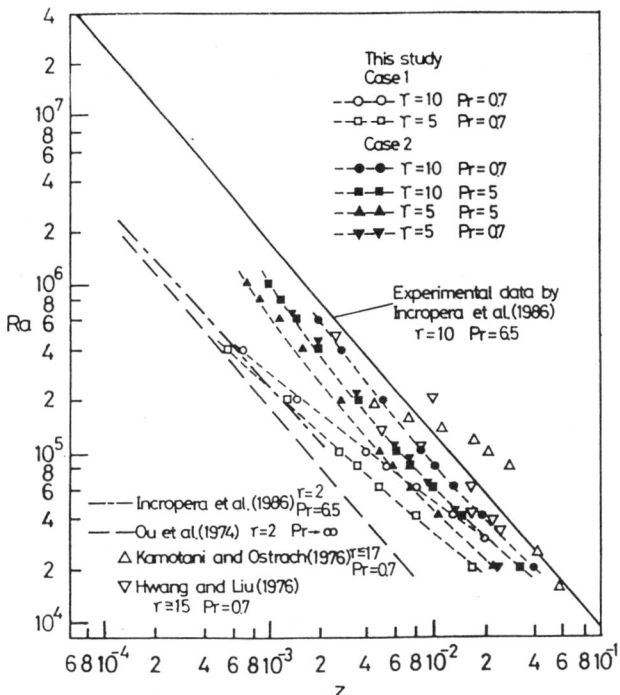

Fig.-5. Comparison of the present and previous results for onset of convective instability in a horizontal rectangular channel

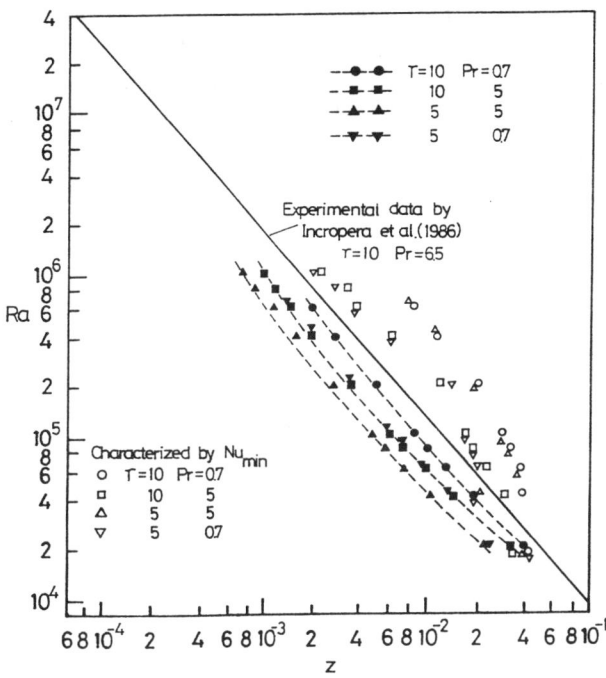

Fig.-6. Comparison between the results of onset of convective instability based on two different cribteria

a weak secondary flow in the corner and is also associated with the departure of Nu from the forced convection result. It can be seen that the present predictions of onset of convective instability based on the 2% deviation from the classical Graetz theory show a good agreement with the previous experimental results.

To ascertain the effects of Prandtl number, aspect ratio and thermal boundary condition on convective instability, it is of interest to compare the results of the present and previous predictions for the cases of Pr=0.7, 5 and ∞, γ=2, 5 and 10, and the isothermal channels and that heated from below. It is shown that the onset of convective instability is always slightly earlier for =5 than for γ=10 under fixed Prandtl number and thermal boundary condition. The results for case 1 with =2 by Ou et al.(1974) show a significantly earlier onset. This phenomenon is believed to be caused by the heated or adiabatic side wall effect which is more significant when aspect ratio is smaller. The side walls induced a strong secondary flow in the region near the side walls (as shown in Fig. 2(a)) and make the onset of instability occur earlier. In the bottom heated situation(case 2), the results for Pr= 5 show earlier onset of instability than that of Pr=0.7 for a fixed aspect ratio. This phenomenon that air is slightly more stable than water was also suggested by Maughan and Incropera(1987). Furthermore, the comparison between the results of cases 1 and 2 shows that the onset of instability is earlier for case 1 than case 2. This effect is more pronounce at higher Rayleigh numbers. This phenomenon is due to the fact that the strength

of the buoyancy effect increases more quickly with Rayleigh number near the heated side wall as compared to the adiabatic wall. Thus we may conclude that the combined effects of aspect ratio, Prandtl number and heated side wall are the reason why the results of Ou et al.(1974) for the case of γ =2 and Pr→∞ in an isothermal channel show a much earlier onset of instability when compared with the experiments of air flow in bottom-heated and large aspect ratio channels.

To study the effect of the selection of criterion on the analytical prediction of onset of convective instability, another criterion, based on the location of minimum local Nusselt number, is also considered for comparison. The predictions of the onset of convective instability based on the forgoing two criteria are shown in Fig.6. It is seen that the predictions based on 2% deviation from that for pure forced convection are all below the curve of correlation equation by Incropera et al.(1986), and that based on the location of minimum local Nusselt number are almost above the curve by Incropera et al.. The predictions based on the two different criteria agree well when $Ra \leq 4 \times 10^4$. The deviations between the predictions based on the two criteria are seen to increase with the increase of Rayleigh number when $4 \times 10^4 \leq Ra \leq 2 \times 10^5$, but the deviations are seen to decrease with the increase of Rayleigh number when $Ra \geq 2 \times 10^5$.

4.CONCLUDING REMARKS

Previous analytical results always showed an underprediction of the onset of convective instability as compared to the experimental

335

results. Since the experiments were usually conducted by flow visualization, the experimental results of onset of convective instability are characterized by a finite amplitude buoyancy effect. Thus a criterion based on 2% departure of local Nusselt from that for pure forced convection is used in the present work. This analytical prediction of onset point of convective instability for the bottom heated cases is compared with the previous experimental results with good agreement.

Furthermore, it can be seen that both reducing the aspect ratio and increasing the Prandtl number will induce earlier onset of convective instability. And because the strength of the buoyancy effect near the heated side wall increases more quickly than that near the adiabatic wall when the Rayleigh number increases, therefore the higher the Rayleigh number is, the earlier the onset of instability is for the isothermal channels compared to that for the bottom heated channels.

The predictions of onset of instability by the criterion based on the minimum local Nusselt number are also shown for comparison. It is seen that the predictions by the two different criteria agree well when $Ra \leq 4 \times 10^4$. The deviations between them increase with the increase of Rayleigh number when $4 \times 10^4 \leq Ra \leq 2 \times 10^5$, and decrease with the increase of Rayleigh number when $Ra \geq 2 \times 10^5$.

Acknowledgement- The authors would like to thank the National Science Council of Republic of China for its support of the present work through project NSC78-0401-E008-01.

REFERENCES

Abou-Ellail, M.M.M., and Morcos, S.M., 1983, "Buoyancy Effects in the Entrance Region of Horizontal Rectangular Channels," ASME J. Heat Transfer, Vol. 105, pp.152-159.

Akiyama, M., Hwang, G.J. and Cheng, K.C., 1971, " Experiments on the Onset of Longitudinal Vortices in Laminar Forced Convection Between Horizontal Plates," ASME J. Heat Transfer, Vol. 93, pp.335-341.

Cheng, K.C., Hong, S.W., and Hwang, G.J., 1972, " Buoyancy Effect on Laminar Heat Transfer in the Thermal Entrance Region of Rectangular Channels with Uniform Wall Heat Flux for Large Prandtl Number Fluids," Int, J. Heat Mass Transfer, Vol. 17, pp.1819-1836.

Chou, F.C. and Hwang, G.J., 1987, "Vorticity-Velocity Method for Graetz Problem with the Effect of Natural Convection in a horizontal Rectangular Channel with Uniform Wall Heat Flux," ASME J. Heat Transfer, Vol. 109, pp.704-710.

Coutier, J.P. and Greif, R., 1985, "An Investigation of Laminar Mixed Convection Inside a Horizontal Tube With Isothermal Wall Conditions", Int. J. Heat Mass Transfer, Vol. 28, pp. 1293-1305.

Hwang, G.J., and Cheng, K.C., 1973, " Convective Instability in the Thermal Entrance Region of a Horizontal Parallel Plate Channel Heated From Below," ASME J. Heat Transfer, Vol. 95, pp. 72-77.

Hwang, G.J., and Liu, C.L., 1976, " An Experiment Study of Convective Instability in the Thermal Entrance Region of a Horizontal Parallel-Plate Channel Heated From Below," Candian J. Chem. Eng. Vol. 54, pp. 521-525.

Incropera, F.P., and Schutt, J.A., 1985, " Numerical Simulation of Laminar Mixed Convection in the Thermal Entrance Region of Horizontal Rectangular Ducts," Numerical Heat Transfer, Vol. 8, pp.707-729.

Incropera, F.P., Knox, A.L. and Schutt, J.A., 1986, "Onset of Thermally Driven Secondary Flow in Horizontal Rectangular Ducts, " Procs. 8th Int. Heat Transfer Conf. pp.1395-1400.

Kamotani, Y., and Ostrach, S., 1976, "Effect of Thermal Instability on Thermally Developing Laminar Channel Flow," ASME J.Heat Transfer, Vol. 98, pp. 62-66.

Kamotani,Y., Ostrach, S., and Miao, H., 1979, "Convective Heat Transfer Augmentation By Means of Thermal Instability, "ASME J. Heat Transfer, Vol. 101, pp.222-226.

Mahaney, H.V., Incropera F.P., and Ramadhyani, S., 1987, "Development of Laminar Mixed Convection in a Horizontal Rectangular Duct with Uniform Bottom Heating," Numerical Heat Transfer, Vol. 12, pp.137-155.

Maughan, J.R., and Incropera, F.P., 1987, " Experiments on Mixed Convection Heat Transfer for Airflow in a Horizontal and Inclined Channel," Int. J. Heat Mass Transfer, Vol. 30, pp. 1307-1318.

Mori, Y., and Uchida, Y., 1966, " Forced Convective Heat Transfer Between Horizontal Flat Plates," Int. J. Heat Mass Transfer Vol. 9, pp. 803-817.

Nakayama, W., Hwang, G.J., and Cheng, K.C., 1970, "Thermal Instability in Plane Poiseuille Flow," ASME J. Heat Transfer, Vol. 92, pp. 61-68.

Osborne, D.G., and Incropera, F.P., 1985, " Laminar Mixed Convective Heat Transfer for Flow Between Horizontal Parallel Plates with Asymmetric Heating," Int. J. Heat Mass Transfer, Vol. 28, pp. 207-217.

Ou, J.W., Cheng, K.C., and Lin, R.C., 1974 "Natural Convection Effect on Graetz Problem in Horizontal Rectangular Channels with Uniform Wall Temperature for Large Pr," Int. J. Heat Mass Transfer, Vol. 17, pp.835-843.

Patankar, S.V. and Spalding, D.B., 1972, "A Calculation Procedure for Heat, Mass and Momentum Transfer in Three-Dimensional Parabolic Flows," Int. J. Heat Mass Transfer, Vol. 15, pp.1787-1806.

Ramakrishna, K,, Rubin, S.G, and Khosla, P.K., 1982, "Laminar Natural Convection Along Vertical Square Ducts," Numerical Heat Transfer, Vol. 5, pp.59-79.

Wu, R.S., and Cheng, K.C., 1976, " Thermal Instability of Blasius Flow Along Horizontal Plates," Int. J. Heat Mass Transfer, Vol.19, pp.907-913.

1989 National Heat Transfer Conference
HTD-Vol. 107, Heat Transfer in Convective Flows

HEAT TRANSFER AUGMENTATION THROUGH
WALL SHAPE INDUCED FLOW DESTABILIZATION

M. Greiner, R.-F. Chen, and R. A. Wirtz
Mechanical Engineering Department
University of Nevada-Reno
Reno, Nevada

ABSTRACT

Experiments on heat transfer augmentation in a rectangular cross section water channel are reported. The channel geometry is designed to excite normally damped Tollmien-Schlichting modes in order to enhance mixing. In this experiment, a hydrodynamically fully developed flow encounters a test section where one channel boundary is a saw-tooth series of periodic, transverse grooves. Free shear layers span the groove openings, separating the main channel flow from the recirculating vorticies contained within each cavity. The periodicity length of the grooves is equal to one-half of the expected wavelength of the most unstable mode. The remaining channel walls are flat, and the channel has an aspect ratio of 10:1. Experiments are performed over the Reynolds number range of 200 to 15000.

Streak line flow visualization shows that the flow is steady at the entrance, but becomes oscillatory downstream of an onset location. This location moves upstream with increasing Reynolds numbers. Initially formed traveling waves are two-dimensional with a wave length equal to the predicted most unstable Tollmien-Schlichting mode. Waves become three-dimensional with increasing Reynolds number and distance from onset. Some evidence of wave motion persists into the turbulent flow regime.

Heat Transfer measurements along the smooth channel boundary opposite the grooved wall show augmentation (65%) over the equivalent flat channel in the Reynolds number range 1200 to 4800. The degree of enhancement obtained is shown to depend on the channel Reynolds number, and increases with the distance from the onset location.

NOTATION

a	groove depth
b	groove length
D_h	hydraulic diameter
E	enhancement ratio, $\mathrm{Nu}(grooved)/\mathrm{Nu}(flat)$
Gz^{-1}	inverse Graetz number, $(x/D_h)/RePr$
h	heat transfer coefficient
H	channel height
k	fluid thermal conductivity
L	test section length
Nu	Nusselt number, hD_h/k
Pr	fluid Prandtl number
q''	heat flux
Re	Reynolds number, VD_h/v
Re_c	critical Reynolds number
T	temperature

T_m	mixed mean temperature
T_o	lower wall temperature
T_s	heated wall surface temperature
V	average velocity
W	channel span
x	axial coordinate
λ	flow periodicity length
v	kinematic viscosity

INTRODUCTION AND PROBLEM DEFINITION

In heat exchanger applications involving low density fluids (gases), it is well known that mechanical pumping power requirements may be comparable to heat transfer rates if flow velocities are not limited to small values. As a result, "compact" exchangers involving passages with small hydraulic diameter have been developed where gas side velocities are kept low, and surface area is maximized in order to make up for the relatively low heat transfer coefficients. Kays and London [1964] point out that another way to minimize friction power requirements is to select enhanced surfaces which "...plot 'high' on a heat transfer-friction power plot...".

A recent review [Bergles and Webb, 1985] shows an exponential growth in the heat transfer augmentation literature during the past thirty years, underscoring the importance of the problem which, if current interest is any indication, has resisted a solution that is satisfactory for all applications. Enhancement schemes are classified as active (those requiring external energy input) or passive (which are inherently more reliable). Passive schemes generally involve an increase in the complexity of the heat transfer surface. For example, offset-strip fins, louvers and flutes are added to plate fin exchangers in order to promote mixing in the plate-side stream, and to interrupt the thermal boundary layer along the plate-fin, giving rise to an increase in heat transfer coefficient [Mori and Nakayama, 1980; Joshi and Webb, 1987]. Since leading edges also cause very high surface shear stresses, these mechanisms also lead to significant increases in friction power requirements.

Due to the generally small passage hydraulic diameters and low face velocities employed in compact heat exchangers, gas side passage Reynolds numbers are relatively low (approximately 2000 or less) giving rise to laminar, or at best transitional flows. Therefore, a successful augmentation scheme must be operable in a flow with strong viscous damping characteristics.

The present investigation involves a passive scheme which enhances mixing in the fluid stream, not by turbulation, but by promotion of a secondary flow through careful design of the channel contour. The

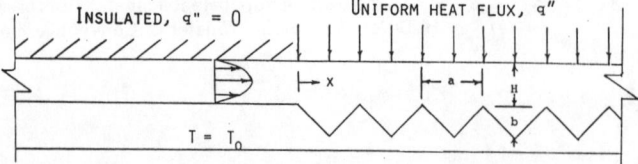

Figure 1. Geometric and thermal boundary conditions.

flow is inherently laminar, but is more complex than flat channel counterparts. The secondary flow is designed to increase convective mixing normal to the heat transfer surface, thus promoting transport augmentation.

Since the complexity of the flow is increased, it is expected that for a given Reynolds number, pumping power requirements for the present geometry will be greater than flow in a flat channel. However, increases in pumping power requirements come about because of organized mixing of the flow; not because of a direct manipulation of the surface shear stress (as is done with periodically interrupted surfaces), or through the introduction of chaotic motion (as in flow turbulation). Some evidence [Kozlu et al., 1988; Karniadakis et al., 1988] indicates that the pumping power cost of the current technique is favorable compared to more conventional methods. While this issue is not addressed in the present work, it is the subject of current investigation.

The focus of the present paper is to extend concepts which have been developed numerically [Ghaddar et al., 1986], using experimental techniques. In that work rectangular shaped grooves were studied under periodically fully developed flow conditions. A new cavity geometry which shows promise as an efficient heat transfer enhancement device is investigated. Transport measurements are made in a thermally developing flow along an adjacent flat surface in order that direct and meaningful comparisons may be made with the performance of a plane channel flow.

The specific geometry considered in this work is indicated in Figure 1. A fully developed flow is discharged from a parallel-wall flow development channel whose upper wall is insulated and whose bottom wall is maintained at a uniform temperature, T_0. At the position $x = 0$, this flow enters a test section, of height H, whose upper surface dissipates a uniform heat flux, q'', and whose lower surface is maintained at a constant temperature, T_0. Heat transfer measurements are performed at the upper wall under two sets of geometric conditions of the bottom surface: a flat surface, and a saw-toothed profile. The first condition provides a baseline against which enhancement, caused by a grooved lower surface, can be compared. It also serves to certify the validity of the measurement system.

The grooved surface is specifically designed to excite Tollmien-Schlichting waves in the channel flow which under flat wall conditions are damped at Reynolds numbers less than a critical value, $Re_c = 15400$ [Drazen and Reid, 1981]. The critical Reynolds number is reduced by introducing spatially periodic disturbances whose periodicity length is compatible with the most unstable Tollmien-Schlichting wave length. At supercritical conditions, traveling waves whose amplitude increase with $Re - Re_c$, promote transport perpendicular to the channel walls.

A groove aspect ratio of $a/b = 2$ is used so that grooves act as "open" cavities (whose openings are completely spanned by free shear layers) for the Reynolds number range considered [Yee, 1986]. These shear layers cause velocity profile inflection points to form in the channel. At Reynolds numbers above the critical value, Kelvin-Helmholtz instabilities of these layers have been shown numerically [Ghaddar et al., 1986] and experimentally [Greiner, 1986] to destabilized the normally-damped Tollmien-Schlichting waves. The destabilized flow exhibits traveling waves which alternately pump core fluid to and away from the walls.

The channel periodicity length, a, is chosen to destabilize the most unstable Tollmein-Schlichting wavelength, λ. Since this wavelength is Reynolds number dependent, and the present investigation considers the range $200 \leq Re \leq 15000$, a representative value of $Re = 1400$ is selected. At this Reynolds number $\lambda = 2.4H$ [Ghaddar et al.], and the channel periodicity length, a is chosen so that $\lambda = a/2$.

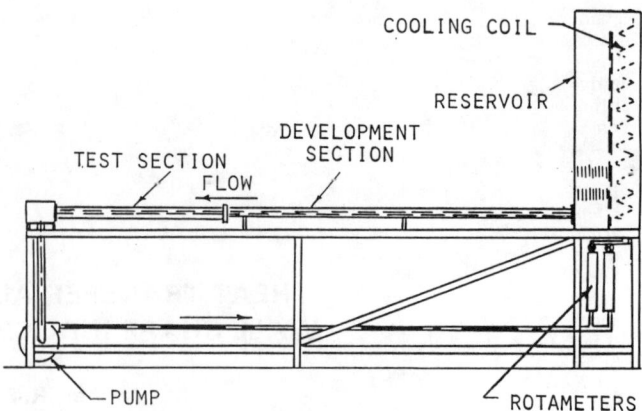

Figure 2. Closed loop test apparatus.

The critical Reynolds number for the onset of natural oscillations has been shown to decrease with spacing between the grooves [Greiner, 1986]. In the current design, a "saw-tooth" profile is selected since it minimizes cavity separation, and the critical Reynolds number in fully developed flow is predicted to be roughly $Re_c = 330$ [Greiner et al., 1988]. Furthermore, this geometry is expected to produce an efficient cavity flow, and be easily manufactured.

Convective heat transfer measurements are performed for the Reynolds number range, $200 \leq Re \leq 15000$, where $Re = VD_h/\nu$ with V the average velocity, D_h the channel hydraulic diameter, and ν the fluid kinematic viscosity. The local Nusselt number, $Nu = hD_h/k$, where h is the local heat transfer coefficient and k is the fluid thermal conductivity, is reported at five x-locations along the upper flat surface. The Nusselt number dependence on Reynolds number, channel location and channel geometry is determined experimentally.

EXPERIMENTAL APPARATUS

Measurements were made using the temperature-controlled recirculating water channel shown in Figure 2. A centrifugal pump delivers distilled water through a bank of rotameters and control valves to the right side of a partitioned reservoir which contains a cooling coil for temperature control. The left side of the tank is fed via a weir at the top of the partition and can supply up to a 1.2 meter head to the test channel. The flow passes two honeycomb sections and enters the channel flow development section through a "soda straw" flow straightener. The flow development section has height, H = 20 mm, and width, W = 203 mm giving a hydraulic diameter, $D_h = 36.4$ mm. Velocimetry measurements show that the flow development length of $67 D_h$ is sufficient to assure fully developed conditions at the test section inlet (x = 0 in Figure 1). The bottom and side walls of the flow development section are aluminum, and the top surface is Plexiglas. The temperature at x = -50 mm is monitored by a digital thermometer and maintained at $T_0 = 29.4 \pm 0.6$ C, which is within 3 C of the laboratory room temperature.

The test section is 30.3 hydraulic diameters long, and includes forty-six V-shaped grooves which span the lower surface. These are constructed by mounting right-triangular aluminum ribs, 12 mm high and 24 mm at their base, to a 12 mm thick aluminum base plate, which is in turn backed by a water jacket for temperature control. The bottom wall is maintained isothermal to within ± 0.2 C while the overall driving temperature difference for experiments ranges from 2 to 10 C. A 12 mm thick aluminum plate is substituted in place of the triangular elements for the base-line flat channel experiments.

Two different upper surfaces are employed in the test section; one for flow visualization experiments, the other for heat transfer measurements. Both are fabricated from 25 mm thick Plexiglas. Flow visualizations are performed by injecting colored tracer into the flow field and recording the resulting patterns on video tape. A variable volume flow rate syringe pump is used to inject the pigment at channel center-height and mid-span through an L-shaped tube (1.0 mm O.D.) inserted through the channel ceiling and bent downstream. For each Reynolds number, the tracer flow rate is adjusted so that its velocity at the injector tip is the same as the average fluid velocity, V, thus minimizing disturbances to the channel flow. For the applicable channel

Reynolds number range, the injector Reynolds number is less than 40, thus causing negligible flow disturbance. The pigment is a solution of red food coloring, diluted with roughly seventeen parts water. While the dye density is slightly greater than that of the working fluid, its settling velocity is much less than that of the center channel speed, even at the lowest Reynolds numbers considered.

The heat transfer surface has six custom heater/thermocouple/heat flux gage assemblies bonded to its surface. The heat flux passing to the fluid from each assembly is monitored by a 102 mm by 102 mm thermopile-type heat flux gage (accuracy ± 1%) located at its center. The assemblies contain copper-constantan thermocouple junctions located 0.28 mm beneath the wetted surface at eighteen equally spaced points along the channel center-line. These thermocouples are referenced to a junction located at the center of the inlet channel at x = -50 mm. After corrections are made for the conduction temperature drop between the wall thermocouples and the wetted wall surface, and the local fluid mixed mean temperature is computed using an energy balance, the local temperature difference between the wall and the mixed mean fluid temperature, $\Delta T = T_s - T_m$, is used with the local heat flux to determine the local heat transfer coefficient, h.

Each of the six combination gage's heaters are wired in series with a trimming rheostat, and these subsystems are wired in parallel to a regulated DC power supply. During an experimental run, the trimming rheostats are adjusted so that the indicated heat flux through each combination gage is the same, resulting in a heat flux uniformity of ± 1% of the average heat flux input.

Following the test section the flow passes another "soda straw" flow straightener and enters a small plenum chamber with a return line to the pump. Edge walls of the test section are 6mm thick Plexiglas, and all metallic surfaces are coated with baked epoxy paint. The apparatus is covered with 50 mm thick expanded foam insulation during data collection.

RESULTS

Flow Visualizations

Figures 3 and 4 show a series of streakline patterns which document the onset of natural flow oscillations and the subsequent breakdown to turbulence. These visualizations are produced by injecting colored tracer at the leading edge of the thirty-eighth groove ($x/D_h = 24.4$). The streakline pattern shown in Figure 3a is typical of subcritical flows. The flow is steady and the grooves contain slowly turning voriticies. The outer channel streaklines move parallel to the flat wall, much like the flow in ungrooved (flat) channels. At Re = 600, small amplitude laminar waves are intermittently observed between long periods of steady flow. At a Reynolds number of 700 (Figure 3b), the flow is almost continuously oscillatory, with occasional steady periods. A traveling wave structure develops with a regular wavelength, which Figure 3b shows to be roughly equal to two groove lengths. The channel geometry is designed so the wavelength of the most unstable Tollmien-Schlichting mode is equal to two groove lengths, and it is not surprising that this mode is the first to be excited. These streaklines are "smooth" and views from the top of the tank indicate that they are mostly two-dimensional. As the Reynolds number increases, however, the patterns become more irregular and three-dimensional. Small scale structures are superimpose on the long wavelength Tollmien-Schlichting waves in Figure 4a for Re = 1000. As the Reynolds number increases, the length scales of the smaller structures decrease and the three-dimensionality of the flow increases, as seen in Figure 4b.

The dye injection visualization technique loses its effectiveness at demonstrating the smallest scale motion at higher Reynolds numbers where the tracer rapidly dissipates. The dye fan envelope does appear to experience periodic large scale oscillatory motion when viewed in real time, suggesting that the groove spanning free shear layer may be capable of effecting fully turbulent flows in channels. This visualization technique, however, is severely limited at this turbulence level.

A graph showing the fraction of time the flow exhibits an oscillatory behavior, as a function of Reynolds number, is presented in Figure 5. These data are determined by viewing a flow pattern for a given time period and measuring the fraction of this time the flow is "oscillatory". These measurements are somewhat qualitative because the oscillatory amplitude varies continuously, and does not exhibit an "on/off" behavior. Multiple data points are reported for Reynolds numbers at which more than one observation is made, indicating the imprecision of this technique. Figure 5 shows that the oscillatory flow

a)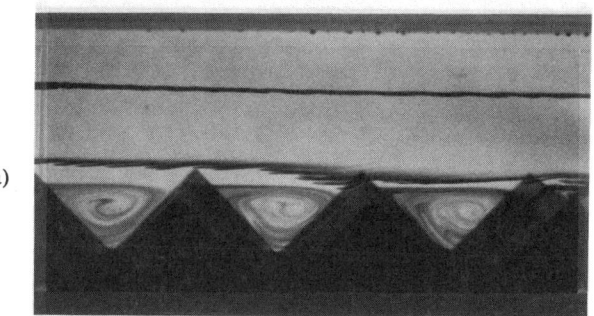

b)

Figure 3. Streakline flow visualizations at $x/D_h = 24.4$. The flow is from left to right. a) Re = 600, steady flow. b) Re = 700, traveling wave structure.

a)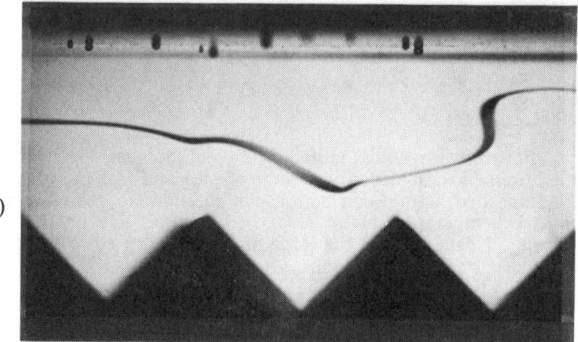

b)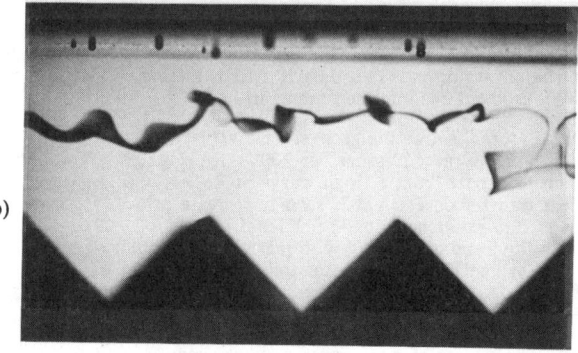

Figure 4. Streakline flow visualization showing three-dimensional wave structure. a) Re = 1000. b) Re = 2000.

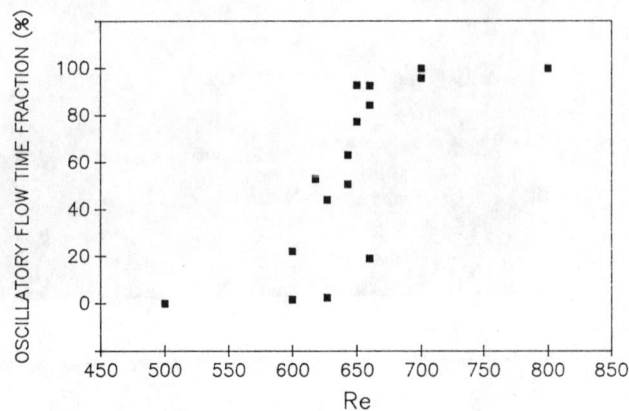

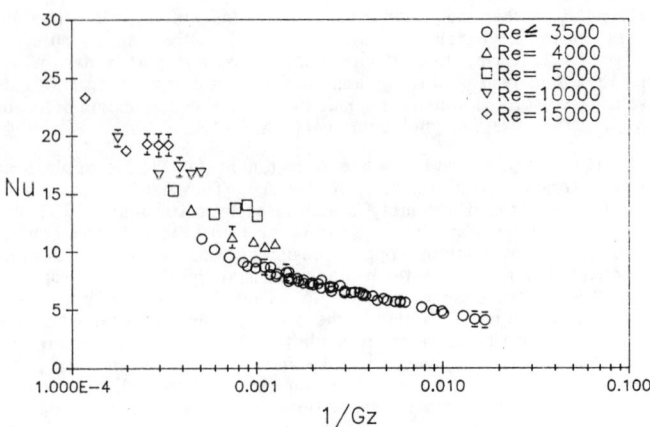

Figure 5. Oscillatory flow time fraction versus Reynolds number. Two-dimensional transition occurs in the range 600 < Re < 660.

Figure 6. Measured flat channel Nusselt number versus inverse Graetz number for 300 < Re < 15,000. Error bars are used to indicate 99.7% confidence band due to random errors when this band is larger than the symbol size.

time-fraction increases sharply in the Reynolds number range 600 to 660. As the Reynolds number increases beyond 700, the flow is observed to be continuously oscillatory.

If the observed critical Reynolds number, Re_c, is defined (arbitrarily) as the value at which the flow is oscillatory fifty percent of the time, then for this channel location, $Re_c = 630 \pm 20$. While oscillatory modes are known to be susceptible to forced flow rate modulation [Greiner et al., 1986], it is not known what effect any flow rate oscillations inherent in the apparatus has on the value of Re_c. The dependence of the observed value of Re_c on location is discussed in the following section in connection with the onset of heat transfer enhancement. Reference to Figure 10 (triangular symbols) shows that Re_c decreases with distance downstream. However, for the present apparatus an asymptotic value is not achieved. Work in progress with a test section which is 60 D_h long shows Re_c approaches a value of 350 after about 35 D_h, in good agreement with previous predictions [Greiner et al., 1988].

Heat Transfer

For each Reynolds number and axial location, local temperature measurements are made for a range of wall heat flux in order to detect any effect of natural convection, and to eliminate the effects of systematic temperature offset errors. A straight line is fit to each q″ versus ΔT data set, and the slope of this line is used to determine the local heat transfer coefficient, $h = dq''/d\Delta T$. The relation is found to be highly linear, indicating the absence of significant buoyancy effects. In the following heat transfer plots, error bars representing the 99.7% confidence level due to random errors are indicated where the bar size exceeds data point symbol size. In many cases the error bar size is smaller than the symbol size, and is not shown.

Figure 6 shows the local Nusselt number as a function of inverse Graetz number for a flat channel flow. Data for Re < 3500 collapse to a single line, while that for Re $\geq$ 4000 show the onset of significant turbulent transport. This data is used as the base-line for comparison with grooved channel measurements.

Figure 7 shows the grooved channel Nusselt number as a function of inverse Graetz number for 300 $\leq$ Re $\leq$ 15,000. The data is now seen to depart from a single curve at Re $\geq$ 1200. At Re = 1200, the heat transfer coefficient is seen to increase at $Gz^{-1} \approx 0.003$, and the departure point has moved upstream to $Gz^{-1} \approx 0.0008$ at Re = 2000, with the trend continuing with increasing Reynolds number. The heat transfer coefficient is greater than corresponding flat channel values for Re $\geq$ 1200, and it is actually less than the flat channel values for Re $\leq$ 1000, as explained below.

The degree of heat transfer enhancement obtained is more easily seen in Figure 8, which is for a super-critical Reynolds number of 3000. Part (a) of the figure compares the local Nusselt number for the grooved channel flow with that obtained with the flat configuration, and part (b) shows the enhancement factor, E = Nu(grooved)/Nu(flat), both plotted against dimensionless axial location. As expected, flat channel values show a steady decrease in the stream-wise direction,

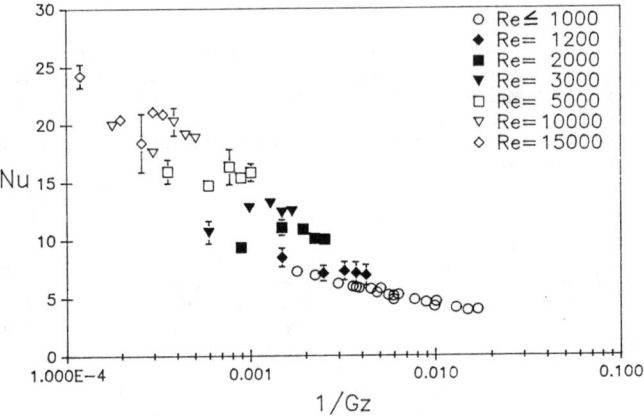

Figure 7. Measured grooved channel Nusselt number versus inverse Graetz number for 300 < Re < 15,000. Error bars are used to indicate 99.7% confidence band due to random errors when this band is larger than the symbol size.

consistent with a thermally developing flow. On the other hand, the grooved channel values show enhancement at $x/D_h \approx 6$, with an augmentation of approximately 65% further down stream.

Figure 9 shows E as a function of x/D_h for subcritical (Re = 300), supercritical (3000), and turbulent (5000) flows. For a subcritical Reynolds number of 300, the groove geometry is seen to actually decrease the heat transfer coefficient along the upper wall by about 10% relative to the ungrooved geometry. Groove spanning free shear layers relax the no slip condition along the lower wall, causing the velocity maximum to shift downward, resulting in reduced transport along the upper wall. As a consequence the enhancement ratio is less than unity over the entire length of the channel. For Re = 1200 (not shown in Figure 9), traveling waves are visually observed for $x/D_h >$ 10. The grooved channel heat transfer is seen to exceed the ungrooved values for $x/D_h > 16$, resulting in enhancement ratios of approximately 1.1.

As the Reynolds number is further increased, the break even point (E = 1.0) rapidly moves upstream, and the magnitude of the local enhancement ratio increases. For Re = 3000, visual observations of the flow show remnants of a traveling wave structure blurred by turbulent mixing. The break even point shifts to $x/D_h = 7$, and E > 1.5 for $x/D_h > 16$ (Figures 8b and 9).

As Reynolds number is increased beyond 3000, the grooved channel local heat transfer coefficient continues to increase. However, the

a)

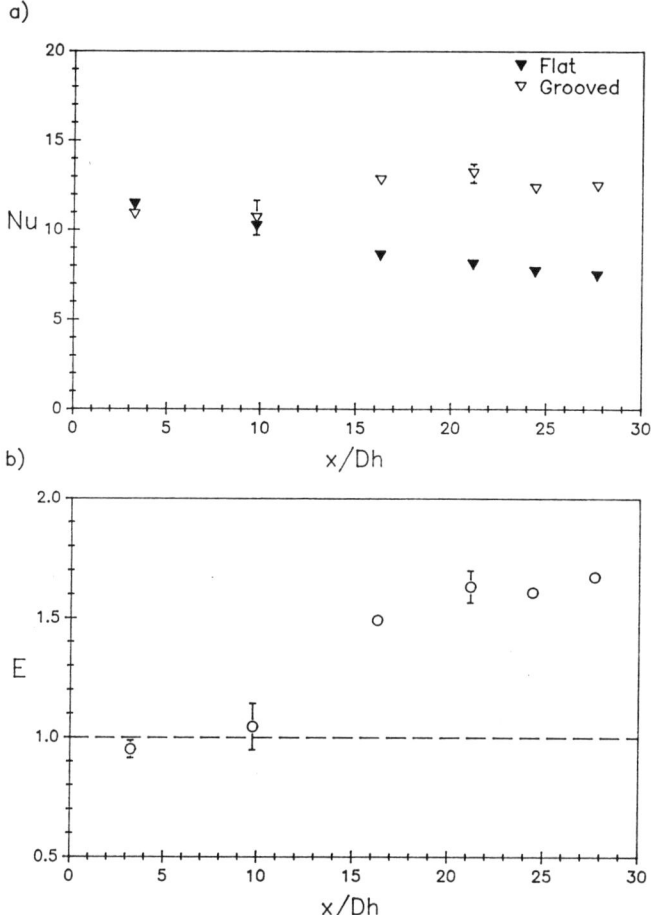

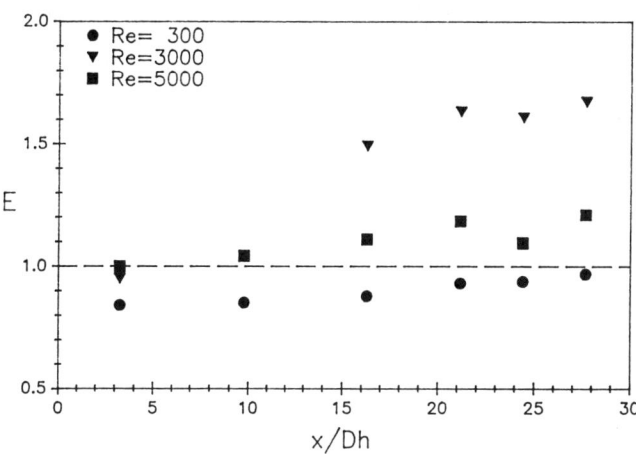

Figure 9. Local grooved channel heat transfer enhancement for Re = 300 (subcritical flow, heat transfer degradation), Re = 3000 (maximum enhancement), and Re = 5000 (typical of high Reynolds numbers).

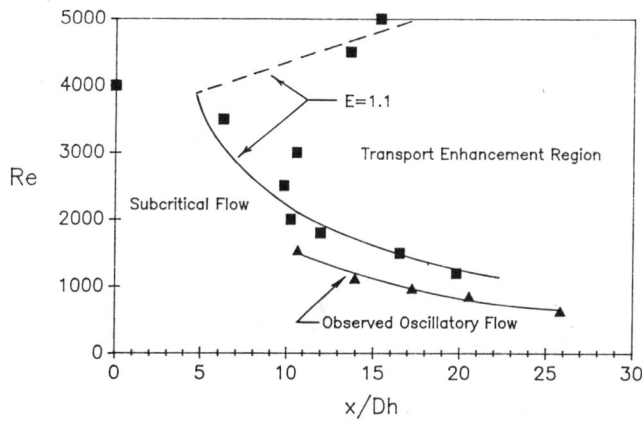

Figure 8. Local heat transfer measurements at Re = 3000. a) Nusselt number. b) Enhancement ratio.

Figure 10. Location of ten percent heat transfer enhancement (squares) as a function of Reynolds number. The onset location of oscillatory flow (triangles) is closely correlated with enhancement.

ungrooved heat transfer coefficient increases at a faster rate due to increased turbulent mixing. As a consequence, the local enhancement ratio is reduced as shown in the figure for Re = 5000. This result is effectively the same for all $5000 \leq Re \leq 15000$.

Figure 10 includes a Re versus x/D_h map of groove-induced heat transfer enhancement. The figure shows the locus of points where enhancement exceeds 10% (E = 1.1). Also shown are wave onset locations determined from analysis of video records of flow visualization experiments. The figure indicates that the occurrence of groove-induced oscillations is closely correlated with effective heat transfer augmentation. Points to the right of the data band of the figure experience heat transfer augmentation via this mechanism. Points to the left of the band experience no enhancement, or a degradation in performance. Enhancement is seen to move upstream with increasing Reynolds number until Re = 4000, and move down stream for higher Reynolds numbers. It is currently thought that for Re > 4000, turbulent mixing mechanisms become sufficiently strong that they dominate the large scale wavy structures, reducing the difference between grooved and ungrooved channel flows.

Enhancement data as a function of Reynolds number for five measurement stations of the current apparatus is condensed in Figure 11. For the current channel configuration, the figure shows that significant, and spatially constant (i.e. fully developed) enhancement, is obtained for $x/D_h > 16$. Maximum heat transfer enhancement of approximately 65% is obtained over the range, $2000 \leq Re \leq 4000$. At higher Reynolds numbers turbulence appears to overwhelm the natural oscillations, leading to a reduction in E.

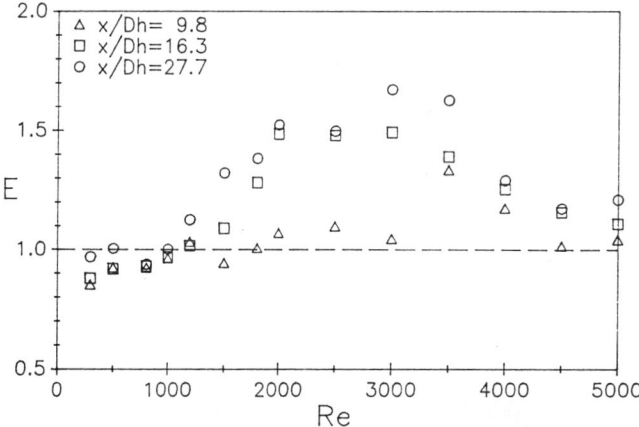

Figure 11. Local grooved channel enhancement versus Reynolds number for $9.8 \leq x/D_h \leq 27.7$. Enhancement factor is fully developed for $x/D_h > 16$.

341

CONCLUSIONS

Groove induced flow oscillations occur at super-critical Reynolds numbers. The onset location for such oscillations is Reynolds number dependent. Oscillation intensity and onset location rapidly move upstream with increases in Reynolds number. For the current channel configuration, oscillatory flows are observed for Re $\geq$ 630, and persist to Re = 4800 (approx.).

The oscillatory flow mechanism is responsible for augmentation of heat transfer. For the current channel configuration, enhancement in excess of 10% extends over the range, 1200 $\leq$ Re $\leq$ 4800, and x/D_h > 16. Maximum enhancement of 65% occurs at Re = 3000 ± 1000. Turbulent mixing at higher Reynolds numbers degrades the effectiveness of the mechanism.

ACKNOWLEDGEMENT

This work is supported by the Gas Research Institute under contract number 5087-260-1562.

REFERENCES

Bergles, A.E. And Webb, R.L., (1985), "A guide to the literature on convective heat transfer augmentation", *Advances in Enhanced Heat Transfer*, S.M. Shenkman, et al., eds. ASME HTD-Vol. 43.

Drazin, P.G. and Reid, W.H., (1981), *Hydrodynamic Stability*, Cambridge University Press, Cambridge.

Ghaddar, N.K., Korczak, Mikic, B.B., And Patera, A.T., (1986), "Numerical investigation of incompressible flow in grooved channels. Part 1: stability and self-sustained oscillations", *J. Fluid Mech.* 163, 99-128.

Greiner, M. (1986), "Experimental investigation of resonance and heat transfer enhancement in grooved channels", Ph.D. thesis, Massachusetts Institute of Technology, Cambridge, Mass.

Greiner, M., Ghaddar, N.K., Mikic, B.B. and Patera, A.T., (1986), "Resonant convective heat transfer in grooved channels", *Proc. Eighth Int. Heat Trans. Conf.*, 6, 2867-2872.

Greiner, M., Karniadakis, G.E., Mikic, B.B. and Patera, A.T., (1988), "Heat transfer augmentation and hydrodynamic stability theory: Understanding and exploitation", *Heat Transfer: Korea-US Seminar on Thermal Engineering and High Technology*, J.H. Kim, S.T. Ro, and T.S. Lee eds., Hemisphere Publishing Corp., New York, 31-50.

Joshi, H.M. And Webb, R.L., (1987), "Heat transfer and friction in the offset strip-fin heat exchanger", *Int. J. Heat Mass Trans.*, 30, pp. 69-84.

Kays, W.M. And London, A.L., (1964), *Compact Heat Exchangers*, McGraw-Hill, New York.

Karniadakis, G.E., Mikic, B.B. and Patera, A.T., (1988), "Minimum-dissipation transport enhancement by flow destabilization: Reynolds' analogy revisited", *J. Fluid Mechanics*, 192, 365-391.

Kozlu, H., Mikic, B.B. and Patera, A.T., (1988), "Minimum-dissipation heat removal by scale-matched flow destabilization", *Int. J. Heat Mass Trans.*, 31, 2023-2032.

Mori, Y. And Nakayama, W. (1980), "Recent advances in compact heat exchangers in Japan", *Compact Heat Exchangers-History, Technological Advancement and Mechanical Design Problems*, R.K. Shah, et al., eds. ASME HTD-Vol. 10.

Yee, E.C. (1986), "The effect of geometry on hydrodynamic resonance in grooved channels", S.B. Thesis, Massachusetts Institute of Technology, Cambridge, Mass.

1989 National Heat Transfer Conference
HTD-Vol. 107, Heat Transfer in Convective Flows

HYSTERESIS EFFECTS IN THREE-DIMENSIONAL
NATURAL CONVECTION IN A
TILTED POROUS MEDIUM

S. J. Pien and M. Sen
Department of Aerospace and Mechanical Engineering
University of Notre Dame
Notre Dame, Indiana

ABSTRACT

Steady state natural convection in a tilted three-dimensional porous material is numerically calculated. For simplicity the Darcy law is used as a governing equation. Two opposing isothermal faces are kept at different temperatures, while the rest are adiabatic. A time marching numerical method is used to determine the velocity and temperature fields. The aspect ratios are chosen such that the convective motion is unicellular. The inclination is also chosen so as to change the axis of the cell through 90° as the angle is increased. The change is found to be at different angles depending on whether the inclination is decreasing or increasing. This hysteresis effect is observed in the magnitude of the Nusselt number as well as the direction of the velocity vector potential.

NOMENCLATURE

A	$= L/W$, aspect ratio
B	$= L/H$, aspect ratio
g	gravity vector
K	permeability of porous medium
L,W,H	dimensions of parallelopiped
Nu	Nusselt number
p	pressure
Ra	$= \rho_0 g \beta K L \alpha (T_H - T_C)/\mu$, Rayleigh number
t	time
T	temperature
T_0	reference temperature
T_H, T_C	temperature of isothermal hot and cold faces respectively
V	velocity vector
x,y,z	Cartesian coordinates

Greek letters

α	effective thermal diffusivity
β	coefficient of volumetric expansion
θ	inclination angle of parallelopiped
λ	effective thermal conductivity of medium
μ	viscosity of fluid
ρ	density of fluid
ρ_0	density of fluid at temperature T_0
σ	ratio of heat capacity of solid matrix and of fluid
ψ	velocity vector potential
ψ_x, ψ_y, ψ_z	components of velocity vector potential

INTRODUCTION

Heat transfer by convection in porous media is of interest in many areas, particularly those relating to geophysical and thermal insulation applications. Many of the natural convection features are in common with those in fluid-filled enclosures. Because of its nonlinearity the problem is also interesting from a fundamental point of view. Multiple steady and unsteady states exist and the transition between them is of considerable importance. A recent review of the subject of convection in porous materials has been carried out by Bejan (1987).

Many different three-dimensional geometries for natural convection in porous media have been studied. There are some results available for rectangular parallelopipeds without inclination. Beck (1972) studied the instability modes of a bottom heated geometry with varying aspect ratios. Zebib and Kassoy (1977) first calculated the effect of viscosity variation and then (Zebib and Kassoy, 1978) used a weakly nonlinear analysis for possible two- and three-dimensional convection patterns. Straus and Schubert (1978) analyzed the stability of finite amplitude two-dimensional convection to determine the dimensions of boxes for which the motion is necessarily unsteady or steady and three-dimensional. Horne (1979) used finite difference numerical results to confirm that three-dimensional convective motions do exist even when the boundary conditions are not three-dimensional. In fact it was demonstrated that there exist more than one mode of convection for any particular configuration and Rayleigh number. Straus and Schubert (1979) used a Galerkin technique to compute convective motion in a cubic geometry. They also compared (Schubert and Straus,1979) three-dimensional and two-dimensional motion in a cube and square respectively. This numerical method has been used by Caltagirone et al. (1981) for a horizontal porous layer. The stability of the possible modes of convection was investigated by Straus and Schubert (1981). The aspect of pattern selection related to all possible initial conditions was addressed by Steen (1983). Some special problems such as those derived from internal heat generation (Beukema et al., 1983) or a maximum in the fluid density (Altimir, 1984) have also been treated.

There has been some work on geometries corresponding to inclined porous materials. Holst and Aziz (1972) carried out a finite difference analysis of natural convection in a confined porous medium. Bories and Monferran (1972) looked for a criterion for the transition between two types of motion observed in inclined layers. This work was extended by Bories et al. (1972). Bories and Combarnous (1973) have carried out a linear stability analysis to determine the critical conditions for the transitions between unicellular and polyhedral cells. Caltagirone and Bories (1980, 1985) used a three-dimensional numerical model based on the Galerkin spectral method. They also analyzed the linear stability of these flows.

Some of the results for the inclined layer have been summarized by Combarnous and Bories (1975). When the Rayleigh number is not too high, the convective motion is unicellular and two-dimensional. Otherwise, two different types of motion have been observed. For small inclinations the convection is in the form of polyhedral cells, while for large inclinations rolls with axes along the inclined coordinate appear.

In the present work we look at a three-dimensional block of material, its plane surfaces being inclined to the gravity vector. Two opposite faces are each isothermal, but kept at different temperatures. Numerical solutions are obtained for the temperature and velocity fields. Since there are many different steady and unsteady solutions possible for different parameter ranges, we restrict ourselves to two unicellular motions and analyze the transitions between them. Our interest is in the occurrence of the transition and the hysteresis involved with respect to increasing or decreasing inclinations.

MATHEMATICAL MODEL

Consider the porous material shown in Fig. 1. The Cartesian coordinate directions are along the sides of the rectangular parallelopiped. The inclination θ from the horizontal is a tilt of the xy plane around the x-axis. L, W and H are the physical dimensions in the x, y and z directions respectively. The lower face is at temperature T_H and the upper one at T_C; the rest are adiabatic.

For purposes of simplicity we make the Boussinesq approximation as well as the assumptions necessary for the Darcy law for flow through porous media to be valid. The mass conservation, Darcy law and energy equations are

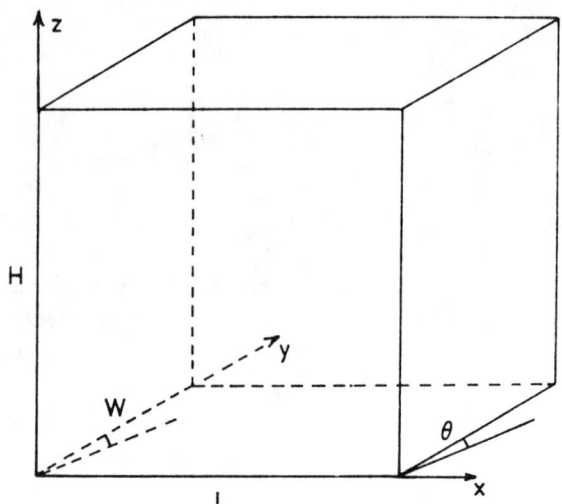

Fig. 1:   Schematic of porous material.

$$\nabla \cdot \mathbf{V} = 0 \qquad (1)$$

$$\mathbf{V} = \frac{K}{\mu}\left[-\nabla p + \rho \mathbf{g}\right] \qquad (2)$$

$$\sigma \frac{\partial T}{\partial t} + \mathbf{V} \cdot \nabla T = \alpha \nabla^2 T \qquad (3)$$

where $\mathbf{V}$, p and T are the filtration velocity vector, the pressure and the temperature. The properties K, μ, $\mathbf{g}$, σ and α are the permeability of the medium, the viscosity of the fluid, the gravity vector, the ratio of the heat capacity of the solid matrix to that of the fluid, and the effective thermal diffusivity of the medium, respectively. We also assume the density change to be of the form

$$\rho = \rho_0[1 - \beta(T - T_0)] \qquad (4)$$

where ρ_0 is the density at a reference temperature T_0. β is the coefficient of volumetric expansion. For computational ease it is convenient to introduce a vector potential ψ defined by $\mathbf{V} = \nabla \times \psi$. The continuity equation (1) is satisfied and equations (2) and (4) can be reduced to (Beukema et al., 1983)

$$\nabla^2 \psi = \frac{K}{\mu}\rho_0 \beta \, \nabla \times (T\mathbf{g}) \qquad (5)$$

In terms of the Cartesian components ψ_x, ψ_y and ψ_z, we have

$$\nabla^2 \psi_x = -\frac{K}{\mu}\rho_0 g\beta \left(\cos\theta \frac{\partial T}{\partial y} - \sin\theta \frac{\partial T}{\partial z}\right) \qquad (6)$$

$$\nabla^2 \psi_y = \frac{K}{\mu}\rho_0 g\beta \cos\theta \frac{\partial T}{\partial x} \qquad (7)$$

$$\nabla^2 \psi_z = -\frac{K}{\mu}\rho_0 g\beta \sin\theta \frac{\partial T}{\partial x} \qquad (8)$$

These have to be solved along with the energy equation (3). Using the following nondimensional variables

$$t^* = \frac{t\alpha}{\sigma L^2}, \quad x^* = \frac{x}{L}, \quad y^* = \frac{y}{W}, \quad z^* = \frac{z}{H},$$

$$T^* = \frac{T - T_C}{T_H - T_C}, \quad \psi_x^* = \frac{\psi_x}{\alpha}\frac{L^2}{HW}, \quad \psi_y^* = \frac{\psi_y}{\alpha}\frac{L}{H}, \quad \psi_z^* = \frac{\psi_z}{\alpha}\frac{L}{W} \quad (9)$$

we get

$$\nabla^2 \psi_x = -AB\,Ra\left(A\cos\theta\frac{\partial T}{\partial y} - B\sin\theta\frac{\partial T}{\partial z}\right) \quad (10)$$

$$\nabla^2 \psi_y = B\,Ra\cos\theta\frac{\partial T}{\partial x} \quad (11)$$

$$\nabla^2 \psi_z = -A\,Ra\sin\theta\frac{\partial T}{\partial x} \quad (12)$$

$$\frac{\partial T}{\partial t} + \left(\frac{\partial\psi_z}{\partial y} - \frac{\partial\psi_y}{\partial z}\right)\frac{\partial T}{\partial x} + \left(\frac{\partial\psi_x}{\partial z} - \frac{\partial\psi_z}{\partial x}\right)\frac{\partial T}{\partial y}$$
$$+ \left(\frac{\partial\psi_y}{\partial x} - \frac{\partial\psi_x}{\partial y}\right)\frac{\partial T}{\partial z} = \frac{\partial^2 T}{\partial x^2} + A^2\frac{\partial^2 T}{\partial y^2} + B^2\frac{\partial^2 T}{\partial z^2} \quad (13)$$

where the asterisks have been dropped. The controlling parameters are the two aspects ratios $A = L/W$ and $B = L/H$, the inclination angle θ, and the Rayleigh number $Ra = \rho_0 g\beta KL\alpha(T_H - T_C)/\mu$.

At each one of the boundary surfaces, the normal velocity is zero. This is equivalent to the conditions

$$\frac{\partial\psi_x}{\partial x} = 0, \psi_y = 0, \psi_z = 0 \text{ at } x = 0 \text{ and } x = 1$$

$$\psi_x = 0, \frac{\psi_y}{\partial y} = 0, \psi_z = 0 \text{ at } y = 0 \text{ and } y = 1 \quad (14)$$

$$\psi_x = 0, \psi_y = 0, \frac{\partial\psi_z}{\partial z} = 0 \text{ at } z = 0 \text{ and } z = 1$$

The two opposite faces $z = 0$ and $z = 1$ are maintained at nondimensional temperatures zero and one respectively. All other faces have zero normal derivative of the temperature. The Nusselt number Nu at the isothermal faces is defined as

$$Nu = \int_0^1 \int_0^1 \left|\left(\frac{\partial T}{\partial z}\right)_{z=0 \text{ or } 1}\right| dx\,dy \quad (15)$$

NUMERICAL PROCEDURE

Equations (10)-(13) are solved numerically to obtain the solutions for the vector function ψ and temperature T. The solution methods follow that of Holst and Aziz (1972). The equations of motion (10)-(12), are in the forms of elliptic type and therefore are solved by using the method of Successive Over-Relaxation (SOR). The energy equation (13) is solved by using the Alternating Direction Implicit (ADI) method. At every time step, the ψ values are first solved with the field variables set at their values at the previous time step. The temperature field is then solved by using the updated ψ values. The newly obtained ψ and T values are then used for the next run of iteration. The iterations proceed until satisfactory convergence is achieved. A time step is chosen and the above solution procedure provides the solutions at the new time. Steady-state solutions are obtained until no significant changes in the field variables are observed.

The solution is considered steady when the relative changes between successive iterations are less than 10^{-6} of the magnitude of the variables. A global check on the energy conservation is also performed by comparing the Nusselt numbers calculated on both the hot and cold surfaces and obtaining values within $10^{-3}\%$ of each other. Comparison tests were initially made with a cubic geometry A = B = 1, and $\theta = 0°$; grid sizes of 20×20×20, 40×40×40 and 50×50×50 in the x, y and z directions were used to check for convergence and grid independence. The variation in Nusselt number between the results of the last two grids was within 1%. The Nusselt numbers compared favorably (to within 1%) with those given by Straus and Schubert (1979). Taking into account the aspect ratios actually considered in the computations, the grid size used was chosen to be 40×20×40, and the time step chosen was 0.01.

For increasing inclination angle, the computation starts at $\theta = 0°$ with an initial guess for the field variables. The steady solution at $\theta = 0°$ is then used as the initial guess for the solution at the next angle. This is continued all the way up to 180°. To show hysteresis, the procedure is repeated with decreasing inclinations. The computation is started at 90° with each initial guess again set at the solution previously obtained. This proceeds until $\theta = 0°$ is reached. A 5° step in angle is typical throughout most of the computations. However, close to the transition points where fluid motion patterns alter, smaller steps in the angle are used, sometimes 2° or even 0.5°. This is necessary to avoid jumping to the "other" branch of the hysteresis curve.

RESULTS

On choosing A = 2 and B = 1, the prediction of linear stability theory of the onset of convective motion with $\theta = 0°$ is that the motion will appear as a single roll with axis parallel to the y-direction (Beck, 1972). The critical Rayleigh number above which this will happen is $4\pi^2$. However, as θ is increased, there will come a point at which the convection roll will turn through 90° in order to have its axis parallel to the x-direction. On this basis, and to study this transition, we have chosen the aspects ratios and Rayleigh numbers for numerical analysis.

Numerical results have been obtained with Ra = 50 and 100, for θ varying between zero and 180°. At $\theta = 0°$, the predicted convection begins as a roll with its axis parallel to the y-direction. This corresponds to $\psi_x = \psi_z = 0$ with ψ_y nonzero. The ψ values plotted in the figures are the maximum within the parallelopiped. Throughout these runs, the z component ψ_z was close to zero and is not plotted. The $\theta = 0°$ unicell was used as initial condition for the next inclination. Figure 2 shows the variation of this ψ_x over the entire inclination range with increasing angle. Over most of this range ψ_x can be taken to be representative of the intensity of the convective motion. The fluid velocities obviously rise with the Rayleigh number. The curves show some sharp corners for small angles, indicating the possibility of hysteresis.

Figure 3 provides more detailed information for Ra = 50 for a smaller range of inclination. Also, the maximum ψ_y values are plotted. Both increasing and decreasing inclinations are now indicated, the former being the continuous lines and the latter the broken lines. It is observed that at $\theta = 0°$, $\psi_x = 0$ and ψ_y is nonzero. As the angle increases, this is reversed indicating the change in the roll axis from the y- to the x-direction. The bend in the ψ_x curve near $\theta = 10°$ is caused by the disappearance of ψ_y. After that the roll is exclusively with axis in the x-direction. Considerable quantitative, though not qualitative, difference is seen between the increasing and decreasing inclination curves. There is thus

345

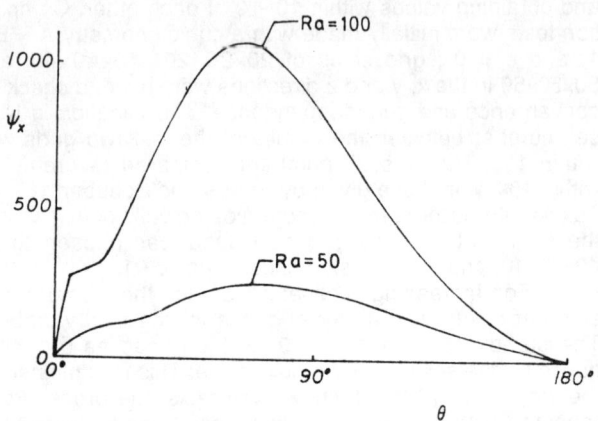

Fig. 2: Variation of the component of the vector potential ψ_x with inclination θ for Ra = 50 and 100. Increasing inclination only.

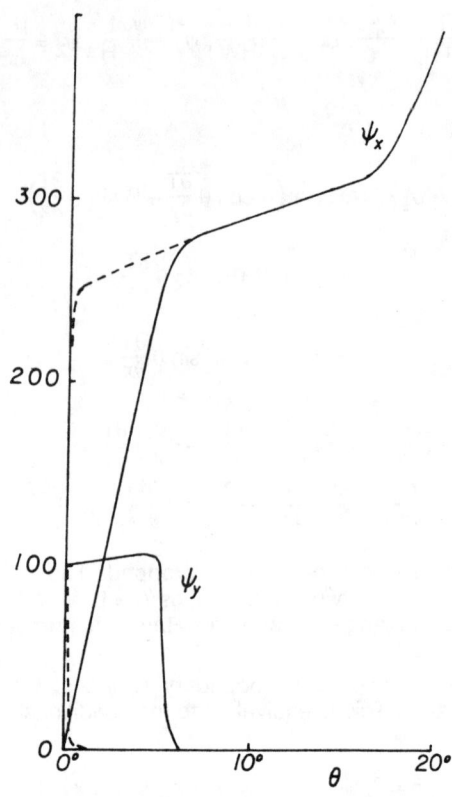

Fig. 4: Variation of the components of the vector potential ψ_x and ψ_y with inclination θ for Ra = 100. Increasing inclination (continuous lines) and decreasing inclination (broken lines).

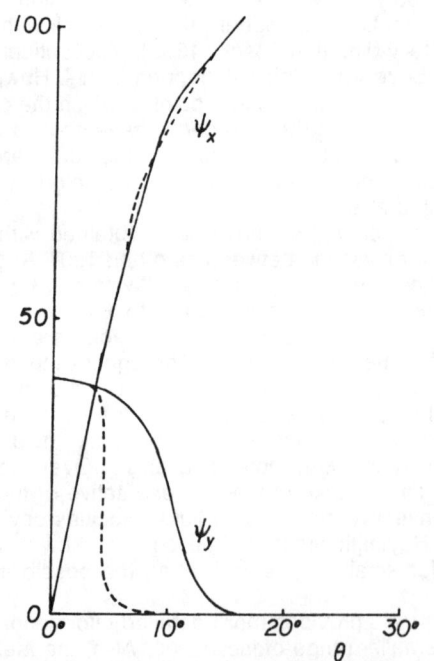

Fig. 3: Variation of the components of the vector potential ψ_x and ψ_y with inclination θ for Ra = 50. Increasing inclination (continuous lines) and decreasing inclination (broken lines).

a range of angle between about 4° and 10°, in which ψ_y may be zero or nonzero, depending on the initial conditions for the computation.

Similar information for Ra = 100 is plotted in Fig. 4. Once again the sudden appearance or disappearance of ψ_y is seen to correspond to the bends in the ψ_x curve. Hysteresis is now observed in the 0° to 5° range of inclination.

Changes in the flow structure have a strong effect of the heat flux and consequently the Nusselt number. Figures 5 and 6 show the variation of Nu with inclination for Ra = 50 and 100 respectively. Once again the continuous lines are for increasing and the broken lines for decreasing inclinations. In the range of angles for which two steady states exist, the difference between the Nusselt numbers of these two flow patterns is quite high.

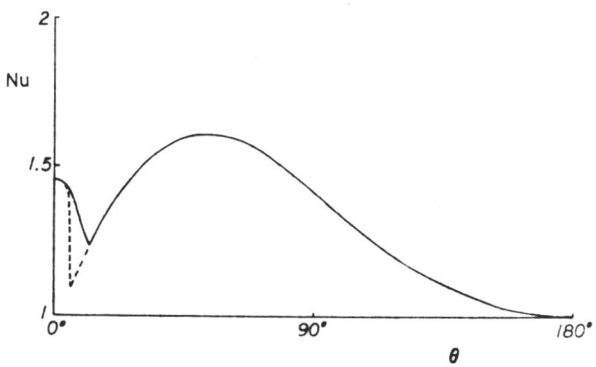

Fig. 5: Variation of the Nusselt number Nu with inclination θ for Ra = 50. Increasing inclination (continuous lines) and decreasing inclination (broken lines).

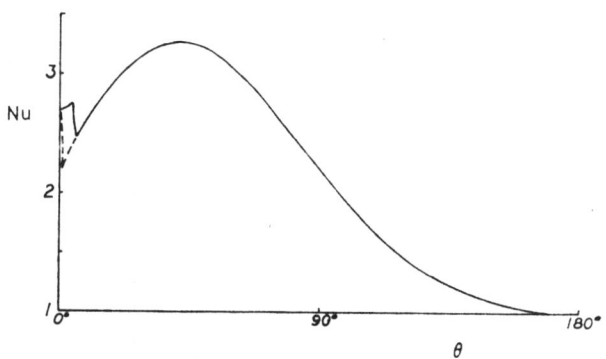

Fig. 6: Variation of the Nusselt number Nu with inclination θ for Ra = 100. Increasing inclination (continuous lines) and decreasing inclination (broken lines).

CONCLUSION

Many different linearly stable flow patterns may co-exist for a given set of parameters, each with its own basin of attraction of initial conditions. As the parameters change, the basins change in size, and if one disappears its corresponding solution loses stability and the resulting steady state must change to one of the other stable flows. Since the study of multiple stable patterns is complex, we have decided here to analyze the transition between two flow patterns represented by unicellular convective rolls with axes at right angles to one another. As the porous material is tilted, the first roll with axis in the y-direction loses stability and a roll with axis in the x-direction appears. On decreasing inclination, the latter loses stability at a different point, hence the observed hysteresis. This hysteresis in the transition of the flow pattern is reflected on the values of velocity vector potential components ψ_x and ψ_y as well as on the Nusselt number.

Convection problems are often known to have solutions which depend strongly on initial conditions. However, the possibilities for initial conditions are infinite, and one can only attempt to use a small subset of these. The results that have been obtained so far are for continuously increasing or decreasing angles of inclination, giving two different solution branches. It is as yet unknown what effect a decrease in inclination would have with initial conditions on the increasing branch, and vice versa, or of using initial conditions away from either of these branches.

Transitions between unicellular convective rolls with axes in different directions have been documented by Yang et al. (1987, 1988) for an inclined fluid-filled cavity, although there the flow is much more complicated than in the Darcy porous medium discussed here. The change in direction of the roll axis is seen to correspond to the local minimum of the Nusselt number, which is also the case here. It would be interesting to find out whether similar hysteresis effects appears in that situation also.

REFERENCES

Altimir, I., 1984, Convection Naturelle Tridimensionelle en Milieu Poreux Sature par un Fluide Presentant un Maximum de Densité, *Int. J. Heat Mass Transfer,* vol. 27, pp. 1813-1824

Beck, J.L., 1972, Convection in a Box of Porous material Saturated with Fluid, *Phys. Fluids,* vol. 15, pp. 1377-1383

Bejan, A., 1987, Convective Heat Transfer in Porous Media, in *Handbook of Single-Phase Convective Heat Transfer,* S. Kakac, R.K. Shah and W. Aung (eds.), Chap. 16, John Wiley and Sons, New York

Beukema, K.J., Bruin, S. and Schenk, J., 1983, Three-Dimensional Natural Convection in a Confined Porous Medium with Internal Heat Generation, *Int. J. Heat Mass Transfer,* vol. 26, pp. 451-458

Bories, S. and Monferran, L., 1972, Condition de Stabilité et Échange Thermique par Convection Naturelle dans une Couche Poreuse Inclinée de Grande Extension, *C.R. Acad. Sc. Paris,* vol. B274, pp. 4-7

Bories, S., Combarnous, M. and Jaffrennou, J.-Y., 1972, Observations des Différentes Formes d'Écoulements Thermoconvectifs dans une Couche Poreuse Inclinée, *C.R. Acad. Sc. Paris,* vol. A275, pp. 857-860

Bories, S.A. and Combarnous, M.A., 1973, Natural Convection in a Sloping Porous Layer, *J. Fluid Mech.,* vol. 57, pp. 63-79

Caltagirone, J.-P. and Bories, S., 1980, Solutions Numériques Bidimensionelles et Tridimensionelles de l'Écoulement de Convection Naturelle dans une Couche Poreuse Inclinée, *C.R. Acad. Sc. Paris,* vol. B291, pp. 197-200

Caltagirone, J.-P. and Bories, S., 1985, Solutions and Stability Criteria of Natural Convective Flow in an Inclined Porous Layer, *J. Fluid Mech.,* vol. 155, pp. 267-287

Caltagirone, J.-P., Meyer, G. and Mojtabi, A., 1981, Structurations Thermoconvectives Tridimensionnelles dans une Couche Poreuse Horizontale, *J. Méc.,* vol. 20, pp. 219-232

Combarnous, M.A. and Bories, S.A., 1975, Hydrothermal Convection in Saturated Porous Media, *Adv. Hydrosciences,* vol. 10, pp. 231-307

Holst, P.H. and Aziz, K., 1972, Transient Three-Dimensional Natural Convection in Confined Porous Media, *Int. J. Heat Mass Transfer,* vol. 15, pp. 73-90

Horne, R., 1979, Three-Dimensional Natural Convection in a Confined Porous Medium Heated from Below, *J. Fluid Mech.,* vol. 92, pp. 751-766

Schubert, G. and Straus, J.M., 1979, Three-Dimensional and Multicellular Steady and Unsteady Convection in a Fluid-Saturated Porous Media at High Rayleigh Numbers, *J. Fluid Mech.,* vol. 94, pp. 25-38

Steen, P.H., 1983, Pattern Selection for Finite-Amplitude Convection States in Boxes of Porous Media, *J. Fluid Mech.,* vol. 136, pp. 219-241

Straus, J.M. and Schubert, G., 1978, On the Existence of Three-Dimensional Convection in a Rectangular Box Containing Fluid-Saturated Porous Material, *J. Fluid Mech.,* vol. 87, pp. 385-394

Straus, J.M. and Schubert, G., 1979, Three-Dimensional Convection in a Cubic Box of Fluid-Saturated Porous Material, *J. Fluid Mech.,* vol. 91, pp. 155-165

Straus, J.M. and Schubert, G., 1981, Modes of Finite-Amplitude Three-Dimensional Convection in Rectangular Boxes of Fluid Saturated Porous Material, *J. Fluid Mech.,* vol. 103, pp. 23-32

Yang, H.Q., Yang, K.T. and Lloyd, J.R., 1987, Laminar Natural-Convection Flow Transitions in Tilted Three-Dimensional Longitudinal Rectangular Enclosures, *Int. J. Heat Mass Transfer* **30**, 1637-1644

Yang, H.Q., Yang, K.T. and Lloyd, J.R., 1988, Three-Dimensional Bimodal Buoyant Flow Transitions in Tilted Enclosures, *Int. J. Heat Fluid Flow* **9**, 90-97

Zebib, A. and Kassoy, D.R., 1977, Onset of natural Convection in a Box of Water-Saturated Porous Media with Large Temperature Variation, *Phys. Fluids,* vol. 20, pp. 4-9

Zebib, A. and Kassoy, D.R., 1978, Three-Dimensional Natural Convection Motion in a Confined Porous Medium, *Phys. Fluids,* vol. 21, pp. 1-3

1989 National Heat Transfer Conference
HTD-Vol. 107, Heat Transfer in Convective Flows

STABILITY OF THREE DIMENSIONAL FLOWS IN A HORIZONTAL ANNULUS WITH A HEATED ROTATING INNER CIRCULAR CYLINDER

L. Yang and B. Farouk
Department of Mechanical Engineering and Mechanics
Drexel University
Philadelphia, Pennsylvania

Abstract

The study of flows in the annulus between two concentric cylinders with one or both cylinders rotating has been a subject of interest to many researchers. The existence of hydrodynamic instabilities leads to the formation of Taylor vortices in these flows. The study of heat transfer within the rotating enclosure is also of considerable interest. For horizontal configuration, the buoyancy and the centrifugal effects (created by the heated rotating cylinder) are orthogonal and give rise to fully three dimensional flows. The results of a three dimensional numerical analysis of flows and heat transfer in a horizontal annulus with a heated rotating inner circular cylinder are presented. Past studies of the problem considered two dimensional formulation, thus precluding the consideration of Taylor vortices in the flow field. Solutions are presented over a wide range of the rotational Reynolds (Taylor) number and the Grashof number. The aspect ratio and diameter ratio for the annulus are 6 and 2.6 respectively for the results presented. The effect of the centrifugal instability on the heat transfer is examined.

Nomenclature

d	gap width $= R_o - R_i$
Gr	Grashof number $= g\beta d^3 (T_i - T_o)/\mu^2$
H	height of annulus
k	thermal conductivity
Nu	Nussel Number $= q\, d/k(T_i - T_o)$
p	pressure
Pr	Prandtl number $= C_p\mu/k$
q	heat flux
r	radius
R	radius of inner or outer cylinder
Re	Reynolds number $= \omega R_i d/\nu$
t	time
T	temperature
u	radial velocity component
v	angular velocity component
w	axial velocity component
z	axial location

Greek letters

θ	circumferential location
σ	densiometric Froude number $= Gr/Re^2$
Γ	aspect ratio $= H/d$
η	radius ratio $= R_o/R_i$
ω	angular speed of rotation of inner cylinder
μ	molecular viscosity
ν	kinematic viscocity

Subscripts

i	inner cylinder
o	outer cylinder

Introduction

The study of heat transfer in rotating bodies has a variety of practical applications in industry. These include cooling of turbine rotors or electrical motor shaft, cooling of high speed gas bearings, rotating condensers for sea water distillation , etc. The flow fields in such systems are complex due to interactions of the inertia, buoyancy and the centrifugal effects. In a heated rotating system the buoyancy and the centrifugal forces are of importance. The resultant combination of them determines the flow pattern and the heat transfer mechanism. Two dimensional natural convection in a horizontal concentric annulus has been intensely studied numerically and experimentally in both laminar and turbulent regimes [Liu et al., 1961; Lis, 1966; Kuehn and Goldstein, 1976;, Hodnet, 1973, Farouk and Guceri, 1982]. The forced flow due to an unheated rotating inner cylinder, in which only the centrifugal force is considered, will lead to the Taylor vortices because of the existence of hydrodynamic instability when the Reynolds (Taylor) number reaches a critical value [Taylor, 1923; Coles, 1965]. A comprehensive review of the analytical and experimental investigations for the annulus with a rotating inner cylinder is given by DiPrima and Swinney [1985]. Ball and Farouk [1987] and Ball [1987] undertook a detailed study on the development of Taylor vortices and the distribution of heat transfer in a vertical annulus with a heated rotating inner cylinder. For the vertical orientation (for moderate speeds of rotation), the flow field generated by the centrifugal and the buoyancy effects are both axisymmetric. For the horizontal configuration, however, the buoyancy and the centrifugal effects will give rise to fully three-dimensional flows when the centrifugal force is strong enough to trigger the formation of the Taylor cells. Fusegi et al. [1986] presented numerical results for two dimensional (r - θ) mixed convection in the annulus between horizontal concentric cylinders with a heated rotating inner cylinder. The study was limited to slow rotational speed of the inner cylinder so that the appearance of Taylor cells was precluded. When the rotational Reynolds number is increased beyond a critical value, the flow will become unstable hydrodynamically and will then lead to the formation of Taylor vortices that necessitates a three dimensional analysis.

This paper presents the results of three-dimensional mixed convection in a horizontal rotating annulus. The main objective of the paper is to quantify the interaction of the buoyancy and centrifugal forces and determine the effects of secondary flow structures (due to the Taylor cells) on the heat transfer. From the heat transfer distributions on the surfaces, the structure of convective flow is evaluated. The inner cylinder is considered to be rotating at a uniform speed while the outer cylinder and end-plates are held stationary. Both the inner and the outer cylinders are isothermal with the inner cylinder being hotter than the outer one. The no-slip conditions are applied for all enclosure surfaces. Thermally insulated flat end plates are considered. The geometry of the problem is shown in Figure 1 where the angle θ is measured from the bottom vertical line .

Mathematical Formulation

The geometry is specified by the radius ratio $\eta = R_O/R_i$ and the aspect ratio $\Gamma = H/d$, where d denotes the gap width (R_O - R_i). The aspect ratio is held fixed at 6.0 and the radius ratio is set equal to 2.6. The aspect ratio was chosen such that more than one pair of counter rotating Taylor cells are formed in the annulus . A very long annulus was not considered as it would have increased the computing costs. Air is considered as the medium, with the Prandtl number being equal to 0.72.

Governing equation and boundary conditions

Three dimensional incompressible Navier-Stokes and energy equations were used to describe the problem. By introducing the following dimensionless variables (an over bar means a dimensional quantity)

$$r = \frac{\bar{r}}{d}, \quad z = \frac{\bar{z}}{d}, \quad u = \frac{\bar{u}}{u_0}, \quad v = \frac{\bar{v}}{u_0}, \quad w = \frac{\bar{w}}{u_0}, \quad t = \frac{\bar{t}\, u_0}{d}$$

$$p = \frac{\bar{p}}{u_0^2 \rho_0}, \quad \text{and} \quad T = \frac{\bar{T} - \bar{T}_0}{\bar{T}_i - \bar{T}_0}, \quad \text{where} \quad u_0 = \frac{v}{d}$$

the dimensionless time-dependent equations of fluid flow and heat transfer in cylindrical coordinates are given by:

$$\frac{1}{r}\frac{\partial}{\partial r}(ru) + \frac{1}{r}\frac{\partial v}{\partial \theta} + \frac{\partial w}{\partial z} = 0$$

$$\frac{\partial u}{\partial t} + u\frac{\partial u}{\partial r} + \frac{v}{r}\frac{\partial u}{\partial \theta} + w\frac{\partial w}{\partial z} - \frac{v^2}{r} = -\frac{\partial p}{\partial r} + (\nabla^2 u - \frac{u}{r^2} - \frac{2}{r^2}\frac{\partial v}{\partial \theta}) - Gr\, T \cos(\theta)$$

$$\frac{\partial v}{\partial t} + u\frac{\partial v}{\partial r} + \frac{v}{r}\frac{\partial v}{\partial \theta} + \frac{uv}{r} + w\frac{\partial v}{\partial z} = -\frac{1}{r}\frac{\partial p}{\partial \theta} + (\nabla^2 v - \frac{v}{r^2} + \frac{2}{r^2}\frac{\partial u}{\partial \theta}) + Gr\, T \sin(\theta)$$

$$\frac{\partial w}{\partial t} + u\frac{\partial w}{\partial r} + \frac{v}{r}\frac{\partial w}{\partial \theta} + w\frac{\partial w}{\partial z} = -\frac{\partial p}{\partial z} + \nabla^2 w$$

and

$$\frac{\partial T}{\partial t} + u\frac{\partial T}{\partial r} + \frac{v}{r}\frac{\partial T}{\partial \theta} + w\frac{\partial T}{\partial z} = \frac{1}{Pr}\nabla^2 T$$

The Boussinesq approximation is applied in the above formulation. The coupled set of equations are numerically integrated with the following boundary conditions:

Along the inner cylinder: (r = 0.625)
$$u = 0, \ v = Re, \ w = 0, \ T = 1$$

Along the outer cylinder: (r = 1.625)
$$u = 0, \ v = 0, \ w = 0, \ T = 0$$

Along the end- plates: (z = 0, and z = 6)

$$u = 0, \ v = 0, \ w = 0, \ \frac{\partial T}{\partial z} = 0$$

For the present formulations and the non-dimensional parameters used, the rotational effects enter via the boundary conditions at the inner cylinder. It can be easily shown that the relative strength of the buoyancy and the centrifugal forces in the problem is given by the ratio σ (densiometric Froude number) = Gr/Re^2. For isothermal flows (Gr = 0.0), the rotating inner cylinder induces a Couette flow for slow speeds of rotation. The rotational instability (Taylor vortex flow) is triggered when the Reynolds number exceeds a critical value (often described as the critical Taylor number in the literature). When a temperature gradient exists in the problem, the buoyancy induced flow interacts with the rotational flow which can delay the onset of the Taylor vortex type flow in the annulus.

The Nusselt number Nu (z,θ) for the inner and the outer cylinders are defined as

$$Nu \ (z,\theta)_i = q(z,\theta)_i \ d/ \ k \ (T_i - T_o) \quad \text{and}$$

$$Nu \ (z,\theta)_o = q(z,\theta)_o \ d/ \ k \ (T_i - T_o) \ \text{respectively.}$$

For convenience in presenting the results, we define the circumferentially averaged Nusselt number $\overline{Nu}_\theta$ for the inner and the outer cylinders as

$$\overline{Nu}_\theta = \frac{1}{2\pi} \int_0^{2\pi} Nu \ (\theta, z) \ d\theta$$

Similarly, the longitudinally averaged Nusselt number $\overline{Nu}_z$ is defined as follows:

$$\overline{Nu}_z = \frac{1}{H} \int_0^{H} Nu \ (\theta, z) \ dz$$

where H is the length of the annulus. The global mean Nusselt number (for the inner or the outer cylinder) then can be defined as:

$$\overline{\overline{Nu}} = \frac{1}{2\pi H} \int_0^{2\pi} \int_0^{H} Nu \ (\theta, z) \ dz \ d\theta$$

Numerical Method

A staggered mesh system is adopted for the derivations and solution of the finite difference approximation to the differential equations. The SIMPLE scheme of Patankar [1980] is used to solve the finite difference equations resulting from the discretization. For specified Grashof and Reynolds numbers, the resulting finite difference equations are solved in a time marching manner until a steady state condition is achieved. The computations employ a uniform mesh system with 11(r) x 24(θ) x 30(z) grid points. For the range of parameters considered and the aspect and radius ratios of the problem geometry, the above grid was found to be adequate. Sample calculations were performed with denser grids which did not produce appreciable differences in the flow structure or the heat transfer characteristics. All computations were performed on a CRAY X-MP supercomputer at the Pittsburgh Supercomputing Center. The typical CPU time used for one complete case is about 600 seconds.

Results and Discussions

Results were obtained for isothermal (Gr = 0.0), non-rotating (Re = 0.0) and mixed convection cases for the finite annular geometry. The maximum allowable values of Gr and Re were sufficiently low so that only laminar flows were encountered. For the isothermal case, the flow field is essentially one dimensional (except near the end plates) when the rotational speed of the inner cylinder is below the critical Reynolds number. From linear stability theory, the critical Reynolds number for the onset of the Taylor vortices (for a wide gap case) is equal to 64. This was verified by our computations also. When Re > 64, the flow field is characterized by toroidal vortices which appear in the form of counter-rotating cells. The cells usually occur in pairs and for moderate speeds of rotation of the inner cylinder, the flow field remains axisymmetric. Figure 2a shows the streamlines (based on the radial and axial components of velocities) along any arbitrary θ location for Re = 100. To illustrate the relative magnitudes of all three components of velocity in the flow field, we show the axial distribution of the velocity components at r = $(r_o - r_i)/2$ for any arbitrary θ location. The base flow (v distribution) is found to be strongly affected by the secondary (u - w) flow field.

The free and mixed convection computations were carried out in two sets. In one set, the Reynolds number was fixed at 100 while the Grashof number was varied from 138 to 6944. In the other set the Grashof number is fixed at 2777 and the Reynolds number changes from 0 (free convection) to 130. Figures 3-a and 3-b show the relief plots of local Nusselt number Nu (z,θ) distribution at the inner and outer cylinder for Re = 100 and Gr 138 with σ = 0.014. The densiometric Froude number is quite small, which means that the centrifugal force dominates. The flow pattern is characterized by the appearance of Taylor cells which strongly affect the heat transfer characteristics at the inner cylinder. A strong jet-like flow exists at the demarcation of two counter rotating cells which results in increased heat transfer at the inner and outer cylinder surfaces [Ball, 1987]. The variation of the Nusselt number is wavelike in the axial direction due to the regular formation of the Taylor cells. The local heat transfer is however, almost constant along the circumferential direction . Because the surface area of the inner cylinder is larger than that of the outer cylinder, the magnitude of the average Nusselt number on the inner cylinder is larger than that of the outer cylinder. Energy balance calculations resulted in excellent agreement between the inner and the outer cylinder heat transfers. For the same Re and σ = 0.14 , the wavy nature of the Nusselt number distribution along the axial direction persists, indicating the continued presence of the Taylor cells. Figures 4-a and 4-b show the results where Re =100, Gr = 6944, and σ = 0.7. It is clear from these figures that at this value of Grashof number, the buoyancy induced field is strong and the effects of the Taylor cells on the heat transfer is subdued. At this high Grashof number, the critical Reynolds number for the onset of the Taylor cells is higher than 100. We did not attempt to evaluate the value by simulation because at the higher Re, steady laminar flow may not be present. For the inner cylinder, the local heat transfer varies along the circumferential direction only. The shape of the plots look similar to those found in typical natural convection flow fields in horizontal annulus. For the outer cylinder, the Nusselt number peaks near θ = π radians. The flow field near the outer cylinder is vigorous and the effects of two stationary end walls is prominent on the heat transfer.

Figures 5-a and 5-b show the distribution of circumferentially averaged Nusselt number $\overline{Nu}_\theta$ (for the inner and the outer cylinders) along the axial direction for the cases where the Re = 100 and σ = 0.014, 0.14 and 0.7 respectively. Figures 6-a and 6-b show the

$\overline{Nu}_z$ distribution along θ direction (for the inner and the outer cylinders) for the above cases. The circumferential averaging provides the general information along the axial direction while $\overline{Nu}_z$ shows the overall heat transfer variation along the θ direction. It could be seen from Figs. 5-a and 5-b that for σ = 0.014, which is the rotation dominated case, that $\overline{Nu}_\theta$ fluctuates along the axial direction according to the distribution of the Taylor vortices. In Figs. 6-a and 6-b, $\overline{Nu}_z$ appears as a flat line for σ = 0.014. For σ = 0.69, which is a buoyancy dominated case, the $\overline{Nu}_z$ changes significantly with θ while the $\overline{Nu}_\theta$ variations in Figs. 5-a and 5-b look rather flat. It is thus evident that the presence of rotational instability in a mixed convection case is easily recognized by the heat transfer distribution on the cylinder surfaces.

Figure 7 and Fig. 8 present the distribution of averaged Nusselt numbers $\overline{Nu}_\theta$ and $\overline{Nu}_z$ along the axial and circumferential directions for the inner cylinder where the Gr number is held fixed at 2777 and the rotational speed of the inner cylinder is varied (σ = 0.164, 0.43 and 1.1). The corresponding results for the outer cylinder are not shown for the sake of brevity. Figure 7 shows that when σ is small (Re is high) , the presence of the Taylor cells considerably augment the heat transfer. From Fig. 8 it is seen that the azimuthal variation becomes important when the buoyancy effects dominate. The peak heat transfer in the stationary inner cylinder occurs at the 0^0 location. The location shifts towards higher θ values as the rotational speed of the inner cylinder increases.

As seen in Fig. 7, when Gr = 2777 and Re=80 , the flow is still found to be buoyancy dominated. From our calculations, at this Gr, the onset of rotational instability occurs somewhere between Re = 80 and 130. As discussed earlier, from linear stability theory, the critical Reynolds number for the onset of the Taylor vortices for the isothermal problem is 64. The natural convection flow field (which is well developed at Gr =2777), thus inhibits and delays the formation of the Taylor cells.

In Fig. 9-a, the variation of the mean global heat transfer at the inner cylinder for several densiometric Froude numbers is presented for Re = 100. It could be seen from the figure that when σ approaches zero (for reduced buoyancy effects), the mean heat transfer reaches an asymptotic value. When σ is gradually increased , mean heat transfer reaches a minimum value around σ = 0.1 and then goes up rapidly. Near σ = 0.1, the Taylor cells and the developing buoyancy induced flow fields appear to have a cumulative destructive effect on the heat transfer. A steady increase is shown in heat transfer with increasing values of σ when the flow becomes buoyancy dominated. Figure 9-b shows similar relation between mean heat transfer and 1/σ for the cases where Gr is held fixed at 2777 and the Reynolds number is varied. When 1/σ is equal to zero, mean heat transfer value is that of pure natural convection. As Re is increased, mean heat transfer first decreases, then increases rapidly. The minima point is around σ = 0.4 (Gr = 2777, Re = 83) in Fig. 9-b as opposed to σ = 0.35 (Gr = 3504, Re = 100) in Fig. 9-a. It is expected that if simulations are carried out for different Re (for Fig. 9-a) and Gr (for Fig. 9-b) similar behavior of the mean heat transfer with σ will be observed.

Conclusions

In this paper, the interaction of centrifugal and buoyancy forces has been studied in a finite horizontal annulus with a heated rotating inner cylinder. The resulting flows are fully three dimensional for cases where the rotational instability triggers the formation of Taylor cells. Due to the existence of the buoyancy force, the critical Reynolds (Taylor) number value will be higher . The mean heat transfer in some intermediate value of σ, the densiometric Froude number can be actually smaller than both the natural convection and the rotation induced flow regimes.

Acknowledgements

The authors gratefully acknowledge the support from the National Science Foundation   (Grant No. CBT-8712218) and the computational resources provided by the Pittsburgh Supercomputing Center (Grant No. 880027P) .

References

Ball, K. S., and Farouk, B., 1987, "On the Development of Taylor Vortices in a Vertical Annulus With a Heated Rotating Inner Cylinder," Int. J. Numerical Methods In Fluids, Vol.7, pp.857-867.

Ball, K. S.,1987 "Mixed Convection Heat Transfer in Rotating System," The thesis of Ph.D.,Dept. of Mechanical Engng. and Mechanics,Drexel University.

Coles, D., 1965, "Transition in Circular Couette Flow," Journal of Fluid Mechanics, Vol.21, pp.385-425.

DiPrima, R. C., and H.L.Swinney, 1985, Hydrodynamic Instabilities and the Transition to Turbulence, pp.139.

Farouk, B. and Guceri, S.I. , 1982,"Laminar and turbulent Natural Convection in the Annulus Between Horizontal Concentric Cylinders," ASME Journal of Heat Transfer,Vol. 104, pp.631-636.

Fusegi, T, Farouk, B and Ball, K. S., 1986, "Mixed Convection Flows within a Horizontal Concentric Annulus with a Heated Rotating Inner Cylinder", Numerical Heat Transfer, Vol. 9, pp. 591-604

Fusegi, T. and Farouk, B., 1986, "A Three -dimensional Study of Natural Convection in The Annulus in Between Horizontal Concentric Cylinders," Eighth Int. Heat Transfer Conf., San Fransico,CA, pp.1575-1579.

Hodnett, P. F., 1973, "Natural Convection Between Horizontal Heated concentric Circular Cylinders," J. Appl. Math. Phys., Vol. 24, pp.507-516.

Kuehn, T. H., and Goldstein, R. J., 1976, "An experimental and Theoretical Study of Natural Convection in the Annulus Between Horizontal Concentric Annulus," Journal of Fluid Mechanics, Vol. 75, pp.695-719.

Liu, C. Y., Mueller, W.K., and Laudis, F., 1961,"Numerical Convection Heat Transfer in Long horizontal Annuli," International Developments in Heat Transfer, ASME, pp.976-984.

Lis, J., 1966, "Experiment Investigation of Natural Convection Heat Transfer in Simple and Obstructed Horizontal Annuli," Proceedings of the 3rd International Heat Transfer Conference, Vol. 2, pp.196-204.

Taylor, G. I., 1923, " Stability of Viscous Liquid Contained Between Two Rotating Cylinders," Philos. Trans. Roy. Soc. A., Vol.223, pp.289-343.

Patankar, S. V., 1980, Numerical Heat Transfer and Fluid

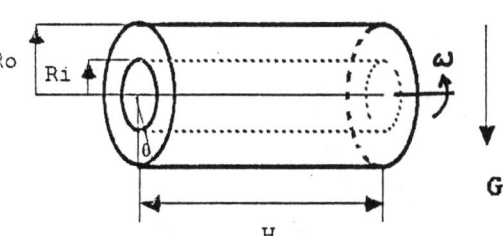

Figure 1   The three dimensional annular geometry

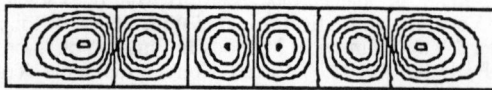

outer

inner

Figure 2a Streamlines for axisymmetric flow in the finite annulus at
any arbitrary θ location Gr = 0.0, Re = 100   Δψ = 1.1

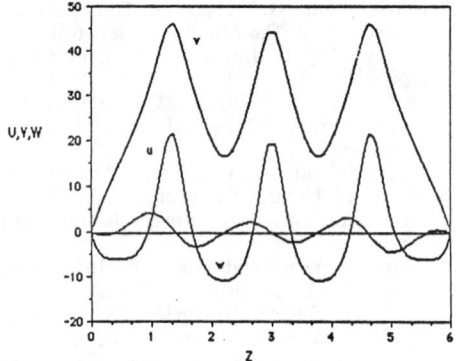

Figure 2b Velocity distributions along the length of the annulus at
r = (r_o - r_i)/2 at any arbitrary θ location. Gr = 0.0,
Re = 100

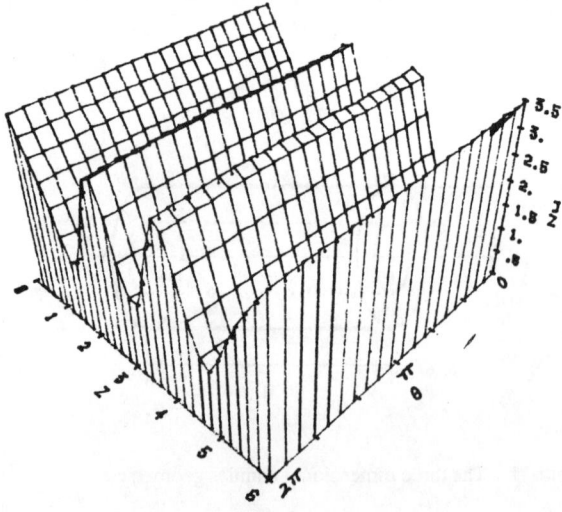

Figure 3a Local Nusselt number distribution on the inner cylinder.
Re = 100, Gr = 138, σ = 0.014

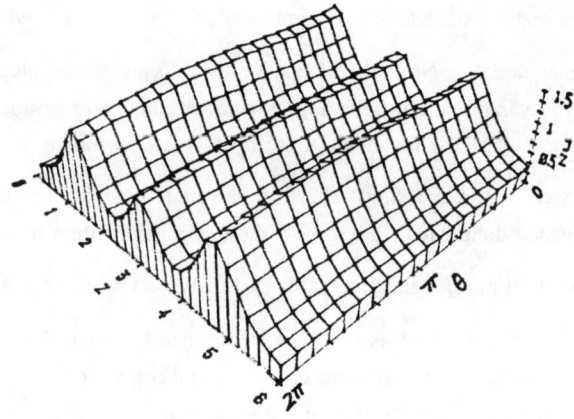

Figure 3b Local Nusselt number distribution on the outer cylinder.
Re = 100, Gr = 138, σ = 0.014

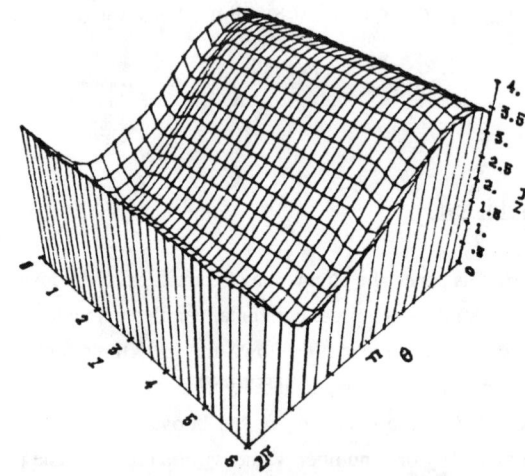

Figure 4a Local Nusselt number distribution on the inner cylinder.
Re = 100, Gr = 6944, σ = 0.7

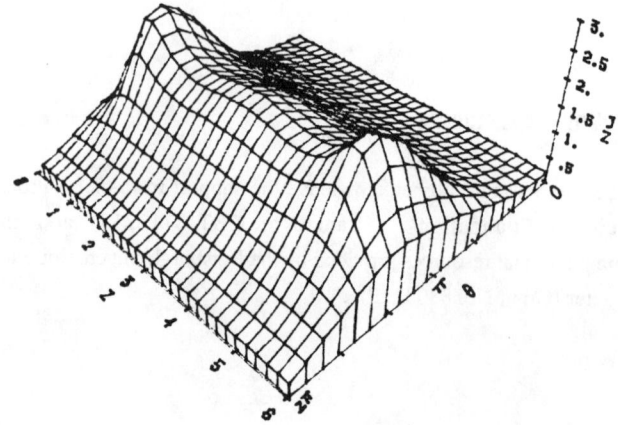

Figure 4b Local Nusselt number distribution on the outer cylinder.
Re = 100, Gr = 6944, σ = 0.7

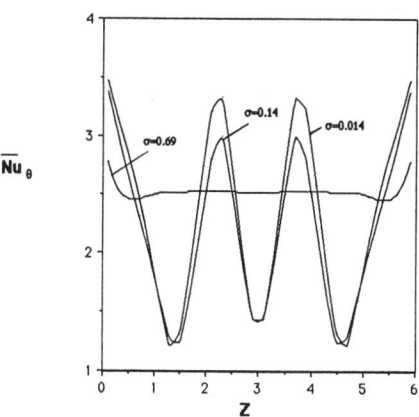

Figure 5a Circumferentially averaged Nusselt number, Nu$_\theta$ distribution for the inner cylinder Re = 100, Gr = 138, 1388, 6944

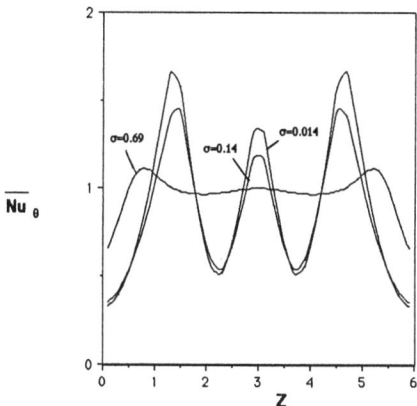

Figure 5b Circumferentially averaged Nusselt number, Nu$_\theta$ distribution for the outer cylinder Re = 100, Gr = 138, 1388, 6944

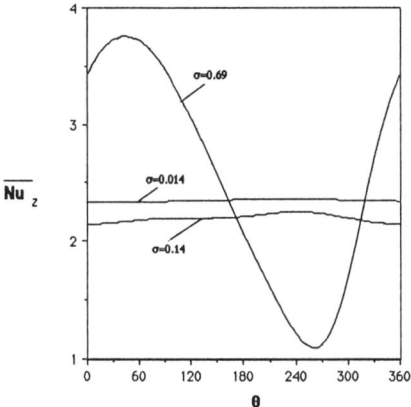

Figure 6a Axially averaged Nusselt number, Nu$_z$ distribution for the inner cylinder Re = 100, Gr = 138, 1388, 6944

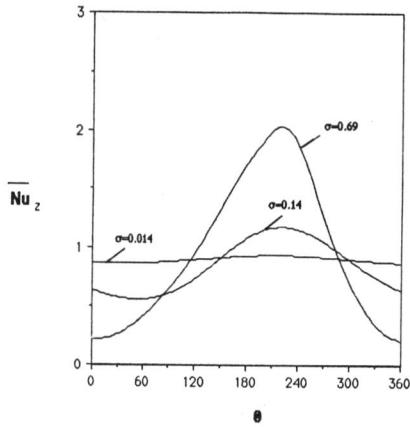

Figure 6b Axially averaged Nusselt number, Nu$_z$ distribution for the outer cylinder Re = 100, Gr = 138, 1388, 6944

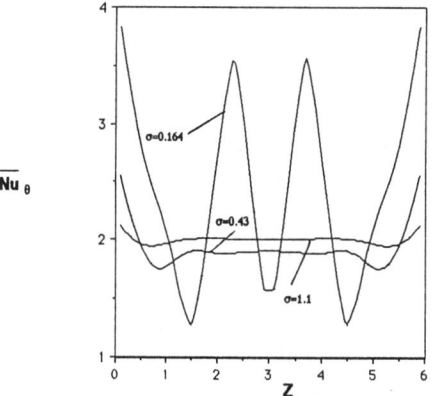

Figure 7 Circumferentially averaged Nusselt number, Nu$_\theta$ distribution for the inner cylinder Gr = 2777, Re = 50, 80, 130

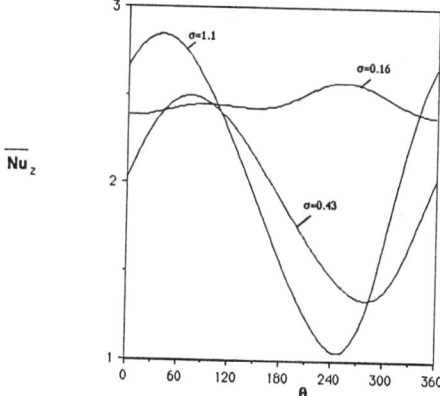

Figure 8 Axially averaged Nusselt number, Nu$_z$ distribution for the inner cylinder Gr = 2777, Re = 50, 80, 130

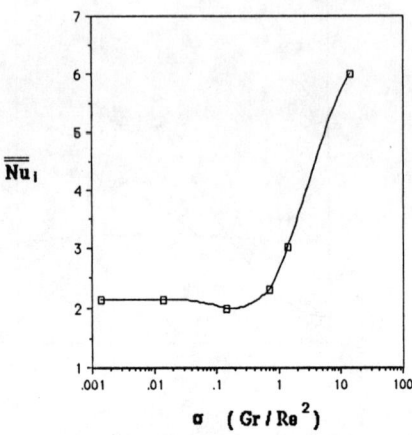

Figure 9a Global mean Nusselt number of inner cylinder as
a function of σ (Gr/Re^2) for Re = 100

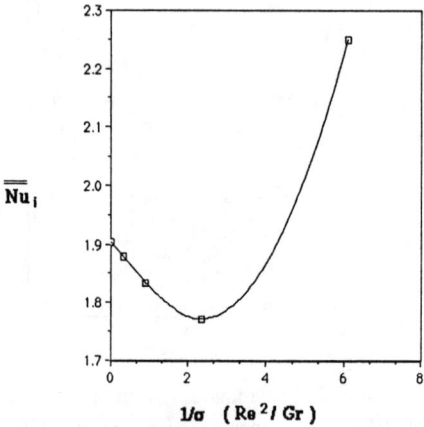

Figure 9b Global mean Nusselt number of inner cylinder as
a function of $1/\sigma$ for Gr = 2777

A NUMERICAL STUDY OF THE INSTABILITY OF DOUBLE-DIFFUSIVE CONVECTION IN A SQUARE ENCLOSURE WITH HORIZONTAL TEMPERATURE AND CONCENTRATION GRADIENTS

R. Krishnan
Vigyan Research Associates, Incorporated
Hampton, Virginia

ABSTRACT

In this work a numerical study of the steady and time-dependent convective motion of a double-diffusive fluid contained in a square enclosure subjected to horizontal temperature and concentration gradients is reported. The partial differential equations governing two-dimensional thermosolutal convection are solved. The case with both gradients in the same direction (opposing effects) and equal thermal and solutal Rayleigh numbers, Ra, is considered. Results are presented for Ra $\leqslant$ 3x10^5. The Prandtl and Schmidt numbers are chosen as 1 and 3.162, respectively.

The transition from the conductive to a steady convective regime is through a finite amplitude instability at Ra = 4.65x10^3. In the convective regime, for values of Ra < 6.25x10^4, a steady solution is obtained. With increasing Rayleigh number, successive bifurcations lead to oscillatory motions containing one frequency and its harmonics, two and then three incommensurate frequencies. Thus, the fluid motion considered in the present study follows a quasi-periodic route to chaos, although fully chaotic motions are not obtained in the range of Ra number considered. Contours of the temperature and solute concentration field are presented as are velocity vector plots. (Differences from the case with only a lateral thermal gradient are noted.) The bifurcation which leads to time dependent motion is found to be an oscillatory blob-type instability. A simple connection between subsequent bifurcations and changing spatial patterns is not as easily evident.

Mesh refinement studies are in progress. Partial results obtained indicate that the mesh used gives reliable results at the moderately large Raleigh numbers considered.

1. INTRODUCTION

This work is concerned with the steady and oscillatory motions of a double-diffusive fluid contained in a square enclosure and subjected to horizontal gradients of temperature and solute concentration. Whereas the case with both gradients vertically oriented and that where a lateral temperature gradient is applied to a stable vertical salinity gradient have received a great deal of attention, little work has been done on the configuration of interest in the present study. Ostrach (1983) and Kamotani, et al. (1985) present experimental studies of this case. Complex time-dependent flows with spatial structure different from the case where only a lateral thermal gradient is applied are observed. As noted in Ostrach (1983) a study of this configuration may be of interest in some crystal growth techniques, e.g., horizontal Bridgman. No numerical study has been reported in the literature.

Turner (1985) gives the following broad definition of multi-component convection, of which double-diffusive convection is a special case: "The fluid must contain at least two components with different molecular diffusivities, each of which affects the density of the fluid. Differential diffusion can produce convective motions that are associated with a decrease in the gravitational potential energy of the system." There have been several reviews of the subject and its possible wide ranging application in fluid motion encountered in engineering, oceanography, geology, crystal growth, solar ponds, liquid natural gas storage, and astrophysics (Turner (1973, 1974, 1985), Huppert and Turner (1981), Chen and Johnson (1984)). Convection in double-diffusive fluids is now recognized as a classic problem of fluid mechanics exhibiting oscillatory motions.

There are several related problems on which analytical, numerical, and in some cases, experimental studies have been done. Of these, the first is with both gradients in the vertical direction. Of particular interest have been the following two situations: a rectangular layer of fluid which is heated and salted from below (the double-diffusive Benard problem), and the case where the upper part of the layer is hotter and more saline. The second problem is the case where a stable, vertical stratification of solute is heated laterally. Two configurations in single component convection are also pertinent. These are the situations where only a lateral thermal gradient is applied and heating a fluid layer from below--the thermal Benard problem. It is of interest to note the differences from the known results of the first situation that the second component in the present study will cause. Also, as will be discussed later, the physical mechanism causing the oscillatory instability arising in the present study has also been observed in studies of the thermal Benard problem.

The double diffusive Benard problem--convection in a fluid layer containing a bottom heavy distribution of solute heated from below--has been extensively studied. Veronis (1968), Huppert and Moore (1976), Moore, et al. (1983), Knobloch, et al. (1986) have numerically studied the finite amplitude oscillatory motions of the case where the top and bottom boundaries are dynamically free (see Chang, et al. (1982) for the case with rigid boundaries). Stern (1960), Veronis (1965), and Baines and Gill (1969) have considered a linear stability analysis of the same problem. Overstable, i.e. growing oscillations, are characteristic of the instability. The physical mechanism causing these oscillations has been elucidated by a simple argument involving the vertical displacment of a parcel of fluid (Stern (1960), Moore and Spiegel (1966)). A transition to aperiodic oscillation takes place through a sequence of instabilities. Period doubling (one of the routes to chaos) was noticed in numerical solutions of the double-diffusive Benard problem by Moore, et al. (1983)--the first instance of solutions of a fluid dynamic partial differential equation system exhibiting such behavior. Unlike this, the problem studied in this paper follows a quasi-periodic route where successive bifurcations result in oscillatory motion containing one, two and three linearly distinct frequencies, with harmonics and several linear combinations also present.

When a vertical, stable stratification of solute is heated from the side, horizontal cells are known to form resulting in a layered structure for the temperature and concentration. Thorpe, et al. (1969) present an experimental study along with a linear stability analysis, while Wirtz, et al. (1972) report experimental and numerical work. For aspect ratio (height to width) $A \gg 1$, and when heat and salt are the two components, they show that the instability arising in this situation is different from the overstable oscillations that result when the imposed gradients are vertical. Termed sideways diffusive instability, it is non-oscillatory and results in convection rolls. The instability encountered in the present study is not of this type. Growing oscillations suggest it is of the overstable kind. No horizontal cells or a layered structure are noticed for the range of Rayleigh numbers considered.

The type of convective flow observed (e.g., single cell or multi-cellular, stationary or time-dependent instability) that arises when only a thermal gradient is imposed laterally on a fluid contained in an enclosure, is dependent on the aspect ratio and Prandtl number. There have been several experimental, analytical and numerical studies of this problem (see e.g., the extensive summary in Lee and Korpela (1983)). Concentrating only on the case with aspect ratio A=1, (Prandtl number, Pr= 0.72), numerical

studies indicate the following (see e.g., deVahl Davis (1983)). At low Rayleigh numbers, the steady convective state consists of a single cell motion. Steady states have been obtained up to (and even beyond) Ra = 10^6. The solution at Ra=10^5, exhibits two secondary cells in the core and at Ra=10^6, three are noticed. These are the result of stationary instabilities. In the double-diffusive case studied here, it will be seen in Section 3 that even at low Rayleigh numbers in the convective regime, the steady state is multicellular. Also, unlike the purely thermal case, the instabilities in the convective regime give rise to time-dependent states.

For the purely thermal Benard problem, it is a well established fact that temporal oscillations may arise from a blob instability--fluid elements which are hotter or colder than their surroundings circulating within the convection rolls. For example, Berge and Dubois (1979) and Moore and Weiss (1973) report experimental and numerical observations, respectively, of this phenomenon. From the results of Section 3, it will be seen that the instability resulting in oscillatory motions in the present study is attributable to the presence of a migrating solutal blob. For higher values of the Rayleigh numbers considered, a thermal blob is also present. Studies of the double-diffusive Benard problem (e.g., Huppert and Moore (1976), Moore, et al. (1983), and Knobloch, et al. (1986)) do not report such a phenomenon nor is it noticed in the problem where convective motions are due only to a lateral temperature gradient.

It is noted that even though only lateral thermal and solutal gradients are applied externally, convective motions induce vertical gradients away from the two horizontal boundaries. The vertical density gradient that results from the combination of these temperature and salinity gradients is conjectured to be the cause of the oscillatory blob instability.

Finally, the following difference in this work from some of the computational efforts discussed above is noted. It can be shown that the equations and boundary conditions of this study allow solutions with the centro-symmetry property (Gill (1966)). This involves a ´reflection´ about the center of the enclosure. However, such a symmetry condition is not imposed in the computational procedure (unlike, for example, the work of Knobloch, et al. (1986) on the double-diffusive Benard problem). This is because it is well known that a bifurcation encountered may be of the symmetry breaking kind. Indeed, for the problem considered in the present study, the bifurcation at Ra = 6.25×10^4 which results in time dependent motion is of the symmetry breaking kind. While solutions in the steady, convective regime for Ra < 6.25×10^4 are centro-symmetric, time-frozen solutions for Ra > 6.25×10^4 are not.

There are five parameters that could be varied in this study. These are the thermal and solutal Rayleigh numbers, the Prandtl and Schmidt numbers (all defined in Section 2) and the aspect ratio of the enclosure. Furthermore, the horizontal gradients in temperature and concentration that are imposed could have effects that oppose (same direction) or augment. An exhaustive study involving a variation of all parameters and situations will be a considerable task. Hence, the study is limited to one aspect ratio (unity) and the case where both the gradients are in the same direction with equal thermal and solutal

Rayleigh numbers. Following Veronis (1968) and Huppert and Moore (1976) the Prandtl number is chosen as 1 and the Schmidt number as 3.162. The solutal boundary layer is then thicker and easier to resolve numerically than when the Schmidt number is $O(10^3)$-- the value quoted in the experiments of Kamotani, et al. (1985). This precludes any comparison with their experiments.

In the next section, the governing partial differential equations and the boundary conditions are first stated. Only a brief outline of the numerical method used in this work and the effort to validate the code is given in Section 2. Details will be reported in Krishnan (1989). In Section 3, first the steady state solutions are discussed. Next, the time-dependent regime is considered. Then the variation in time of the spatial patterns of temperature, solute concentration and velocity fields are studied. Finally, results obtained are summarized.

2. FORMULATION OF THE PROBLEM AND NUMERICAL METHOD

2.1 Governing Equations

The fluid motion studied in this work is governed by the unsteady incompressible Navier-Stokes equations (where the Boussinesq approximation will be made) along with the thermal convection-diffusion equation and an analogous solutal convection-diffusion equation. These equations are

$$\rho_o \left[\frac{\partial \underset{\sim}{u}}{\partial t} + (\underset{\sim}{u} \cdot \underset{\sim}{\nabla}) \underset{\sim}{u} \right] = -\underset{\sim}{\nabla} p + \rho \underset{\sim}{g} + \rho_o \nu \nabla^2 \underset{\sim}{u} , \quad (2.1)$$

$$\frac{\partial T}{\partial t} - \kappa_T \nabla^2 T + (\underset{\sim}{u} \cdot \underset{\sim}{\nabla}) T = 0, \quad (2.2)$$

$$\frac{\partial S}{\partial t} - \kappa_S \nabla^2 S + (\underset{\sim}{u} \cdot \underset{\sim}{\nabla}) S = 0, \quad (2.3)$$

$$\underset{\sim}{\nabla} \cdot \underset{\sim}{u} = 0. \quad (2.4)$$

In the above equations, ρ, ν, κ_T and κ_S are the density, viscosity, and the thermal and solutal diffusivities, respectively and $g = -g\underset{\sim}{j}$ is the acceleration due to gravity, $\underset{\sim}{j}$ being the unit vector in the upward direction. The density is related to the temperature and solute concentration by a linear equation of state

$$\rho(T,S) = \rho_o \left[1 - \alpha(T-T_o) + \beta(S-S_o) \right]. \quad (2.5)$$

Here, ρ_o is the density at $T=T_o$, $S=S_o$, and α, β, the thermal and solutal expansion coefficients, respectively, are defined by

$$\alpha = -\frac{1}{\rho_o} \left(\frac{\partial \rho}{\partial T} \right)_{S_o}, \qquad \beta = \frac{1}{\rho_o} \left(\frac{\partial \rho}{\partial S} \right)_{T_o}.$$

Lengths are non-dimensionalized by the enclosure dimension d and time by d^2/κ_T. The following scales are used for velocity, temperature, solute concentration, and pressure: $u´ = (u/\sqrt{\alpha g \Delta T d})$, $T´ = (T-T_o)/\Delta T$, $S´ = (S-S_o)/\Delta T$, and $p´ = p/(\rho_o \alpha g \Delta T d)$. Then the governing equations in non-dimensional form (with the primes dropped) are

$$\frac{1}{Pr} \frac{\partial \underset{\sim}{u}}{\partial t} - \nabla^2 \underset{\sim}{u} + \sqrt{\frac{Ra_T}{Pr}} (\underset{\sim}{u} \cdot \underset{\sim}{\nabla}) + \sqrt{\frac{Ra_T}{Pr}} \underset{\sim}{\nabla} p$$

$$- \sqrt{\frac{Ra_T}{Pr}} T\underset{\sim}{j} + \frac{Ra_S}{\sqrt{Ra_T Pr}} S\underset{\sim}{j} = \underset{\sim}{0} , \quad (2.6)$$

$$\frac{1}{Pr} \frac{\partial T}{\partial t} - \frac{1}{Pr} \nabla^2 T + \sqrt{\frac{Ra_T}{Pr}} (\underset{\sim}{u} \cdot \underset{\sim}{\nabla}) T = 0 , \quad (2.7)$$

$$\frac{1}{Pr} \frac{\partial S}{\partial t} - \frac{1}{Sc} \nabla^2 S + \sqrt{\frac{Ra_T}{Pr}} (\underset{\sim}{u} \cdot \underset{\sim}{\nabla}) S = 0, \quad (2.8)$$

$$\underset{\sim}{\nabla} \cdot \underset{\sim}{u} = 0. \quad (2.9)$$

In the above equations, four non-dimensional parameters appear. They are the Prandtl number: ν/κ_T, the Schmidt number: ν/κ_S, the thermal Rayleigh number: $(\alpha g \Delta T d^3)/(\kappa_T \nu)$, and the solutal Rayleigh number: $(\beta g \Delta S d^3)/(\kappa_T \nu)$. The Rayleigh numbers are a measure of the externally applied forcing due to temperature and solute gradients. The Prandtl and Schmidt numbers which are a relative (to momentum) measure of the thermal and solute diffusivities are solely dependent on fluid properties.

On the lateral boundaries temperature and solute concentration are specified. The horizontal top and bottom boundaries are insulated against heat and solutal fluxes. All the boundaries being rigid, the no-slip condition is enforced. Thus, as shown in Fig. 1, $T=S=0$ on $x=1$, $T=S=1$ on $x=0$, $\partial T/\partial y = \partial S/\partial y = 0$ on $y=0$ and $y=1$, and $\underset{\sim}{u}=0$ on $x=0,1$ and $y=0,1$.

2.2 Numerical Method Used

The spatial discretization of the equations is done by a Galerkin finite element method. A method proposed by Gresho, et al. (1979, 1980) is used for the time integration of the ordinary differential equations arising from such a spatial discretization. Gresho, et al. found this an accurate and stable technique for solving the two-dimensional, unsteady, incompressible equations of motion. One of the example problems they considered was the computation of the motion induced in square enclosure by a lateral gradient in temperature alone. Except for the changes that are needed for the double diffusive case, the method closely follows Gresho et al. (1979). Hence, only a brief outline is given below. Details are given in Krishnan (1989).

For the spatial discretization of the velocity, temperature and solute concentration field the 9-node biquadratic element is chosen. A 4-node bilinear element is used for the pressure. This choice ensures numerical stability of the spatial discretization (e.g., see Bercovier and Pironneau (1979)). The alternative of a penalty finite element formulation with the 9-node element and a 4-point reduced integration scheme for the pressures would eliminate pressure unknowns. But this scheme is notorious for its pressure oscillations (spatial) and may not be suitable for a problem where as we shall note later, the physical instability is triggered by small numerical perturbations. Furthermore, because the number of unknowns at each node is now u,v,T, and S, the saving incurred by eliminating pressure unknowns is not very great. For example, for the mesh used in the present study, the number of pressure unknowns is a little more than 6% of the total number of unknowns. This will be even smaller for the finer mesh being used in the mesh refinement study in progress.

The time integration of the system of ODE's resulting from the spatial discretization is done by a predictor/corrector scheme (explicit second-order Adams-Bashforth/Crank-Nicholson), together with an automatic time step selector. The time integration error is preset to 10^{-4}.

The amplitude of solutions can be characterized by the vertically averaged thermal Nusselt number, Nu_T, and solutal Nusselt number, Nu_S, computed on the vertical boundary at x=0. For example, Nu_T is defined as the ratio of the net horizontal heat flux at the boundary to that due to conduction alone. To compute the fluxes at the boundary, the so called ´consistent flux´ method is used (see Marshall (1978), Gresho, et al. (1981, 1987), Thornton (1982) among others). This method uses the integral statement of the thermal and solutal convection-diffusion equations. It has been shown to compute far more accurate values for derived boundary quantities than the standard basis function derivative method (see, e.g., Gresho, et al. (1987)). Details of its use in the present study are given in Krishnan (1989).

The finite element mesh used in this work is shown in Fig. 1 where the nodal spacing is also given. Computations with a finer mesh (more nodes in the boundary layers as well as the interior) are in progress. Preliminary results indicate good agreement with results presented in Section 3 using the mesh of Fig. 1.

2.3 Code Validation

The code developed for this work was validated by reproducing selected results for two of the related problems discussed in the introduction. These problems are: (1) the case where only a lateral thermal gradient is applied and, (2) the double-diffusive Benard problem. For the first problem, the average thermal Nusselt number over-approximates the values published (e.g., deVahl Davis (1983)) by 0.2% and 1.7% at $Ra_T=10^4$ and 10^6 respectively. Noting that 10^6 is much larger than the highest Rayleigh number used in the present study (3×10^5) gives confidence in the working of at least part of the code.

For the second problem, both steady and time-dependent results of Veronis (1968) and Huppert and Moore (1976) are reproduced very well. Details are given in Krishnan (1989).

3. RESULTS AND DISCUSSION

3.1 Steady State Solutions

For Ra < 6.25×10^4, a steady state is obtainable. A steady state is assumed to have been reached when (for at least six consecutive timesteps) the ratio of the difference to the mean of the Nusselt numbers at successive timesteps is no more than the preset time-step error. Both the thermal and solutal fields are required to satisfy this criterion. A similar test is also used to ascertain that the Nusselt numbers at the right and left end walls are equal in a steady state (which is required by the global conservation of energy equation and its analog for the solute concentration).

The first instability is the transition from a conductive regime to a steady convective regime. Computations indicate that this occurs at Ra = 4650. In the conductive regime, the lateral variation of temperature and solute concentration across the enclosure is linear and the velocities are very small. In Fig. 2, the variation of Nu_S with Ra is plotted (Nu_T varies similarly). At Ra=4600, Nu_T and Nu_S are equal to unity, while at Ra=4700 they have the values 1.32 and 2.02, respectively. This discontinuous change implies that the conductive regime is unstable to finite amplitude disturbances.

For $Ra=10^4$, Fig. 3 shows the variation in time of Nu_S and Nu_T at the left end wall. The initial condition at this Rayleigh number is a quiescent fluid with the left end wall temperature and solute concentration raised to the final value, unity, at the instant t=0. Because the thermal diffusivity is larger than the solutal diffusivity, the temperature field reaches a steady state earlier than the solute concentration. The temperature and solute concentration fields, and the velocity vector plot are similar to those for Ra = 6×10^4, presented in Fig. 5 and discussed below (although the convective motions are weaker and the thermal and solutal boundary layers thicker than in that case). The same steady state is reached when two other initial conditions are used. These are the steady states of the problems where (1) only the lateral thermal gradient is present and (2) only the lateral solute gradient is present. The steady state values of Nu_T and Nu_S are 1.719 and 2.596, respectively. For convection in the presence of the lateral thermal gradient only, Nu_T is 2.24. The lower value of Nu_T is expected because of the effect of the solutal gradient in the configuration of interest here--convective motion induced by the thermal gradient is inhibited by the solute gradient.

The Rayleigh number is then increased in steps of 10^4 with the steady state of the previous value used as the initial condition. This was possible until Ra= 6×10^4. For Ra = 6×10^4, the variation of Nu_S on the left vertical wall is shown in Fig. 4 (Nu_T varies similarly). As will be noted in the next subsection, the first instability in the convective regime is at Ra = 6.25×10^4. At Ra = 6×10^4, which is close to but less than this critical value, oscillations in Nu_S are clearly decaying and a steady state is eventually reached. This steady state´s solute concentration, temperature, and velocity fields are shown in Figs. 5(a), (b) and (c). Well defined thermal and solutal boundary layers are present on the vertical boundaries. The solutal boundary layer is thinner than its thermal counterpart. This is because κ_S is smaller than κ_T and therefore, a steeper gradient is required to sustain a concentration flux at the boundary.

In the solely thermal gradient case, convective motion consists of a single cell. One circuit is complete when the fluid rises up the hot wall, traverses the length of the enclosure to the cold wall to which it transfers heat and descends before moving along the lower part of the enclosure to the hot wall. In the double-diffusive case, the steady state is multicellular. It consists of a primary cell in which convective motions are strongest in that it has velocities of larger magnitude than those found in the secondary cells. In the primary motion, along the vertical boundary which is hotter and more saline, the fluid rises and descends along the colder, less saline wall. At the top of the hotter vertical boundary, there is a weak secondary circulation with sense of rotation opposite to that of the primary cell. In Fig. 5(d), the motion in this corner is shown magnified several times. The largest velocities in this secondary cell are orders-of-magnitude smaller than those in the primary cell. (This will not be the case at higher Rayleigh numbers.) Thus, their contribution to the convective heat or solutal fluxes is insignificant. A similar secondary circulation is set up at the diagonally opposite corner.

These secondary cells are formed because of the presence of horizontal boundaries and the opposing buoyancy force that the solutal gradient exerts against the motion induced by the thermal gradient. Consider the corner at the top of the hotter, more saline boundary. (A similar argument holds for the other secondary motion also.) As the fluid rises up this wall, the presence of the horizontal boundaries decelerates its vertical motion before it is turned at the corner. Near the corners, when the fluid is sufficiently decelerated, the solutal gradient induced buoyancy force is greater than that due to the thermal gradient and this causes the fluid to descend. A comparison of Figs. 5(a) and 5(b) shows that in this nearly stagnant region, the non-dimensional solute concentration is greater than the non-dimensional temperature. From eq. (3.1) below, since $\beta \Delta S > 0$, it follows that ρ is greater than ρ_o. Note that this argument requires the fluid to be (or nearly be) static. It cannot hold where well-developed convective motion is present (e.g., away from the corners). At such points, the solute gradient only serves to decelerate the convective motion. The size and strength of these secondary motions can be varied by keeping the thermal Rayleigh number fixed and varying the solutal Rayleigh number. The effects of differing Ra_T and Ra_S will not be considered in this study.

At the other two corners, even smaller and weaker secondary motions are present. More importantly, unlike the secondary motions at the top corner of the hotter, more saline wall and its diagonally opposite corner, these cells do not grow in size or strength at higher Ra--in fact, they diminish. It must be noted that a mesh refinement study which is in progress indicates that these are not an artifact of the spatial resolution of the numerical scheme used.

In Fig. 6, the vertical density gradient on the vertical mid-plane is plotted. Using the non-dimensionalizations of Section 2 and since $Ra_T = Ra_S$, the linear equation of state (2.5) becomes

$$\frac{\rho}{\rho_o} = 1 + \beta \Delta S (S´ - T´) \qquad (3.1)$$

That is, letting $\rho´ = \frac{\rho}{\rho_o}$ and dropping primes

$$\frac{1}{\beta \Delta S} \frac{\partial \rho}{\partial y} = \frac{\partial S}{\partial y} - \frac{\partial T}{\partial y} .$$

This quantity is plotted in Fig. 6. While the density stratification is gravitationally stable (negative) in the core, it is unstable (positive) near the upper and lower boundaries. The density gradient is nearly constant in the core. This results from the well mixed state that obtains for the solute concentration

field (see Fig. 5) and the almost linear variation (in y) of the temperature in this region.

Finally, it is noted that all the convective steady state solutions presented in this study exhibit the centro-symmetry property. With the onset of time dependency after the bifurcation at $Ra = 6.25 \times 10^4$, this property is lost. This will be noted when the spatial patterns of the time-dependent solution fields are presented in Section 3.3.

3.2 Time-Dependent Solutions

From studies of several problems in thermal convection, it is well known that as the Rayleigh number increases, the flow undergoes a sequence of instabilities which leads finally to temporal chaos. Gollub and Benson (1980) have identified four distinct routes to chaos in their experiments on the thermal Benard problem. One of the routes they noticed has a time independent regime followed by bifurcation to a state where the flow oscillates with a single frequency and its harmonics. Two further bifurcations lead to states with incommensurate second and third frequencies. A typical power spectrum before broad-band noise begins to grow contains three linearly distinct frequencies and several linear combinations of these. The fluid motion considered in this study follows this route to chaos.

Time dependency begins with a bifurcation from the steady state discussed earlier to an oscillatory periodic flow. In Fig. 7, the time history of Nu_S at the left end wall is shown for $Ra = 6.5 \times 10^4$. (Nu_T varies similarly.) The oscillations triggered by time integration errors are clearly growing in amplitude unlike the case of $Ra = 6.0 \times 10^4$ (Fig. 4) where they decay. This indicates the presence of the first in a sequence of bifurcations in the convective regime. The best estimate of the critical Rayleigh number from this study is 6.25×10^4. It is difficult to determine more precisely by this method (checking whether oscillations grow or decay) the exact location of the bifurcation point. This is because the closer one gets to the point, the oscillations decay or grow very slowly. Hence, long runs would be necessary to ascertain whether Ra is greater than or less than Ra_{crit}. The aim of this study is not to provide a precise quantitative value for the critical Rayleigh number, but a delineation of the various regimes as the flow progresses towards aperiodicity. If integrated far enough in time, the oscillations in Fig. 7 would grow in amplitude until a state of periodic, finite amplitude oscillatory motion is reached. Such a motion is shown in Fig. 8(a) for $Ra = 7 \times 10^4$. In both cases the initial condition is the steady state at 6×10^4. Subsequent computations are with increments of 10^4 in the Rayleigh number. The solution at an instant of time after transients have died serves as the initial condition at the next Ra.

The trend towards time dependency of greater complexity than a single frequency (and harmonics) oscillation is seen in Figs. 8 through 13 where the time variation of Nu_S and Nu_T and the power spectral density of the Nu_S time series are presented. For $Ra = 7 \times 10^4$, a fundamental frequency f_1 and its harmonics are found--Figs. 8(a,b). The oscillations remain singly periodic until $Ra = 2.4 \times 10^5$ --Figs. 9(a,b). With increasing Ra, nonlinear effects become prominent. The nearly sinusoidal oscillations near the first bifurcation point develop into relaxation oscillations--regions of sharp variation with time followed by slow variation; in this case narrow, spiky peaks and broad valleys. The growth of higher harmonics, a characteristic of relaxation oscillations, is noticed in the power spectrum. These effects are noticeable at lower values of Ra also, although they are not as pronounced. The next bifurcation occurs between Ra= 2.4×10^5 and 2.5×10^5. There is an obvious change in the time dependency of Nu_S and Nu_T shown in Fig. 10(a) for Ra=2.5×10^5. The power spectrum in Fig. 10(b) indicates that the solution now oscillates with two incommensurate frequencies, f_1 and f_2, and several linear combinations of these. The solution at Ra= 2.6×10^5 is similar in nature--the power content of the subfrequencies is greater than at 2.5×10^5. The power spectrum for Ra=2.8×10^5 in Fig. 11(b) indicates that the flow is still quasi-periodic with two incommensurate frequencies. A new linear combination, $f_1 - 3f_2$ is now present. The frequency modulation that this causes in the time signal is apparent from Fig. 11(a) which shows approximately two cycles in the oscillation. A similar oscillation with larger period is noticed at Ra = 2.7×10^5 also. Between 2.9×10^5 and 3.0×10^5 the flow undergoes one more bifurcation to another quasi-periodic state with three incommensurate frequencies. For Ra=2.9×10^5, the oscillation in Fig. 12(a) has the power spectrum in Fig. 12(b). The two frequencies marked are incommensurate. All others appearing have been verified as being linear combinations of these two. The power spectrum shown in Fig. 13(b) is for the oscillations in Fig. 13(a) at Ra=3×10^5. This appears to contain a third incommensurate frequency. Computations with Ra greater than 3×10^5 are in progress.

As stated earlier, in the double-diffusive Benard problem (Moore, et al. (1983), Knobloch, et al. (1986)) the transition to aperiodicity is through a succession of period-doubling bifurcations. In the configuration of interest to the present study, from the above it is clear that a different route--the quasi-periodic one--is followed.

Returning to the first bifurcation point at Ra= 6.25×10^5, it is noted that the oscillations following it are temporally asymmetric. The temporal asymmetry (which may not be obvious from Fig. 8(a)) is clearly seen in Fig. 14 where a time series of the temperature and concentration at a fixed location x=(0.22, 0.22) for Ra=7×10^4 is shown. A phase plot of $\widetilde{Nu}_S$ versus $|u(0.22, 0.22)|$ at this Rayleigh number (Fig. 15) also shows this asymmetry. It is a slightly deformed ellipse. With increasing Ra, this asymmetry becomes more pronounced. (See Krishnan (1989) for these details.) As expected, the pointwise temperature, solute concentration, and velocity all oscillate in time with the same frequencies (power contents differ) and pass through the same instabilities as in the discussion above. Also noticed clearly in Fig. 14 is the fact that the concentration field lags behind the temperature. This again is because the solutal diffusivity is smaller than the thermal diffusivity. This lag is also present in the plots of Nu_S and Nu_T versus time, but is barely discernable.

3.3 Spatial Structures and Cause of Oscillations

It was noted in the introduction that one of the well known physical mechanisms producing time-dependent behavior in thermal convection is the presence of one or more blobs of fluid hotter or colder than their surroundings circulating within the convection cell. For example, the time variation of quantity like the Nusselt number (heat transfer at the boundary) would be influenced by the interaction of such fluid elements with the boundary layer. The time frozen temperature and solutal fields that are considered in this section indicate that the instability in the present study also arises from the presence of such blobs.

In Fig. 16, for Ra = 7×10^4, the solute concentration and temperature fields, and velocity vector plot are presented at the instant of time shown in the cycle of oscillation. For brevity, plots at only one such instant are presented here. A sequence of such figures over the complete cycle is presented in Krishnan (1989). Clearly, the centro-symmetry of the steady state has been broken by the first bifurcation. At this value of Ra, which is not much greater than the critical Rayleigh number 6.25×10^4, a small solutal blob is pinched off--see the contour S=0.5. These blobs appear only over a part of the time cycle. Over the remainder of the oscillation, they are dissipated away by the effects of solutal diffusion. Fluctuations over the cycle in solutal contours are much more pronounced than the temperature contours. Also, no thermal blobs appear at this Rayleigh number. This is because the thermal diffusivity is larger than solutal diffusivity. The velocity vector plots indicate that over a cycle, small oscillations in the shape of the primary cell are present. The maximum velocity in the enclosure changes, but not by much. The change with time in the position of the primary vortex center is clearly noticeable. The strength of the secondary cells also undergoes oscillations. All these effects are much more pronounced at higher Ra.

With increasing Rayleigh number, the solutal blob becomes bigger. When Ra is sufficiently large, it is present throughout the time cyle. Figures 17(a) through 17(h) are for Ra=1.8×10^5. At this value of Ra, the flow still oscillates with a single frequency and its harmonics. Figures 17(c) indicates the presence of two solutal blobs. Therefore, the appearance of a second incommensurate frequency (at Ra=2.45×10^5) cannot be attributed to the appearance of a second solutal blob. These two blobs have coalesced into one in Fig. 17(d) It should be remarked here that the second solutal blob appears in the other half of the oscillation cycle also, but not at the instants where the contours are presented. These instants are chosen to be (or as best as can be) symmetrically situated about the peak of one cycle in Nu_S versus time, but the solute and temperature fields are temporally asymmetric. This asymmetry is also noticeable at Ra=7×10^4, but barely. Fluctuations in the temperature and solute contours, shape of the primary cell and the position of its vortex center are much larger than at Ra=7×10^4. In Fig. 18 the secondary cell at the top left corner is shown at the instant of time corresponding to Fig. 17(d). This clearly shows that at certain points in the oscillation, the velocities in the secondary cells are no longer negligibly small compared to those in the primary cell. Thus, their contribution to the convective heat and solutal fluxes will not be insignificant at these high Rayleigh numbers.

Further increase in Rayleigh number produces changes in the spatial structure before the next bifurcation. At Ra=2.4×10^5, the spatial structures are quite similar to those at Ra=1.8×10^5. One exception is that now a thermal blob is also present over a part of the time cycle. Evidence of this is seen in Fig. 19 which is at a position in the time

cycle similar to Fig. 17(c) for Ra=1.8×10⁵. Such a
thermal blob is also present at Ra=2.2×10⁵. Recall
that the next bifurcation to an oscillation with two
incommensurate frequencies does not occur until Ra=
2.45×10⁵. Since it appears long before this critical
Rayleigh number, the thermal blob may not be the cause
of this instability. (This must be clarified by the
mesh refinement study in progress.) Another
interesting phenomenon noticed is that the boundary
layers on the horizontal walls separate. This is
clearly evident in Fig. 20 where at one instant in a
cycle, the velocity profile in the horizontal boundary
layer is plotted at a point where separation is noted
to occur. At other instants in the cycle, separation
is noticed on the upper boundary also. (The finer
mesh calculations that are in progress should indicate
whether or not this is a truly physical phenomenon or
merely an artifact of numerical resolution.) Once
again, it is noted that the separated horizontal
boundary layer is mildly apparent even at Ra=2.0×10⁵
and cannot be the cause for the instability at
Ra=2.45×10⁵.

Finally, in the transition from a single frequency to a two incommensurate frequency oscillation,
no new spatial structures are noticed. Thus, temperature and solute concentration contours at Ra=2.6×10⁵
do not show more than a slight modification over those
discussed earlier. These are presented in Krishnan
(1989). This is also the case with the appearance of
the third incommensurate frequency. The physical
mechanisms causing the second and third instabilities
in the convective regime are not clearly evident from
a study of the spatial patterns.

4. CONCLUSIONS

The instability of double-diffusive convection in
a square enclosure due to the opposing effects of
horizontal gradients in temperature and solute
concentration has been studied numerically. The
conductive regime is unstable to a finite amplitude
instability which results in steady, multicellular
convective motion at Ra=4.65×10³. This steady convection becomes unstable to oscillatory convection
with a single frequency and its harmonics at Ra=
6.25×10⁴. The cause of this instability is a small
solutal blob of greater/ lesser concentration than its
immediate surroundings that is present over parts of a
time cycle. With increasing Ra, this blob becomes
larger and is present throughout the oscillation. A
thermal blob also appears. The boundary layers on the
horizontal walls separate. At Ra=2.45×10⁵ the next
instability produces oscillations with two incommensurate frequencies. One more instability at Ra=2.95×
10⁵ which results in a third linearly distinct
frequency is present in the range of Rayleigh numbers
considered. Computations are being repeated on a
finer mesh. This should ascertain the quantitative
and qualitative validity of the results presented.

Acknowledgment

The author would like to thank Barbara Kraft for
her expert typing.

REFERENCES

Baines, P. G. and Gill, A. E., On Thermohaline
Convection with Linear Gradients, Journal of Fluid
Mechanics, Vol. 37, 1969, pp. 289-306

Bercovier, M. and Pironneau, O., Error Estimates
for Finite Element Method Solution of the Stokes
Problem in the Primitive Variables, Numerische
Mathematik, Vol. 33, 1979, pp. 211-224.

Berge, P. and DuBois, M., Study of Unsteady
convection through Simultaneous Velocity and
Interferometric Measurements, Le Journal de Physique-
Lettres, Vol. 40, 1979, pp. L505-509.

Chang, S.-M., Korpela, S. A., and Lee, Y., Double
Diffusive Convection in the Diffusive Regime, Applied
Scientific Research, Vol. 39, 1982, pp. 301-319.

Chen, C. F., and Johnson, D. H., Double-Diffusive
Convection: A Report on an Engineering Foundation
Conference, Journal of Fluid Mechanics, Vol. 138,
1984, pp. 405-416.

de Vahl Davis, G., Natural Convection of Air in a
Square Cavity: A Bench Mark Numerical Solution,
International Journal for Numerical Methods in Fluids,
Vol. 3, 1983, pp. 249-264.

Gill, A. E., The Boundary-Layer Regime for
Convection in a Rectangular Cavity, Journal of Fluid
Mechanics, Vol. 26, 1966, pp. 515-536.

Gollub, J. P. and Benson, S. V., Many Routes to
Turbulent Convection, Journal of Fluid Mechanics, Vol.
100, 1980, pp. 449-470.

Gresho, P. M., Lee, R. L., Chan, S. T., and Sani,
R. L.: Lawrence Livermore Laboratory Preprint UCRL-
82899, 1979.

Gresho, P. M., Lee, R. L., and Sani, R. L., On the
Time-Dependent Solution of the Incompressible Navier-
Stokes Equations in Two and Three Dimensions, Recent
Advances in Numerical Methods in Fluids, Vol. 1,
Taylor, C. and Morgan, K. (eds), Pineridge Press,
Swansea, UK, 1980, pp. 27-79.

Gresho, P. M., Lee, R. L., and Sani, R. L., The
Consistent Method for Computing Derived Boundary
Quantities when the Galerkin FEM is Used to Solve
Thermal and/or Fluids Problems, R. W. Lewis et al.
(eds), Proceedings, 2nd International Conference on
Numerical Methods in Thermal Problems, Pineridge
Press, 1981, pp. 663-675.

Gresho, P. M., Lee, R. L., Sani, R. L., Maslanik,
M. K., Eaton, B. E., The Consistent Galerkin FEM for
Computing Derived Boundary Quantities in Thermal
and/or Fluid Problems, International Journal for
Numerical Methods in Fluids, Vol. 7, 1987, pp. 371-
394.

Huppert, H. E. and Moore, D. R., Nonlinear Double-
Diffusive Convection, Journal of Fluid Mechanics, Vol.
78, 1976, pp. 821-854.

Huppert, H. E. and Turner, J. S., Double-Diffusive
Convection, Journal of Fluid Mechanics, Vol. 106,
1981, pp. 299-329.

Kamotani, Y., Wang, L. W., Ostrach, S., and Jiang,
H. D., Experimental Study of Natural Convection in
Shallow Enclosures with Horizontal Temperature and
Concentration Gradients, International Journal of Heat
& Mass Transfer, Vol. 28, No. 1, 1985, pp. 165-173.

Knobloch, E., Moore, D. R., and Toomre, J., and
Weiss, N. O., Transitions to Chaos in Two-Dimensional
Double-Diffusive Convection, Journal of Fluid
Mechanics, Vol. 166, 1986, pp. 409-448.

Krishnan, R., in preparation, 1989.

Lee, Y. and Korpela, Multicellular Natural
Convection in a Vertical Slot, Journal of Fluid
Mechanics, Vol. 126, 1983, pp. 91-121.

Marshall, R. S., Heinrich, J. C., and Zienkiewicz,
O. C., Natural Convection in a Square Enclosure by a
Finite-Element, Penalty Function Method Using
Primitive Fluid Variables, Numerical Heat Transfer,
Vol. 1, 1978, pp. 315-330.

Moore, D. R. and Weiss, N. O., Two-Dimensional Rayleigh-Benard Convection, _Journal of Fluid Mechanics_, Vol. 58, 1973, pp. 289-312.

Moore, D. R., Toomre, J., Knobloch, E., and Weiss, N. O., Period Doubling and Chaos in Partial Differential Equations for Thermosolutall Convection, _Nature_, Vol. 303, 1983, pp. 663-667.

Moore, D. W. and Spiegel, E. A., A Thermally Excited Non-linear Oscillator, _Astrophysical Journal_, Vol. 143, 1966, pp. 871-887

Ostrach, S., Fluid Mechanics in Crystal Growth - The 1982 Freeman Scholar Lecture, _Journal of Fluids Engineering_, Vol. 105, No. 5, 1983, pp. 5-20.

Stern, M. E., The "Salt-Fountain" and Thermohaline Convection, _Tellus_, Vol. 12, 1960, pp. 172-175.

Thornton, E. A., Computation of Consistent Boundary Quantities in Finite Element Thermal-Fluid Solutions, in _Finite Element Flow Analysis_, T. Kawai, (ed.), University of Tokyo Press, Tokyo, Japan, 1982, pp. 263-270.

Thorpe, S. A., Hutt, P. K., and Soulsby, R., The Effect of Horizontal Gradients on Thermohaline Convection, _Journal of Fluid Mechanics_, Vol. 38, Part 2, 1969, pp. 375-400.

Turner, J. S.: _Buoyancy Effects in Fluids_, Cambridge University Press, 1973.

Turner, J. S., Double-Diffusive Phenomena, _Annual Reviews of Fluid Mechanics_, Vol. 6, 1974, pp. 37-56.

Turner, J. S., Multicomponent Convection, _Annual Reviews of Fluid Mechanics_, Vol. 17, 1985, pp. 11-44.

Veronis, G., On Finite Amplitude Instability in Thermohaline Convection, _Journal of Marine Research_, Vol. 23, 1965, pp. 1-17.

Veronis, G., Effect of a Stabilizing Gradient of Solute on Thermal Convection, _Journal of Fluid Mechanics_, Vol. 34, Part 2, 1968, pp. 315-336.

Wirtz, R. A., Briggs, D. G., and Chen, C. F., Physical and Numerical Experiments on Layered Convection in a Density-Stratified Fluid, _Geophysical Fluid Dynamics_, Vol. 3, 1972, pp. 265-288.

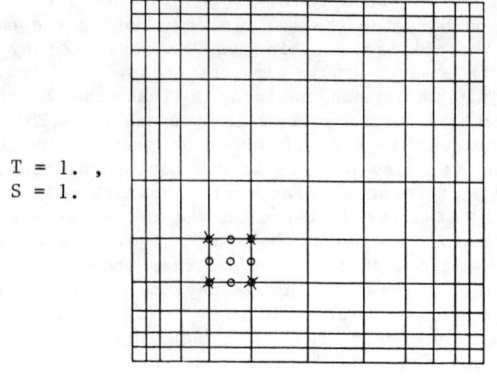

$$\frac{\partial T}{\partial y} = 0. \quad , \quad \frac{\partial S}{\partial y} = 0.$$

$T = 1.$, $S = 1.$

$T = 0.$, $S = 0.$

$$\frac{\partial T}{\partial y} = 0. \quad , \quad \frac{\partial S}{\partial y} = 0.$$

Fig. 1. Problem domain showing finite-element mesh and boundary conditions. On a typical element, nodes where velocity, solute concentration, and temperature are unknowns are marked as o. Nodes where pressure is an additional unknown are marked x. The spacing of the nodes in both x and y directions is 0.0, 0.02, 0.04, 0.06, 0.08, 0.11, 0.14, 0.18, 0.22, 0.28, 0.34, 0.42, and 0.50 (symmetrical about midplanes).

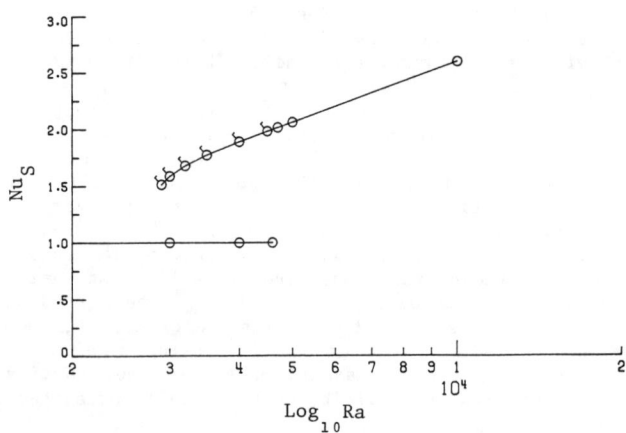

Fig. 2. Plot of Nu_S as a function of Ra. Flags on the symbols are to indicate that those solutions are obtained when the solution at the immediately higher Ra is used as the initial condition.

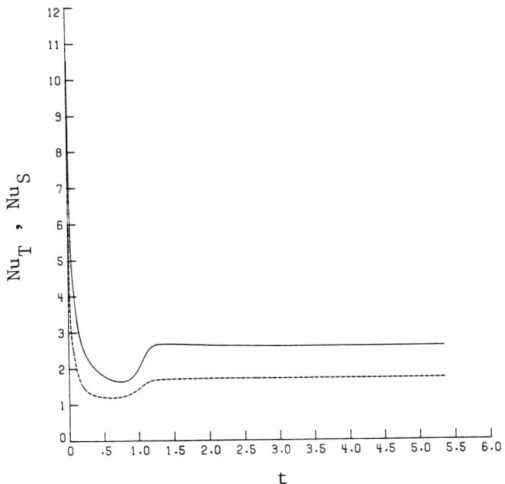

Fig. 3. Time history of Nu_S (——) and Nu_T (- - -) on the left end wall for $Ra = 10^4$.

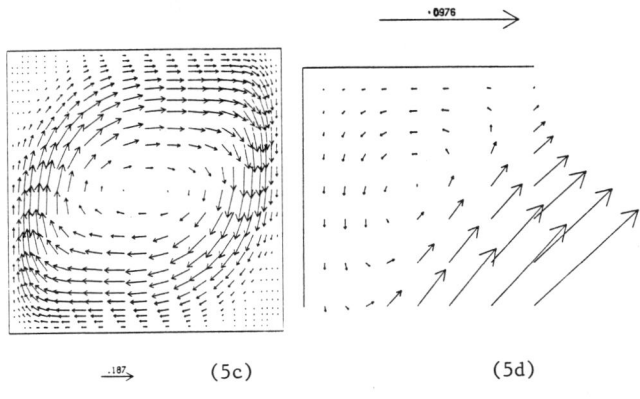

(5c)   (5d)

Fig. 5. The steady state at $Ra = 6 \times 10^4$; (a) solute concentration field, (b) temperature field, (c) the velocity vector field with the magnification of the secondary cell in the top left corner shown in 5(d).

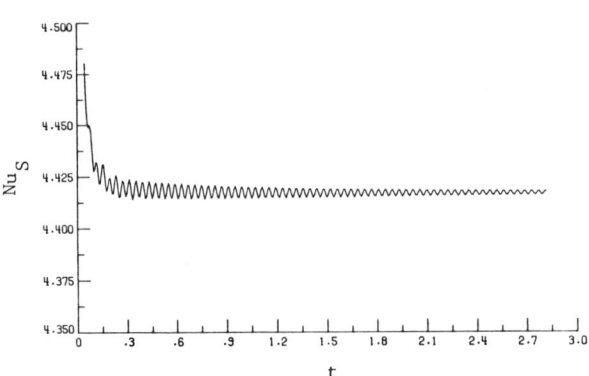

Fig. 4. Time variation of Nu_S at $Ra = 6 \times 10^4$.

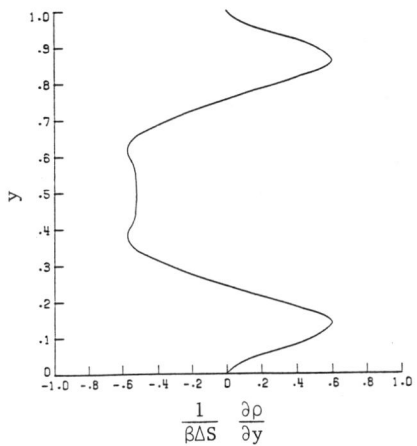

Fig. 6. Variation of the density gradient on the vertical midplane.

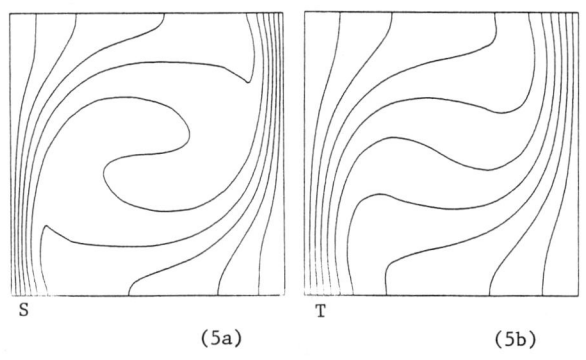

S   T
(5a)   (5b)

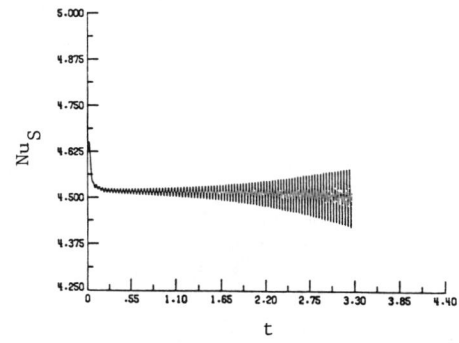

Fig. 7. Time history of Nu_S at $Ra = 6.5 \times 10^4$.

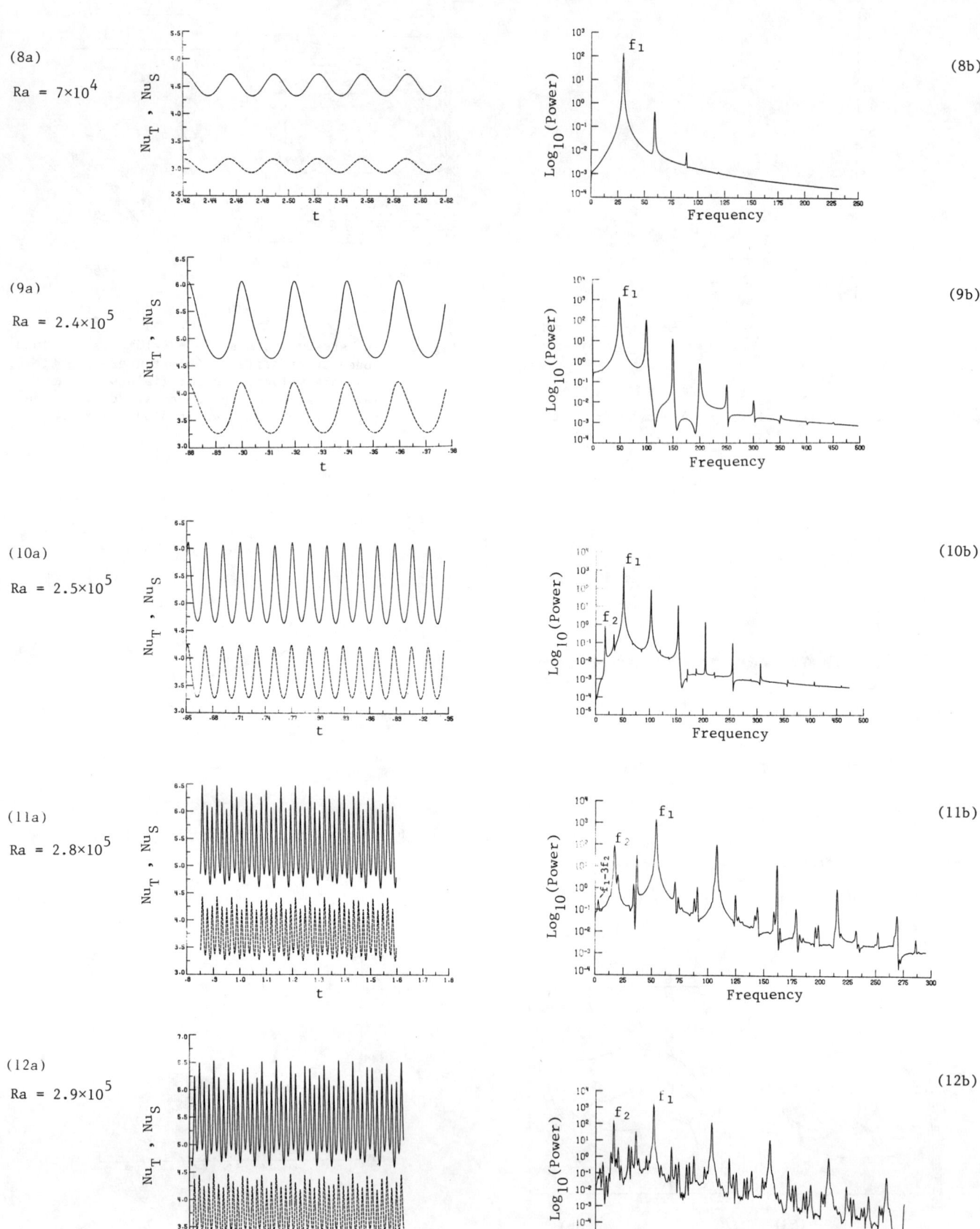

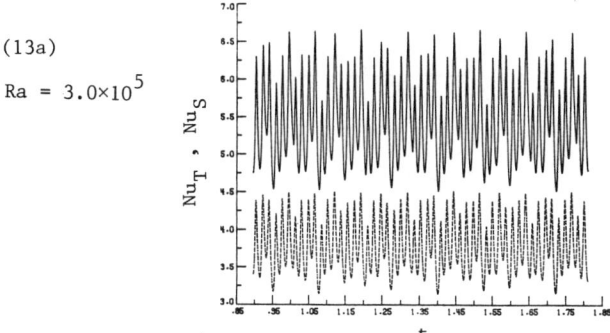

(13a)

Ra = 3.0×10^5

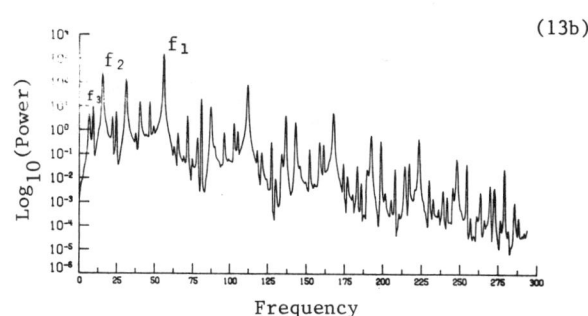

(13b)

Figs. 8-13. These figures contain: (a) oscillations of Nu_S (——) and Nu_T (---), (b) the power spectrum of the Nu_S time series. The Rayleigh numbers are as shown in Figs. 8(a) - 13(a).

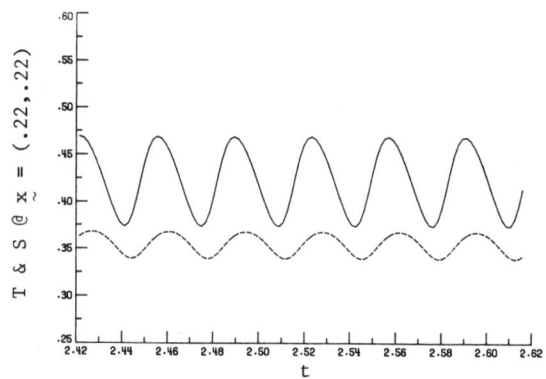

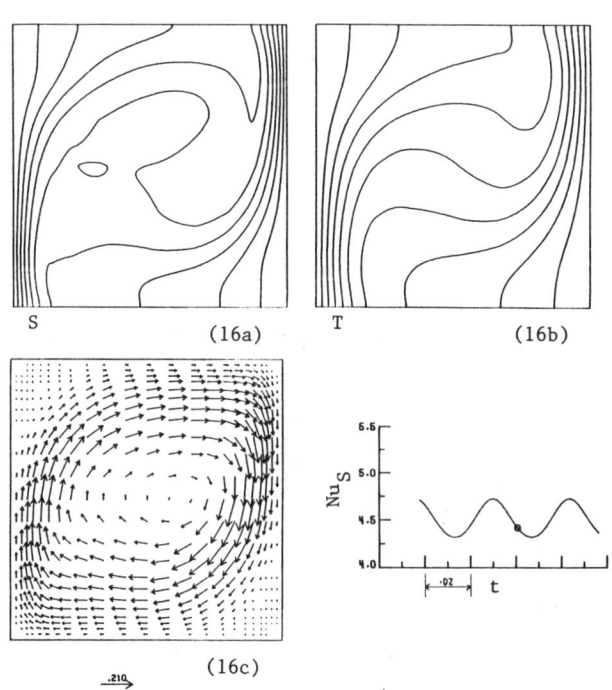

S (16a)   T (16b)

(16c)

Fig. 16. The spatial fields of solute concentration (S), temperature (T), and velocity. The instant of time in one cycle at which these fields are presented is also shown. The Rayleigh number is 7×10^4.

Fig. 14. Oscillations in pointwise solute concentration (——) and temperature (---) at $\underset{\sim}{x} = (0.22, 0.22)$ for Ra = 7×10^4.

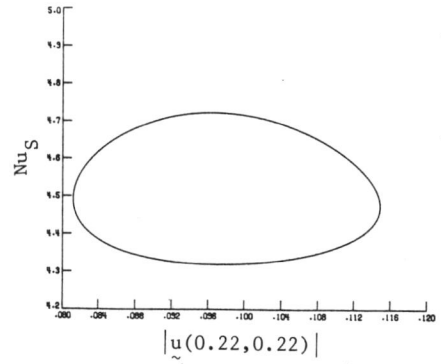

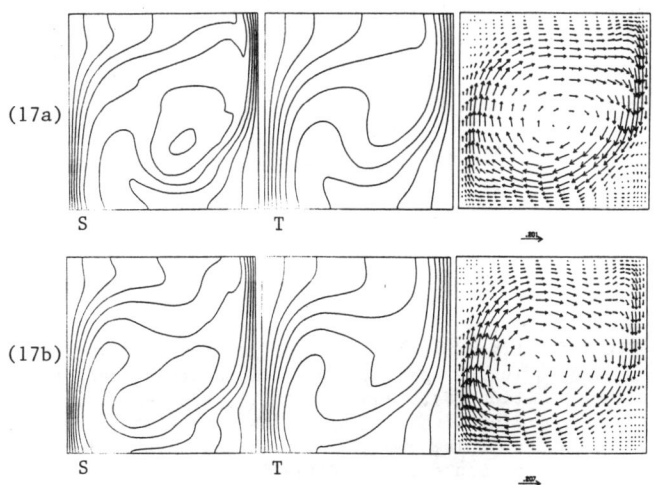

(17a)

S   T

(17b)

S   T

Fig. 15. Limit cycle of the oscillation at Ra = 7×10^4 projected onto the $|\underset{\sim}{u}(0.22, 0.22)| - Nu_S$ plane.

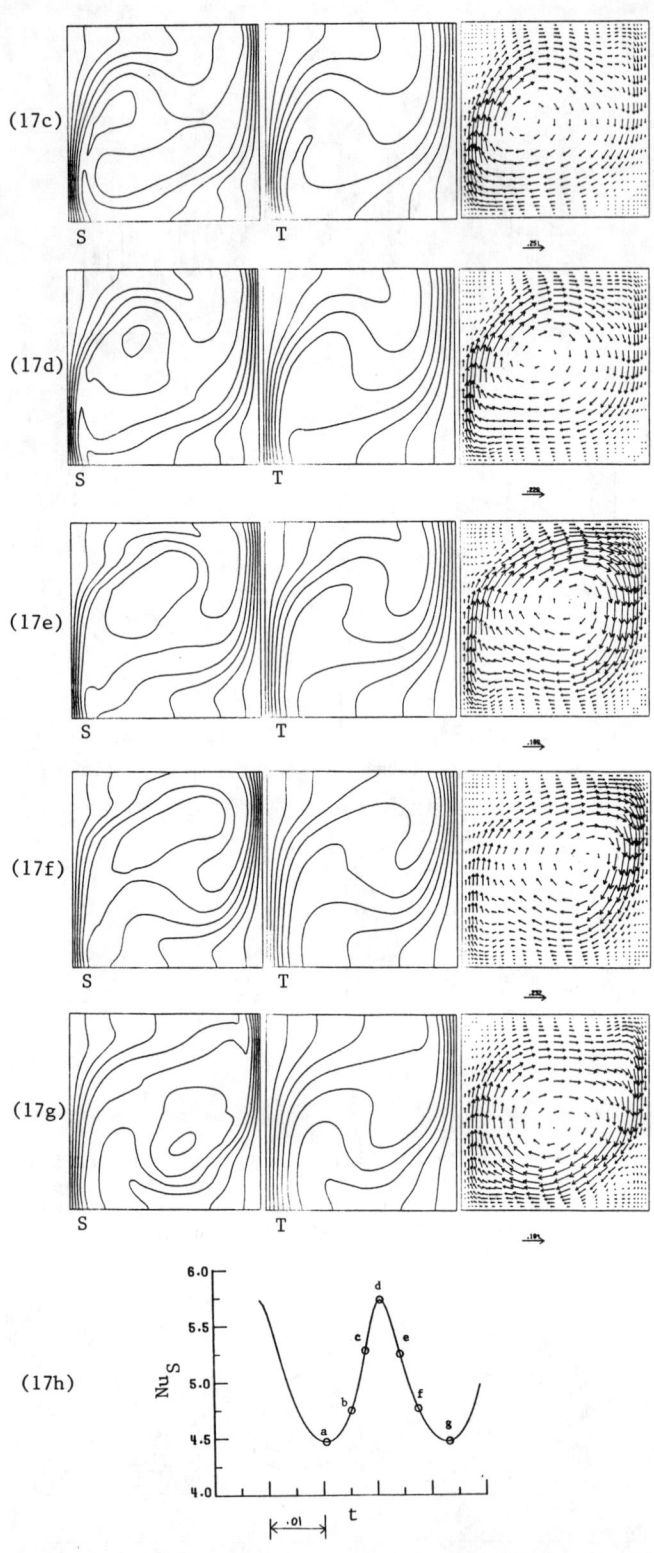

(17c)

S   T

(17d)

S   T

(17e)

S   T

(17f)

S   T

(17g)

S   T

(17h)

Fig. 17. Figs. 17(a) through 17(g) show oscillations in the spatial fields of solute concentration (S), temperature (T), and velocity. Fig. 17(h) shows the instants of time spanning one cycle at which these fields are presented. The Rayleigh number 1.8×10^5.

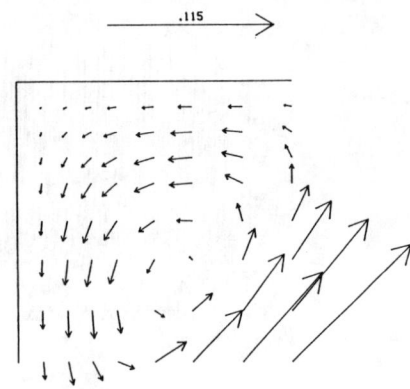

Fig. 18. The velocity field in the secondary cell at the top left corner at the instant of time corresponding to Fig. 17(d). This is for $Ra = 1.8 \times 10^5$.

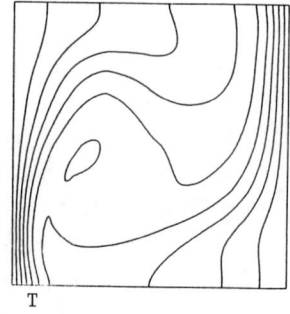

T

Fig. 19. Temperature field at an instant of time (similar to Fig. 17(c)) showing the presence of a thermal blob. The Rayleigh number is 2.4×10^5.

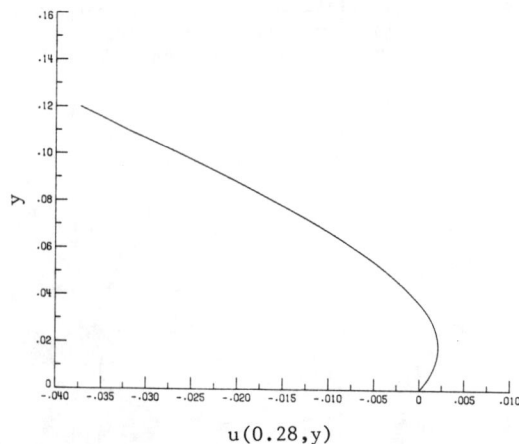

$u(0.28, y)$

Fig. 20. Velocity profile in the boundary layer on the lower horizontal surface. This is along the vertical at $x = 0.28$. The reverse flow in this region of separation is apparent. The Raleigh number is 2.4×10^5.

1989 National Heat Transfer Conference
HTD-Vol. 107, Heat Transfer in Convective Flows

CONVECTIVE INSTABILITIES IN HORIZONTAL POROUS LAYERS HEATED FROM BELOW: EFFECTS OF GRAIN SIZE AND ITS PROPERTIES

N. Kladias and V. Prasad
Department of Mechanical Engineering
Columbia University
New York, New York

ABSTRACT

Non-Darcy effects on convective instabilities and heat transfer in horizontal porous layers heated from below are examined numerically. The consideration of porosity variation in the wall region results in a lower critical Rayleigh number for the onset of convection compared to that for a uniform porosity media. The effect is, however, opposite on the onset of oscillatory flows. The range of stable convective state is therefore substantially increased when the porosity variation is considered. In the stable convection regime, the heat transfer rate is greatly enhanced by the wall channeling. The effects of porosity and specific heat ratio are significant only in the oscillatory flow regime. The highly complex flow structures obtained for periodic and random oscillations show that all the governing parameters strongly affect the evolution of instabilities and bifurcation.

NOMENCLATURE

A   aspect ratio, W/H

b   porous matrix structure property associated with the inertia term, m

c   isobaric specific heat, J/kg-K

C   constant in equation (10), C = 0.55 for $\varepsilon = 0.4$

d   characteristic particle dimension, m

Da   Darcy number, K/H^2

Fs   Forchheimer number, b/H

$\vec{g}$ body force vector, m/s^2

H   height of porous layer, m

k   thermal conductivity, W/m-K

k_m stagnant thermal conductivity of fluid-saturated porous medium, W/m-K

K   permeability of porous medium, m^2

Nu   local Nusselt number, equation (13)

$\overline{Nu}$ overall Nusselt number, equation (14)

$\overline{Nu}_f$ overall Nusselt number based on k_f ($= \overline{Nu}/\lambda$)

p   dimensionless pressure

Pr* modified Prandtl number, ν/α_m

Pr_f fluid Prandtl number, ν/α_f

r   stretching parameter, equation (12)

Ra_f fluid Rayleigh number, $g\beta H^3(T_h-T_c)/\nu\alpha_f$

Ra* modified Rayleigh number, $g\beta KH(T_h-T_c)/\nu\alpha_m$

S   specific heat ratio, $(\rho c)_m/(\rho c)_f$

t   dimensionless time

T   temperature, K

T_h temperature at heated surface, K

T_c temperature at cooled surface, K

u,v dimensionless velocity in x and y-directions

$\vec{V}$ filtration velocity vector, m/s

V   dimensionless velocity vector, $\vec{V}/(\alpha_f/H)$

W   width of porous layer, m

X,Y dimensionless distance in x and y-direction, x/H and y/H

KEYWORDS: Porous Media, Natural convection, Flow instability.

Greek Symbols

α thermal diffusivity, $k/\rho c$, m^2/s

α_m thermal diffusivity of porous medium, $k_m/(\rho c)_f$, m^2/s

β isobaric coefficient of thermal expansion of fluid, $1/K$

γ dimensionless particle diameter, d/H

ε porosity

ζ dimensionless vorticity, $\partial v/\partial X - \partial u/\partial Y$

θ dimensionless temperature, $(T - T_c)/(T_h - T_c)$

λ conductivity ratio, k_f/k_m

Λ viscosity ratio, μ'/μ

μ dynamic viscosity of fluid, kg/m-s

μ' apparent viscosity for Brinkman's viscous drag term, kg/m-s

ν kinematic viscosity of fluid, m^2/s

ρ density, kg/m^3

τ period of oscillation

ψ dimensionless stream function

Subscripts

c   critical value for the onset of convection

f   fluid

m   fluid-saturated porous medium

s   solid

w   wall

∞ asymptotic (bulk) value

INTRODUCTION

Convective instability in a horizontal porous layer subjected to a destabilizing temperature gradient has been investigated extensively by several authors in the past(see Combarnous and Bories, 1975; Cheng, 1978; Catton, 1985). It has been reported that the onset of convection takes place at a critical modified Rayleigh number, $Ra^*_c = 4\pi^2$. When Ra* is increased beyond a value between 240 and 280, the convective motion changes in character and becomes oscillatory (Combarnous and Bories, 1975). Caltagirone et al. (1971) argued that the fluctuations are caused by an increase in local values of Ra* at the base of the descending column of the unicellular flow above the value at which a convection cell can appear.

Caltagirone (1975) predicted the critical Ra* for the onset of fluctuations as 384 for unicellular flow in a square cross-section. Horne and O'Sullivan (1974) observed that the flow displays a periodically oscillatory behavior, generating 'tongues' of fluid in the descending and ascending columns of the flow. They further noted that at low Ra* the convective flow begins in the most favorable mode for that value of Ra* which is unicellular, and this mode dominates the later development of the flow until fluctuations occur as an attempt of the system to form a more favorable steady multicellular pattern which is repressed by the dominant circulation. In a later study (Horne and O'Sullivan, 1978), they investigated the significance of cyclic 'triggering' in the evolution of fluctuating convection in a rectangular porous cell by comparing the behavior of confined and unconfined flows. The similarity of the flows observed suggests that the oscillatory flow arises from the instability of the thermal boundary layer.

Recently, Schubert and Straus (1982) have shown that there exist four transitions in time-dependent convection in porous media. At the onset (Ra* between 380 and 400), the oscillatory flow is periodic with a single basic frequency. However, as Ra* increases the periodic state evolves to a quasi-periodic situation (at Ra* between 480 and 500) with two basic frequencies. Then, with increasing Ra*, a reverse transition occurs to a periodic state at a value somewhere between 530 to 650. In a follow up study, Kimura et al. (1986) have observed that the second periodic flow evolves to a chaotic non-periodic flow regime at a modified Rayleigh number which lies between 850-1000. The time-dependence of the chaotic state arises from the random generation of tongue-like disturbances within the horizontal thermal boundary layers, and the transition to the chaotic regime destroys the centro-symmetry property of convection.

It is important to note that all the above-mentioned studies have been based on the Darcy flow model where the modified Rayleigh number appears as the only governing parameter. However, it is now widely realized that the Darcy model is applicable only under special circumstances, and a generalized model for accurate prediction of convection in porous media must include the Forchheimer's inertia term and Brinkman's viscous diffusion term.

Rudraiah (1984) was probably the first to consider the Darcy-Brinkman model to study Benard convection in porous media, and recognize the importance of Darcy number other than Ra* in producing the instability. Georgiadis and Catton (1986) obtained solutions for the Darcy-Brinkman-Forchheimer (DBF) model, and noted that the divergence in heat transfer results is primarily due to inertia effects. Their numerical solutions compared reasonably well with the experimental data of Jonsson and Catton (1987).

We have also conducted a series of studies on Benard convection in porous media by employing the DBF model (Kladias and Prasad, 1987, 1988a,b), and have argued strongly in favor of considering the fluid Rayleigh and Prandtl numbers as independent parameters to analyze the non-Darcy convection. Our numerical solutions, for the unicellular flow initiated in a square cell, show that the critical Rayleigh number, $Ra_{f,c}$, for the onset of convection is a strong function of the fluid Prandtl number, and the porous matrix structure properties (Da and Fs) and the conductivity ratio (Kladias and Prasad, 1988a). In the stable convection regime, the heat transfer rate increases with the Rayleigh number, the Prandtl number, and the Darcy number while the conductivity ratio has an opposite effect. These results demonstrate the existance of an asymptotic convection regime where the porous media solutions are independent of the permeability of the porous matrix or Da.

We have demonstrated also that the onset of fluctuating convection takes place at lower fluid Rayleigh numbers as the Darcy number and/or the conductivity ratio increases (Kladias and Prasad, 1988b). Initially, the oscillatory convection is highly periodic. However, at a high Rayleigh and/or Darcy number, this gives rise to the random fluctuations where the flow structure and the temperature field are highly complex, a phenomenon earlier observed by Schubert and Straus (1982), and Kimura et al. (1986) for the Darcy convection. Consequently, the overall heat transfer rate varies significantly with time.

It should be noted that the above studies on non--Darcy Benard convection are for a porous medium of uniform porosity. However, the convective flows in bounded porous media, generally, become highly complex due to a large-scale variation in porosity, and hence, in permeability in the wall region, causing flow maldistribution and channeling in the boundary layer.

Channeling, which refers to the occurrence of a maximum velocity in a region close to an external boundary, has been reported by several investigators. However, only very recently, the investigators have considered the effect of wall-porosity variation on buoyancy-induced flow and heat transfer for a flat plate (Hong et al., 1987; Cheng, 1987) and a vertical cavity (Nishimura et al., 1984; David et al., 1988). These studies indicate that the heat transfer rate increases significantly with the wall-channeling effect.

The purpose of the present work is to examine the effects of the grain size and its thermophysical properties on the convective instability in a horizontal porous layer heated from below. Special attention is given to the effects of the governing parameters and the wall channeling on the evolution and structure of the oscillatory flows as an extension to our earlier work (Kladias and Prasad, 1988b).

MATHEMATICAL MODEL AND NUMERICAL ANALYSIS

The conservation equations for mass, momentum and energy based on the Brinkman-Forchheimer extended Darcy equation of motion for a two-dimensional porous layer may be written in the dimensionless form as (Kladias and Prasad, 1988a):

$$\frac{\partial u}{\partial X} + \frac{\partial v}{\partial Y} = 0 \qquad (1)$$

$$\frac{1}{\varepsilon}\frac{\partial u}{\partial t} + \frac{1}{\varepsilon^2}(u\frac{\partial u}{\partial X} + v\frac{\partial u}{\partial Y}) = -\frac{1}{\rho_f}\frac{H^2}{a_f}\frac{\partial p}{\partial X} - \frac{\nu}{K}\frac{H^2}{a_f}u - \frac{b}{K}H|V|u +$$
$$\frac{\mu'}{a_f \rho_f}\frac{1}{\varepsilon}(\frac{\partial^2 u}{\partial X^2} + \frac{\partial^2 u}{\partial Y^2}) \qquad (2)$$

$$\frac{1}{\varepsilon}\frac{\partial v}{\partial t} + \frac{1}{\varepsilon^2}(u\frac{\partial v}{\partial X} + v\frac{\partial v}{\partial Y}) = -\frac{1}{\rho_f}\frac{H^2}{a_f}\frac{\partial p}{\partial Y} + \beta(T_h - T_c)g\frac{H^2}{a_f}\theta - \frac{\nu}{K}\frac{H^2}{a_f}v -$$
$$\frac{b}{K}H|V|v + \frac{\mu'}{a_f \rho_f}\frac{1}{\varepsilon}(\frac{\partial^2 v}{\partial X^2} + \frac{\partial^2 v}{\partial Y^2}) \qquad (3)$$

$$\frac{(\rho c)_m}{(\rho c)_f}\frac{\partial \theta}{\partial t} + (u\frac{\partial \theta}{\partial X} + v\frac{\partial \theta}{\partial Y}) = \frac{k_m}{k_f}(\frac{\partial^2 \theta}{\partial X^2} + \frac{\partial^2 \theta}{\partial Y^2}) \qquad (4)$$

where H, $(T_h - T_c)$, α_f/H, H^2/α_f, and $\alpha_f\mu/K$ have been used as scales for length, temperature, velocity, time and pressure, respectively. Defining the stream function and vorticity in the usual way, the above equations may be reduced to,

$$\nabla^2 \psi = \zeta \qquad (5)$$

$$\frac{1}{Pr_f}[\frac{1}{\varepsilon}\frac{\partial \zeta}{\partial t} + \frac{1}{\varepsilon^2}(u\frac{\partial \zeta}{\partial X} + v\frac{\partial \zeta}{\partial Y}) + \frac{1}{\varepsilon^3}[\frac{\partial \varepsilon}{\partial Y}(u\frac{\partial u}{\partial X} + v\frac{\partial u}{\partial Y}) - \frac{\partial \varepsilon}{\partial X}(u\frac{\partial v}{\partial X} + v\frac{\partial v}{\partial Y})]] =$$

$$Ra_f\frac{\partial \theta}{\partial X} + \frac{1}{Da^2}[\frac{Fs}{Pr_f}|V| - 1](u\frac{\partial Da}{\partial Y} - v\frac{\partial Da}{\partial X}) + \Lambda(\frac{\partial^2 \zeta}{\partial X^2} + \frac{\partial^2 \zeta}{\partial Y^2}) - \frac{1}{Da}\zeta +$$

$$\frac{1}{Pr_f}\frac{1}{Da}|V|(u\frac{\partial Fs}{\partial Y} - v\frac{\partial Fs}{\partial X}) - \frac{Fs}{Pr_f}\frac{1}{Da}[\frac{\partial}{\partial X}(|V|v) - \frac{\partial}{\partial Y}(|V|u)] \qquad (6)$$

$$S\frac{\partial \theta}{\partial t} + (u\frac{\partial \theta}{\partial X} + v\frac{\partial \theta}{\partial Y}) = \frac{1}{\lambda}(\frac{\partial^2 \theta}{\partial X^2} + \frac{\partial^2 \theta}{\partial Y^2}) \qquad (7)$$

where the porosity, ε, the Darcy number, Da, the Forchheimer number, Fs, the specific heat ratio, S, and the conductivity ratio, λ have been considered to vary spatially in the case of a nonuniform porosity medium. Otherwise, they can be treated as constant and their derivatives may be dropped.

In the case of a nonuniform porosity medium, a variation in ε with distance from the bounding walls is assumed as follows (Roblee et al., 1958; Benenati and Brosilow, 1962):

$$\varepsilon = \varepsilon_\infty[1 + C_1 exp(-N_1 Y/\gamma)] \qquad (8)$$

where, ε_∞ is the porosity in the bulk of the porous medium and γ is the ratio of the particle diameter to the height of the bed, d/H. Equation (8) neglects the small oscillations of porosity, which are considered to be secondary (Vortmeyer and Schuster, 1983). The emphasis here is on the decay of porosity from the external surface, which has the primary effect. The empirical constants, C_1 and N_1, determine the value of the porosity on the wall and the distance over which the porosity varies, respectively, and their choice generally depends on the dimensionless particle size, γ.

For the present study, the Ergun model has been employed to describe the porous matrix structure which yields Darcy and Forchheimer numbers as functions of the porosity and the dimensionless particle size:

$$Da = \frac{\varepsilon^3}{150(1-\varepsilon)^2}\gamma^2 \qquad (9)$$

$$Fs = \frac{1.75}{150(1-\varepsilon)}\gamma = C(\varepsilon)\sqrt{Da} \qquad (10)$$

The geometric configuration considered for the present study is a two-dimensional square cavity (A=1), which is isothermally heated from below and cooled at the top, the side walls being adiabatic. The relevant hydrodynamic and thermal boundary conditions are:
$$u = v = \psi = 0 \text{ on all boundaries} \qquad (11a)$$

$$\theta = 1, Y = 0; \ \theta = 0, Y = 1; \ and \ \partial\theta/\partial X = 0, X = 0 \ and \ A \qquad (11b)$$

The Thom's first-order vorticity boundary condition has been used for the vorticity. It should be noted that a higher-order representation of ζ_w is inappropriate for the present problem since the velocity peaks lie very close to the wall particularly when ε is varying in the wall-region and/or the Darcy number is not very high.

To obtain numerical solutions, an ADI procedure, described in Kladias and Prasad (1988a), has been employed to perform the time integration for the vorticity and temperature equations (6) and (7) while the stream function equation (5) is solved by the Gauss-Seidel SOR iterative scheme at each time step. In the present study, a nonvarying grid distribution is used in the case of uniform porosity media, a 31x31 grid field being the best compromise between accuracy, convergence and economy. On the other hand, a nonuniform grid distribution is more appropriate for the case of variable porosity media since a large number of nodes are required near the walls to predict accurately at high Rayleigh and Darcy numbers. The requirement for very fine mesh in the wall region arises from the fact that the porosity varies exponentially from a high value at the wall to its asymptotic value (Eq. 8) within a thin layer. The fluid flow is observed to be very strong in this region while the velocities in the core are small. Generally, a 61x61 grid field has been employed in the case of variable porosity media.

To obtain a nonuniform grid field, a hyperbolic tangential function, has been used in the present work. Employing a stretching parameter, r, we are able to distribute the grids as per requirement. For example, the X-direction grid points are determined as:

$$X_i = \frac{1}{2}[1 - \frac{tanh(r(1 - i/N))}{tanh r}] \quad 1 \le i \le N \qquad (12)$$

where, there are 2N + 1 nodes in the X-direction and the distribution is symmetric about X = 0.5. For the same number of nodes an increase in r results in finer grids in the wall region.

An initial sinusoidal perturbation of temperature

field is introduced to drive the solution (Caltagirone, 1975; Kladias and Prasad, 1988a,b). The iterative convergence of the stream function equation (5) is checked by obtaining the fractional difference between the values of ψ obtained from two successive iterations, less than 10^{-3}. Solutions are considered to be steady when the fractional changes in ζ and θ between two time-steps are both less than 10^{-4} at all nodes in the computational domain. Generally, very small time steps (of the order of 10^{-3}-10^{-5}) are used to achieve the desired accuracy and the numerical stability. Hence, the present scheme is formally second-order accurate at least for the stable convection solutions, and does not suffer from the artificial diffusion errors.

The average Nusselt number at any Y-location is obtained as:

$$\overline{Nu} = (1/A) \int_0^A Nu(X,Y)\,dX \qquad (13)$$

where the local Nusselt number,

$$Nu = \lambda v\theta - \frac{\partial \theta}{\partial Y} \qquad (14)$$

The details of this numerical scheme as well as an excellent agreement with the Darcy flow solutions (within 0.55 percent) of Caltagirone (1975), and Schubert and Straus (1979), and non-Darcy results (within 3.7 percent) of Georgiadis and Catton (1986) are discussed in a separate paper (Kladias and Prasad, 1988a), and will be omitted here for brevity.

CONVECTIVE FLOW REGIMES

As noted earlier the present numerical solutions indicate also the existence of four flow regimes: (i) conduction, (ii) stable convection, (iii) periodic oscillatory convection, and (iv) random (non-periodic) oscillatory convection. The onset of convection takes place at a critical Rayleigh number, $Ra_{f,c}$ which is equal to $4\pi^2/\lambda Da$ (Darcy model) only under special circumstances. Otherwise, $Ra_{f,c}$ strongly depends on Da, Fs, Pr_f and λ (Kladias and Prasad, 1988a). Beyond $Ra_f > Ra_{f,c}$, the initial perturbation yields a stable solution for the convective flow, where an unicellular flow is observed in the square cell.

Monocellular convection in a square cavity starts oscillating when the Rayleigh number increases beyond a second critical value. Then, irrespective of the number of time-steps in the calculation, both the temperature and flow fields, and the heat transfer rate fluctuate with time. At the onset, the oscillatory convection has a strong periodic nature (Schubert and Straus, 1982; Kimura et al., 1986; Kladias and Prasad, 1988b). It is seen that the periodic modulations of the depth of the thermal boundary layers are responsible for the oscillatory behavior (Howard instability).

As the Rayleigh number increases a transition to multiple frequency solutions is observed where the well-defined periodic nature of the oscillations vanishes, and highly random fluctuations in the local and global variables are observed. Indeed, the monocellular flow gives rise to four cells flow pattern in a square cavity, whose shape and size change with time. A similar four-cells pattern was also reported by Caltagirone (1975) for Darcy porous media (Ra* > 800), and by Muller (1982) for a fluid layer. This sequence of transitions reported here is characterized by a monotonically increasing disorder similar to what has been observed in the fluid layer heated from below. Indeed, the experiments of Gollub and Benson (1980) in a fluid-filled cavity show four paths from a state of rest to turbulence depending on the aspect ratio and the Prandtl number. All of them begin with a bifurcation from a steady-state to an oscillatory periodic flow and gradually toward

more complex time-dependent flows which end in a chaotic state. On the other hand, Kimura et al. (1986) reported a reverse transition from a more-disordered to a less-disordered state which takes place before the transition to non-periodic state. This discrepancy between our findings and the observations of Kimura et al. (1986) may be due to the fact that we have considered the inertial effects while the work of Kimura et al. is based on the Darcy flow model.

RESULTS AND DISCUSSION

To study the effect of the grain size, γ which is related to the Darcy and Forchheimer numbers by equations (9) and (10), respectively, the computations have been performed in two stages. In the first part, the results have been obtained for a porous medium of $\varepsilon_\infty = 0.4$ with the porosity variation in the wall region and have been compared with the uniform porosity solutions reported earlier (Kladias and Prasad, 1988a,b). In the second part of the paper, the effects of the porosity, ε, the specific heat ratio, S, and the conductivity ratio, λ are considered. For the present computations, the fluid Prandtl number has been fixed to $Pr_f = 1$, and the viscosity ratio has been taken as $\Lambda = 1/\varepsilon$.

Effect of Grain Size: Variable Porosity Media

To consider the variation in porosity according to Eq. (8), the empirical constants C_1 and N_1 have been assigned the values of 0.98 and 5.0, respectively (Vafai, 1984). The porosity at the bulk of the porous medium has been fixed to $\varepsilon_\infty = 0.4$ which is generally true for a randomly packed sphere bed. This fixes the wall porosity at $\varepsilon_w \simeq 0.8$ while the asymptotic value of porosity is reached within a distance of five particle-diameter from the wall. Also, the conductivity ratio has been fixed to $\lambda = 1 (k_s = k_f)$ and no variation in k_m has been considered; the fluid Prandtl number, Pr_f, and the specific heat ratio, S, are taken as unity. On the other hand, the viscosity ratio, Λ, changes according to the variation in porosity ($\Lambda = 1/\varepsilon$).

Onset of Convection. The effect of wall channeling on critical Rayleigh number for the onset of convection is presented in Fig. 1. It is clearly seen that for variable porosity media with $Da_\infty > 10^{-6}$, the stable convection regime initiates at much lower Ra_f compared to the homogeneous porous medium with the same asymptotic Darcy number (or particle size) and conductivity ratio. For example, the critical Rayleigh number for a uniform porosity medium with $Da = 10^{-4}$ is $Ra_{f,c} = 4\times10^5$ while the convective flow for a variable ε with the same grain size (or bulk Darcy number) initiates at $Ra_{f,c} \simeq 5\times10^4$.

Stable Convective State. To study the effects of flow maldistribution and wall channeling, the numerical results have been obtained for $Da_\infty = 10^{-6}, 10^{-5}$ and 10^{-4}. It should be noted, that it has been extremely difficult to obtain numerical solutions for $Da_\infty > 10^{-3}$. Due to the large gradients in equation (6) we had to use very small time steps ($\Delta t < 10^{-5}$) in order to achieve the desired numerical stability, and consequently, the computational cost became prohibitive. However, the present range covers completely the Darcy numbers generally encountered in laboratory experiments with spherical beads as well as in most of the practical applications.

Figure 2 and table 1 show that the effects of wall channeling are quite pronounced for large particle size (or Da_∞) and/or at high Rayleigh numbers. Indeed, the enhancement in heat transfere can be more than 100 percent for variable porosity media of large grain size ($Da_\infty > 10^{-5}$). This agrees qualitatively with the findings of David et al. (1988) for a vertical cavity. For example, the above authors have reported an enhancement of the

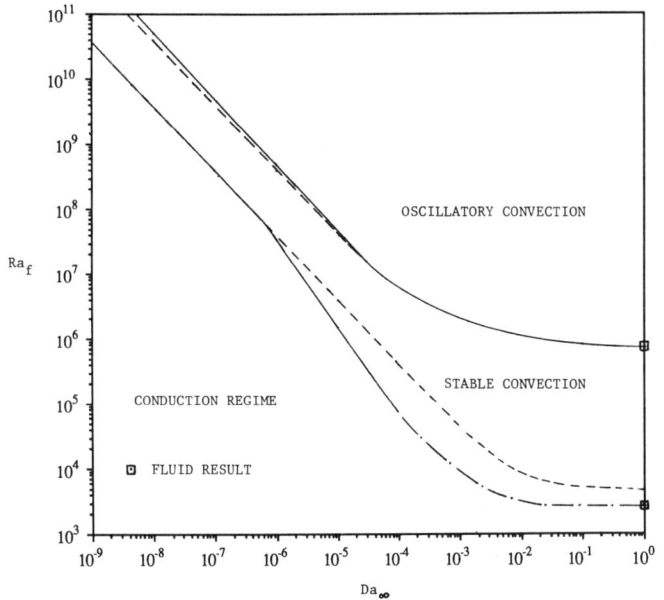

Fig. 1 Criteria for the onsets of stable and fluctuating convection regimes predicted by the Darcy-Brinkman-Forchheimer formulation (A = 1, $Pr_f = 1$, $\lambda = 1$), --: uniform porosity, -: variable porosity, -.-: expected solution for variable porosity at $Da_\infty > 10^{-4}$).

Table 1. Comparison of overall Nusselt number and maximum stream function for uniform and variable porosity media (A = 1, $C_1 = 0.98$, $N_1 = 5.0$).

Da	Ra_f	Darcy Solution $\overline{Nu}$	ψ_{max}	Uniform Porosity $\overline{Nu}$	ψ_{max}	Variable Porosity $\overline{Nu}$	ψ_{max}
	4×10^7	1.029	0.430	1.019	0.398	1.022	0.427
	5×10^7	1.450	2.112	1.409	2.057	1.423	2.065
	10^8	2.651	5.377	2.530	5.311	2.644	5.352
10^{-6}	2×10^8	3.813	8.942	3.556	8.841	3.808	8.912
	3×10^8	4.523	11.405	4.131	11.241	4.516	11.372
	4×10^8	-	-	-	-	9.054	21.230
	4×10^6			1.009	0.326	2.367	3.254
	5×10^6			1.390	2.006	3.129	4.499
10^{-5}	10^7			2.489	5.214	6.208	9.628
	2×10^7			3.487	8.651	8.880	16.936
	4×10^5			1.000	0	3.291	4.863
	5×10^5			1.297	1.728	3.944	6.136
	10^6			2.312	4.809	5.742	10.585
10^{-4}	2×10^6			3.202	7.952	6.963	16.238
	3×10^6			3.690	10.024	8.122	20.290

(*Da $= 10^{-6}$, $\gamma = 0.029$; Da $= 10^{-5}$, $\gamma = 0.092$; Da $= 10^{-4}$, $\gamma = 0.291$.)

order of 100-110 percent for $Da_\infty = 2.93\times10^{-5}$ when $Ra_f > 10^7$.

The vertical velocity profiles, which are displayed in Figs. 3a-c reveal that the increase in $\overline{Nu}$ is primarily due to the increase in convective velocities near the boundary. These plots show that for small Darcy numbers ($Da_\infty \leq 10^{-6}$), the wall channeling is confined within a thin layer, the thickness of which increases with Da_∞ (Figs. 3a-b). Indeed, the consideration of variable ε has resulted in a very marked peak of the vertical velocity near the end walls. However, the trend is reversed in the core of the cavity, where the velocities for uniform porosity media are higher than that for the variable porosity media. Again, the effect

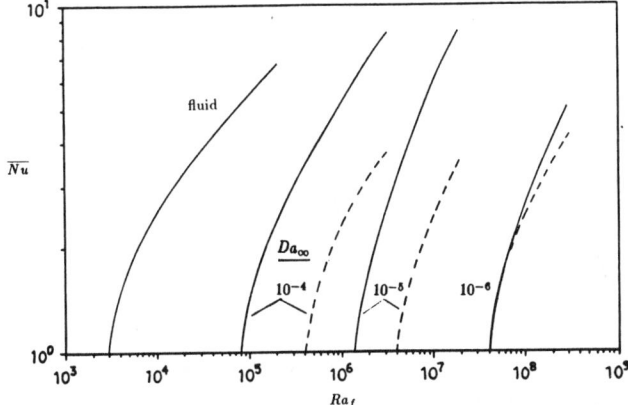

Fig. 2 Heat transfer rates for horizontal porous layers with and without porosity variation together with the fluid results (- -:uniform porosity, -:variable porosity).

is more pronounced at higher Da_∞. The effect of Rayleigh number, Ra_f, on these velocity profiles can be visualized from Fig. 3c. Clearly, the higher the Rayleigh number the stronger is the influence of variable porosity on predicted velocities. The above variation in the velocity field is also displayed by the streamline patterns in Figs. 4(a)-(c) and Figs. 5(a)-(b) for $Da_\infty = 10^{-6}$ and 10^{-4}, respectively. Note the substantial change in the flow structure because of the strong flows along the boundaries.

The temperature fields change accordingly (Figs. 4 and 5). The isotherms now move closer to the wall region indicating a thermally stratified core with thermal boundary layers along the horizontal walls (compare Figs. 5b and 5c). As a result, the sharper temperature gradients on the boundaries increase the Nusselt number. The effect is more pronounced at higher Ra_f or Da; the effect of Darcy number being much more stronger. For example, Table 1 and Figure 2 show that for $Da_\infty = 10^{-6}$ the wall channeling increases $\overline{Nu}$ by 9.3 percent ($Ra_f = 3\times10^8$) while for $Da_\infty = 10^{-4}$ $\overline{Nu}$ increases by 120 percent, although the fluid Rayleigh number in this case is smaller ($Ra_f = 3 \times 10^6$).

Another interesting aspect, which is revealed from Fig. 2, is that the asymptotic convection regime is reached much earlier (in terms of Da_∞) if the variation in ε is considered. Indeed, the heat transfer curves for $Da_\infty = 10^{-5}$ and $Da_\infty = 10^{-4}$ have different slopes and are non-parallel. Clearly, for nonhomogeneous porous media the effect of Darcy number diminishes with an increase in Da_∞ beyond 10^{-5} whereas, the same is true for homogeneous porous media only when Da $> 10^{-4}$ (Kladias and Prasad, 1988a). The same asymptotic trend has been demonstrated by the experimental data for

373

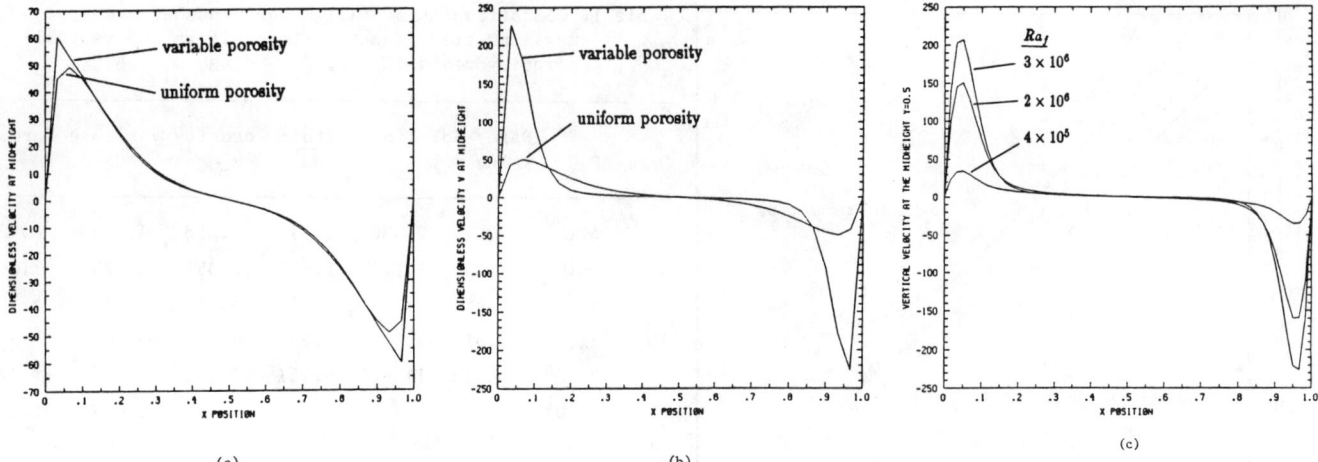

(a)　　　　　　　　　　(b)　　　　　　　　　　(c)

Fig. 3　Effect of wall channeling on vertical velocity profiles at the midheight; (a) $Ra_f = 3 \times 10^8$, $Da_\infty = 10^{-6}$, (b) $Ra_f = 3 \times 10^6$, $Da_\infty = 10^{-4}$, and (c) $Da_\infty = 10^{-4}$ (effect of Ra_f).

15mm ($Da_\infty = 7.78 \times 10^{-5}$) and 25mm diameter ($Da_\infty = 1.85 \times 10^{-4}$) solid beads, when the saturating fluid is the same (Kladias, 1988).

The above reported enhancement in heat transfer due to the consideration of wall channeling effect is in agreement with the findings of Hong et al. (1984) for a vertical flat plate, David et al. (1988) for a differentially heated vertical porous cavity, and of Vafai (1984) for forced convection in variable porosity media. However, the present results do not agree with the conclusion of Nishimura et al. (1984) that the Nusselt number in the variable porosity case is smaller than that

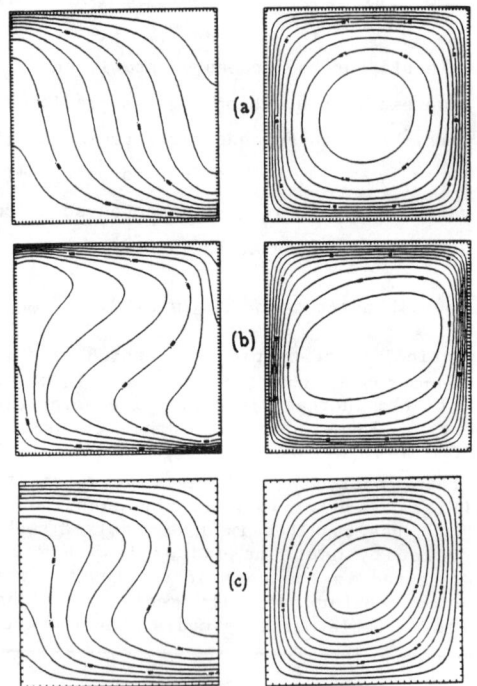

Fig. 5　Isotherms ($\Delta\theta = 0.1$) and streamlines for variable porosity media,

$Da_\infty = 10^{-4}$; (a) $Ra_f = 4 \times 10^5$ ($\Delta\psi = 0.486$), (b) $Ra_f = 2 \times 10^6$ ($= 1.623$), (c) $Ra_f = 2 \times 10^6$ ($\Delta\psi = 0.722$), uniform porosity.

for the uniform porosity media. This may be due to the fact that the above authors did not consider the variation in Darcy number with the distance from the wall. Also, the present results do not support the assumption of Georgiadis and Catton (1988), that is, the effect of neglecting the variation in porosity is balanced by the elimination of Brinkman's viscous diffusion term.

It is also interesting to note that the DBF solutions with variable porosity ($Da_\infty = 10^{-6}$) are within 2 percent of the Darcy solutions for $4 \times 10^7 \leq Ra_f \leq 3 \times 10^8$ (Table 1). However, this is true only for $Pr_f = 1$, $\lambda = 1$. As can be expected from the numerical results presented in Kladias and Prasad (1988a), the effect of

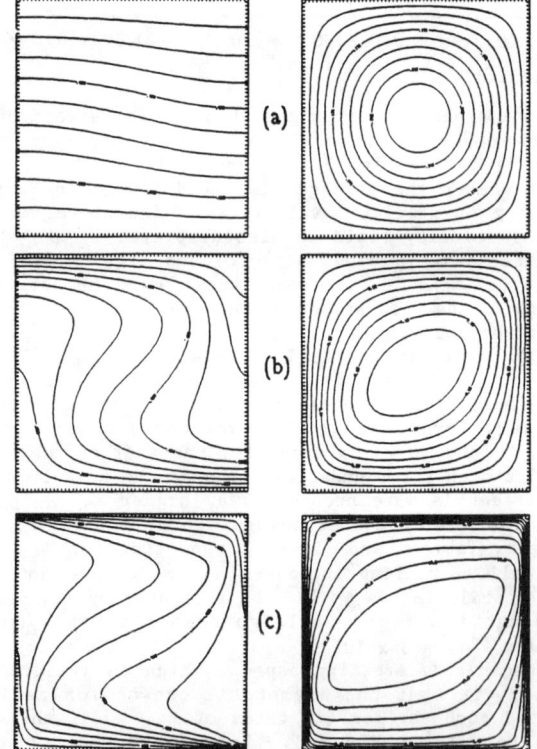

Fig. 4　Isotherms ($\Delta\theta = 0.1$) and streamlines for variable porosity media,

$Da_\infty = 10^{-6}$, (a) $Ra_f = 4 \times 10^7$ ($\Delta\psi = 0.427$), (b) $Ra_f = 2 \times 10^8$ ($\Delta\psi = 0.891$), (c) $Ra_f = 4 \times 10^8$ ($\Delta\psi = 2.129$).

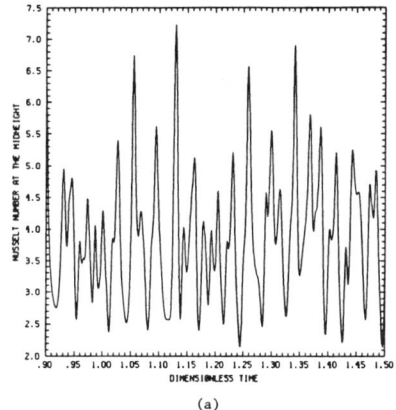
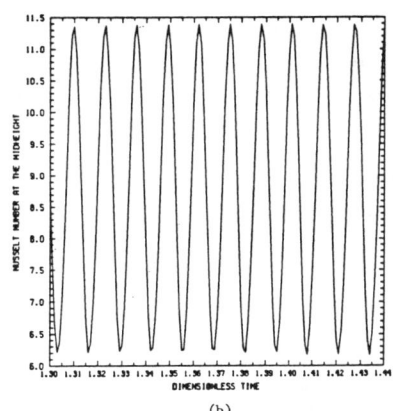
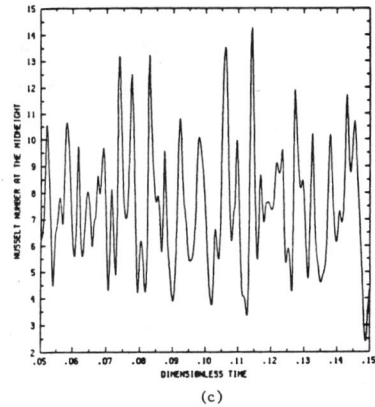

(a)   (b)   (c)

Fig. 6 Fluctuating Nusselt number at the midheight for variable porosity media, $Da_\infty = 10^{-4}$; (a) Ra_f = 5×10^8, uniform porosity, (b) Ra_f = 5×10^8, variable porosity, (c) $Ra_f = 10^9$, variable porosity.

Prandtl number and the conductivity ratio on the heat transfer results will be significant especially at high Ra_f and Da_∞. Therefore, a more realistic description of transport processes in porous media is obtained by employing the DBF formulation with variable porosity in the wall-region rather than by employing the Darcy model.

Oscillatory Convective State. To examine the effect of wall channeling on the evolution and structure of oscillatory flows, results have been obtained for $Da_\infty = 10^{-6}$, $Ra_f = 4 \times 10^8 - 10^9$ and $Da_\infty \leq 10^{-4}$, $Ra_f = 5 \times 10^6 - 10^7$. The highest maximum and lowest minimum Nusselt numbers observed during the extent of calculations together with the range over which the period varies, are listed in Table 2.

The effect of variation in porosity on the critical Rayleigh number for the onset of periodic and random oscillations is demonstrated in Fig. 1 and Table 1. It can be seen that, for a fixed Da, the bifurcations to periodic and random oscillatory flows are delayed (in terms of Ra_f) by the consideration of variable porosity. For example, the periodic oscillatory flow for a uniform porosity medium ($\varepsilon = 0.4$, $Da = 10^{-6}$, $Ra_f = 4 \times 10^8$) reverts to stable convective motion at the same Da_∞ and Ra_f when ε is variable. Similarly, the random oscillation for $Ra_f = 5 \times 10^8$, $Da = 10^{-6}$ (Fig. 6a) reverts to periodic when the effect of wall channeling is considered (Fig. 6b). This is primarily due to the fact that the core of the cavity becomes more or less stagnant, whereas the thermal activity and fluid motion is confined within thin boundary layers along the walls. However, the oscillations for $Da_\infty = 10^{-6}$ become random at $Ra_f = 10^9$ (Fig. 6c).

Effect of Porosity

Stable Convective State. The effect of porosity on the overall heat transfer rate in the stable convection regime has been found to be insignificant. This is clearly shown in Table 3 where $\overline{Nu}$ and ψ_{max} are presented for $\varepsilon = 0.2, 0.4$, and 0.6, $Ra_f = 2 \times 10^8$ and $Da = 10^{-6}$. Although in the steady-state, stable convection regime, the contribution of the first term in equation (6) is zero, a small variation in the solutions has been possible because $\Lambda = 1/\varepsilon$ which reduces, the contribution of the viscous term as ε is increased.

Oscillatory Convective State. To study the effect of ε in the oscillatory convective regime, results have been obtained for $Da = 10^{-6}$, $Ra_f = 4 \times 10^7 - 5 \times 10^7$, and $\varepsilon = 0.2$, 0.4, 0.6. The quantitative results of this study are presented in Table 4, where the largest maximum and smallest minimum Nusselt numbers observed during the extent of the calculations are listed together with the range over which the period varies. Also, the Nusselt number at the midheight as a function of the dimensionless time is plotted in Figs. 7(a)-(c). It is seen that the periodic oscillation for $Da = 10^{-6}$, $Ra_f = 4 \times 10^7$, $\varepsilon = 0.4$ (Kladias and Prasad, 1986b) reverts to random fluctuations for $\varepsilon = 0.6$ (Fig. 7a). On the other hand, a decrease in ε stabilizes the flow (Table 4). Generally, the amplitude of fluctuation increases with ε. This may be a result of the relative increase of bouyancy and inertia terms in equation (6).

Effect of Specific Heat Ratio

Stable Convective State. Since the specific heat ratio appears only in the time derivative term in equation

Table 2. Maximum and minimum values of Nusselt number and periods of oscillations in the fluctuating convection regime - Variable porosity media.

γ	Da_∞	Ra_f	Uniform Porosity		Variable Porosity	
			$\overline{Nu}$ max, min	τ Period	$\overline{Nu}$ max, min	τ Period
0.029	10^{-6}	5×10^8	7.23, 2.14	0.005 - 0.035	11.40, 6.20	0.013
		10^9	13.92, 4.60	0.008 - 0.028	14.27, 2.37	0.00016 - 0.00058
0.291	10^{-4}	5×10^6	4.16 4.15	0.023	8.921, 8.768	0.01
		10^7	8.12 2.62	0.009 - 0.016	11.94, 4.22	0.0022 - 0.00044

375

Table 3. Effect of porosity on Nusselt number and maximum stream function in the stable convection regime (uniform porosity), $Ra_f = 2\times10^8$, $Da = 10^{-6}$.

ε	$\overline{Nu}$	ψ_{max}
0.2	3.519	8.7216
0.4	3.556	8.8409
0.6	3.565	8.8697

(7), no effect of this parameter on the steady-state solution in the stable convection regime is expected. Oscillatory Convective State. Table 5 presents the effect of the specific heat ratio. It is observed that the amplitude of fluctuation is minimum for S = 1 and is seen to increase as S increases or decreases from unity. On the other hand, the period of oscillation increases as S increases. This is obvious from a comparison

Table 4 Maximum and minimum values of Nusselt number and periods of oscillations in the fluctuating convection regime – Effect of Porosity (Da = 10^{-6}).

Ra_f	ε	$\overline{Nu}_{max, min}$	period τ
4×10^8	0.6	7.15, 1.61	0.005-0.049
	0.4	4.43, 4.38	0.016
	0.2	5.39	–
5×10^8	0.6	10.57, 1.70	0.012-0.024
	0.4	7.23, 2.14	0.005-0.035
	0.2	6.71, 2.32	0.007-0.025

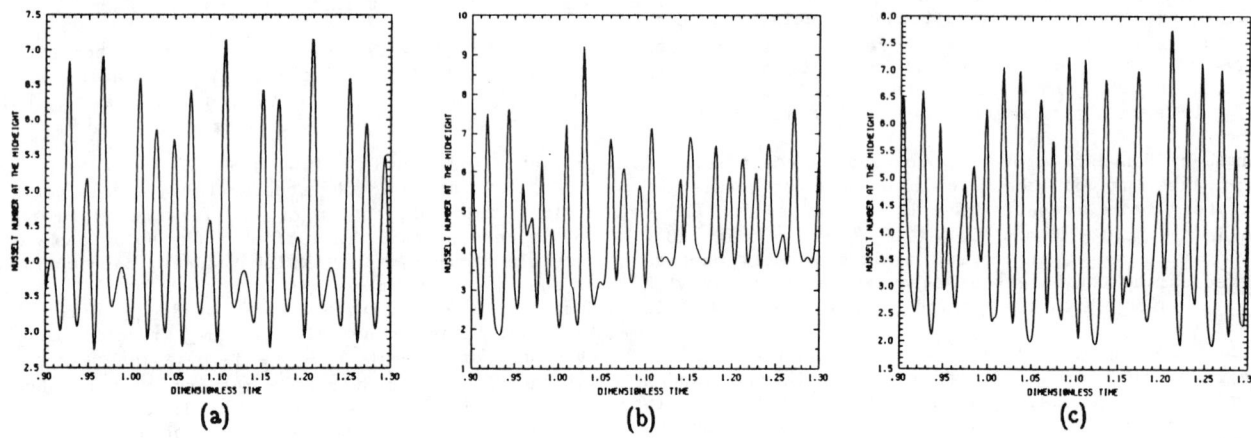

Fig. 7 Effect of porosity on fluctuating Nusselt number; (a) Ra$_f$ = 4x10^8, Da = 10^{-6}, ε = 0.6, (b) Ra$_f$ = 5x10^8, Da = 10^{-6}, ε = 0.6, and (c) Ra$_f$ = 5x10^8, Da = 10^{-6}, ε = 0.2

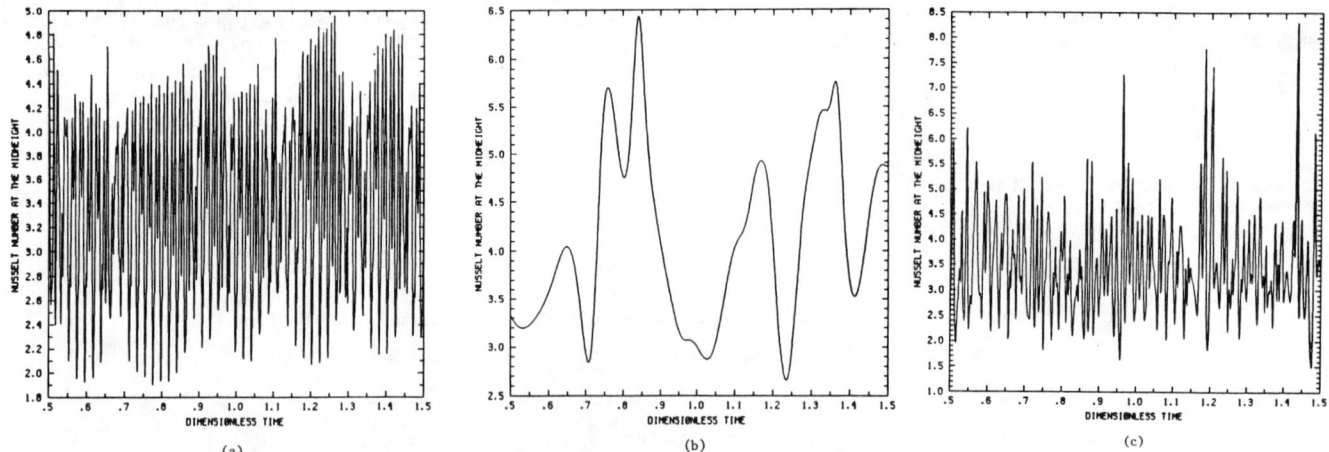

Fig. 8 Effect of specific heat ratio on fluctuating Nusselt number at midheight for $Ra_f = 5\times10^6$, $Da = 10^{-4}$, (a) S=0.5, and (b) S=10, and (c) $Ra_f = 4\times10^8$, $Da = 10^{-6}$, S = 0.5.

between Figs. 8a (S = 0.5) and 8b (S = 10). Another interesting aspect of the specific heat ratio effect is that the periodic fluctuation in $\overline{Nu}$ for S = 1, always reverts to random oscillations as S decreases to 0.5 (a value which is applicable to geothermal systems). This becomes clear from a comparison between Figs. 8a, 8c and those presented in Kladias and Prasad (1988b).

Furthermore, it is interesting to observe the effect of S on transient evolution of the oscillatory flow pattern in Figs. 9(a)-(i) (Ra$_f$ = 10^9, Da = 10^{-6}, S = 100). Figures 9(a)-(c) (t = 0.05-0.63) show that as time increases the flow is intensified and the temperature gradients become steeper. Then, the local instabilities develop in the areas of large temperature gradient (Fig. 9d, t = 0.8) to cause the appearance of two microvortices (Fig. 9e, t = 0.88). As time passes, the microvortices grow, and the flow structure becomes highly complex (Fig. 9f, t = 1.024), finally reverting to a three cells swing from left to right (Figs. 9h, t = 1.8 and 9i, t = 20).

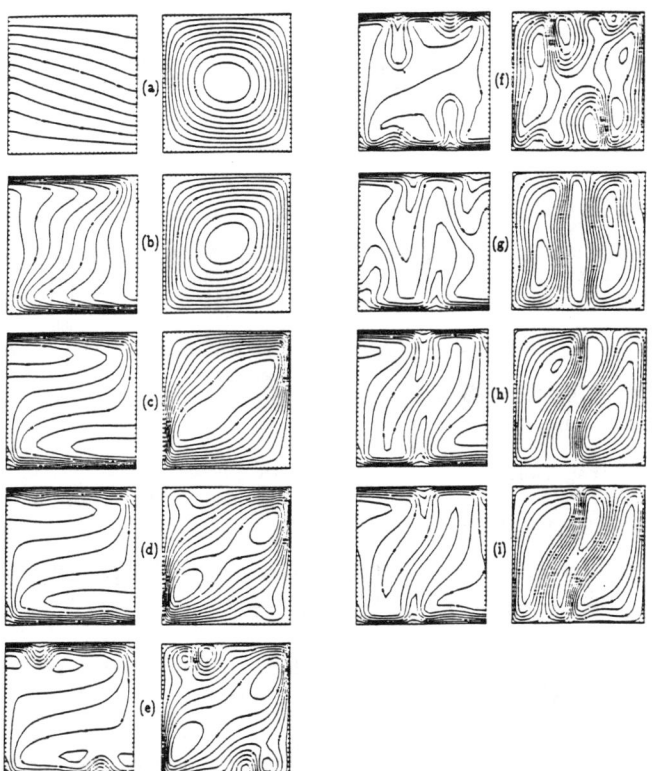

Fig. 9   Effect of specific heat ratio on fluctuating Nusselt number at midheight for Ra$_f$ = 10^7, Da = 10^{-4}, (a) S=0.5, and (b)S=10, Fig. 10 Effect of specific heat ratio on the evolution and structure of oscillatory flows-Isotherms ($\Delta\theta$ = 0.1) and streamlines for Ra$_f$=10^9, Da=10^{-6}, S=100;(a) t=0.05, $\overline{Nu}$ = 2.68, (b) t = 0.375, $\overline{Nu}$ = 43.08, (c) t = 0.630, $\overline{Nu}$ = 2.69, (d) t=0.80,$\overline{Nu}$=1.45, (e) t = 0.88, $\overline{Nu}$ = 1.38, (f) t=1.024, $\overline{Nu}$ = 1.14, (g) t = 1.418, $\overline{Nu}$ = 8.18, (h) t=1.80, $\overline{Nu}$ = 8.47, and (i) t = 2.0, $\overline{Nu}$ = 8.82.

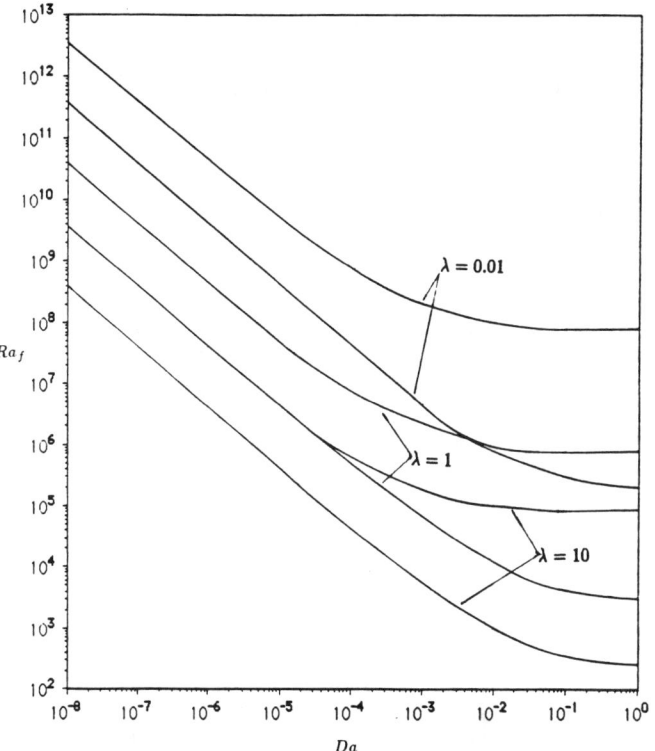

Fig. 10   Effect of conductivity ratio on critical values of Rayleigh and Darcy numbers for the onsets of stable and fluctuating convection regimes (for every λ the upper curve delimits the stableand fluctuating convection regimes while the lower curve delimits the conduction and stable convection egimes).

Effect of Thermal Conductivity Ratio

To present the effect of the thermal conductivity ratio, λ , on the convective heat transfer, numerical results have been obtained for Da = 10^{-6}-10^{-1}.
Onset of Convection.   The effect of conductivity ratio on Ra$_{f,c}$ can be seen in Fig. 10. It is clear that the smaller the conductivity ratio, the higher is the Rayleigh number required for the onset of convection. The expression λ= k$_f$/k$_m$, and the fact that k$_m$ increases with k$_s$ imply that the onset of convection for a given saturating fluid and a porous matrix of fixed Da, will be delayed with an increase in the thermal conductivity of the solid particles. This is primarily because the high conductivity solid particles have stabilizing effect.

An interesting aspect of Fig. 10 is that the delimiting curves corresponding to different values of λ are seen to be parallel. This trend can be explained if we nondimensionalize the governing equations (1)-(4) using α_m instead of α_f. Then λ will be eliminated from the energy equation (7) and appear, on the other hand, in the vorticity transport equation (6):

$$\frac{Da}{\lambda Pr_f}[\frac{1}{\varepsilon}\frac{\partial\zeta}{\partial t} + \frac{1}{\varepsilon^2}(\dot{u}\frac{\partial\zeta}{\partial X} + v\frac{\partial\zeta}{\partial Y})] = \lambda Ra_f Da\frac{\partial\theta}{\partial X} - \frac{Fs}{\lambda Pr_f}(v\frac{\partial|V|}{\partial X} - u\frac{\partial|V|}{\partial Y}) -$$

$$\frac{Fs}{\lambda Pr_f}|V|\zeta - \zeta + \Lambda(\frac{\partial^2\zeta}{\partial X^2} + \frac{\partial^2\zeta}{\partial Y^2}) \quad (15)$$

Equation (16) and Fig. 10 show that it is possible to use a more general Rayleigh number, λRa_f, as parameter of interest, instead of Ra$_f$. Thus, solutions for Ra$_f$ (λ=1), reported in the previous paragraphs, are valid for corresponding values of λRa_f, which makes them more general. However, in that case λPr_f has to be used as the Prandtl number rather than Pr_f.

Stable Convective State. As reported earlier (Kladias and Prasad, 1988a) the heat transfer rate for fixed values of Ra_f and Da always increase with an increase in thermal conductivity of the solid particles, k_s, and therefore, with a reduction in the conductivity ratio, λ.

Oscillatory Convective State. The effect of conductivity ratio on critical Rayleigh number for the onset of oscillating convection can also be seen in Fig. 10. It is clear that the onset of oscillations is delayed (in terms of Ra_f) with a decrease in the conductivity ratio (or an increase in the conductivity of solid particles). Again, this is due to the stabilizing effect of high conductivity solid particles. On the other hand, the effect of λ on the structure of oscillatory flows and the periods of oscillations can be studied if, as was made clear before, we present the previous results in terms of λRa_f, instead of Ra_f.

CONCLUSIONS

The Darcy-Brinkman-Forchheimer solutions support the existence of four flow regimes in the case of free convection in horizontal porous layers heated from below; conduction, stable convection, periodic oscillatory, and random oscillatory convection regimes. The critical Rayleigh number for the onset of convection decreases from a value predicted by the Darcy flow model, $Ra_{f,c} = 4\pi^2/\lambda Da$, as the particle diameter (or Da) and/or the conductivity ratio is increased.

The numerical solutions show that a porous medium can possibly transport larger amounts of energy compared to the fluid alone if the porous matrix highly permeable and the thermal conductivity of solid particles is higher than that for the fluid, $k_s > k_f$. On the other hand, the effects of porosity, ε, and specific heat ratio, S, have been shown to be insignificant in the stable convective regime. The effects of specific heat ratio and porosity are, however, seen to be significant in the unstable convective flow regime. The highly complex flow structures, obtained in the fluctuating regime, show that all the parameters under consideration strongly affect the evolution of instabilities and emphasize the need to consider the three-dimensional effects.

The solutions for variable porosity media show that the convective heat transfer is greatly enhanced when the wall channeling, due to the variation in porosity, is considered. The enhancement in heat transfer is primarily due to the increase in convective velocities near the boundaries. The thickness of boundary layer within which the wall channeling is confined, increases with the particle diameter. The effect of variation in porosity is, thus, stronger as γ and/or Ra_f increases.

Moreover, for $\gamma > 0.092$ ($Da_\infty > 10^{-5}$), the critical Rayleigh number for the onset of stable convective flow in a variable porosity medium is much lower than that for the onset of stable convection in a uniform porosity medium of the same particle size (or permeability). However, the wall channeling has an opposite effect on the critical value of Ra_f for the onset of oscillatory flows. Therefore the range of stable convective state is substantialy broadened when the variation in porosity is considered. Also, the nonhomogeneous porous media solutions approach the fluid results for $Da_\infty > 10^{-4}$ which is much lower than the corresponding Darcy number ($Da = 10^{-2}$) in the case of a homogeneous porous layer.

Table 5 Maximum and minimum values of Nusselt number and periods of oscillations in the fluctuating convection regime – Effect of Specific Heat Ratio, S.

Da	Ra_f	S	$\overline{Nu}_{max, min}$	period τ
10^{-6}	4×10^8	0.5	8.31, 1.48	0.005-0.023
		1	4.43, 4.38	0.016
		10	4.69, 4.15	0.098
		100	16.88, 1.62	1.763
	10^9	0.5	13.88, 3.04	0.0006-0.003
		1	13.92, 4.60	0.008-0.028
		10	16.56, 2.81	0.011-0.085
		100	43.08, 1.13	0.649
10^{-4}	5×10^6	0.5	4.88, 1.92	0.005-0.013
		1	4.16, 4.15	0.023
		10	6.63, 2.87	0.005 - 0.183
		100	15.99, 1.04	1.129
	10^7	0.5	13.22, 3.90	0.002-0.01
		1	8.12, 2.62	0.009-0.016
		10	8.37, 2.35	0.0215-0.112
		100	30.53, 1.10	1.287

REFERENCES

Benenati, R. F., and Brosilow, C. B., 1962, "Void Fraction Distribution in Packed Beds," AIChE Journal, Vol. 8, pp. 359-361.

Caltagirone, J. P., Cloupeau, M. and Combarnous, M. A., 1971, "Convection Naturelle Fluctuante dans une Couche Poreuse Horizontale," C. R. Academy of Sciences, B, Vol. 273, pp. 833-836.

Caltagirone, J. P., 1975, "Thermoconvective Instabilities in a Horizontal Porous Layer," Journal of Fluid Mechanics, Vol. 72, pp. 269-287.

Catton, I., 1985, "Natural Convection in Porous Media," Natural Convection Fundamentals and Applications, Eds. S. Kakac, W. Aung, R. Viskanta, pp. 514-547.

Cheng, P., 1978, "Heat Transfer in Geothermal Systems," Advances in Heat Transfer, Vol. 14, pp. 1-105.

Cheng, P., 1987, "Wall Effects on Fluid Flow and Heat Transfer in Porous Media," Proceedings ASME/JSME Joint Thermal Engineering Conference, pp. 297-303.

Combarnous, M. and Bories, S., 1975, "Hydrothermal Convection in Saturated Porous Media," Advances in Hydroscience, Vol.10, pp. 231-307.

David, E., Lauriat, G. and Cheng, P., 1988, "Natural Convection in Rectangular Cavities with Variable Porosity Media," ASME HTD-Vol. 96, pp. 605-612.

Georgiadis, J. G. and Catton, I., 1986, "Prandtl Number Effect on Benard Convection in Porous Media," Journal of Heat Transfer, Vol. 108, pp. 284-290.

Georgiadis, J. G. and Catton, I., 1988, "Dispersion in Cellular Thermal Convection in Porous Layers," International Journal of Heat and Mass Transfer, Vol. 31, pp. 1081-1091.

Gollub, J. P. and Benson, S. V., 1980, "Many Routes to Turbulent Convection," Journal of Fluid Mechanics, Vol. 100, pp. 449-470.

Hong, J. T., Yamada, Y. and Tien C. L., 1987, "Effects of Non-Darcian and Nonuniform Porosity on Vertical-Plate Natural Convection in Porous Media," Journal of Heat Transfer, Vol. 109, pp. 356-362.

Horne, R. N. and O'Sullivan, M. J., 1974, "Oscillatory Convection in a Porous Medium Heated from Below," Journal of Fluid Mechanics, Vol. 66, pp. 339-352.

Horne, R. N., and O'Sullivan, M. J., 1978, "Origin of Oscillatory Convection in a Porous Medium Heated from Below," Physics of Fluids, Vol. 21, pp. 1260-1264.

Jonsson, T. and Catton, I., 1987, "Prandtl Number Dependence of Natural Convection in Porous Media," Journal of Heat Transfer, Vol. 109, pp. 371-377.

Kimura, S., Schubert, G. and Straus, J. M., 1986, "Route to Chaos in Porous-Medium Thermal Convection," Journal of Fluid Mechanics, Vol.166, pp. 305-324.

Kladias, N. and Prasad, V., 1987, "Numerical Study for Inertia and Viscous Diffusion effects on Benard Convection in Porous Media," Proc. Fifth International Conference on Numerical Methods for Thermal Problems, Vol. (1). pp. 797-810.

Kladias, N. and Prasad, V., 1988a, "Natural Convection in Horizontal Porous Layers: Effects of Darcy and Prandtl Numbers," ASME HTD-Vol. 96, pp. 593-604; Journal of Heat Transfer (in Press).

Kladias, N. and Prasad, V., 1988b "Non-Darcy Oscillating Convection in Horizontal Porous Layers Heated from Below," Proceedings, 1st National Fluid Dynamics Congress, Part 3, pp. 1757-1764.

Kladias, N., 1988, "Non-Darcy Free Convection in Horizontal Porous Layers," Ph.D. Dissertation, Columbia University.

Muller, V., 1982, "Benard Convection in Gaps and Cavities," Convective Transport and Instability Phenomena, Eds. J. Zievep, H. Oerted, Jr., pp. 71-100.

Nishimura, T., Kawamura, Y., Takumi, T. and Ozoe, H., 1984, "Analysis of Natural Convection Heat Transfer at the Walls of Packed Beds with Voidage Variations," Kagaku Kogaku Ronbushu, Vol. 10, pp. 648-652.

Roblee, L. H.S. et al., 1958, "Radial Porosity Variation Packed Beds," AIChE Journal, Vol. 4, pp. 450-464.

Rudraiah, N., 1984, "Non-Linear Convection in a Porous Medium with Convective Acceleration and Viscous Force," Arabian Journal of Science and Engineering., Vol. 9, pp. 153-167.

Schubert, G. and Straus, J. M., 1979, "Three-dimensional and Multicellular Steady and Unsteady Convection in Fluid-Saturated Porous Media at High Rayleigh Numbers," Journal of Fluid Mechanics, Vol. 94, pp. 25-38.

Vafai, K., 1984, "Convective Flow and Heat Transfer in Variable-Porosity Media," Journal of Fluid Mechanics, Vol. 147, pp. 233-259.

Vortmeyer, D. and Schuster, J., 1983, "Evaluation of Steady Flow Profiles in Rectangular and Circular Packed Beds by a Variational Method," Chemical Engineering Science, Vol. 38, pp. 1691-1699.

1989 National Heat Transfer Conference
HTD-Vol. 107, Heat Transfer in Convective Flows

THE STABILITY ANALYSIS OF CLOSED-LOOP
THERMOSYPHON SYSTEM

C. C. Hwang, S. H. Yin,
J. T. Teng, and M. J. Tsai
Department of Mechanical Engineering
Chung Yuan Christian University
Chung Li, Taiwan

ABSTRACT

The onset and oscillatory instabilities of natural convection flows in a single-phase rectangular closed-loop thermosyphon are investigated using the one-dimensional model. In this study the axial conduction term in the energy equation is included and it is found that the effect of axial conduction stabilizes the thermosyphon systems. The critical Grashof number of onset and oscillatory instabilities are found to increase with increasing the values of shape ratio. The range of the Grashof numbers evaluated from this study can provide essential information on the various parameters for the design of a thermosyphon system.

NOMENCLATURE

A	aspect ratio, A = W/H
a,b	coefficient and exponent in the expression for the friction coefficient, equation (7)
C	specific heat
C_i	variables determined in Appendix 1
F	friction parameter, equation (8f)
f	friction coefficient
Gr	Grashof number, equation (8e)
Grc	critical onset Grashof number
Groc	critical oscillatory Grashof number
Groc'	critical oscillatory Grashof number with axial conduction effect neglected, Fig. 3
g	gravitational acceleration
H	twice the height of thermosyphon, Fig. 1
h	convective heat transfer coefficient
K	complex time factor
m,n	coefficient and exponent in the expression for heat transfer coefficient, equation (6)
p	pressure
Pem	modified Peclet number
Re	Reynolds number, vr/ν
r	radius of flow channel Fig. 1
SR	shape ratio, equation (8j)
S^*	coordinate along the loop
T^*	temperature
Tw^*	temperature at the wall
t^*	time
V^*	mean fluid velocity in the loop

V_{ref}	reference velocity, equation (8c)
W	twice the width of thermosyphon, Fig. 1
ω	angular frequency
β	thermal expansion coefficient
θ	dimensionless temperature, equation (8b)
θ_w	dimensionless temperature at the wall
k	thermal conductivity
ν	kinematic viscosity
τ_w	wall shear stress
ϕ	angle between tube and the vertical

Subscripts

c	heat sink conditions
h	heat source conditions

Superscripts

$^-$	steady state
$'$	perturbed quantities
$\sim$	spatial variation of perturbed quantities

INTRODUCTION

The problem of heat transfer and stability for flow in natural convection loops (thermosyphons) has stimulated considerable interest in the past due to its important applications in oceanic and geophysical systems as well as in other practical engineering systems such as those for nuclear reactors and solar heaters.

Keller (1966) and Welander (1967) considered a simple rectangular thermosyphon consisting of two insulated vertical branches with point heat source and sink at the center of the lower and upper horizontal segments, respectively. It was point out that when a certain parameter which physically represents the ratio of the buoyancy force to the friction force exceeds a critical value, an oscillatory motion in the fluid will occur. Dumerell and Schoenhals (1975) studied the oscillatory instability of toriodal loops. The work aforementioned did not include the axial conduction term. Zvirin (1985) used a simple geometric configuration to study the onset motion. It is shown that there exits a critical modified Rayleigh number, Rac, below which the rest state is stable and any flow perturbation will decay, while for

Ra > Rac the rest state of the loop is always unstable. Chen (1983) also examined thoroughly the influence of the loop configuration on the instability. However, the axial conduction term was not included in his analysis.

The present work deals with the onset motion, steady-state analysis and the oscillatory instability of the flow in a rectangular loop accounting for the axial conduction effects. Critical conditions are obtained for different aspect ratio, A, shape ratio, SR, and the friction parameter, F, of the system. The result from this study can provide vital information about choosing suitable fluid media and geometrical configurations under the circumstances of fixed temperature difference for the design of a thermosyphon system.

ANALYSIS

The closed loop thermosyphon to be analyzed is shown in Fig. 1. This thermosyphon is made of two vertically insulated pipes of equal diameter and joined together at the top and bottom by two horizontal pipes. Heat is received from a constant-temperature, Th, heat source over the bottom horizontal pipe and rejected to a constant-temperature, Tc, heat sink over the top horizontal pipe.

1. Derivation of Governing Equations:

The working fluid completely fills up the system and is treated as single-phase flow. Fluid properties are considered constant except for the effect of density variations in producing buoyancy. Thus the Boussinesq assumption is adopted with $\rho = \rho_0[1-\beta(T-T_0)]$, where ρ_0 is the density at the reference temperature T_0.

For simplicity, a one-dimensional analysis is performed; and the continuity equation and the equation of motion for the flow in the loop can be written as

$$\frac{\partial V^*}{\partial S^*} = 0 \quad , \quad \text{and} \tag{1}$$

$$\rho_0 \, \pi r^2 \oint \frac{\partial V^*}{\partial t^*} dS^* = - \pi r^2 \oint dp -$$

$$g \, \pi r^2 \oint \rho \cos\phi \, dS^* - 2\pi r \oint \mathcal{T}_\omega \, dS^* \quad , \tag{2}$$

where S^* is the arc length measured from the beginning of the heated section in a counter-closewise manner, and ϕ is the angle between the tube and the vertical. The mean velocity, V^*, in the pipe is only a function of time. The energy equation, with the viscous heating term neglected and the axial conduction effect included can be written as

$$\rho_0 \, C\pi r^2 \left(\frac{\partial T^*}{\partial t^*} + V^* \frac{\partial T^*}{\partial S^*} \right) = 2\pi r \, h \, (T_w^* - T^*) +$$

$$k \, \pi \, r^2 \frac{d^2 T^*}{dS^{*2}} \quad . \tag{3}$$

Zvirin (1985) mentioned that the heat conduction term was neglected in most analyses of thermosyphons, since this term is generally small compared to the convection term. But in the study of the problem of the onset of fluid motion, the heat conduction term must be included.

Because the energy equation is a second order one, thus two boundary conditions are needed. The temperatures and the axial temperature gradients are continuous at the four corners of the loop and can be expressed as

$$T^*_1 = T^*_2 \, , \, \frac{dT^*_1}{dS^*} = \frac{dT^*_2}{dS^*} \quad \text{at } S^* = W/2 \, ,$$

$$T^*_2 = T^*_3 \, , \, \frac{dT^*_2}{dS^*} = \frac{dT^*_3}{dS^*} \quad \text{at } S^* = (W+H)/2 \, ,$$
$$\tag{4}$$

$$T^*_3 = T^*_4 \, , \, \frac{dT^*_3}{dS^*} = \frac{dT^*_4}{dS^*} \quad \text{at } S^* = W + H/2 \, , \text{ and}$$

$$T^*_4 = T^*_1 \, , \, \frac{dT^*_4}{dS^*} = \frac{dT^*_1}{dS^*} \quad \text{at } S^* = W + H \, .$$

It should be noted that T^*_i represents the temperature at the ith-section. The section numbers are shown in Fig. 1.

The wall shear stress, $\mathcal{T}_\omega$, and the convective heat transfer coefficient, h, are related to the mean velocity by the following relationships:

$$\mathcal{T}_\omega = \frac{1}{8} \, \rho_0 \, V^{*2} \, f \quad , \tag{5}$$

and

$$h = m \, \frac{k}{2r} \, (2 \, Re)^n \quad , \tag{6}$$

where f is the friction factor and usually can be correlated by

$$f = \frac{a}{Re^b} \quad , \tag{7}$$

Re in equations (6) and (7) is the Reynolds number based on the pipe radius and a, b, m and n are the empirical constants. These constants can be obtained experimentally for the specific thermosyphon loop under consideration. However, for this study, thease constants were obtained from those for the straight tubes.

Substituting equations (5), (6) and (7) into equations (2) and (3) and introducing the following dimensionless parameters:

$$S = \frac{S^*}{H} \quad , \tag{8a}$$

$$\theta = \frac{T^* - \left(\dfrac{T^*h + T^*c}{2} \right)}{T^*h - T^*c} \quad , \tag{8b}$$

$$V = \frac{V^*}{V_{ref}} , \qquad (8c)$$

where $V_{ref} = (\frac{m \, k}{\nu \rho_o C})^{1/(1-n)} (\frac{\nu}{2H})$, and

$$t = \frac{t^* \, V_{ref}}{H} , \qquad (8d)$$

$$Grm = \frac{4g\beta(\, T^*h - T^*c)H^{1+b} \, V_{ref}^{b-2}}{a \, \nu^b} , \qquad (8e)$$

$$F = \frac{a \, \nu^b}{V_{ref}^b \, H^b} , \qquad (8f)$$

$$K = \frac{K^* \, H}{V_{ref}} , \qquad (8g)$$

$$A = \frac{W}{H} , \qquad (8h)$$

$$Pem = \frac{H \, V_{ref}}{\alpha} , \quad \text{and} \qquad (8i)$$

$$SR = \frac{H}{r} . \qquad (8j)$$

Equations 1 through 4 will now take the following forms:

$$\frac{4(1+A)}{F \, Gr} \frac{\partial V}{\partial t} \qquad (9)$$

$$= (\int_{0.5A}^{0.5(1+A)} \theta \, dS - \int_{0.5+A}^{1+A} \theta \, dS) - \frac{(1+A)}{Gr} V^{(2-b)} SR^{(b+1)}$$

$$\frac{\partial \theta}{\partial t} + V \frac{\partial \theta}{\partial S} = 2 \, V^n \, SR^{(2-n)} \, (\theta_w - \theta) + \frac{1}{Pem} \frac{\partial^2 \theta}{\partial S^2}$$

for $0 < S < 0.5A$, $0.5(1+A) < S < 0.5 + A$.

$$(10)$$

$$\frac{\partial \theta}{\partial t} + V \frac{\partial \theta}{\partial S} = \frac{1}{Pem} \frac{\partial^2 \theta}{\partial S^2}$$

for $0.5A < S < 0.5(1+A)$, $0.5 + A < S < 1 + A$.

$$\theta_1 = \theta_2 , \quad \frac{d\theta_1}{dS} = \frac{d\theta_2}{dS} \quad \text{at } S = 0.5A , \qquad (11)$$

$$\theta_2 = \theta_3 , \quad \frac{d\theta_2}{dS} = \frac{d\theta_3}{dS} \quad \text{at } S = 0.5 + 0.5A , \qquad (12)$$

$$\theta_3 = \theta_4 , \quad \frac{d\theta_3}{dS} = \frac{d\theta_4}{dS} \quad \text{at } S = 0.5 + A , \qquad (13)$$

$$\theta_4 = \theta_1 , \quad \frac{d\theta_4}{dS} = \frac{d\theta_1}{dS} \quad \text{at } S = 1 + A . \qquad (14)$$

In this paper, the onset of motion is studied first, then the steady state behavior and the stability of the steady flow are studied next.

2. Evaluation of Flow Stability Associated with the Onset of Motion from Rest

For this rest state, we can introduce $\overline{V}=0$ into equations (9) and (10) and determine the steady-state temperature distribution first with the transient effect neglected. The following solution can be obtained.

$$\overline{\theta_1} = C_1 \exp(r_1 S) + C_2 \exp(- r_1 S) + 0.5 \qquad (15)$$

$$\overline{\theta_2} = C_3 (S-0.5A) + C_4 \qquad (16)$$

$$\overline{\theta_3} = - C_1 \exp[r_1 (S-(1+A)/2)] - C_2 \exp[- r_1 (S-(1+A)/2)] - 0.5 \qquad (17)$$

$$\overline{\theta_4} = -C_3 [S-(0.5+A)] - C_4 \qquad (18)$$

where $r_1 = (2 \, Pem \, SR^2)^{1/2}$, C_i's (see Appendix 1) can be obtained from the boundary conditions.

Linearized stability analysis is used to analyze the onset problem. Then the small perturbation of the rest state is now introduced which takes the following form.

$$V(t) = \widetilde{V} \exp(Kt) \qquad (19)$$

$$\theta(S, t) = \overline{\theta}(S) + \widetilde{\theta} \exp(Kt) \qquad (20)$$

Where $\widetilde{V}$ and $\widetilde{\theta}$ are the amplitudes of the velocity and temperature disturbances, and K is the growth rate. The system will be unstable if K has a positive real part.

Equations (19) and (20) are substituted into equations (9) and (10) (where $\overline{V}=0$), with the rest state deleted and the nonlinear term neglected. The perturbation equations and boundary conditions can now be written as:

383

$$\frac{4(1+A)}{F \ Gr} \ K \ \tilde{V} = (\int_{0.5A}^{0.5(1+A)} \tilde{\theta} \ dS - \int_{0.5+A}^{1+A} \tilde{\theta} \ dS) \tag{21}$$

$$- \frac{(1+A)}{Gr} \ \tilde{V} \ SR^2$$

$$K \ \tilde{\theta} + \tilde{V} \ \frac{\partial \bar{\theta}}{\partial S} = -2 \ SR^2 \ \tilde{\theta} + \frac{1}{Pem} \ \frac{\partial^2 \tilde{\theta}}{\partial S^2}$$

for $0 < S < 0.5A$, $0.5(1+A) < S < 0.5 + A$. (22)

$$K \ \tilde{\theta} + \tilde{V} \ \frac{\partial \bar{\theta}}{\partial S} = \frac{1}{Pem} \ \frac{\partial^2 \tilde{\theta}}{\partial S^2}$$

for $0.5A < S < 0.5(1+A)$, $0.5 + A < S < 1 + A$.

$$\tilde{\theta}_1 = \tilde{\theta}_2 , \quad \frac{d\tilde{\theta}_1}{dS} = \frac{d\tilde{\theta}_2}{dS} \quad \text{at } S = 0.5A , \tag{23}$$

$$\tilde{\theta}_2 = \tilde{\theta}_3 , \quad \frac{d\tilde{\theta}_2}{dS} = \frac{d\tilde{\theta}_3}{dS} \quad \text{at } S = 0.5(1+A) , \tag{24}$$

$$\tilde{\theta}_3 = \tilde{\theta}_4 , \quad \frac{d\tilde{\theta}_3}{dS} = \frac{d\tilde{\theta}_4}{dS} \quad \text{at } S = 0.5 + A , \tag{25}$$

$$\tilde{\theta}_4 = \tilde{\theta}_1 , \quad \frac{d\tilde{\theta}_4}{dS} = \frac{d\tilde{\theta}_1}{dS} \quad \text{at } S = 1 + A . \tag{26}$$

The perturbation temperature distribution can be determined first and the resulting $\tilde{\theta}$ is substituted into the buoyancy integral in equation (21), then the characteristic equation for the K can be expressed implicitly as

$$Gr = \frac{\frac{4(1+A)}{F} K + (1+A) SR^2}{2\{\frac{C_7}{r_3} [\exp(r_3/2)-1] - \frac{C_8}{r_3} [\exp(-r_3/2)-1] - \frac{C_3}{2K}\}} , \tag{27}$$

where $r_3 = (k \ Pem)^{1/2}$, Ci's (see Appendix 1) can be obtained from the boundary conditions.

Zvirin (1985) mentioned that the instability associated with the initiation of motion from rest is monotonic. Expanding equation (27) for small real values of K and taking the limit as $K \rightarrow 0$, the critical onset Grashof number can be evaluated.

When appropriate working fluid is chosen, then the parameter F will be determined. Here the parameter F can be treated as a constant. The critical onset Grashof number, Grc, is a function of shape ratio, SR, but not a function of F as discussed in Zvirin (1985). It should be noted here that in the present work Grc is also the function of aspect ratio, A.

3. Determination of Steady-State Solution

To analyze the oscillatory instability, we must determine the steady-state solution. All time derivatives in equations (9) and (10) vanish. We first determine the temperature distribution from the energy equation, assuming a constant value of velocity $\bar{V}$ which is to be determined. The solutions are

$$\bar{\theta}_1 = C_9 \exp(r_4 S) + C_{10} \exp(r_5 S) + 0.5$$

$$\text{at} \quad 0 < S < 0.5A , \tag{28}$$

$$\bar{\theta}_2 = C_{11} \exp[r_6 (S-0.5A)] + C_{12}$$

$$\text{at} \quad 0.5A < S < 0.5(1+A) , \tag{29}$$

$$\bar{\theta}_3 = - C_9 \exp\{r_4 [S-0.5(1+A)]\} - C_{10} \exp\{r_5 [S-0.5(1+A)]\} - 0.5$$

$$\text{at} \quad 0.5(1+A) < S < 0.5 + A , \text{ and} \tag{30}$$

$$\bar{\theta}_4 = - C_{11} \exp[r_6 (S-(0.5+A))] - C_{12}$$

$$\text{at} \quad 0.5A < S < 0.5(1+A) , \tag{31}$$

where

$$r_4 = \frac{Pem + [(Pem \ \bar{V})^2 + 8 SR^2 Pem]^{1/2}}{2} ,$$

$$r_5 = \frac{Pem - [(Pem \ \bar{V})^2 + 8 SR^2 Pem]^{1/2}}{2} ,$$

$$r_6 = Pem \ \bar{V} , \quad \text{and}$$

Ci's (see Appendix 1) can be obtained from the boundary conditions.

Substituting equations (29) and (31) into the momentum equation (9), finally we obtain the result of $\bar{V}$ in the following algebraic form.

$$\frac{(1+A) \ \bar{V} \ SR^2}{Gr} = 2 \{\frac{1}{r_6} C_{11} [\exp(r_6/2)-1] + \frac{C_{12}}{2} \} \tag{32}$$

In general, this equation must be solved numerically to obtain $\bar{v}$ which is a function of A, Gr and SR. The solution procedure is quite simple, because equation (32) can be rearranged to find Gr (as a function of $\bar{V}$, A, SR) by straightforward computation. $\bar{V}$ obtained from equation (32) can be introduced into equations (28) through (31) in order to find the temperature distributions.

It is found that $-\bar{V}$ (here, the minus sign represents the clockwise direction) is also the solution of equation (32). On the other hand, Torrance (1980) mentioned that a pair of symmetry may be taking place (i.e., the flow may be moving in either the counter clockwise or clockwise direction), but only the state with undirectional flow through the loop was observed in the experiments.

4. Analysis of Stability of the Steady-State Solution

Using the same procedure as the previous analysis about the onset motion, the time-dependent velocity and temperature can be written as

$$V(t) = \bar{V} + V'(t) = \bar{V} + \tilde{V} \exp(Kt) \qquad (33)$$

$$\theta(S, t) = \bar{\theta}(S) + \tilde{\theta} \exp(Kt) \qquad (34)$$

Substituting these two equations into equations (9) and (10), we obtain the small perturbation equation with a growth rate of K as the following:

$$\frac{4(1+A)}{F\ Gr} K\ \tilde{V} = \left(\int_{0.5A}^{0.5(1+A)} \tilde{\theta}\ dS - \int_{0.5+A}^{1+A} \tilde{\theta}\ dS \right)$$
$$- \frac{(1+A)}{Gr} \tilde{V}\ SR^2 \qquad (35)$$

$$K\ \tilde{\theta} + \bar{V}\ \frac{\partial \tilde{\theta}}{\partial S} = -\tilde{V}\ \frac{\partial \bar{\theta}}{\partial S} - 2\ SR^2\ \tilde{\theta} + \frac{1}{Pem}\ \frac{\partial^2 \tilde{\theta}}{\partial S^2}$$

for $0 < S < A/2$, $0.5(1+A) < S < 0.5 + A$, and $\quad (36a)$

$$K\ \tilde{\theta} + \bar{V}\ \frac{\partial \tilde{\theta}}{\partial S} = -\tilde{V}\ \frac{\partial \bar{\theta}}{\partial S} + \frac{1}{Pem}\ \frac{\partial^2 \tilde{\theta}}{\partial S^2}$$

for $0.5A < S < 0.5(1+A)$, $0.5 + A < S < 1 + A$. $\quad (36b)$

With $\bar{\theta}$ and $\bar{V}$ determined from steady-state analysis, the perturbed energy equation can be solved analytically. The results are:

$$\tilde{\theta}_1 = C_{13}\exp(r_7 S) + C_{14}\exp(r_8 S) + R_2\exp(r_4 S) \qquad (37)$$
$$+ R_3\exp(r_5 S)$$

$$\tilde{\theta}_2 = C_{15}\exp[r_9(S-A/2)] + C_{16}\exp[r_{10}(S-A/2)] \qquad (38)$$
$$+ R_4\exp[r_6(S-A/2)]$$

$$\tilde{\theta}_3 = -C_{13}\exp[r_7(S-(1+A)/2)] - C_{14}\exp[r_8(S-(1+A)/2)]$$
$$- R_2\exp[r_4(S-(1+A)/2)] - R_3\exp[r_5(S-(1+A)/2)]$$
$$\qquad (39)$$

$$\tilde{\theta}_4 = -C_{15}\exp[r_9(S-(1/2+A))] - C_{16}\exp[r_{10}(S-(1/2+A))]$$
$$- R_4\exp[r_6(S-(1/2+A))] \qquad (40)$$

where

$$r_7 = \frac{Pem\ \bar{V} + [(Pem\ \bar{V})^2 + 4 Pem(2\ SR^2+K)]^{1/2}}{2} ,$$

$$r_8 = \frac{Pem\ \bar{V} - [(Pem\ \bar{V})^2 + 4\ Pem(2\ SR^2+K)]^{1/2}}{2} ,$$

$$r_9 = \frac{Pem\ \bar{V} + [(Pem\ \bar{V})^2 + 4\ Pem\ K]^{1/2}}{2} ,$$

$$r_{10} = \frac{Pem\ \bar{V} - [(Pem\ \bar{V})^2 + 4\ Pem\ K]^{1/2}}{2} ,$$

$$R_2 = \frac{(Pem\ r_4\ C_5)\ \tilde{V}}{r_4^2 - Pem\ \bar{V}\ r_4 - Pem(2\ SR^2+K)} ,$$

$$R_3 = \frac{(Pem\ r_5\ C_6)\ \tilde{V}}{r_5^2 - Pem\ \bar{V}\ r_5 - Pem(2\ SR^2+K)} , \text{ and}$$

$$R_4 = \frac{Pem\ r_6\ C_7\ \tilde{V}}{r_6^2 - Pem\ \bar{V}\ r_6 - Pem\ K} .$$

Substituting the calculated $\tilde{\theta}$ into the equation (35) then the characteristic equation for the growth rate, K, can be obtained:

$$\frac{4(1+A)}{F\ Gr} K = 2\ \{\frac{C_{15}}{r_9}[\exp(r_9/2)-1] - \frac{C_{16}}{r_{10}}[\exp(r_{10}/2)-1]$$

$$\qquad\qquad (41)$$

$$\frac{R_4}{r_6}[\exp(r_6/2)-1] - \frac{C_3}{2K}\} - \frac{(1+A)}{Gr} SR^2 .$$

Equation (41) is rearranged in the form of $F = F(K, A, SR, Gr)$, where $K = i\omega$ and F is a complex function of ω. At a point of marginal stability for parameters A, SR and Gr, $Im[F(\omega)] = 0$ because the parameter F is real. Thus solving this equation algebraically, we obtain the frequency ω for the critical disturbance. The value of ω is then used to calculate $K = Re[K(\omega)]$, giving a point on the marginal stability line for the parameters A, SR and Gr specified. Finally the critical oscillatory Grashof number, Groc can be determined by choosing appropriate parameters A, SR and F. When the magnitude of the Grashof number is greater than the value of Groc, the thermosyphon system will be unstable.

RESULT AND DISCUSSION

The height of the loop and the temperatures for the heat source and sink of the loop in practical thermosyphon operations are usually fixed. The radii of pipes and the lengths of the cooled and the heated sections are the variables to be specified in the design of the thermosyphon systems. When an aspect ratio is chosen, a corresponding value of the critical onset Grashof number, Grc, can be obtained from Fig. 2 and it is found that Grc is increased with an increased value of the shape ratio SR. The present work indicates that the critical onset Grashof number is on the order of $10^3 \sim 10^4$, depending on the aspect ratio. If $Gr < Grc$, then the fluid in the loop system is stationary and no convective heat transfer can be achieved by the system, and if $Gr > Grc$, then the fluid will be in motion and convective heat transfer will be steadily developed in the loop.

The results are shown in Fig. 3 for the two cases. As can be seen, the effect of axial conduction is prominent at the onset and oscillatory critical conditions. The two curves will overlap when Gr reaches some specific value. However the predicted of oscillatory critical Grashof number for the case with the effect of axial conduction neglected is less than (about 9%) that predicted by the current study. When the Grashof number is greater than the critical oscillatory Grashof number, the system will be unstable. Under such circumstances, a more complicated nonlinear model will be needed to trace them.

Figures 4 and 5 show the results for the steady velocity, $\bar{V}$, as a function of Gr, with the shape ratio, SR, and the aspect ratio, A, as the parameters. It is found that $\bar{V}$ increases with an increased value of the Grashof number. It should be noted that all the flow solutions are pairs of $\bar{V}$ and $-\bar{V}$. With a fixed aspect ratio, if SR decreases, the difference between the highest and the lowest temperatures around the loop becomes less pronounced as shown in Fig. 6. This is due to an increase in the axial conduction effect, and the associated decrease in the driving force resulting from a decrease in the SR value. On the contrary, the friction effect will decrease. Thus the value of the velocity depends on the Grashof number. While for a fixed SR, large aspect ratio may not correspond to a small velocity and the results also depend on the Grashof number as shown in Fig. 5.

The oscillatory stability of the steady flow has been determined by the Grashof number and parameter F. The neutral curve is shown in Figures 7 and 8 for different aspect ratio and shape ratio. For a fixed F, there exists a Groc value. Steady flow will become unstable only when Gr > Groc. It is shown in Fig. 7 that for a specified SR value the flow regime is stable below the curve and unstable if the flow regime is above the curve.

Fig. 9 shows that for each shape ratio, SR, there exists a minimum critical Grashof number of oscillatory instability. It is noted that at a fixed A value, Groc increases with an increased SR value.

CONCLUSION

A simplified one-dimensional model is used to study the onset motion and instability of the steady flow. It is found that for different aspect ratio, A, shape ratio, SR, and friction parameter, F, there exist two values Grc and Groc, where the value of Groc is always greater than the value of Grc. If Gr < Grc the fluid in the loop system is stationary and no steady flow will develop, while for Gr > Groc the steady flow will be oscillatorily unstable. The results show that the effect of axial conduction is strong at critical condition.

For each SR investigated, there exists a minimum critical onset Grashof number, Grc, and critical oscillatory Grashof number, Groc as a function of the aspect ratio. But for a fixed aspect ratio, the values of Grc and Groc are monotonically increasing with SR. The result also shows that the flow is always stable if Gr associated with the flow lies below the the Gr verse F curve.

In the engineering design, the limits of static stability and oscillatory stability of the steady flow must be determined first. Based on the two limits, we choose appropriate geometrical configuration and working fluid to assure that the system is in stable motion. It is hoped that the present study on the onset and oscillatory motion of the working fluid in the thermosyphon loop can be extended in the future to include a study on the efficiency of a thermosyphon system which operates at suitable aspect ratio and shape ratio for a specific fluid under consideration.

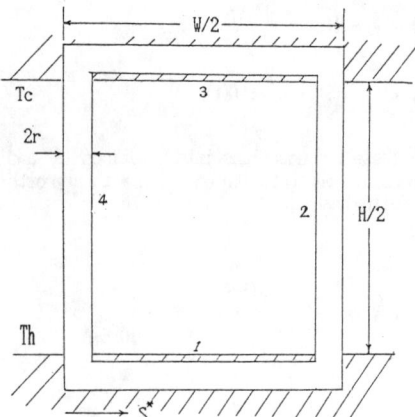

Fig.1 Rectangular natural convection loop

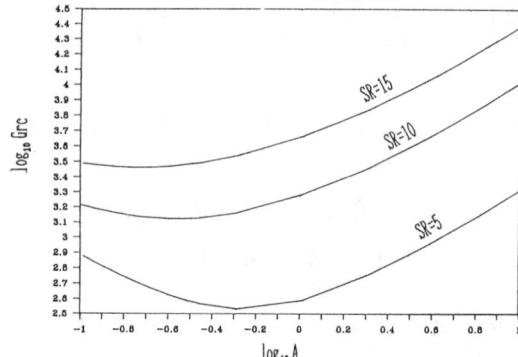

Fig. 2 Relationship of critical onset Grashof number, Grc with aspect ratio, A at different values of shape ratio, SR.

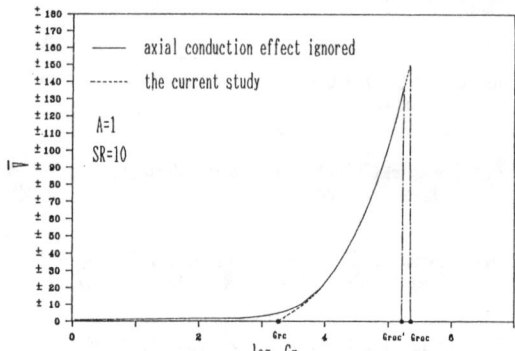

Fig. 3 Relationship of dimensionless velocity, V with Grashof numbers, Gr.

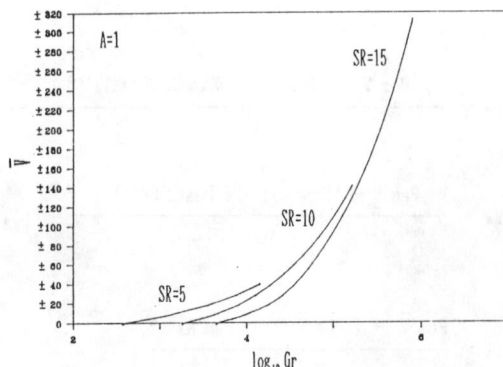

Fig. 4 Relationship of dimensionless velocity, V with Grashof numbers, Gr at different values of shape ratio, SR.

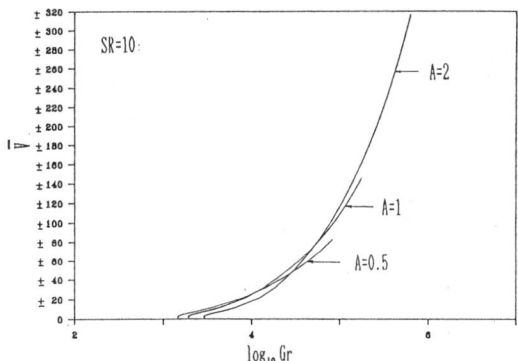

Fig. 5 Relationship of dimensionless velocity, V with Grashof numbers, Gr at different values of aspect ratio, A.

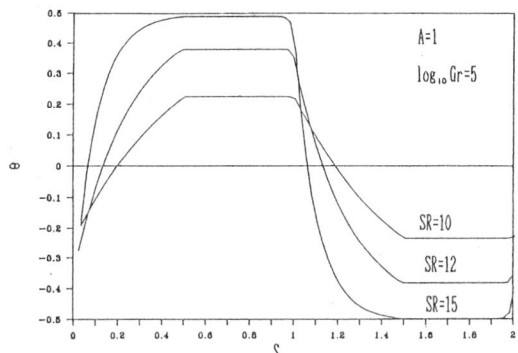

Fig. 6 Axial temperature distribution for different values of shape ratio, SR.

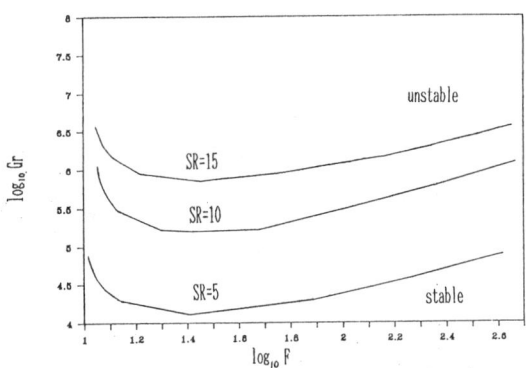

Fig. 7 Neutral stability curves for different values of shape ratio, SR.

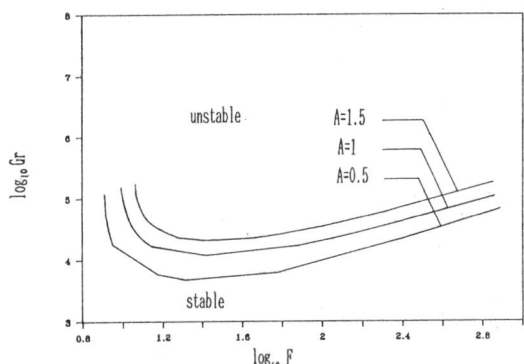

Fig. 8 Neutral stability curves for different values of aspect ratio, A.

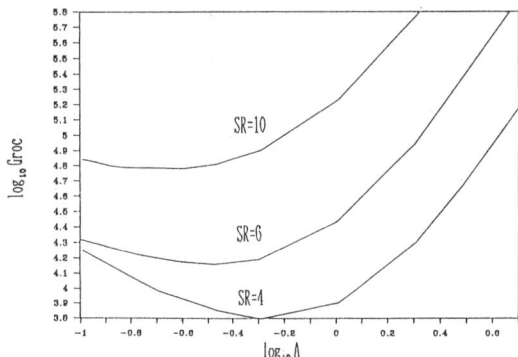

Fig. 9 Relationship of critical oscillatory Grashof number, Groc with aspect ratio, Λ at different values of shape ratio, SR.

ACKNOWLEDGMENT

The authors wish to acknowledge with appreciation the financial support(Grant No. NSC78-0401-E033-01)provided by the National Science Council of the Republic of China.

REFERENCES

Bau, H. H., and Torrance, K. E., 1980, "Transient and Steady Behavior Of an Open, Symmetrically-Heated, Free Convection Loop," Int.J. Heat & Mass Transfer, Vol. 24, No. 4, pp. 597-609.

Chen, K., 1982, "The Influence of Loop Configuration on Closed-loop Thermosyphons," ASME Paper No 82-WA/HT-63.

Chen, k., 1983, "On The Oscillatory Instability Of Closed-Loop Thermosyphons," ASME Paper No 83-WA/HT-95.

Creveling, H. F., Depaz, J. F., Baladi, J. Y., Schoenhals, R.J., 1975, "Stability Characteristics of a Single-Phase Free Convection Loop," J. Fluid Mech, Vol. 67, part 1, pp. 65-84.

Japikes, D., 1972, "Advances in Thermosyphon Technology," in Advances in Heat Transfer, edited by Irvine, T. F., Jr and Hartnett, J. P., Vol, 9, pp. 1-111, Academic Press, New York.

Keller, J. B., 1966, "Periodic Oscillations in a Model of Thermal Convection" J. Fluid Mech., Vol. 26, part 3, pp. 599-606.

Kays, W. M., 1976, "Convective Heat and Mass Transfer," 1st ed., pp. 110 and 173. McGraw-Hill Book Co. ,New York.

Mertol, A., R. Greif, 1986, "Natural Circulation Loops: Background and Selected Current Studies and Application," Heat Transfer in Thermal Systems Seminar-Phase II, National Cheng Kung University, Tainan, January 13-14.

Welander, P., 1967 "On The Oscillatory Instability of Differentially Heated Fluid Loop," J. Fluid Mech., Vol. 29, Part 1, pp. 17-30.

Zvirin, Y., 1985, "The Instability Associated With The Onset of Motion in a Thermosyphon," Int. J. Heat & Mass Transfer , Vol. 28, No. 11, pp. 2105-2111.

Zvirin, Y., 1986, "Instability in a Double-Diffusive Thermosyphon" Int. J. Heat & Mass Transfer, Vol. 30, No. 7, pp. 1319-1329.

APPENDIX 1

The C_i's discussed in equations (15), (16), (17), (18), (27), (28), (29), (30), (31), (32), (37), (38), (39), (40) and (41) can be determined from the following matrices:

$$\begin{bmatrix} \exp(\frac{A}{2}r_1) & \exp(\frac{-A}{2}r_1) & 0 & -1 \\ r_1\exp(\frac{A}{2}r_1) & -r_1\exp(\frac{A}{2}r_1) & -1 & 0 \\ 1 & 1 & 1/2 & 1 \\ r_1 & -r_1 & 1 & 0 \end{bmatrix} \begin{bmatrix} C_1 \\ C_2 \\ C_3 \\ C_4 \end{bmatrix} = \begin{bmatrix} -1/2 \\ 0 \\ -1/2 \\ 0 \end{bmatrix}$$

$$\begin{bmatrix} \exp(\frac{A}{2}r_2) & \exp(\frac{-A}{2}r_2) & -1 & -1 \\ r_2\exp(\frac{A}{2}r_2) & r_2\exp(\frac{-A}{2}r_2) & -r_3 & r_3 \\ 1 & 1 & \exp(r_3/2) & \exp(r_3/2) \\ -r_2 & r_2 & -r_3\exp(\frac{A}{2}r_3) & r_3\exp(\frac{-A}{2}r_3) \end{bmatrix} \begin{bmatrix} C_5 \\ C_6 \\ C_7 \\ C_8 \end{bmatrix} = \begin{bmatrix} S_1 \\ S_2 \\ S_3 \\ S_4 \end{bmatrix}$$

$$S_1 = -R_1\exp(\frac{A}{2}r_1) + R_1\exp(\frac{-A}{2}r_1) - C_3/K \quad,$$

$$S_2 = -r_1R_1\exp(\frac{A}{2}r_1) - r_1R_1\exp(\frac{-A}{2}r_1) \quad,$$

$$S_3 = C_3/K \quad,$$

$$S_4 = 2r_1R_1 \quad,$$

$$r_2 = [(K + 2 SR^2)Pem]^{1/2} \quad,$$

$$r_3 = (K Pem)^{1/2} \quad,$$

$$R_1 = \frac{Pem\ C_1\ r_1\ \tilde{V}}{r_1^2 - (K + 2 SR^2)Pem} \quad,$$

$$\begin{bmatrix} \exp(\frac{A}{2}r_4) & \exp(\frac{A}{2}r_5) & -1 & -1 \\ r_4\exp(\frac{A}{2}r_4) & r_5\exp(\frac{A}{2}r_5) & -r_6 & 0 \\ 1 & 1 & \exp(\frac{r_6}{2}) & 1 \\ r_4 & r_5 & r_6\exp(\frac{r_6}{2}) & 0 \end{bmatrix} \begin{bmatrix} C_9 \\ C_{10} \\ C_{11} \\ C_{12} \end{bmatrix} = \begin{bmatrix} -1/2 \\ 0 \\ -1/2 \\ 0 \end{bmatrix}$$

$$\begin{bmatrix} \exp(\frac{A}{2}r_7) & \exp(\frac{A}{2}r_8) & -1 & -1 \\ r_7\exp(\frac{A}{2}r_7) & r_8\exp(\frac{A}{2}r_8) & -r_9 & -r_{10} \\ 1 & 1 & \exp(\frac{A}{2}r_9) & \exp(\frac{A}{2}r_{10}) \\ r_7 & r_8 & r_9\exp(\frac{A}{2}r_9) & r_{10}\exp(\frac{A}{2}r_{10}) \end{bmatrix} \begin{bmatrix} C_{13} \\ C_{14} \\ C_{15} \\ C_{16} \end{bmatrix} = \begin{bmatrix} T_1 \\ T_2 \\ T_3 \\ T_4 \end{bmatrix}$$

$$T_1 = -R_2\exp(\frac{A}{2}r_4) - R_3\exp(\frac{A}{2}r_5) + R_4 \quad,$$

$$T_2 = -r_4R_2\exp(\frac{A}{2}r_4) - r_5R_3\exp(\frac{A}{2}r_5) + r_6 R_4 \quad,$$

$$T_3 = -R_2 - R_3 - R_4\exp(\frac{1}{2}r_6) \quad\text{and}$$

$$T_4 = -r_4R_2 - r_5R_3 - r_6R_4\exp(\frac{1}{2}r_6) \quad.$$

ONSET OF THERMOCAPILLARY CONVECTION
IN HYDROMAGNETIC AND RADIATING FLUIDS

T. T. Lam and S. C. Lee
Spacecraft Thermal Department
The Aerospace Corporation
El Segundo, California

ABSTRACT

The criteria for the onset of thermocapillary convection in a thermally radiating fluid layer under the influence of magnetic field in a microgravity environment are determined. The hydrodynamic boundary conditions for the layer include a heat conducting free upper surface and an isothermal rigid lower surface. The radiative boundaries of black-black, diffuse-diffuse, and black-diffuse are considered. Linear stability analysis is performed on the continuity, momentum, energy, hydromagnetic, Maxwell, and radiative transfer equations. The resulting linearized equations for the thermal instability problem are solved by the sequential gradient-restoration algorithm, a numerical optimization technique. The results are presented in terms of the Hartmann number, the Biot number, the Crispation number, the critical Marangoni number, and the optical thickness for a wide range of thermal and radiative properties. It is shown that the magnetic field, coupled with thermal radiation, suppresses thermocapillary convection considerably in a microgravity environment.

NOMENCLATURE

a	= horizontal wave number
Bi	= Biot number, hd/k
Cr	= Crispation number, $\upsilon_o \alpha / S_o d$
C_v	= specific heat at constant volume
d	= thickness of the fluid layer
D	= differential operator, d/dZ
E_{bo}	= blackbody emissive power of the fluid
f(Z)	= initial steady state temperature distribution
h	= convective heat transfer coefficient
H	= magnetic field intensity vector
H_i, H_j	= magnetic field intensities
H_o, H_z	= magnetic field intensity in the z-direction
I	= radiative intensity
j	= first moment of radiative intensity, $\int_\Omega I d\Omega$
J	= amplitude function of the first moment of radiative intensity, $j/12\sigma T_o^3$
k	= thermal conductivity
Ma	= Marangoni number, $(-\partial S/\partial T)_o (\Delta T d/\mu_o \alpha)$
P	= pressure
Po	= Planck number, $\alpha_M k/4\sigma T_o^3$
Q	= Hartmann number, $\mu_m^2 H_o^2 d^2 \sigma_e / \rho \upsilon_o$
S	= surface tension
t	= time
T	= temperature
$T\infty$	= temperature of the free space
u_i, u_j	= fluid velocity vectors
w	= fluid velocity in the z-direction
W	= amplitude function of the velocity w
x_i, x_j	= spatial coordinates
Z	= dimensionless depth, z/d

Greek Symbols:

α	= thermal diffusivity
α_M	= root mean square absorption coefficient, $\sqrt{(\alpha_P \alpha_R)}$
α_P	= Planck mean absorption coefficient

α_R = Rosseland mean absorption coefficient

β = thermal expansion coefficient

ε = diffuse emissivity of boundary, $1/(0.5 + 1/4\lambda)$

η = degree of nongrayness of fluid, $\sqrt{(\alpha_P/\alpha_R)}$

η_m = electrical resistivity, $1/4\pi\mu_m\sigma_e$

Θ = amplitude function of the disturbance temperature

λ = 0 for diffusely reflecting surface

= 0.5 for black-rigid or -free surfaces

μ = dynamic viscosity

μ_m = magnetic permeability

υ = kinematic viscosity, μ/ρ

ξ = dimensionless interfacial deflection, ξ^*/d

ξ^* = local deflection of the free surface from the mean

ρ = density

σ = Stefan-Boltzmann constant

σ_e = electrical conductivity

τ = optical thickness, $\sigma_M d$

χ = η/Po

Ω = solid angle

Subscripts:

o = reference state or lower boundary

1 = upper boundary

c = critical

i = spatial indices

j = spatial indices

INTRODUCTION

The study of surface-tension gradient flows has received a great deal of interest in recent years because of its technological importance in materials science (Napolitano, 1984; Monti et al., 1988; Saghir, 1988) and in many space experiments involving materials processing (Malmejac, et al., 1981; Napolitano et al., 1984; Regel', 1988). In a reduced-gravity environment, the buoyancy effect diminishes and surface tension becomes a dominant force affecting the transport processes in fluids. Configurations of interest in materials processing often involve interfacial transport investigations. The presence of temperature and electrical potential variations along the fluid interfaces can generate surface-tension gradients. Surface-tension gradients act like shear stresses applied by the interface on the adjoining bulk fluids and induce interfacial tractions, thereby generating convective instabilities and macroscopic flows. It is well known that surface tension, and not buoyancy, is the dominant mechanism in convection processes in many of these space experiments. Surface-tension-driven flows can often influence the material composition and lead to undesirable compositional changes in the end product. Thus,

a better understanding of controlled surface-tension-driven convection is important in order to improve the techniques of materials processing in space.

The onset of buoyancy-driven convection in a thermally radiating fluid layer, but neglecting the magnetic field has received considerable attention in the past. By using a variational method, Goody (1956) studied the thermal instability of a radiative fluid layer which has very large and very small absorption coefficients and bounded by free surfaces. Spiegel (1960) considered the same stability problem for rigid boundaries but neglected the effect of conduction. Following Goody's (1956) approach, Khosla and Murgai (1963) investigated the combined effects of thermal radiation and rotation on the Benard problem. Arpaci and Gozum (1973) studied the thermal instability of fluid layers by using the Eddington approximation for radiation. In a recent investigation, Bayazitoglu and Lam (1987) studied the thermocapillary convection in fluid layers with thermal radiation.

Marangoni convection in a magnetic field, but neglecting thermal radiation has been the subject of theoretical investigation by several authors, such as Nield (1966), Chun and Wuest (1978), Sarma (1978, 1987), Rudraiah et al. (1985) and Maekawa and Tanasawa (1988). Much of these previous works focused only on the effect of a magnetic field on thermocapillary convection. The combined effect of thermal radiation and magnetic field on convection was studied by Murgai and Khosla (1962) for an ionized fluid layer under the influence of gravity. Maekawa and Tanasawa (1988) investigated the onset of thermocapillary convection in a horizontal layer of an electrically conducting liquid with a vertical temperature gradient and a magnetic field, but neglecting radiation. Their results showed that only the vertical component of the magnetic field affected the critical Marangoni number.

In this study, we examine the coupled effects of a vertical magnetic field and thermal radiation on thermocapillary convection in a microgravity environment. A brief description of the physical problem, governing equations, and boundary conditions is first presented. The sequential gradient-restoration algorithm (SGRA) is used to find the eigenvalues for the thermal instability problem. The results are presented in terms of the Hartmann number, the Biot number, the Crispation number, the critical Marangoni number and optical thickness for a wide range of thermal and radiative properties, including the Planck number, nongrayness of the fluid, and the emissivity of the boundaries.

MATHEMATICAL FORMULATION

Consider an electrically conducting and incompressible fluid layer of infinite horizontal extent which is heated from below (Fig. 1). The fluid layer is assumed to be an absorbing, emitting and nongray medium. The origin of the Cartesian coordinate system is affixed on the lower boundary of the fluid layer and the z-axis is directed vertically upwards. A nonlinear temperature gradient exists across the fluid layer due to internal thermal radiation. A uniform vertical magnetic field H_0 is applied. The viscosity, magnetic permeability, and thermal and electrical conductivities are assumed to be constant.

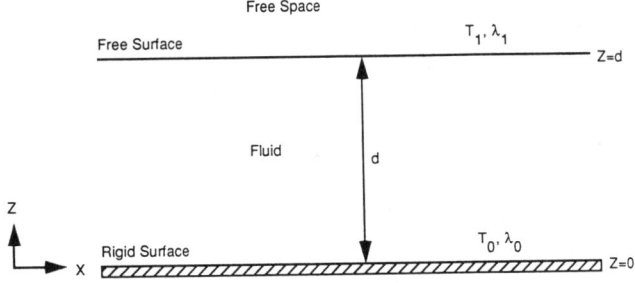

Fig. 1.   Schematic diagram of the physical system.

The fluid layer is confined between the region $0 \le z \le d+\xi^*$, where d is the mean thickness and $\xi^*(x,y,t)$ is the local deflection of the free surface from the mean. The lower boundary is a perfectly heat and electric conducting flat solid surface. The upper surface is adjacent to an electrically non-conducting medium, which exchanges heat with the free space above. The governing equations of fluid motion are:

Continuity

$$\frac{\partial u_i}{\partial x_i} = 0, \qquad (1)$$

Momentum

$$\frac{\partial u_i}{\partial t} + u_j \frac{\partial u_i}{\partial x_j} = -\frac{\partial}{\partial x_i}\left(\frac{P}{\rho_0} + \frac{\mu|H|^2}{8\pi\rho_0}\right) +$$

$$\frac{\mu_m}{4\pi\rho_0}H_j\frac{\partial H_i}{\partial x_j} + \upsilon\nabla^2 u_i, \qquad (2)$$

Energy

$$\frac{\partial T}{\partial t} + u_j\frac{\partial T}{\partial x_j} = \alpha\nabla^2 T + \left(\frac{\alpha_P}{\rho_0 C_v}\right)(j - 4\sigma T^4), \qquad (3)$$

Radiative transfer

$$\nabla^2 j - 3\alpha_P\alpha_R j = -12\alpha_P\alpha_R\sigma T^4, \qquad (4)$$

Hydromagnetic

$$\frac{\partial H_i}{\partial t} + u_j\frac{\partial H_i}{\partial x_j} = H_j\frac{\partial u_i}{\partial x_j} + \eta_m\nabla^2 H_i, \qquad (5)$$

Magnetic field

$$\frac{\partial H_i}{\partial x_i} = 0. \qquad (6)$$

In a microgravity environment, the surface tension is dominant compared to buoyancy. Therefore, the body force term due to gravity effect has been neglected in the formulation. The contributions of viscous dissipation and radiative stress are assumed to be negligible, whereas the Lorentz force resulting from the magnetic field is included in the momentum equation. The thermal energy equation has been modified by relating the blackbody radiation to the temperature field. Thermal radiation inside the fluid is described by the radiative transfer equation with the Eddington (P-1) approximation as developed by Traugott (1963) for an absorbing and emitting medium.

The general equations of motion (1) - (6) are to be solved subject to the appropriate boundary conditions. The relevant velocity, thermal and radiative boundary conditions are as follows.

At the lower rigid surface z=0:
(1)   No slip condition

$$u = v = w = 0, \qquad (7)$$

$$\frac{\partial u}{\partial x} = \frac{\partial v}{\partial y} = \frac{\partial w}{\partial z} = 0. \qquad (8)$$

(2)   Constant temperature

$$T = T_0. \qquad (9)$$

(3)   Perfect electric conductor

$$\frac{\partial H_z}{\partial z} = 0 \qquad (10)$$

(4)   Radiative boundary

$$\frac{\partial j}{\partial z} = 0, \qquad (11a)$$

for a perfectly reflecting diffuse surface or

$$j = 4E_{bo} + \left(\frac{\eta}{3\lambda_0\tau}\right)\frac{\partial j}{\partial z}, \qquad (11b)$$

for a black-rigid surface ($\lambda_0 = 0.5$).

391

At the deformable upper free surface $z=d+\xi^*(x,y,t)$

(1) Kinematic condition

$$w = \frac{\partial \xi^*}{\partial t}. \tag{12}$$

(2) No normal velocity component

$$w = 0. \tag{13}$$

(3) Interfacial heat transfer

$$-k\frac{\partial T}{\partial z} = h(T - T_\infty). \tag{14}$$

(4) Perfect electric insulator

$$H_z = 0. \tag{15}$$

(5) Continuity of tangential stress

$$\mu_0(\frac{\partial w}{\partial x} + \frac{\partial u}{\partial z}) = \frac{\partial S}{\partial x}, \quad \text{in the x-direction;} \tag{16a}$$

$$\mu_0(\frac{\partial w}{\partial y} + \frac{\partial v}{\partial z}) = \frac{\partial S}{\partial y}, \quad \text{in the y-direction.} \tag{16b}$$

(6) Continuity of normal stress

$$-(P-P_0) + 2\mu_0\frac{\partial w}{\partial z} = S_0(\frac{\partial^2 \xi^*}{\partial x^2} + \frac{\partial^2 \xi^*}{\partial y^2}). \tag{17}$$

(7) Radiative boundary

$$\frac{\partial j}{\partial z} = 0, \tag{18a}$$

for a perfectly reflecting diffuse surface or

$$j = 4E_{bo} + (\frac{\eta}{3\lambda_1\tau})\frac{\partial j}{\partial z}, \tag{18b}$$

for a black-rigid surface ($\lambda_1 = 0.5$).

On the upper fluid - free space interface, the differential temperature generates surface tension, and thus induces conventional convection flows or unstable celluar flows. In the present study, the surface tension S is assumed to vary with the temperature according to

$$S = S_0 + (\frac{\partial S}{\partial T})_0(T-T_0). \tag{19}$$

The Linear stability theory, based on Lord Rayleigh's (1916) fundamental study, assumes that the field variables undergo infinitesimal disturbances and investigates the reaction of the system to these small perturbations. In the mathematical analysis, all nonlinear terms in the governing equations are neglected. The governing equations and boundary conditions can be made dimensionless using d, d^2/α, α/d, ΔT, $\rho\upsilon_0/\alpha d^2$, σT_0^3 and H_0 as length, time, velocity, temperature, pressure, radiative intensity and magnetic field scales, respectively. The normal and tangential stress balance conditions at the upper free surface can be expanded about the mean interface (z=d) using the Taylor series. Murgai and Khosla (1962) have shown that the exchange of stabilities is valid for any optical thickness in a system with free boundaries. The exchange of stabilities is assumed to be also applicable to the present configuration. By using the standard procedure of the linear stability theory as outlined by Chandrasekhar (1961), the nondimensional perturbation equations governing the marginal state for stationary instability are

$$(D^2 - a^2)^2 W = QD^2 W, \tag{20}$$

$$(D^2 - a^2 - 4\chi\tau^2)\Theta + 3\chi\tau^2 J = -f(Z)W, \tag{21}$$

$$(D^2 - a^2 - 3\tau^2)J = -4\tau^2\Theta. \tag{22}$$

The boundary conditions for convective instability at Z=0 are

$$W = 0, \tag{23}$$

$$DW = 0, \tag{24}$$

$$\Theta = 0, \tag{25}$$

$$DJ = 0, \tag{26a}$$

for a perfectly reflecting diffuse surface or

$$J - (\eta/3\lambda_0\tau)DJ = 0, \tag{26b}$$

for a black-rigid surface ($\lambda_0 = 0.5$).

and at Z = 1 are

$$W = 0, \tag{27}$$

$$D\Theta + Bi[\Theta - f(1)\xi] = 0, \tag{28}$$

$$(D^2 + a^2)W + a^2Ma[\Theta - f(1)\xi] = 0, \tag{29}$$

$$D^3W - 3a^2DW - a^4\xi/Cr = 0, \tag{30}$$

$$DJ = 0, \tag{31a}$$

for a perfectly reflecting diffuse surface or

$$J + (\eta/3\lambda_1\tau)DJ = 0, \tag{31b}$$

for a black-free surface ($\lambda_1 = 0.5$).

In the above equations, the dimensionless parameters Bi, Cr, Ma, and Q are the Biot number, Crispation number, Marangoni number, and Hartmann number, respectively. The Biot number is the ratio of the convective heat transfer to the internal thermal conduction across the fluid layer. It represents the heat transfer condition at the free surface. The degree of deformability of the upper free surface is characterized by the Crispation number. The Marangoni number, defined as the ratio of the surface tension to heat diffusion and the viscous force, is a measure of both the surface tension gradient along the free surface and the temperature difference across the layer. The critical Marangoni number, Ma_c, denotes the minimum temperature difference at which instability will occur. The Hartmann number denotes the relative effects of the magnetic field intensity and the viscous force.

The boundary conditions require that the the normal components of the velocities given by equations (23) and (27) vanish at both boundaries, Z=0 and 1. Equation (24) represents the no-slip, impermeability conditions at the bounding lower rigid surface. Boundary condition (25) states that the surface at Z=0 is maintained at constant temperature. Equations (26) and (31) account for the effect of radiation. The thermal condition at the interface is described by equation (28), which represents the heat convection to the environment. The surface-tension effects at the free surface, represented by equations (29) and (30), designate the continuity of the tangential and normal stresses at the interface.

In order to solve the system of Eqns. (20) - (31), the initial temperature gradient for the fluid is needed. Based on the linearized radiation energy equation, Arpaci and Gozum (1973) obtained the initial temperature gradient as

$$f(Z) = C_3 + C_4\sinh[\varphi\tau(Z-0.5)] + C_5\cosh[\varphi\tau(Z-0.5)], \qquad (32)$$

where

$$\varphi = \sqrt{(3+4\chi)}, \qquad \chi = \frac{\eta}{Po}, \qquad Po = \frac{\alpha_M k}{4\sigma T_0^3},$$

$\lambda_0, \lambda_1 = 0$,   for perfectly reflecting diffuse surfaces

         = 0.5, for black-rigid or -free surfaces

$$C_0 = 2\sinh(\varphi\tau/2) + (\varphi/\eta)(\lambda_0+\lambda_1)\cosh(\varphi\tau/2),$$

$$C_1 = \{[1 + \lambda_0\lambda_1(\varphi/\eta)^2]\sinh(\varphi\tau) + (\varphi/\eta)(\lambda_0+\lambda_1)\cosh(\varphi\tau)\}/C_0,$$

$$C_2 = C_1 + (8\chi/3\varphi\tau)\sinh(\varphi\tau/2),$$

$$C_3 = C_1/C_2,$$

$$C_4 = (\lambda_0-\lambda_1)(4\chi\varphi/3\eta)\sinh(\varphi\tau/2) / (C_0C_2),$$

$$C_5 = (4\chi/3) / C_2.$$

Equations (20) - (31) are the required perturbation equations governing marginal stability for the problem under consideration. A numerical optimization technique, the sequential gradient-restoration algorithm (SGRA) is used to solved the resulting governing equations (Eqns. 20 - 22) of the convective instability problem. The SGRA has been successfully applied to the study of a convective instability problem by Lam and Bayazitoglu (1986). In the present paper, we employ the same algorithm to study the critical conditions for thermocapillary convection in a hydromagnetic and radiating fluid layer.

RESULTS AND DISCUSSION

The conditions leading to the onset of thermocapillary convection in a horizontal hydromagnetic and radiating fluid layer have been obtained by using the SGRA. The critical Marangoni number which defines the threshold for the onset of convective instability has been determined as a function of the Biot number, the Crispation number, the Hartmann number, and the radiative properties, including the degree of nongrayness of the fluid, the Planck number, and the emissivity of the boundaries.

The accuracy of the SGRA numerical procedure is first verfied by comparing the results from this numerical scheme with those reported by Nield (1966), Rudraiah, et al. (1985) and Maekawa, et al. (1988) for the conditions of Bi=0 and a non-radiating fluid. As shown in Table 1, the critical Marangoni number and the corresponding wave number obtained by SGRA are in good agreement with those reported in the literature.

Table 1. Comparison of the critical Marangoni number and corresponding wave number for various values of Q for f(Z)=1, Bi=0 and Cr=0.

Investigator(s)	Rudraiah et al.		Nield		Maekawa et al.		Present Analysis	
Method of Solution	One-term Galerkin Procedure		Fourier Series		Analytical Solution		SGRA	
Q	Ma_c	a_c	Ma_c	a_c	Ma_c	a_c	Ma_c	a_c
0.0	78.44	2.43	79.61	1.99	79.61	1.99	79.61	1.99
1.0					82.17	2.02	82.17	2.02
2.5	82.06	2.46	85.97	2.05			85.97	2.05
5.0					92.18	2.09	92.18	2.09
10.0					104.22	2.18	104.22	2.18
12.5	92.26	2.58	110.08	2.22			110.08	2.22
20.0					127.11	2.33	127.11	2.33
25.0	113.58	2.71	138.09	2.39			138.09	2.39
50.0	147.18	2.90	189.87	2.63	189.87	2.63	189.87	2.63
100.0					284.22	2.96	284.22	2.96

The influence of the Biot number on the critical Marangoni number and the corresponding wave number is shown in Figs. 2a and 2b for the conditions of $Cr=10^{-6}$, $Po=1.0$, $\eta=1.0$, $\tau=0.5$, $\varepsilon_0=1$ and $\varepsilon_1=0$. The Biot number designates the relative magnitudes of the external heat transfer by convection to the internal thermal conduction across the fluid layer. As $Bi \to \infty$, the free surface cannot support temperature gradients, and the thermocapillary stress therefore vanishes, leading to stabilization. However, if $Bi=0$, the free surface is insulated and energy is retained within the fluid. This causes temperature fluctuations to increase and hence the surface tension, thus resulting in a less stable system. The critical Marangoni number therefore increases with the Biot number. The results are in accord with the conclusions of Pearson (1958) and Nield (1966).

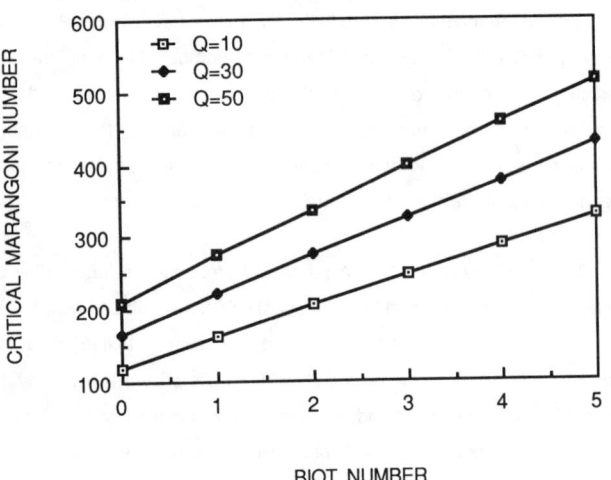

Fig. 2a. Influence of interface heat transfer on critical Marangoni number for the case $Cr=10^{-6}$, $Po=1.0$, $\eta=1.0$, $\tau=0.5$, $\varepsilon_0=1$ and $\varepsilon_1=0$.

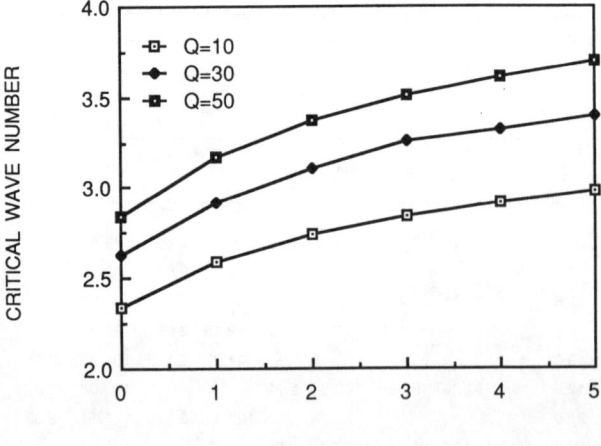

Fig. 2b. Influence of interface heat transfer on critical wave number for the case $Cr=10^{-6}$, $Po=1.0$, $\eta=1.0$, $\tau=0.5$, $\varepsilon_0=1$ and $\varepsilon_1=0$.

The Crispation number represents the degree of deformability of the free surface. As defined, the Crispation number is inversely proportional to the mean surface tension. Figure 3 shows that the critical Marangoni number decreases with increasing Crispation number. It indicates that less energy is required to generate convection when the mean surface tension is decreased from an infinite to a small finite value. Hence, the stability limit is lowered with increasing Crispation number. The figure also reveals that little variation occurs in the critical Marangoni number for Crispation numbers less than 10^{-4}. Below this value, the critical Marangoni numbers increase slightly. The same conclusion was reached by Scriven and Sternling (1964), Davis and Homsy (1980), and Cloot and Lebon (1985).

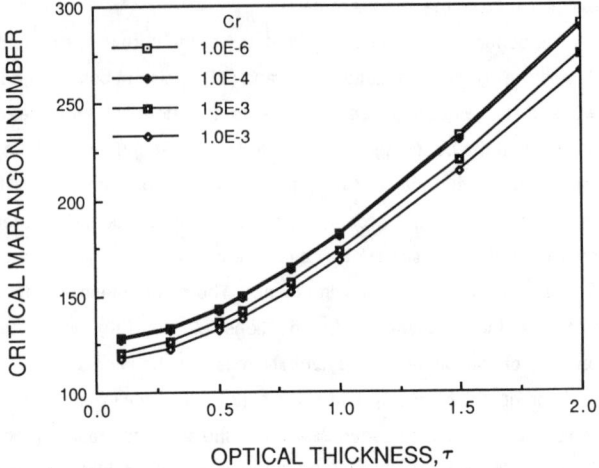

Fig. 3. Influence of surface tension for the case $Q=20$, $Bi=0$, $Po=0.1$, $\eta=0.1$, $\varepsilon_0=1$ and $\varepsilon_1=1$.

Landau and Lifshitz (1966) have pointed out that when a magnetic field is imposed on an electrically conducting fluid, the liquid motion is suppressed because of the interaction between the induced electric current and the external magnetic field. The Lorentz force due to electromagnetic effects is considered to be a stabilizing factor in thermocapillary convection. Since the Hartmann number denotes the relative effects due to the magnetic field intensity and the viscous force, the influence of the magnetic field on convection can be revealed by examining the dependence of the critical Marangoni number on the Hartmann number. In Figs 4a and 4b, the critical Marangoni number and the corresponding critical wave number are presented against the Hartmann number for the case $Cr=10^{-6}$, $Po=1.0$, $\eta=1.0$, $\tau=0.5$, $\varepsilon_0=1$ and $\varepsilon_1=0$. The results clearly demonstrate that the onset of thermocapillary convection is suppressed by the presence of a magnetic field in an electrically conductive fluid.

The effect of nongrayness of the fluid with upper perfectly reflecting diffuse boundary and lower black-rigid boundary is shown in Fig. 5 for the case $Cr=10^{-6}$, $Bi=0.0$, $Po=0.5$, $\tau=0.5$, $\varepsilon_0=1$ and $\varepsilon_1=0$. The nongrayness η is defined as the ratio of the optical depths based on the Rosseland

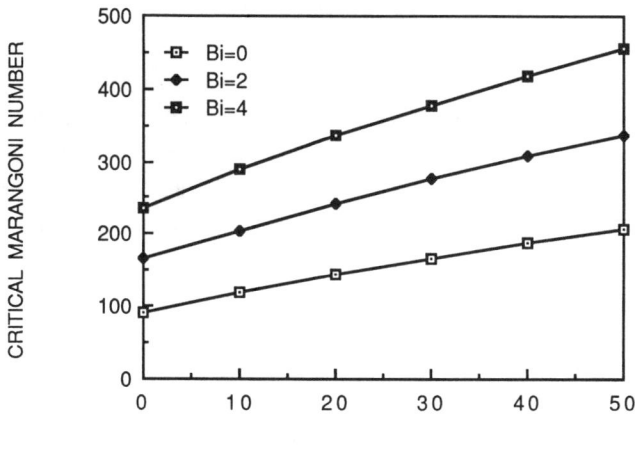

CRITICAL MARANGONI NUMBER

HARTMANN NUMBER

Fig. 4a. Influence of magnetic field on critical Marangoni number for the case $Cr=10^{-6}$, $Po=1.0$, $\eta=1.0$, $\tau=0.5$, $\varepsilon_0=1$ and $\varepsilon_1=0$.

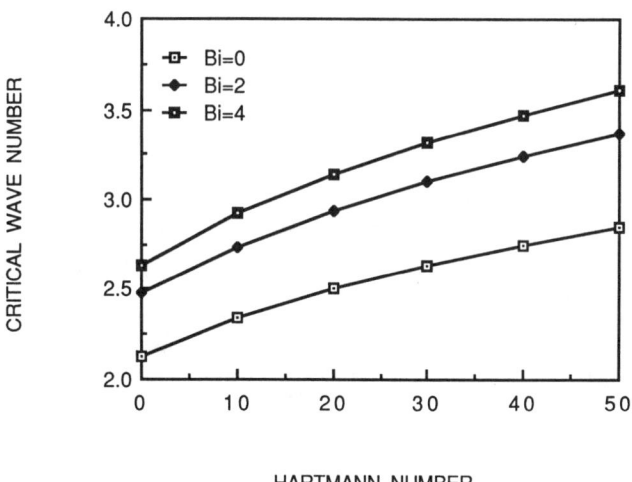

CRITICAL WAVE NUMBER

HARTMANN NUMBER

Fig. 4b. Influence of magnetic field on critical wave number for the case $Cr=10^{-6}$, $Po=1.0$, $\eta=1.0$, $\tau=0.5$, $\varepsilon_0=1$ and $\varepsilon_1=0$.

mean and the Planck mean absorption coefficients. The results indicates that the critical Marangoni number increases with increasing η. The effect of η on stability can be explained by examining its effect on the initial nonlinear temperature distribution. As we increase the degree of nongrayness of the fluid, the initial nonlinear temperature gradient is deformed and the temperature variations are concentrated in boundary layers at the horizontal boundaries. This means that, the core of the fluid is sujected to little temeperature gradient and thus causes stabilization.

Figure 6 demonstrates the effect of thermal radiation as characterized by the Planck number Po for the condition of $Cr=10^{-6}$, $Bi=0.0$, $\eta=0.6$, $\tau=0.5$, $\varepsilon_0=1$ and $\varepsilon_1=0$. The Planck number Po is defined as the ratio of conduction to radiation heat transfer. Thermal radiation becomes more

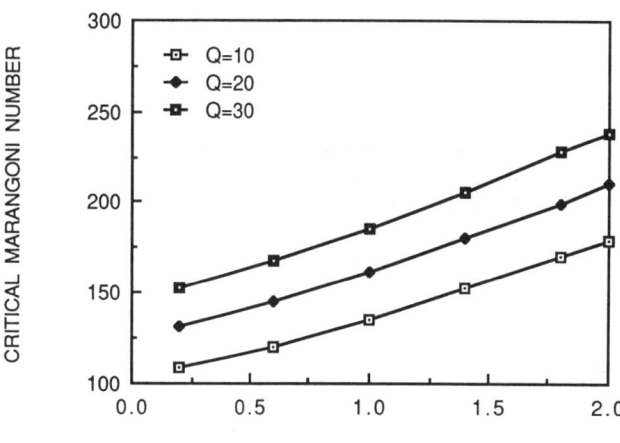

CRITICAL MARANGONI NUMBER

DEGREE OF NONGRAYNESS OF FLUID, η

Fig. 5. Influence of nongrayness of the fluid for the case $Cr=10^{-6}$, $Bi=0.0$, $Po=0.5$, $\tau=0.5$, $\varepsilon_0=1$ and $\varepsilon_1=0$.

dominant for Po < 1. It provides an extra heat transport mechanism in addition to the molecular diffusion to damp out any convective motions. As expected, the critical Marangoni number increases with decreasing Planck number, indicating that radiation enhances the stability of the fluid, as does the magnetic field.

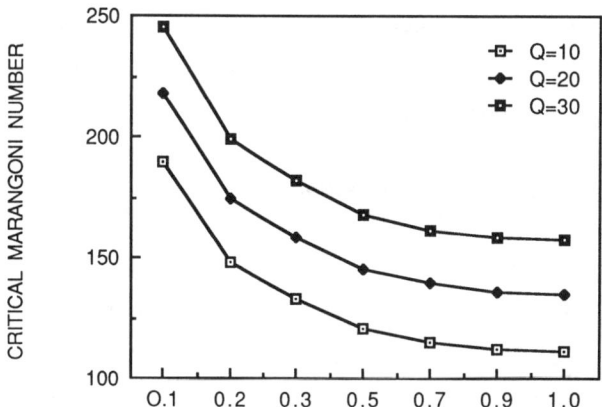

CRITICAL MARANGONI NUMBER

PLANCK NUMBER, Po

Fig. 6. Effect of thermal radiation for the case $Cr=10^{-6}$, $Bi=0$, $\eta=0.6$, $\tau=0.5$, $\varepsilon_0=1$ and $\varepsilon_1=0$.

The influence of the radiative boundaries of the fluid layer is illustrated in Fig. 7 for the case $Cr=10^{-6}$, $Bi=0.0$, $Q=20$, $Po=0.1$ and $\eta=0.1$ for the black-black, black-diffuse, diffuse-black and diffuse-diffuse surface conditions. The critical Marangoni number is plotted against the optical thickness for various emissivities of the boundaries. For a perfectly reflecting diffuse lower boundary, $\varepsilon_0=0$, temperature gradient is steep at the boundary and flat near the core, thus supplying a stabilizing mechanism. On the other hand, the fluid layer becomes less stable if the rigid lower boundary is black since it absorbs and retains energy within the system.

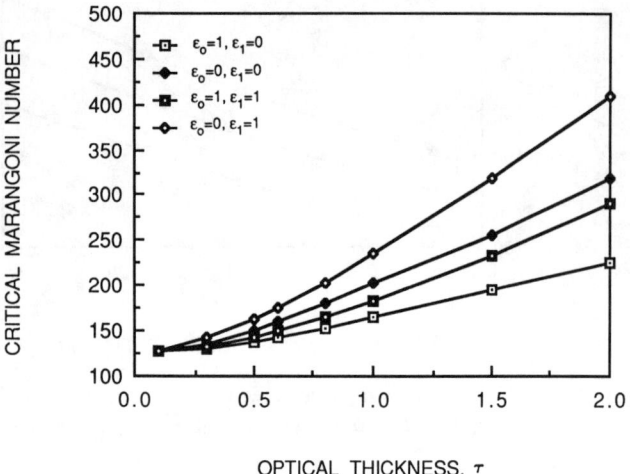

Fig. 7. Influence of radiative boundaries for the case Cr=10^{-6}, Bi=0.0, Q=20, Po=0.1 and η=0.1.

The combined effects of radiation and magnetic field in convective instability is shown in Table 2 for various radiative boundaries. It reveals the same effects as mentioned above. For the upper surface two cases, ε_1=1 (black) and ε_1=0 (diffuse) are considered. The perfectly reflecting diffuse lower surface, ε_0=0, stabilizes the fluid layer, and the black lower surface ε_0=1 destabilizes the fluid layer. Both the magnetic field and thermal radiation, therefore, are considered to be an effective means for suppressing the onset of convective motion.

Table 2. The effects of radiation and magnetic field on critical Marangoni number for the case Bi=0.0, Po=0.1, Cr=10^{-6} and η=0.1.

Boundary Conditions	ε_0=1, ε_1=0		ε_0=0, ε_1=0		ε_0=1, ε_1=1		ε_0=0, ε_1=1	
τ	Rad. Only	Rad. Q=20	Rad. Only	Rad. Q=20	Rad. Only	Rad. Q=20	Rad. Only	Rad. Q=20
0.1	80.07	127.62	80.32	128.17	80.22	127.92	80.69	128.71
0.3	83.58	131.59	86.32	136.34	84.47	133.44	88.91	140.80
0.5	89.65	138.64	97.04	151.21	92.16	143.38	103.64	162.21
0.6	93.32	143.03	103.68	160.48	97.07	149.71	112.81	175.36
0.8	101.50	153.14	118.47	181.34	108.63	164.59	133.57	204.73
1.0	110.35	164.61	133.99	203.66	122.08	181.83	156.40	236.44
1.5	132.68	196.05	168.76	256.50	161.59	232.40	217.17	318.88
2.0	151.70	225.96	192.10	318.92	206.79	290.43	278.96	410.05

CONCLUDING REMARKS

The coupled effects of thermal radiation and magnetic field on thermocapillary convection in a horizontal layer of fluid in a microgravity environment have been investigated. The fluid is electrically conducting and heated from below. The conditions leading to the onset of convective motions in the fluid are determined by using a normal mode linear stability analysis. The perturbation equations governing the marginal state for stationary instability are solved by the SGRA numerically. The accuracy of the SGRA is first verified by comparing the results with those reported in the literature for several special cases. The SGRA is then applied to analyze the influences of the Biot number, Crispation number, Hartmann number, optical thickness, Planck number, degree of nongrayness of the fluid, and radiative properties of the boundaries on convective instability. The results indicate that the critical Marangoni number increases with thermal radiation and magnetic field. The two agencies acting independently are unconditionally stabilizing. Hence, both of these factors play an important role in enhancing the stability of the fluid.

REFERENCES

Arpaci, V.S., and Gozum, D., 1973, "Thermal Stability of Radiating Fluids: The Benard Problem," *The Physics of Fluids*, Vol. 16, pp. 581-588.

Bayazitoglu, Y., and Lam, T.T., 1987, "Marangoni Convection in Radiating Fluids," *ASME Journal of Heat Transfer*, Vol. 109, pp. 717-722.

Chandrasekhar, S.,1961,*Hydrodynamic and Hydromagnetic Stability*, Clarendon Press, London.

Chun, C.H., and Wuest, W., 1978, "Flow Phenomena in Gravitationless Melting Zones in the Presence of Electromagnetic Fields," *COSPAR: Space Research*, Vol. 18, pp. 523-526.

Cloot, A., and Lebon, G., 1985, "Marangoni Instability in a Fluid Layer with Variable Viscosity and Free Interface, in Microgravity," *PCH PhysicoChemical Hydrodynamics*, Vol. 6, pp. 453-462.

Davis, S.H., and Homsy, G., 1980, "Energy Stability Theory for Free-Surface Problems: Buoyancy-Thermocapillary Layers," *Journal of Fluid Mechanics*, Vol. 98, pp. 527-553.

Goody, R.M., 1956, "The Influence of Radiative Transfer on Cellular Convection," *Journal of Fluid Mechanics*, Vol. 2, pp. 424-435.

Khosla, R.K., and Murgai, M.P., 1963, "A Study of the Combined Effects of Thermal Radiative Transfer and Rotation on the Gravitational Stability of a Hot Fluid," *Journal of Fluid Mechanics*, Vol. 16, pp. 97-107.

Lam, T.T., and Bayazitoglu, Y., 1988, "Marangoni Instability With Non-Uniform Volumetric Energy Sources due to Incident Radiation," *Acta Astronautica*, Vol. 17, pp. 31-38.

Landau, L.D., and Lifshitz, E.M., 1966, *Electrodynamics of Continuous Media: Course of Theoretical Physics*, Vol. 8, Chapter 8, Pergamon Press, London.

Lord Rayleigh, 1916, "On Convection Currents in a Horizontal Layer of Fluid, When the Higher Temperature is on the Under Side," *Philosophical Magazine*, Series 6, Vol. 32, pp. 529-546.

Maekawa, T., and Tanasawa, I., 1988, "Effect of Magnetic Field on Onset of Marangoni Convection," *International Journal of Heat and Mass Transfer*, Vol. 31, pp. 285-293.

Malmejac, Y., Bewersdorff, A, Da-Riva, I., and Napolitano, L.G., 1981, "Challenges and Prospectives of Microgravity Research in Space," ESA Report Br. 05, Paris.

Monti, R., Fortezza, R., and Mannara, G., 1988, "Results of the TEXUS 14-B Flight Experiment on a Floating Zone. First Approach Towards Telescience in Fluid Science," *Acta Astronautica*, Vol. 17, pp. 1221-1228.

Murgai, M.P., and Khosla, P.K., 1962, "A Study of the Combined Effect of Thermal Radiative Transfer and a Magnetic Field on the Gravitational Convection of an Ionized Fluid," *Journal of Fluid Mechanics*, Vol. 14, pp. 433-451.

Napolitano, L.G., 1984, "Marangoni Convection in Space Microgravity Environments," *Science*, Vol. 225, pp. 197-198.

Napolitano, L.G., Monti, R., and Russo, G., 1984, "Some Results of the Marangoni Free Convection Experiment," *Proceedings of the 5th European Symposium on Materials Science under Microgravity*, ESA Report SP-222, Schloss Elmau, Germany, pp. 15-22.

Nield, D.A., 1966, "Surface Tension and Buoyancy Effects in the Cellular Convection of an Electrically Conducting Liquid in a Magnetic Field," *Zeitschrift fur Angewandte Mathematik und Physik*, Vol. 17, pp. 131-139.

Pearson, J.R.A., 1958, "On Convection Cells Induced by Surface Tension," *Journal of Fluid Mechanics*, Vol. 4, pp. 489-500.

Regel', L.L., 1988, "Investigations of Gravity Effect on Crystal Growth. Achievements and Prospects," *Acta Astronautica*, Vol. 17, pp. 1241-1244.

Rudraiah, N., Ramachandramurthy, V., and Chandna, O.P., 1985, "Effects of Magnetic Field and Non-Uniform Temperature Gradient on Marangoni Convection," *International Journal of Heat and Mass Transfer*, Vol. 28, pp. 1621-1624.

Saghir, M.Z., 1988, "The Study of Marangoni Convection and the Solid/Liquid Interface Shape on a Germanium Float Zone in Microgravity," *Acta Astronautica*, Vol. 17, pp. 1211-1219.

Sarma, G.S.R., 1978, "Marangoni Convection in a Fluid Layer Under the Action of a Transverse Magnetic Field," *COSPAR: Space Research*, Vol. 19, pp. 575-578.

Sarma, G.S.R., 1987, "Interaction of Surface-Tension and Buoyancy Mechanisms in Horizontal Liquid Layers," *Journal of Thermophysics and Heat Transfer*, Vol. 1, pp. 129-135.

Scriven, L.E., and Sternling, C.V., 1964, "On Cellular Convection Driven by Surface Tension Gradients: Effects of Mean Surface Tension and Surface Viscosity," *Journal of Fluid Mechanics*, Vol. 19, pp. 320-340.

Spiegel, E.A., 1960, "The Convective Instability of a Radiating Fluid Layer," *Astrophysical Journal*, Vol. 132, pp. 716-728.

Traugott, S.C., 1963, "A Differential Approximation for Radiative Transfer with Application to Normal Shock Structure," in *Proceedings of the 1963 Heat Transfer and Fluid Mechanics Institute*, edited by A. Roshko, B. Sturtevant and D.R. Bartz, Stanford University Press, Stanford, Calif., pp. 1-13.

1989 National Heat Transfer Conference
HTD-Vol. 107, Heat Transfer in Convective Flows

OBSERVATIONS OF SECONDARY MOTIONS IN NATURAL CONVECTION THROUGH INCLINED CHANNELS HEATED FROM BELOW

L. F. A. Azevedo and M. J. Kaskus
Department of Mechanical Engineering
Pontifícia Universidade Católica – RJ
Rio de Janeiro, Brazil

ABSTRACT

Flow visualization experiments were conducted in one-sided heated, open-ended inclined channels in natural convection. The studies were aimed at revealing the structure of secondary flows developed when the channel is heated from below. Previous studies demonstrated that the presence of such secondary motions, in the form of longitudinal vortices, influences the heat transfer from the heated wall. In the present research the longitudinal vortices were visualized, seemingly for the first time, by long-time-exposure photographs of illuminated, neutrally-buoyant solid particles suspended in water. It was observed that, as the wall-to-ambient temperature difference increased, the flow patterns changed from predominantly axial flow to a regime where secondary motions were present. The photographs obtained suggested that, above a certain value of the temperature difference, vortices of different sizes coexist in the channel.

NOMENCLATURE

g – acceleration of gravity
H – channel height, Figure 1
Pr – Prandtl number
Ra_S – Rayleigh number
S – interwall spacing
T_w – average wall temperature
T_∞ – ambient temperature
W – plate width
x – axial coordinate
β – coefficient of thermal expansion
θ – angle of inclination, Figure 1
λ – wave length of a vortex pair, Figure 4
ν – kinematic viscosity

INTRODUCTION

The vast majority of the published work dealing with natural convection through one-sided heated channels has been concentrated on the vertical orientation of the channel, with very little attention being directed to the inclined geometry. In 1942, Elenbaas investigated the effects of inclination on natural convection of air in a symmetrically heated channel with uniform wall temperatures. Forty-one years later, Kennedy and Kanehl published what seems to be the only other experimental investigation of natural convection in inclined channels to that date. In

that research however, the highly specific nature of the thermal boundary conditions utilized did not allow a generalization of the results obtained.

In his pioneer work, Elenbaas presented the heat transfer results as a function of a modified Rayleigh number calculated with the streamwise component of gravity. This dimensionless presentation was suggested by an analysis of the two-dimensional boundary layer form of the governing equations. The heat transfer data for the vertical orientation of the channel were very well correlated by the suggested dimensionless parameter. However, considerable spread of the data was verified for the inclined channel configuration.

This outcome indicates that the inability of a dimensionless parameter, obtained from a two dimensional analysis, to correlate the data, might be due to the presence of secondary motions in the channel, i.e., three dimensional flow.

It is known that heating from below can have a destabilizing effect on fluid layers due to the existence of a top-heavy situation. Indeed, Sparrow and Husar (1969) have demonstrated that secondary motions in the form of longitudinal vortices were present in natural convection flows over inclined flat plates heated from below. The same flow pattern was revealed by visualization experiments conducted on laminar mixed convection through horizontal ducts of rectangular cross sections (Akiyama et al., 1971).

In a recently published paper, Azevedo and Sparrow (1985) studied inclined channels in natural convection. This study investigated the effects on the average heat transfer from the channel walls due to interwall spacing, wall-to-ambient temperature difference, angle of inclination, and mode of heating (i.e., heating from below, heating from above, and both walls heated). Perfect correlation of the heat transfer data were reported by using the modified Rayleigh number suggested by Elenbaas, when the channel was heated from above. Heating from below, however, produced data displaying a separate dependence on interwall spacing, wall-to-ambient temperature difference, and angle of inclination. For a fixed spacing and angle of inclination the heat transfer data progressively lift off the vertical channel results as the Rayleigh number (i.e., the dimensionless wall-to-ambient temperature difference) increases. The heat transfer augmentation was shown to reach as much as twenty percent beyond the vertical channel values. The same general trends, although not so pronounced, were reported for the channels with both walls heated.

With the objective of explaining the just described behavior of the heat transfer results, Azevedo and Sparrow (1985) reported flow visualization experiments employing the thymol blue visualization method (Baker, 1966). This is an electrochemical technique in which a change in fluid color is produced by changes in pH brought about by an imposed d.c. voltage. The main feature of this method is to produce a neutrally-buoyant tracer fluid which faithfully follows the natural-convection-induced motions. The heating surface of the channel was employed as the tracer producing electrode, so that, when the d.c. voltage was imposed, the whole surface of the heating wall (i.e., the lower wall) was covered with tracer fluid which was carried along with the flow passing through the channel.

The flow visualization results revealed that, up to a certain value of the wall-to-ambient temperature difference (for fixed values of the interwall spacing and angle of inclination), the tracer fluid emerged from the channel exit opening as a continuous sheet, indicating the existence of flow strictly in the streamwise direction. Beyond a critical value of the temperature difference, this pattern was replaced by an array of more or less regularly spaced streaks adjacent to the bottom wall. For the same angle of inclination, the spacing between two adjacent streaks was shown to be a function of the temperature difference. In addition, when the upper plate was used as the tracer producing electrode, a continuous sheet of tracer fluid was observed at the channel exit, for the whole range of parameters investigated.

The formation of the longitudinal streaks was attributed to the presence in the channel of a system of counterrotating longitudinal vortices, which sweep the layer of tracer fluid adjacent to the heated wall in the transverse direction, accumulating it at the boundary line between two vortices. The observation of a continuous layer of tracer fluid adjacent to the upper wall was interpreted as being an indication that the vortex system does not occupy the entire cross section of the channel. An illustration of the vortex system just described is presented in the inset of Figure 4, which was reproduced from Azevedo and Sparrow, (1985). Kennedy and Kanehl, (1983) have previously reported three dimensional longitudinal vortices producing heat transfer enhancement for angles of inclination less than 70°. However, no results of the smoke flow visualization experiments conducted were published.

The aforementioned visualization experiments of Azevedo and Sparrow, (1985) were unable to capture the complete structure of the longitudinal vortex system described, since it only documented the effects of the secondary motions on a layer adjacent to the heating wall (i.e., the longitudinal streaks). It is the purpose of the present research to implement a visualization technique capable of revealing the details of the whole flow field, thereby verifying the correctness of the flow structure inferred from the thymol blue visualizations.

The choice of the visualization technique to be employed is constrained by the nature of the flow under investigation. Unlike in the case of forced convection and, to some extent, mixed convection, natural convection flows can be disturbed by density differences between the tracer fluid and the working fluid. For this reason, the techniques based on the injection of a tracer fluid, widely used for mixed convection flows were discarded.

The flow visualization method employed in the present research is based on long time exposure photographs of small, neutrally-buoyant solid particles evenly distributed in the working fluid. The flow is made visible by illuminating a particular channel cross section with a plane of light. This way, two-dimensional slices of the three dimensional flow can be obtained and used to visualize it.

Finally, it should be mentioned that the photographs to be presented shortly are, to the best knowledge of the authors, the first documentation of the flow structure of natural convection in open-ended inclined channels heated from below.

EXPERIMENTS

The Test Section. The visualization experiments were conducted in water utilizing the apparatus which will now be described.

The main component of the experimental setup is the test section shown schematically in Figure 1. As can be seen in the figure, the channel geometry is formed by two main walls, one heated and one unheated, and by two side walls.

The heated wall was a 12-mm-thick copper plate with height H of 150 mm and width W of 100 mm. In order to simulate a constant temperature condition at the heating surface, the plate was electrically heated by means of three independently controlled heating circuits. This heating-circuit layout was designed based on previous knowledge of the behaviour of the rate of heat transfer from the plate, which decreases along its height.

Three regulated d.c. sources supplied power to the heating circuits. The voltage drop across each heater was carefully adjusted with the objective of attaining temperature uniformity at the heating surface. This adjusting process was guided by the readings of eight pre-calibrated thermocouples distributed in the plate. The thermocouples were made of 0.127-mm-diameter, Teflon-coated, chromel-constantan wires, and were installed in holes drilled through the back side of the plate. The junctions of the thermocouples were positioned about 0.5 mm from the front face of the plate. A digital voltmeter with an accuracy of $1\,\mu V$ was used for the readings of the thermocouples.

Extraneous heat losses from the heated plate were minimized by affixing to its rear face a 40-mm-thick block of closed-pore water-tolerante polystyrene.

The unheated wall was a plexiglass plate having the same dimensions as the copper plate. As in the case of the unheated wall, a block of polystyrene insulation was affixed to the rear face of the unheated wall (not shown in Fig. 1).

Heat losses from the test section to the water environment were estimated by simplified heat conduction models which considered losses through the back insulation and through the bundled wire cable (thermocouples and power lead wires). Typically, the heat losses amounted to 0.5% of the total heat input to the plate, being always less than 1%. The low levels of the heat losses encountered are explained by the relatively high heat transfer coefficients obtained in water.

As will be seen shortly in the Results Section, the flow visualization results will be presented in terms of the Rayleigh number. An estimate of the uncertainty levels in the experimentally determined Rayleigh numbers was obtained by an uncertainty analysis performed according to Moffat, (1982), which lead to a value of ±2.5%.

To complete the channel geometry, two 1.5-mm-thick plexiglass side walls were fabricated. These walls served a dual purpose. Firstly, they were used as spacers to maintain the desired interwall spacing S. For this reason a pair of such plates was available for each interwall spacing investigated. Secondly, as will be explained shortly, they were also used to partially block the incident light during the visualization runs, in order to provide an illuminated plane of fairly constant thickness. To this end, the inner face of both side walls were covered with black tape normally used for electrical insulation. Then, a 1-mm-thick slit was cut horizontally on the tape, along the width of the side walls. This slit allowed for the illumination of the channel. Care was taken to guarantee that the slits in both side walls were cut at the same axial position.

The two main walls and the side walls were mounted in a plexiglass supporting frame (not shown in Figure 1). This supporting frame was designed so as not to obstruct or deflect the flow passing in or out the channel openings. A special mechanism was provided in the supporting frame to allow the setting of the inclination angle θ.

The Test Environment. The test environment for the visualization runs was provided by a water-filled plexiglass tank with dimensions $420\times450\times720$ mm (height×width× length), in which the supporting frame carrying the test section was positioned. The temperature of the water in the test tank was monitored by three thermocouples (0.127-mm-dia., chromel-constantan) located approximately 300 mm away from the test section. The thermocouples were fixed on a vertical rod at, respectively, 40, 205, and 370 mm from the floor of the tank.

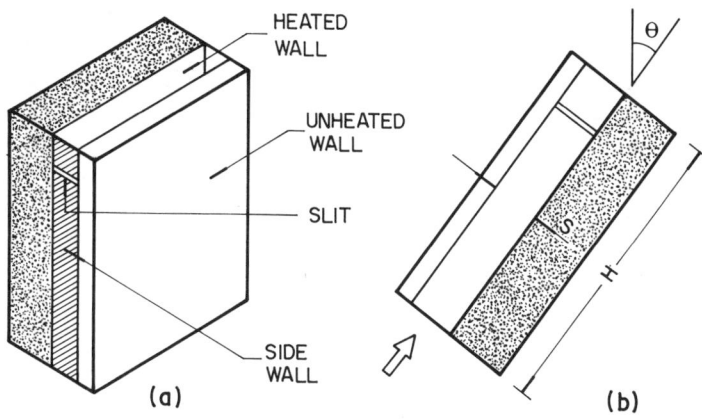

Fig. 1 – Test section

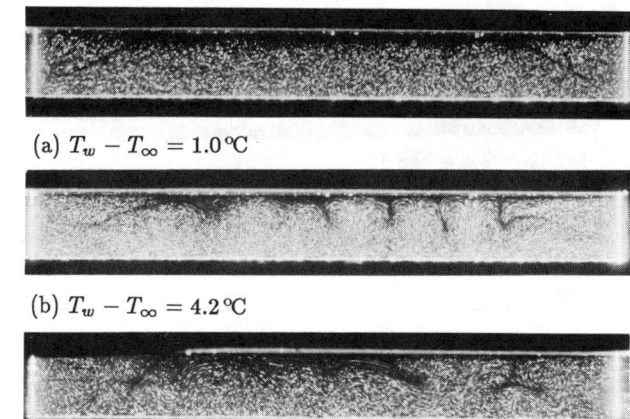

(a) $T_w - T_\infty = 1.0\,°C$

(b) $T_w - T_\infty = 4.2\,°C$

(c) $T_w - T_\infty = 7.6\,°C$

Fig. 2 – Flow visualization for $S/H = 0.066$ and $\theta = 45°$

Visualization Technique. The flow visualization technique employed in the present research relied on low intensity light reflected by small illuminated plastic particles carried along the flow. For this reason, all possible extraneous sources of light should be eliminated, otherwise the quality of the photographs obtained would be affected. With this objective, the inner surfaces of the tank walls were covered with plastic panels painted with flat black paint. Slits were provided in two opposing panels to allow for the illumination of the test section. In addition, during the photographic sessions, the water tank was covered with a black sheet of fabric. Small openings in the sheet were provided for illumination of the test section and for the camera.

The photographs of the channel cross sections were taken through the water surface, maintaining the camera in a plane parallel to the exit plane of the channel. As it would be expected in this situation, refraction of the light rays will distort the image of the channel. To avoid this effect, a plexiglass box with dimensions $200 \times 200 \times 150$ mm was constructed and partially submerged in water, with its base parallel to the exit plane of the channel. The photographs were taken through the base of the plexiglass box.

Photographs were obtained with a 35-mm Pentax camera utilizing a 50-mm macro lens, and 400-ASA Kodak TRI-X black and white film.

Uniform illumination of any particular channel cross section was obtained by utilizing two Kodak slide projectors. The projectors were situated in opposite sides of the tank. The plane of light necessary for the visualization was obtained by inserting in the projectors slides with 1-mm-thick slits. The plane of light from the projectors reached the interior of the channel after passing through the openings in the fabric sheet, through the slits in the panels covering the tank walls, and through the slits in the side walls of the channel. This multiple slit arrangement produced a plane of light in the channel with a fairly constant thickness of approximately 1 mm.

Experimental Procedure. The experiments consisted of the photographic documentation of the flow patterns prevailing in the channel for different operating conditions. Each data run was characterized by a value of the interwall spacing S, the angle of inclination of the channel, and the wall-to-ambient temperature difference.

Visualization of the flow patterns were conducted utilizing time exposure photographs of illuminated solid particles dispersed in the water. The particles utilized were 50-μm Pliolite resin made by the Goodyear Rubber Co.. These particles are well suited for visualization experiments in water since they are nearly neutrally buoyant (specific gravity of 1.02).

The optimum particle concentration is a function of the flow characteristics being observed. In the present research several test runs were conducted leading to an optimum concentration in the range of 0.5-0.8 grams per 100 liters of water.

The particles were mixed with water in a separate container.

Vigorous mixing and some drops of liquid soap were necessary to obtain a good dispersion and avoid the formation of cluster of particles. This solution was added to the tank where the visualization experiments were to be carried out. Stirring of the tank was necessary to evenly distribute the particles and to eliminate possible temperature gradients in the water. This condition was verified by monitoring the three thermocouples deployed in the tank. An hour waiting period was allowed to guarantee that all motions generated by the stirring operation had died out.

After the waiting period, power was applied to the three heating circuits, and the attainment of the steady-state condition was verified by monitoring the plate thermocouples. At steady-state, the wall and fluid temperatures and the heaters voltages and currents were recorded.

At this point, the projectors were switched on while the lights of the laboratory were switched off. Several photographs with different exposure times were taken. Then, the new voltage settings for the next run were dialed in and the power switched off. The water in the tank was stirred up and, after an hour's wait, a new data run was performed.

RESULTS AND DISCUSSION

The experiments to be reported in the present paper were conducted with the objective of implementing a visualization technique capable of revealing secondary motions within the channel. They were not intended to be a parametric study of the flow conditions in the channel. This latter objective is being pursued in an experimental research program in course. However, even with a limited number of experiments, some interesting observations could be made, as will be described shortly.

Heat transfer data in the form of Nusselt numbers were obtained and compared with the available literature (Azevedo and Sparrow, 1985) with the objective of validating the experimental procedure and the apparatus constructed. In all runs performed the average Nusselt numbers agreed to within five percent with the data from the literature.

Visualization experiments were performed for a fixed angle of inclination equal to 45 degrees, and for dimensionless spacings S/H of 0.066 and 0.100. For each dimensionless spacing three levels of the wall-to-ambient temperature difference were investigated.

Figure 2 presents photographs of the cross section of the channel for three different levels of the wall-to-ambient temperature difference, for a channel with S/H equal to 0.066. The cross section photographed was situated at a dimensionless axial distance from the inlet of the channel, x/H, equal to 0.95. This particular position was chosen for being in the same region of observation as that for the thymol blue experiments reported in

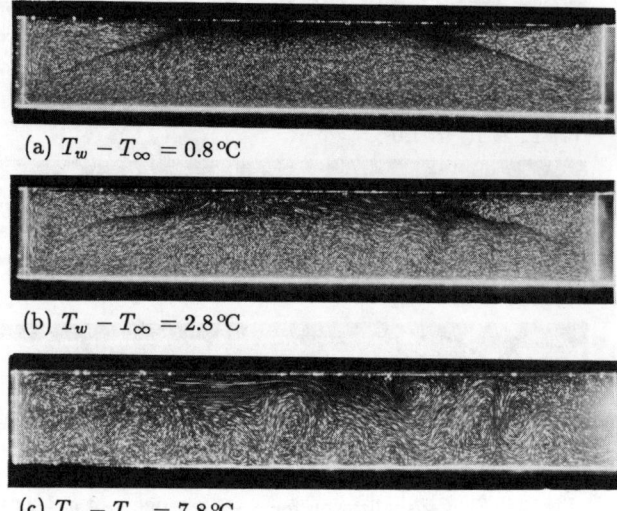

(a) $T_w - T_\infty = 0.8\,^\circ\text{C}$

(b) $T_w - T_\infty = 2.8\,^\circ\text{C}$

(c) $T_w - T_\infty = 7.8\,^\circ\text{C}$

Fig. 3 – Flow visualization for $S/H = 0.100$ and $\theta = 45^\circ$

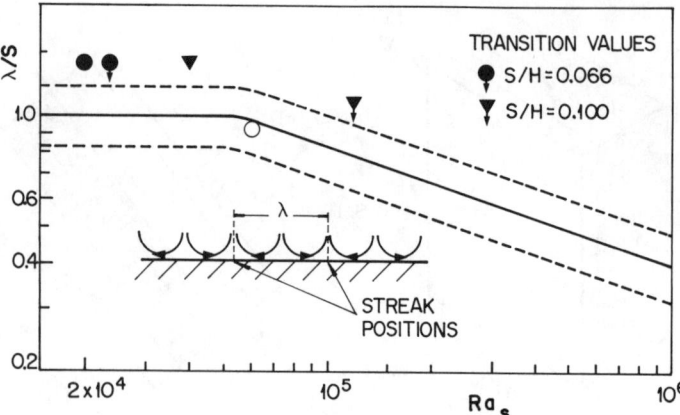

Fig. 4 – Wavelength for pairs of vortices
(Azevedo and Sparrow, 1985)

the literature. The exposure time for all three photographs was equal to 40 seconds.

An overall observation of Figure 2(a) shows that for the case of low temperature differences $(T_w - T_\infty = 1\,^\circ\text{C})$ the flow is strictly in the axial direction. A careful observation along a vertical line crossing, for example, the mid-plane of the cross section, shows a gradient in the illumination of the picture. Close to the top (unheated) wall there exists a dark region which slowly becomes more illuminated as one moves towards the bottom (heated) wall. A brighter region means that more light-reflecting particles have crossed the illuminated plane in a fixed time interval (40 seconds in this particular case). Thus, a brighter region can be associated with higher axial velocities. Following this line of argument, it can be concluded that the axial velocity profiles are not symmetric, displaying higher velocities in the lower half of the cross section. A similar velocity distribution has been numerically predicted for mixed convection flows through horizontal ducts heated from below (Incropera and Schutt, 1985). It can also be observed in Figure 2(a) the existence of one cell of secondary motion adjacent to each side wall. This is attributed to free convective motions induced by heat losses from these non-insulated walls.

Figure 2(b) displays a flow structure totally distinct from the one previously described. This situation corresponds to a wall-to-ambient temperature difference of 4.2 °C. The secondary motions are clearly visible in the central region of the cross section, although, as predicted by Azevedo and Sparrow (1985), these cells do not seem to touch the upper wall. It should be noted that the cells adjacent to the side walls have grown into the cross section of the channel.

As the temperature difference is increased (Figure 2(c), $T_w - T_\infty = 7.6\,^\circ\text{C}$), the well organized structures of Figure 2(b) tend to disappear. Vortices of different sizes occupy the channel cross section.

The thymol visualization experiments (Azevedo and Sparrow, 1985) were not able to reveal this new flow structure, yielding the same qualitative results for the cases corresponding to Figures 2(b) and 2(c), i.e., regularly spaced longitudinal streaks adjacent to the heated wall. This outcome suggests that there should exist a smaller vortex system acting closer to the heated wall and responsible for the formation of the longitudinal streaks shown by the thymol blue visualizations. These smaller vortices would coexist with the larger structures revealed by the visualization experiments of the present research.

Figure 3 presents photographs of a cross section of the channel positioned at a dimensionless axial distance from the inlet equal to 0.95, for a channel with $S/H = 0.100$, inclined at 45

degrees. As can be seen, the qualitative results are similar to the case described in Figure 2, i.e., axial flow with natural convection cells close to the side walls for small temperature differences, and the development of secondary motions as the temperature difference is increased. Figure 3(c) shows clearly the presence of vortices of different sizes, which corroborates the arguments of the preceding paragraph.

The information extracted from the visualization experiments will now be compared with the results from Azevedo and Sparrow, (1985), as far as the wavelengths of the longitudinal vortices are concerned. The inset in Figure 4 helps define the wavelength for a pair of counterrotating vortices, which was computed from the distance between two adjacent longitudinal streaks in the thymol blue visualization studies. This figure was adapted from Azevedo and Sparrow (1985), and displays the dimensionless wavelength λ/S as a function of the Rayleigh number, given by $Ra_S = \{g\beta(T_w - T_\infty)S^3/\nu^2\}Pr$. The data points presented in the original paper were replaced here by a curve fitted line sided by two dashed lines which represent the uncertainty levels associated with the experiments. Figure 4 also presents information regarding the transition of the flow from a pure streamwise motion to one in which secondary flows are present. The transition values are represented in the figure by the arrow-attached symbols.

Figures 2(a) and 3(a) represent situations of predominantly axial flow. These two cases are plotted in Figure 4, respectively as a dark circle and a dark diamond. As it can be verified in the figure, these data points are in agreement with the measured transition values, i.e., they lay to the left of each transition value.

Results related to the wavelength of the vortices were very difficult to be extracted from the photographs presented in Figures 2 and 3. This was due to the complex nature of the secondary flows observed which, as already mentioned, presented different vortex sizes. However, specifically for Figure 2(b), the better organized flow structures allowed the measurements of the wavelength of the vortex pair. This result is plotted as an open circle in Figure 4, and it is seen to agree very well with the solid line representing the data from Azevedo and Sparrow.

Additional information regarding the secondary flow structures in the channel can be obtained by examining the results presented in Figure 5. In this figure, two different channel cross sections were photographed for the same set of operating conditions as those for the case of Figure 2(b), i.e., $S/H = 0.066$, $\theta = 45^\circ$ and $T_w - T_\infty = 4.2\,^\circ\text{C}$. The cross sections observed were situated at dimensionless axial distances from the inlet of the channel, x/H, equal to 0.15 and 0.50 which are presented, respectively, in Figures 5(a) and 5(b). A marked vertical motion can be noticed in the region adjacent to the heated (bottom) wall in Figure 5(a). This type of motion, already documented for mixed convection flows in horizontal ducts heated from below (e.g., Incropera et al., 1987), represent the early stages of the formation of the lon-

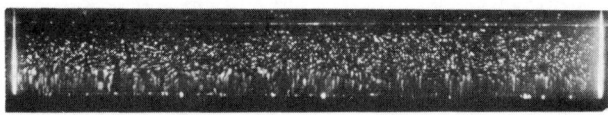

(a) $x/H = 0.15$

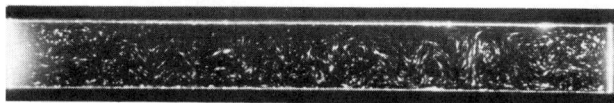

(b) $x/H = 0.50$

Fig. 5 – Flow visualization for $S/H = 0.066$, $\theta = 45°$, and $T_w - T_\infty = 4.2\,°C$

gitudinal vortex system. At channel mid-length, Figure 5(b), the vertical motions are no longer dominant, while the circular patterns which eventually will lead to the flow structure displayed in Figure 2(b) prevail.

CONCLUSIONS

A visualization technique based on the illumination of neutrally-buoyant solid particles suspended in water was implemented to study the flow patterns prevailing in natural convection in open-ended inclined channels. It was observed that, as the temperature difference was increased, the flow patterns changed from predominantly axial flow to a regime where secondary motions were present. The results obtained were compared with previously published data. They were found to be in good agreement with those data regarding the transition from the axial flow regime to the secondary flow regime.

Based on the results obtained with the visualization technique employed, it was suggested that distinct flow structures represented by different vortex sizes coexist in the channel, the smaller ones being responsible for the longitudinal streaks revealed by the thymol blue experiments reported in the literature.

REFERENCES

Akiyama, M., Hwang, G. J., and Cheng, K. C., 1971, "Experiments on the Onset of Longitudinal Vortices in Laminar Forced Convection Between Horizontal Plates," *ASME J. Heat Transfer*, Vol. 93, pp. 335–341.

Aung, W., Fletcher, L. S., and Sernas, V., 1972, "Developing Laminar Free Convection Between Vertical Flat Plates with Asymmetric Heating," *Int. J. Heat and Mass Transfer*, Vol. 15, pp. 2293–2308.

Azevedo, L. F. A. and Sparrow, E. M., 1985, "Natural Convection in Open-Ended Inclined Channels," *ASME Journal of Heat Transfer*, Vol. 107, pp. 893–901.

Baker, D. J., 1966, "A Technique for the Precise Measurement of Small Fluid Velocities," *J. Fluid Mechanics*, Vol. 26, pp. 573–575.

Elenbaas, W., 1942, "Heat Dissipation of Parallel Plates by Free Convection," *Physica*, Vol. 9, pp. 1–28.

Incropera, F. P. and Schutt, J. A., 1985, "Numerical Simulation of Laminar Mixed Convection in the Entrance Region of Horizontal Rectangular Ducts," *Numerical Heat Transfer*, Vol. 8, pp. 707–729.

Incropera, F. P., Knox, A. L., and Maughan, J. R., 1987, "Mixed-Convection Flow and Heat Transfer in the Entry Region of a Horizontal Rectangular Duct," *ASME J. Heat Transfer*, Vol. 109, pp. 434–439.

Kennedy, K. J. and Kanehl, J., 1983, "Free Convection in Tilted Enclosures," in *Heat Transfer in Electronic Equipment*, ASME HTD – Vol. 28, ed. S. Oktay and A. Bar-Cohen, pp. 43–47.

Moffat, R. J., 1982, "Contributions to the Theory of Single–Sample Uncertainty Analysis," *ASME J. of Fluids Engineering*, Vol. 104, pp. 250–260.

Sparrow, E. M., and Husar, R. B., 1969, "Longitudinal Vortices in Natural Convection Flow on Inclined Plates," *J. Fluid Mechanics*, Vol. 37, pp. 251–255.

1989 National Heat Transfer Conference
HTD-Vol. 107, Heat Transfer in Convective Flows

INFLUENCE OF SURFACE VISCOSITY ON MARANGONI-BENARD INSTABILITIES

P. Queeckers, J. C. Dupin, and J. C. Legros
Université Libre de Bruxelles
Service de Chimie-Physique E.P.
Brussels, Belgium

ABSTRACT

We studied the influence of the surface viscosity on the Marangoni-Bénard Instabilities (MBI) when heating from top, layers of aqueous solutions of long chain alcohols (C_7 - C_{10} - C_{12}). These solutions have surface tension variations with temperature which presents a minimum. We worked at temperature for which the temperature coefficient of surface tension is positive. The linear stability analysis performed leads to the definition of a modified critical Marangoni number in which the surface viscosity appears.

We have measured the surface tension and the surface viscosity of these solutions.

By heating from the top, we observed Marangoni Bénard convection in heptanol and decanol. For the dodecanol solution, no convection has been detected due to surface viscosity effect.

For increasing values of the imposed heat flux at the lower plate, at the onset of convection, we observed an abrupt decrease of the resulting temperature difference. This is explained by the existence of an inverse bifurcation at the critical point.

This demonstrates the important role played by the surface viscosity in the Marangoni-Bénard instability problem.

NOMENCLATURE

c_p = heat capacity
d = thickness of the layer
g = gravity acceleration
H = heat-transfer coefficient at the free surface
K = surface dilational viscosity
k_c = heat conductivity
Q = heat loss rate per surface unit at the interface
T = temperature
v = velocity along the z-axis
α = thermal volume expansion

β = adverse temperature gradient
ε = surface shear viscosity
θ = temperature perturbation of vertical uniforn temperature gradient
κ = heat diffusivity
μ = dynamic viscosity
ν = μ/ρ kinematic viscosity
ρ = density
σ = surface tension

$$Cr = \frac{\mu \, k_c}{\sigma \, d} \qquad \text{Crispation number}$$

$$Ma = \frac{-\frac{d\sigma}{dT} \beta \, d^2}{\mu \, k_c} \qquad \text{Marangoni number}$$

$$Nu = \frac{H \, d}{\rho \, c_p \, k_c} \qquad \text{Nusselt number}$$

$$Pr = \frac{\nu}{k_c} \qquad \text{Prandtl number}$$

$$Vi = \frac{K + \varepsilon}{\mu \, d} \qquad \text{Surface viscosity group}$$

INTRODUCTION

When a horizontal fluid layer limited on the upper side by a liquid/gas interface is heated from below, it has been demonstrated that a critical temperature gradient is existing beyond which the rest state of the liquid is unstable. Convective motions start and are organised in prismatic cell patterns which are generally hexagonals. This problem has been studied first in 1901 by Bénard [1].

The onset of this instability can be explained as follow : at this critical point, the fluid "cannot" anymore dissipate the fluctuations by diffusing momentum (ν : kinematic viscosity) and heat (κ : heat diffusivity). These fluctuations are thus growing, become macroscopic and motions are starting. In 1916, Rayleigh [2] explained the experimental observations of Bénard by the buoyancy forces. The linear stability analysis that he performed did not take into account the surface tension and the surface viscosity. In that case, motions were induced, when a dimensionless number called Rayleigh Number (Ra) was greater than a critical value.

In 1958, the onset of the motions was explained by Pearson [3] by introducing in his analysis surface tension stress at the liquid/gas interface. He did not take into account gravity and the dimensionless parameter was called Marangoni number (Ma). These two mechanisms inducing instabilities are both present in real situations and were present in the Nield analysis [4]. The gravity is the leading force in thick layers (some millimeters), the surface tension forces are the dominant parameter for thin layers (< 1 mm). In experiments, it is very difficult to separate these two destabilising factors. The convective phenomena related to the surface tension are studied from years, but the experimental approach is very difficult and there are not many results :

- thin layers have to be used in order to decrease the relative importance of buoyancy with respect to surface forces.

- the study of fluid motions through a window limiting the vapour phase is disturbed by condensation of liquid arising from the heated liquid bulk phase.

- very small amount of impurities adsorbed at the interface changes completely the measured results.

From this last point we thought that intermediate situations can exist for which intentionally adsorbed molecules modify the properties of the interface and particularly its viscosity. This could deeply modify the hydrodynamical stability of the interface because in the Marangoni-Bénard Instability (MBI) problem, the dissipation of fluctuations of surface tension has to take place in the interface. Thus this phenomenon is directly related to the viscosity in the interface instead of the viscosity of the bulk phase. In a following paragraph we shall indicate by a stability analysis that this behaviour is possible. The viscosity contribution introduced through the boundary condition at the liquid/gas interface can be different than in the bulk phase.

We have measured the critical temperature gradient for the onset of convection in diluted aqueous solutions of long chain alcohols. The molecules adsorbed at the interface lead to a saturated film with a viscosity which intuitively has to depend on the length of the carbon chain. We have measured the surface viscosities for heptanol, decanol and dodecanol solutions.

On the other hand, these alcohol solutions have the interesting property that their surface tensions have a minimum as a function of the temperature. This means that beyond a temperature T_{min}, the surface tension is increasing with the temperature and allows us to study the onset of convective motions in horizontal layers heated from the top. This yields :

- to avoid the condensation problem on the surface located near the liquid/gas interface in order to control its temperature.

- the gravity is not any more a motor to induce convection, it plays a stabilising role and the observed convective motions are only induced by the surface tension gradients.

ANALYSIS

Generally when the stability analysis of the MBI problem is performed, the surface viscosity is supposed zero, except in the work of Scriven and Sternling [5] in which they used an equation describing the interface dynamic. The 2D obtained equation [6] is similar to the Navier-Stockes equation with the assumption of a newtonian interface of isotropic fluid. Taking into account the fact that the fluxes in the interface are connected to the fluxes in higher and lower phases, they have linked the motion equation in the interface to the constraints, the velocities and the accelerations in the bulk phases. Using the same method as Pearson (thus without any contribution of gravity), the steady marginal stability has been studied. The stability curves, depending on the following dimensionless parameters have been obtained :

Crispation group : $Cr = \dfrac{\mu\, \kappa}{\sigma\, d}$

Surface viscosity group : $Vi = \dfrac{\kappa + \varepsilon}{\mu\, d}$

Marangoni number : $Ma = \dfrac{-\frac{d\sigma}{dT}\beta\, d^2}{\mu\, \kappa}$

Nusselt number : $Nu = \dfrac{H\, d}{\rho\, c_p\, \kappa}$

In this analysis, when Vi increases, the critical Marangoni number for the onset of convection increases too.

We have studied the influence of the surface viscosity when both gravity and surface tension gradients contribute to convection onset.

Let us make the following assumptions :

- the horizontal layer of a Newtonian liquid with a depth d is limited by a undeformable horizontal free surface and is submitted to a vertical temperature gradient.

- the kinematic viscosity of the liquid phase is ν and the surface viscosity is $\nu*$

- the classical Boussinesq approximation [7] is valid.

- $\vec{1}_z$ is vertical upwards, $\vec{1}_x$ and $\vec{1}_y$ are horizontal.

In order to define how the viscosities ν and ν^* operate in the definition of the critical Marangoni number, we shall use the same approach as Nield [4].

Starting from the linearized Navier Stokes and heat equations for the perturbations, gravity is the only external force

$$\frac{\partial}{\partial t} \nabla^2 v = g\alpha \left(\frac{\partial^2 \theta}{\partial x^2} + \frac{\partial^2 \theta}{\partial y^2} \right) + \nu \nabla^4 v \qquad (1)$$

$$\frac{\partial \theta}{\partial t} = \beta v + \kappa \nabla^2 \theta \qquad (2)$$

in which : v = velocity along the z-axis
θ = temperature perturbation of vertical uniform temperature gradient.
g = gravity acceleration
α = thermal volume expansion
T = $T_0 - \beta z + \theta$; β is positive when the layer is heated from below.

The surface tension can be written as

$$\sigma = \sigma_0 - \gamma \theta_s$$

where $\gamma = -d\sigma/dT$ with θ_s is the temperature perturbation at the free surface.
The heat loss rate per surface unit at the interface is

$$Q = Q_0 + q_0 \theta_s$$

with $q_0 = \frac{\partial Q}{\partial T}$

The expression of the shear stress at the interface leads to

$$\mu \left(\frac{\partial^2 v}{\partial x^2} + \frac{\partial^2 v}{\partial y^2} - \frac{\partial^2 v}{\partial z^2} \right) = -\gamma \left(\frac{\partial^2 \theta}{\partial x^2} + \frac{\partial^2 \theta}{\partial y^2} \right) \qquad (3)$$

and the heat flux conservation through the interface is written as :

$$- k_c \frac{\partial \theta}{\partial z} = q_0 \theta \qquad (4)$$

with k_c = heat conductivity of the liquid.
The conditions on the vertical velocity at the undeformable interface is v = 0 at z = d.
The boundary conditions at the lower conductive solid plate are :

$$\theta = v = \frac{\partial v}{\partial z} = 0 \qquad \text{at} \quad z = 0$$

Classical normal modes are used :

$$v = V(z) \exp[i(k_x x + k_y y) + pt]$$
$$\theta = \Theta(z) \exp[i(k_x x + k_y y) + pt]$$

with $k = \sqrt{k_x^2 + k_y^2}$, p can be complex.
Eqs. (1) and (2) become :

$$p\left(\frac{d^2}{dz^2} - k^2 \right) V = -g\alpha k^2 \Theta + \nu \left(\frac{d^2}{dz^2} - k^2 \right)^2 V \qquad (5)$$

$$p \Theta = \beta V + \kappa \left(\frac{d^2}{dz^2} - k^2 \right) \Theta \qquad (6)$$

with corresponding boundary conditions
for z = 0 $\qquad V = dv/dz = \Theta = 0$
and for z = d $\qquad V = 0$
$$q_0 \Theta = -k_c \frac{d\Theta}{dz}$$

$$\mu^* \frac{d^2 V}{dz^2} = -\gamma k^2 \Theta \qquad (7)$$

In Eq. 7 the viscosity which appears in the l.h.s. is related to the interface and is thus written μ^*.

In order to write the equations in dimensionless forms, the following units are used for length : d/π

time : $d^2/\pi^2 \nu$

velocity : $\pi \nu / d$

temperature : $\frac{\beta d}{\pi} \frac{\nu}{\kappa}$

The following groups are defined.

$$b = \frac{kd}{\pi}$$

$$\sigma_1 = \frac{p d^2}{\pi^2 \nu}$$

$$V_1 = \frac{V d}{\pi \nu}$$

$$\Theta_1 = \frac{\Theta \pi \kappa}{\beta d \nu}$$

$$Ra = \frac{g \alpha \beta d^4}{\kappa \nu}$$

$$Pr = \frac{\nu}{\kappa}$$

In these groups, ν is the kinematic viscosity of the liquid. After these transformations, the Eqs. 5, 6 are :

$$\left(D^2 - b^2 \right) \left(D^2 - b^2 - \sigma_1 \right) V_1 = \frac{Ra}{\pi^4} b^2 \Theta_1$$

$$\left(D^2 - b^2 - Pr \sigma_1 \right) \Theta_1 = -V_1 \qquad (8)$$

D is the undimensional derivation along z and the boundary conditions write as

$$V_1 = DV_1 = \Theta_1 = 0 \qquad \text{at} \quad z = 0$$
and $\quad V_1 = 0 \ ; \ D\Theta_1 = -\frac{L}{\pi} \Theta_1 \ , \ D^2 V_1 = -\frac{Ma^*}{\pi^2} b^2 \Theta_1$
with $L = \frac{q_0 d}{k_c}$ $\qquad$ at $\quad z = \pi$
and $Ma^* = \frac{\gamma \beta c d^2}{\rho \nu^* \kappa}$ $\qquad$ is a modified Marangoni number.

Let us remark that ν^*, the surface viscosity, appears only in the modified Marangoni number and not in the conservation equations. The classical results obtained for Ma^{cr} are thus still valid for Ma^{*cr}, only the numerical value of the critical temperature difference ΔT^{cr} for the onset of convection will be modified. The approximate results express the coupling with gravity forces.

$$\frac{Ra}{Ra^{cr}} + \frac{Ma^*}{Ma^{*cr}} = 1 \qquad (9)$$

Difficulties arise from the fact that ν^* has not the dimensions of a surface viscosity, a characteristic depth δ of the surface layer has to be introduced : $\nu_s = \delta \nu^*$.
If the surface viscosity is classically defined as an excess quantity, it is vanishing for a pure compound, then $\nu^* = \nu$ and the viscosity which appears in Ma is equal to the viscosity of the liquid. For an adsorbed phase with $\nu^* > \nu$ then the critical value of the temperature difference will increase.

EXPERIMENTAL RESULTS

We have experimentally demonstrated the role of the surface viscosity on the hydrodynamical stability of a layer at rest and submitted to a vertical temperature gradient.

We used aqueous solutions of long chain alcohols : n-heptanol $6.298 \, 10^{-3}$ m, n-decanol $1.386 \, 10^{-4}$ m and a n-dodecanol saturated at 20°C and diluted afterward by the addition of 20 % of water. The surface tensions of these solutions as a function of the temperature present a minimum. In the Marangoni-Bénard experiments that we performed with these solutions the temperatures of the interface were choosen in order that the surface tension is increasing with the temperature and the layer was heated from top. A great care was taken in order to avoid interface contamination and to clean the experimental set-up.

Surface Tension Measurements

The surface tensions of the three solutions used were measured by the Wilhelmy plate technique. The values obtained are given on Figs. 1, 2, 3. The minimums of surface tension were respectively around 39, 25 and 10°C.

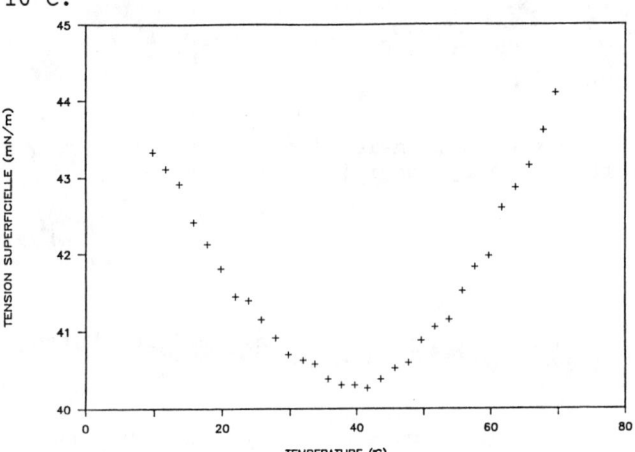

Figure 1. Surface tension of aqueous solution of n-heptanol $6.298 \, 10^{-3}$ m as a function of temperature.

Surface Viscosity Measurements

We have also determined the surface viscosity using a channel viscometer [8, 9, 10] which allows to determine these parameters from measurements of the surface velocity induced by the motions of the bottom of the circular channel.

Classical theory performed by Burton and Mannheimer [8] shows that

$$\mu_S = \frac{4 \, y_o \, \mu_b}{\pi^2 \sinh \pi D} \, \frac{v_d}{v_m} - \frac{y_o \, \mu_b}{\pi} \coth \pi D \qquad (10)$$

$$\text{with } D > \frac{2}{\pi}$$

where μ_S = surface viscosity
μ_l = dynamic viscosity of liquid
v_d = linear velocity at the bottom of the channel
v_m = velocity of the surface

y_o = width of the channel

$D = \dfrac{x_o}{y_o}$ with x_o = thickness of the liquid layer

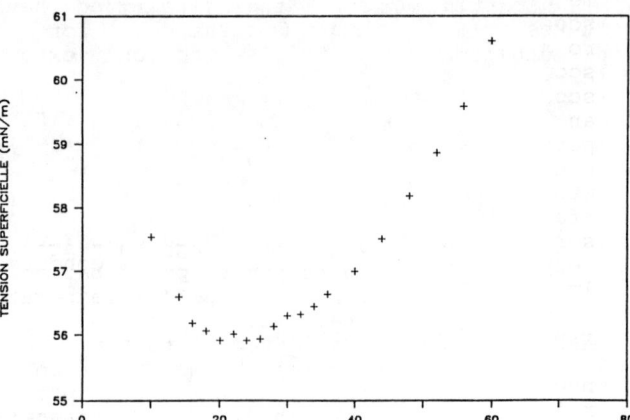

Figure 2. Surface tension of aqueous solution of n-decanol $1.386 \, 10^{-4}$ m as a function of temperature.

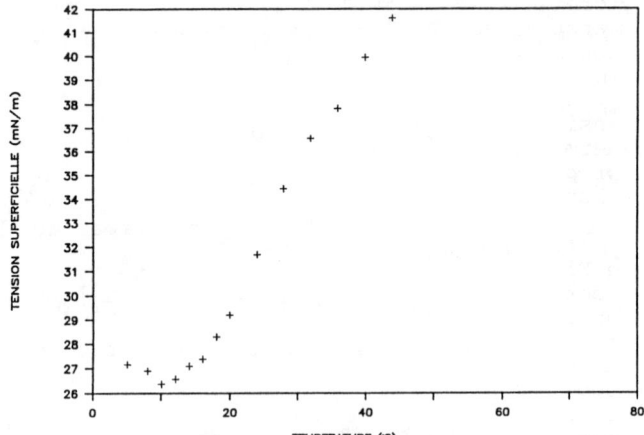

Figure 3. Surface tension of aqueous solution of n-dodecanol as a function of temperature. The concentration is 80 % of saturation at 20°C.

The inaccuracy is of the order of ± 7 %, mainly due to the fact that the effects of the curvature of the liquid interface are neglected and to the inaccuracy of geometry. We measured the surface viscosities of pure water and of the alcohol solutions. The results are given in table 1.

	μ_S (sp)
Water	0.0005 ± 0.0003
n-heptanol	0.0008 ± 0.0004
n-decanol	0.0007 ± 0.0003
n-dodecanol	± 0.01

Table 1.
The results obtained for pure water were around 0.0005 sp[+]. This non zero value could

+ sp = surface poice = poise cm.

be due to approximations used to derive Eq.10, or to slight contamination of the water surface. Table 1 shows that the values obtained for heptanol and decanol are of the same order as for water. The (excess) surface viscosities of these solutions are close to zero and cannot be measured with this viscometer. On the other hand, the surface viscosity of dodecanol is twenty times larger than water and far larger than the experimental error. The obtained values depending on the rotation speed show the non newtonian behaviour of the dodecanol solution surface. This can be related to the possible description of heptanol and decanol solutions as expanded films and dodecanol as condensed film. [11]

Marangoni-Bénard Experiments

In order to determine the onset of convection, without the use of tracer particles which could contaminate the interface, we have measured the heat flux across the liquid layer as a function of the temperature difference imposed at the boundaries. The convective motions will increase the heat transport over the diffusive transport of the layer at rest. The change of slope of $\Delta T = \Delta T (W)$ indicates the critical temperature. In order to derive Ma, Figs. 1, 2, 3 are used to determine $\Delta \sigma / \Delta T$, the density, the kinematic viscosity and the thermal diffusivity are taken in literature for pure water because the solutions are very diluted.

MBI cell. A sketch of the cell is given on Fig. 4, it is constituted by two copper blocks. Between which a Teflon plate is inserted. The temperature drop at the limit of this plate allows to evaluate the heat flux pumped by the four Peltier elements. These copper blocks are chromium coated in order to be cleaned using sulfochromic acid. The thickness of the liquid layer was 1 mm, the air layer was 3 mm thick and the diameter was 80 mm. The interface temperature is kept constant through the experimental runs by controlling the temperatures of the thermoregulated water flowing on top of the

cell depending on the heat fluxes which are pumped by the Peltier elements. This is adjusted iteratively.

n-heptanol solution. Using Fig. 1, the temperature of the interface was chosen in order that $d\sigma/dT = 6,67 \ 10^{-3}$ mN/m°C, this temperature was 39.5°C and leads to $\Delta T^{cr} = 1$°C.

The knowledge of the thickness and the heat conductivity of the air layer (at rest) and of the Pyrex plate allows to determine the temperature of the flowing thermoregulated water.

To start the measurements, we impose a potential difference to the Peltier elements and the temperature of the flowing water. But this temperature has to be adjusted as a function of the heat flux in order to keep the interface temperature constant. Taking into account that the heat flux pumped depends also on the temperatures on the walls of the Peltier elements, this delicate regulation has to be adjusted iteratively because changing one element modifies the other parameters. The obtained results are given on Fig. 5. For heat flows smaller than 3.9 Watts, the experimental points are on a straightline going through the origin. At the critical point, T is decreasing when the heat flux is increased. This behaviour is compatible with the existence of an inverse bifurcation.

The onset of convection is observed for T (between Pyrex and copper) equal to 3.4°C.

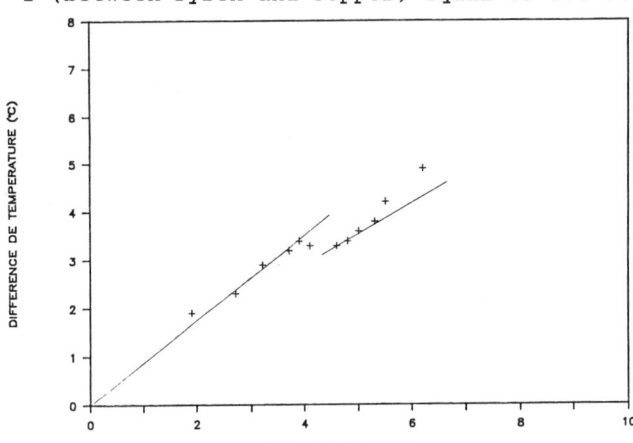

Figure 5.   Schmidt-Milverton plot for the n-heptanol system

n-decanol solution. The experimental procedure is the same as for the n-heptanol solution. The temperature of the interface corresponding to the same value of $d\sigma/dT$ as for the n-heptanol solution is 26.3°C (see Fig. 2). The obtained results are given on Fig. 6. The experimental points for heat flux lower than 4 watts are on a line going through the origin and the behaviour is the same as for n-heptanol. The ΔT is decreasing after the critical point corresponding to a critical value equal to 5.1°C.

n-dodecanol solution. On Fig. 3, we determine that the interface has to be kept at

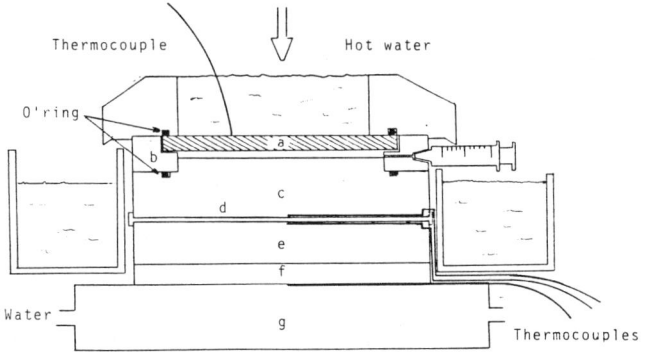

Figure 4.   Sketch of the experimental cell

a : Pyrex window   e : Copper block
b : Teflon walls   f : Four Peltier
c : Copper block   elements
d : Teflon plate   g : Cooling water

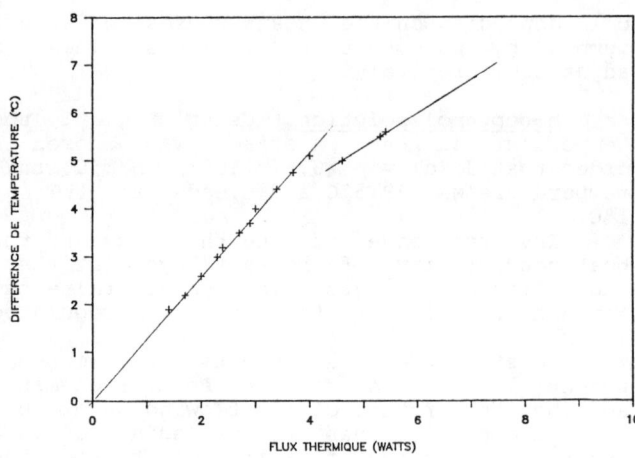

Figure 6. Schmidt-Milverton plot for the
n-decanol system

10.3°C to have the same value of dσ/dT as the
two other systems.
The procedure was identical, but the onset of
convection was not observed for heat flux
higher than 8 watts which was the limit of the
present experimental hardware.

Discussion Of The Experimental Results
The measured value for the surface
viscosity of the n-dodecanol solution was
about 100 10^{-4} sp.This value is significant,
it is twenty times larger than the dispersal
of the measured results. On the other hand we
measured the surface viscosity for the n-
heptanol, the n-decanol solutions and water
which were very small and considered as zero.
This has to lead to classical critical values
of the Marangoni number ($\approx$ 80) for these three
last systems (and thus calculated using the $\check{v}$
value of the bulk phase) and a critical
temperature difference at least six time
larger (using [5]). The obtained results are
in agreement with these conclusions.
These investigation results will soon be
improved by

- increasing the accuracy on the measurement
  of the liquid volume (3.85 ml) in order to
  have a better determination of the liquid
  thickness
- replace the Pyrex windows by a sapphire
  plate to decrease the temperature drop
- improve the measurement of the heat flux.

CONCLUSIONS

We have shown the role played by the
surface viscosity on the hydrodynamical
instability of Marangoni-Bénard. We believe
that the transport coefficients in the
interface are of importance regarding
fluctuation relaxations located in the
interface, where they are the motors for the
onset of convection.
We demonstrated that the viscosity
coefficient introduced in the Marangoni-Bénard
number arises from the boundary condition at
the liquid/gas interface and can be related to
a surface viscosity (defined as an excess

quantity). This surface viscosity does not
appear elsewhere. Thus the classical linear
stability analysis results remain valid. Only
the couple ΔT^* - ν^* value changes at the
critical value.
Experimental investigations performed on
aqueous solution of long chain alcohols allow
the determination of the critical onset of
convection when the layer is heated from the
top. This eliminates the destabilizing role of
gravity and the condensation problem on the
plate limiting the gas phase.

ACKNOWLEDGEMENTS

We are very grateful to Professor A.
Jaumotte for his constant interest in this
work.
This investigation was performed under
ARC contract of SPPS and financial support of
FRFC.
The surface viscosity measurements were
conducted on the equipment of Professor P.
Joos, UIA, that we thank.

BIBLIOGRAPHY

[1] Bénard, H., "Les tourbillons cellulaires
dans une nappe liquide", Revue Générale des
Sciences Pures et Appliquées, 1901, pp. 1261-
1271 and pp. 1309- 1328.

[2] Rayleigh, L., "On the convection currents
in a horizontal layer of fluid when the higher
temperature is on the underside ", Phil. Mag.,
Vol 32, 1916, pp. 529-46.

[3] Pearson, J.R.A., "On convection cells
induced by surface tension", Journal of Fluid
Mechanics, Vol 4, 1958, pp. 489-500.

[4] Nield, D.A., "Surface tension and
buoyancy effects in cellular convection"
Journal of Fluid Mechanics, Vol 19, 1964, pp
341-352.

[5] Scriven, L.E., Sternling, C.V., "On
cellular convection driven by surface tension
gradients : effects of mean surface tension
and surface viscocity", Journal of Fluid
Mechanics, Vol 19, 1964, pp. 321-340.

[6] Scriven, L.E., "Dynamics of a fluid
interface", Chemical Engineering Science, Vol
12, 1960, pp. 9-108.

Aris, R., "Vectors, Tensors and the Basic
Equations of Fluid Mechanics", Prentice-Hall,
inc., 1962.

[7] Chandrasekhar, S., "Hydrodynamic and
Hydromagnetic Stability", Clarendon Press,
Oxford, 1961.

[8] Burton, R.A. and Mannheimer, R.J.,
"Analysis and Apparatus for Surface
Rheological Measurements", Advances in
Chemistry Series, Vol 63, pp. 315-328 (1967)

[9] Mannheimer, R.J. and Schechter, R.S., "An
Improved Apparatus and Analysis for Surface

Rheological Measurements", <u>Journal of Colloïd and Interface Science</u>, Vol 32, N° 2, 1970.

[10] Mannheimer, R.J. and Schechter, R.S., "The theory of Interfacial Viscoelastic Measurement by the Viscous-Traction Method", <u>Journal of Colloïd and Interface Science</u>, Vol 32, N° 2, 1970.

[11] Gaines, G.L., "<u>Insoluble monolayers at liquid-gas interfaces</u>", Interscience publishers, 1966.